Collection of Simulated XRD Powder Patterns for Zeolites

Collection of Simulated XRD Powder Patterns for Zeolites

M.M.J. Treacy
and
J.B. Higgins

Published on behalf of the Stucture Commision of the International Zeolite Association

Fourth Revised Edition

ELSEVIER

Amsterdam - London - New York - Oxford - Paris - Shannon - Tokyo

ELSEVIER SCIENCE B.V.
Sara Burgerhartstraat 25
P.O. Box 211, 1000 AE Amsterdam, The Netherlands

Fourth revised edition, first impression 2001.
Fourth revised edition, second impression 2001.

Library of Congress Cataloging in Publication Data
A catalog record from the Library of Congress has been applied for.

ISBN: 0-444-50702-7

∞ The paper used in this publication meets the requirements of ANSI/NISO Z39.48-1992 (Permanence of Paper).
Printed in The Netherlands.

TABLE OF CONTENTS

PREFACE

The synthesis and characterization of new zeolite materials continues unabated. The IZA Structure Commission has recognized thirty-five new topologies since the 3rd edition of this *Collection* was prepared in 1995. The total number of zeolite structure refinements has surpassed 3000 with over 1000 new refinements since 1995. Not surprisingly, the most scrutinized framework topology is that of faujasite, **FAU**, with about 600 structural studies. Probably no other inorganic host structure has received such attention. With these numbers it is apparent that this *Collection* is not comprehensive. The scope of the *Collection* has been broadly defined to include materials of interest to zeolite scientists, following the policies established at recent IZA conferences. We have attempted to be as inclusive as possible to give the users of this publication the maximum information. The structures of the porous solids that we considered comprise corner-sharing tetrahedra, and are not limited to a specific chemical composition. Materials such as metal phosphates and silica polymorphs are included.

The present *Collection* serves as a source of reference patterns for pure zeolite phases. The data will be helpful in establishing the structural purity of experimental phases and in indexing their diffraction patterns. The data will also aid in the determination of changes in the lattice parameters with changing composition, assessing preferred orientation effects, background evaluation, and line broadening. We have also included diffraction patterns of several common dense silicate phases to facilitate their detection in mixed phase syntheses.

The numerical data comprises 2θ values for CuKα radiation ($\lambda = 1.5418$ Å), d-spacings, relative intensities, hkl Miller indices and multiplicity, M_{hkl}. Data representing 133 framework topologies have been included in this *Collection*. In most cases, X-ray or neutron refinements of hydrated or as-synthesized forms are used.

This edition differs significantly from the 1996 3rd edition which included atomic coordinates. Because of space constraints coordinate data has not been included in this edition, but are available in electronic form, along with the complete contents of this edition, at:

http://www.iza-structure.org/databases/

This web site also contains an interactive powder pattern calculator that allows the user to change the input and variables for a powder pattern calculation.

This book was typeset using LaTeX with the standard computer modern typeface. The LaTeX file, and associated postscript plots, were generated using a C program written by M. M. J. Treacy.

We wish to acknowledge the assistance and collaboration of the members of the IZA Structure Commission in proofreading the manuscript and for providing additional information. We are indebted to our companies (NEC Research Institute, Inc., and Air Products and Chemicals, Inc.) for support of this project. We are grateful to Peggy Bisher for keeping the reference data files impeccably organized. Finally, we acknowledge the patience and support of our wives Laura and Carol.

Michael M. J. Treacy, Princeton, NJ
John B. Higgins, Bad Dog Ridge, PA
January 2001

STRUCTURE COMMISSION

MEMBERS

Additional IZA publications:

Atlas of Zeolite Framework Types, 5th revised edition (2001), Ch. Baerlocher, W. M. Meier and D. H. Olson.

Compilation of Extra Framework Sites in Zeolites (1982), W. J. Mortier. (out of print)

Verified Syntheses of Zeolitic Materials, 2nd revised edition (2001), H. Robson, editor; and K. P. Lillerud, XRD patterns.

See also: http://www.iza-online.org/

EXPLANATORY NOTES

The numerical data and the simulated powder patterns presented in this *Collection* are to a great extent self-explanatory. In order to facilitate the use of these reference patterns some pertinent remarks regarding the keywords used in the data section are summarized below. The input structural data have been deposited on the worldwide web at:

http://www.iza-structure.org/databases/

This *Collection*, including updates, will be accessible at the above address.

ZEOLITE FRAMEWORK TYPES

The three-letter framework type codes, recognized by the IUPAC Commission on Zeolite Nomenclature, have been used to organize the entries in this publication. The powder diffraction data and simulated patterns for the reference structures are listed alphabetically according to the respective framework type code. An index of material names, and associated three-letter codes, is included in the companion volume, the *Atlas of Zeolite Framework Types* (Baerlocher, Meier and Olson (2001)).

COMPOSITION

Compositions are expressed in terms of the full unit cell content. Two compositions are provided. The CHEMICAL COMPOSITION is the nominal material composition provided in the original reference, and is usually obtained from chemical analysis. The REFINED COMPOSITION is derived from the structure refinement. Because of the complexities of structure refinements, the chemical and refined compositions do not always concur. When available, refinements of hydrated zeolites were used to calculate the patterns. For synthetic zeolites, if the zeolite had been synthesized in the presence of organic material, those refinements of the uncalcined products that contained the occluded organic molecules were chosen. The sample locality is given in the case of natural zeolites.

CRYSTAL DATA

Crystal data includes lattice parameters and space group information from the International Tables for Crystallography, 4th revised edition 1995, and incorporates the new e 'double-glide' plane. Consequently, some space group symbols will differ from those listed in the original references. Two entries in this *Collection* are affected by this change; EU-1 (**EUO**) which has space group symbol $Cmme$ (formerly $Cmma$), and gottardiite (**NES**) which has space group symbol $Cmce$ (formerly $Cmca$). The type of refinement, along with the final R-values, is listed with the unit cell parameters.

REFERENCE

Reference cites the literature from which the crystal data, atomic coordinates, and temperature factors were obtained. In many cases there are multiple refinements of the same zeolitic material, but because of space limitations not all refinements could be included. We would be appreciative if authors and users would inform us of any errors or omissions. A listing of the references for isotypic species can be found in the *Atlas of Zeolite Framework Types* (Baerlocher, Meier and Olson (2001)). A list of references to structure analyses of zeolites with different cations, up to 1982, is given in the *Compilation of Extra Framework Sites in Zeolites*, W. J. Mortier (1982).

POWDER PATTERN IDENTIFICATION TABLE

A table is provided to assist in the identification of powder patterns of unknown materials. The 2θ (°) values of the three most pronounced low-angle reflections are listed. Usually, these reflections are simply the three strongest peaks. In many instances, a pronounced low-angle reflection will be included, even if it is not among the most intense. In the 3rd edition (1996) of the *Collection* this table was assembled by visual inspection of the computed powder patterns. For this fourth edition (2001) of the *Collection*, the table was generated automatically from the computed diffraction patterns. To achieve good correspondence with the hand-generated table of the 3rd (1996) edition of the *Collection*, a peak intensity weighting function $W(2\theta)$ was used

$$W(2\theta) = 1 + A\exp\left[-(2\theta)^2/2\sigma^2\right]. \quad (1)$$

The parameters were set to $A = 10$, and $\sigma = 7°$. This function is strongly weighted towards the low angle peaks, particularly those below about 10°, which, even if relatively weak, tend to offer a more characteristic fingerprint of a material compared to the abundance of strong peaks that tend to cluster around 25° in most zeolitic materials.

The data for all the materials presented in this work are sorted by increasing 2θ value in the table. To identify an unknown material, measure the 2θ values of the three most pronounced peaks (assigning strong weighting to any pronounced low-angle peaks, particularly those below about 10°) and find those materials with corresponding reflections at those values. This provides a starting point for a more detailed comparison of the experimental and calculated patterns.

CALCULATED POWDER DIFFRACTION DATA

The powder diffraction data include the 2θ-values for CuKα radiation, d-spacings, relative intensities $I_{\rm rel}$, Miller indices hkl and multiplicity M_{hkl}, for the strongest 135 reflections with an integrated intensity $I_{\rm rel}$ greater than 0.1. The strongest reflection is set to $I_{\rm rel} = 100$. The scattering factors used for the framework T- and O-atoms in the structure factor computations were those selected by the authors of the original work. If none had been specified, we chose atomic (neutral) scattering factors. No adsorption corrections were applied to the data, and anisotropic temperature factors were converted to isotropic temperature factors $B_{\rm iso}$ (Å^2) using equations reported by W. C. Hamilton, *Acta Cryst.*, **12** 609–610 (1959).

Materials with the same framework type code (i.e. framework topology) may have very different diffraction patterns, so for some framework type codes several different reference materials have been included. Examples listed under **FAU**, **GIS**, **MFI** and **NAT** illustrate the extent of the differences observed in the diffraction patterns of materials with identical framework topologies but variations in composition and/or symmetry.

SIMULATED POWDER PATTERNS

The powder patterns for CuKα radiation are reproduced from 0° to 50° 2θ. The patterns, and associated tables, were calculated using a custom C program written by M. M. J. Treacy. The intensity scale is kept variable to accommodate extreme situations. The intensity range is usually plotted between 0 and $I_{\max} = 100$. If only one or a few peaks are very intense, and the rest of the pattern consists of low-intensity reflections, $I_{\max}$ is set at a lower value (see ordinate) to show sufficient detail in the low-intensity region of the pattern. The scale of the ordinate is always calibrated in percent relative intensity of the maximum peak height. Frequently, the peak intensities in the plots will not be identical to the integrated intensities listed in the corresponding tables, because of the possibility of overlapping diffraction peaks. A Lorentzian profile was assumed for the calculation of the pattern. The full width at half maximum (FWHM) of all peaks was set to be 0.08° 2θ. Many real samples will in general give diffraction patterns with broader lines due to instrumental broadening, disorder, or small crystallite size. However, synchrotron powder diffraction data of some zeolite materials exhibit FWHM of less than 0.04° 2θ.

Because hydrated forms of natural zeolites or as-synthesized forms were used whenever refined atomic parameters were available, the plots should be easily comparable to experimental patterns. In some cases, only structure refinements of dehydrated or calcined forms were available. Significant differences in the intensities of low-angle peaks may be found when comparing the calculated pattern to experimental patterns of hydrated or as-synthesized materials.

Finally, a note of caution. The patterns are useful in helping to establish the structural purity of a zeolite phase, yet they may not always allow one to readily and unambiguously determine the framework type of the sample. This assignment is often not straightforward and may require more sophisticated analyses. W. J. Rohrbaugh and E. W. Wu review the factors affecting the diffraction characteristics of zeolite materials (ACS Symposium Series **411** 279–302 (1989)).

POWDER PATTERN SIMULATIONS OF DISORDERED INTERGROWTHS

A number of zeolitic materials crystallize as disordered planar intergrowths of "end-member" frameworks. Two of the better-studied series are the **FAU/EMT** intergrowths (such as ZSM-2, ZSM-3, ZSM-20, ECR-30, CSZ-1 and CSZ-3) and the intergrown zeolite beta and its natural analog tschernichite. Powder patterns for five of the more commonly observed coherently intergrown series were calculated using the DIFFaX computer program (M. M. J. Treacy, J. M. Newsam and M. W. Deem, *Proc. R. Soc. Lond.* A **433** 499–520 (1991)). Random stacking faults are assumed. The five series are; the beta family; SSZ-33/SSZ-26; **FAU/EMT**; **MFI/MEL**; and **OFF/ERI**. The stacking fault probability is incremented from 0 to 1 in steps of 0.1. Stacking fault probabilities of 0 and 1 represent the unfaulted end-members.

Additional information on disordered intergrowths can be found in the *Catalog of Disordered Zeolite Structures*, which is available electronically at http://www.iza-structure.org/databases/.

CHANGES SINCE THE THIRD REVISED EDITION

- 35 new framework types have been approved since the third revised edition, and are included in this fourth revised edition. The new framework types are:

ACO	**AEN**	**AFN**	**ASV**	**AWO**	**CFI**	**CGF**
CGS	**CZP**	**DFT**	**DON**	**ESV**	**FRA**	**GON**
IFR	**ISV**	**ITE**	**MSO**	**MTF**	**MWW**	**OSI**
OSO	**SAO**	**SAS**	**SAT**	**SAV**	**SBE**	**SBS**
SBT	**SFE**	**SFF**	**STF**	**STT**	**TER**	**TSC**

- Due to space constraints, the refined structure coordinates are not listed in the printed version of this *Collection*. An electronic PDF version of this *Collection*, complete with coordinates, can be found at:

  http://www.iza-structure.org/databases/

- Errors in the space group settings of Maricopaite (**MOR**) and chiavennite (**-CHI**) have been fixed. These corrections change the diffraction patterns and the listed peak intensities.
- The distance least squares refinement of liottite (**LIO**), given in the third revised edition, has been replaced by a single crystal x-ray refinement.
- The entry for Na-exchanged K-F (**EDI**) has been removed due to inconsistencies in the refinement.
- A typographic error in the published coordinates for perlialite (**LTL**) has been corrected. This correction has changed the diffraction pattern noticeably.
- An error in the cation occupancies for the mineral faujasite (**FAU**) has been corrected. This correction has changed the diffraction pattern noticeably.
- The refinements of MAPO–39 (**ATN**) and offretite (**OFF**) have been replaced.
- A refinement for low silica zeolite-X (**FAU**) has been added.
- A small number of typographic errors in atom coordinates have been corrected.

POWDER PATTERN IDENTIFICATION TABLE

Powder Pattern Identification Table

$2\theta^\circ$	Material	Code
3.33	Franzinite	**FRA**
3.42	Cloverite	**-CLO**
3.56	Paulingite	**PAU**
4.14	NaZ-21	**LTN**
4.56	Decamethonium DAF-1	**DFO**
4.80	VPI-9	**VNI**
4.83	Cloverite	**-CLO**
4.84	Tschörtnerite	**TSC**
5.04	Paulingite	**PAU**
5.31	AlPO-8	**AET**
5.38	VPI-5	**VFI**
5.50	Perlialite	**LTL**
5.55	Linde Type L	**LTL**
5.55	Mazzite	**MAZ**
5.59	Tschörtnerite	**TSC**
5.63	UCSB-8Co	**SBE**
5.64	Chiavennite	**-CHI**
5.72	UCSB-6GaCo	**SBS**
5.76	MCM-35	**MTF**
5.87	EMC-2, Calcined	**EMT**
5.87	EMC-2, Partially Dehydrated	**EMT**
5.92	Cloverite	**-CLO**
5.96	Li-LSX	**FAU**
6.02	UCSB-10GaZn	**SBT**
6.04	UTD-1	**DON**
6.10	Na-X, Dehydrated	**FAU**
6.12	Na-X, Hydrated	**FAU**
6.19	Faujasite	**FAU**
6.23	EMC-2, Partially Dehydrated	**EMT**
6.24	EMC-2, Calcined	**EMT**
6.31	Na-Y, Siliceous	**FAU**
6.32	UCSB-10GaZn	**SBT**
6.33	Ultrastable Y, Dehydrated Dealuminated	**FAU**
6.39	Maricopaite	**MOR**
6.41	UCSB-8Co	**SBE**
6.48	1-aminoadamantane Deca-Dodecasil 3R	**DDR**
6.50	UCSB-6GaCo	**SBS**
6.51	Mordenite	**MOR**
6.57	AlPO-8	**AET**
6.57	Di-n-propylamine MAPSO-46	**AFS**
6.58	UCSB-6GaCo	**SBS**
6.58	SSZ-48, Calcined	**SFE**
6.65	EMC-2, Calcined	**EMT**
6.65	EMC-2, Partially Dehydrated	**EMT**
6.69	(Cs,K) ZK5, Dehydrated	**KFI**
6.77	NaZ-21	**LTN**
6.78	VPI-8	**VET**
6.82	UiO-6, Calcined	**OSI**
6.82	Roggianite	**-RON**
6.85	AlPO-41, Calcined	**AFO**
6.88	ITQ-7, Siliceous, Calcined	**ISV**
6.93	CIT-5	**CFI**
6.96	GUS-1	**GON**
6.98	Beta, Polymorph A SiO_2 Framework	***BEA**
7.01	ITQ-7, Siliceous, Calcined	**ISV**
7.04	UCSB-10GaZn	**SBT**
7.05	NU-87	**NES**
7.09	ITQ-1, Calcined (Siliceous MCM-22)	**MWW**
7.10	Beryllophosphate-H	**BPH**
7.18	ITQ-1, Calcined (Siliceous MCM-22)	**MWW**

$2\theta^\circ$	Material	Code
7.18	Linde Type A, Hydrated	**LTA**
7.20	Linde Type A, Dehydrated	**LTA**
7.33	CIT-5	**CFI**
7.40	Terranovaite	**TER**
7.41	N_2, Piperidine Dodecasil 1H	**DOH**
7.42	SAPO-56	**AFX**
7.42	Gmelinite	**GME**
7.43	Boggsite	**BOG**
7.43	Tetrapropylammonium Fluoride AlPO-5	**AFI**
7.47	ZSM-12, Calcined	**MTW**
7.49	MCM-61	**MSO**
7.55	UTD-1	**DON**
7.61	Tetrapropylammonium SAPO-40	**AFR**
7.63	ZSM-12, Calcined	**MTW**
7.66	Offretite	**OFF**
7.66	CIT-1	**CON**
7.67	STA-1, Magnesium Aluminophosphate	**SAO**
7.68	1-aminoadamantane Deca-Dodecasil 3R	**DDR**
7.69	Erionite	**ERI**
7.71	Bellbergite	**EAB**
7.71	Piperidine AlPO-17	**ERI**
7.72	ITQ-7, Siliceous, Calcined	**ISV**
7.72	Di-n-propylamine MAPSO-46	**AFS**
7.72	SSZ-44, Calcined	**SFF**
7.73	ERS-7 Framework	**ESV**
7.74	Beta, Polymorph A SiO_2 Framework	***BEA**
7.75	ZSM-18, SiO_2 Framework	**MEI**
7.78	Ferrierite	**FER**
7.80	Gottardiite	**NES**
7.82	ZSM-57, SiO_2 Framework	**MFS**
7.84	Boggsite	**BOG**
7.84	CIT-1	**CON**
7.88	MAPO-36, Calcined	**ATS**
7.89	Dodecasil 3C	**MTN**
7.90	NU-87	**NES**
7.91	Decamethonium DAF-1	**DFO**
7.92	EU-1, Calcined, Rehydrated	**EUO**
7.93	ZSM-11, Calcined	**MEL**
7.94	NaZ-21	**LTN**
7.94	ZSM-5, Calcined	**MFI**
7.94	Ammonium Fluoride ZSM-23	**MTT**
7.95	Tetrapropylammonium ZSM-5	**MFI**
7.95	2-aminopentane Nonasil	**NON**
7.99	SSZ-35, Calcined	**STF**
8.01	ZSM-5, Calcined	**MFI**
8.01	Di-n-propylamine CoAPO-50	**AFY**
8.01	ITQ-1, Calcined (Siliceous MCM-22)	**MWW**
8.02	AlPO-52, Calcined, Rehydrated	**AFT**
8.06	Tetrapropylammonium SAPO-40	**AFR**
8.06	SSZ-44, Calcined	**SFF**
8.07	SSZ-48, Calcined	**SFE**
8.09	SSZ-35, Calcined	**STF**
8.09	Di-isopropylamine MnAPO-11	**AEL**
8.10	AlPO-11, Calcined	**AEL**
8.11	ERS-7 Framework	**ESV**
8.11	SSZ-23	**STT**
8.11	Beryllophosphate-H	**BPH**
8.15	Decamethonium DAF-1	**DFO**
8.15	ITQ-4, Calcined	**IFR**
8.15	Diethylamine Theta-1 (Silica ZSM-22)	**TON**

Powder Pattern Identification Table *continued*

$2\theta^\circ$	Material	Code
8.17	STA-1, Magnesium Aluminophosphate	**SAO**
8.20	Cobalt Gallium Phosphate-6	**CGS**
8.20	MAPO–36, Calcined	**ATS**
8.21	Partheite	**-PAR**
8.22	Ammonium Fluoride ZSM-23	**MTT**
8.32	Gottardiite	**NES**
8.32	Rho, Hydrated	**RHO**
8.37	NU-87	**NES**
8.49	Di-n-propylamine SAPO–31	**ATO**
8.56	Levyne	**LEV**
8.58	ITQ–3, Calcined	**ITE**
8.62	Terranovaite	**TER**
8.64	SAPO–56	**AFX**
8.64	RUB-13	**RTH**
8.68	1-aminoadamantane NU-3	**LEV**
8.71	SSZ-44, Calcined	**SFF**
8.72	RUB-17	**RSN**
8.73	STA-6	**SAS**
8.75	EU-1, Calcined, Rehydrated	**EUO**
8.81	Terranovaite	**TER**
8.81	ZSM-11, Calcined	**MEL**
8.82	GUS-1	**GON**
8.84	Tetrapropylammonium ZSM-5	**MFI**
8.87	VPI-7	**VSV**
8.88	Tetrapropylammonium ZSM-5	**MFI**
8.90	ZSM-5, Calcined	**MFI**
8.91	ZSM-12, Calcined	**MTW**
8.93	Rho, Deuterated Berylloarsenate	**RHO**
8.94	Lovdarite	**LOV**
8.95	Ammonium Fluoride ZSM-23	**MTT**
8.98	Maricopaite	**MOR**
9.01	ITQ–3, Calcined	**ITE**
9.01	1-aminoadamantane Sigma-2	**SGT**
9.05	RUB-13	**RTH**
9.07	Pahasapaite	**RHO**
9.08	AlPO–14, Calcined	**AFN**
9.10	CIT-1	**CON**
9.14	AlPO–C, Hydrated	**APC**
9.16	RUB-3	**RTE**
9.18	Quinuclidine AlPO–22	**AWW**
9.23	Ferrierite	**FER**
9.23	Cobalt Gallophosphate	**LAU**
9.24	Laumontite	**LAU**
9.28	UCSB–8Co	**SBE**
9.30	Tetramethylammonium OH AlPO–12	**ATT**
9.36	Magnesium STA-7	**SAV**
9.36	AlPO–25	**ATV**
9.38	Leonhardite	**LAU**
9.40	Chabazite	**CHA**
9.42	ZSM-57, SiO_2 Framework	**MFS**
9.42	Magnesium STA-7	**SAV**
9.43	SSZ–48, Calcined	**SFE**
9.45	Ferrierite, Siliceous	**FER**
9.45	Di-isopropylamine MnAPO–11	**AEL**
9.47	MAPO–39	**ATN**
9.47	(Cs,K) ZK5, Dehydrated	**KFI**
9.48	Methylbutylamine SAPO–47	**CHA**
9.49	AlPO–18, Calcined	**AEI**
9.49	Bellbergite	**EAB**
9.50	SSZ-23	**STT**

$2\theta^\circ$	Material	Code
9.54	AlPO–52, Calcined, Rehydrated	**AFT**
9.54	Perlialite	**LTL**
9.55	ZSM-18, SiO_2 Framework	**MEI**
9.57	AlPO–11, Calcined	**AEL**
9.59	VPI-8	**VET**
9.61	2-aminopentane Nonasil	**NON**
9.62	Mazzite	**MAZ**
9.64	(Na, Tetramethylammonium)-E	**EAB**
9.65	UiO-6, Calcined	**OSI**
9.65	Roggianite	**-RON**
9.68	Erionite	**ERI**
9.68	Beta, Polymorph A SiO_2 Framework	***BEA**
9.69	Tschörtnerite	**TSC**
9.69	ITQ-4, Calcined	**IFR**
9.70	Stilbite	**STI**
9.71	Stellerite	**STI**
9.72	Barrerite	**STI**
9.72	AlPO–41, Calcined	**AFO**
9.73	SSZ-35, Calcined	**STF**
9.74	Li-LSX	**FAU**
9.75	MCM-35	**MTF**
9.75	Chiral Sodium Zincophosphate, $P6_122$	**CZP**
9.77	Mordenite	**MOR**
9.77	Piperidine AlPO–17	**ERI**
9.81	Chiral Sodium Zincophosphate, $P6_522$	**CZP**
9.81	Di-n-propylamine CoAPO–50	**AFY**
9.82	ERS-7 Framework	**ESV**
9.85	Heulandite	**HEU**
9.88	Clinoptilolite	**HEU**
9.90	Stellerite	**STI**
9.91	Barrerite	**STI**
9.95	Dachiardite	**DAC**
9.96	Stilbite	**STI**
9.97	Epistilbite	**EPI**
9.97	Na-X, Dehydrated	**FAU**
9.99	AlPO–8	**AET**
10.00	Cobalt Gallium Phosphate-5	**CGF**
10.00	Na-X, Hydrated	**FAU**
10.06	RUB-10, SiO_2 Framework	**RUT**
10.06	ITQ–3, Calcined	**ITE**
10.07	ASU-7	**ASV**
10.08	Tetramethylammonium ZAPO-M1	**ZON**
10.10	AlPO–21	**AWO**
10.12	OSB-1	**OSO**
10.14	AlPO-EN3	**AEN**
10.16	VPI-9	**VNI**
10.16	Diethylamine Theta-1 (Silica ZSM-22)	**TON**
10.17	Linde Type A, Hydrated	**LTA**
10.18	RUB-13	**RTH**
10.19	Linde Type A, Dehydrated	**LTA**
10.27	MCM-61	**MSO**
10.31	Na-Y, Siliceous	**FAU**
10.34	Ultrastable Y, Dehydrated Dealuminated	**FAU**
10.47	AlPO – H2	**AHT**
10.48	Magnesium STA-7	**SAV**
10.50	STA-6	**SAS**
10.58	SSZ-23	**STT**
10.61	AlPO–18, Calcined	**AEI**
10.70	Paulingite	**PAU**
10.75	Nepheline Hydrate	**JBW**

Powder Pattern Identification Table *continued*

2θ°	Material	Code
10.77	RUB-10, SiO_2 Framework	**RUT**
10.77	VPI-5	**VFI**
10.85	Levyne	**LEV**
10.90	Harmotome	**PHI**
10.90	Partheite	**-PAR**
11.07	Heulandite	**HEU**
11.08	1-aminoadamantane NU-3	**LEV**
11.11	Mazzite	**MAZ**
11.17	ZSM-18, SiO_2 Framework	**MEI**
11.19	Clinoptilolite	**HEU**
11.29	Bikitaite	**BIK**
11.30	Chiavennite	**-CHI**
11.33	Cobalt Gallium Phosphate-5	**CGF**
11.38	Chiral Sodium Zincophosphate, $P6_122$	**CZP**
11.41	Chiral Sodium Zincophosphate, $P6_522$	**CZP**
11.42	Li-LSX	**FAU**
11.43	Barium Chloroaluminosilicate	**MER**
11.45	Quinuclidine AlPO–16	**AST**
11.49	Cobalt Gallium Phosphate-6	**CGS**
11.52	Gmelinite	**GME**
11.53	Levyne	**LEV**
11.54	Losod	**LOS**
11.57	AlPO–14, Calcined	**AFN**
11.57	SAPO–56	**AFX**
11.64	AlPO–52, Calcined, Rehydrated	**AFT**
11.65	Offretite	**OFF**
11.67	Quinuclidinium fluoride octadecasil	**AST**
11.69	Na-X, Dehydrated	**FAU**
11.73	Na-X, Hydrated	**FAU**
11.77	Linde Type L	**LTL**
11.79	Nepheline Hydrate	**JBW**
11.86	Faujasite	**FAU**
11.86	Wenkite	**-WEN**
11.95	Magnesium Phosphate UiO-20	**DFT**
12.10	Na-Y, Siliceous	**FAU**
12.13	Amicite	**GIS**
12.14	Ultrastable Y, Dehydrated Dealuminated	**FAU**
12.15	Gismondine	**GIS**
12.22	AlPO–21	**AWO**
12.22	ACP-1	**ACO**
12.32	Cobalt Gallium Phosphate-6	**CGS**
12.35	Gobbinsite	**GIS**
12.36	Phillipsite	**PHI**
12.38	Garronite	**GIS**
12.38	Di-n-propylamine MAPO–43	**GIS**
12.40	Harmotome	**PHI**
12.41	Goosecreekite	**GOO**
12.42	Tetrapropylammonium SAPO–40	**AFR**
12.44	Merlinoite	**MER**
12.44	Tetramethylammonium ZAPO-M1	**ZON**
12.46	Na-P1	**GIS**
12.46	Linde Type A, Hydrated	**LTA**
12.48	Phillipsite	**PHI**
12.49	Linde Type A, Dehydrated	**LTA**
12.54	Merlinoite	**MER**
12.55	AlPO–C, Dehydrated	**APC**
12.57	Gobbinsite	**GIS**
12.58	Ferrierite, Siliceous	**FER**
12.59	Goosecreekite	**GOO**
12.60	ACP-1	**ACO**

2θ°	Material	Code
12.65	Tetramethylammonium OH AlPO–12	**ATT**
12.71	Laumontite	**LAU**
12.73	MCM-61	**MSO**
12.76	Franzinite	**FRA**
12.78	ZSM-57, SiO_2 Framework	**MFS**
12.78	Dachiardite	**DAC**
12.80	Maricopaite	**MOR**
12.83	Ferrierite, Siliceous	**FER**
12.84	Epistilbite	**EPI**
12.88	RUB-17	**RSN**
12.89	Methylbutylamine SAPO–47	**CHA**
12.90	AlPO–C, Hydrated	**APC**
12.91	Dodecasil 3C	**MTN**
12.91	AlPO–D	**APD**
12.91	AlPO–18, Calcined	**AEI**
12.99	AlPO-EN3	**AEN**
13.00	Leonhardite	**LAU**
13.07	Brewsterite	**BRE**
13.15	AlPO–14, Calcined	**AFN**
13.16	Lovdarite	**LOV**
13.17	Di-n-propylamine MAPSO–46	**AFS**
13.33	(Na, Tetramethylammonium)-E	**EAB**
13.34	Erionite	**ERI**
13.37	Scolecite	**NAT**
13.38	Piperidine AlPO–17	**ERI**
13.39	Thomsonite	**THO**
13.41	Montesommaite	**MON**
13.41	MAPO–39	**ATN**
13.41	Gonnardite	**NAT**
13.44	Mesolite	**NAT**
13.45	Mordenite	**MOR**
13.50	Bellbergite	**EAB**
13.56	Natrolite	**NAT**
13.60	Edingtonite	**EDI**
13.64	AlPO–C, Hydrated	**APC**
13.72	ITQ-4, Calcined	**IFR**
13.72	Losod	**LOS**
13.73	AlPO–41, Calcined	**AFO**
13.76	Tetramethylammonium ZAPO-M1	**ZON**
13.80	Li-ABW	**ABW**
13.88	RUB-17	**RSN**
13.89	STA-1, Magnesium Aluminophosphate	**SAO**
13.89	VPI-7	**VSV**
13.90	Harmotome	**PHI**
14.00	AlPO–C, Dehydrated	**APC**
14.02	Cancrinite	**CAN**
14.02	Brewsterite	**BRE**
14.14	RUB-3	**RTE**
14.16	Sodalite Octahydrate	**SOD**
14.31	Tugtupite	**SOD**
14.41	RUB-10, SiO_2 Framework	**RUT**
14.50	Roggianite	**-RON**
14.56	Partheite	**-PAR**
14.71	Linde Type L	**LTL**
14.74	Melanophlogite	**MEP**
14.80	AlPO–21	**AWO**
14.96	Gonnardite	**NAT**
14.96	AlPO–25	**ATV**
14.97	Weinebeneite	**WEI**
15.00	(Cs,K) ZK5, Dehydrated	**KFI**

Powder Pattern Identification Table *continued*

$2\theta^\circ$	Material	Code
15.04	Natrolite	NAT
15.10	Scolecite	NAT
15.10	N_2, Piperidine Dodecasil 1H	DOH
15.11	Mesolite	NAT
15.14	Dodecasil 3C	MTN
15.20	Chiavennite	-CHI
15.23	Ferrierite	FER
15.26	Chiral Sodium Zincophosphate, $P6_522$	CZP
15.27	Chiral Sodium Zincophosphate, $P6_122$	CZP
15.30	Yugawaralite	YUG
15.42	OSB-1	OSO
15.45	Weinebeneite	WEI
15.61	Faujasite	FAU
15.81	Analcime	ANA
16.16	Melanophlogite	MEP
16.24	STA-6	SAS
16.34	STA-2	SAT
16.42	STA-2	SAT
16.57	Merlinoite	MER
16.84	N_2, Piperidine Dodecasil 1H	DOH
17.13	1-aminoadamantane Deca-Dodecasil 3R	DDR
17.35	Amicite	GIS
17.65	Gmelinite	GME
17.82	Lovdarite	LOV
17.87	RUB-3	RTE
17.87	Garronite	GIS
17.90	AlPO–C, Dehydrated	APC
18.04	Gismondine	GIS
18.35	Liottite	LIO
18.40	Afghanite	AFG
18.54	Quinuclidine AlPO–22	AWW
18.61	Losod	LOS
18.67	Rho, Hydrated	RHO
18.75	Quinuclidine AlPO–16	AST
18.89	Edingtonite	EDI
19.06	MAPO–36, Calcined	ATS
19.15	Cancrinite	CAN
19.16	Yugawaralite	YUG
19.21	Thomsonite	THO
19.56	1-aminoadamantane Sigma-2	SGT
19.76	1-aminoadamantane Sigma-2	SGT
19.82	Goosecreekite	GOO
20.01	Moganite	–
20.08	CIT-5	CFI
20.09	VPI-8	VET
20.18	Di-n-propylamine SAPO–31	ATO
20.34	Gottardiite	NES
20.36	Diethylamine Theta-1 (Silica ZSM-22)	TON
20.45	Chabazite	CHA
20.52	Tridymite	–
20.53	EU-1, Calcined, Rehydrated	EUO
20.58	Tridymite	–
20.61	Quinuclidine AlPO–22	AWW
20.62	2-aminopentane Nonasil	NON
20.62	AlPO–D	APD
20.62	Methylbutylamine SAPO–47	CHA
20.64	UiO-6, Calcined	OSI
20.66	GUS-1	GON
20.71	AlPO–D	APD
20.82	Gismondine	GIS

$2\theta^\circ$	Material	Code
20.86	Alpha Quartz	–
20.96	UTD-1	DON
20.97	Tetrapropylammonium Fluoride AlPO-5	AFI
21.03	Amicite	GIS
21.04	Di-isopropylamine MnAPO–11	AEL
21.10	Barium Chloroaluminosilicate	MER
21.13	AlPO–25	ATV
21.18	Cobalt Gallophosphate	LAU
21.22	Bikitaite	BIK
21.23	AlPO–11, Calcined	AEL
21.52	Di-n-propylamine MAPO–43	GIS
21.52	STA-2	SAT
21.57	Gobbinsite	GIS
21.60	Cs-Aluminosilicate	CAS
21.67	Na-P1	GIS
21.68	Tridymite	–
21.87	AlPO – H2	AHT
21.89	Stilbite	STI
21.90	Stellerite	STI
21.92	Barrerite	STI
22.00	Alpha Cristobalite	–
22.03	1-aminoadamantane NU-3	LEV
22.03	Quinuclidine AlPO–16	AST
22.13	Quinuclidinium fluoride octadecasil	AST
22.22	Heulandite	HEU
22.38	Tetrapropylammonium Fluoride AlPO-5	AFI
22.49	Clinoptilolite	HEU
22.52	Perlialite	LTL
22.58	Di-n-propylamine SAPO–31	ATO
22.59	MAPO–39	ATN
22.60	VPI-5	VFI
22.62	Quinuclidinium fluoride octadecasil	AST
22.65	ASU-7	ASV
22.96	Epistilbite	EPI
23.05	Boggsite	BOG
23.11	Cobalt Gallium Phosphate-5	CGF
23.13	ZSM-11, Calcined	MEL
23.24	Magnesium Phosphate UiO-20	DFT
23.48	ASU-7	ASV
23.53	Offretite	OFF
23.57	(Na, Tetramethylammonium)-E	EAB
23.78	Rho, Deuterated Berylloarsenate	RHO
23.86	Keatite	–
23.86	Franzinite	FRA
23.95	Liottite	LIO
23.96	AlPO-EN3	AEN
24.08	Afghanite	AFG
24.16	Pahasapaite	RHO
24.18	Di-n-propylamine CoAPO–50	AFY
24.65	Sodalite Octahydrate	SOD
24.68	Cs-Aluminosilicate	CAS
24.71	Bicchulite	SOD
24.79	Edingtonite	EDI
24.79	Melanophlogite	MEP
24.80	Tugtupite	SOD
24.83	Cs-Aluminosilicate	CAS
25.14	Tugtupite	SOD
25.14	Rho, Hydrated	RHO
25.18	Laumontite	LAU
25.19	Cobalt Gallophosphate	LAU

Powder Pattern Identification Table *continued*

$2\theta^\circ$	Material	Code
25.37	Leonhardite	**LAU**
25.42	Thomsonite	**THO**
25.59	MCM-35	**MTF**
25.73	Dachiardite	**DAC**
25.76	Bikitaite	**BIK**
25.77	Wenkite	**-WEN**
25.96	Analcime	**ANA**
26.03	Keatite	–
26.07	Magnesium Phosphate UiO-20	**DFT**
26.10	Nepheline Hydrate	**JBW**
26.29	Moganite	–
26.49	Tiptopite	**CAN**
26.65	Alpha Quartz	–
26.73	Moganite	–
26.73	Keatite	–
26.91	Liottite	**LIO**
27.01	Montesommaite	**MON**
27.02	Afghanite	**AFG**
27.02	Rho, Deuterated Berylloarsenate	**RHO**
27.43	VPI-7	**VSV**
27.46	Pahasapaite	**RHO**
27.54	Tetramethylammonium OH AlPO–12	**ATT**
27.66	Di-n-propylamine MAPO–43	**GIS**
27.72	ACP-1	**ACO**
27.74	Cancrinite	**CAN**
27.89	Phillipsite	**PHI**
28.10	Na-P1	**GIS**

$2\theta^\circ$	Material	Code
28.15	Li-ABW	**ABW**
28.33	Garronite	**GIS**
28.37	Wenkite	**-WEN**
28.42	Montesommaite	**MON**
29.03	AlPO – H2	**AHT**
29.10	VPI-9	**VNI**
29.29	Yugawaralite	**YUG**
29.36	OSB-1	**OSO**
29.50	Li-ABW	**ABW**
30.19	Tiptopite	**CAN**
30.40	Chabazite	**CHA**
30.50	Brewsterite	**BRE**
30.54	Analcime	**ANA**
30.96	Scolecite	**NAT**
30.98	Mesolite	**NAT**
31.01	Gonnardite	**NAT**
31.22	Natrolite	**NAT**
31.44	Alpha Cristobalite	–
31.99	Sodalite Octahydrate	**SOD**
32.07	Bicchulite	**SOD**
32.08	Beryllophosphate-H	**BPH**
32.32	Barium Chloroaluminosilicate	**MER**
35.58	Tiptopite	**CAN**
35.73	Weinebeneite	**WEI**
36.15	Alpha Cristobalite	–
39.49	Alpha Quartz	–
43.51	Bicchulite	**SOD**

POWDER PATTERNS

ABW Li-ABW

CHEMICAL COMPOSITION: $|Li_4(H_2O)_4|$ $[Si_4Al_4O_{16}]$

REFINED COMPOSITION: $|Li_4(H_2O)_4|$ $[Si_4Al_4O_{16}]$

CRYSTAL DATA: $Pna2_1$ (No. 33)
$a = 10.313$ Å $b = 8.194$ Å $c = 4.993$ Å
$\alpha = 90°$ $\beta = 90°$ $\gamma = 90°$
X-ray single crystal refinement, $R_w = 0.081$

REFERENCE: E. Krogh Andersen and G. Ploug-Sørensen, *Z. Kristallogr.* **176** 67–73 (1986).

h	k	l	2θ	d	M	I_{rel}	h	k	l	2θ	d	M	I_{rel}	h	k	l	2θ	d	M	I_{rel}
1	1	0	13.80	6.416	4	88.5	0	0	2	35.97	2.497	2	34.7	3	3	0	42.26	2.139	4	0.5
2	0	0	17.20	5.156	2	20.6	4	1	0	36.53	2.459	4	3.3	1	2	2	43.34	2.088	8	0.7
2	1	0	20.35	4.364	4	17.5	2	3	0	37.25	2.414	4	0.1	0	4	0	44.21	2.049	2	6.6
0	1	1	20.83	4.264	4	59.3	0	3	1	37.53	2.396	4	9.0	1	4	0	45.12	2.009	4	0.4
0	2	0	21.69	4.097	2	4.3	1	3	1	38.57	2.334	8	5.5	5	1	0	45.34	2.000	4	1.9
2	1	1	27.14	3.286	8	5.9	3	2	1	38.65	2.329	8	11.3	4	2	1	45.36	2.000	8	0.4
2	2	0	27.81	3.208	4	0.1	1	1	2	38.70	2.327	8	6.0	2	2	2	46.07	1.970	8	0.2
3	1	0	28.15	3.170	4	94.6	4	0	1	39.33	2.291	4	0.3	3	3	1	46.18	1.966	8	2.2
1	2	1	29.50	3.028	8	100.0	2	0	2	40.13	2.247	4	5.0	3	1	2	46.29	1.961	8	9.3
2	2	1	33.19	2.699	8	2.7	4	1	1	40.90	2.206	8	3.7	2	4	0	47.77	1.904	4	0.7
3	1	1	33.48	2.676	8	0.5	4	2	0	41.38	2.182	4	2.5	4	3	0	48.56	1.875	4	1.2
1	3	0	33.95	2.640	4	4.1	2	3	1	41.56	2.173	8	7.9	1	4	1	48.86	1.864	8	0.9
3	2	0	34.04	2.633	4	1.1	2	1	2	41.68	2.167	8	1.3	5	2	0	49.47	1.842	4	1.3

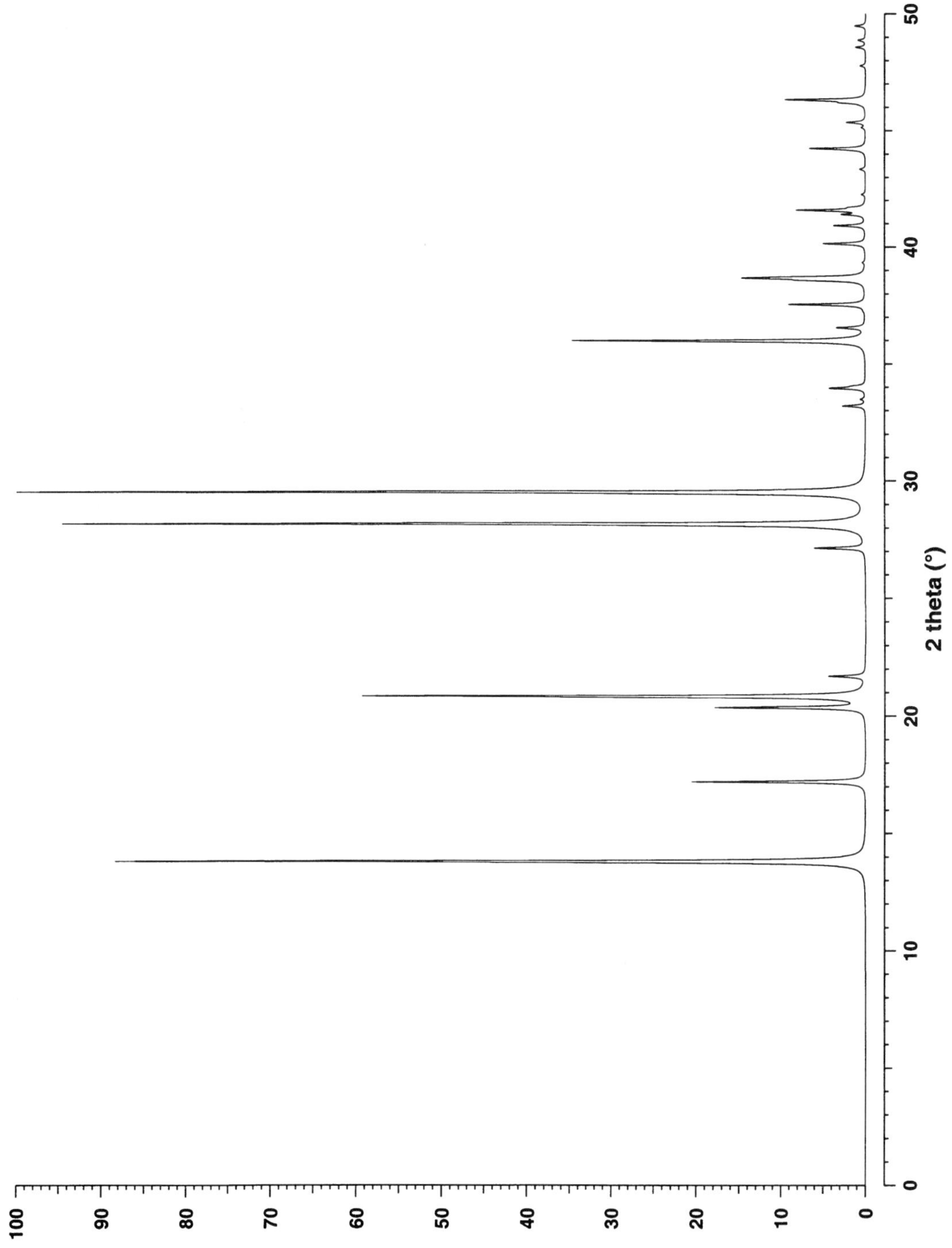
2 theta (°)
0
10
20
30
40
50
0
10
20
30
40
50
60
70
80
90
100

ACO ACP-1

CHEMICAL COMPOSITION: $|(NH_2(CH_2)_2NH_2)_4(H_2O)_2|$ $[Al_{0.88}Co_{7.12}P_8O_{32}]$
$NH_2(CH_2)_2NH_2$ = ethylenediamine

REFINED COMPOSITION: $|C_8N_8O_2|$ $[Co_8P_8O_{32}]$

CRYSTAL DATA: $I\bar{4}2m$ (No. 121)
$a = 10.240$ Å $b = 10.240$ Å $c = 9.652$ Å
$\alpha = 90.0°$ $\beta = 90.0°$ $\gamma = 90.0°$
X-ray single crystal refinement, $R_p = 0.0775$

REFERENCE: P. Feng, X. Bu and G. D. Stucky,
Nature **388** 735–741 (1997).

h	k	l	2θ	d	M	I_{rel}
1	1	0	12.22	7.241	4	43.6
1	0	1	12.60	7.024	8	100.0
2	0	0	17.32	5.120	4	15.6
0	0	2	18.38	4.826	2	7.4
2	1	1	21.48	4.137	16	0.2
1	1	2	22.14	4.016	8	16.5
2	2	0	24.59	3.620	4	1.1
2	0	2	25.36	3.512	8	6.6
3	1	0	27.54	3.238	8	36.4
3	0	1	27.72	3.218	8	46.5
1	0	3	29.09	3.069	8	26.4
2	2	2	30.88	2.896	8	10.6
3	2	1	32.87	2.725	16	8.4
3	1	2	33.32	2.689	16	8.7
2	1	3	34.06	2.633	16	14.7
4	0	0	35.05	2.560	4	1.0
3	3	0	37.25	2.414	4	7.1
4	1	1	37.39	2.405	16	4.5
3	0	3	38.45	2.341	8	5.2
1	1	4	39.36	2.289	8	1.0
4	0	2	39.86	2.262	8	0.3
3	3	2	41.85	2.159	8	2.0
3	2	3	42.45	2.129	16	2.9
4	2	2	43.76	2.069	16	0.4
2	2	4	45.16	2.008	8	0.5
4	3	1	45.26	2.003	16	2.8
4	1	3	46.17	1.966	16	4.8
3	1	4	46.96	1.935	16	3.5
5	2	1	48.81	1.866	16	1.1
5	1	2	49.14	1.854	16	2.3

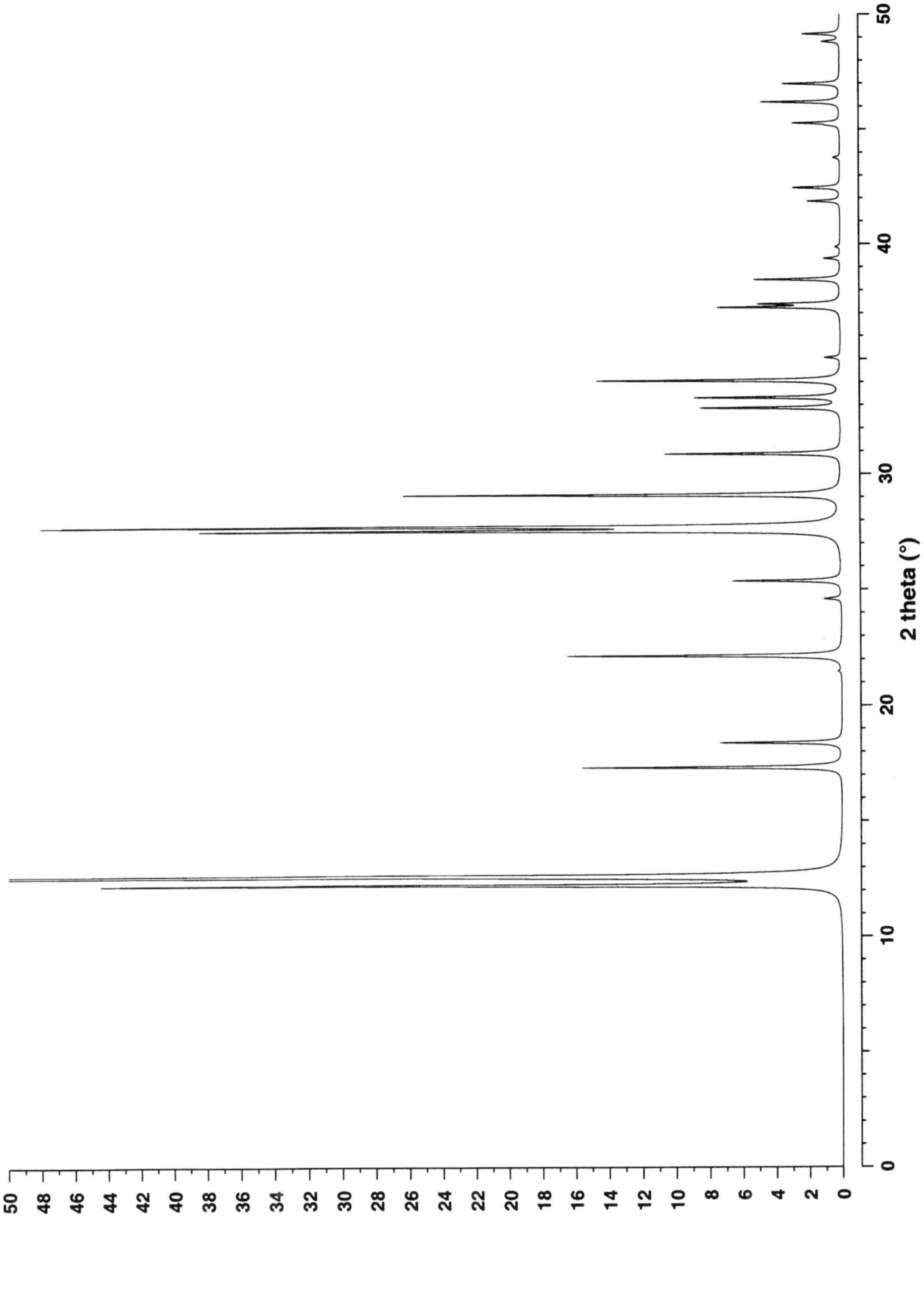
2 theta (°)

AEI AlPO–18, Calcined

CHEMICAL COMPOSITION: $[Al_{24}P_{24}O_{96}]$

REFINED COMPOSITION: $[Al_{24}P_{24}O_{96}]$

CRYSTAL DATA: $C1\,2/c\,1$ (No. 15) unique axis **b**, cell choice 1
a = 13.7114 Å b = 12.7315 Å c = 18.5706 Å
$\alpha = 90°$ $\beta = 90.01°$ $\gamma = 90°$
X-ray Rietveld refinement, $R_{exp} = 0.047$, $R_{wp} = 0.108$, $R_F = 0.032$

REFERENCE: A. Simmen, L. B. McCusker, Ch. Baerlocher and W. M. Meier, *Zeolites* **11** 654–661 (1991). Ch. Baerlocher, Private communication.

h	k	l	2θ	d	M	I_{rel}
1	1	0	9.48	9.330	4	100.0
0	0	2	9.52	9.285	2	57.9
-1	1	1	10.61	8.337	4	13.5
1	1	1	10.61	8.336	4	14.6
2	0	0	12.91	6.856	2	13.5
-1	1	2	13.45	6.582	4	2.4
1	1	2	13.45	6.581	4	2.8
0	2	0	13.91	6.366	2	1.9
0	2	1	14.71	6.022	4	0.4
-2	0	2	16.07	5.516	2	6.5
2	0	2	16.07	5.515	2	6.0
0	2	2	16.89	5.250	4	19.2
-1	1	3	17.19	5.158	4	9.3
1	1	3	17.19	5.158	4	9.5
2	2	0	19.02	4.665	4	3.0
0	0	4	19.12	4.643	2	0.9
-2	2	1	19.62	4.524	4	1.6
2	2	1	19.62	4.524	4	2.0
0	2	3	20.01	4.438	4	6.6
3	1	0	20.65	4.302	4	11.8
3	1	1	21.20	4.191	4	3.0
-2	2	2	21.31	4.169	4	0.9
2	2	2	21.32	4.168	4	5.6
-1	1	4	21.38	4.157	4	4.2
1	1	4	21.38	4.156	4	0.5
1	3	0	21.92	4.054	4	0.7
1	3	1	22.45	3.961	4	0.4
-3	1	2	22.78	3.903	4	0.3
-2	0	4	23.14	3.844	2	1.5
0	2	4	23.72	3.751	4	1.2
-2	2	3	23.88	3.726	4	4.4
2	2	3	23.89	3.725	4	4.6
-1	3	2	23.95	3.715	4	1.6
1	3	2	23.95	3.715	4	3.6
-3	1	3	25.21	3.533	4	0.4
3	1	3	25.21	3.532	4	0.9
-1	1	5	25.82	3.451	4	1.2
1	1	5	25.82	3.451	4	0.9
4	0	0	25.99	3.428	2	3.8
-1	3	3	26.28	3.392	4	3.3
1	3	3	26.28	3.391	4	2.8
-2	2	4	27.10	3.291	4	0.4
2	2	4	27.10	3.290	4	0.4
-4	0	2	27.74	3.216	2	0.2
4	0	2	27.74	3.216	2	0.4

h	k	l	2θ	d	M	I_{rel}
0	2	5	27.81	3.208	4	5.9
0	4	0	28.03	3.183	2	0.8
-1	3	4	29.24	3.054	4	0.6
1	3	4	29.25	3.054	4	0.5
4	2	0	29.60	3.018	4	1.0
0	4	2	29.67	3.011	4	0.4
-4	2	1	29.99	2.979	4	1.4
4	2	1	30.00	2.979	4	2.2
-3	3	2	30.31	2.949	4	1.4
3	3	2	30.31	2.949	4	0.6
-1	1	6	30.43	2.938	4	1.3
1	1	6	30.43	2.938	4	1.9
-2	2	5	30.77	2.906	4	1.5
2	2	5	30.77	2.905	4	1.2
2	4	0	30.98	2.887	4	2.6
-4	2	2	31.16	2.870	4	5.1
4	2	2	31.16	2.870	4	3.7
2	4	1	31.36	2.853	4	0.2
0	4	3	31.61	2.831	4	1.8
2	0	6	31.72	2.821	2	0.4
0	2	6	32.16	2.784	4	4.2
-3	3	3	32.21	2.779	4	1.0
3	3	3	32.21	2.779	4	0.9
-4	0	4	32.46	2.758	2	0.5
4	0	4	32.47	2.757	2	0.4
-2	4	2	32.48	2.757	4	0.5
2	4	2	32.48	2.757	4	0.5
-1	3	5	32.70	2.739	4	1.7
1	3	5	32.70	2.739	4	2.5
-4	2	3	33.02	2.713	4	0.9
4	2	3	33.02	2.713	4	0.9
-5	1	1	33.78	2.653	4	0.5
5	1	1	33.78	2.653	4	0.5
-2	4	3	34.27	2.616	4	0.4
2	4	3	34.27	2.616	4	0.2
3	3	4	34.72	2.584	4	0.4
-2	2	6	34.78	2.579	4	0.3
5	1	2	34.83	2.575	4	0.4
-4	2	4	35.47	2.531	4	0.4
1	5	0	35.87	2.503	4	0.9
-1	5	1	36.20	2.481	4	0.2
1	5	1	36.20	2.481	4	0.3
-5	1	3	36.52	2.460	4	0.5
5	1	3	36.53	2.460	4	0.5
0	2	7	36.70	2.449	4	0.3

h	k	l	2θ	d	M	I_{rel}
-3	3	5	37.72	2.385	4	0.2
4	4	0	38.60	2.332	4	0.3
0	0	8	38.79	2.321	2	0.3
-4	4	1	38.91	2.314	4	0.3
4	4	1	38.92	2.314	4	0.2
-2	2	7	39.06	2.306	4	0.5
5	3	0	39.11	2.303	4	0.3
-5	3	1	39.42	2.286	4	0.5
5	3	1	39.42	2.286	4	0.3
-5	3	2	40.34	2.236	4	0.2
5	3	2	40.35	2.235	4	0.2
-3	3	6	41.14	2.194	4	0.4
3	3	6	41.15	2.194	4	0.6
0	2	8	41.40	2.181	4	1.0
-4	2	6	41.80	2.161	4	0.2
2	4	6	42.84	2.111	4	0.2
0	6	1	42.90	2.108	4	0.4
-6	2	2	43.17	2.095	4	0.5
-3	5	3	43.22	2.093	4	0.7
3	5	3	43.22	2.093	4	0.5
1	5	5	43.60	2.076	4	0.3
5	3	4	43.88	2.063	4	0.2
-3	1	8	44.34	2.043	4	0.3
3	1	8	44.34	2.043	4	0.2
5	1	6	44.72	2.026	4	0.2
1	3	8	45.00	2.014	4	0.2
-5	3	5	46.38	1.958	4	0.2
5	3	5	46.39	1.957	4	0.2
-6	2	4	46.53	1.952	4	0.2
7	1	0	46.93	1.936	4	0.2
7	1	1	47.20	1.926	4	0.3
7	1	2	48.00	1.895	4	0.2
5	5	0	48.80	1.866	4	0.9
4	4	6	48.90	1.863	4	0.5
-3	3	8	48.96	1.860	4	0.2
0	0	10	49.05	1.857	2	1.1
-5	5	1	49.07	1.857	4	0.4
5	5	1	49.07	1.857	4	0.5
-5	3	6	49.31	1.848	4	0.4
-7	1	3	49.32	1.848	4	0.3
7	1	3	49.32	1.848	4	0.3
5	3	6	49.32	1.848	4	0.2
0	6	5	49.47	1.842	4	0.3
-5	5	2	49.84	1.829	4	0.2
5	5	2	49.85	1.829	4	0.4

AlPO–18, Calcined

AEI

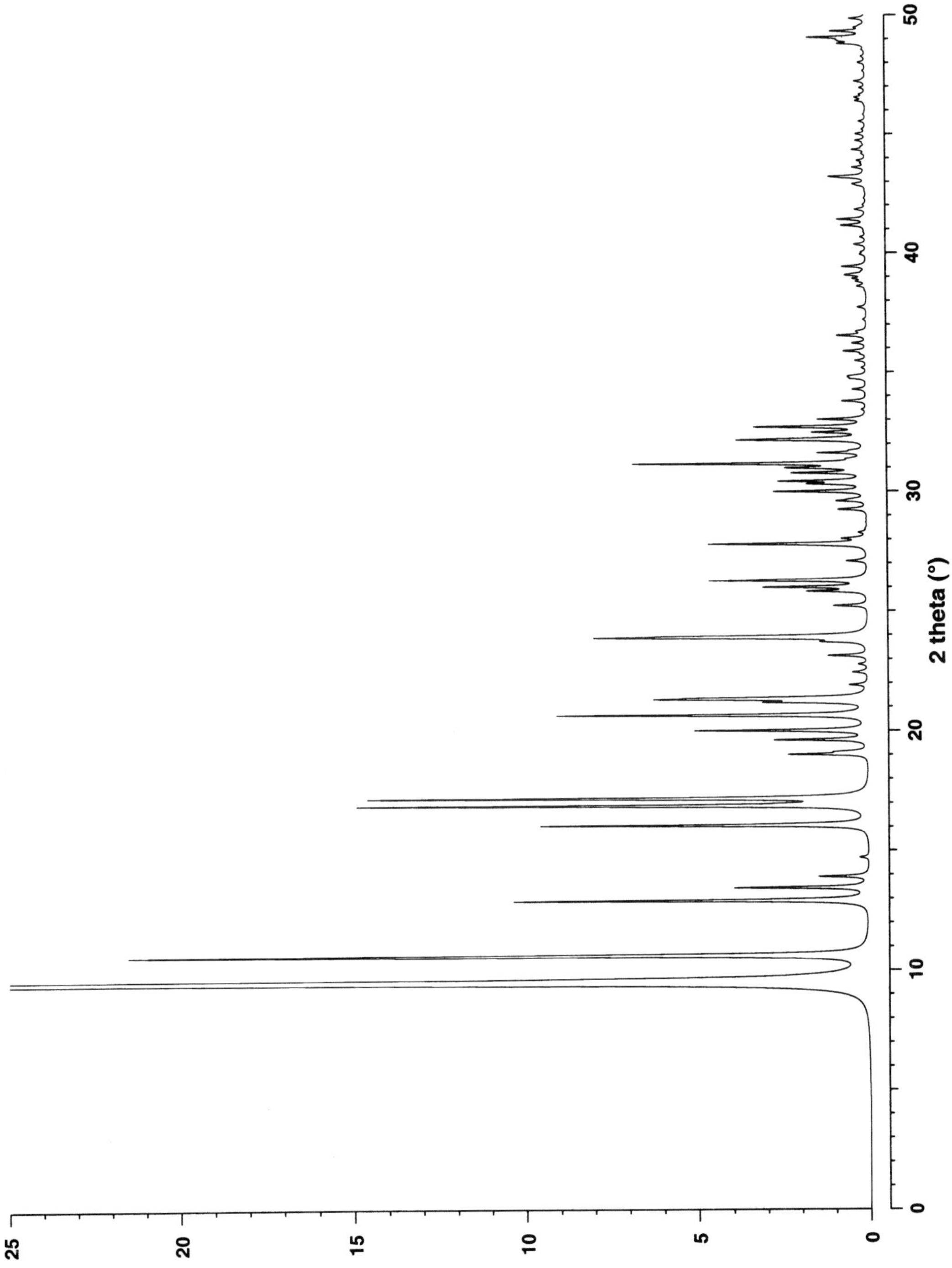

AEL AlPO–11, Calcined

CHEMICAL COMPOSITION: $[Al_{20}P_{20}O_{80}]$

REFINED COMPOSITION: $[Al_{20}P_{20}O_{80}]$

CRYSTAL DATA: *Ibm*2 (No. 46) **ba$\bar{\mathbf{c}}$** setting
$a = 13.5336$ Å $b = 18.4821$ Å $c = 8.3703$ Å
$\alpha = 90°$ $\beta = 90°$ $\gamma = 90°$
Neutron Rietveld refinement, $R_{wp} = 0.027$, $R_{F^2} = 0.058$

REFERENCE: J. W. Richardson, Jr., J. J. Pluth and J. V. Smith, *Acta Cryst.* **B44** 367–373 (1988).

h	k	l	2θ	d	M	I_{rel}
1	1	0	8.10	10.919	4	100.0
0	2	0	9.57	9.241	2	82.1
2	0	0	13.08	6.767	2	18.9
1	3	0	15.80	5.607	4	35.6
2	2	0	16.23	5.460	4	1.3
0	4	0	19.21	4.621	2	0.6
3	1	0	20.26	4.383	4	17.7
0	0	2	21.23	4.185	2	71.4
2	3	1	22.22	4.001	8	26.2
3	0	1	22.39	3.971	4	18.9
1	1	2	22.75	3.908	8	5.0
1	4	1	22.95	3.876	8	23.9
2	4	0	23.31	3.816	4	26.6
0	2	2	23.33	3.812	4	6.2
3	2	1	24.40	3.649	8	1.7
3	3	0	24.46	3.640	4	1.7
1	5	0	24.97	3.566	4	0.9
2	0	2	25.02	3.559	4	1.9
4	0	0	26.34	3.383	2	11.3
1	3	2	26.58	3.354	8	5.9
2	2	2	26.84	3.322	8	0.2
4	2	0	28.08	3.177	4	2.0
0	4	2	28.78	3.102	4	0.2
4	1	1	28.87	3.093	8	2.3
0	6	0	28.99	3.080	2	7.0
3	1	2	29.51	3.027	8	2.7
2	5	1	29.53	3.025	8	0.3
3	4	1	29.66	3.012	8	1.2
3	5	0	31.28	2.859	4	0.5
1	6	1	31.65	2.827	8	1.2
2	4	2	31.73	2.820	8	5.6
2	6	0	31.92	2.804	4	0.9
4	3	1	32.02	2.795	8	0.6
3	3	2	32.60	2.746	8	0.2
4	4	0	32.81	2.730	4	10.6
4	0	2	34.07	2.631	4	2.6
1	7	0	34.61	2.591	4	4.6
5	0	1	34.83	2.575	4	0.1
4	2	2	35.47	2.531	8	0.8
0	6	2	36.21	2.481	4	1.8
5	3	0	36.25	2.478	4	4.3
3	6	1	36.93	2.434	8	0.7
4	5	1	37.61	2.392	8	0.1
2	3	3	37.81	2.379	8	4.8
3	0	3	37.92	2.373	4	4.5
3	5	2	38.12	2.361	8	0.1
2	7	1	38.13	2.360	8	0.9
1	4	3	38.26	2.352	8	5.7
3	2	3	39.20	2.298	8	0.4
4	4	2	39.41	2.286	8	2.8
3	7	0	39.55	2.279	4	0.6
4	6	0	39.56	2.278	4	0.8
6	0	0	39.97	2.256	2	0.2
5	4	1	40.08	2.250	8	0.2
1	7	2	40.96	2.203	8	0.3
5	5	0	41.34	2.184	4	0.3
6	1	1	41.76	2.163	8	0.7
4	1	3	42.27	2.138	8	0.8
3	4	3	42.84	2.111	8	0.8
0	0	4	43.23	2.093	2	3.0
1	1	4	44.06	2.055	8	0.1
1	6	3	44.31	2.044	8	0.7
0	2	4	44.39	2.041	4	0.4
6	4	0	44.71	2.027	4	2.3
4	7	1	44.87	2.020	8	2.0
3	7	2	45.31	2.001	8	0.3
4	6	2	45.33	2.001	8	1.0
2	0	4	45.36	1.999	4	0.1
3	8	1	45.42	1.997	8	0.6
6	0	2	45.69	1.986	4	0.5
1	3	4	46.31	1.960	8	0.4
5	0	3	46.76	1.943	4	0.1
5	5	2	46.93	1.936	8	0.4
5	2	3	47.84	1.901	8	0.1
5	7	0	48.14	1.890	4	0.3
3	1	4	48.19	1.888	8	0.1
7	0	1	48.31	1.884	4	1.2
3	6	3	48.42	1.880	8	0.3
6	5	1	48.51	1.876	8	0.1
3	9	0	48.72	1.869	4	0.3
7	2	1	49.37	1.846	8	0.1
2	7	3	49.39	1.845	8	1.0
7	3	0	49.41	1.845	4	0.6
2	4	4	49.69	1.835	8	0.3

AlPO–11, Calcined AEL

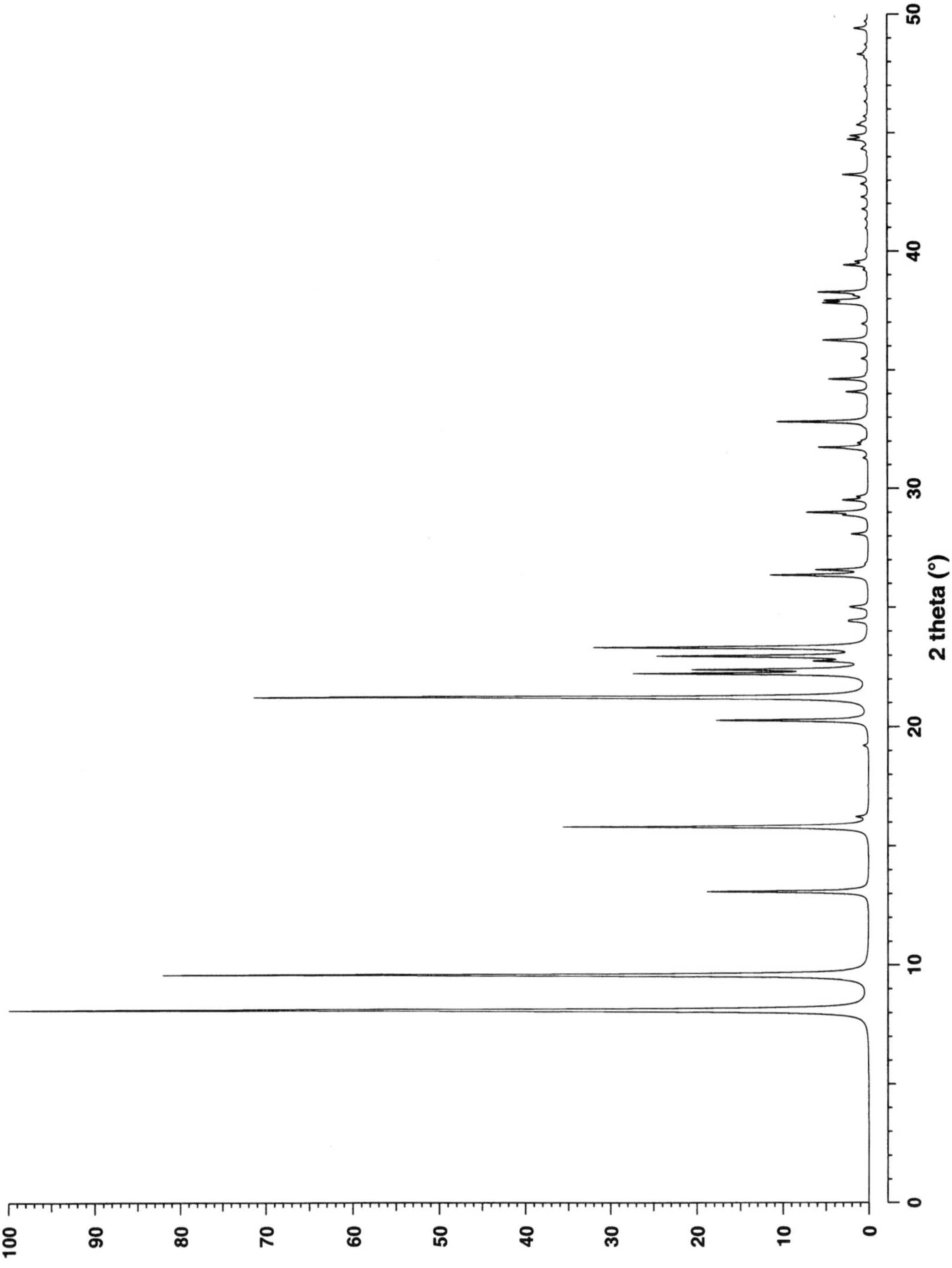

AEL Di-isopropylamine MnAPO-11

CHEMICAL COMPOSITION: $|(C_6H_{16}N)_2|$ $[Mn_2Al_{18}P_{20}O_{80}]$
$C_6H_{16}N$ = di-isopropylamine

REFINED COMPOSITION: $|C_{15.5}N_{2.58}|$ $[Al_{20.54}P_{18.22}O_{80}]$

CRYSTAL DATA: $Ibm2$ (No. 46) **ba$\bar{\textbf{c}}$** setting
$a = 13.472$ Å $b = 18.712$ Å $c = 8.4431$ Å
$\alpha = 90°$ $\beta = 90°$ $\gamma = 90°$
X-ray single crystal refinement, $R_w = 0.042$

REFERENCE: J. J. Pluth, J. V. Smith and J. W. Richardson, Jr., *J. Phys. Chem.* **92** 2734–2738 (1988).

h	k	l	2θ	d	M	I_{rel}
1	1	0	8.09	10.933	4	29.6
0	2	0	9.45	9.356	2	41.6
2	0	0	13.14	6.736	2	14.9
1	3	0	15.66	5.660	4	31.0
2	2	0	16.21	5.467	4	3.7
2	1	1	17.50	5.069	8	0.2
0	4	0	18.97	4.678	2	5.2
3	1	0	20.34	4.367	4	36.2
0	0	2	21.04	4.222	2	100.0
2	3	1	22.09	4.024	8	44.4
3	0	1	22.42	3.965	4	31.6
1	1	2	22.58	3.938	8	11.8
1	4	1	22.71	3.915	8	41.0
0	2	2	23.11	3.848	4	12.5
2	4	0	23.15	3.842	4	60.4
3	2	1	24.38	3.651	8	3.2
3	3	0	24.42	3.644	4	0.7
1	5	0	24.69	3.606	4	7.3
2	0	2	24.89	3.577	4	4.0
1	3	2	26.34	3.384	8	14.2
4	0	0	26.46	3.368	2	16.5
2	2	2	26.68	3.341	8	0.3
4	2	0	28.16	3.169	4	2.8
0	4	2	28.48	3.134	4	0.2
0	6	0	28.62	3.119	2	13.7
4	1	1	28.94	3.085	8	3.2
2	5	1	29.28	3.050	8	0.5
3	1	2	29.43	3.035	8	6.4
3	4	1	29.53	3.025	8	1.3
3	5	0	31.11	2.875	4	0.4
1	6	1	31.29	2.859	8	0.9
2	4	2	31.48	2.842	8	7.8
2	6	0	31.61	2.830	4	0.4
4	3	1	32.01	2.796	8	0.6
3	3	2	32.45	2.759	8	0.1
4	4	0	32.76	2.733	4	17.1
5	1	0	33.60	2.667	4	1.0
4	0	2	34.05	2.633	4	2.8
1	7	0	34.20	2.622	4	6.4
5	0	1	34.95	2.567	4	0.3
4	2	2	35.42	2.534	8	1.5
0	6	2	35.80	2.508	4	2.1
5	2	1	36.29	2.475	8	0.1
5	3	0	36.32	2.473	4	6.0
3	6	1	36.66	2.451	8	1.1
4	5	1	37.47	2.400	8	0.5
2	3	3	37.52	2.397	8	9.8
3	0	3	37.72	2.385	4	6.5
2	7	1	37.74	2.384	8	2.0
1	4	3	37.90	2.374	8	8.9
2	6	2	38.29	2.351	8	0.2
0	8	0	38.49	2.339	2	0.3
3	2	3	38.97	2.311	8	0.9
3	7	0	39.22	2.297	4	0.6
4	4	2	39.27	2.294	8	2.4
4	6	0	39.37	2.288	4	0.3
5	4	1	40.07	2.250	8	0.1
6	0	0	40.16	2.245	2	0.2
1	7	2	40.50	2.227	8	0.9
1	8	1	40.58	2.223	8	0.5
2	8	0	40.84	2.210	4	0.4
5	5	0	41.29	2.187	4	0.3
6	1	1	41.91	2.155	8	1.0
4	1	3	42.12	2.145	8	0.9
2	5	3	42.36	2.134	8	0.2
3	4	3	42.55	2.125	8	0.5
0	0	4	42.84	2.111	2	4.7
1	1	4	43.67	2.073	8	0.1
1	6	3	43.85	2.065	8	0.7
0	2	4	43.97	2.059	4	0.3
0	8	2	44.27	2.046	4	0.4
4	3	3	44.39	2.041	8	0.3
4	7	1	44.58	2.032	8	3.0
6	4	0	44.77	2.024	4	3.1
3	7	2	44.92	2.018	8	0.6
3	8	1	45.00	2.015	8	0.6
2	0	4	45.01	2.014	4	0.1
4	6	2	45.06	2.012	8	0.6
6	0	2	45.77	1.982	4	1.2
1	3	4	45.88	1.978	8	0.4
2	8	2	46.38	1.958	8	0.1
5	0	3	46.67	1.946	4	0.1
5	5	2	46.79	1.942	8	0.9
6	2	2	46.85	1.939	8	0.1
2	9	1	46.99	1.934	8	0.2
3	1	4	47.86	1.900	8	0.2
5	7	0	47.94	1.898	4	0.5
3	6	3	48.03	1.894	8	0.6
3	9	0	48.23	1.887	4	0.1
6	5	1	48.49	1.877	8	0.3
7	0	1	48.51	1.876	4	1.8
0	10	0	48.66	1.871	2	0.1
4	5	3	48.68	1.871	8	0.2
2	7	3	48.90	1.863	8	1.3
2	4	4	49.25	1.850	8	0.5
7	2	1	49.54	1.840	8	0.2
7	3	0	49.57	1.839	4	0.9
6	4	2	49.97	1.825	8	0.7

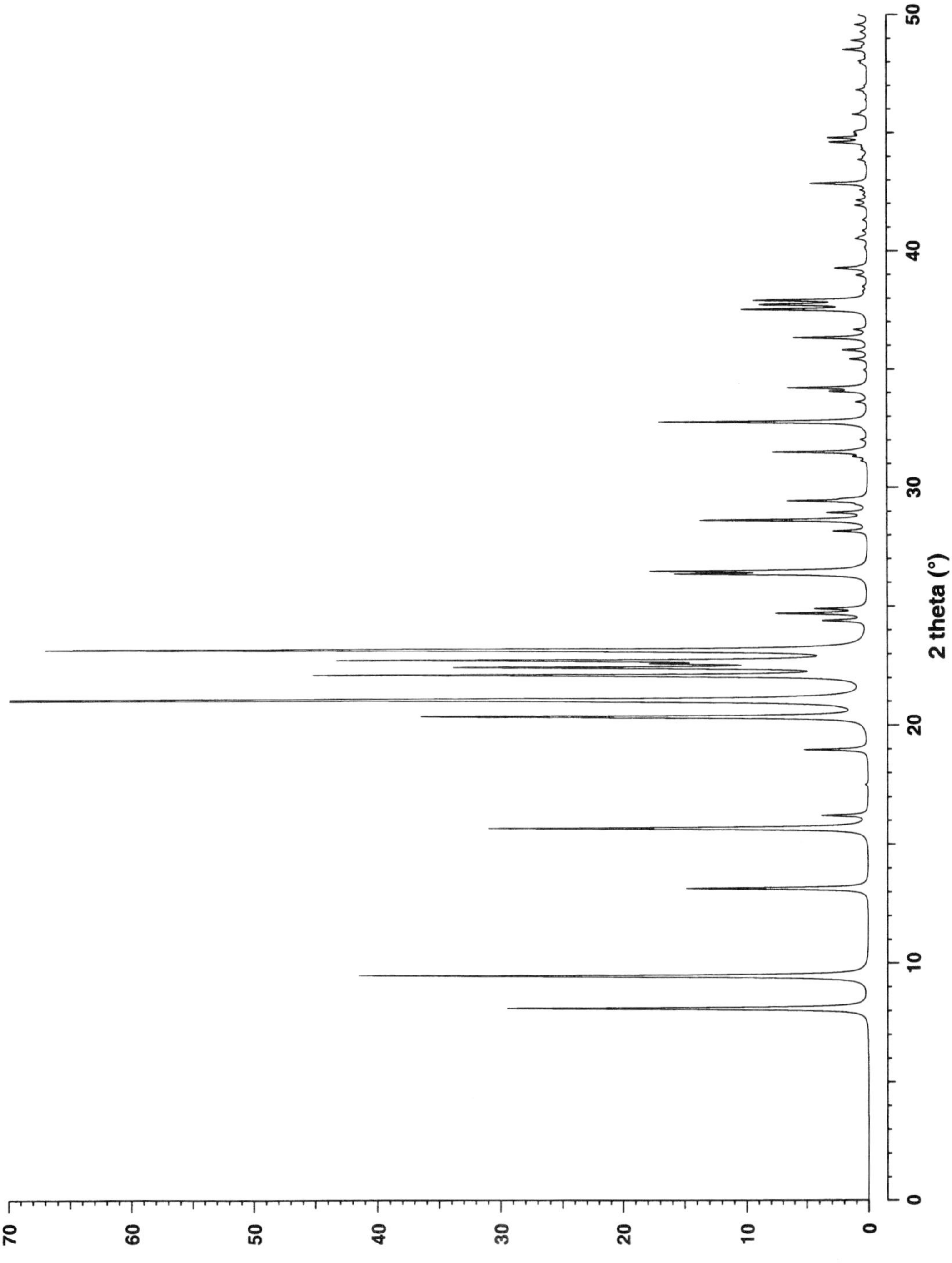
2 theta (°)
0
10
20
30
40
50
0
10
20
30
40
50
60
70

AEN

AlPO-EN3

CHEMICAL COMPOSITION: $|(C_2H_8N_2)_4(H_2O)_{16}|\ [Al_{24}P_{24}O_{96}]$
$C_2H_8N_2$ = ethylene diamine

REFINED COMPOSITION: $|(C_8N_8O_8)(H_2O)_8|\ [P_{24}Al_{24}O_{96}]$

CRYSTAL DATA: $P2_12_12_1$ (No. 19)
$a = 10.321$ Å $b = 13.631$ Å $c = 17.454$ Å
$\alpha = 90°$ $\beta = 90°$ $\gamma = 90°$
X-ray single crystal refinement. $R_p = 0.140$, $R_{wp} = 0.182$, $R_{F^2} = 0.103$

REFERENCE: R. M. Kirchner, R. W. Grosse-Kunstleve, J. J. Pluth, S. T. Wilson, R. W. Broach and J. V. Smith, *Microporous and Mesoporous Materials* **39** 319–332 (2000).

h	k	l	2θ	d	M	I_{rel}
0	1	1	8.23	10.743	4	0.5
0	0	2	10.14	8.727	2	100.0
1	1	0	10.75	8.228	4	2.0
1	1	1	11.89	7.443	8	14.8
0	1	2	12.04	7.350	4	1.3
0	2	0	12.99	6.815	2	49.0
1	1	2	14.80	5.987	8	1.2
1	2	0	15.58	5.687	4	13.0
1	2	1	16.39	5.408	8	1.1
0	2	2	16.50	5.372	4	11.3
2	0	0	17.18	5.161	2	4.6
1	2	2	18.62	4.765	8	22.1
1	1	3	18.68	4.750	8	16.3
2	1	1	19.08	4.652	8	68.2
2	0	2	19.99	4.442	4	2.1
0	0	4	20.35	4.364	2	0.5
0	1	4	21.38	4.156	4	0.7
0	3	2	22.06	4.030	4	2.7
2	2	1	22.20	4.004	8	4.2
2	0	3	23.04	3.861	4	10.6
1	1	4	23.07	3.855	8	1.2
1	3	2	23.70	3.754	8	0.9
2	2	2	23.91	3.721	8	3.3
2	1	3	23.96	3.715	8	98.8
0	2	4	24.22	3.675	4	35.8
1	2	4	25.73	3.462	8	1.8
0	4	0	26.15	3.408	2	44.7
1	3	3	26.34	3.383	8	10.2
2	2	3	26.53	3.359	8	0.9
2	3	1	26.63	3.347	8	23.2
3	1	1	27.22	3.276	8	3.1
2	1	4	27.56	3.237	8	0.4
1	4	0	27.56	3.236	4	3.5
1	1	5	27.76	3.214	8	30.0
1	4	1	28.04	3.182	8	4.4
0	4	2	28.11	3.174	4	47.5
3	1	2	28.65	3.116	8	4.2
3	2	0	29.07	3.071	4	4.9
1	4	2	29.44	3.034	8	26.7
3	2	1	29.53	3.025	8	0.5
1	3	4	29.68	3.010	8	1.4
1	2	5	30.04	2.975	8	2.9
2	3	3	30.38	2.942	8	2.6
0	0	6	30.73	2.909	2	0.8
3	2	2	30.86	2.897	8	5.3
3	1	3	30.90	2.894	8	16.5
0	1	6	31.44	2.845	4	0.6
2	1	5	31.63	2.828	8	4.2
2	4	1	31.88	2.807	8	1.1
1	1	6	32.65	2.743	8	5.9
3	3	1	33.06	2.710	8	1.7
2	3	4	33.35	2.687	8	2.3
0	4	4	33.36	2.686	4	1.0
0	2	6	33.49	2.675	4	2.8
2	2	5	33.67	2.662	8	0.9
3	1	4	33.82	2.650	8	10.4
1	5	0	34.01	2.636	4	0.3
3	3	2	34.27	2.617	8	1.0
1	4	4	34.51	2.599	8	1.0
1	2	6	34.63	2.590	8	10.2
4	0	0	34.77	2.580	2	16.1
2	4	3	35.12	2.555	8	0.3
4	0	1	35.16	2.553	4	0.9
3	2	4	35.75	2.512	8	0.3
4	1	1	35.79	2.509	8	0.6
2	1	6	36.05	2.491	8	0.5
4	0	2	36.31	2.474	4	3.3
2	3	5	36.85	2.439	8	6.7
3	1	5	37.28	2.412	8	1.8
1	5	3	37.46	2.401	8	0.4
4	2	1	37.63	2.390	8	1.2
2	5	1	37.67	2.388	8	3.5
1	1	7	37.70	2.386	8	8.4
2	2	6	37.88	2.375	8	3.0
1	4	5	37.91	2.373	8	0.7
2	5	2	38.75	2.324	8	0.9
3	3	4	38.78	2.322	8	1.8
3	2	5	39.06	2.306	8	1.1
0	6	0	39.67	2.272	2	0.5
1	5	4	39.96	2.256	8	0.8
4	2	3	40.47	2.229	8	0.3
2	5	3	40.51	2.227	8	4.4
4	3	1	40.54	2.225	8	1.6
4	0	4	40.62	2.221	4	0.7
1	6	0	40.66	2.219	4	1.9
2	1	7	40.73	2.215	8	0.9
2	3	6	40.77	2.213	8	1.3
0	4	6	40.78	2.212	4	1.8
1	6	1	41.01	2.201	8	0.6
0	6	2	41.05	2.199	4	0.8
4	3	2	41.56	2.173	8	0.6
1	4	6	41.75	2.163	8	0.9
1	3	7	42.26	2.138	8	5.0
3	5	1	42.63	2.121	8	4.2
3	4	4	42.71	2.117	8	1.2
4	2	4	42.82	2.112	8	2.5
1	1	8	42.88	2.109	8	2.3
3	5	2	43.61	2.075	8	0.4
1	6	3	43.66	2.073	8	0.5
4	1	5	44.15	2.051	8	1.7
1	2	8	44.48	2.037	8	2.9
2	4	6	44.56	2.033	8	0.4
5	1	1	44.70	2.027	8	3.8
2	6	2	44.81	2.023	8	0.7
0	6	4	44.99	2.015	4	3.1
2	3	7	45.04	2.013	8	1.1
3	5	3	45.21	2.006	8	0.4
4	4	2	45.29	2.002	8	4.9
3	1	7	45.41	1.997	8	1.1
3	4	5	45.60	1.989	8	1.0
0	5	6	45.60	1.989	4	0.3
2	1	8	45.63	1.988	8	0.6
5	1	2	45.65	1.987	8	2.3
4	2	5	45.71	1.985	8	0.4
2	5	5	45.74	1.984	8	2.5
1	6	4	45.88	1.978	8	3.3
5	2	1	46.25	1.963	8	0.4
1	5	6	46.49	1.953	8	1.3
3	2	7	46.94	1.936	8	0.5
4	0	6	47.08	1.930	4	1.6
2	2	8	47.15	1.927	8	0.8
5	2	2	47.17	1.927	8	1.7
5	1	3	47.19	1.926	8	2.4
3	5	4	47.37	1.919	8	0.5
1	7	1	47.82	1.902	8	1.6
1	1	9	48.21	1.888	8	3.0
4	3	5	48.21	1.887	8	0.8
2	6	4	48.50	1.877	8	1.0
1	6	5	48.62	1.873	8	1.2
3	4	6	48.95	1.861	8	2.3
4	4	4	48.95	1.861	8	3.2
4	2	6	49.05	1.857	8	1.9
3	6	2	49.18	1.853	8	0.3
0	4	8	49.61	1.837	4	1.2
1	2	9	49.67	1.836	8	0.8

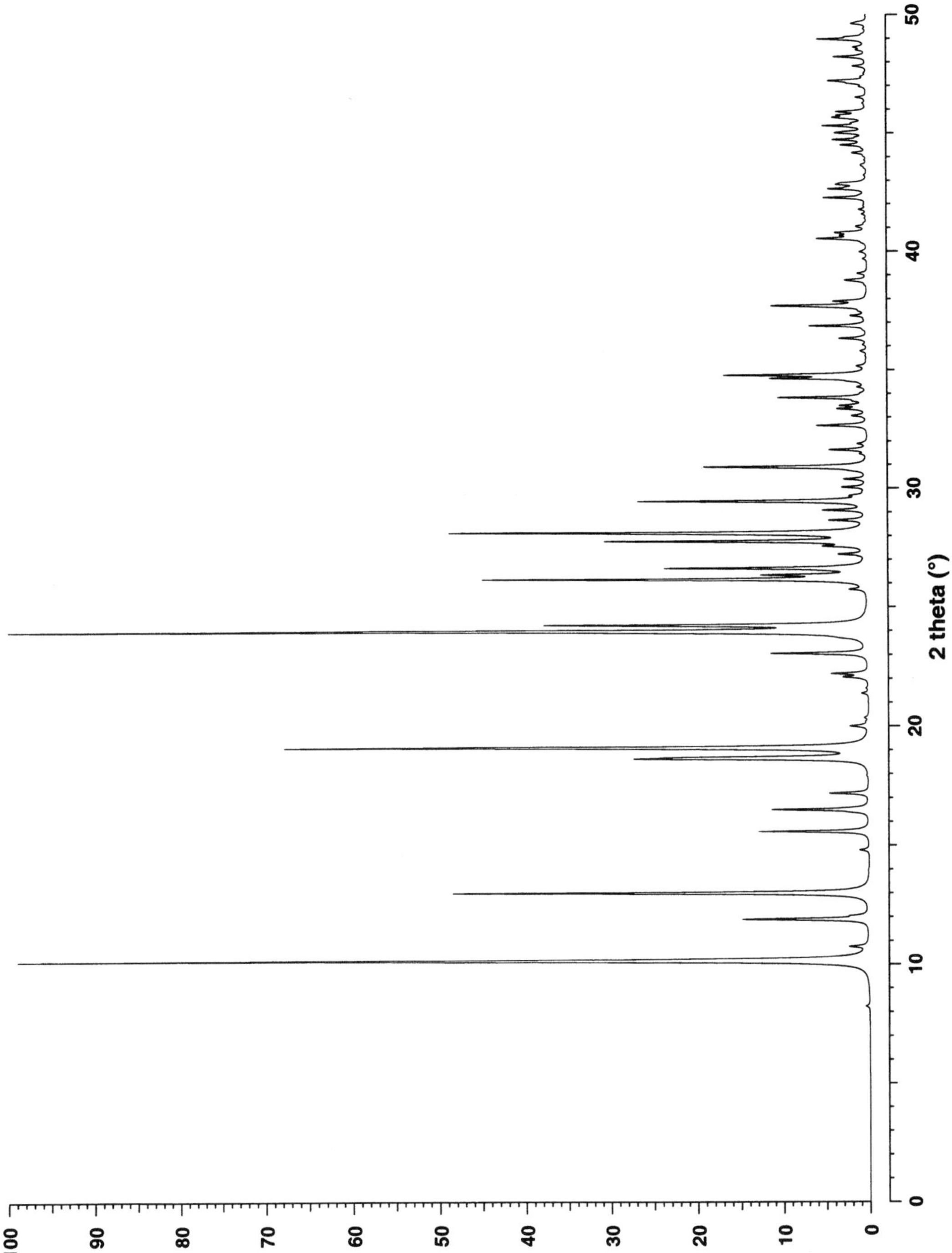

0
10
20
30
40
50
2 theta (°)
100
90
80
70
60
50
40
30
20
10
0

AET

AlPO-8

CHEMICAL COMPOSITION: $[Al_{36}P_{36}O_{144}]$

REFINED COMPOSITION: $[Si_{72}O_{144}]$

CRYSTAL DATA: *Cmcm* (No. 63)
a = 33.2900 Å b = 14.7036 Å c = 8.3863 Å
α = 90° β = 90° γ = 90°
Neutron Rietveld refinement, R_{wp} = 0.043, R_{F^2} = 0.077

REFERENCE: J. W. Richardson, Jr. and E. T. C. Vogt, *Zeolites* **12** 13–19 (1992).

h	k	l	2θ	d	M	I_{rel}
2	0	0	5.31	16.645	2	100.0
1	1	0	6.57	13.450	4	94.7
3	1	0	9.99	8.857	4	17.1
0	2	0	12.04	7.352	2	0.2
2	2	0	13.16	6.725	4	0.1
3	1	1	14.55	6.090	8	0.1
5	1	0	14.60	6.065	4	6.9
6	0	0	15.97	5.548	2	0.5
4	2	0	16.09	5.510	4	7.6
1	3	0	18.30	4.849	4	4.7
4	2	1	19.27	4.605	8	0.3
7	1	0	19.62	4.525	4	2.9
6	2	0	20.05	4.429	4	6.0
1	3	1	21.16	4.198	8	4.3
0	0	2	21.19	4.193	2	27.9
8	0	0	21.35	4.161	2	0.5
2	0	2	21.86	4.066	4	2.6
1	1	2	22.21	4.003	8	5.0
7	1	1	22.32	3.982	8	5.7
3	3	1	22.49	3.954	8	7.3
5	3	0	22.53	3.947	4	0.2
6	2	1	22.71	3.916	8	10.1
3	1	2	23.47	3.790	8	0.9
0	4	0	24.21	3.676	2	4.4
0	2	2	24.44	3.642	4	0.2
8	2	0	24.58	3.621	4	0.8
9	1	0	24.82	3.587	4	5.9
5	3	1	24.93	3.571	8	0.2
2	2	2	25.03	3.558	8	0.1
5	1	2	25.83	3.449	8	0.7
0	4	1	26.47	3.367	4	0.2
4	4	0	26.51	3.363	4	3.7
4	2	2	26.72	3.337	8	1.2
10	0	0	26.78	3.329	2	0.4
8	2	1	26.81	3.325	8	0.4
9	1	1	27.03	3.298	8	0.3
1	3	2	28.13	3.172	8	1.1
7	3	1	28.23	3.161	8	0.7
4	4	1	28.60	3.121	8	0.2
7	1	2	29.03	3.076	8	0.8
6	4	0	29.14	3.064	4	2.5
6	2	2	29.33	3.045	8	1.3
10	2	0	29.45	3.033	4	3.0
11	1	0	30.15	2.964	4	0.2
8	0	2	30.26	2.954	4	0.4
6	4	1	31.07	2.878	8	0.3
3	5	0	31.47	2.843	4	2.1
11	1	1	32.02	2.795	8	0.2
12	0	0	32.27	2.774	2	0.8
0	4	2	32.39	2.764	4	0.8
1	1	3	32.72	2.737	8	0.1
9	1	2	32.86	2.726	8	1.4
3	5	1	33.28	2.692	8	0.2
5	5	0	33.31	2.690	4	1.0
4	4	2	34.18	2.623	8	0.6
8	4	1	34.26	2.617	8	0.3
10	0	2	34.40	2.607	4	0.6
2	2	3	34.75	2.581	8	0.2
13	1	0	35.59	2.523	4	1.5
7	5	0	35.90	2.501	4	0.2
6	4	2	36.31	2.474	8	0.7
10	4	0	36.41	2.468	4	0.4
10	2	2	36.57	2.457	8	0.8
0	6	0	36.67	2.451	2	0.6
2	6	0	37.08	2.424	4	0.7
1	3	3	37.12	2.422	8	1.2
7	5	1	37.52	2.397	8	0.2
7	1	3	37.83	2.378	8	1.6
3	3	3	37.93	2.372	8	1.9
10	4	1	38.01	2.367	8	0.2
6	2	3	38.07	2.364	8	1.6
3	5	2	38.25	2.353	8	0.3
0	6	1	38.26	2.352	4	0.2
9	5	0	39.13	2.302	4	0.1
5	3	3	39.50	2.281	8	0.6
13	3	0	39.71	2.270	4	0.3
5	5	2	39.81	2.264	8	0.2
6	6	0	40.23	2.242	4	0.2
0	4	3	40.54	2.225	4	0.2
9	1	3	40.93	2.205	8	0.7
15	1	0	41.13	2.194	4	0.1
13	3	1	41.20	2.191	8	0.3
7	3	3	41.77	2.163	8	1.0
13	1	2	41.79	2.162	8	0.1
4	4	3	42.03	2.150	8	0.1
12	4	1	42.21	2.141	8	0.2
8	6	0	42.82	2.112	4	0.1
2	6	2	43.10	2.099	8	0.2
0	0	4	43.15	2.097	2	1.7
16	0	0	43.49	2.081	2	0.3
1	1	4	43.69	2.072	8	0.4
3	7	0	43.87	2.064	4	0.2
4	6	2	44.17	2.051	8	0.3
11	5	1	44.28	2.045	8	0.1
3	1	4	44.40	2.040	8	0.1
9	5	2	44.92	2.018	8	0.2
5	7	0	45.27	2.003	4	0.3
13	3	2	45.44	1.996	8	0.8
6	6	2	45.90	1.977	8	0.4
15	3	1	46.19	1.965	8	0.2
8	4	3	46.27	1.962	8	0.2
12	4	2	46.37	1.958	8	0.4
16	2	1	46.64	1.947	8	0.3
14	4	1	46.77	1.942	8	0.2
17	1	0	46.80	1.941	4	0.3
13	5	0	47.05	1.931	4	0.8
1	3	4	47.23	1.924	8	0.1
7	7	0	47.31	1.921	4	0.1
10	6	1	47.32	1.921	8	0.1
7	1	4	47.81	1.902	8	0.1
7	7	1	48.61	1.873	8	0.4
13	1	3	48.61	1.873	8	0.1
7	5	3	48.86	1.864	8	0.2
16	0	2	48.86	1.864	4	0.3
3	7	2	49.20	1.852	8	0.1
10	4	3	49.26	1.850	8	0.3
0	6	3	49.46	1.843	4	0.3
12	6	0	49.64	1.837	4	0.3

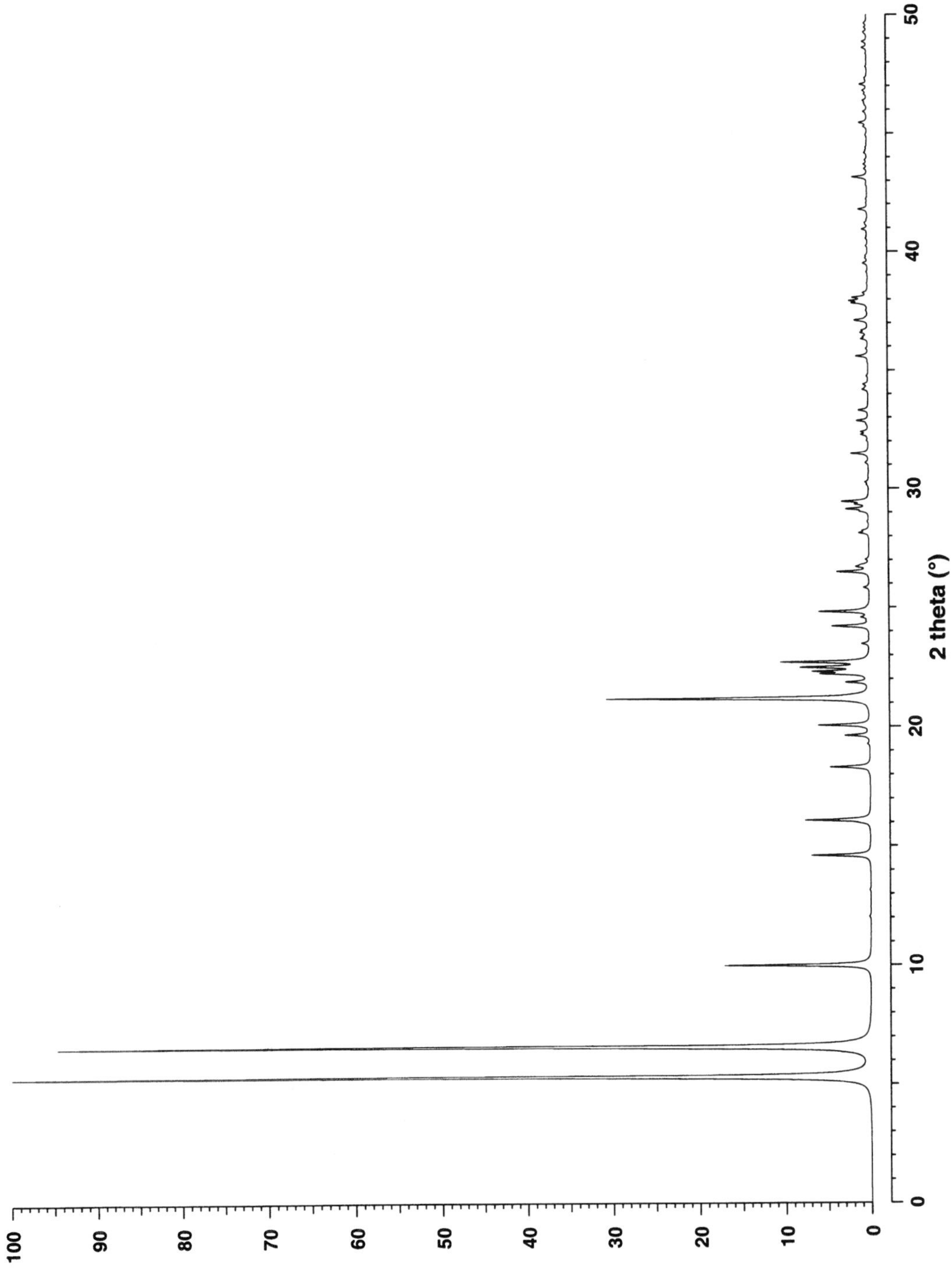
2 theta (°)
0
10
20
30
40
50
0
10
20
30
40
50
60
70
80
90
100

AFG Afghanite

CHEMICAL COMPOSITION: $|(Na_{17.28}K_{4.72}Ca_{10})(SO_4)_6Cl_{5.8}F_{0.2}|$ $[Si_{24}Al_{24}O_{96}]$
Pitigliano, Tuscany, Italy

REFINED COMPOSITION: $|(Na_{10.134}K_{5.304}Ca_{9.332})((SO_4)_6Cl_{5.8}F_{0.18})|$ $[Si_{24}Al_{24}O_{96}]$

CRYSTAL DATA: $P31c$ (No. 159)
$a = 12.801$ Å $b = 12.801$ Å $c = 21.412$ Å
$\alpha = 90°$ $\beta = 90°$ $\gamma = 120°$
X-ray single crystal refinement, $R_p = 0.045$, $R_{wp} = 0.102$

REFERENCE: P. Ballirano, E. Bonaccorsi, A. Maras and S. Merlino, *Eur. J. Mineral.* **9** 21–30 (1997).

h	k	l	2θ	d	M	I_{rel}
0	0	2	8.26	10.706	2	2.1
1	0	1	8.98	9.845	12	5.5
1	0	2	11.49	7.701	12	1.3
1	1	0	13.84	6.400	6	0.1
1	0	3	14.76	6.001	12	9.3
2	-1	2	16.13	5.494	6	0.2
1	1	2	16.13	5.494	6	0.2
2	0	1	16.52	5.366	12	2.4
0	0	4	16.56	5.353	2	1.7
2	0	2	18.02	4.922	12	0.7
1	0	4	18.40	4.820	12	48.5
2	0	3	20.28	4.378	12	1.1
2	1	0	21.20	4.190	12	2.6
3	-1	1	21.61	4.112	12	2.1
2	1	1	21.61	4.112	12	2.2
2	-1	4	21.64	4.106	6	1.2
1	1	4	21.64	4.106	6	1.4
1	0	5	22.25	3.995	12	26.1
2	1	2	22.79	3.902	12	0.5
3	-1	2	22.79	3.902	12	0.5
2	0	4	23.10	3.851	12	4.4
3	0	0	24.08	3.695	6	100.0
3	0	1	24.44	3.641	12	0.6
3	-1	3	24.64	3.613	12	2.2
2	1	3	24.64	3.613	12	2.2
2	0	5	26.30	3.389	12	1.9
3	-1	4	27.02	3.299	12	36.4
2	1	4	27.02	3.299	12	35.7
3	0	3	27.17	3.282	12	1.0
2	2	0	27.88	3.200	6	1.5
2	-1	6	28.64	3.117	6	0.5
1	1	6	28.64	3.117	6	0.5
3	1	0	29.04	3.075	12	0.2
4	-2	2	29.12	3.066	6	0.2
2	2	2	29.12	3.066	6	0.1
3	1	1	29.35	3.043	12	2.6
4	-1	1	29.35	3.043	12	2.7
3	0	4	29.37	3.041	12	1.7
2	0	6	29.77	3.001	12	0.3
1	0	7	30.31	2.949	12	5.6
4	-1	3	31.69	2.824	12	3.1
3	1	3	31.69	2.824	12	2.9
3	0	5	31.99	2.798	12	2.3
4	0	0	32.30	2.771	6	6.7
4	0	1	32.58	2.749	12	8.1
4	-2	4	32.60	2.747	6	1.4
2	2	4	32.60	2.747	6	1.1
3	-1	6	32.97	2.717	12	0.1
2	1	6	32.97	2.717	12	0.1
2	0	7	33.46	2.678	12	0.2
0	0	8	33.48	2.677	2	26.0
3	1	4	33.61	2.666	12	4.7
4	-1	4	33.61	2.666	12	4.6
1	0	8	34.47	2.602	12	3.1
4	0	3	34.72	2.584	12	7.7
3	0	6	34.95	2.567	12	0.1
4	-1	5	35.96	2.498	12	3.2
3	1	5	35.96	2.498	12	3.1
1	1	8	36.38	2.469	6	0.6
2	-1	8	36.38	2.469	6	0.5
4	0	4	36.51	2.461	12	18.3
4	1	0	37.16	2.419	12	6.3
2	0	8	37.31	2.410	12	0.3
3	0	7	38.19	2.356	12	0.7
4	-1	6	38.65	2.329	12	0.1
4	0	5	38.70	2.327	12	4.6
1	0	9	38.71	2.326	12	1.1
5	-2	4	39.22	2.297	12	0.3
3	2	4	39.22	2.297	12	0.2
5	-1	3	39.32	2.291	12	0.3
4	1	3	39.32	2.291	12	0.4
2	1	8	39.97	2.256	12	0.1
3	-1	8	39.97	2.256	12	0.1
5	0	0	40.69	2.217	6	0.8
5	0	1	40.92	2.205	12	0.4
4	0	6	41.24	2.189	12	0.2
5	-2	5	41.28	2.187	12	0.3
3	2	5	41.28	2.187	12	0.3
2	0	9	41.29	2.186	12	0.2
3	1	7	41.65	2.169	12	1.5
4	-1	7	41.65	2.169	12	1.6
3	0	8	41.67	2.168	12	1.8
0	0	10	42.20	2.141	2	0.4
3	3	0	42.36	2.133	6	26.5
5	0	3	42.70	2.117	12	0.1
1	0	10	43.02	2.102	12	0.2
4	2	0	43.18	2.095	12	1.4
3	-1	9	43.75	2.069	12	0.5
2	1	9	43.75	2.069	12	0.5
6	-2	2	44.04	2.056	12	0.2
4	2	2	44.04	2.056	12	0.1
4	0	7	44.09	2.054	12	2.9
4	-2	8	44.11	2.053	6	3.4
2	2	8	44.11	2.053	6	3.2
5	0	4	44.21	2.048	12	0.2
1	1	10	44.62	2.031	6	0.2
2	-1	10	44.62	2.031	6	0.2
3	1	8	44.90	2.019	12	0.1
2	0	10	45.41	1.997	12	0.3
5	1	0	45.56	1.991	12	0.9
3	3	4	45.78	1.982	6	0.2
6	-3	4	45.78	1.982	6	0.2
5	0	5	46.10	1.969	12	0.3
3	2	7	46.43	1.956	12	0.2
5	-2	7	46.43	1.956	12	0.1
4	2	4	46.55	1.951	12	0.5
6	-2	4	46.55	1.951	12	0.5
4	0	8	47.21	1.925	12	7.4
1	0	11	47.42	1.917	12	0.4
4	2	5	48.36	1.882	12	0.1
3	1	9	48.37	1.882	12	0.5
4	-1	9	48.37	1.882	12	0.5
5	1	4	48.80	1.866	12	2.0
6	-1	4	48.80	1.866	12	2.3
3	0	10	49.18	1.853	12	0.2
6	0	0	49.32	1.848	6	1.3
6	0	1	49.52	1.841	12	0.3

Afghanite AFG

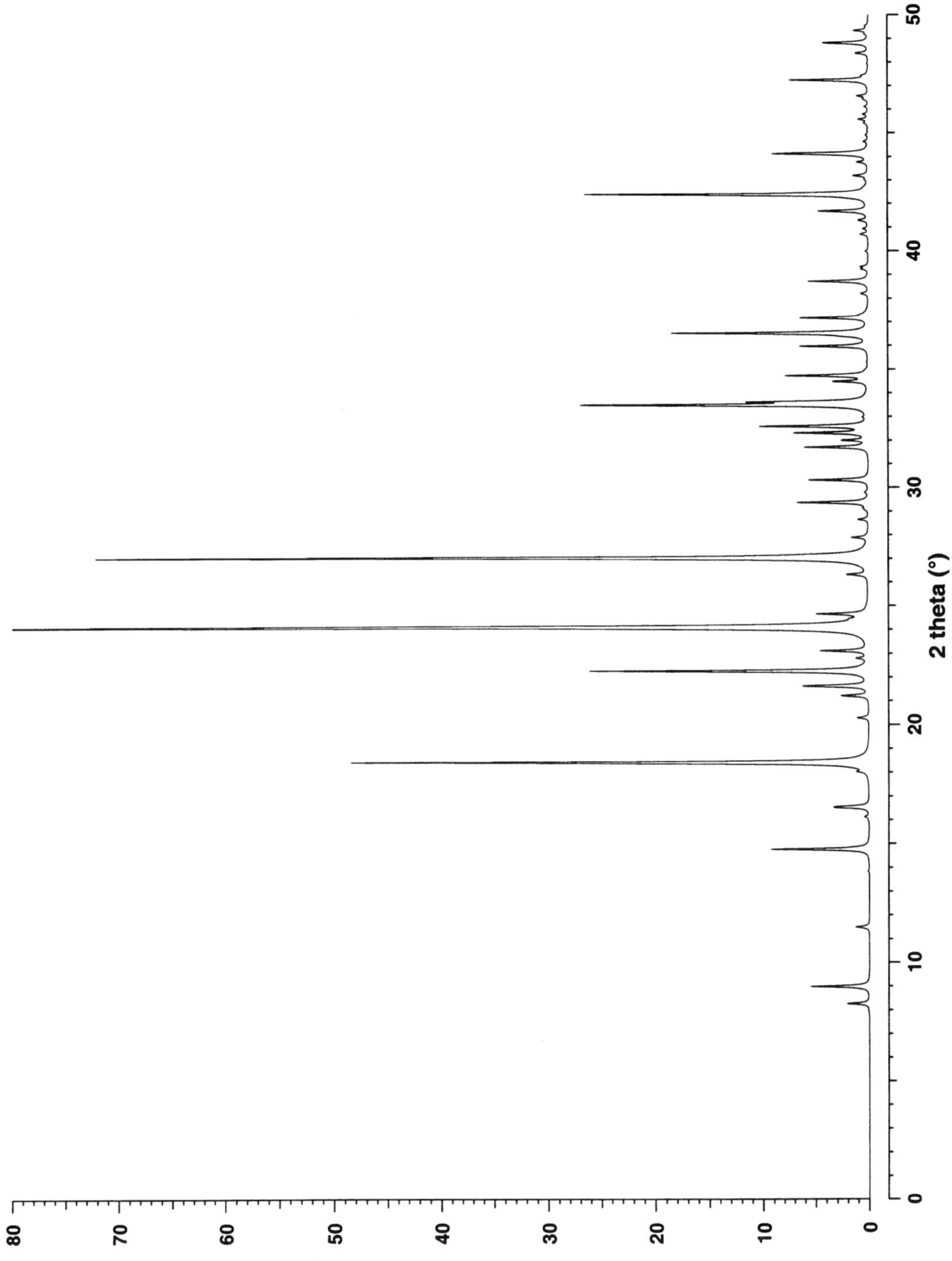

AFI Tetrapropylammonium Fluoride AlPO-5

CHEMICAL COMPOSITION: $|(C_3H_7)_4NF|$ $[Al_{12}P_{12}O_{48}]$
$(C_3H_7)_4NF$ = tetrapropylammonium fluoride

REFINED COMPOSITION: $|C_{12}NF|$ $[Al_{12}P_{12}O_{48}]$

CRYSTAL DATA: *P6cc* (No. 184)
$a = 13.740$ Å $b = 13.740$ Å $c = 8.474$ Å
$\alpha = 90°$ $\beta = 90°$ $\gamma = 120°$
X-ray single crystal refinement, $R_w = 0.042$

REFERENCE: S. Qiu, W. Pang, H. Kessler and J. L. Guth, *Zeolites* **9** 440–444 (1989).

h	k	l	2θ	d	M	I_{rel}	h	k	l	2θ	d	M	I_{rel}	h	k	l	2θ	d	M	I_{rel}
1	0	0	7.43	11.899	6	100.0	4	0	0	30.04	2.975	6	11.4	5	1	0	42.29	2.137	12	1.9
1	1	0	12.89	6.870	6	6.3	2	2	2	33.59	2.668	12	3.7	0	0	4	42.68	2.118	2	3.0
2	0	0	14.89	5.950	6	20.6	3	1	2	34.45	2.604	24	0.4	1	0	4	43.38	2.086	12	0.6
2	1	0	19.74	4.497	12	42.4	3	2	1	34.52	2.598	24	1.0	5	0	2	43.62	2.075	12	1.8
0	0	2	20.97	4.237	2	45.3	4	1	0	34.54	2.597	12	10.3	5	1	1	43.68	2.072	24	0.2
1	0	2	22.27	3.992	12	12.5	4	1	1	36.18	2.483	24	0.2	1	1	4	44.77	2.024	12	0.2
2	1	1	22.38	3.973	24	55.3	4	0	2	36.92	2.435	12	2.8	3	3	2	45.00	2.015	12	1.5
3	0	0	22.41	3.966	6	1.3	2	1	3	37.60	2.392	24	12.1	6	0	0	45.75	1.983	6	0.8
1	1	2	24.69	3.606	12	2.5	5	0	0	37.80	2.380	6	0.2	3	2	3	46.25	1.963	24	0.6
2	0	2	25.81	3.451	12	0.5	3	3	0	39.34	2.290	6	0.2	2	1	4	47.44	1.917	24	0.8
2	2	0	25.94	3.435	6	19.6	4	1	2	40.75	2.214	24	0.8	5	1	2	47.66	1.908	24	0.4
2	1	2	28.95	3.084	24	8.2	4	2	1	41.55	2.174	24	2.6	4	3	1	47.71	1.906	24	2.4
3	1	1	29.04	3.075	24	5.6	3	1	3	42.11	2.146	24	1.9	5	2	0	47.73	1.905	12	2.4

Tetrapropylammonium Fluoride AlPO-5 AFI

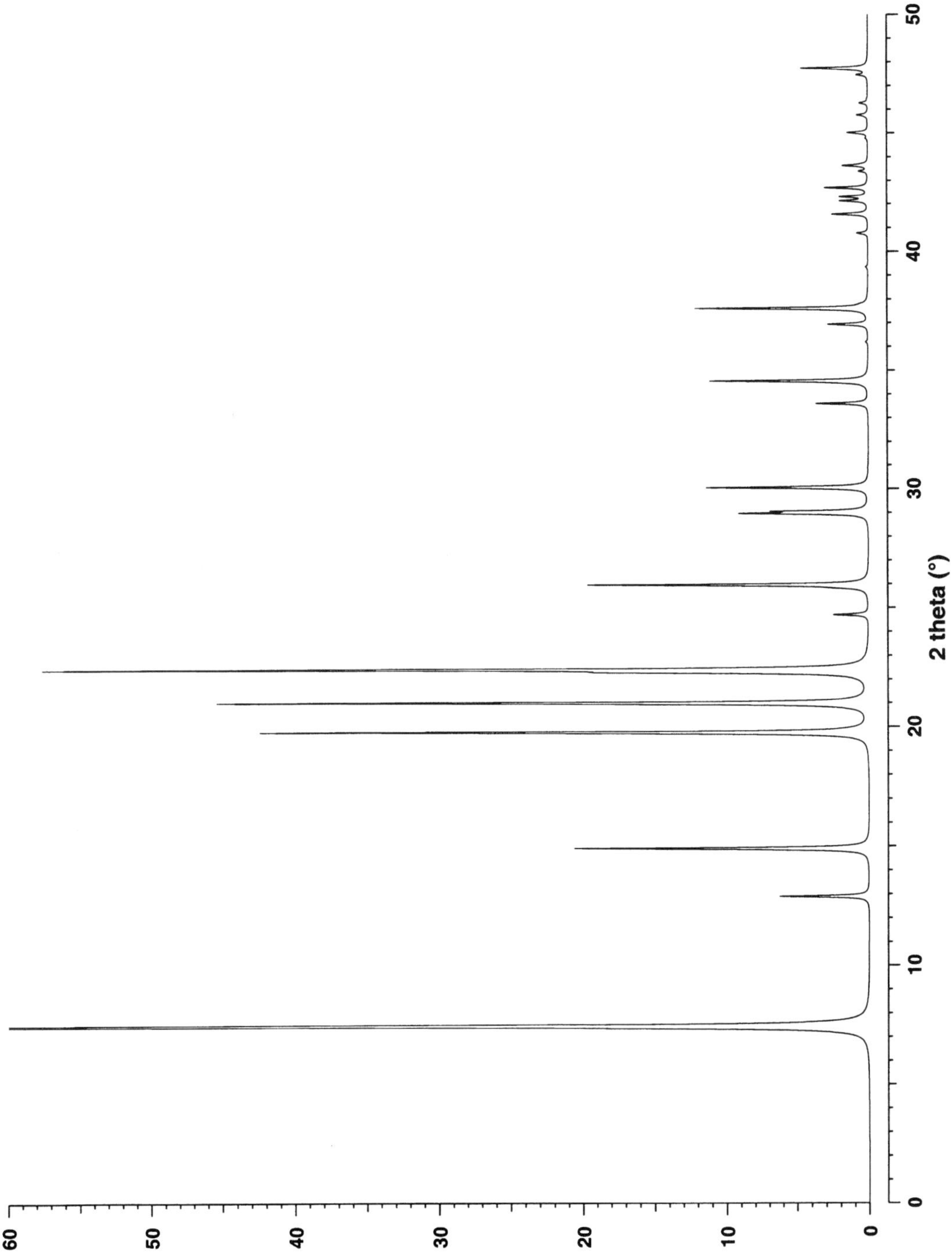

AFN

AlPO–14, Calcined

CHEMICAL COMPOSITION: $[Al_8P_8O_{32}]$

REFINED COMPOSITION: $[Al_8P_8O_{32}]$

CRYSTAL DATA: $P\bar{1}$ (No. 2)
$a = 9.7041$ Å $b = 9.7361$ Å $c = 10.202$ Å
$\alpha = 77.811°$ $\beta = 77.504°$ $\gamma = 87.691°$
X-ray Rietveld refinement, $R_F = 0.059$, $R_P = 0.114$, $R_{wP} = 0.135$

REFERENCE: R. W. Broach, S. T. Wilson and R. M. Kirchner, in *Proceedings of the 12th International Zeolite Conference*, Ed. by M. M. J. Treacy, B. K. Marcus, J. B. Higgins and M. E. Bisher (Materials Research Society: Warrendale) vol. **III** 1715–1722 (1999).

h	k	l	2θ	d	M	I_{rel}
0	0	1	9.08	9.744	2	100.0
1	0	0	9.33	9.474	2	4.9
1	0	1	11.56	7.655	2	36.1
0	1	1	11.57	7.647	2	53.6
1	-1	0	13.15	6.733	2	66.3
1	1	0	13.22	6.695	2	2.5
1	1	1	13.65	6.488	2	37.5
0	-1	1	14.30	6.196	2	7.1
-1	0	1	14.36	6.168	2	6.5
1	-1	1	15.97	5.551	2	13.0
-1	1	1	16.03	5.528	2	1.7
0	0	2	18.21	4.872	2	26.7
1	0	2	18.62	4.764	2	3.0
0	2	0	18.65	4.758	2	1.8
0	1	2	18.66	4.755	2	4.5
2	0	0	18.73	4.737	2	0.4
0	2	1	18.98	4.675	2	12.8
2	0	1	19.01	4.669	2	9.2
2	1	1	20.39	4.356	2	0.9
1	2	0	20.94	4.242	2	6.3
2	1	0	21.00	4.231	2	7.1
2	-1	1	21.98	4.044	2	28.1
-1	2	1	22.00	4.041	2	8.9
0	-1	2	22.16	4.012	2	6.6
-1	0	2	22.22	4.000	2	4.8
0	-2	1	22.43	3.964	2	0.6
1	-1	2	22.48	3.955	2	1.0
-1	1	2	22.57	3.939	2	13.2
2	0	2	23.24	3.828	2	2.9
0	2	2	23.26	3.824	2	1.7
1	2	2	23.63	3.764	2	4.4
2	1	2	23.64	3.764	2	1.6
-2	1	1	23.65	3.763	2	2.0
-1	-2	1	25.13	3.543	2	3.9
-2	-1	1	25.20	3.534	2	4.3
2	2	1	25.44	3.501	2	0.8
-1	-1	2	25.60	3.479	2	1.6
2	-2	0	26.47	3.367	2	5.6
-1	2	2	26.52	3.361	2	1.2
2	2	0	26.63	3.348	2	12.7
1	1	3	26.72	3.336	2	2.9
1	0	3	27.06	3.295	2	8.4
0	1	3	27.11	3.289	2	9.4
0	0	3	27.46	3.248	2	4.2
2	2	2	27.49	3.244	2	1.0
0	3	1	27.71	3.219	2	0.9
3	0	1	27.78	3.211	2	0.4
0	3	0	28.13	3.172	2	1.2
3	0	0	28.26	3.158	2	2.2
3	1	1	28.78	3.102	2	1.1
1	2	3	29.58	3.020	2	0.4
1	-3	0	29.65	3.013	2	0.4
1	3	0	29.75	3.003	2	6.8
3	1	0	29.86	2.992	2	2.7
2	0	3	29.86	2.992	2	1.0
-1	3	1	29.88	2.990	2	15.3
3	-1	1	29.92	2.987	2	6.7
0	2	3	29.92	2.987	2	0.8
0	3	2	30.26	2.953	2	1.8
3	0	2	30.27	2.952	2	0.5
1	-1	3	30.50	2.930	2	3.0
1	3	2	30.57	2.924	2	4.0
3	1	2	30.61	2.921	2	1.5
-1	1	3	30.61	2.920	2	3.0
-2	-2	1	30.70	2.912	2	5.2
0	-1	3	30.88	2.896	2	1.1
-1	0	3	30.95	2.889	2	0.4
-3	0	1	31.59	2.833	2	0.4
-1	-2	2	31.61	2.830	2	0.4
2	2	3	32.21	2.779	2	1.3
2	3	1	32.60	2.746	2	3.9
3	2	1	32.65	2.743	2	3.2
-1	3	2	32.86	2.725	2	1.8
2	-1	3	33.02	2.712	2	2.6
-1	2	3	33.14	2.703	2	3.4
-1	-3	1	33.46	2.678	2	2.7
-3	-1	1	33.60	2.668	2	0.4
-1	-1	3	34.06	2.632	2	1.0
2	3	0	34.10	2.629	2	1.5
3	2	0	34.16	2.624	2	0.8
3	1	3	34.93	2.569	2	0.8
1	3	3	34.93	2.568	2	1.4
1	1	4	35.28	2.544	2	1.4
2	-3	1	35.68	2.516	2	2.0
-3	2	1	35.82	2.507	2	2.9
-2	-2	2	36.81	2.442	2	1.3
0	-3	2	36.88	2.437	2	0.5
0	0	4	36.90	2.436	2	1.3
1	2	4	37.05	2.426	2	0.4
4	0	1	37.10	2.423	2	1.2
-3	1	2	37.27	2.413	2	0.9
1	4	1	37.78	2.381	2	1.2
4	1	1	37.90	2.374	2	0.6
2	-2	3	38.43	2.343	2	0.8
0	4	2	38.51	2.338	2	0.4
-2	2	3	38.60	2.332	2	0.7
4	1	2	38.85	2.318	2	0.4
3	3	2	39.10	2.304	2	1.6
-1	1	4	39.39	2.287	2	0.9
2	-3	2	39.67	2.272	2	0.9
-3	2	2	39.87	2.261	2	1.1
-1	4	2	40.63	2.220	2	0.7
4	-1	2	40.65	2.220	2	1.2
2	-1	4	40.85	2.209	2	0.5
1	-4	1	41.49	2.176	2	0.4
3	0	4	41.73	2.164	2	0.8
2	-4	0	42.42	2.131	2	0.4
4	-2	0	42.54	2.125	2	0.5
-2	4	1	42.61	2.122	2	1.2
2	4	0	42.62	2.121	2	0.4
4	-2	1	42.64	2.120	2	0.4
-1	-1	4	43.12	2.098	2	1.0
4	2	3	43.99	2.058	2	0.4
-2	-2	3	44.24	2.047	2	3.4
-3	-3	1	44.24	2.047	2	1.5
1	1	5	44.40	2.040	2	2.0
-3	3	2	44.52	2.035	2	0.4
2	-2	4	45.88	1.978	2	0.6
3	4	2	46.02	1.972	2	0.8
3	4	1	46.03	1.972	2	0.6
4	3	2	46.06	1.971	2	1.0
-1	-3	3	46.28	1.962	2	0.8
-3	-1	3	46.43	1.956	2	0.5
2	2	5	46.45	1.955	2	0.5
1	5	1	47.35	1.920	2	0.8
4	0	4	47.51	1.914	2	0.5
5	1	1	47.52	1.913	2	0.5
0	4	4	47.56	1.912	2	0.4
3	-4	0	47.73	1.905	2	0.5
4	-3	0	47.79	1.903	2	0.5
5	-1	1	48.23	1.887	2	0.5
-1	1	5	48.70	1.870	2	1.4
3	-4	1	49.06	1.857	2	1.0
-4	3	1	49.21	1.852	2	0.5
5	-1	2	49.41	1.845	2	0.5

AlPO–14, Calcined

AFN

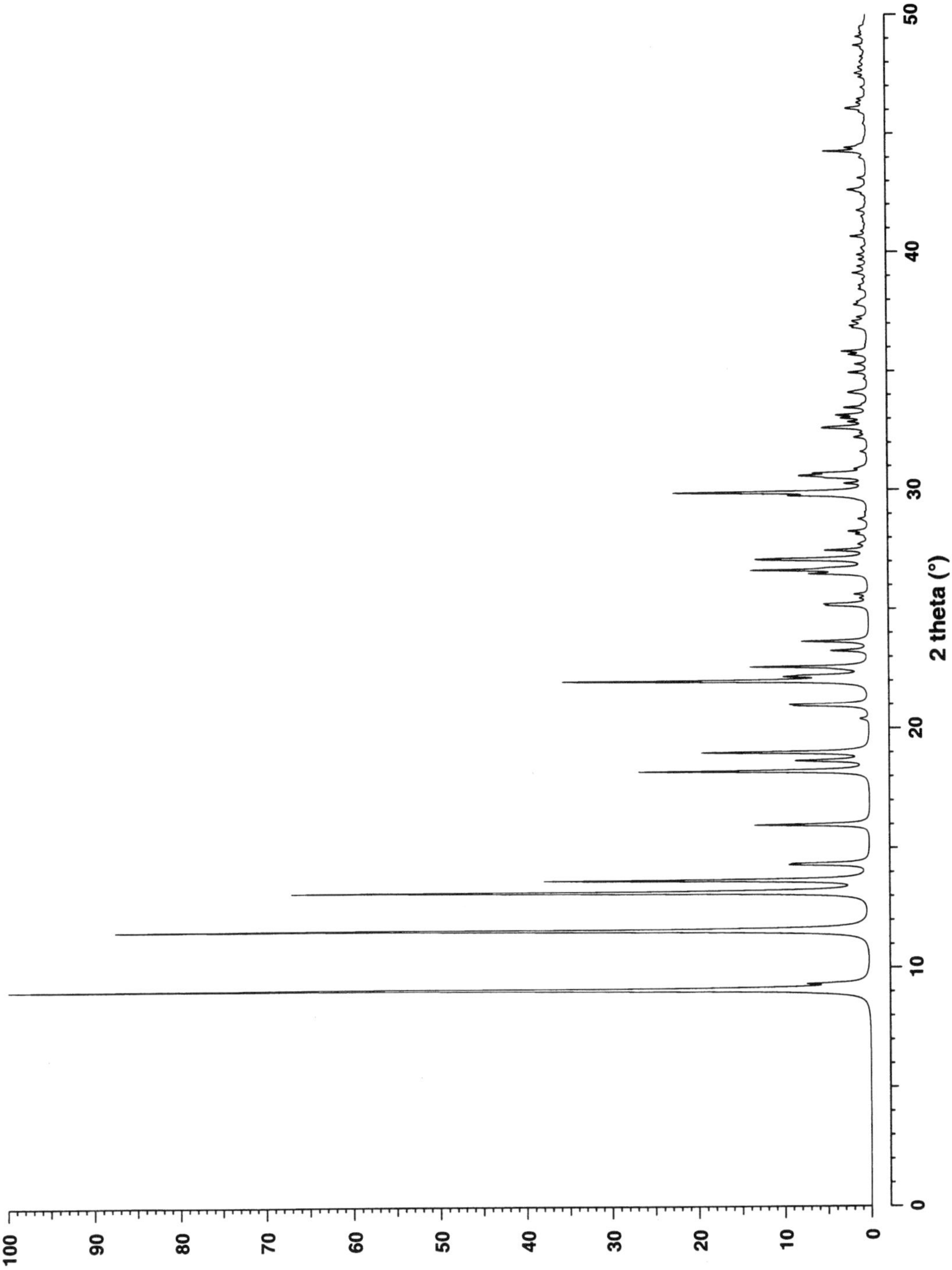

AFO AlPO–41, Calcined

CHEMICAL COMPOSITION: $[Al_{10}P_{10}O_{40}]$

REFINED COMPOSITION: $[Al_{10}P_{10}O_{40}]$

CRYSTAL DATA: $P112_1$ (No. 4) unique axis **c**
$a = 9.7179$ Å $b = 13.7915$ Å $c = 8.3591$ Å
$\alpha = 90°$ $\beta = 90°$ $\gamma = 110.6°$
X-ray Rietveld refinement, $R_{exp} = 0.023$, $R_{wp} = 0.164$, $R_F = 0.074$

REFERENCE: R. M. Kirchner and J. M. Bennett, *Zeolites* **14** 523–528 (1994).

h	k	l	2θ	d	M	I_{rel}
0	1	0	6.85	12.910	2	100.0
1	0	0	9.72	9.097	2	42.8
1	-1	0	9.73	9.094	2	35.6
0	1	1	12.62	7.017	4	0.6
0	2	0	13.72	6.455	2	41.9
1	1	0	13.74	6.445	2	10.1
1	-2	0	13.75	6.441	2	8.6
1	-1	1	14.39	6.154	4	0.2
0	2	1	17.36	5.109	4	0.6
1	1	1	17.37	5.104	4	0.3
1	-2	1	17.38	5.102	4	0.4
2	-1	0	18.26	4.859	2	10.6
1	-3	0	19.47	4.559	2	0.5
2	0	0	19.52	4.548	2	0.8
2	-2	0	19.52	4.547	2	0.9
0	3	0	20.64	4.303	2	3.1
2	-1	1	21.15	4.201	4	0.3
0	0	2	21.26	4.180	2	68.3
1	2	1	22.20	4.004	4	10.5
1	-3	1	22.21	4.002	4	9.7
2	0	1	22.25	3.995	4	19.5
2	-2	1	22.26	3.994	4	23.0
0	1	2	22.36	3.976	4	3.3
2	1	0	22.90	3.883	2	18.0
2	-3	0	22.91	3.881	2	15.9
0	3	1	23.25	3.826	4	13.5
1	0	2	23.42	3.798	4	1.6
1	-1	2	23.42	3.798	4	4.7
2	1	1	25.29	3.522	4	1.7
2	-3	1	25.30	3.520	4	1.8
0	2	2	25.39	3.508	4	2.8
1	1	2	25.40	3.507	4	3.0
1	-2	2	25.40	3.506	4	2.1
1	3	0	25.83	3.449	2	10.7
1	-4	0	25.84	3.448	2	11.5
2	2	0	27.68	3.222	2	1.6
2	-4	0	27.70	3.221	2	0.7
3	-2	0	27.76	3.214	2	1.7
1	3	1	27.98	3.188	4	0.6
1	-4	1	27.99	3.187	4	0.3
2	-1	2	28.16	3.169	4	2.1
1	2	2	28.98	3.081	4	0.9
1	-3	2	28.98	3.081	4	0.7
2	0	2	29.01	3.077	4	0.3
2	-2	2	29.02	3.077	4	0.3
3	0	0	29.46	3.032	2	11.6
3	-3	0	29.47	3.031	2	5.1
0	4	1	29.67	3.011	4	0.9
3	-1	1	29.78	3.000	4	1.4
3	-2	1	29.78	3.000	4	1.0
0	3	2	29.80	2.998	4	1.4
3	0	1	31.38	2.850	4	0.4
2	1	2	31.45	2.845	4	2.8
2	-3	2	31.45	2.844	4	4.4
0	1	3	32.88	2.724	4	0.3
2	3	0	33.32	2.689	2	1.4
2	-5	0	33.33	2.688	2	4.6
1	-1	3	33.64	2.664	4	0.8
1	3	2	33.69	2.660	4	4.2
1	-4	2	33.70	2.660	4	0.6
1	4	1	34.31	2.614	4	0.5
1	-5	1	34.32	2.613	4	0.4
3	1	1	34.39	2.608	4	0.2
3	-4	1	34.40	2.607	4	1.0
0	5	0	34.74	2.582	2	2.8
2	3	1	35.05	2.560	4	0.5
2	-5	1	35.07	2.559	4	0.5
0	2	3	35.08	2.558	4	0.3
1	1	3	35.09	2.558	4	0.3
1	-2	3	35.09	2.557	4	0.3
2	2	2	35.17	2.552	4	0.3
3	-2	2	35.23	2.548	4	0.3
0	5	1	36.42	2.467	4	0.5
3	0	2	36.61	2.454	4	1.4
3	-3	2	36.62	2.454	4	2.2
3	2	0	36.92	2.435	2	1.2
3	-5	0	36.94	2.433	2	1.1
4	-2	0	37.00	2.429	2	2.6
2	-1	3	37.20	2.417	4	0.3
4	-1	0	37.67	2.388	2	0.8
1	-3	3	37.84	2.377	4	2.9
2	0	3	37.87	2.376	4	6.2
2	-2	3	37.87	2.376	4	2.2
0	3	3	38.49	2.339	4	4.5
4	-3	1	39.25	2.296	4	0.6
1	-6	0	39.50	2.281	2	0.2
4	0	0	39.63	2.274	2	1.4
2	1	3	39.82	2.264	4	0.2
2	-3	3	39.83	2.263	4	0.8
2	3	2	39.86	2.261	4	0.4
2	-5	2	39.88	2.261	4	0.7
1	5	1	41.00	2.201	4	0.2
2	-6	1	41.04	2.199	4	0.4
0	6	0	41.99	2.152	2	0.3
3	3	0	42.06	2.148	2	0.9
0	4	3	42.88	2.109	4	0.3
2	-4	3	42.92	2.107	4	0.3
3	-1	3	42.96	2.105	4	0.4
3	-5	2	43.01	2.103	4	0.3
0	0	4	43.29	2.090	2	3.2
3	3	1	43.50	2.081	4	0.4
3	-6	1	43.52	2.080	4	0.4
0	1	4	43.89	2.063	4	0.3
3	0	3	44.14	2.052	4	0.3
4	1	1	44.15	2.051	4	0.9
4	-5	1	44.17	2.051	4	0.8
1	-1	4	44.48	2.037	4	0.5
1	5	2	45.28	2.003	4	0.7
2	4	2	45.30	2.002	4	0.2
4	0	2	45.40	1.998	4	0.8
4	-4	2	45.41	1.997	4	0.9
1	1	4	45.64	1.988	4	0.2
1	-2	4	45.64	1.988	4	0.4
1	4	3	46.37	1.958	4	0.2
1	-5	3	46.38	1.958	4	0.3
3	-4	3	46.45	1.955	4	0.6
5	-2	0	46.87	1.938	2	1.0
5	-3	0	46.88	1.938	2	0.7
2	-5	3	46.97	1.934	4	0.3
2	-1	4	47.35	1.920	4	0.3
2	5	1	47.47	1.915	4	0.9
2	-7	1	47.48	1.915	4	1.6
0	6	2	47.53	1.913	4	0.3
3	3	2	47.59	1.911	4	0.4
3	4	0	47.84	1.901	2	0.9
3	-7	0	47.87	1.900	2	0.4
1	-3	4	47.88	1.900	4	0.3
1	6	1	48.00	1.895	4	0.3
1	-7	1	48.01	1.895	4	0.4
0	5	3	48.04	1.894	4	0.5
5	-2	1	48.20	1.888	4	0.4
0	3	4	48.42	1.880	4	0.4
5	-4	1	49.29	1.849	4	0.2
0	7	0	49.42	1.844	2	1.9
2	-3	4	49.54	1.840	4	0.5

AlPO-41, Calcined

AFO

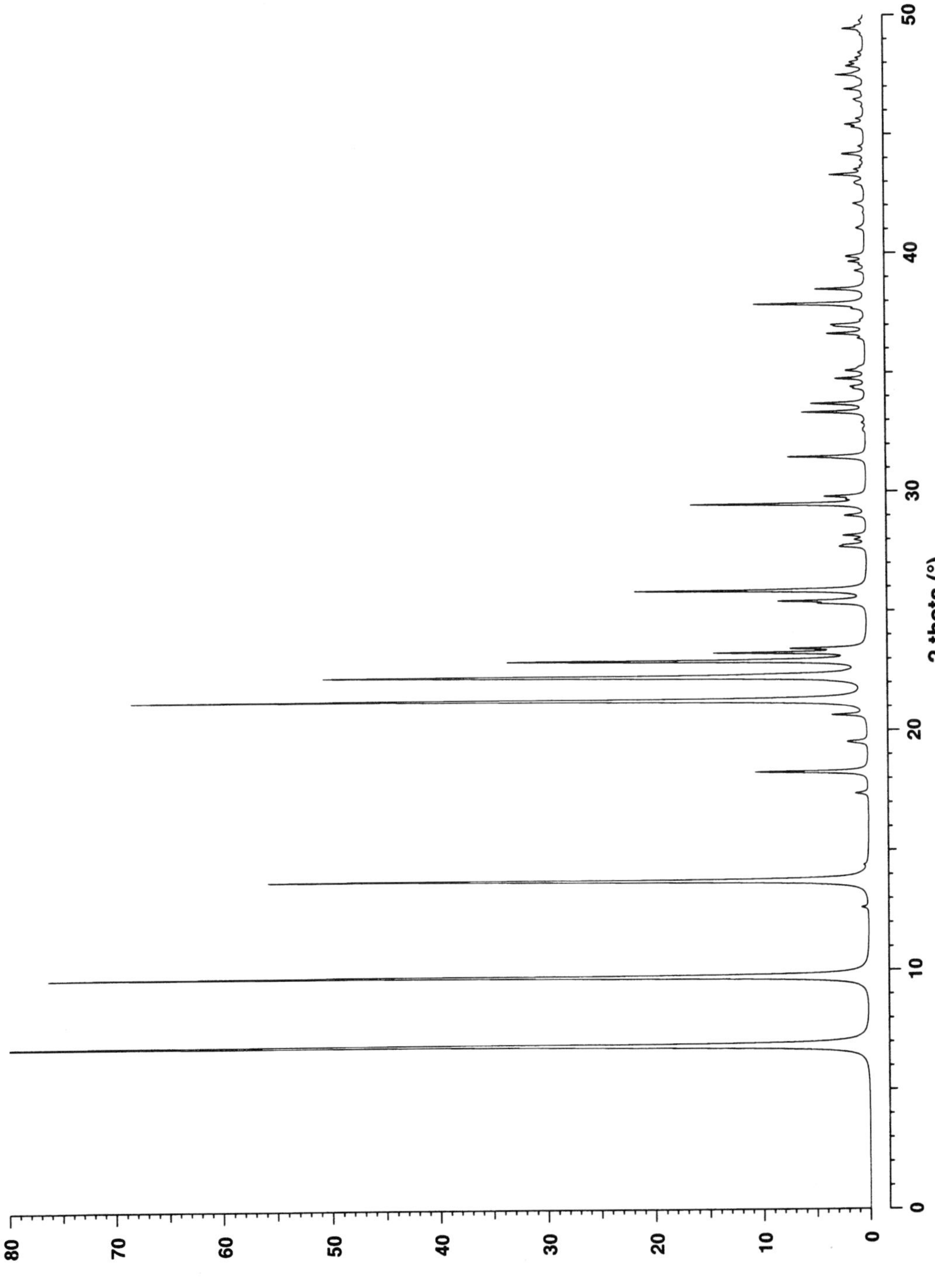

AFR Tetrapropylammonium SAPO–40

CHEMICAL COMPOSITION: $|((C_3H_7)_4NOH)_4|$ $[Si_7Al_{29}P_{28}O_{128}]$
$(C_3H_7)_4NOH$ = tetrapropylammonium hydroxide

REFINED COMPOSITION: $|((C_3H_7)_4N)_4|$ $[Al_{32}P_{32}O_{128}]$

CRYSTAL DATA: *Pccn* (No. 56)
$a = 21.9443$ Å $b = 13.6911$ Å $c = 14.2486$ Å
$\alpha = 90°$ $\beta = 90°$ $\gamma = 90°$
X-ray Rietveld refinement, $R_{exp} = 0.149$, $R_{wp} = 0.161$, $R_F = 0.066$

REFERENCE: L. B. McCusker and Ch. Baerlocher,
Microporous Materials **6** 51–54 (1996).

h	*k*	*l*	2θ	*d*	*M*	I_{rel}
1	1	0	7.61	11.616	4	100.0
2	0	0	8.06	10.972	2	83.4
0	0	2	12.42	7.124	2	83.9
1	0	2	13.07	6.776	4	5.2
3	1	0	13.73	6.452	4	80.7
0	1	2	14.01	6.320	4	75.2
2	0	2	14.83	5.975	4	0.7
2	2	0	15.26	5.808	4	5.7
2	1	2	16.18	5.476	8	1.8
3	0	2	17.38	5.104	4	15.1
0	2	2	17.97	4.936	4	0.5
1	2	2	18.42	4.816	8	34.7
4	1	1	18.50	4.795	8	0.8
3	1	2	18.55	4.782	8	35.6
2	2	2	19.72	4.502	8	10.4
1	3	0	19.87	4.468	4	62.3
1	1	3	20.20	4.396	8	0.7
4	0	2	20.43	4.347	4	21.9
4	2	0	20.75	4.281	4	3.6
1	3	1	20.83	4.263	8	1.7
5	1	0	21.26	4.179	4	4.2
2	1	3	21.39	4.153	8	2.9
4	1	2	21.45	4.143	8	21.7
4	2	1	21.68	4.100	8	1.9
3	2	2	21.72	4.092	8	22.1
5	1	1	22.17	4.010	8	2.4
3	3	0	22.97	3.872	4	25.8
0	3	2	23.14	3.843	4	10.4
1	2	3	23.15	3.842	8	3.3
1	3	2	23.50	3.785	8	7.0
5	0	2	23.81	3.737	4	32.8
4	2	2	24.25	3.669	8	17.4
6	0	0	24.34	3.657	2	0.7
2	3	2	24.54	3.627	8	3.6
5	1	2	24.70	3.605	8	23.5
5	2	1	24.90	3.576	8	0.6
0	0	4	25.00	3.562	2	1.8
0	1	4	25.84	3.447	4	4.4
0	4	0	26.03	3.423	2	33.2
4	3	1	26.16	3.407	8	0.7
1	1	4	26.17	3.406	8	5.4
3	3	2	26.19	3.402	8	7.4
2	0	4	26.30	3.388	4	1.3
2	1	4	27.11	3.289	8	0.5
2	4	0	27.29	3.267	4	6.4

h	*k*	*l*	2θ	*d*	*M*	I_{rel}
6	0	2	27.41	3.254	4	29.6
6	2	0	27.65	3.226	4	11.1
3	0	4	27.86	3.203	4	55.7
6	1	2	28.19	3.166	8	7.1
0	2	4	28.24	3.160	4	1.7
4	3	2	28.35	3.147	8	3.2
1	2	4	28.54	3.128	8	1.7
3	1	4	28.62	3.118	8	2.9
0	4	2	28.94	3.085	4	1.8
1	4	2	29.23	3.055	8	6.5
4	0	4	29.91	2.988	4	0.5
6	2	2	30.42	2.939	8	6.4
4	1	4	30.63	2.919	8	4.8
4	4	0	30.79	2.904	4	3.3
3	2	4	30.82	2.901	8	2.3
7	0	2	31.17	2.869	4	2.5
3	4	2	31.47	2.843	8	1.4
7	1	2	31.86	2.808	8	5.2
0	3	4	31.87	2.808	4	1.6
5	0	4	32.37	2.766	4	8.4
8	0	0	32.64	2.743	2	0.9
2	3	4	32.92	2.720	8	1.6
4	4	2	33.32	2.689	8	5.3
6	3	2	33.83	2.649	8	4.9
7	2	2	33.87	2.646	8	1.0
3	3	4	34.20	2.622	8	0.5
5	2	4	34.99	2.564	8	1.8
8	0	2	35.05	2.560	4	0.7
0	5	2	35.11	2.556	4	16.4
8	1	2	35.68	2.516	8	0.5
6	1	4	35.79	2.509	8	4.6
4	3	4	35.93	2.500	8	3.5
6	4	0	35.93	2.499	4	6.1
2	5	2	36.08	2.489	8	2.8
7	3	2	37.01	2.429	8	0.7
3	5	2	37.26	2.413	8	1.1
2	4	4	37.34	2.408	8	1.4
9	1	0	37.46	2.400	4	2.3
6	2	4	37.62	2.391	8	1.8
0	0	6	37.89	2.375	2	2.0
6	4	2	38.16	2.358	8	9.4
0	1	6	38.47	2.340	4	4.0
3	4	4	38.50	2.339	8	3.8
1	1	6	38.70	2.327	8	0.5
9	0	2	39.04	2.307	4	0.6

h	*k*	*l*	2θ	*d*	*M*	I_{rel}
0	6	0	39.49	2.282	2	2.5
9	1	2	39.62	2.275	8	0.6
3	0	6	39.91	2.259	4	2.8
5	2	5	39.95	2.257	8	0.5
8	3	2	40.40	2.233	8	1.8
1	2	6	40.41	2.232	8	2.7
7	2	4	40.53	2.226	8	0.5
10	0	0	41.13	2.194	2	0.5
9	2	2	41.30	2.186	8	6.0
0	6	2	41.56	2.173	4	0.7
5	4	4	42.00	2.151	8	1.4
3	2	6	42.13	2.145	8	1.3
2	5	4	42.44	2.130	8	1.7
4	6	0	42.93	2.107	4	2.1
10	0	2	43.13	2.097	4	1.2
5	0	6	43.32	2.089	4	1.5
3	6	2	43.44	2.083	8	2.9
7	5	0	43.90	2.062	4	1.3
9	3	2	43.98	2.059	8	1.8
8	4	2	44.18	2.050	8	0.9
6	4	4	44.27	2.046	8	0.7
3	3	6	44.77	2.024	8	1.2
4	5	4	44.90	2.019	8	0.7
9	0	4	45.06	2.012	4	1.7
6	0	6	45.54	1.992	4	1.5
9	1	4	45.57	1.991	8	2.0
7	5	2	45.80	1.981	8	1.1
11	1	0	45.97	1.974	4	1.1
4	3	6	46.16	1.967	8	0.5
8	3	4	46.27	1.962	8	0.6
1	7	0	46.62	1.948	4	1.7
5	5	4	46.68	1.946	8	1.8
1	4	6	46.74	1.943	8	0.8
6	6	0	46.93	1.936	4	0.7
9	2	4	47.07	1.930	8	2.0
2	4	6	47.32	1.921	8	1.0
6	2	6	47.54	1.912	8	1.3
0	7	2	48.25	1.886	4	3.2
7	1	6	48.55	1.875	8	2.6
6	5	4	48.78	1.867	8	4.3
11	2	2	49.26	1.850	8	1.1
8	4	4	49.69	1.835	8	0.7
12	0	0	49.87	1.829	2	4.8
3	7	2	49.93	1.826	8	1.0
6	3	6	49.96	1.825	8	0.5

Tetrapropylammonium SAPO–40 AFR

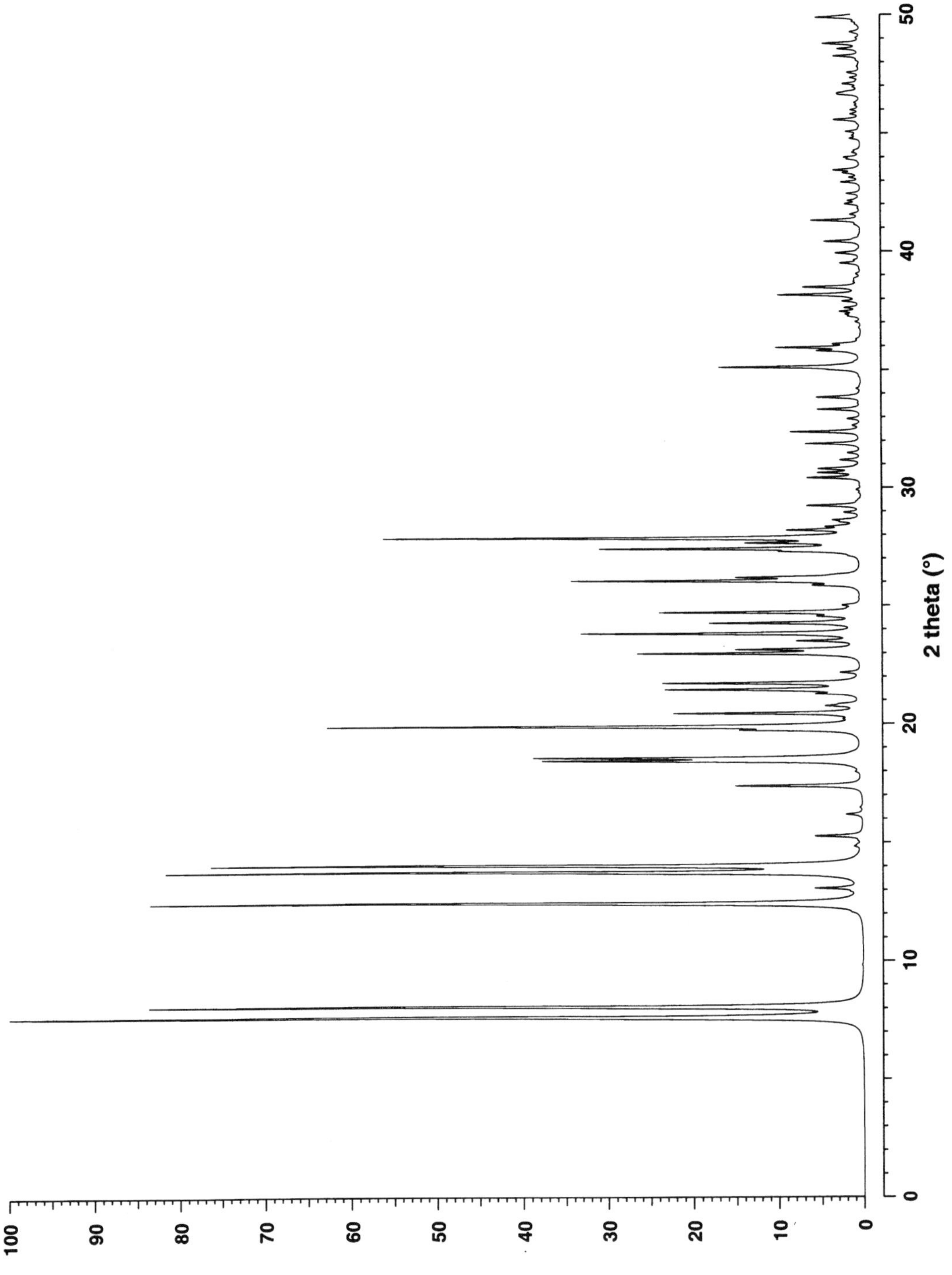

AFS Di-n-propylamine MAPSO–46

CHEMICAL COMPOSITION: $|((C_3H_7)_2NH_2)_8(H_2O)_{14}|$ $[Mg_6Al_{22}P_{26}Si_2O_{112}]$
$(C_3H_7)_2NH_2$ = di-n-propylamine

REFINED COMPOSITION: $|(C_4N)_6|$ $[Al_{28}P_{28}O_{112}]$

CRYSTAL DATA: *P*3*c*1 (No. 158)
$a = 13.2251$ Å $b = 13.2251$ Å $c = 26.8922$ Å
$\alpha = 90°$ $\beta = 90°$ $\gamma = 120°$
X-ray single crystal refinement, $R_w = 0.119$

REFERENCE: J. M. Bennett and B. K. Marcus,
in *Innovations in Zeolite Material Science* (*Stud. Surf. Sci. Catal. No. 37*)
Ed. by P. J. Grobet, W. J. Mortier, E. F. Vansant and G. Schulz-Ekloff
(Elsevier: Amsterdam) 269–279 (1988).
And J. M. Bennett, Private communication.

h	*k*	*l*	2*θ*	*d*	*M*	I_{rel}
0	0	2	6.57	13.446	2	13.6
1	0	0	7.72	11.453	6	100.0
0	1	2	10.15	8.719	6	0.1
1	0	2	10.15	8.719	6	0.2
0	0	4	13.17	6.723	2	7.8
1	1	0	13.39	6.613	6	0.2
1	1	1	13.79	6.421	12	3.4
1	1	2	14.93	5.934	12	5.4
1	0	4	15.28	5.798	6	2.2
0	1	4	15.28	5.798	6	1.6
1	1	3	16.66	5.321	12	9.2
1	1	4	18.82	4.714	12	0.1
0	0	6	19.81	4.482	2	4.4
0	2	4	20.37	4.359	6	0.5
2	1	0	20.52	4.329	12	2.5
1	2	1	20.78	4.274	12	0.1
2	1	1	20.78	4.274	12	0.6
1	0	6	21.29	4.174	6	1.0
1	1	5	21.29	4.173	12	2.0
2	1	2	21.57	4.121	12	6.7
1	2	2	21.57	4.121	12	3.6
2	1	3	22.81	3.898	12	1.4
1	2	3	22.81	3.898	12	5.3
1	1	6	23.99	3.710	12	3.2
3	0	2	24.23	3.673	6	2.1
0	3	2	24.23	3.673	6	1.4
2	1	4	24.46	3.640	12	0.6
2	0	6	25.23	3.530	6	0.2
0	2	6	25.23	3.530	6	0.5
1	2	5	26.43	3.372	12	0.5
2	1	5	26.43	3.372	12	0.3
0	0	8	26.52	3.362	2	0.5
1	1	7	26.84	3.322	12	5.7
0	3	4	26.85	3.320	6	0.7
3	0	4	26.85	3.320	6	0.3
2	2	0	26.97	3.306	6	0.9
2	2	1	27.17	3.282	12	0.2
2	2	2	27.79	3.211	12	2.2
1	3	1	28.29	3.155	12	0.6
3	1	1	28.29	3.155	12	0.5
1	2	6	28.67	3.114	12	0.6
2	1	6	28.67	3.114	12	0.9
2	2	3	28.78	3.102	12	4.4
3	1	2	28.88	3.091	12	0.1
1	3	2	28.88	3.091	12	0.3
1	1	8	29.82	2.997	12	0.6
1	3	3	29.84	2.994	12	0.3
3	1	3	29.84	2.994	12	0.8
2	2	4	30.12	2.967	12	1.0
3	0	6	30.76	2.906	6	0.2
0	3	6	30.76	2.906	6	0.3
0	2	8	30.84	2.899	6	0.5
2	0	8	30.84	2.899	6	0.5
2	1	7	31.12	2.873	12	0.7
1	2	7	31.12	2.873	12	0.7
3	1	4	31.14	2.872	12	0.3
1	3	4	31.14	2.872	12	0.6
4	0	0	31.24	2.863	6	1.1
2	2	5	31.77	2.817	12	0.8
4	0	2	31.96	2.801	6	0.1
1	1	9	32.89	2.723	12	1.8
2	2	6	33.68	2.661	12	0.2
0	4	4	34.03	2.634	6	0.2
4	0	4	34.03	2.634	6	0.3
3	2	0	34.12	2.628	12	0.2
2	3	1	34.29	2.615	12	0.6
3	2	1	34.29	2.615	12	0.7
3	1	6	34.61	2.592	12	0.1
2	2	7	35.83	2.506	12	0.4
4	1	0	35.93	2.499	12	1.7
1	1	10	36.05	2.491	12	0.1
1	2	9	36.54	2.459	12	0.5
2	1	9	36.54	2.459	12	0.7
2	0	10	36.93	2.434	6	0.1
4	0	6	37.26	2.413	6	0.2
1	4	3	37.35	2.407	12	0.1
5	0	0	39.33	2.291	6	0.1
3	2	6	39.76	2.267	12	0.1
2	3	6	39.76	2.267	12	0.1
4	1	5	39.77	2.267	12	0.2
0	0	12	40.24	2.241	2	0.1
2	2	9	40.70	2.217	12	0.7
3	3	1	41.09	2.197	12	0.5
1	3	9	41.49	2.176	12	0.1
4	2	2	42.29	2.137	12	0.1
2	4	2	42.29	2.137	12	0.2
1	1	12	42.60	2.122	12	0.2
4	1	7	43.18	2.095	12	0.1
1	4	7	43.18	2.095	12	0.1
2	0	12	43.36	2.087	6	0.1
2	2	10	43.37	2.086	12	0.1
2	3	8	43.73	2.070	12	0.2
3	2	8	43.73	2.070	12	0.3
5	1	0	44.02	2.057	12	0.8
1	5	1	44.15	2.051	12	0.2
5	1	1	44.15	2.051	12	0.2
3	3	5	44.42	2.040	12	0.2
3	3	6	45.88	1.978	12	0.2
1	1	13	45.97	1.974	12	0.6
3	1	11	46.89	1.937	12	0.1
3	3	7	47.56	1.912	12	0.1
6	0	0	47.64	1.909	6	0.5
1	2	13	48.79	1.866	12	0.2
2	1	13	48.79	1.866	12	0.3
1	1	14	49.41	1.845	12	0.2
3	3	8	49.45	1.843	12	0.2

Di-n-propylamine MAPSO–46 AFS

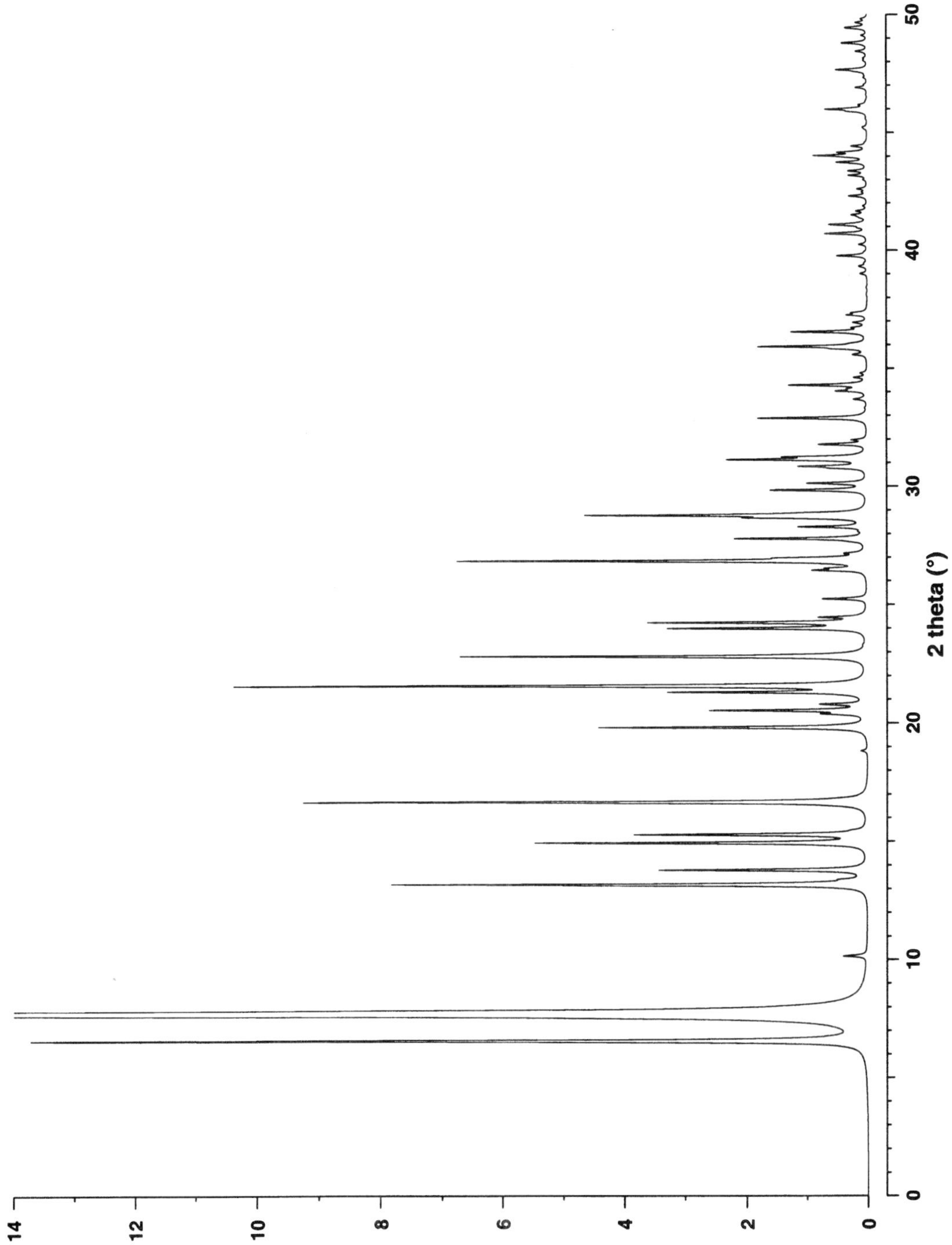

AFT AlPO–52, Calcined, Rehydrated

CHEMICAL COMPOSITION: $|(H_2O)_x|$ $[Al_{36}P_{36}O_{144}]$
Partially Rehydrated.

REFINED COMPOSITION: $[Al_{36}P_{36}O_{144}]$

CRYSTAL DATA: $P\bar{3}1c$ (No. 163)
$a = 13.715$ Å $b = 13.715$ Å $c = 29.676$ Å
$\alpha = 90°$ $\beta = 90°$ $\gamma = 120°$
X-ray Rietveld refinement, $R_{exp} = 0.130$, $R_{wp} = 0.197$, $R_p = 0.173$

REFERENCE: N. K. McGuire, C. A. Bateman, C. S. Blackwell, S. T. Wilson and R. M. Kirchner, *Zeolites* **15** 460–469 (1995).

h	*k*	*l*	2*θ*	*d*	*M*	I_{rel}	*h*	*k*	*l*	2*θ*	*d*	*M*	I_{rel}	*h*	*k*	*l*	2*θ*	*d*	*M*	I_{rel}
1	0	1	8.02	11.027	12	69.6	2	0	8	28.37	3.146	12	0.9	2	1	10	36.29	2.476	12	0.8
1	0	2	9.54	9.273	12	100.0	4	-1	3	28.56	3.125	12	0.9	4	0	7	36.96	2.432	12	0.1
1	0	3	11.64	7.601	12	36.5	3	1	3	28.56	3.125	12	0.5	5	-1	5	37.88	2.375	12	0.1
1	1	0	12.91	6.857	6	41.0	3	0	6	28.89	3.091	12	0.2	5	0	2	38.37	2.346	12	0.4
1	0	4	14.07	6.292	12	3.3	2	1	7	28.97	3.082	12	0.5	4	0	8	38.85	2.318	12	0.3
2	0	1	15.21	5.823	12	4.8	3	-1	7	28.97	3.082	12	1.2	3	-1	11	38.95	2.312	12	0.1
2	0	2	16.07	5.514	12	11.2	4	-1	4	29.67	3.011	12	0.5	5	0	3	38.99	2.310	12	1.0
1	0	5	16.70	5.309	12	0.2	4	0	1	30.25	2.955	12	5.3	3	3	0	39.42	2.286	6	0.3
2	0	3	17.42	5.092	12	5.8	4	0	2	30.71	2.912	12	9.5	5	0	4	39.85	2.262	12	1.3
0	0	6	17.93	4.946	2	17.7	2	0	9	31.02	2.883	12	0.7	4	0	9	40.90	2.207	12	0.6
2	0	4	19.14	4.636	12	1.2	4	-1	5	31.05	2.880	12	0.1	5	0	5	40.92	2.205	12	0.4
3	-1	1	20.00	4.439	12	4.9	1	0	10	31.06	2.879	12	3.1	4	2	4	42.05	2.148	12	0.1
2	1	1	20.00	4.439	12	7.2	3	-1	8	31.28	2.860	12	3.3	6	-2	4	42.05	2.148	12	0.1
2	1	2	20.67	4.297	12	15.5	2	1	8	31.28	2.860	12	1.4	4	2	5	43.08	2.100	12	0.1
3	-1	2	20.67	4.297	12	6.9	4	0	3	31.45	2.844	12	5.5	4	0	10	43.09	2.099	12	0.5
2	1	3	21.74	4.088	12	9.7	2	2	6	31.75	2.818	6	0.3	3	0	12	43.13	2.097	12	1.2
3	-1	3	21.74	4.088	12	3.1	4	-2	6	31.75	2.818	6	1.0	6	-3	6	43.62	2.075	6	1.8
2	-1	6	22.16	4.011	6	2.1	4	0	4	32.48	2.757	12	1.2	3	3	6	43.62	2.075	6	1.0
1	0	7	22.26	3.993	12	2.5	5	-2	2	33.43	2.680	12	0.2	5	0	8	45.33	2.000	12	0.2
3	0	0	22.46	3.959	6	1.6	3	2	2	33.43	2.680	12	0.5	4	0	11	45.42	1.997	12	0.3
2	1	4	23.16	3.841	12	1.9	3	-1	9	33.73	2.658	12	3.2	6	0	0	45.84	1.980	6	0.2
3	-1	4	23.16	3.841	12	0.1	2	1	9	33.73	2.658	12	1.9	1	0	15	46.54	1.952	12	0.6
3	-1	5	24.87	3.580	12	0.3	2	0	10	33.76	2.655	12	0.6	4	3	1	46.61	1.948	12	0.2
1	0	8	25.15	3.541	12	7.2	1	0	11	34.08	2.631	12	0.4	7	-3	2	46.93	1.936	12	0.1
2	0	7	25.82	3.450	12	0.1	5	-2	3	34.13	2.627	12	0.4	4	3	2	46.93	1.936	12	0.6
2	2	0	25.99	3.429	6	13.6	3	2	3	34.13	2.627	12	0.4	5	0	9	47.15	1.927	12	1.3
3	0	5	27.07	3.294	12	0.1	4	1	0	34.61	2.592	12	5.4	6	-2	8	47.33	1.920	12	0.1
3	1	1	27.24	3.274	12	0.4	4	1	1	34.74	2.582	12	0.1	5	2	0	47.82	1.902	12	2.0
4	-1	2	27.74	3.216	12	0.3	5	-2	4	35.08	2.558	12	0.2	2	0	15	48.50	1.877	12	0.1
3	1	2	27.74	3.216	12	0.9	5	-2	5	36.28	2.476	12	0.1	5	0	10	49.13	1.855	12	1.6
1	0	9	28.08	3.177	12	6.9	3	-1	10	36.29	2.476	12	1.1	1	0	16	49.75	1.833	12	0.4

AlPO–52, Calcined, Rehydrated **AFT**

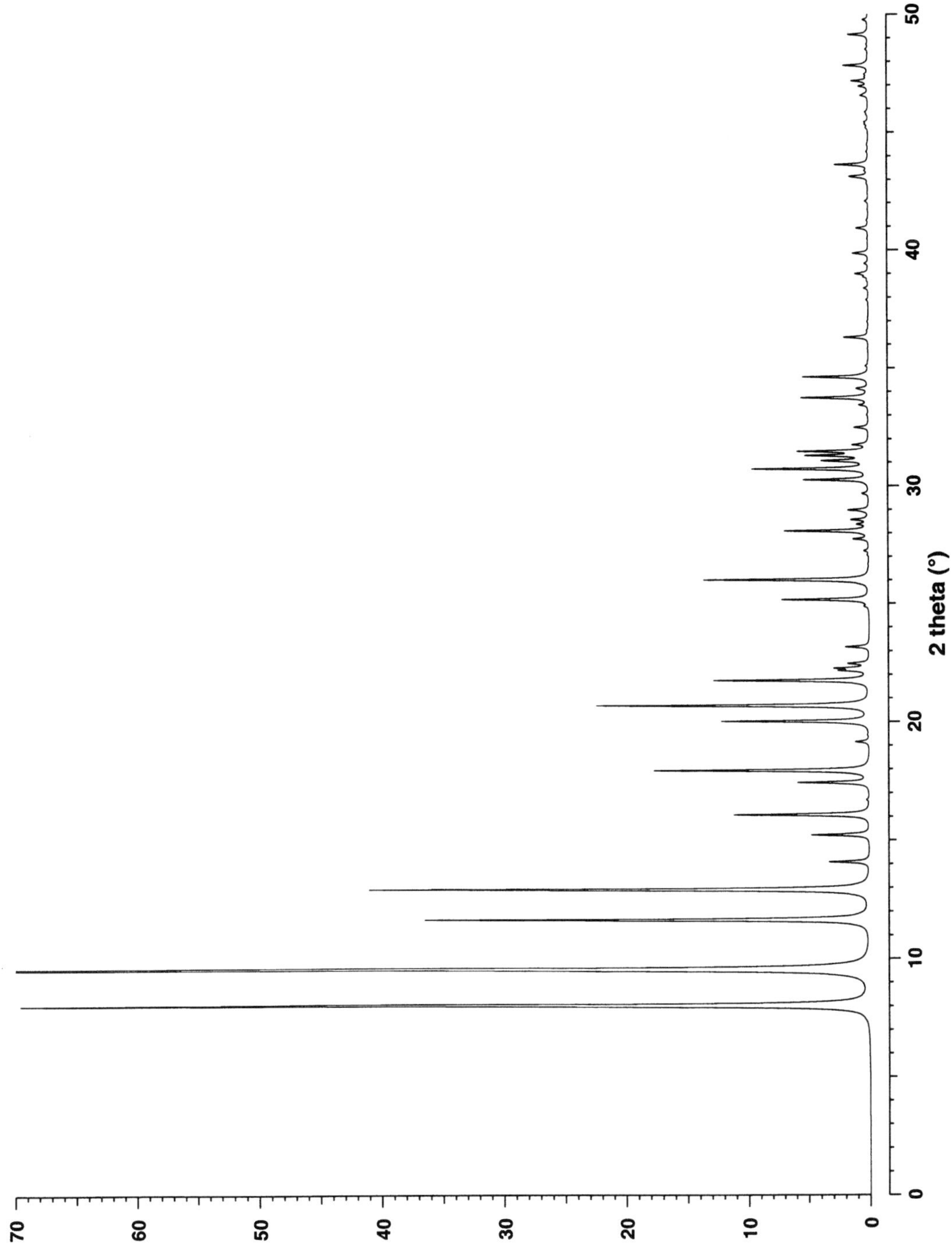

AFX SAPO–56

CHEMICAL COMPOSITION: $[Si_5Al_{23}P_{20}O_{96}]$

REFINED COMPOSITION: $[Al_{24}P_{24}O_{96}]$

CRYSTAL DATA: $P\overline{3}1c$ (No. 163)
$a = 13.7617$ Å $b = 13.7617$ Å $c = 19.949$ Å
$\alpha = 90°$ $\beta = 90°$ $\gamma = 120°$
X-ray Rietveld refinement, $R_{exp} = 0.068$, $R_{wp} = 0.219$

REFERENCE: N. K. McGuire, C. S. Blackwell, C. A. Bateman, S. T. Wilson and R. M. Kirchner, *Microporous and Mesoporous Materials* **28** 125–137 (1999).

h	k	l	2θ	d	M	I_{rel}
1	0	0	7.42	11.918	6	30.0
1	0	1	8.64	10.231	12	100.0
1	0	2	11.57	7.649	12	54.0
1	1	0	12.87	6.881	6	43.2
2	0	0	14.87	5.959	6	1.4
1	0	3	15.26	5.807	12	0.2
2	0	1	15.52	5.710	12	8.0
2	0	2	17.33	5.116	12	8.7
0	0	4	17.78	4.987	2	19.1
1	0	4	19.29	4.601	12	1.5
2	1	0	19.71	4.505	12	4.4
2	0	3	20.01	4.438	12	0.8
2	1	1	20.21	4.394	12	7.3
3	-1	1	20.21	4.394	12	14.8
3	-1	2	21.65	4.105	12	15.7
2	1	2	21.65	4.105	12	3.8
1	1	4	22.01	4.038	6	1.5
3	0	0	22.38	3.973	6	2.5
1	0	5	23.51	3.783	12	4.7
2	1	3	23.86	3.729	12	0.1
3	-1	3	23.86	3.729	12	0.3
2	2	0	25.90	3.440	6	12.6
2	1	4	26.67	3.343	12	0.8
3	-1	4	26.67	3.343	12	0.4
2	0	5	26.89	3.315	12	0.3
4	-1	1	27.35	3.261	12	0.7
1	0	6	27.86	3.203	12	11.9
4	-1	2	28.45	3.138	12	0.9
3	1	2	28.45	3.138	12	1.2
3	0	4	28.73	3.107	12	0.2
3	-1	5	29.92	2.987	12	1.1
2	1	5	29.92	2.987	12	2.6
4	0	0	29.99	2.979	6	2.0
3	1	3	30.19	2.960	12	0.5
4	-1	3	30.19	2.960	12	0.1
4	0	1	30.33	2.947	12	10.3
2	0	6	30.79	2.903	12	1.8
4	0	2	31.33	2.855	12	9.0
4	-2	4	31.59	2.832	6	0.2
2	2	4	31.59	2.832	6	0.9
1	0	7	32.30	2.772	12	1.4
4	0	3	32.94	2.719	12	0.3
3	-1	6	33.50	2.675	12	3.9
2	1	6	33.50	2.675	12	4.0
5	-2	2	34.00	2.637	12	0.2
3	2	2	34.00	2.637	12	1.0
4	1	0	34.49	2.601	12	4.6
2	0	7	34.90	2.571	12	0.4
5	-2	3	35.50	2.529	12	0.2
3	2	3	35.50	2.529	12	0.5
4	1	3	37.12	2.422	12	0.2
2	1	7	37.34	2.408	12	0.6
3	-1	7	37.34	2.408	12	0.2
4	0	5	37.68	2.387	12	0.2
5	0	1	38.02	2.367	12	0.1
2	-1	8	38.39	2.344	6	0.3
5	0	2	38.84	2.318	12	1.7
3	3	0	39.28	2.294	6	0.4
5	0	3	40.19	2.244	12	1.1
6	-2	1	40.30	2.238	12	0.2
4	0	6	40.66	2.219	12	1.2
6	-2	2	41.08	2.197	12	0.2
4	2	2	41.08	2.197	12	0.1
5	-1	5	41.44	2.179	12	0.2
4	1	5	41.44	2.179	12	0.1
3	0	8	42.82	2.112	12	1.1
6	-1	2	43.23	2.093	12	0.3
6	-3	4	43.42	2.084	6	1.0
3	3	4	43.42	2.084	6	1.6
4	0	7	43.96	2.059	12	0.4
4	-2	8	44.89	2.019	6	0.2
6	0	0	45.67	1.986	6	0.1
1	0	10	46.13	1.968	12	0.9
7	-3	1	46.58	1.950	12	0.4
5	0	6	46.90	1.937	12	2.4
7	-3	2	47.28	1.923	12	0.1
4	3	2	47.28	1.923	12	0.3
5	-1	7	47.32	1.921	12	0.2
5	2	0	47.65	1.908	12	2.6
2	0	10	48.10	1.892	12	0.2
5	0	7	49.88	1.828	12	1.7

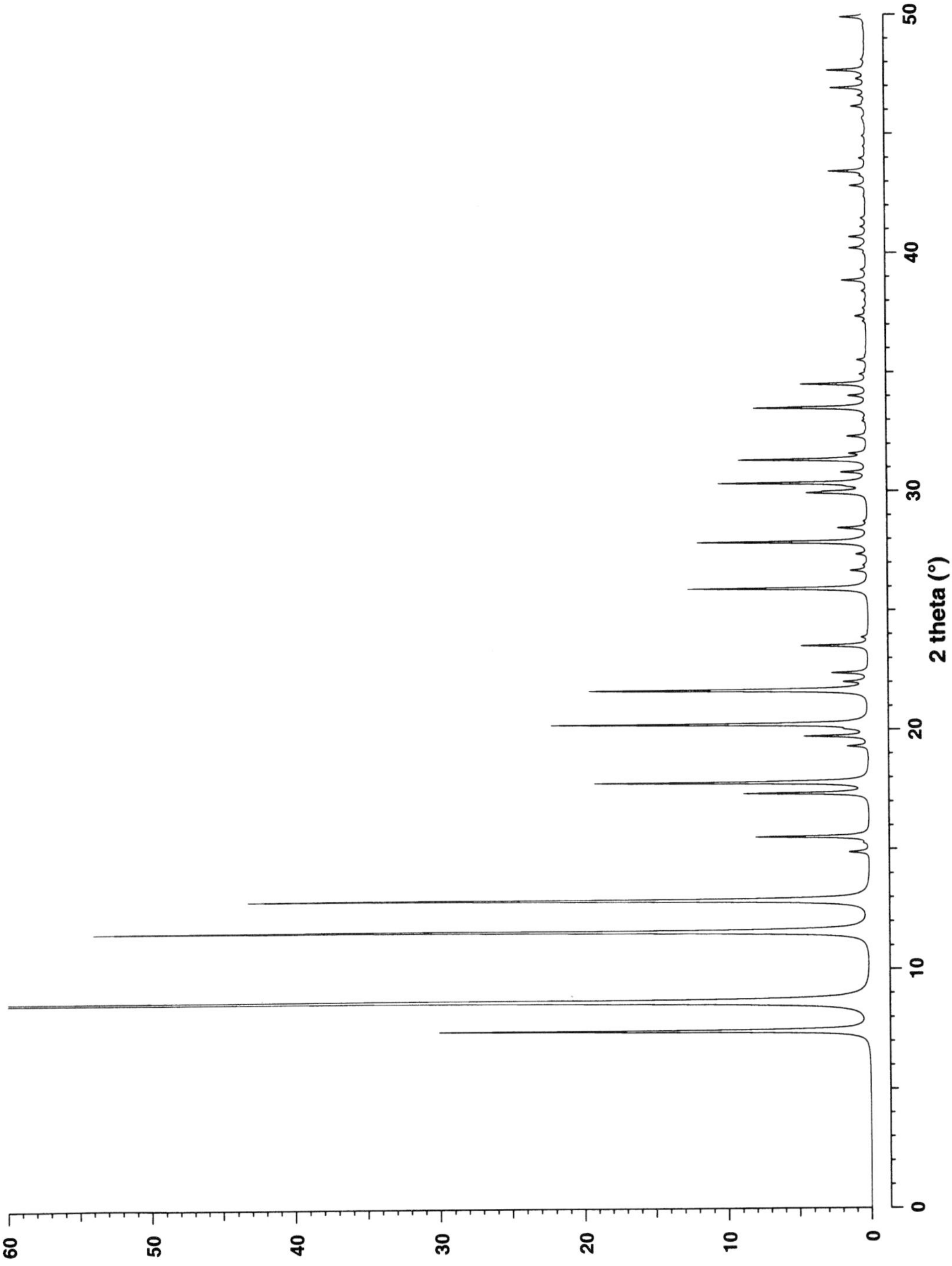
2 theta (°)
0
10
20
30
40
50
60

AFY Di-n-propylamine CoAPO-50

CHEMICAL COMPOSITION: $|((C_3H_7)_2NH)_3(H_2O)_{6.7}|$ $[Co_3Al_5P_8O_{32}]$
$(C_3H_7)_2NH$ = di-n-propylamine

REFINED COMPOSITION: $|(C_4N)_6|$ $[Al_6Co_2P_8O_{32}]$

CRYSTAL DATA: $P\overline{3}$ (No. 147)
a = 12.7468 Å b = 12.7468 Å c = 9.0153 Å
α = 90° β = 90° γ = 120°
X-ray single crystal refinement, R_w = 0.071

REFERENCE: J. M. Bennett and B. K. Marcus,
in *Innovations in Zeolite Material Science* (*Stud. Surf. Sci. Catal. No. 37*)
Ed. by P. J. Grobet, W. J. Mortier, E. F. Vansant and G. Schulz-Ekloff
(Elsevier: Amsterdam) 269–279 (1988).
and J. M. Bennett, Private communication.

h	k	l	2θ	d	M	I_{rel}
0	1	0	8.01	11.039	6	100.0
0	0	1	9.81	9.015	2	8.8
0	1	1	12.68	6.983	6	5.1
1	1	0	13.89	6.373	6	2.8
0	2	0	16.06	5.520	6	1.2
2	-1	1	17.04	5.204	6	3.1
2	1	0	21.29	4.172	6	0.5
1	2	0	21.29	4.172	6	0.2
1	2	1	23.49	3.787	6	0.3
3	-1	1	23.49	3.787	6	12.5
3	-2	1	23.49	3.787	6	9.5
1	1	2	24.18	3.680	6	1.3
2	-1	2	24.18	3.680	6	19.4
0	3	0	24.19	3.680	6	3.6
0	2	2	25.51	3.491	6	0.5
2	0	2	25.51	3.491	6	3.1
3	0	1	26.16	3.407	6	4.4
0	3	1	26.16	3.407	6	3.0
3	-2	2	29.16	3.062	6	3.6
3	-1	2	29.16	3.062	6	0.4
1	3	0	29.17	3.062	6	0.7
3	1	0	29.17	3.062	6	0.4
0	0	3	29.73	3.005	2	0.3
4	-2	1	29.73	3.005	6	9.2
0	1	3	30.84	2.900	6	0.2
4	-3	1	30.84	2.899	6	2.1
4	-1	1	30.84	2.899	6	1.5
0	3	2	31.38	2.851	6	1.6
3	0	2	31.38	2.851	6	1.7
0	4	0	32.44	2.760	6	0.2
1	1	3	32.95	2.718	6	3.2
2	-1	3	32.95	2.718	6	0.1
0	2	3	33.97	2.639	6	0.9
2	0	3	33.97	2.639	6	0.3
0	4	1	33.97	2.639	6	1.1
4	0	1	33.97	2.639	6	0.1
4	-2	2	34.47	2.602	6	0.4
2	2	2	34.47	2.602	6	0.6
4	-3	2	35.44	2.533	6	0.3
4	-1	2	35.44	2.533	6	0.8
2	3	0	35.44	2.533	6	0.2
1	2	3	36.86	2.438	6	0.7
2	1	3	36.86	2.438	6	0.4
3	2	1	36.86	2.438	6	0.2
1	4	0	37.33	2.409	6	1.3
4	1	0	37.33	2.409	6	1.9
4	0	2	38.24	2.354	6	0.3
1	4	1	38.69	2.327	6	0.2
5	-1	1	38.69	2.327	6	0.1
5	-4	1	38.69	2.327	6	0.2
4	1	1	38.69	2.327	6	0.3
0	1	4	40.86	2.208	6	0.3
5	-2	2	40.87	2.208	6	0.1
0	5	0	40.87	2.208	6	0.1
4	-2	3	41.29	2.186	6	0.2
2	2	3	41.29	2.186	6	1.7
0	5	1	42.14	2.144	6	0.3
1	1	4	42.54	2.125	6	0.5
2	-1	4	42.54	2.125	6	0.2
5	-4	2	42.55	2.125	6	0.3
3	3	0	42.55	2.124	6	0.5
2	0	4	43.36	2.087	6	0.3
4	2	0	43.37	2.086	6	0.5
4	0	3	44.58	2.033	6	0.1
4	2	1	44.58	2.032	6	0.2
6	-4	1	44.58	2.032	6	0.5
1	2	4	45.75	1.983	6	0.5
5	0	2	45.76	1.983	6	0.3
0	5	2	45.76	1.983	6	0.2
1	5	0	45.76	1.983	6	0.4
5	1	0	45.76	1.983	6	0.3
5	-3	3	46.92	1.937	6	0.8
5	-2	3	46.92	1.937	6	0.5
3	2	3	46.92	1.937	6	0.2
3	3	2	47.30	1.922	6	0.6
6	-3	2	47.30	1.922	6	0.5
5	-1	3	48.43	1.880	6	0.4
2	2	4	49.54	1.840	6	0.8
0	6	0	49.54	1.840	6	2.2

Di-n-propylamine CoAPO–50 AFY

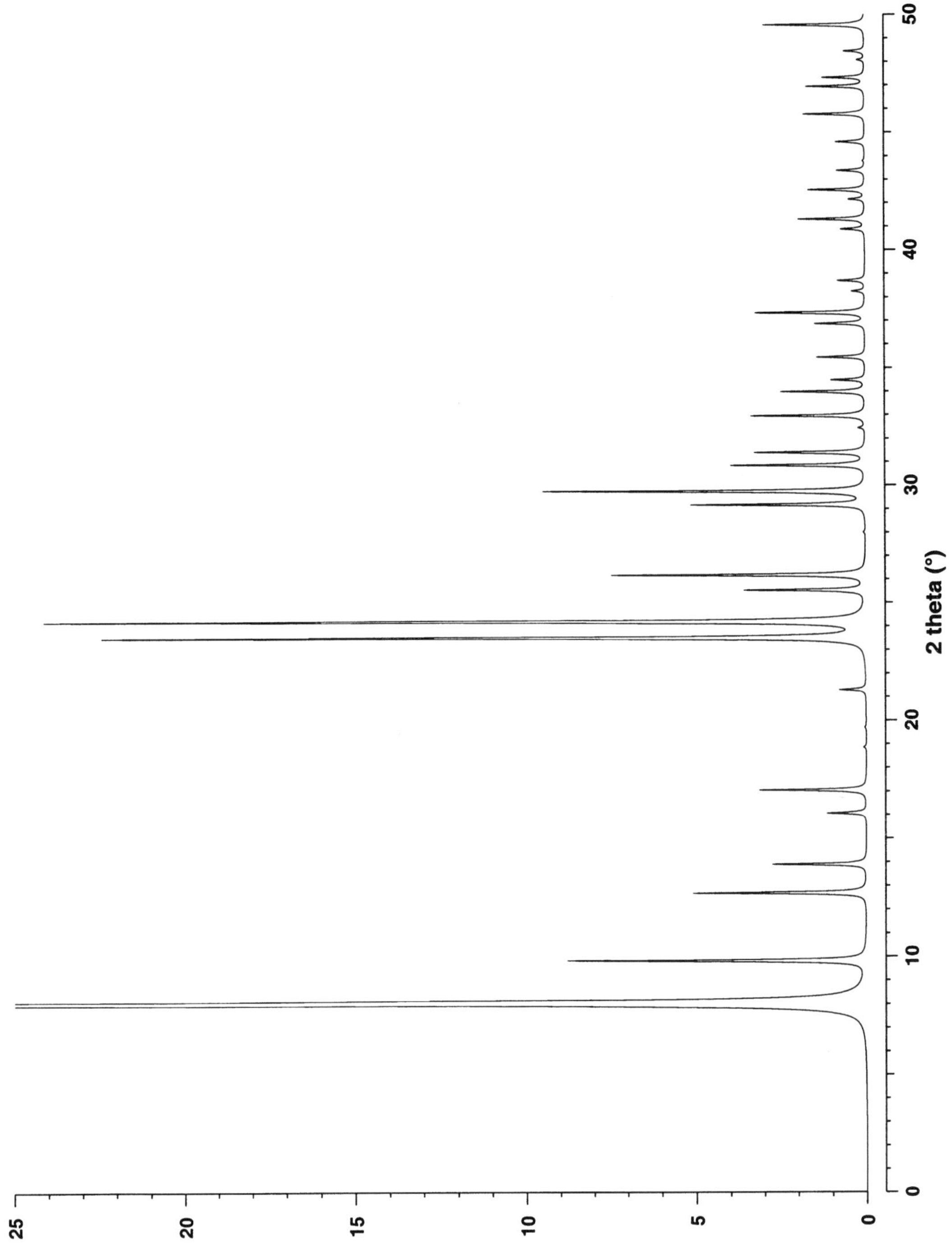

AHT

AlPO - H2

CHEMICAL COMPOSITION: $|(H_2O)_4|\ [Al_6P_6O_{24}]$

REFINED COMPOSITION: $|(H_2O)_4|\ [Al_6P_6O_{24}]$

CRYSTAL DATA: $P112_1$ (No. 4) unique axis **c**
$a = 9.486$ Å $b = 9.914$ Å $c = 8.126$ Å
$\alpha = 90°$ $\beta = 90°$ $\gamma = 121.49°$
X-ray Rietveld refinement, $R_{wp} = 0.127$

REFERENCE: J. B. Higgins, R. M. Dessau, H-X. Li, M. E. Davis and J.M. Newsam, *American Crystallographic Association Abstracts* **K005** 39 (1993).

h	k	l	2θ	d	M	I_{rel}
0	1	0	10.46	8.454	2	87.1
1	-1	0	10.47	8.452	2	100.0
1	0	0	10.94	8.089	2	4.1
0	1	1	15.12	5.858	4	4.7
1	-1	1	15.12	5.858	4	3.9
1	0	1	15.46	5.733	4	0.2
1	-2	0	17.89	4.957	2	14.8
1	1	0	18.73	4.738	2	5.7
2	-1	0	18.73	4.737	2	3.1
1	-2	1	20.99	4.232	4	3.8
0	2	0	21.02	4.227	2	0.5
2	-2	0	21.02	4.226	2	0.2
1	1	1	21.71	4.093	4	13.4
2	-1	1	21.72	4.092	4	11.4
0	0	2	21.87	4.063	2	56.5
2	0	0	21.98	4.045	2	6.1
0	2	1	23.73	3.750	4	15.1
2	-2	1	23.73	3.749	4	13.6
0	1	2	24.30	3.662	4	12.2
1	-1	2	24.31	3.662	4	10.3
1	0	2	24.52	3.631	4	4.3
2	0	1	24.59	3.621	4	0.2
1	-3	0	27.55	3.238	2	8.7
2	-3	0	27.55	3.238	2	12.4
1	-2	2	28.40	3.142	4	8.1
3	-2	0	28.49	3.133	2	0.2
1	1	2	28.95	3.084	4	5.4
2	-1	2	28.95	3.084	4	5.8
2	1	0	29.03	3.076	2	15.6
3	-1	0	29.03	3.076	2	13.1
1	-3	1	29.70	3.008	4	2.1
2	-3	1	29.70	3.008	4	6.1
0	2	2	30.52	2.929	4	2.8
2	-2	2	30.52	2.929	4	2.5
1	2	1	30.57	2.924	4	1.7
3	-2	1	30.58	2.924	4	1.5
2	1	1	31.09	2.877	4	1.7
3	-1	1	31.09	2.877	4	2.5
2	0	2	31.20	2.866	4	3.7
0	3	0	31.75	2.818	2	4.3
3	-3	0	31.76	2.817	2	6.0
3	0	0	33.23	2.696	2	11.3
0	3	1	33.66	2.662	4	2.3
3	-3	1	33.67	2.662	4	2.1
0	1	3	34.78	2.579	4	0.4
1	-1	3	34.78	2.579	4	0.6
1	0	3	34.93	2.568	4	0.2
3	0	1	35.06	2.559	4	0.2
1	-3	2	35.45	2.532	4	1.6
2	-3	2	35.45	2.532	4	2.2
1	2	2	36.20	2.482	4	0.3
3	-2	2	36.20	2.481	4	0.6
2	-4	0	36.24	2.478	2	1.1
2	1	2	36.64	2.453	4	1.5
3	-1	2	36.64	2.452	4	2.8
1	-2	3	37.85	2.377	4	0.9
2	-4	1	37.95	2.371	4	0.9
1	-4	0	37.97	2.370	2	0.2
4	-2	0	37.99	2.368	2	0.2
1	1	3	38.27	2.351	4	3.2
2	-1	3	38.28	2.351	4	0.9
0	3	2	38.89	2.316	4	0.2
3	-3	2	38.90	2.315	4	1.2
1	3	0	38.95	2.312	2	0.7
4	-3	0	38.96	2.312	2	1.0
2	-2	3	39.52	2.280	4	2.6
1	-4	1	39.61	2.275	4	0.4
3	-4	1	39.62	2.275	4	0.6
2	2	1	39.63	2.274	4	0.6
4	-2	1	39.64	2.274	4	0.6
2	0	3	40.06	2.251	4	0.5
3	0	2	40.14	2.247	4	0.3
1	3	1	40.56	2.224	4	0.3
4	-3	1	40.57	2.223	4	0.3
3	1	1	41.63	2.169	4	1.1
4	-1	1	41.64	2.169	4	0.6
2	-4	2	42.73	2.116	4	0.3
4	-4	0	42.79	2.113	2	0.3
1	-3	3	43.56	2.078	4	1.0
2	-3	3	43.56	2.077	4	0.5
1	2	3	44.19	2.049	4	0.4
3	-2	3	44.20	2.049	4	0.1
2	2	2	44.26	2.046	4	0.4
4	-2	2	44.27	2.046	4	0.5
0	4	1	44.28	2.045	4	0.3
4	-4	1	44.29	2.045	4	0.1
2	1	3	44.57	2.033	4	0.2
3	-1	3	44.57	2.033	4	0.3
0	0	4	44.60	2.032	2	3.9
1	3	2	45.11	2.010	4	0.3
0	1	4	45.94	1.975	4	0.8
1	-1	4	45.94	1.975	4	0.3
1	0	4	46.07	1.970	4	0.8
3	1	2	46.10	1.969	4	0.3
4	-1	2	46.10	1.969	4	0.4
0	3	3	46.50	1.953	4	0.4
3	-3	3	46.51	1.953	4	0.2
2	-5	1	47.53	1.913	4	1.0
3	-5	1	47.54	1.913	4	0.9
3	0	3	47.58	1.911	4	0.2
5	-3	0	48.04	1.894	2	0.2
1	-2	4	48.42	1.880	4	1.7
0	4	2	48.55	1.875	4	0.3
4	-4	2	48.56	1.875	4	0.9
1	1	4	48.77	1.867	4	1.3
2	-1	4	48.77	1.867	4	1.4
2	3	1	49.40	1.845	4	2.9
5	-3	1	49.42	1.844	4	0.6
0	2	4	49.80	1.831	4	1.8
2	-2	4	49.80	1.831	4	1.2
2	-4	3	49.87	1.829	4	0.1
1	4	0	49.98	1.825	2	2.4
3	2	1	49.98	1.825	4	1.8
5	-4	0	49.99	1.824	2	1.6
5	-2	1	49.99	1.824	4	1.0

AlPO - H2 **AHT**

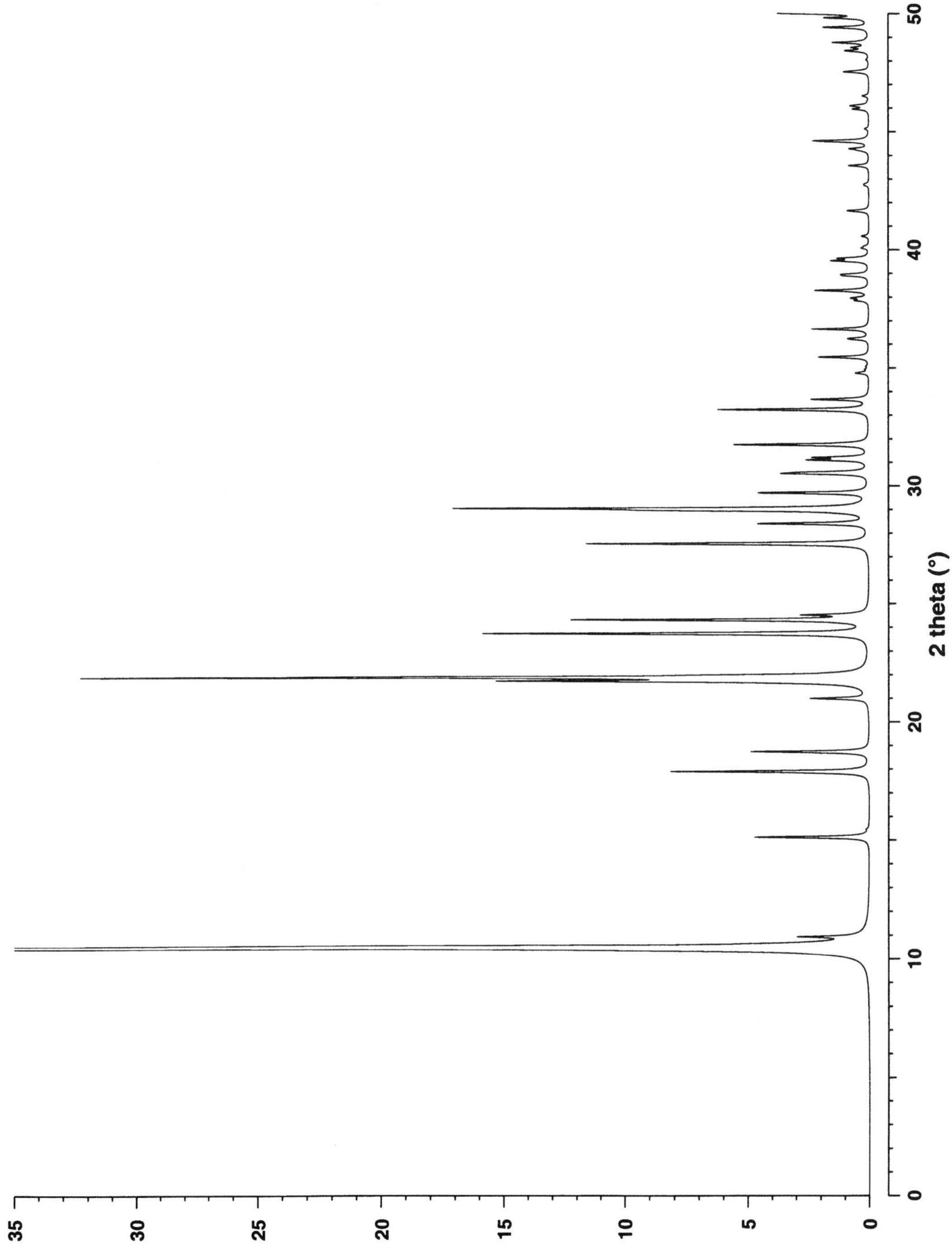

ANA Analcime

CHEMICAL COMPOSITION: $|Na_{16}(H_2O)_{16}|$ $[Si_{32}Al_{16}O_{96}]$
Cyclopean Islands, Greece

REFINED COMPOSITION: $|Na_{16}(H_2O)_{16}|$ $[Si_{32}Al_{16}O_{96}]$

CRYSTAL DATA: $Ia\bar{3}d$ (No. 230)
$a = 13.73$ Å $b = 13.73$ Å $c = 13.73$ Å
$\alpha = 90°$ $\beta = 90°$ $\gamma = 90°$
X-ray single crystal refinement, $R = 0.04$

REFERENCE: G. Ferraris, D. W. Jones and J. Yerkess, *Z. Kristallogr.* **135** 240–252 (1972).

h	k	l	2θ	d	M	I_{rel}	h	k	l	2θ	d	M	I_{rel}	h	k	l	2θ	d	M	I_{rel}
2	1	1	15.81	5.605	24	60.2	5	2	1	35.82	2.507	48	11.9	4	4	4	45.78	1.982	8	0.4
2	2	0	18.28	4.854	12	14.1	4	4	0	37.04	2.427	12	7.7	5	4	3	46.78	1.942	48	0.7
3	2	1	24.25	3.669	48	5.4	6	1	1	40.50	2.227	24	2.8	6	4	0	47.77	1.904	24	14.9
4	0	0	25.96	3.433	6	100.0	5	3	2	40.50	2.227	48	5.5	5	5	2	48.74	1.868	24	0.2
3	3	2	30.54	2.927	24	51.3	6	2	0	41.60	2.171	24	1.0	6	3	3	48.74	1.868	24	6.9
4	2	2	31.93	2.803	24	5.5	5	4	1	42.68	2.119	48	0.9	7	2	1	48.74	1.868	48	2.3
4	3	1	33.27	2.693	48	14.8	6	3	1	44.77	2.024	48	1.3	6	4	2	49.69	1.835	48	0.4

Analcime ANA

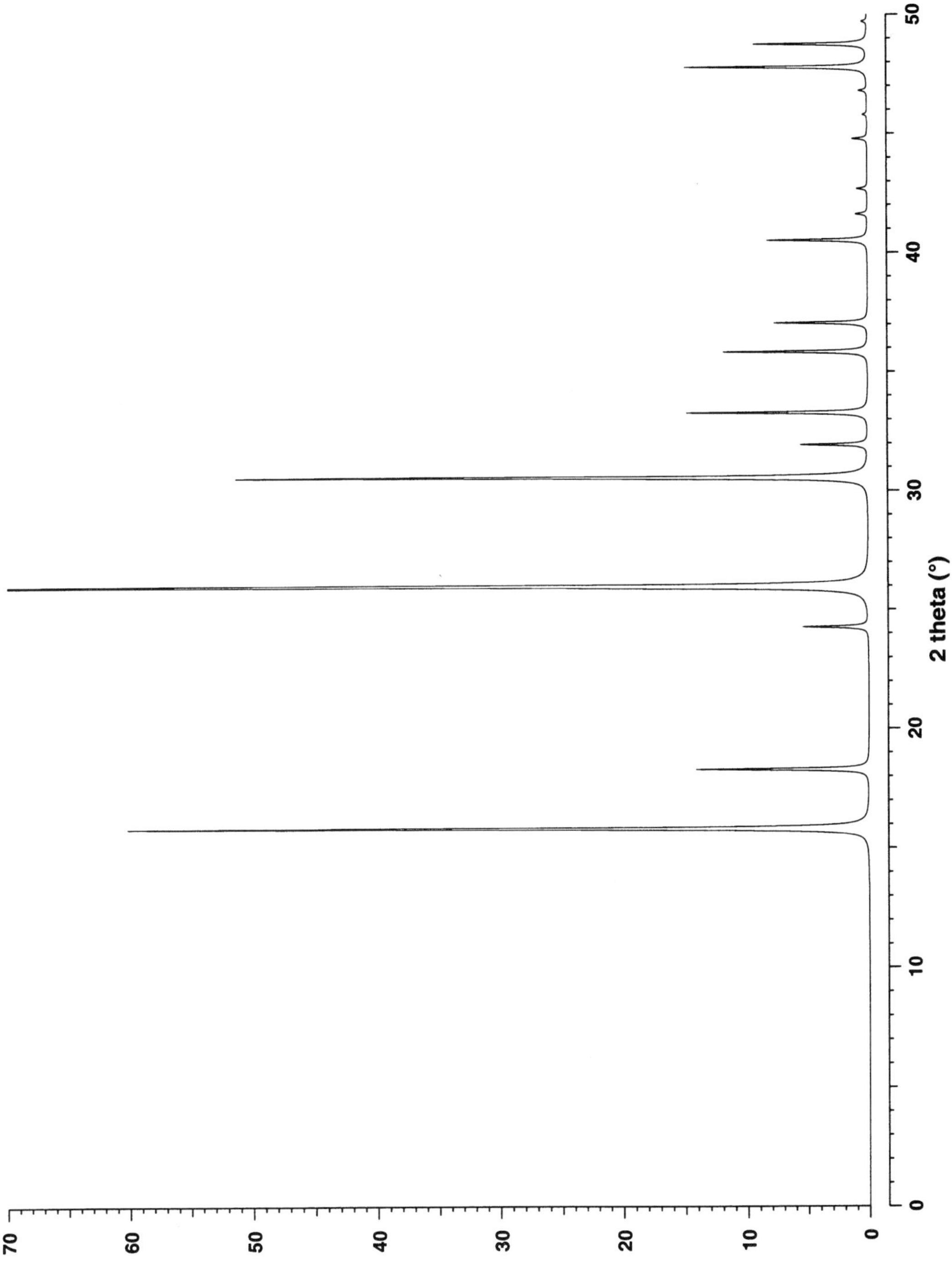

APC AlPO–C, Hydrated

CHEMICAL COMPOSITION: $|(H_2O)_{24}|$ $[Al_{16}P_{16}O_{64}]$

REFINED COMPOSITION: $|(H_2O)_{16}O_8|$ $[Al_{16}P_{16}O_{64}]$

CRYSTAL DATA: *Pbca* (No. 61)
$a = 19.3525$ Å $b = 9.7272$ Å $c = 9.7621$ Å
$\alpha = 90°$ $\beta = 90°$ $\gamma = 90°$
X-ray single crystal refinement, $R = 0.033$

REFERENCE: J. J. Pluth and J. V. Smith,
Acta Cryst. **C42** 1118–1120 (1986).

h	*k*	*l*	2θ	*d*	*M*	I_{rel}
2	0	0	9.14	9.676	2	45.5
2	1	0	12.90	6.860	4	100.0
1	1	1	13.64	6.491	8	79.8
2	1	1	15.79	5.613	8	7.1
0	0	2	18.17	4.881	2	30.8
0	2	0	18.24	4.864	2	0.1
4	0	0	18.34	4.838	2	42.4
1	0	2	18.75	4.733	4	0.2
3	1	1	18.84	4.709	8	0.7
2	0	2	20.38	4.358	4	0.7
0	2	1	20.40	4.353	4	0.2
2	2	0	20.44	4.346	4	0.8
4	1	0	20.50	4.332	4	3.3
1	1	2	20.87	4.256	8	21.0
1	2	1	20.92	4.247	8	54.6
2	1	2	22.35	3.977	8	1.9
4	1	1	22.45	3.960	8	2.0
3	1	2	24.63	3.614	8	3.5
3	2	1	24.67	3.608	8	6.4
0	2	2	25.86	3.445	4	5.4
4	0	2	25.93	3.436	4	1.3
4	2	0	25.98	3.430	4	0.2
1	2	2	26.27	3.392	8	34.8
5	1	1	26.41	3.375	8	5.9
2	2	2	27.48	3.246	8	0.4
4	1	2	27.53	3.240	8	0.2
6	0	0	27.66	3.225	2	0.3
2	3	0	29.04	3.074	4	10.2
6	1	0	29.17	3.061	4	59.8
1	1	3	29.31	3.047	8	24.1
1	3	1	29.39	3.039	8	1.1
5	0	2	29.45	3.033	4	0.6
2	1	3	30.40	2.940	8	0.8
2	3	1	30.48	2.932	8	11.8
6	1	1	30.60	2.921	8	5.7
5	1	2	30.88	2.895	8	2.3
5	2	1	30.91	2.893	8	12.7
3	1	3	32.15	2.784	8	1.6
3	3	1	32.23	2.777	8	10.2
0	2	3	33.12	2.705	4	0.1
4	3	0	33.26	2.693	4	0.6
6	0	2	33.29	2.691	4	0.2
6	2	0	33.33	2.688	4	1.0
1	2	3	33.45	2.679	8	35.8
2	2	3	34.43	2.605	8	0.4
4	1	3	34.47	2.602	8	1.8
2	3	2	34.48	2.601	8	1.1
4	3	1	34.54	2.596	8	0.1
6	1	2	34.58	2.594	8	0.5
5	2	2	34.86	2.573	8	1.9
7	1	1	34.97	2.566	8	4.9
3	2	3	36.01	2.494	8	0.8
3	3	2	36.05	2.491	8	0.8
0	0	4	36.83	2.441	2	1.9
0	4	0	36.96	2.432	2	0.9
1	0	4	37.13	2.421	4	3.4
8	0	0	37.17	2.419	2	0.5
5	1	3	37.26	2.413	8	3.4
5	3	1	37.33	2.409	8	0.6
7	0	2	37.38	2.406	4	2.5
2	0	4	38.02	2.366	4	0.1
4	2	3	38.12	2.361	8	0.1
0	4	1	38.14	2.360	4	2.5
2	4	0	38.16	2.358	4	2.7
4	3	2	38.16	2.358	8	0.7
6	2	2	38.22	2.355	8	0.4
1	1	4	38.31	2.350	8	1.8
8	1	0	38.34	2.348	4	0.4
1	4	1	38.43	2.342	8	0.7
7	2	1	38.58	2.334	8	0.2
2	1	4	39.18	2.299	8	0.1
2	4	1	39.30	2.293	8	1.9
6	3	0	39.40	2.287	4	3.5
3	0	4	39.48	2.283	4	0.8
8	1	1	39.48	2.282	8	0.2
1	3	3	39.51	2.281	8	0.4
2	3	3	40.36	2.235	8	0.2
6	1	3	40.45	2.230	8	0.2
6	3	1	40.52	2.226	8	0.9
3	1	4	40.60	2.222	8	1.7
5	2	3	40.70	2.217	8	0.7
3	4	1	40.71	2.216	8	2.0
5	3	2	40.74	2.215	8	0.5
0	2	4	41.39	2.181	4	1.4
4	0	4	41.44	2.179	4	1.6
0	4	2	41.49	2.177	4	0.8
4	4	0	41.56	2.173	4	0.4
1	2	4	41.67	2.168	8	1.3
8	0	2	41.67	2.167	4	1.8
8	2	0	41.70	2.166	4	4.6
3	3	3	41.74	2.164	8	1.5
1	4	2	41.76	2.163	8	0.9
7	2	2	41.90	2.156	8	3.8
2	2	4	42.48	2.128	8	1.3
4	1	4	42.51	2.126	8	1.6
2	4	2	42.57	2.124	8	0.6
8	2	1	42.76	2.115	8	1.0
4	3	3	43.62	2.075	8	0.8
3	2	4	43.81	2.066	8	1.2
7	1	3	43.97	2.059	8	1.5
7	3	1	44.03	2.057	8	1.0
9	1	1	44.12	2.053	8	0.4
5	1	4	44.88	2.019	8	0.3
4	2	4	45.62	1.989	8	0.5
4	4	2	45.71	1.985	8	1.0
8	2	2	45.83	1.980	8	0.1
5	3	3	45.94	1.975	8	0.7
0	4	3	46.63	1.948	4	1.9
6	4	0	46.78	1.942	4	0.4
1	3	4	46.83	1.940	8	0.7
1	4	3	46.88	1.938	8	1.2
7	2	3	47.00	1.933	8	0.3
7	3	2	47.04	1.932	8	3.1
9	1	2	47.12	1.929	8	0.7
9	2	1	47.14	1.928	8	1.0
2	3	4	47.57	1.911	8	0.8
2	4	3	47.62	1.910	8	0.3
6	1	4	47.65	1.908	8	0.5
1	1	5	47.74	1.905	8	3.4
6	4	1	47.76	1.904	8	0.8
8	1	3	47.77	1.904	8	0.2
8	3	1	47.83	1.902	8	0.4
5	2	4	47.87	1.900	8	0.7
1	5	1	47.91	1.899	8	0.2
10	1	0	47.93	1.898	4	4.8
5	4	2	47.95	1.897	8	0.1
2	5	1	48.64	1.872	8	0.3
6	3	3	48.67	1.871	8	1.0
3	4	3	48.84	1.865	8	3.2
3	1	5	49.68	1.835	8	0.1
3	5	1	49.84	1.830	8	0.3

AlPO-C, Hydrated

APC

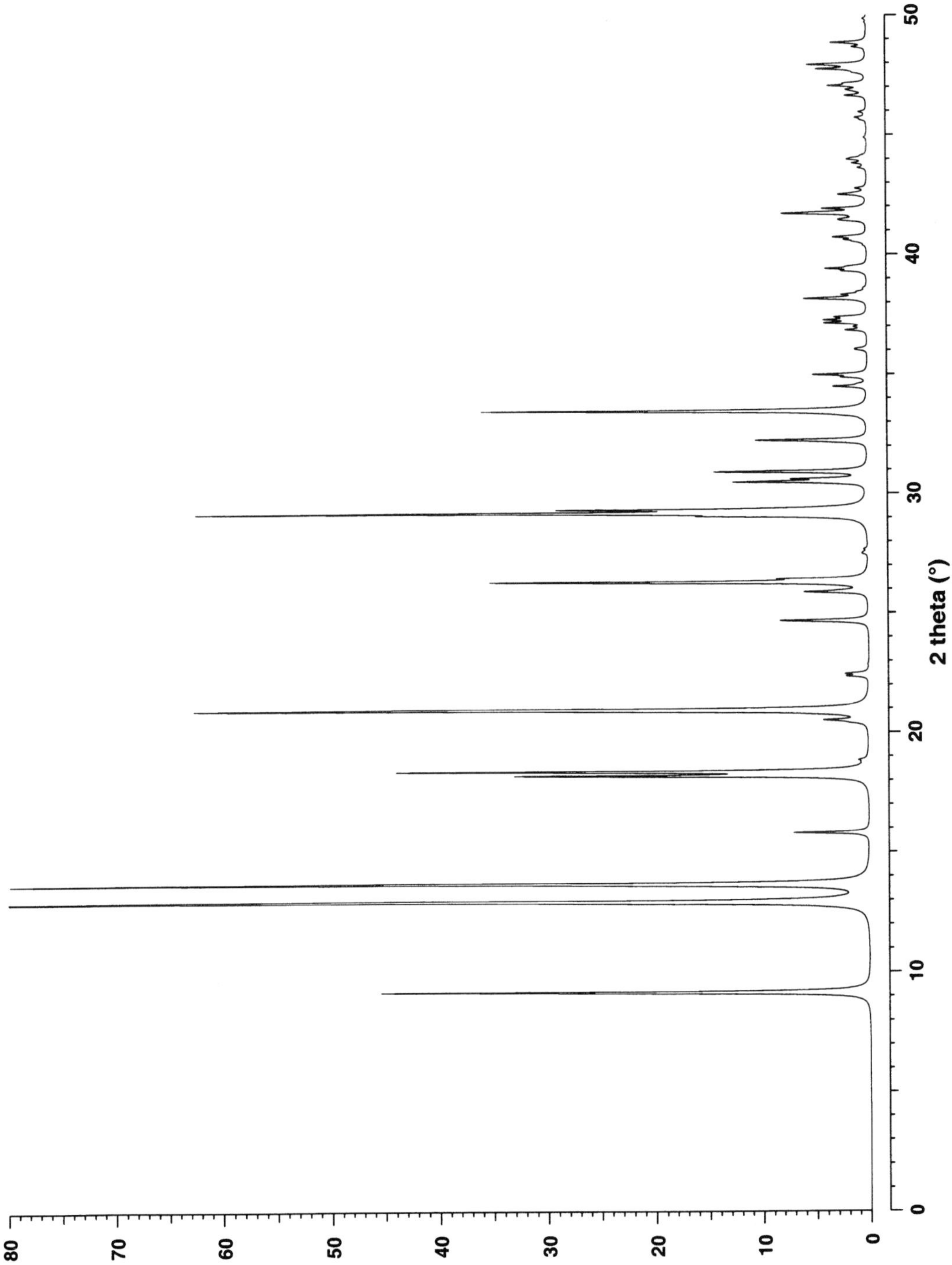

APC AlPO–C, Dehydrated

CHEMICAL COMPOSITION: $[Al_{16}P_{16}O_{64}]$

REFINED COMPOSITION: $[Al_{16}P_{16}O_{64}]$

CRYSTAL DATA: *Pbca* (No. 61)
$a = 19.816$ Å $b = 10.047$ Å $c = 8.935$ Å
$\alpha = 90°$ $\beta = 90°$ $\gamma = 90°$
X-ray Rietveld refinement, $R_{wp} = 0.17$, $R_F = 0.051$; At $T = 100°C$.

REFERENCE: E. B. Keller, W. M. Meier and R. M. Kirchner, *Solid State Ionics* **43** 93–102 (1990).

h	k	l	2θ	d	M	I_{rel}
2	1	0	12.55	7.055	4	100.0
1	1	1	14.00	6.327	8	37.6
2	1	1	16.01	5.537	8	8.8
0	2	0	17.65	5.023	2	3.0
4	0	0	17.90	4.954	2	40.4
3	1	1	18.90	4.696	8	0.8
0	0	2	19.87	4.468	2	39.2
1	0	2	20.38	4.358	4	4.0
1	2	1	20.77	4.276	8	27.4
2	2	1	22.19	4.005	8	4.2
1	1	2	22.23	3.998	8	8.0
4	1	1	22.35	3.978	8	1.9
2	1	2	23.57	3.774	8	5.9
3	1	2	25.65	3.473	8	0.1
5	1	1	26.15	3.408	8	4.9
0	2	2	26.70	3.338	4	4.2
4	0	2	26.87	3.318	4	2.9
1	2	2	27.09	3.292	8	6.3
4	2	1	27.18	3.281	8	0.2
2	3	0	28.13	3.173	4	8.1
2	2	2	28.21	3.164	8	1.6
4	1	2	28.33	3.150	8	0.3
6	1	0	28.45	3.137	4	26.1
1	3	1	28.82	3.097	8	3.3
2	3	1	29.88	2.990	8	5.2
5	0	2	30.14	2.965	4	0.4
6	1	1	30.19	2.960	8	2.8
5	2	1	30.42	2.938	8	5.5
5	1	2	31.46	2.844	8	2.3
1	1	3	31.66	2.826	8	5.1
4	2	2	32.34	2.768	8	0.2
2	1	3	32.63	2.744	8	2.0
1	3	2	33.75	2.655	8	0.8
6	2	1	34.00	2.637	8	0.1
3	1	3	34.21	2.621	8	0.8
7	1	1	34.41	2.606	8	2.3
2	3	2	34.68	2.587	8	1.9
6	1	2	34.94	2.568	8	1.4
0	2	3	35.02	2.562	4	0.5
5	2	2	35.15	2.553	8	2.3
1	2	3	35.33	2.541	8	3.8
0	4	0	35.75	2.512	2	3.6
3	3	2	36.17	2.483	8	0.1
5	3	1	36.54	2.459	8	0.9
0	4	1	37.18	2.418	4	0.8
7	0	2	37.61	2.391	4	0.1
3	2	3	37.66	2.389	8	0.4
7	2	1	37.84	2.377	8	2.1
4	3	2	38.18	2.357	8	0.2
6	3	0	38.27	2.352	4	2.9
7	1	2	38.71	2.326	8	0.6
5	1	3	38.87	2.317	8	0.5
6	3	1	39.63	2.274	8	0.2
3	4	1	39.69	2.271	8	0.2
4	4	0	40.26	2.240	4	0.5
0	0	4	40.38	2.234	2	0.3
8	2	0	40.61	2.222	4	0.3
5	3	2	40.64	2.220	8	0.5
1	0	4	40.64	2.220	4	0.2
0	4	2	41.23	2.189	4	0.1
2	0	4	41.44	2.179	4	0.1
1	4	2	41.49	2.176	8	0.7
4	4	1	41.56	2.173	8	0.8
1	1	4	41.67	2.167	8	1.5
8	0	2	41.69	2.166	4	0.2
6	1	3	41.82	2.160	8	1.2
7	2	2	41.84	2.159	8	0.8
8	2	1	41.90	2.156	8	0.6
3	3	3	42.88	2.109	8	0.7
7	3	1	43.04	2.101	8	0.9
0	2	4	44.38	2.041	4	1.7
4	0	4	44.49	2.036	4	0.3
1	2	4	44.63	2.030	8	2.2
7	1	3	45.10	2.010	8	0.6
4	4	2	45.28	2.003	8	0.6
2	2	4	45.37	1.999	8	0.2
4	1	4	45.45	1.996	8	0.1
8	2	2	45.60	1.989	8	0.1
2	5	0	46.09	1.969	4	0.2
1	5	1	46.55	1.951	8	0.2
7	3	2	46.67	1.946	8	0.3
10	1	0	46.72	1.944	4	3.2
9	1	2	46.88	1.938	8	0.2
1	4	3	47.58	1.911	8	0.1
5	1	4	47.60	1.910	8	0.7
10	1	1	47.88	1.900	8	0.1
7	2	3	47.89	1.900	8	1.0
4	2	4	48.22	1.887	8	0.6
3	5	1	48.43	1.879	8	0.1
1	3	4	49.25	1.850	8	0.3
6	3	3	49.38	1.846	8	0.4
1	5	2	49.98	1.825	8	1.6

AlPO–C, Dehydrated

APC

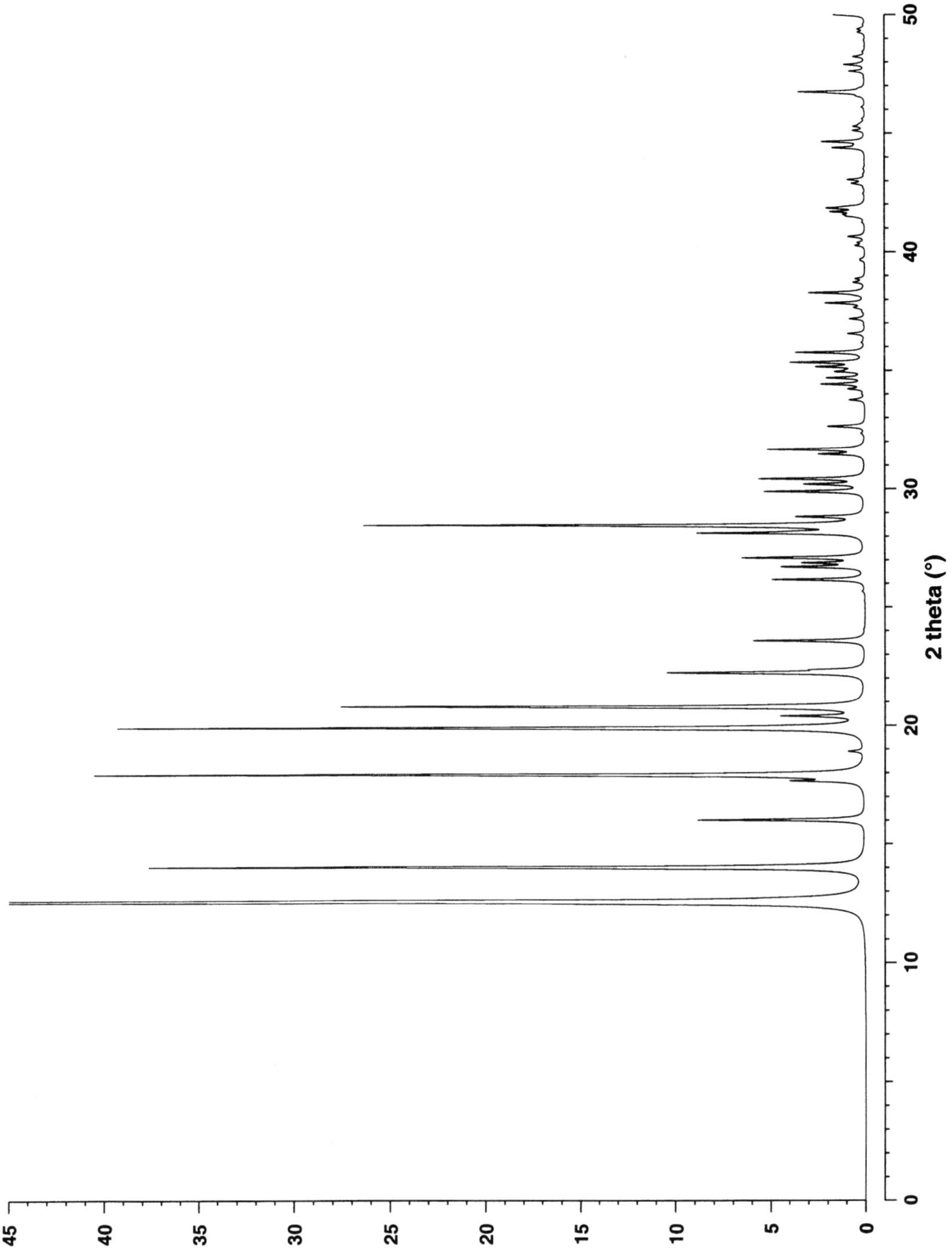

APD AlPO–D

CHEMICAL COMPOSITION: $[Al_{16}P_{16}O_{64}]$

REFINED COMPOSITION: $[Al_{16}P_{16}O_{64}]$

CRYSTAL DATA: $Pca2_1$ (No. 29)
$a = 19.187$ Å $b = 8.576$ Å $c = 9.804$ Å
$\alpha = 90°$ $\beta = 90°$ $\gamma = 90°$
X-ray Rietveld refinement, $R_{wp} = 0.203$, $R_F = 0.115$; At $T = 100°C$.

REFERENCE: E. B. Keller, W. M. Meier and R. M. Kirchner, *Solid State Ionics* **43** 93–102 (1990).

h	k	l	2θ	d	M	I_{rel}
2	0	0	9.22	9.594	2	0.4
1	1	0	11.30	7.829	4	0.4
2	0	1	12.91	6.857	4	100.0
1	1	1	14.48	6.118	8	3.7
2	1	1	16.55	5.355	8	1.0
3	1	0	17.30	5.127	4	4.9
0	0	2	18.10	4.902	2	7.4
4	0	0	18.50	4.797	2	18.8
3	1	1	19.54	4.543	8	17.0
2	0	2	20.34	4.365	4	13.9
4	0	1	20.61	4.309	4	20.9
0	2	0	20.71	4.288	2	89.1
0	1	2	20.87	4.256	4	2.6
4	1	0	21.22	4.186	4	2.8
1	2	0	21.23	4.185	4	0.3
1	1	2	21.39	4.155	8	15.0
2	1	2	22.86	3.890	8	3.8
4	1	1	23.10	3.850	8	3.7
1	2	1	23.11	3.849	8	1.4
2	2	1	24.48	3.636	8	12.1
3	2	0	25.00	3.562	4	0.2
3	1	2	25.13	3.543	8	10.4
5	1	0	25.43	3.503	4	2.1
4	0	2	25.99	3.428	4	11.3
3	2	1	26.63	3.348	8	0.5
5	1	1	27.03	3.299	8	7.3
0	2	2	27.64	3.227	4	2.0
6	0	0	27.90	3.198	2	0.2
4	2	0	27.91	3.197	4	3.7
4	1	2	28.03	3.183	8	0.4
1	2	2	28.03	3.183	8	0.6
2	0	3	28.86	3.093	4	13.7
2	2	2	29.19	3.059	8	4.2
6	0	1	29.38	3.040	4	13.7
4	2	1	29.39	3.039	8	3.4
1	1	3	29.62	3.016	8	1.8
2	1	3	30.72	2.910	8	0.8
3	2	2	31.04	2.881	8	0.2
6	1	1	31.21	2.865	8	0.3
5	1	2	31.39	2.850	8	1.7
3	1	3	32.49	2.756	8	1.7
1	3	1	32.97	2.717	8	0.2
4	0	3	33.17	2.701	4	4.0
6	0	2	33.46	2.678	4	5.0
4	2	2	33.46	2.678	8	3.4
7	1	0	34.35	2.611	4	0.1
3	3	0	34.36	2.610	4	1.0
4	1	3	34.83	2.576	8	0.3
1	2	3	34.83	2.576	8	0.3
6	1	2	35.10	2.557	8	0.1
3	3	1	35.60	2.522	8	4.7
2	2	3	35.79	2.509	8	3.4
6	2	1	36.22	2.480	8	3.0
5	2	2	36.37	2.470	8	0.2
0	3	2	36.38	2.469	4	0.3
4	3	0	36.59	2.456	4	0.3
0	0	4	36.66	2.451	2	3.3
1	3	2	36.69	2.449	8	3.8
3	2	3	37.34	2.408	8	0.2
8	0	0	37.50	2.398	2	1.2
2	3	2	37.61	2.391	8	0.9
4	3	1	37.76	2.382	8	0.8
2	0	4	37.89	2.375	4	1.1
0	1	4	38.19	2.357	4	0.5
1	1	4	38.49	2.339	8	0.2
8	1	0	38.99	2.310	4	0.2
3	3	2	39.10	2.304	8	1.3
5	3	0	39.30	2.292	4	0.5
6	0	3	39.42	2.286	4	0.2
4	2	3	39.43	2.285	8	0.7
6	2	2	39.68	2.272	8	0.1
5	3	1	40.41	2.232	8	2.4
6	1	3	40.86	2.209	8	0.4
4	3	2	41.11	2.196	8	0.2
4	0	4	41.37	2.183	4	1.0
8	0	2	41.93	2.154	4	2.1
5	2	3	41.98	2.152	8	0.1
0	4	0	42.15	2.144	2	5.2
1	3	3	42.27	2.138	8	0.8
1	4	0	42.42	2.131	4	0.4
0	2	4	42.48	2.128	4	1.2
8	2	0	43.22	2.093	4	0.3
8	1	2	43.30	2.089	8	0.1
7	2	2	43.31	2.089	8	0.1
6	3	1	43.45	2.083	8	0.1
1	4	1	43.46	2.082	8	0.8
2	2	4	43.57	2.077	8	0.4
5	3	2	43.58	2.077	8	1.2
9	1	0	43.75	2.069	4	1.6
8	2	1	44.25	2.047	8	1.2
2	4	1	44.26	2.046	8	0.8
7	1	3	44.41	2.040	8	1.0
3	3	3	44.42	2.039	8	0.4
3	4	0	44.57	2.033	4	0.1
9	1	1	44.77	2.024	8	0.6
6	2	3	44.94	2.017	8	1.3
5	1	4	45.15	2.008	8	0.9
3	4	1	45.57	1.990	8	0.2
7	3	0	45.86	1.978	4	0.1
0	4	2	46.21	1.964	4	0.5
4	4	0	46.39	1.957	4	0.3
1	4	2	46.47	1.954	8	0.3
6	0	4	46.69	1.945	4	0.5
4	2	4	46.70	1.945	8	0.2
7	3	1	46.84	1.939	8	0.6
8	0	3	46.99	1.934	4	0.8
8	2	2	47.21	1.925	8	0.6
2	4	2	47.23	1.924	8	0.6
2	0	5	47.32	1.921	4	0.5
4	4	1	47.36	1.919	8	0.3
9	1	2	47.71	1.906	8	0.5
1	1	5	47.82	1.902	8	1.1
6	1	4	47.95	1.897	8	0.1
8	1	3	48.25	1.886	8	0.2
10	0	1	48.33	1.883	4	1.1
2	1	5	48.56	1.875	8	0.1
0	3	4	48.95	1.861	4	0.1
1	3	4	49.20	1.852	8	0.3
3	1	5	49.79	1.831	8	0.1

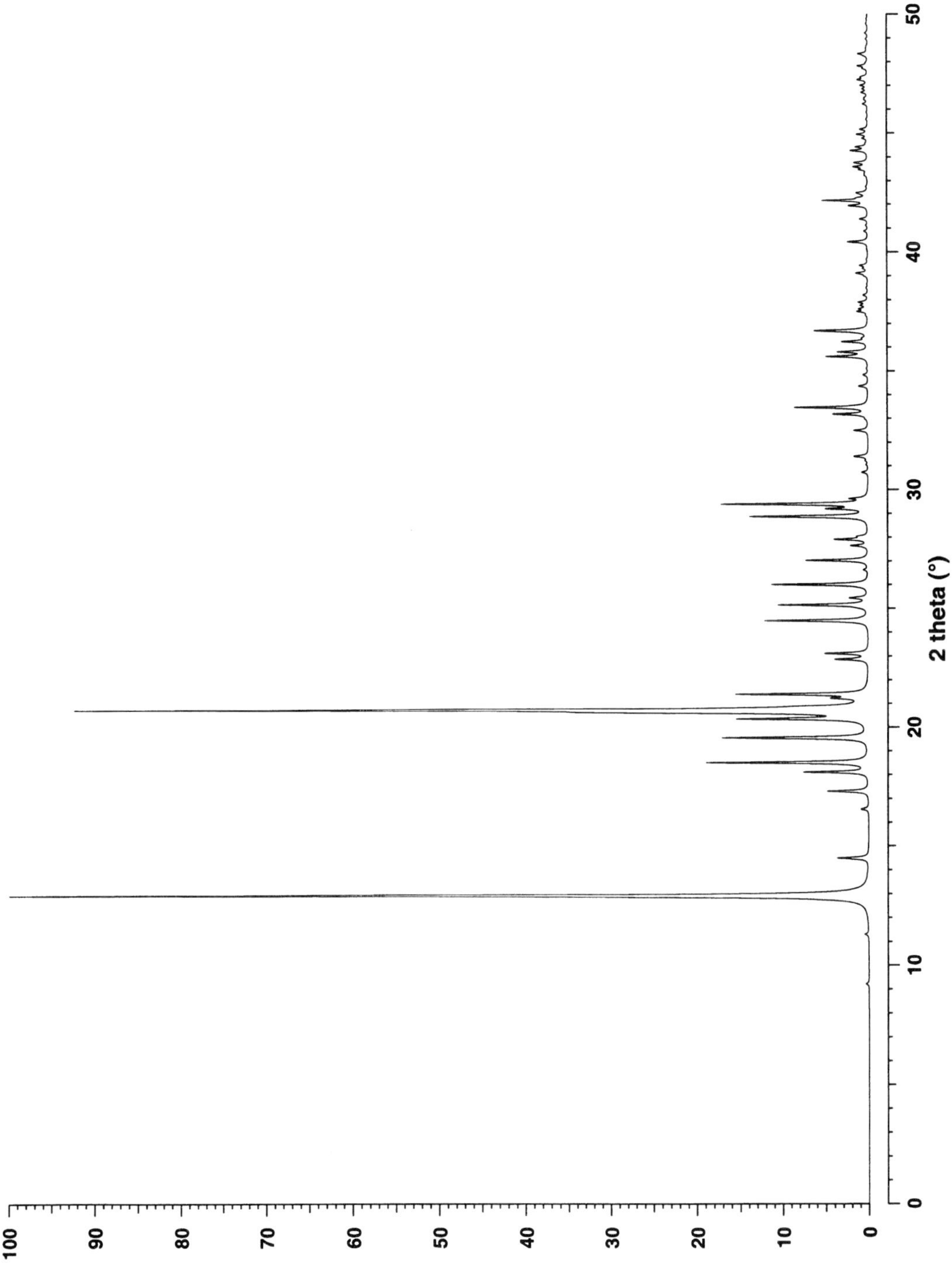
2 theta (°)
0
10
20
30
40
50
100
90
80
70
60
50
40
30
20
10
0

AST Quinuclidine AlPO–16

CHEMICAL COMPOSITION: $|(C_7H_{13}N)_4(H_2O)_{16}|$ $[Al_{20}P_{20}O_{80}]$
$C_7H_{13}N$ = quinuclidine

REFINED COMPOSITION: $|(C_7H_{13}N)_4|$ $[Al_{20}P_{20}O_{80}]$

CRYSTAL DATA: $F23$ (No. 196)
a = 13.3832 Å b = 13.3832 Å c = 13.3832 Å
α = 90° β = 90° γ = 90°
X-ray Rietveld refinement, R_{wp} = 0.192, R_F = 0.106

REFERENCE: J. M. Bennett and R. M. Kirchner,
Zeolites **11** 502–506 (1991).

h	k	l	2θ	d	M	I_{rel}
1	1	1	11.45	7.727	8	61.9
2	0	0	13.23	6.692	6	0.3
2	2	0	18.75	4.732	12	52.0
3	1	1	22.03	4.035	24	100.0
2	2	2	23.02	3.863	8	9.2
4	0	0	26.64	3.346	6	24.4
3	3	1	29.08	3.070	24	8.4
4	2	0	29.86	2.993	12	9.2
4	0	2	29.86	2.993	12	18.6
4	2	2	32.78	2.732	24	5.0
3	3	3	34.83	2.576	8	0.3
5	1	1	34.83	2.576	24	5.1
4	4	0	38.03	2.366	12	10.3
5	3	1	39.85	2.262	24	2.1
4	4	2	40.44	2.231	24	0.3
6	0	0	40.44	2.231	6	0.5
6	0	2	42.73	2.116	12	0.2
6	2	0	42.73	2.116	12	0.2
5	3	3	44.39	2.041	24	4.6
6	2	2	44.93	2.018	24	0.6
4	4	4	47.04	1.932	8	0.2
7	1	1	48.58	1.874	24	4.8
5	5	1	48.58	1.874	24	3.7
6	0	4	49.09	1.856	12	1.1

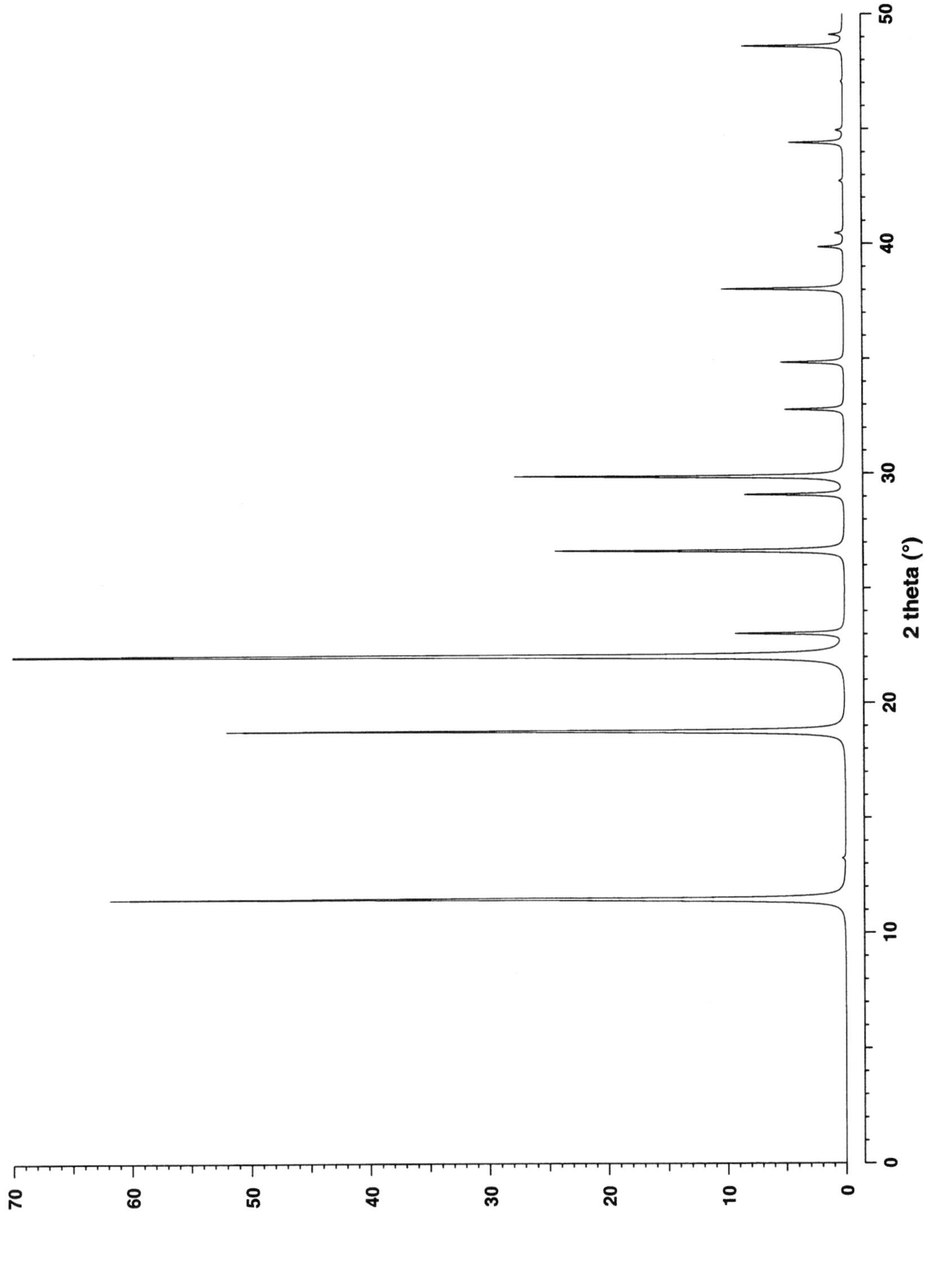
2 theta (°)
0
10
20
30
40
50
0
10
20
30
40
50
60
70

AST Quinuclidinium fluoride octadecasil

CHEMICAL COMPOSITION: $|(C_7H_{14}NF)_2|$ $[Si_{20}O_{40}]$
$C_7H_{14}NF$ = quinuclidinium fluoride

REFINED COMPOSITION: $|C_{6.18}F_{0.07}|$ $[Si_{20}O_{40}]$

CRYSTAL DATA: $I\,4/m$ (No. 87)
$a = 9.194$ Å $b = 9.194$ Å $c = 13.396$ Å
$\alpha = 90°$ $\beta = 90°$ $\gamma = 90°$
X-ray single crystal refinement, $R_w = 0.039$

REFERENCE: P. Caullet, J. L. Guth, J. Hazm, J. M. Lamblin and H. Gies, *Eur. J. Solid State Inorg. Chem.* **28** 345–361 (1991).

h	k	l	2θ	d	M	I_{rel}	h	k	l	2θ	d	M	I_{rel}	h	k	l	2θ	d	M	I_{rel}
1	0	1	11.67	7.580	8	100.0	1	1	4	30.01	2.977	8	6.3	3	2	3	40.75	2.214	8	1.8
0	0	2	13.22	6.698	2	2.9	2	2	2	30.57	2.924	8	5.5	2	3	3	40.75	2.214	8	0.7
1	1	0	13.62	6.501	4	5.1	1	3	0	30.75	2.907	4	8.5	4	1	1	41.03	2.200	8	2.4
1	1	2	19.02	4.665	8	12.0	2	0	4	33.09	2.707	8	1.8	1	4	1	41.03	2.200	8	0.3
2	0	0	19.31	4.597	4	2.7	1	3	2	33.60	2.667	8	0.1	1	3	4	41.11	2.195	8	0.7
1	0	3	22.13	4.017	8	14.9	3	1	2	33.60	2.667	8	2.6	4	0	2	41.54	2.174	8	0.4
2	1	1	22.62	3.931	8	3.6	1	0	5	34.88	2.572	8	1.8	3	3	2	43.91	2.062	8	0.1
1	2	1	22.62	3.931	8	35.4	3	0	3	35.53	2.527	8	0.5	2	4	0	44.05	2.056	4	1.7
2	0	2	23.47	3.790	8	4.3	2	3	1	35.85	2.505	8	0.4	3	0	5	44.94	2.017	8	0.6
0	0	4	26.62	3.349	2	7.5	3	2	1	35.85	2.505	8	1.2	2	0	6	45.14	2.008	8	0.2
2	2	0	27.44	3.251	4	8.2	2	2	4	38.60	2.333	8	6.6	1	4	3	45.46	1.995	8	0.8
2	1	3	29.53	3.025	8	0.3	4	0	0	39.19	2.299	4	0.5	2	4	2	46.19	1.965	8	0.4
1	2	3	29.53	3.025	8	5.0	2	1	5	40.17	2.245	8	0.7	1	0	7	48.59	1.874	8	1.2
3	0	1	29.91	2.987	8	0.4	0	0	6	40.40	2.233	2	0.3	2	3	5	49.34	1.847	8	4.6

Quinuclidinium fluoride octadecasil AST

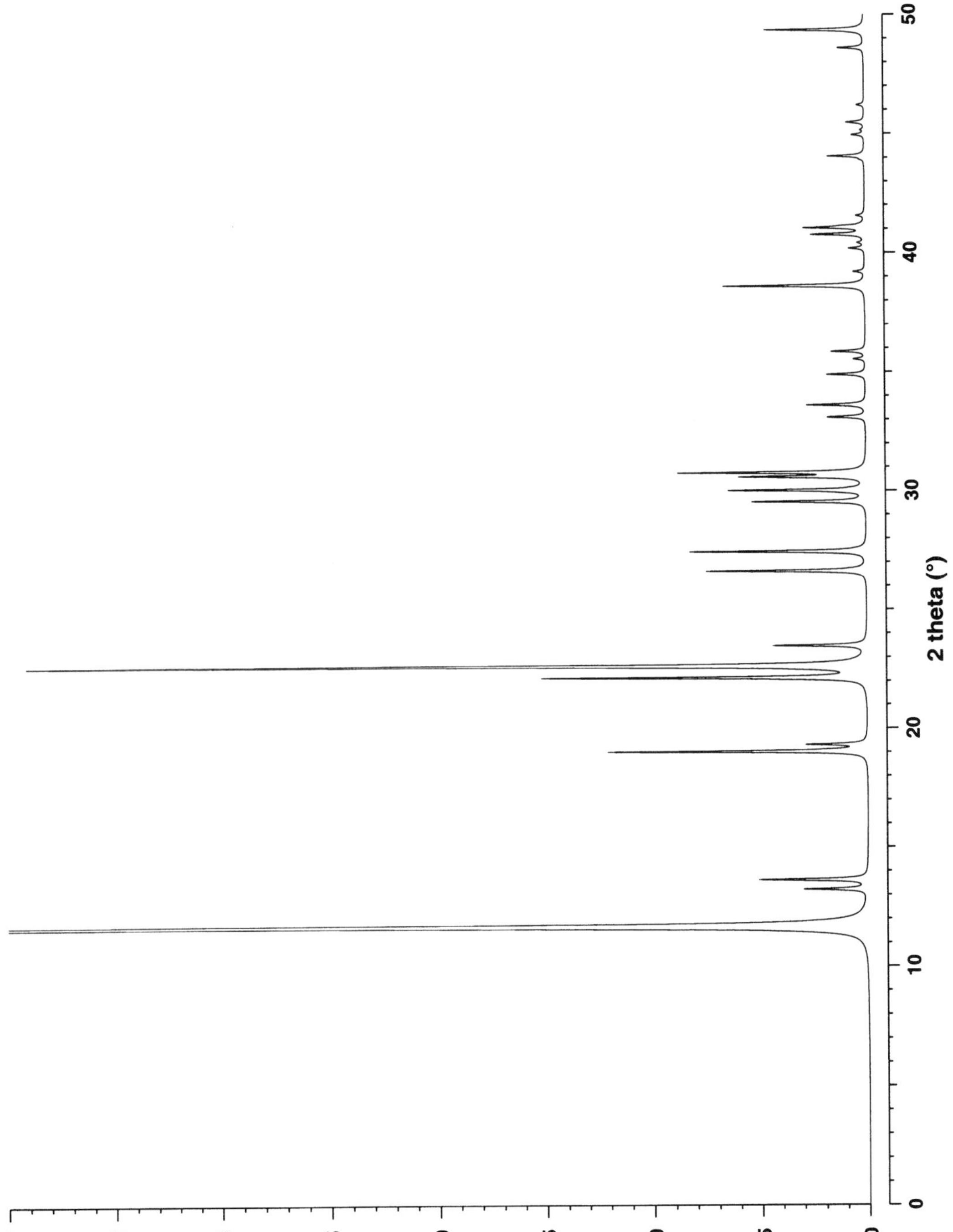

ASV ASU-7

CHEMICAL COMPOSITION: $|((CH_3)_2NH)_2(H_2O)_2|$ $[Ge_{20}O_{40}]$
$(CH_3)_2NH$ = dimethylamine

REFINED COMPOSITION: $|(C_4N_2)(H_2O)_2|$ $[Ge_{20}O_{40}]$

CRYSTAL DATA: $P\,4/m\,cc$ (No. 124)
$a = 8.7795$ Å $b = 8.7795$ Å $c = 14.4696$ Å
$\alpha = 90°$ $\beta = 90°$ $\gamma = 90°$
X-ray single crystal refinement, $R_F = 0.025$, $wR_{F2} = 0.053$; at -75 C.

REFERENCE: H. Li and O. M. Yaghi,
J. Am. Chem. Soc. **120** 10569–10570 (1998).

h	*k*	*l*	2*θ*	*d*	*M*	I_{rel}
1	0	0	10.07	8.780	4	100.0
0	0	2	12.23	7.235	2	1.6
1	1	0	14.27	6.208	4	7.7
1	0	2	15.87	5.583	8	6.5
1	1	2	18.84	4.711	8	12.4
2	0	0	20.23	4.390	4	0.3
2	1	0	22.65	3.926	8	26.5
2	1	1	23.48	3.789	16	25.8
2	0	2	23.71	3.753	8	9.6
0	0	4	24.61	3.617	2	10.9
2	1	2	25.82	3.451	16	0.9
1	0	4	26.65	3.345	8	14.0
1	1	4	28.56	3.126	8	5.0
2	2	0	28.76	3.104	4	2.3
2	1	3	29.33	3.045	16	4.1
3	0	0	30.55	2.926	4	0.3
2	2	2	31.36	2.853	8	5.4
2	0	4	32.06	2.792	8	9.4
3	1	0	32.24	2.776	8	1.1
3	1	1	32.85	2.727	16	18.7
3	0	2	33.02	2.713	8	0.6
2	1	4	33.69	2.660	16	12.4
3	1	2	34.60	2.592	16	3.8
3	2	0	36.91	2.435	8	0.2
0	0	6	37.29	2.412	2	1.2
3	1	3	37.37	2.406	16	1.3
3	2	1	37.45	2.401	16	1.8
2	2	4	38.20	2.356	8	15.0
2	1	5	38.65	2.330	16	1.3
1	0	6	38.72	2.325	8	0.7
3	2	2	39.03	2.308	16	0.2
3	0	4	39.61	2.275	8	0.5
1	1	6	40.11	2.248	8	0.5
3	1	4	40.98	2.202	16	0.4
4	0	0	41.12	2.195	4	1.1
4	1	0	42.45	2.129	8	3.5
2	0	6	42.78	2.114	8	0.2
4	1	1	42.93	2.107	16	3.9
3	3	0	43.74	2.069	4	0.5
2	1	6	44.07	2.055	16	1.8
4	1	2	44.34	2.043	16	0.5
3	2	4	44.87	2.020	16	0.5
3	1	5	45.26	2.003	16	10.7
3	3	2	45.59	1.990	8	1.0
4	2	0	46.24	1.963	8	8.1
4	1	3	46.63	1.948	16	1.1
4	2	2	48.02	1.895	16	0.3
3	2	5	48.88	1.863	16	6.1
4	1	4	49.68	1.835	16	2.8

ASU-7 ASV

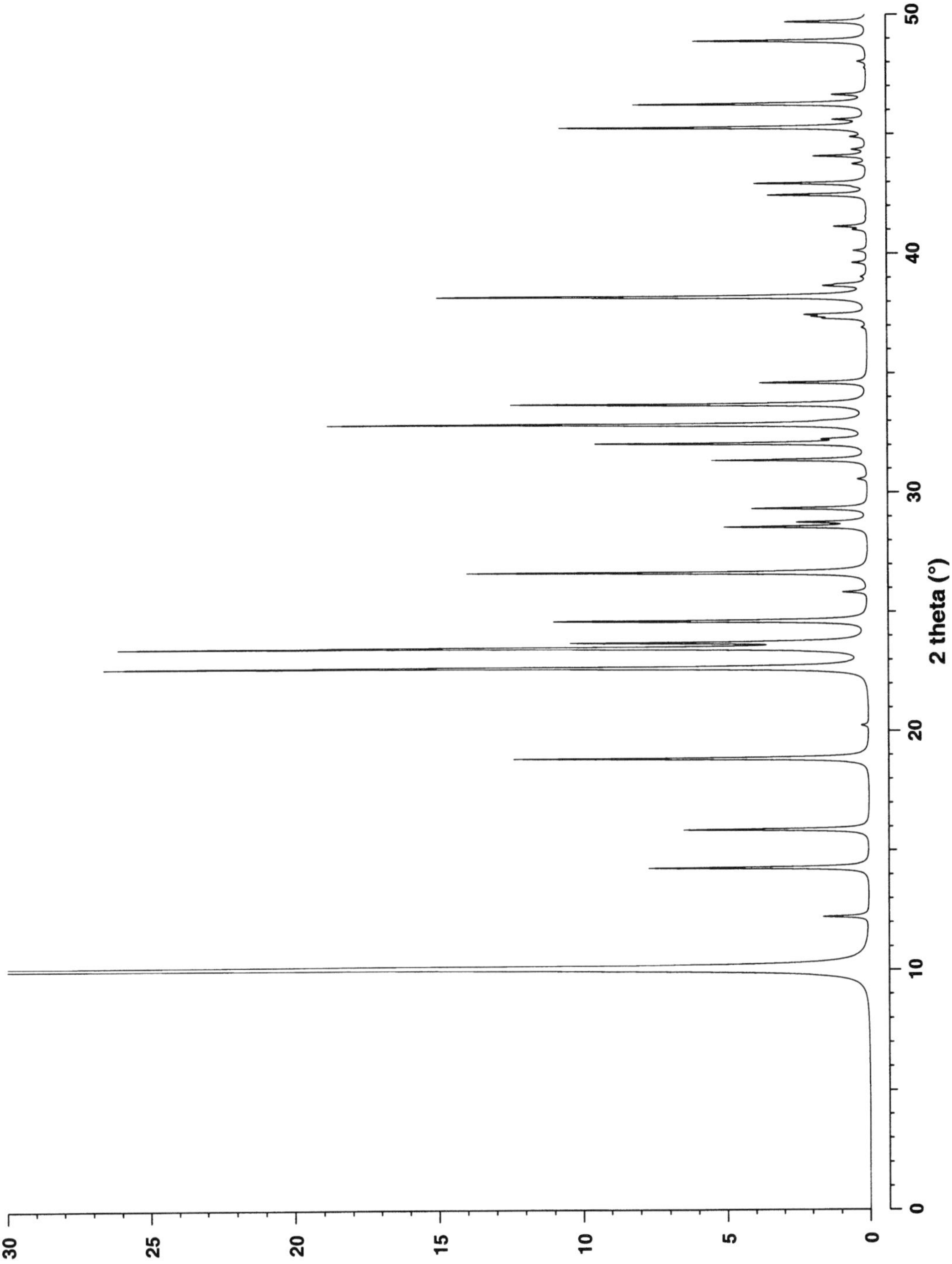

ATN MAPO–39

CHEMICAL COMPOSITION: $|((C_3H_7)_2NH)_x|$ $[Al_7MgP_8O_{32}]$
$(C_3H_7)_2NH$ = di-n-propylamine

REFINED COMPOSITION: $|C_{13.16}O_{0.168}|$ $[Al_{7.512}P_{7.504}O_{32}]$

CRYSTAL DATA: $I\,4/m$ (No. 87)
a = 13.2088 Å b = 13.2088 Å c = 5.2771 Å
α = 90° β = 90° γ = 90°
X-ray single crystal refinement, R_w = 0.02, R_F = 0.093

REFERENCE: W. H. Baur, W. Joswig, D. Kassner, A. Bieniok, G. Finger and J. Kornatowski, *Z. Kristallogr.* **214** 154–159 (1999).

h	k	l	2θ	d	M	I_{rel}
1	1	0	9.47	9.340	4	39.6
2	0	0	13.41	6.604	4	75.6
1	0	1	18.10	4.900	8	69.0
2	2	0	19.00	4.670	4	0.3
1	3	0	21.27	4.177	4	20.8
3	1	0	21.27	4.177	4	72.5
2	1	1	22.59	3.935	8	42.3
1	2	1	22.59	3.935	8	100.0
3	0	1	26.36	3.381	8	1.1
4	0	0	27.00	3.302	4	5.3
3	3	0	28.67	3.113	4	11.9
2	3	1	29.69	3.009	8	6.4
3	2	1	29.69	3.009	8	14.6
2	4	0	30.26	2.954	4	23.9
4	2	0	30.26	2.954	4	15.7
4	1	1	32.70	2.738	8	10.5
1	4	1	32.70	2.738	8	9.1
0	0	2	33.98	2.639	2	16.5
5	1	0	34.63	2.590	4	0.8
2	0	2	36.68	2.450	8	4.4
3	4	1	38.09	2.362	8	1.7
5	0	1	38.09	2.362	8	9.7
4	3	1	38.09	2.362	8	4.3
4	4	0	38.56	2.335	4	0.3
2	2	2	39.22	2.297	8	1.4
3	5	0	39.79	2.265	4	0.4
5	3	0	39.79	2.265	4	0.6
1	3	2	40.43	2.231	8	0.6
2	5	1	40.56	2.224	8	0.1
6	0	0	41.00	2.201	4	3.1
6	2	0	43.32	2.088	4	1.1
2	6	0	43.32	2.088	4	1.5
4	0	2	43.92	2.061	8	2.6
3	3	2	45.04	2.013	8	1.0
6	1	1	45.15	2.008	8	0.1
4	2	2	46.13	1.968	8	3.1
2	4	2	46.13	1.968	8	1.6
5	4	1	47.31	1.921	8	1.3
4	5	1	47.31	1.921	8	0.9
5	5	0	48.75	1.868	4	4.6
1	7	0	48.75	1.868	4	0.2
5	1	2	49.30	1.848	8	0.8
1	5	2	49.30	1.848	8	0.4
3	6	1	49.40	1.845	8	1.2
6	3	1	49.40	1.845	8	1.1
4	6	0	49.78	1.832	4	0.3
6	4	0	49.78	1.832	4	0.2

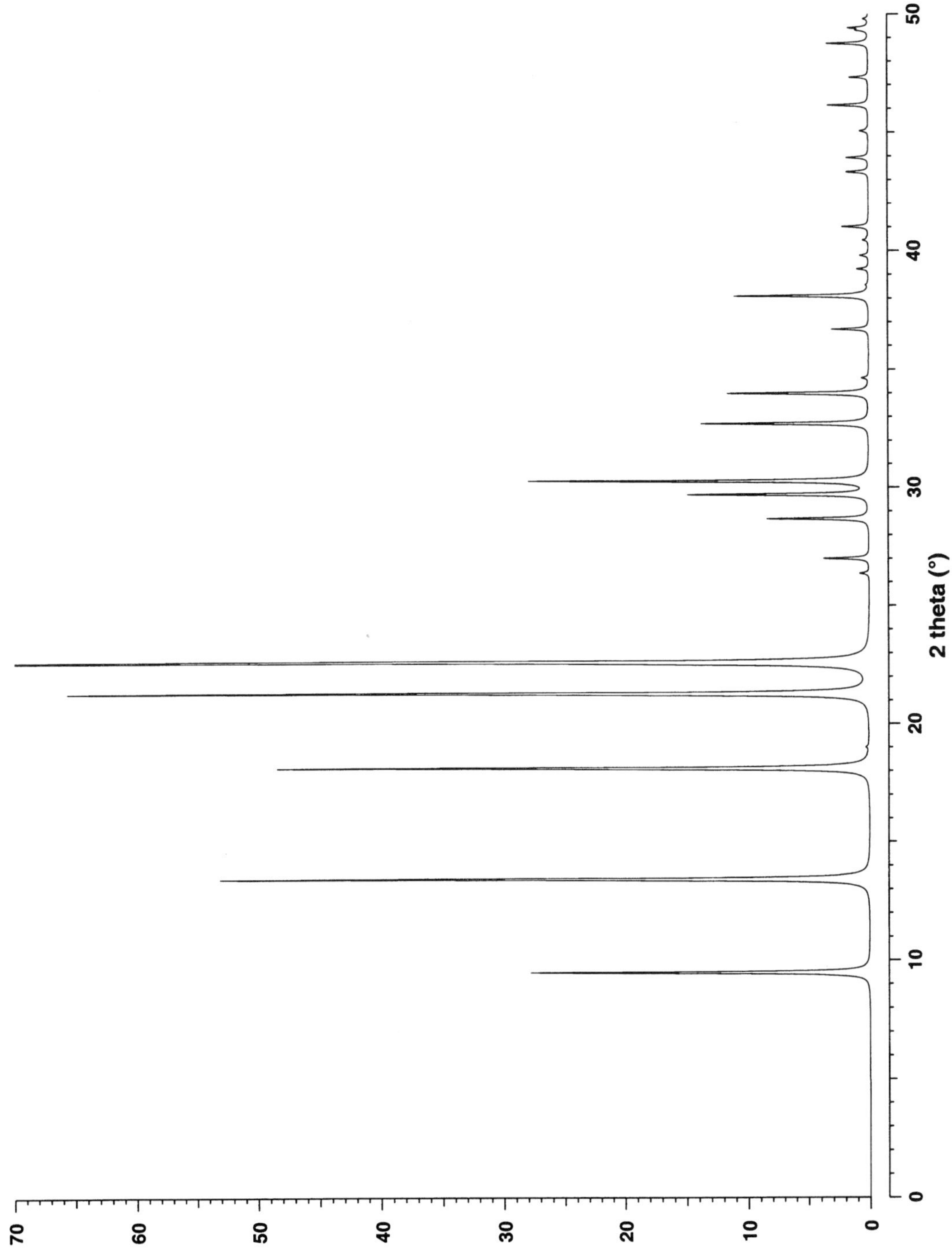
0
10
20
30
40
50
2 theta (°)
0
10
20
30
40
50
60
70

ATO Di-n-propylamine SAPO–31

CHEMICAL COMPOSITION: $|((C_3H_7)_2NH)_2|$ $[Al_{18}P_{17}SiO_{72}]$
$(C_3H_7)_2NH$ = di-n-propylamine

REFINED COMPOSITION: $|C_{14.76}|$ $[Al_{17.71}P_{16.67}O_{72}]$

CRYSTAL DATA: $R\overline{3}$ (No. 148) hexagonal setting
a = 20.839 Å b = 20.839 Å c = 5.041 Å
α = 90° β = 90° γ = 120°
X-ray single crystal refinement, R_{wF} = 0.018, R_F = 0.051

REFERENCE: W. H. Baur, W. Joswig, D. Kassner and J. Kornatowski, *Acta Cryst.* **B50** 290–294 (1994).

h	k	l	2θ	d	M	I_{rel}
1	1	0	8.49	10.419	6	100.0
0	3	0	14.73	6.016	6	2.8
2	2	0	17.02	5.210	6	7.2
1	0	1	18.27	4.855	6	5.0
0	2	1	20.18	4.401	6	91.6
2	1	1	21.92	4.054	6	8.8
3	-1	1	21.92	4.054	6	46.8
4	1	0	22.58	3.938	6	99.9
1	4	0	22.58	3.938	6	53.5
4	-3	1	25.07	3.552	6	7.7
1	3	1	25.07	3.552	6	0.4
3	3	0	25.65	3.473	6	5.2
5	-2	1	27.88	3.199	6	1.9
3	2	1	27.88	3.199	6	21.1
0	6	0	29.70	3.008	6	9.7
5	2	0	30.94	2.890	6	1.0
6	-4	1	31.67	2.825	6	5.7
2	4	1	31.67	2.825	6	32.5
6	-1	1	32.85	2.726	6	0.3
5	1	1	32.85	2.726	6	0.1
4	4	0	34.43	2.605	6	0.4
4	3	1	35.09	2.557	6	5.4
7	-3	1	35.09	2.557	6	11.6
0	1	2	35.98	2.496	6	5.8
2	0	2	37.03	2.428	6	4.5
7	-6	1	37.22	2.416	6	0.8
1	7	0	37.63	2.390	6	2.0
7	1	0	37.63	2.390	6	0.7
3	-2	2	38.06	2.364	6	2.9
1	2	2	38.06	2.364	6	2.4
8	-5	1	39.25	2.295	6	2.6
3	5	1	39.25	2.295	6	0.3
7	0	1	39.25	2.295	6	3.3
6	3	0	39.64	2.274	6	1.0
3	6	0	39.64	2.274	6	2.0
4	-1	2	40.05	2.251	6	0.6
3	1	2	40.05	2.251	6	3.8
6	2	1	40.23	2.242	6	1.4
2	3	2	41.96	2.153	6	0.4
5	-3	2	41.96	2.153	6	0.2
5	4	1	43.06	2.101	6	0.4
9	-4	1	43.06	2.101	6	0.8
5	5	0	43.42	2.084	6	0.6
5	0	2	43.81	2.067	6	1.9
0	8	1	43.97	2.059	6	0.9
4	2	2	44.71	2.027	6	1.1
6	-2	2	44.71	2.027	6	0.5
9	-7	1	44.87	2.020	6	0.3
2	7	1	44.87	2.020	6	1.0
0	9	0	45.22	2.005	6	3.0
1	5	2	45.59	1.990	6	0.4
2	8	0	46.10	1.969	6	0.8
8	2	0	46.10	1.969	6	1.0
9	-1	1	46.62	1.948	6	0.2
8	1	1	46.62	1.948	6	6.5
3	4	2	47.32	1.921	6	2.3
4	6	1	47.48	1.915	6	1.8
10	-6	1	47.48	1.915	6	0.1
7	3	1	48.32	1.883	6	0.2
10	-3	1	48.32	1.883	6	1.4
7	4	0	48.65	1.871	6	1.5
4	7	0	48.65	1.871	6	2.6
6	1	2	49.01	1.859	6	0.2
7	-1	2	49.01	1.859	6	2.5

Di-n-propylamine SAPO-31 ATO

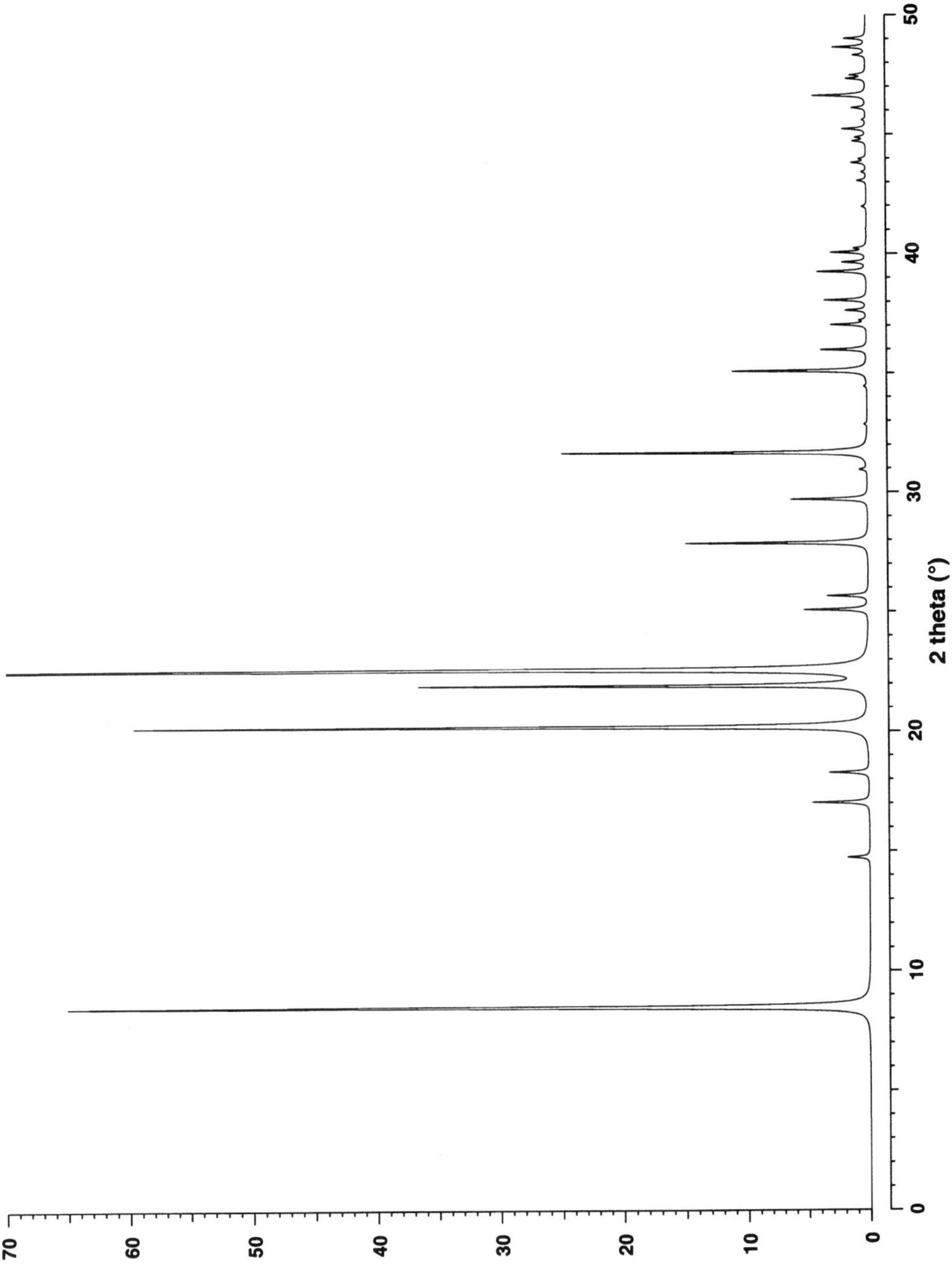

ATS MAPO–36, Calcined

CHEMICAL COMPOSITION: $[(Al_{10.8}Mg)P_{12.2}O_{48}]$

REFINED COMPOSITION: $[Si_{24}O_{48}]$

CRYSTAL DATA: $C1\,2/c\,1$ (No. 15) unique axis **b**, cell choice 1
$a = 13.1483$ Å $b = 21.5771$ Å $c = 5.1639$ Å
$\alpha = 90°$ $\beta = 91.84°$ $\gamma = 90°$
X-ray Rietveld refinement, $R_{exp} = 0.112$, $R_{wp} = 0.135$, $R_F = 0.058$

REFERENCE: J. V. Smith, J. J. Pluth and K. J. Andries, *Zeolites* **13** 166–169 (1993).

h	*k*	*l*	2*θ*	*d*	*M*	I_{rel}	*h*	*k*	*l*	2*θ*	*d*	*M*	I_{rel}	*h*	*k*	*l*	2*θ*	*d*	*M*	I_{rel}
1	1	0	7.88	11.224	4	100.0	-3	3	1	29.09	3.069	4	2.4	-1	5	2	41.14	2.194	4	0.1
0	2	0	8.20	10.789	2	25.3	1	7	0	29.77	3.001	4	0.2	6	0	0	41.22	2.190	2	1.0
2	0	0	13.48	6.571	2	4.9	3	3	1	29.88	2.990	4	0.6	3	1	2	41.39	2.182	4	0.1
1	3	0	14.04	6.309	4	0.3	0	6	1	30.29	2.951	4	2.8	4	6	1	41.51	2.175	4	0.2
2	2	0	15.79	5.612	4	1.3	4	4	0	31.89	2.806	4	3.4	2	4	2	41.53	2.175	4	0.6
0	4	0	16.43	5.394	2	12.2	-2	6	1	33.05	2.710	4	0.2	1	5	2	41.53	2.174	4	0.3
-1	1	1	18.72	4.739	4	0.6	0	8	0	33.22	2.697	2	0.3	1	9	1	42.22	2.141	4	0.1
0	2	1	19.06	4.656	4	19.8	-3	5	1	33.59	2.668	4	0.6	3	3	2	43.13	2.098	4	0.2
1	1	1	19.12	4.641	4	1.0	-1	7	1	34.46	2.603	4	1.1	0	6	2	43.15	2.097	4	0.4
3	1	0	20.69	4.293	4	10.7	1	7	1	34.69	2.586	4	3.6	-5	5	1	43.52	2.079	4	0.4
2	4	0	21.31	4.169	4	2.2	0	0	2	34.76	2.581	2	3.6	2	10	0	44.18	2.050	4	0.7
1	5	0	21.67	4.100	4	3.7	-1	1	2	35.48	2.530	4	0.6	-6	2	1	45.24	2.004	4	0.3
-1	3	1	22.08	4.025	4	8.8	0	2	2	35.78	2.510	4	0.5	5	7	0	45.34	2.000	4	0.4
1	3	1	22.42	3.965	4	9.8	1	1	2	35.92	2.500	4	0.2	-3	5	2	45.37	1.999	4	0.6
-2	2	1	23.09	3.852	4	2.2	2	8	0	35.99	2.495	4	0.4	2	6	2	45.77	1.983	4	0.2
2	2	1	23.74	3.748	4	1.7	-4	4	1	36.01	2.494	4	1.0	-1	7	2	46.23	1.964	4	0.1
3	3	0	23.78	3.741	4	1.1	5	3	0	36.39	2.469	4	0.2	6	2	1	46.32	1.960	4	0.4
0	6	0	24.76	3.596	2	0.2	0	8	1	37.63	2.390	4	0.4	-4	8	1	46.66	1.947	4	0.3
4	0	0	27.14	3.285	2	6.2	2	0	2	37.86	2.376	2	0.3	4	8	1	47.36	1.919	4	0.3
-2	4	1	27.22	3.276	4	2.2	1	3	2	37.87	2.376	4	0.3	-2	10	1	47.56	1.912	4	0.8
3	1	1	27.44	3.251	4	0.7	-5	1	1	38.15	2.359	4	0.5	-6	4	1	47.66	1.908	4	0.1
-1	5	1	27.65	3.226	4	0.2	2	2	2	38.81	2.321	4	0.3	-5	7	1	48.41	1.880	4	0.1
2	4	1	27.78	3.211	4	0.7	5	1	1	39.18	2.299	4	0.3	6	6	0	48.67	1.871	4	0.3
1	5	1	27.93	3.195	4	0.6	-3	7	1	39.49	2.282	4	0.2	7	1	0	48.68	1.870	4	0.3
2	6	0	28.29	3.155	4	1.8	5	5	0	40.17	2.245	4	1.0	6	4	1	48.70	1.870	4	0.2
4	2	0	28.40	3.143	4	0.3	2	8	1	40.34	2.236	4	0.2	0	8	2	48.84	1.865	4	0.1
3	5	0	29.05	3.074	4	2.3	-2	4	2	40.74	2.215	4	0.4	5	7	1	49.26	1.850	4	0.5

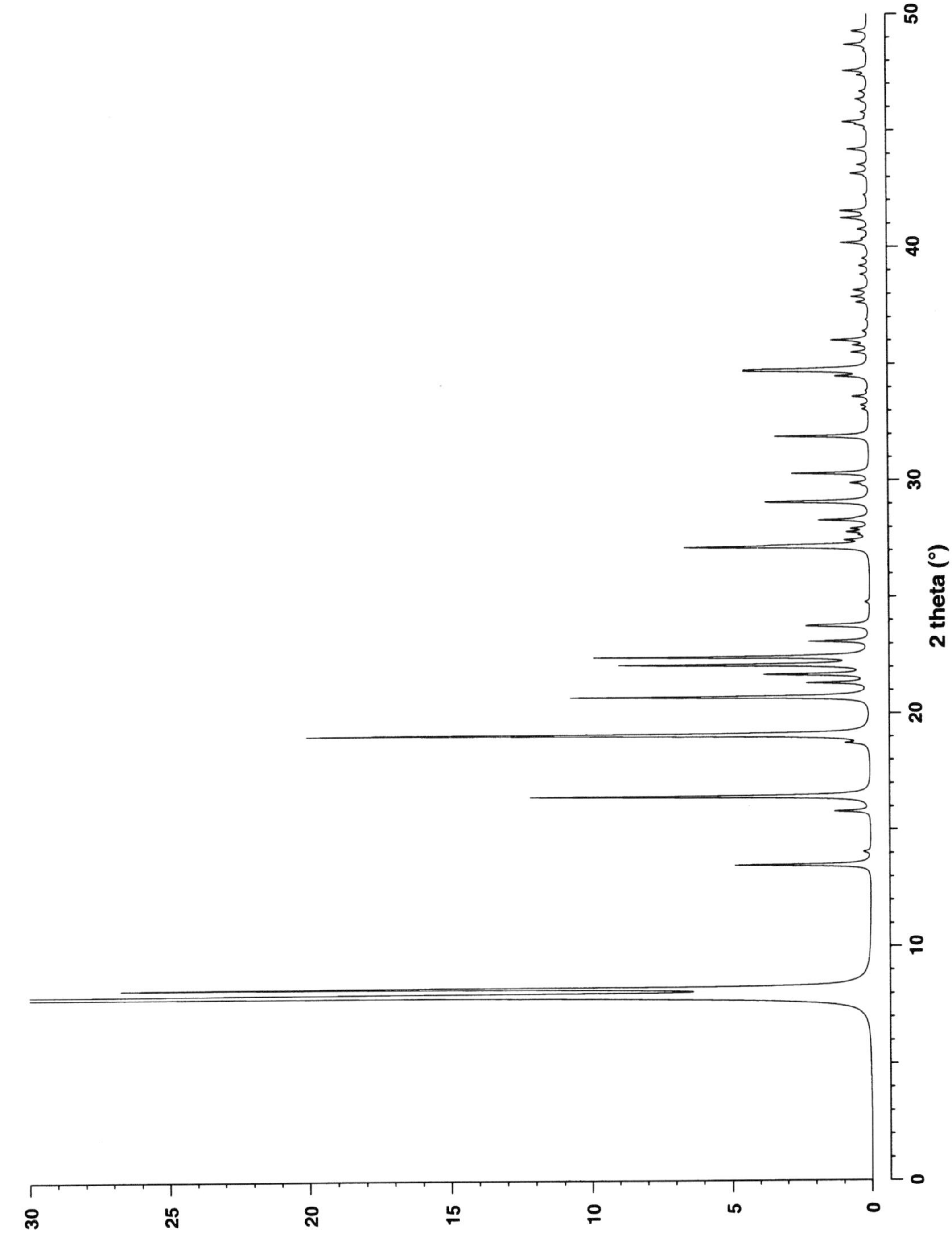
2 theta (°)
0
10
20
30
40
50
0
5
10
15
20
25
30

ATT Tetramethylammonium hydroxide AlPO–12-TAMU

CHEMICAL COMPOSITION: $|((CH_3)_4NOH)_4|\ [Al_{12}P_{12}O_{48}]$
$(CH_3)_4NOH$ = tetramethylammonium hydroxide

REFINED COMPOSITION: $|(C_4NO)_4|\ [Al_{12}P_{12}O_{48}]$

CRYSTAL DATA: $P2_12_12$ (No. 18)
$a = 10.3325$ Å $b = 14.6405$ Å $c = 9.5112$ Å
$\alpha = 90°$ $\beta = 90°$ $\gamma = 90°$
X-ray Rietveld refinement, $R_{wp} = 0.19$, $R_F = 0.17$

REFERENCE: P. R. Rudolf, C. Saldarriaga-Molina and A. Clearfield, *J. Phys. Chem.* **90** 6122–6125 (1986).

h	k	l	2θ	d	M	I_{rel}
0	0	1	9.30	9.511	2	24.3
1	1	0	10.48	8.442	4	0.3
0	1	1	11.09	7.976	4	0.5
0	2	0	12.09	7.320	2	0.9
1	0	1	12.65	6.998	4	100.0
1	1	1	14.03	6.314	8	0.9
1	2	0	14.83	5.973	4	0.4
0	2	1	15.27	5.801	4	26.6
2	0	0	17.16	5.166	2	7.8
1	2	1	17.53	5.058	8	21.7
0	0	2	18.66	4.756	2	0.5
2	0	1	19.55	4.540	4	2.7
0	1	2	19.63	4.523	4	0.5
1	3	0	20.12	4.413	4	1.3
0	3	1	20.45	4.342	4	2.8
2	1	1	20.48	4.336	8	0.7
1	0	2	20.56	4.320	4	19.7
2	2	0	21.05	4.221	4	5.8
1	1	2	21.45	4.143	8	1.3
1	3	1	22.21	4.003	8	4.3
0	2	2	22.29	3.988	4	0.3
2	2	1	23.05	3.858	8	6.9
1	2	2	23.92	3.720	8	30.9
0	4	0	24.32	3.660	2	1.0
2	3	0	25.10	3.548	4	1.3
2	0	2	25.46	3.499	4	0.2
0	4	1	26.09	3.416	4	44.4
0	3	2	26.16	3.406	4	0.1
2	1	2	26.19	3.403	8	0.5
2	3	1	26.82	3.324	8	1.4
1	4	1	27.50	3.243	8	5.3
3	0	1	27.54	3.238	4	51.5
1	3	2	27.58	3.235	8	0.5
0	0	3	28.15	3.170	2	0.4
2	2	2	28.27	3.157	8	0.5
0	1	3	28.81	3.099	4	0.1
1	0	3	29.47	3.031	4	9.7
2	4	0	29.92	2.987	4	1.6
1	1	3	30.11	2.968	8	0.2
0	2	3	30.73	2.909	4	7.6
0	4	2	30.83	2.901	4	0.2
2	4	1	31.39	2.849	8	1.0
2	3	2	31.46	2.844	8	1.0
1	5	0	31.76	2.817	4	0.5
1	2	3	31.96	2.800	8	0.5
1	4	2	32.05	2.793	8	11.1
3	0	2	32.09	2.789	4	1.4
3	1	2	32.68	2.740	8	0.1
1	5	1	33.16	2.701	8	0.3
3	3	1	33.20	2.698	8	1.2
0	3	3	33.71	2.659	4	0.5
3	2	2	34.40	2.607	8	9.2
4	0	0	34.73	2.583	2	0.9
2	2	3	35.41	2.535	8	0.8
2	4	2	35.49	2.529	8	1.0
4	0	1	36.03	2.493	4	0.3
2	5	1	36.52	2.461	8	1.2
4	1	1	36.56	2.457	8	0.2
0	6	0	36.83	2.440	2	1.1
3	4	1	37.07	2.425	8	0.9
1	5	2	37.09	2.424	8	1.3
3	3	2	37.12	2.422	8	0.1
0	0	4	37.84	2.378	2	1.8
2	3	3	38.06	2.364	8	0.2
4	2	1	38.14	2.360	8	0.6
0	1	4	38.35	2.347	4	0.4
1	4	3	38.57	2.334	8	0.3
3	0	3	38.60	2.333	4	0.1
1	0	4	38.86	2.317	4	0.5
1	6	1	39.09	2.304	8	0.9
3	1	3	39.10	2.304	8	0.8
1	1	4	39.37	2.289	8	0.3
4	0	2	39.71	2.270	4	1.7
2	5	2	40.16	2.246	8	0.7
4	1	2	40.20	2.243	8	0.6
3	5	0	40.43	2.231	4	0.4
3	2	3	40.59	2.222	8	0.7
4	3	1	40.64	2.220	8	0.3
3	4	2	40.67	2.219	8	3.1
1	2	4	40.85	2.209	8	0.2
2	4	3	41.54	2.174	8	0.5
3	5	1	41.58	2.172	8	0.1
0	6	2	41.60	2.171	4	0.2
4	2	2	41.66	2.168	8	2.4
2	0	4	41.82	2.160	4	0.3
0	5	3	42.00	2.151	4	1.3
2	6	1	42.04	2.149	8	0.1
0	3	4	42.28	2.138	4	0.3
2	1	4	42.29	2.137	8	1.0
1	6	2	42.55	2.125	8	0.1
4	4	0	42.85	2.110	4	0.1
1	5	3	42.95	2.106	8	0.3
3	3	3	42.98	2.105	8	0.3
1	3	4	43.22	2.093	8	0.2
2	2	4	43.69	2.072	8	0.5
4	3	2	43.99	2.058	8	0.7
1	7	0	44.18	2.050	4	0.2
5	1	0	44.26	2.046	4	0.3
0	7	1	44.34	2.043	4	0.2
5	0	1	44.88	2.019	4	0.5
1	7	1	45.25	2.004	8	0.5
4	0	3	45.28	2.003	4	0.7
2	6	2	45.31	2.001	8	0.3
5	1	1	45.33	2.000	8	0.6
1	4	4	46.38	1.958	8	0.2
3	0	4	46.40	1.957	4	0.2
3	6	1	46.60	1.949	8	0.6
5	2	1	46.66	1.947	8	0.3
3	1	4	46.84	1.940	8	0.2
4	2	3	47.04	1.932	8	2.1
4	4	2	47.11	1.929	8	0.1
0	7	2	47.49	1.915	4	0.8
5	3	0	47.80	1.903	4	0.3
0	0	5	47.81	1.902	2	0.9
1	6	3	47.86	1.901	8	1.9
2	7	1	47.89	1.900	8	0.2
3	2	4	48.13	1.890	8	0.6
5	1	2	48.43	1.880	8	0.4
1	0	5	48.67	1.871	4	0.3
2	4	4	48.96	1.860	8	0.2
1	1	5	49.09	1.856	8	0.1
0	5	4	49.37	1.846	4	0.2
0	2	5	49.51	1.841	4	0.1
3	6	2	49.64	1.837	8	0.2
5	2	2	49.69	1.835	8	0.2
0	8	0	49.83	1.830	2	2.3
3	5	3	49.99	1.824	8	0.2

Tetramethylammonium hydroxide AlPO–12-TAMU ATT

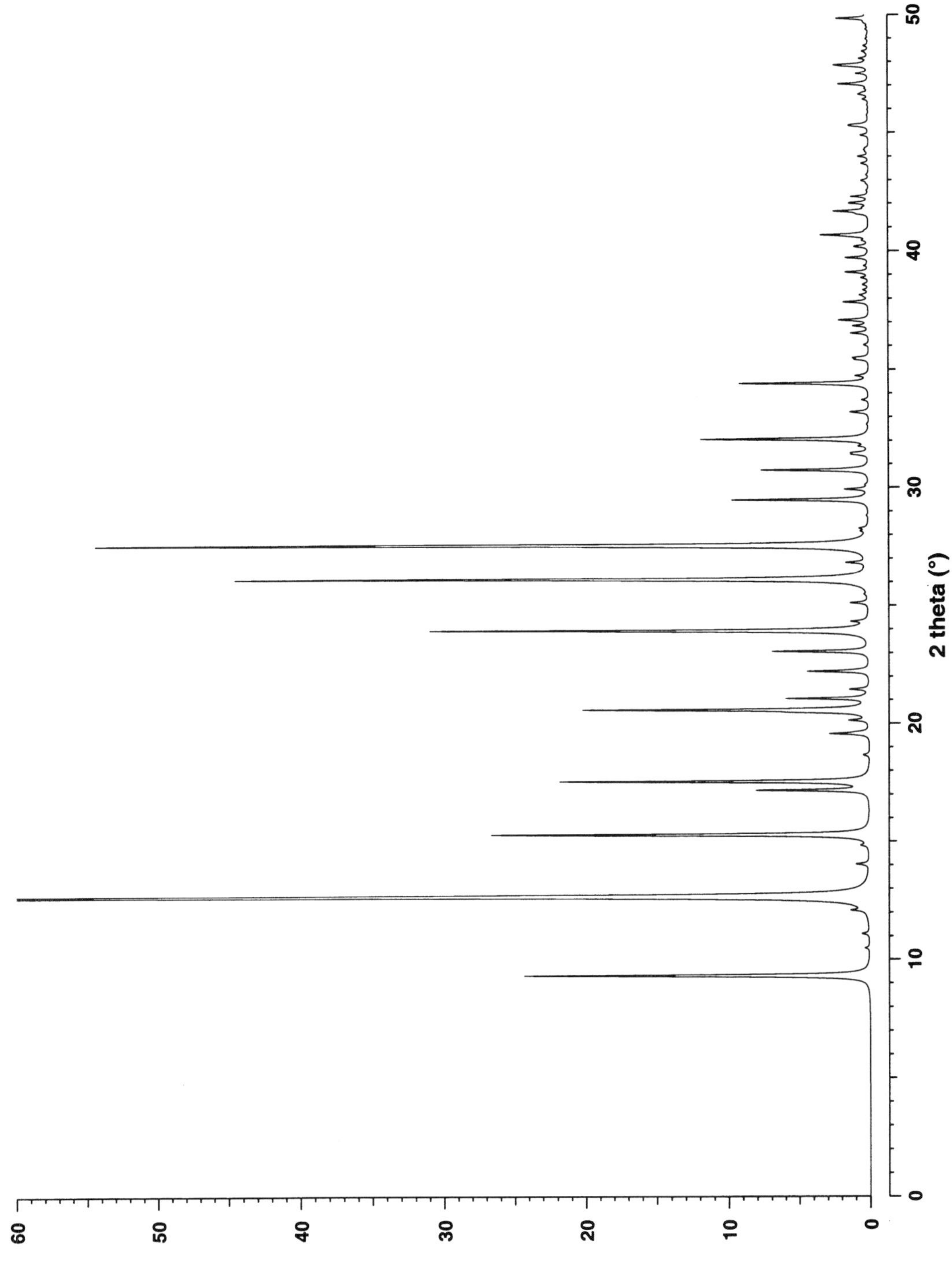

ATV AlPO-25

CHEMICAL COMPOSITION: $[Al_{12}P_{12}O_{48}]$

REFINED COMPOSITION: $[Al_{12}P_{12}O_{48}]$

CRYSTAL DATA: *Acmm* (No. 67) $\bar{\mathbf{c}}$**ba** setting
$a = 9.4489$ Å $b = 15.2028$ Å $c = 8.4084$ Å
$\alpha = 90°$ $\beta = 90°$ $\gamma = 90°$
Neutron Rietveld refinement, $R_{wp} = 0.041$, $R_{F^2} = 0.133$; At $T = 593$ K.

REFERENCE: J. W. Richardson, Jr., J. V. Smith and J. J. Pluth, *J. Phys. Chem.* **94** 3365–3367 (1990).

h	k	l	2θ	d	M	I_{rel}	h	k	l	2θ	d	M	I_{rel}	h	k	l	2θ	d	M	I_{rel}
1	0	0	9.36	9.449	2	48.0	3	1	1	30.88	2.896	8	0.3	0	0	4	43.03	2.102	2	9.0
0	2	0	11.64	7.601	2	12.6	0	4	2	31.74	2.819	4	1.8	4	3	1	43.62	2.075	8	0.8
1	2	0	14.96	5.923	4	95.2	1	5	1	32.72	2.737	8	1.2	3	1	3	43.64	2.074	8	0.3
1	1	1	15.26	5.805	8	0.3	1	4	2	33.16	2.702	8	11.3	4	0	2	43.97	2.059	4	0.3
2	0	0	18.78	4.724	2	29.7	1	1	3	33.88	2.646	8	0.2	1	7	1	44.12	2.053	8	0.3
0	0	2	21.13	4.204	2	100.0	3	3	1	35.21	2.549	8	1.3	1	0	4	44.13	2.052	4	0.7
2	2	0	22.15	4.013	4	0.5	0	6	0	35.43	2.534	2	4.9	1	5	3	45.02	2.013	8	1.0
2	1	1	22.36	3.976	8	33.4	2	5	1	36.74	2.446	8	2.2	4	2	2	45.64	1.988	8	0.8
1	3	1	22.54	3.944	8	44.0	3	4	0	37.07	2.425	4	4.4	1	2	4	45.80	1.981	8	1.4
1	0	2	23.16	3.841	4	6.7	2	4	2	37.13	2.421	8	0.5	3	6	0	45.97	1.974	4	1.1
0	4	0	23.41	3.801	2	5.3	3	2	2	37.59	2.393	8	6.8	2	6	2	46.02	1.972	8	3.0
0	2	2	24.19	3.679	4	1.1	2	1	3	37.79	2.381	8	6.9	3	3	3	46.95	1.935	8	0.7
1	4	0	25.26	3.526	4	47.2	1	3	3	37.90	2.374	8	8.3	2	7	1	47.32	1.921	8	4.4
1	2	2	25.99	3.428	8	11.4	4	0	0	38.09	2.362	2	0.7	2	0	4	47.33	1.921	4	0.9
2	3	1	27.91	3.196	8	1.9	4	2	0	39.97	2.256	4	0.1	0	8	0	47.87	1.900	2	3.6
3	0	0	28.34	3.150	2	3.0	4	1	1	40.09	2.249	8	1.4	2	5	3	48.17	1.889	8	1.1
2	0	2	28.42	3.141	4	7.0	2	6	0	40.39	2.233	4	0.8	2	2	4	48.91	1.862	8	0.3
2	4	0	30.18	2.961	4	0.2	2	3	3	41.48	2.177	8	0.4	0	4	4	49.55	1.839	4	0.7
3	2	0	30.73	2.910	4	21.5	3	5	1	42.71	2.117	8	0.2	5	2	0	49.71	1.834	4	0.9

AlPO-25 ATV

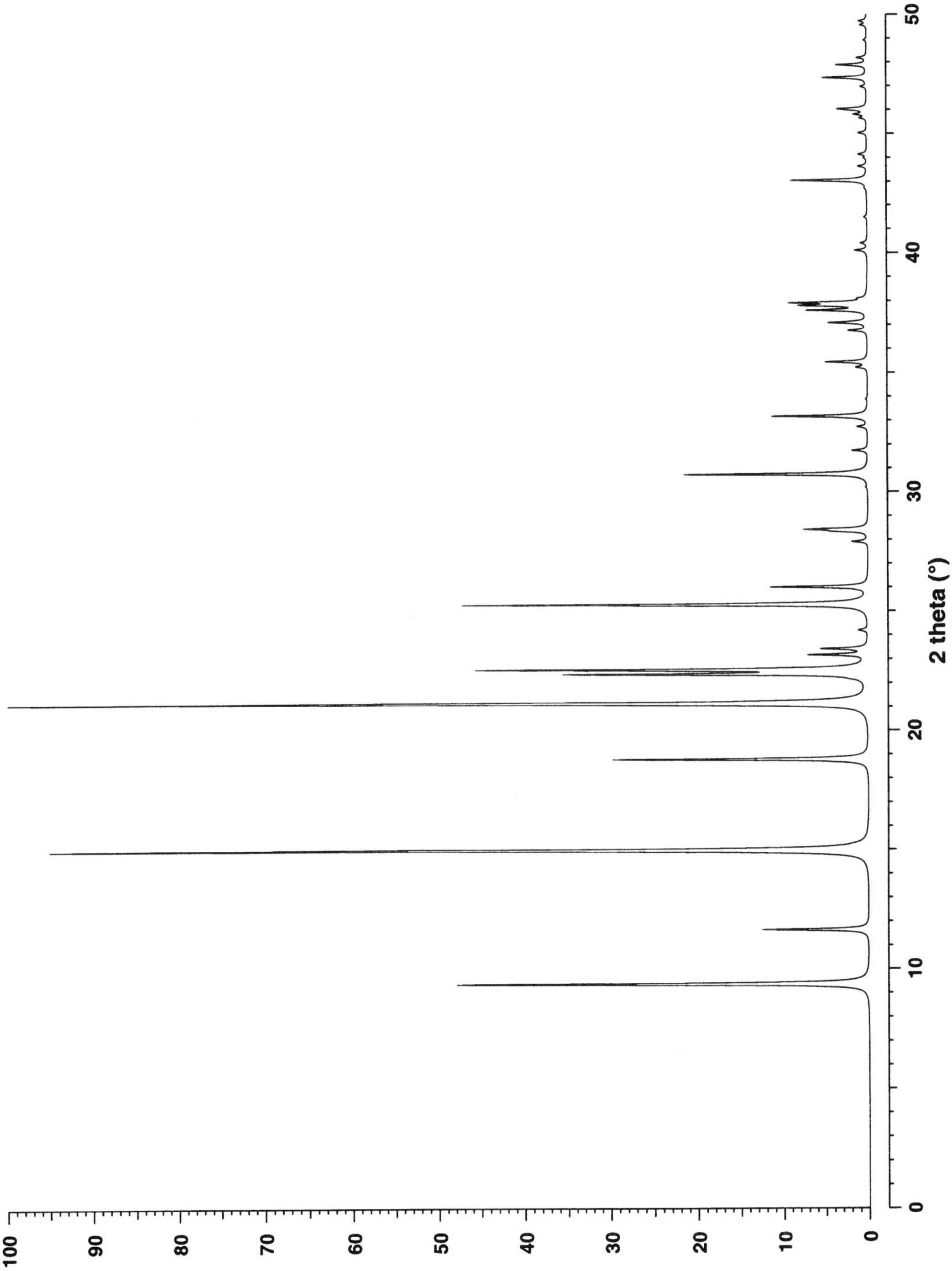

AWO AlPO–21

CHEMICAL COMPOSITION: $|(C_7H_{21}N_2)_{1.33}|\ [Al_{12}P_{12}O_{48}]$
$C_7H_{21}N_2 = N,N,N',N'$-tetramethyl-1,3-propanediamine

REFINED COMPOSITION: $|N_{2.84}C_{13.41}O_4H_4|\ [Al_{12}P_{12}O_{48}]$

CRYSTAL DATA: $P1\,2_1/a\,1$ (No. 14) unique axis **b**, cell choice 3
$a = 10.3307$ Å $b = 17.5241$ Å $c = 8.6757$ Å
$\alpha = 90.0°$ $\beta = 123.369°$ $\gamma = 90.0°$
X-ray single crystal refinement, $R = 0.046$, $R_w = 0.042$

REFERENCE: J. M. Bennett, J. M. Cohen, G. Artioli, J. J. Pluth and J. V. Smith, *Inorg. Chem.* **24** 188–193 (1985).

h	k	l	2θ	d	M	I_{rel}
0	2	0	10.10	8.762	2	100.0
1	1	0	11.43	7.740	4	25.7
-1	1	1	11.92	7.424	4	33.2
0	0	1	12.22	7.245	2	53.3
0	1	1	13.22	6.696	4	9.4
1	2	0	14.41	6.148	4	56.0
-1	2	1	14.80	5.986	4	64.2
0	2	1	15.87	5.584	4	46.5
-2	0	1	17.19	5.158	2	18.4
-2	1	1	17.93	4.948	4	1.7
1	3	0	18.34	4.837	4	3.1
-1	3	1	18.65	4.757	4	7.4
0	3	1	19.52	4.548	4	1.9
-2	2	1	19.98	4.445	4	11.8
0	4	0	20.27	4.381	2	3.2
1	1	1	20.51	4.330	4	31.3
-1	1	2	21.34	4.163	4	28.9
-2	0	2	21.68	4.098	2	10.4
-2	1	2	22.28	3.991	4	1.6
1	2	1	22.33	3.981	4	0.8
1	4	0	22.76	3.906	4	29.1
2	2	0	22.98	3.870	4	18.2
-2	3	1	23.00	3.866	4	37.8
-1	4	1	23.02	3.864	4	69.3
-1	2	2	23.10	3.850	4	1.4
0	4	1	23.73	3.749	4	21.4
0	0	2	24.57	3.623	2	19.5
1	3	1	25.09	3.549	4	3.9
0	1	2	25.10	3.548	4	51.7
0	2	2	26.63	3.348	4	77.9
-2	4	1	26.70	3.339	4	2.8
-3	1	1	26.81	3.326	4	20.5
-3	1	2	27.46	3.248	4	17.2
-1	5	1	27.68	3.223	4	15.2
-3	2	1	28.25	3.159	4	39.2
0	5	1	28.29	3.155	4	8.0
-3	2	2	28.87	3.093	4	17.2
0	3	2	29.00	3.079	4	0.6
-1	4	2	29.15	3.064	4	8.5
2	0	1	29.35	3.043	2	0.8
-3	3	1	30.51	2.930	4	2.1
-2	5	1	30.84	2.899	4	0.8
-3	3	2	31.09	2.877	4	5.4
3	1	0	31.52	2.838	4	8.4
1	1	2	32.03	2.794	4	9.5
0	4	2	32.06	2.792	4	2.2
1	6	0	32.36	2.766	4	3.1
1	5	1	32.47	2.758	4	3.9
-1	6	1	32.54	2.751	4	3.2
-2	2	3	32.61	2.746	4	11.5
3	2	0	32.77	2.732	4	0.6
2	5	0	32.93	2.720	4	18.5
-1	1	3	32.95	2.718	4	1.5
-1	5	2	33.01	2.713	4	1.6
0	6	1	33.07	2.709	4	18.0
1	2	2	33.26	2.693	4	9.3
-3	4	1	33.44	2.680	4	26.8
-2	5	2	33.65	2.664	4	0.9
-1	2	3	34.15	2.625	4	5.6
-3	2	3	34.38	2.608	4	4.8
-2	3	3	34.61	2.592	4	0.6
-4	0	2	34.79	2.579	2	15.3
-4	1	2	35.17	2.551	4	0.9
1	3	2	35.23	2.547	4	1.5
-1	3	3	36.08	2.489	4	1.2
-3	3	3	36.30	2.475	4	1.3
-4	2	2	36.31	2.474	4	3.2
-4	0	1	36.34	2.472	2	0.6
-3	5	1	36.90	2.436	4	3.3
-1	6	2	37.25	2.414	4	0.6
-3	5	2	37.40	2.405	4	0.6
1	7	0	37.40	2.404	4	12.2
3	4	0	37.40	2.404	4	2.5
-1	7	1	37.57	2.394	4	1.3
0	1	3	37.59	2.393	4	2.0
-4	0	3	37.65	2.389	2	1.7
1	4	2	37.84	2.377	4	3.0
-4	3	2	38.15	2.359	4	2.6
-1	4	3	38.64	2.330	4	1.1
2	5	1	39.20	2.298	4	0.9
-4	3	1	39.58	2.277	4	1.3
0	6	2	39.64	2.274	4	2.1
3	1	1	39.91	2.259	4	1.8
0	3	3	40.41	2.232	4	4.4
-2	5	3	40.44	2.231	4	3.4
-3	6	1	40.79	2.212	4	1.3
-4	3	3	40.81	2.211	4	3.4
3	2	1	40.94	2.204	4	0.9
1	5	2	40.99	2.202	4	0.8
1	7	1	41.34	2.184	4	3.1
2	7	0	41.71	2.165	4	0.6
2	2	2	41.72	2.165	4	0.7
4	0	0	41.88	2.157	2	0.6
-3	5	3	41.93	2.155	4	0.9
-3	1	4	42.17	2.143	4	0.8
-2	7	2	42.30	2.136	4	0.9
-1	8	1	42.73	2.116	4	1.1
0	4	3	42.75	2.115	4	1.5
4	2	0	43.19	2.094	4	2.1
-2	2	4	43.47	2.082	4	5.3
-2	6	3	44.07	2.055	4	1.1
3	6	0	44.20	2.049	4	1.4
-5	1	2	44.31	2.044	4	0.9
1	1	3	44.47	2.037	4	1.0
-4	1	4	44.52	2.035	4	0.8
1	6	2	44.58	2.033	4	1.9
4	3	0	44.79	2.023	4	1.0
3	4	1	44.85	2.021	4	3.5
-5	1	3	45.02	2.014	4	0.6
-3	7	1	45.03	2.013	4	0.9
-2	3	4	45.06	2.012	4	0.8
-5	2	2	45.26	2.004	4	1.7
-1	6	3	45.28	2.003	4	1.2
1	2	3	45.41	1.997	4	2.7
-4	2	4	45.46	1.995	4	2.1
-3	6	3	45.46	1.995	4	1.3
-5	2	3	45.95	1.975	4	0.6
-4	5	3	45.97	1.974	4	3.1
1	8	1	46.15	1.967	4	2.0
-1	2	4	46.37	1.958	4	2.6
-1	8	2	46.56	1.951	4	0.6
-4	3	4	46.99	1.934	4	3.6
-4	6	2	47.00	1.933	4	0.9
-2	8	2	47.04	1.932	4	4.4
-5	1	1	47.24	1.924	4	0.6
-5	3	3	47.47	1.915	4	2.0
3	5	1	47.61	1.910	4	1.6
-1	3	4	47.89	1.900	4	1.7
1	9	0	47.89	1.899	4	4.9
-4	6	1	48.23	1.887	4	2.5
2	5	2	48.31	1.884	4	2.5
-5	4	2	48.89	1.863	4	1.3
-1	7	3	49.20	1.852	4	1.1
-4	6	3	49.28	1.849	4	1.0
-2	5	4	49.88	1.828	4	0.9

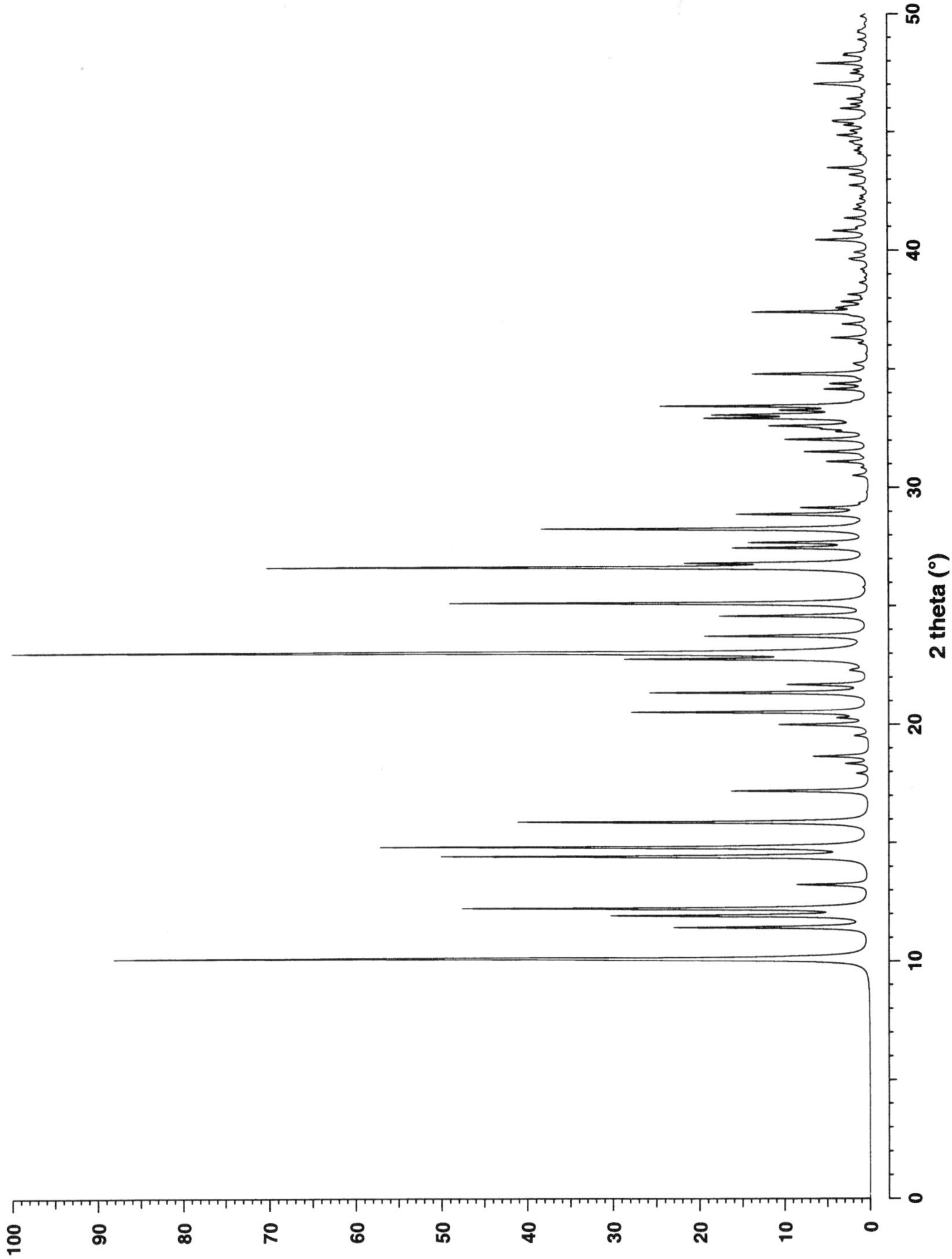
2 theta (°)
0
10
20
30
40
50
0
10
20
30
40
50
60
70
80
90
100

AWW Quinuclidine AlPO–22

CHEMICAL COMPOSITION: $|(C_7H_{13}N)_4(PO_3OH)_2|\ [Al_{24}P_{24}O_{96}]$
$C_7H_{13}N$ = quinuclidine

REFINED COMPOSITION: $|P_2O_{8.48}(C_7N)_4|\ [Al_{24}P_{24}O_{96}]$

CRYSTAL DATA: $P\,4/n\,cc$ (No. 130) origin at centre $\bar{1}$
$a = 13.628$ Å $b = 13.628$ Å $c = 15.463$ Å
$\alpha = 90°$ $\beta = 90°$ $\gamma = 90°$
X-ray single crystal refinement, $R = 0.18$

REFERENCE: J. W. Richardson, Jr., J. J. Pluth and J. V. Smith, *Naturwiss.* **76** 467–469 (1989).
And J. J. Pluth, Private communication (1990).

h	k	l	2θ	d	M	I_{rel}	h	k	l	2θ	d	M	I_{rel}	h	k	l	2θ	d	M	I_{rel}
1	1	0	9.18	9.636	4	92.8	4	2	1	29.88	2.990	16	0.2	6	1	2	41.98	2.152	16	0.4
0	0	2	11.44	7.732	2	2.8	3	3	2	30.13	2.966	8	0.6	5	2	4	42.70	2.117	16	0.4
2	0	0	12.99	6.814	4	6.3	3	1	4	31.08	2.878	16	2.3	6	2	2	43.60	2.076	16	0.5
1	0	2	13.17	6.725	8	21.1	4	2	2	31.56	2.835	16	14.1	6	1	3	44.07	2.055	16	0.2
1	1	2	14.69	6.030	8	3.6	4	1	3	32.17	2.782	16	0.4	5	4	2	44.13	2.052	16	1.1
2	1	1	15.63	5.670	16	0.2	2	1	5	32.46	2.758	16	0.2	4	4	4	44.30	2.045	8	4.1
2	0	2	17.35	5.112	8	31.3	3	2	4	33.15	2.703	16	3.3	4	1	6	44.58	2.032	16	1.1
2	2	0	18.41	4.818	4	1.6	5	1	0	33.53	2.673	8	0.4	5	1	5	44.82	2.022	16	0.4
2	1	2	18.54	4.786	16	100.0	5	1	1	34.04	2.634	16	1.2	3	3	6	45.10	2.010	8	4.6
3	1	0	20.61	4.310	8	63.9	0	0	6	34.81	2.577	2	9.7	5	3	4	45.34	2.000	16	0.3
3	1	1	21.40	4.151	16	1.9	5	0	2	34.90	2.571	8	8.3	4	2	6	46.13	1.968	16	0.3
2	2	2	21.73	4.089	8	2.8	4	3	2	34.90	2.571	16	4.4	6	3	2	46.20	1.965	16	1.9
2	1	3	22.59	3.936	16	3.1	4	0	4	35.11	2.556	8	3.2	6	0	4	46.36	1.958	8	0.3
3	0	2	22.70	3.917	8	5.0	1	0	6	35.45	2.532	8	0.4	6	1	4	46.87	1.938	16	1.6
0	0	4	23.01	3.866	2	3.2	5	1	2	35.54	2.526	16	6.3	0	0	8	47.01	1.933	2	1.3
3	1	2	23.63	3.764	16	16.5	3	1	5	35.73	2.513	16	0.3	5	5	0	47.16	1.927	4	1.0
1	0	4	23.93	3.719	8	18.0	4	1	4	35.74	2.512	16	0.7	1	0	8	47.51	1.914	8	0.1
3	2	1	24.24	3.672	16	3.0	5	2	1	35.96	2.497	16	0.2	7	1	1	47.54	1.912	16	0.1
1	1	4	24.82	3.588	8	10.7	1	1	6	36.08	2.490	8	0.6	1	1	8	48.01	1.895	8	0.4
4	0	0	26.16	3.407	4	9.5	5	2	2	37.39	2.405	16	1.5	6	4	0	48.15	1.890	8	0.2
3	2	2	26.24	3.396	16	6.4	3	2	5	37.58	2.394	16	0.5	7	0	2	48.20	1.888	8	5.2
2	0	4	26.51	3.362	8	9.1	4	2	4	37.58	2.393	16	1.6	6	2	4	48.36	1.882	16	0.2
3	1	3	26.97	3.306	16	0.2	2	1	6	37.90	2.374	16	1.1	5	0	6	48.62	1.873	8	0.4
2	1	4	27.32	3.264	16	25.7	4	4	2	39.17	2.300	8	1.5	7	1	2	48.69	1.870	16	2.1
4	1	1	27.60	3.232	16	0.1	2	2	6	39.66	2.273	8	1.7	5	5	2	48.69	1.870	8	2.6
3	3	0	27.77	3.212	4	5.9	6	0	0	39.68	2.271	4	1.7	5	4	4	48.85	1.864	16	0.3
4	0	2	28.63	3.118	8	8.3	3	0	6	40.23	2.242	8	0.8	2	0	8	48.98	1.860	8	2.0
3	2	3	29.30	3.048	16	1.0	5	3	2	40.31	2.237	16	0.6	7	2	1	49.02	1.858	16	0.1
4	2	0	29.31	3.047	8	18.8	4	3	4	40.49	2.228	16	0.1	5	1	6	49.11	1.855	16	2.8
4	1	2	29.39	3.039	16	10.1	5	1	4	41.06	2.198	16	3.5	2	1	8	49.47	1.842	16	0.3
2	2	4	29.63	3.015	8	0.9	6	2	0	41.93	2.155	8	1.9	6	4	2	49.66	1.836	16	0.2

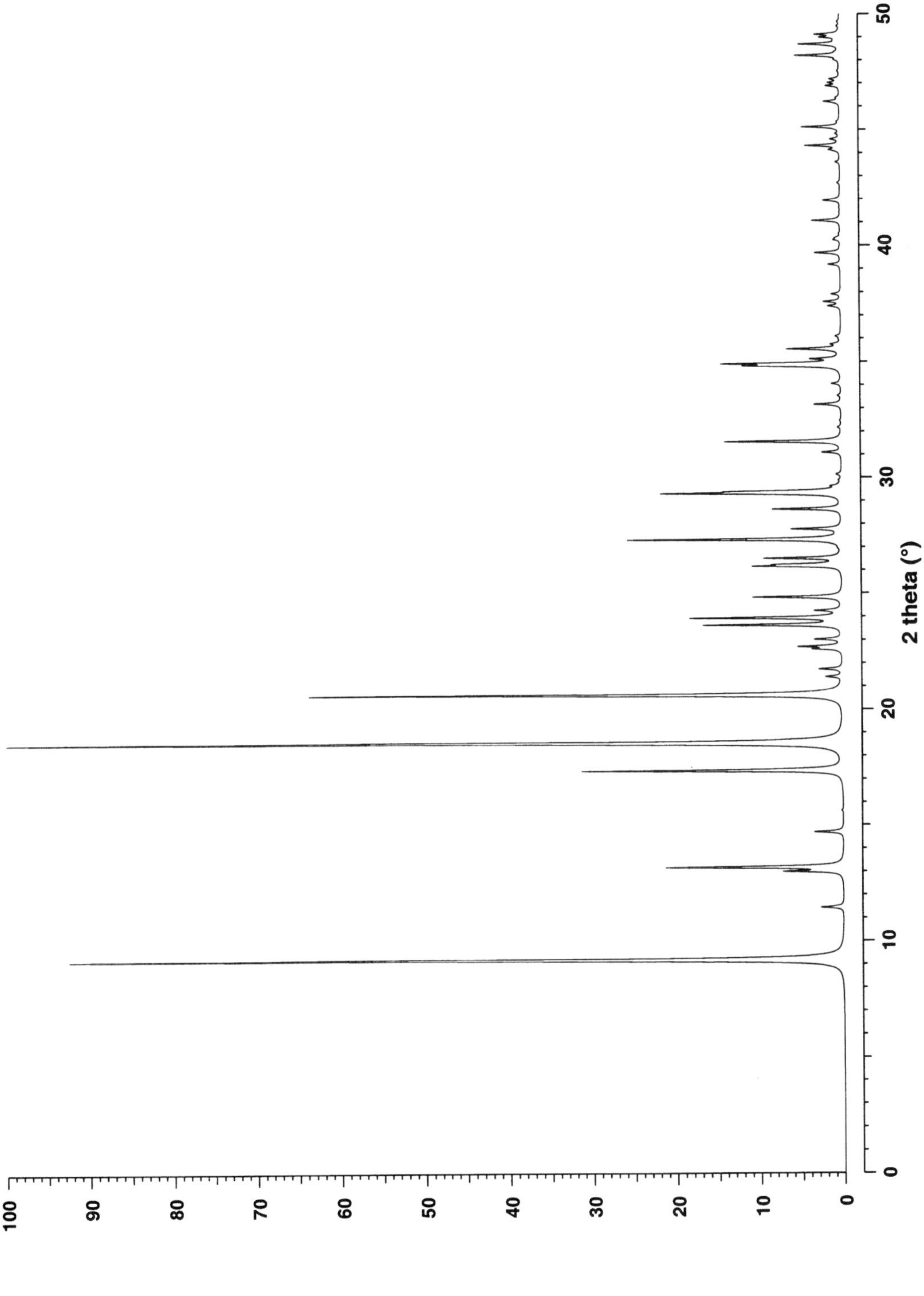
2 theta (°)
0
10
20
30
40
50
100
90
80
70
60
50
40
30
20
10
0

*BEA Beta, Polymorph A SiO_2 Framework

CHEMICAL COMPOSITION: $[Si_{64}O_{128}]$

REFINED COMPOSITION: $[Si_{64}O_{128}]$

CRYSTAL DATA: $P4_122$ (No. 91)
$a = 12.661$ Å $b = 12.661$ Å $c = 26.406$ Å
$\alpha = 90°$ $\beta = 90°$ $\gamma = 90°$
DLS refinement.

REFERENCE: J. M. Newsam, M. M. J. Treacy, W. T. Koetsier and C. B. deGruyter, *Proc. Roy. Soc.* (London) **A420** 375–405 (1988).

h	k	l	2θ	d	M	I_{rel}	h	k	l	2θ	d	M	I_{rel}	h	k	l	2θ	d	M	I_{rel}
1	0	0	6.98	12.661	4	22.8	3	2	2	26.26	3.394	16	0.1	2	1	10	37.58	2.393	16	0.1
1	0	1	7.74	11.417	8	100.0	0	0	8	27.01	3.301	2	2.4	5	1	3	37.64	2.390	16	0.1
1	0	2	9.68	9.138	8	14.7	2	0	7	27.52	3.241	8	1.8	4	2	6	37.78	2.381	16	0.2
1	1	1	10.43	8.479	8	0.3	1	0	8	27.93	3.194	8	0.2	3	1	9	38.02	2.367	16	0.1
1	1	2	11.94	7.410	8	0.3	4	0	0	28.19	3.165	4	0.9	5	0	4	38.06	2.364	8	0.1
1	0	3	12.25	7.227	8	6.3	4	0	1	28.40	3.143	8	2.9	1	0	11	38.16	2.359	8	0.1
0	0	4	13.41	6.602	2	2.9	2	1	7	28.43	3.139	16	0.5	5	1	4	38.74	2.324	16	0.2
2	0	0	13.99	6.331	4	0.3	2	2	6	28.44	3.138	8	0.6	4	0	8	39.44	2.285	8	0.3
2	0	1	14.39	6.156	8	3.8	3	2	4	28.80	3.100	16	0.4	2	2	10	39.63	2.274	8	0.1
1	0	4	15.14	5.854	8	0.3	4	0	2	29.01	3.078	8	0.2	5	2	3	39.68	2.271	16	0.2
2	0	2	15.52	5.708	8	1.0	3	0	6	29.32	3.046	8	2.1	2	0	11	40.17	2.245	8	0.4
2	1	1	16.01	5.536	16	0.2	3	3	0	29.94	2.984	4	0.5	5	2	4	40.74	2.215	16	0.6
2	1	2	17.04	5.204	16	0.1	2	0	8	30.54	2.927	8	1.0	2	1	11	40.83	2.210	16	0.1
1	0	5	18.20	4.874	8	4.2	3	2	5	30.57	2.924	16	1.9	5	3	1	41.74	2.164	16	0.1
2	1	3	18.63	4.762	16	0.5	4	1	3	30.84	2.899	16	0.1	5	1	6	41.77	2.163	16	0.3
2	0	4	19.43	4.569	8	1.1	2	2	7	31.00	2.885	8	0.1	4	0	9	41.99	2.152	8	0.4
1	1	5	19.51	4.549	8	0.1	1	0	9	31.29	2.858	8	0.7	5	2	5	42.07	2.148	16	0.2
2	2	1	20.12	4.413	8	0.5	4	1	4	32.15	2.784	16	0.5	3	2	10	42.85	2.110	16	0.3
2	1	4	20.67	4.298	16	1.7	4	2	2	32.34	2.768	16	0.1	6	0	0	42.86	2.110	4	2.1
2	2	2	20.95	4.239	8	1.4	3	1	7	32.61	2.746	16	1.1	5	3	3	42.90	2.108	16	0.2
3	0	0	21.05	4.220	4	2.8	3	3	4	32.94	2.719	8	1.1	5	0	7	43.02	2.102	8	0.3
1	0	6	21.37	4.157	8	1.0	4	0	5	32.99	2.715	8	0.2	6	1	1	43.62	2.075	16	0.2
2	0	5	21.92	4.055	8	6.9	2	0	9	33.67	2.662	8	0.8	5	2	6	43.65	2.074	16	0.4
3	0	2	22.11	4.020	8	19.2	4	1	5	33.76	2.655	16	0.1	2	1	12	44.15	2.051	16	0.3
3	1	0	22.20	4.004	8	0.7	2	1	9	34.43	2.605	16	0.1	1	0	13	45.21	2.006	8	0.1
2	2	3	22.28	3.990	8	0.3	4	2	4	34.47	2.602	16	0.1	5	2	7	45.46	1.995	16	0.2
3	1	1	22.46	3.959	16	3.7	3	0	8	34.50	2.600	8	0.2	5	1	8	45.72	1.984	16	0.2
1	1	6	22.51	3.950	8	0.4	1	0	10	34.70	2.585	8	0.2	3	2	11	45.78	1.982	16	1.1
2	1	5	23.03	3.862	16	0.7	3	2	7	34.91	2.570	16	0.3	2	0	13	46.98	1.934	8	0.2
3	1	2	23.21	3.831	16	0.7	3	1	8	35.24	2.547	16	0.3	5	0	9	47.42	1.917	8	0.9
3	1	3	24.42	3.644	16	1.3	5	0	0	35.45	2.532	4	0.2	5	2	8	47.48	1.915	16	0.2
1	0	7	24.62	3.615	8	1.0	4	3	1	35.62	2.521	16	0.7	6	0	6	47.80	1.903	8	0.3
2	0	6	24.64	3.614	8	0.3	5	0	1	35.62	2.521	8	0.4	5	1	9	48.00	1.895	16	0.1
3	0	4	25.04	3.556	8	2.3	4	3	2	36.12	2.487	16	0.3	5	4	4	48.03	1.894	16	0.2
3	2	0	25.36	3.512	8	0.1	5	1	1	36.34	2.472	16	0.2	5	3	7	48.37	1.882	16	0.3
3	2	1	25.59	3.481	16	0.8	5	1	2	36.83	2.440	16	0.4	6	2	5	48.64	1.872	16	0.2
2	1	6	25.64	3.475	16	1.6	4	3	3	36.94	2.434	16	0.7	6	3	2	48.74	1.868	16	0.4
3	1	4	26.03	3.423	16	0.3	5	0	3	36.94	2.434	8	0.6	5	2	9	49.69	1.835	16	0.2
2	2	5	26.09	3.415	8	0.2	4	0	7	37.08	2.425	8	0.3	5	0	10	49.90	1.828	8	0.4

Zeolite beta and its mineral analog tschernichite are intergrown materials and *BEA represents the framework of a hypothetical end member. Simulated powder patterns for intergrown variants are presented later in this book.

Beta, Polymorph A SiO_2 Framework *BEA

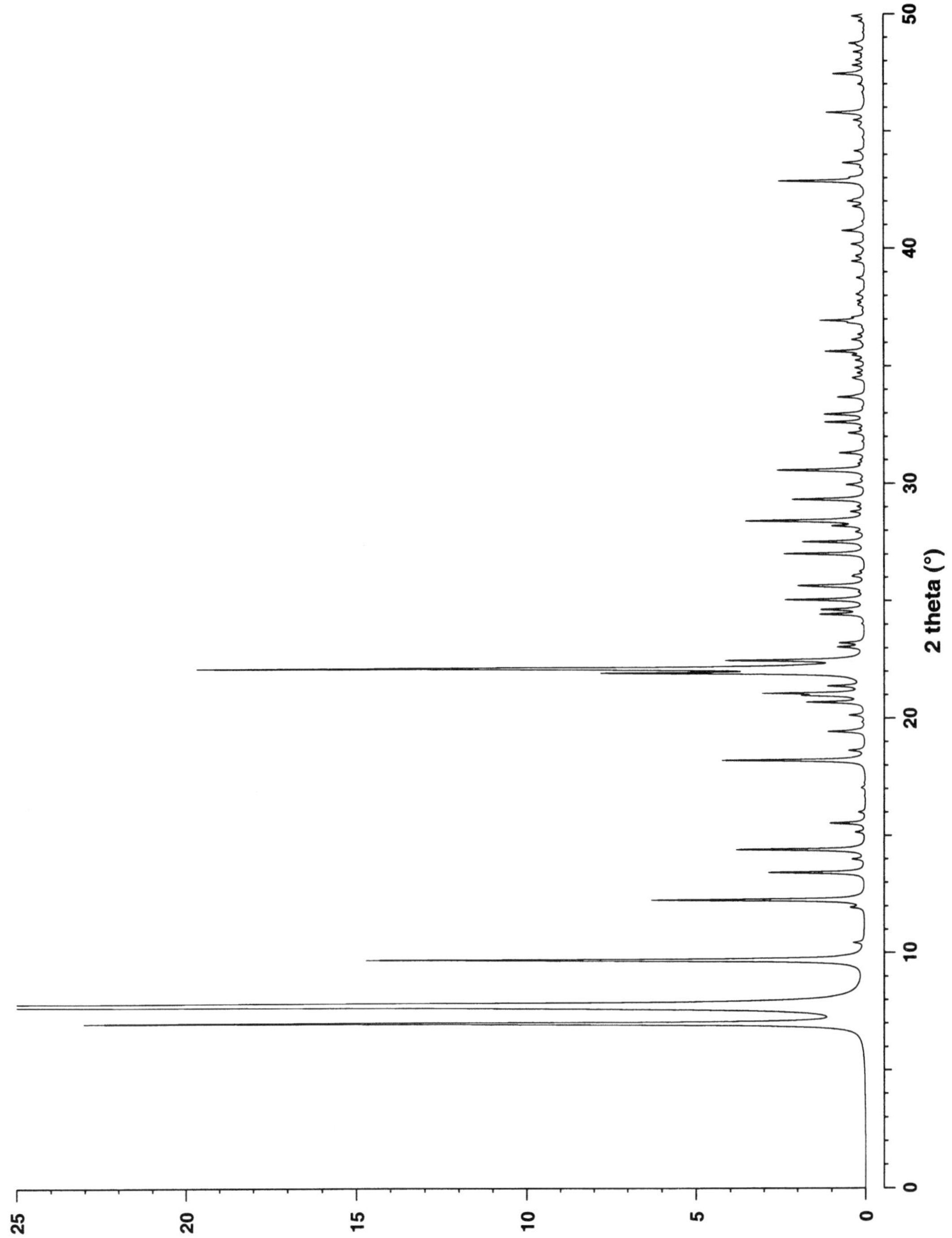

BIK Bikitaite

CHEMICAL COMPOSITION: $|Li_2(H_2O)_2|$ $[Si_4Al_2O_{12}]$
Bikita, Zimbabwe

REFINED COMPOSITION: $|Li_{1.86}(H_2O)_2|$ $[Si_{3.89}Al_{2.13}O_{12}]$

CRYSTAL DATA: $P1$ (No. 1)
$a = 8.6071$ Å $b = 4.9540$ Å $c = 7.5972$ Å
$\alpha = 89.900°$ $\beta = 114.437°$ $\gamma = 89.988°$
Neutron single crystal refinement, $R_w = 0.056$

REFERENCE: K. Ståhl, Å. Kvick and S. Ghose, *Zeolites* **9** 303–311 (1989).

h	k	l	2θ	d	M	I_{rel}
1	0	0	11.29	7.836	2	45.2
0	0	1	12.80	6.917	2	30.6
-1	0	1	13.11	6.754	2	16.1
1	0	1	20.34	4.366	2	31.3
-2	0	1	20.93	4.244	2	9.7
1	1	0	21.21	4.189	2	69.7
1	-1	0	21.23	4.185	2	68.2
0	1	1	22.05	4.031	2	0.6
0	-1	1	22.09	4.024	2	0.5
-1	1	1	22.24	3.997	2	5.4
-1	-1	1	22.27	3.993	2	4.6
2	0	0	22.69	3.918	2	2.4
-1	0	2	23.43	3.797	2	9.3
0	0	2	25.76	3.458	2	100.0
-2	0	2	26.39	3.377	2	78.4
1	1	1	27.20	3.279	2	25.4
1	-1	1	27.25	3.273	2	23.3
-2	1	1	27.67	3.223	2	20.0
-2	-1	1	27.68	3.223	2	20.4
2	1	0	29.04	3.075	2	13.5
2	-1	0	29.07	3.072	2	14.9
-1	1	2	29.62	3.016	2	0.6
-1	-1	2	29.67	3.011	2	1.0
2	0	1	30.52	2.929	2	5.9
-3	0	1	31.20	2.867	2	8.7
0	1	2	31.52	2.838	2	0.7
0	-1	2	31.58	2.833	2	0.7
-2	1	2	32.06	2.792	2	2.7
-2	-1	2	32.09	2.789	2	3.0
1	0	2	32.33	2.769	2	0.1
2	1	1	35.58	2.523	2	8.8
2	-1	1	35.64	2.519	2	9.3
-1	0	3	35.76	2.511	2	0.3
-3	-1	1	36.20	2.482	2	18.6
-3	1	1	36.21	2.481	2	18.4
0	2	0	36.27	2.477	2	43.5
1	1	2	37.17	2.419	2	1.9
1	-1	2	37.23	2.415	2	2.5
1	2	0	38.09	2.363	2	0.5
-3	1	2	38.10	2.362	2	3.9
1	-2	0	38.11	2.361	2	0.5
-3	-1	2	38.12	2.361	2	4.5
0	2	1	38.58	2.333	2	1.5
0	-2	1	38.63	2.330	2	1.3
-1	2	1	38.70	2.326	2	1.5
-1	-2	1	38.73	2.325	2	1.7
3	1	0	38.96	2.312	2	1.1
3	-1	0	39.00	2.310	2	1.7
0	0	3	39.07	2.306	2	1.0
-1	1	3	40.24	2.241	2	1.9
-1	-1	3	40.30	2.238	2	2.0
-2	1	3	40.56	2.224	2	0.4
-2	-1	3	40.61	2.221	2	0.6
3	0	1	41.71	2.165	2	0.1
1	2	1	41.90	2.156	2	0.9
1	-2	1	41.97	2.153	2	1.0
-2	2	1	42.24	2.139	2	1.0
-2	-2	1	42.25	2.139	2	1.3
-4	0	1	42.44	2.130	2	0.1
2	2	0	43.19	2.095	2	0.6
2	-2	0	43.23	2.093	2	0.6
0	1	3	43.25	2.092	2	0.4
0	-1	3	43.32	2.089	2	0.7
-3	1	3	44.17	2.050	2	4.0
-3	-1	3	44.21	2.049	2	4.1
0	2	2	44.97	2.016	2	4.5
0	-2	2	45.06	2.012	2	3.6
1	0	3	45.36	1.999	2	7.5
-2	2	2	45.38	1.998	2	4.9
2	-1	2	45.43	1.996	2	0.4
-2	-2	2	45.43	1.996	2	4.3
3	1	1	45.70	1.985	2	4.3
3	-1	1	45.75	1.983	2	4.4
4	0	0	46.35	1.959	2	7.0
-4	-1	1	46.39	1.957	2	0.9
-4	1	1	46.41	1.956	2	0.8
-4	1	2	46.55	1.951	2	0.7
-4	-1	2	46.56	1.951	2	0.6
-4	0	3	46.81	1.941	2	2.2
-2	0	4	47.92	1.898	2	15.0
2	2	1	48.07	1.893	2	1.8
2	-2	1	48.15	1.890	2	1.7
-3	-2	1	48.57	1.875	2	2.7
-3	2	1	48.58	1.874	2	2.6
-1	0	4	49.03	1.858	2	0.5
1	1	3	49.11	1.855	2	6.9
1	-1	3	49.18	1.853	2	6.0
1	-2	2	49.42	1.844	2	0.1
-3	0	4	49.76	1.832	2	0.4

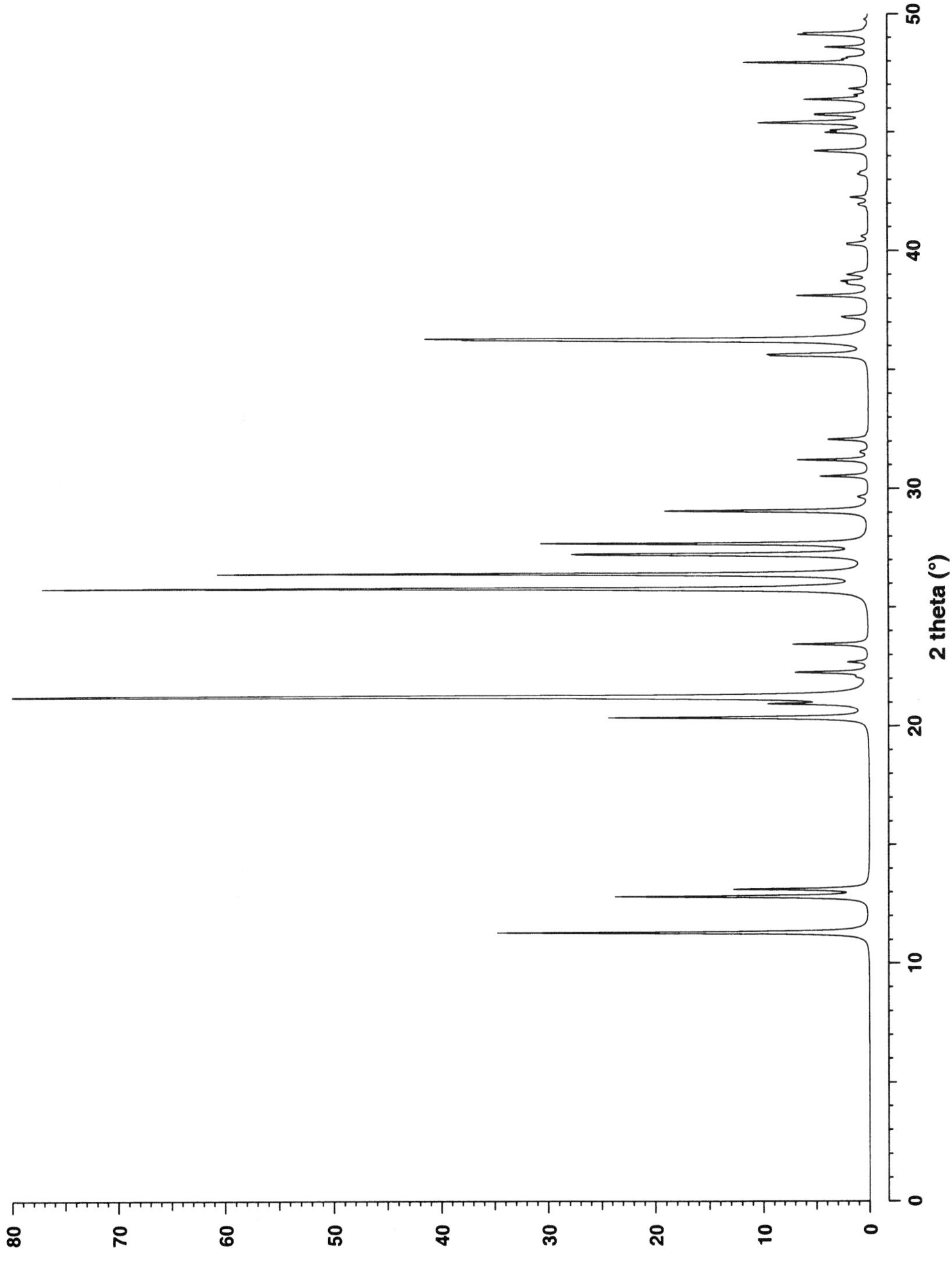
2 theta (°)
0
10
20
30
40
50
0
10
20
30
40
50
60
70
80

BOG Boggsite

CHEMICAL COMPOSITION: $|Na_{2.9}Ca_{7.8}(H_2O)_{70}|\ [Si_{77.5}Al_{18.3}O_{192}]$
Goble area, Oregon, U.S.A.

REFINED COMPOSITION: $|(H_2O)_{137.04}|\ [Si_{77.76}Al_{18.24}O_{192}]$

CRYSTAL DATA: *Imma* (No. 74)
$a = 20.236$ Å $b = 23.798$ Å $c = 12.798$ Å
$\alpha = 90°$ $\beta = 90°$ $\gamma = 90°$
X-ray single crystal refinement, $R_w = 0.094$

REFERENCE: J. J. Pluth and J. V. Smith,
American Mineralogist **75** 501–507 (1990).

h	k	l	2θ	d	M	I_{rel}
0	2	0	7.43	11.899	2	13.6
0	1	1	7.84	11.271	4	68.2
1	0	1	8.17	10.816	4	9.2
2	0	0	8.74	10.118	2	10.8
1	2	1	11.05	8.004	8	8.7
2	2	0	11.48	7.708	4	3.9
2	1	1	11.75	7.529	8	4.6
0	3	1	13.13	6.742	4	11.0
3	0	1	14.85	5.967	4	3.0
0	2	2	15.72	5.636	4	5.7
2	3	1	15.79	5.611	8	1.8
1	4	1	17.01	5.213	8	2.3
2	4	0	17.29	5.129	4	1.3
4	0	0	17.53	5.059	2	0.7
2	2	2	18.02	4.924	8	8.8
4	2	0	19.06	4.656	4	1.0
0	5	1	19.90	4.461	4	29.4
0	4	2	20.38	4.357	4	29.4
3	4	1	21.09	4.213	8	5.6
0	1	3	21.16	4.199	4	2.3
1	0	3	21.28	4.174	4	3.0
3	3	2	22.19	4.007	8	8.4
2	4	2	22.21	4.002	8	0.6
4	0	2	22.40	3.969	4	0.8
0	6	0	22.41	3.966	2	9.0
1	2	3	22.57	3.939	8	1.0
5	0	1	23.05	3.859	4	100.0
4	4	0	23.08	3.854	4	0.5
0	3	3	23.68	3.757	4	15.1
1	6	1	23.90	3.724	8	3.2
2	6	0	24.10	3.693	4	15.7
5	2	1	24.25	3.671	8	12.1
3	0	3	24.69	3.605	4	37.9
2	3	3	25.29	3.522	8	10.1
1	4	3	26.08	3.417	8	4.4
5	1	2	26.32	3.386	8	61.3
0	6	2	26.44	3.371	4	64.6
3	5	2	26.83	3.323	8	2.7
3	6	1	26.99	3.303	8	5.5
4	4	2	27.01	3.301	8	1.0
6	2	0	27.49	3.245	4	2.4
5	4	1	27.55	3.237	8	0.5
6	1	1	27.61	3.231	8	1.5
4	1	3	27.61	3.231	8	0.6
0	0	4	27.88	3.200	2	7.0
2	6	2	27.89	3.198	8	19.0
0	5	3	28.09	3.177	4	1.5
5	3	2	28.42	3.141	8	1.7
4	6	0	28.60	3.121	4	0.5
0	2	4	28.90	3.090	4	0.7
3	4	3	28.96	3.083	8	5.7
2	0	4	29.28	3.051	4	8.0
0	8	0	30.04	2.975	2	15.8
5	0	3	30.44	2.936	4	1.6
1	3	4	30.45	2.936	8	4.4
1	6	3	31.10	2.875	8	11.4
3	1	4	31.17	2.870	8	9.1
5	2	3	31.38	2.851	8	1.8
0	4	4	31.75	2.818	4	4.2
4	6	2	31.90	2.805	8	0.8
5	5	2	32.23	2.778	8	17.2
5	6	1	32.37	2.766	8	2.1
7	2	1	32.63	2.744	8	0.5
3	3	4	32.98	2.716	8	1.0
2	4	4	33.00	2.715	8	3.5
4	5	3	33.30	2.690	8	0.8
3	6	3	33.59	2.668	8	5.5
6	4	2	33.60	2.667	8	0.6
5	4	3	34.05	2.633	8	21.0
1	5	4	34.05	2.633	8	0.7
6	1	3	34.10	2.630	8	0.6
7	1	2	34.24	2.618	8	1.6
0	9	1	34.64	2.590	4	12.9
6	6	0	34.92	2.569	4	0.7
7	4	1	35.22	2.548	8	1.7
0	1	5	35.27	2.545	4	0.6
1	0	5	35.35	2.539	4	1.5
8	0	0	35.49	2.530	2	1.6
6	3	3	35.78	2.510	8	0.6
7	3	2	35.92	2.500	8	3.2
0	6	4	36.07	2.490	4	1.2
3	5	4	36.36	2.471	8	4.4
0	3	5	36.90	2.436	4	4.6
5	7	2	37.29	2.411	8	8.1
7	0	3	37.58	2.393	4	0.8
5	3	4	37.59	2.393	8	2.2
6	6	2	37.73	2.384	8	0.8
0	10	0	37.80	2.380	2	0.9
2	3	5	37.99	2.368	8	3.7
5	8	1	38.20	2.356	8	0.5
3	2	5	38.37	2.346	8	0.5
8	4	0	38.68	2.328	4	1.3
6	5	3	38.95	2.312	8	1.3
7	5	2	39.08	2.305	8	1.1
7	6	1	39.20	2.298	8	4.4
3	8	3	39.26	2.295	8	1.5
0	5	5	39.99	2.254	4	2.7
0	9	3	40.12	2.247	4	1.7
3	4	5	40.63	2.220	8	0.4
5	5	4	40.64	2.220	8	0.5
3	10	1	40.82	2.210	8	0.7
8	1	3	41.68	2.167	8	0.6
5	0	5	41.75	2.163	4	3.2
0	11	1	42.37	2.133	4	2.5
2	8	4	42.44	2.130	8	2.3
8	3	3	43.11	2.098	8	0.9
5	9	2	43.25	2.092	8	0.7
7	3	4	43.72	2.071	8	0.6
4	9	3	44.09	2.054	8	2.1
1	3	6	44.20	2.049	8	1.7
3	6	5	44.20	2.049	8	2.0
1	9	4	44.68	2.028	8	1.0
3	1	6	44.72	2.026	8	0.9
5	10	1	44.74	2.026	8	1.0
10	0	0	44.79	2.024	2	14.2
0	4	6	45.16	2.008	4	1.0
2	7	5	45.24	2.004	8	1.3
8	0	4	45.72	1.984	4	6.4
0	12	0	45.75	1.983	2	5.2
8	5	3	45.86	1.979	8	3.0
3	3	6	46.07	1.970	8	0.8
8	2	4	46.39	1.957	8	1.4
3	9	4	46.55	1.951	8	8.9
9	6	1	46.99	1.934	8	0.6
10	0	2	47.10	1.929	4	0.5
5	6	5	47.90	1.899	8	0.5
0	12	2	48.03	1.894	4	1.5
8	4	4	48.35	1.882	8	1.1
0	6	6	48.46	1.879	4	1.4
3	5	6	48.69	1.870	8	2.7
2	12	2	48.92	1.862	8	0.9
5	3	6	49.66	1.836	8	0.5
10	4	2	49.67	1.835	8	0.5
8	7	3	49.75	1.833	8	0.8
5	11	2	49.87	1.828	8	0.7

Boggsite BOG

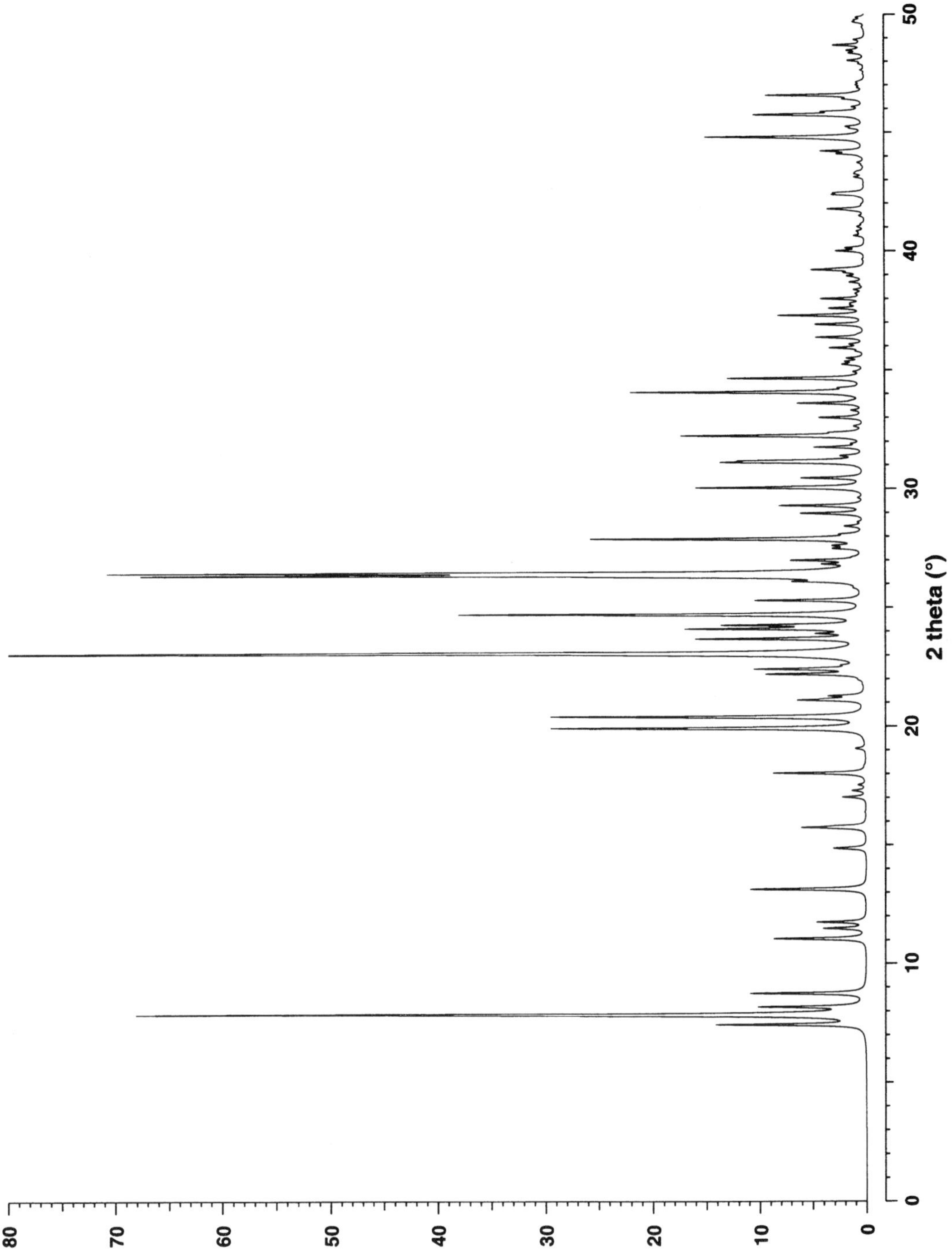

BPH

Beryllophosphate-H

CHEMICAL COMPOSITION: $|Na_7K_7(H_2O)_{20}|$ $[Be_{14}P_{14}O_{56}]$

REFINED COMPOSITION: $|Na_{8.36}K_{5.49}(H_2O)_{30.48}|$ $[Be_{14}P_{14}O_{56}]$

CRYSTAL DATA: $P321$ (No. 150)
$a = 12.5815$ Å $b = 12.5815$ Å $c = 12.4508$ Å
$\alpha = 90°$ $\beta = 90°$ $\gamma = 120°$
X-ray Rietveld refinement, $R_{wp} = 0.192$, $R_F = 0.106$

REFERENCE: G. Harvey, Ch. Baerlocher and T. Wroblewski, *Z. Kristallogr.* **201** 113–123 (1992).

h	k	l	2θ	d	M	I_{rel}
0	0	1	7.10	12.451	2	68.8
1	0	0	8.11	10.896	6	100.0
1	0	1	10.79	8.200	6	0.1
0	1	1	10.79	8.200	6	1.3
1	1	0	14.08	6.291	6	2.6
0	0	2	14.23	6.225	2	2.4
1	1	1	15.78	5.615	12	7.5
2	0	0	16.27	5.448	6	1.1
0	1	2	16.40	5.405	6	0.5
1	0	2	16.40	5.405	6	1.5
2	0	1	17.77	4.991	6	1.3
1	1	2	20.07	4.425	12	1.8
0	0	3	21.41	4.150	2	2.9
2	1	0	21.58	4.118	12	9.1
2	0	2	21.68	4.100	6	2.6
0	2	2	21.68	4.100	6	3.8
2	1	1	22.74	3.910	12	15.5
1	2	1	22.74	3.910	12	24.3
1	0	3	22.93	3.878	6	14.3
0	1	3	22.93	3.878	6	0.9
3	0	1	25.55	3.487	6	4.3
1	1	3	25.72	3.464	12	11.2
2	1	2	25.94	3.435	12	12.6
1	2	2	25.94	3.435	12	6.1
0	2	3	27.01	3.301	6	0.1
2	0	3	27.01	3.301	6	0.4
2	2	0	28.37	3.145	6	6.3
0	3	2	28.45	3.137	6	9.4
3	0	2	28.45	3.137	6	2.1
2	2	1	29.29	3.050	12	20.4
3	1	0	29.56	3.022	12	5.9
1	0	4	29.85	2.993	6	0.2
0	1	4	29.85	2.993	6	0.1
1	3	1	30.44	2.937	12	1.4
3	1	1	30.44	2.937	12	16.1
2	1	3	30.58	2.923	12	10.8
1	2	3	30.58	2.923	12	5.9
2	2	2	31.88	2.807	12	27.3
1	1	4	32.08	2.790	12	47.5
3	0	3	32.77	2.733	6	3.4
0	3	3	32.77	2.733	6	5.6
4	0	0	32.88	2.724	6	10.1
1	3	2	32.95	2.719	12	4.4
3	1	2	32.95	2.719	12	4.3
2	0	4	33.15	2.703	6	1.4
0	2	4	33.15	2.703	6	0.2
4	0	1	33.68	2.661	6	0.4
0	4	1	33.68	2.661	6	1.9
2	2	3	35.82	2.507	12	4.3
3	2	0	35.93	2.500	12	4.4
4	0	2	35.99	2.496	6	0.2
0	0	5	36.07	2.490	2	1.6
1	2	4	36.17	2.483	12	9.1
2	1	4	36.17	2.483	12	5.2
3	2	1	36.67	2.451	12	1.9
2	3	1	36.67	2.451	12	0.8
3	1	3	36.79	2.443	12	1.5
1	3	3	36.79	2.443	12	0.4
0	1	5	37.03	2.428	6	2.9
4	1	0	37.84	2.378	12	1.8
0	3	4	38.07	2.363	6	0.1
1	4	1	38.55	2.335	12	0.4
4	1	1	38.55	2.335	12	0.6
2	3	2	38.82	2.320	12	0.3
3	2	2	38.82	2.320	12	1.5
1	1	5	38.90	2.315	12	1.8
0	4	3	39.57	2.277	6	0.6
2	0	5	39.80	2.265	6	0.5
0	2	5	39.80	2.265	6	0.7
1	4	2	40.62	2.221	12	0.6
2	2	4	40.78	2.212	12	2.1
5	0	0	41.43	2.179	6	9.2
3	1	4	41.65	2.168	12	3.7
1	3	4	41.65	2.168	12	6.6
0	5	1	42.09	2.147	6	0.6
5	0	1	42.09	2.147	6	1.3
2	3	3	42.20	2.141	12	2.1
3	2	3	42.20	2.141	12	1.5
1	2	5	42.42	2.131	12	0.5
2	1	5	42.42	2.131	12	1.0
3	3	0	43.14	2.097	6	4.5
0	0	6	43.62	2.075	2	3.0
3	3	1	43.78	2.068	12	2.7
1	4	3	43.88	2.063	12	1.4
4	1	3	43.88	2.063	12	4.7
0	5	2	44.02	2.057	6	0.7
5	0	2	44.02	2.057	6	1.6
0	3	5	44.09	2.054	6	0.1
4	0	4	44.18	2.050	6	0.3
0	4	4	44.18	2.050	6	0.2
1	0	6	44.44	2.038	6	0.3
2	4	1	44.60	2.032	12	0.2
4	2	1	44.60	2.032	12	1.5
3	3	2	45.65	1.987	12	0.7
1	1	6	46.06	1.971	12	0.8
5	1	0	46.40	1.957	12	4.1
2	4	2	46.45	1.955	12	0.3
4	2	2	46.45	1.955	12	1.4
2	2	5	46.51	1.952	12	0.4
3	2	4	46.60	1.949	12	0.2
2	3	4	46.60	1.949	12	0.2
2	0	6	46.85	1.939	6	1.1
0	2	6	46.85	1.939	6	0.2
5	1	1	47.00	1.933	12	0.6
1	5	1	47.00	1.933	12	1.7
0	5	3	47.10	1.929	6	1.7
5	0	3	47.10	1.929	6	0.4
3	1	5	47.30	1.922	12	1.8
1	3	5	47.30	1.922	12	1.8
4	1	4	48.16	1.889	12	0.5
3	3	3	48.65	1.872	12	0.2
1	5	2	48.78	1.867	12	0.3
5	1	2	48.78	1.867	12	0.2
2	1	6	49.16	1.853	12	0.6
1	2	6	49.16	1.853	12	0.9
4	2	3	49.41	1.845	12	1.0
2	4	3	49.41	1.845	12	0.2
4	0	5	49.60	1.838	6	0.1
0	4	5	49.60	1.838	6	0.2

Beryllophosphate-H BPH

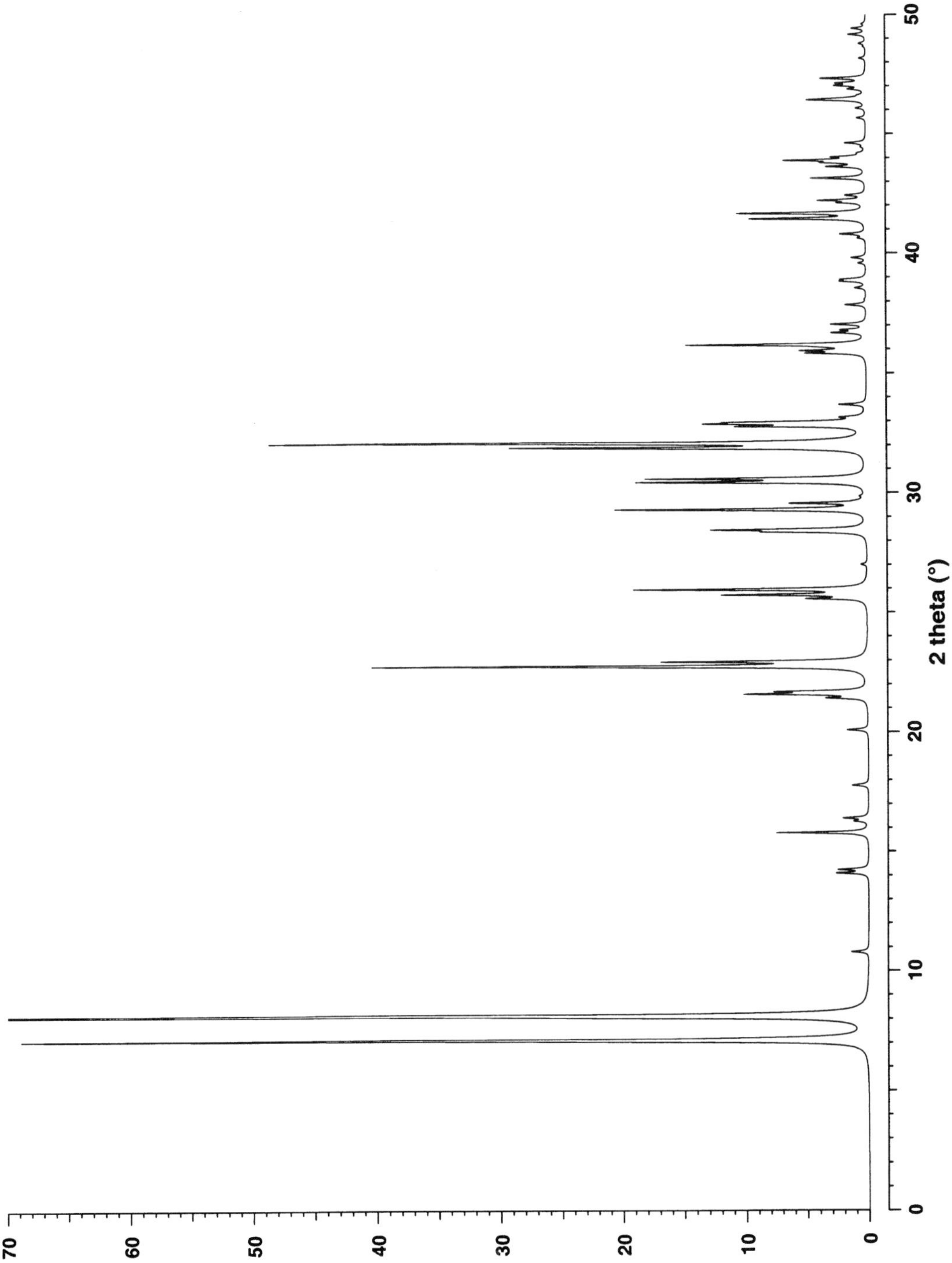

BRE Brewsterite

CHEMICAL COMPOSITION: $|K_{0.02}Ba_{0.48}Sr_{1.42}(H_2O)_{10}|$ $[Si_{12}Al_4O_{32}]$
Strontian, Argyll, Scotland

REFINED COMPOSITION: $|Ba_{0.52}Sr_{1.48}(H_2O)_{10}|$ $[Si_{12}Al_4O_{32}]$

CRYSTAL DATA: $P1\,2_1/m\,1$ (No. 11) unique axis **b**
$a = 6.793$ Å $b = 17.573$ Å $c = 7.759$ Å
$\alpha = 90°$ $\beta = 94.54°$ $\gamma = 90°$
X-ray single crystal refinement, $R_w = 0.06$

REFERENCE: J. L. Schlenker, J. J. Pluth and J. V. Smith, *Acta Cryst.* **B33** 2907–2910 (1977).

h	k	l	2θ	d	M	I_{rel}
0	2	0	10.07	8.786	2	5.3
0	0	1	11.44	7.735	2	5.4
0	1	1	12.50	7.079	4	18.8
1	0	0	13.07	6.772	2	43.0
1	1	0	14.02	6.319	4	100.0
0	2	1	15.26	5.806	4	2.9
-1	0	1	16.70	5.307	2	18.6
-1	1	1	17.45	5.081	4	58.5
1	0	1	18.08	4.906	2	22.5
1	1	1	18.78	4.725	4	31.1
0	3	1	19.00	4.670	4	79.8
-1	2	1	19.54	4.543	4	75.6
1	3	0	20.04	4.430	4	1.1
0	4	0	20.21	4.393	2	1.0
1	2	1	20.74	4.284	4	5.0
-1	3	1	22.61	3.933	4	35.6
0	0	2	23.00	3.867	2	18.9
0	4	1	23.28	3.820	4	10.0
0	1	2	23.55	3.777	4	30.4
1	3	1	23.65	3.761	4	0.7
1	4	0	24.15	3.686	4	13.2
0	2	2	25.16	3.540	4	25.9
-1	0	2	25.60	3.479	2	1.9
-1	1	2	26.11	3.413	4	28.6
2	0	0	26.32	3.386	2	0.6
-1	4	1	26.33	3.384	4	3.2
2	1	0	26.81	3.325	4	2.7
1	4	1	27.25	3.273	4	69.2
1	0	2	27.45	3.249	2	1.3
-1	2	2	27.58	3.235	4	0.6
0	5	1	27.88	3.200	4	5.7
-2	0	1	27.92	3.196	2	10.0
1	1	2	27.92	3.195	4	23.0
2	2	0	28.25	3.159	4	2.1
-2	1	1	28.38	3.144	4	3.0
1	5	0	28.61	3.119	4	28.6
1	2	2	29.30	3.048	4	41.2
2	0	1	29.63	3.015	2	10.4
-2	2	1	29.74	3.003	4	9.3
-1	3	2	29.87	2.991	4	5.9
2	1	1	30.07	2.972	4	18.0
2	3	0	30.49	2.931	4	56.5
-1	5	1	30.51	2.930	4	41.1
0	6	0	30.52	2.929	2	13.8
0	4	2	30.80	2.903	4	7.1
1	5	1	31.31	2.857	4	7.9
1	3	2	31.48	2.841	4	6.4
-2	3	1	31.90	2.806	4	18.3
0	6	1	32.69	2.739	4	13.7
-1	4	2	32.84	2.727	4	23.8
1	6	0	33.33	2.688	4	1.1
2	4	0	33.41	2.682	4	9.9
2	3	1	33.42	2.681	4	10.1
-2	0	2	33.78	2.654	2	0.6
-2	1	2	34.17	2.624	4	14.1
1	4	2	34.33	2.612	4	2.8
0	5	2	34.48	2.601	4	3.5
0	0	3	34.80	2.578	2	22.0
-1	6	1	34.99	2.564	4	0.8
-2	2	2	35.33	2.540	4	5.0
1	6	1	35.70	2.515	4	13.5
2	4	1	36.13	2.486	4	8.6
-1	0	3	36.29	2.476	2	1.4
0	2	3	36.31	2.474	4	4.9
-1	5	2	36.33	2.472	4	3.4
2	0	2	36.63	2.453	2	5.0
-1	1	3	36.66	2.451	4	5.5
2	5	0	36.86	2.438	4	0.6
-2	3	2	37.20	2.417	4	11.2
0	7	1	37.67	2.388	4	1.4
1	5	2	37.70	2.386	4	3.1
-1	2	3	37.75	2.383	4	4.9
-2	5	1	38.06	2.365	4	7.0
2	2	2	38.09	2.363	4	1.8
1	7	0	38.23	2.354	4	16.8
0	6	2	38.56	2.335	4	3.4
1	1	3	38.68	2.328	4	3.5
-1	3	3	39.52	2.280	4	1.6
-2	4	2	39.68	2.271	4	9.5
-1	7	1	39.72	2.269	4	1.4
1	2	3	39.73	2.269	4	1.9
2	3	2	39.84	2.263	4	8.6
3	0	0	39.94	2.257	2	0.7
-1	6	2	40.25	2.241	4	0.7
3	1	0	40.28	2.239	4	3.8
1	7	1	40.36	2.235	4	3.1
0	4	3	40.57	2.224	4	7.6
2	6	0	40.73	2.215	4	8.6
-3	1	1	41.08	2.197	4	0.8
0	8	0	41.09	2.197	2	8.5
3	2	0	41.29	2.186	4	1.4
1	3	3	41.42	2.180	4	3.0
1	6	2	41.51	2.176	4	1.1
-2	6	1	41.83	2.159	4	1.2
-1	4	3	41.89	2.157	4	2.0
-3	2	1	42.08	2.147	4	6.5
2	4	2	42.19	2.142	4	11.7
-2	0	3	42.35	2.134	2	1.6
3	0	1	42.60	2.122	2	7.7
-2	1	3	42.67	2.119	4	2.3
-2	5	2	42.69	2.118	4	4.6
3	3	0	42.94	2.106	4	10.3
0	5	3	43.53	2.079	4	1.7
-3	3	1	43.70	2.071	4	1.9
1	4	3	43.70	2.071	4	7.7
3	2	1	43.89	2.063	4	5.9
-1	7	2	44.50	2.036	4	2.0
-1	8	1	44.64	2.030	4	4.4
-1	5	3	44.78	2.024	4	20.8
-3	0	2	44.86	2.020	2	7.3
1	8	1	45.23	2.005	4	1.1
3	3	1	45.46	1.995	4	0.8
1	7	2	45.67	1.987	4	2.6
-3	4	1	45.89	1.977	4	10.6
2	0	3	45.90	1.977	2	1.5
-2	7	1	45.97	1.974	4	1.6
-3	2	2	46.10	1.969	4	1.7
1	5	3	46.51	1.953	4	14.9
0	6	3	46.95	1.935	4	1.8
0	0	4	46.99	1.934	2	1.1
2	7	1	47.10	1.929	4	2.4
2	2	3	47.11	1.929	4	2.0
0	1	4	47.29	1.922	4	0.6
-2	4	3	47.35	1.920	4	1.8
-3	3	2	47.61	1.910	4	0.8
-1	0	4	47.89	1.899	2	4.8
3	5	0	47.89	1.899	4	2.6
0	9	1	48.06	1.893	4	6.0
3	0	2	48.26	1.886	2	3.2
-1	8	2	49.04	1.857	4	1.4
-1	2	4	49.07	1.857	4	5.5
3	2	2	49.44	1.844	4	1.5
2	8	0	49.46	1.843	4	2.5
0	3	4	49.65	1.836	4	4.2
-3	4	2	49.67	1.836	4	1.0

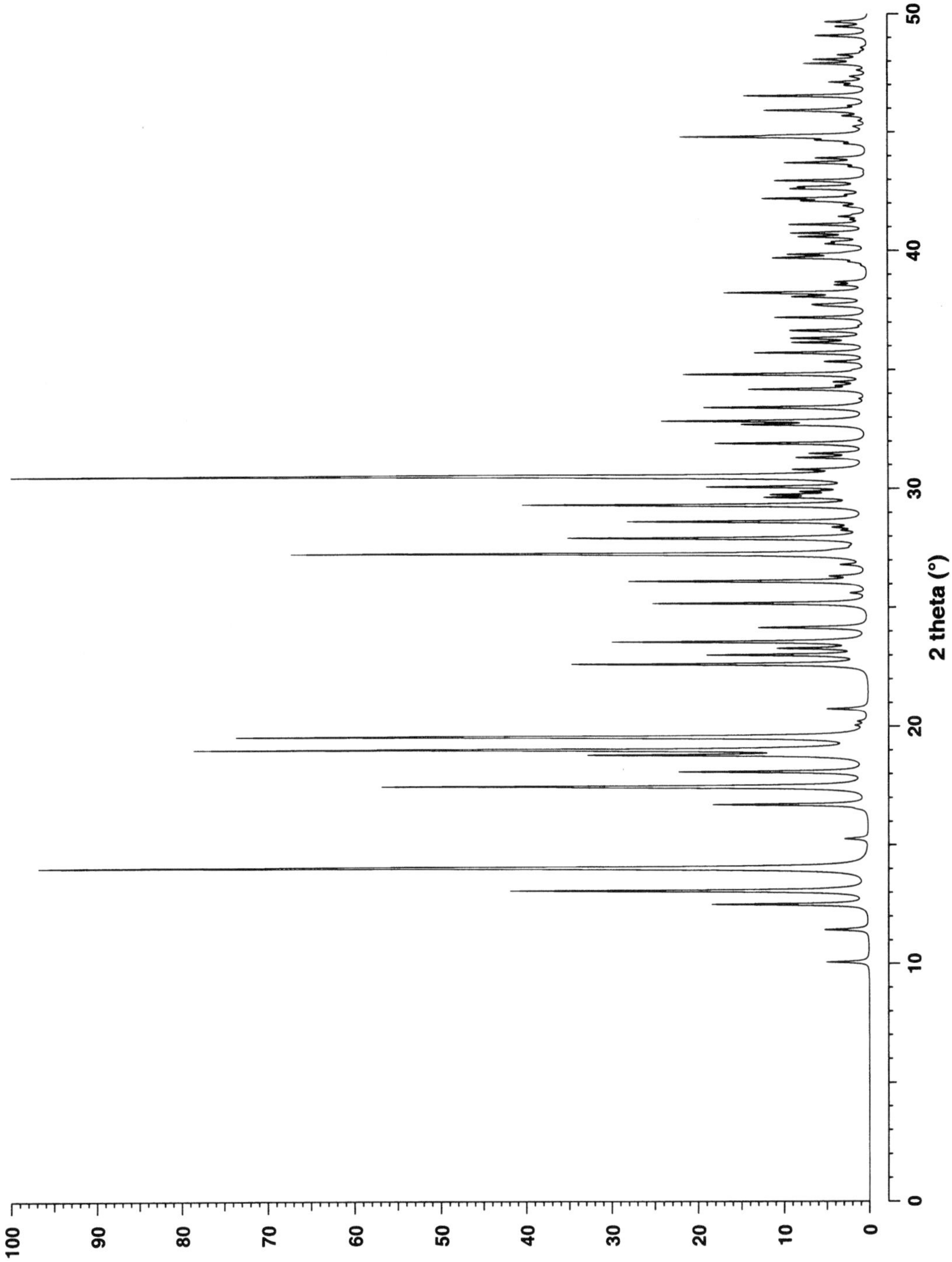
2 theta (°)
0
10
20
30
40
50
100
90
80
70
60
50
40
30
20
10
0

CAN Cancrinite

CHEMICAL COMPOSITION: $|Na_7Ca_{0.9}(CO_3)_{1.4}(H_2O)_{2.1}|$ $[Si_6Al_6O_{24}]$
Synthetic crystal.

REFINED COMPOSITION: $|Na_8(CO_3)_{1.2}(H_2O)_2|$ $[Si_6Al_6O_{24}]$

CRYSTAL DATA: $P6_3$ (No. 173)
$a = 12.635$ Å $b = 12.635$ Å $c = 5.115$ Å
$\alpha = 90°$ $\beta = 90°$ $\gamma = 120°$
X-ray single crystal refinement, $R = 0.04$

REFERENCE: Y. I. Smolin, Y. F. Shepelev, I. K. Butikova and I. B. Kobyakov, *Kristallografiya* **26** 63–66 (1981).

h	k	l	2θ	d	M	I_{rel}	h	k	l	2θ	d	M	I_{rel}	h	k	l	2θ	d	M	I_{rel}
1	0	0	8.08	10.942	6	3.9	2	2	1	33.34	2.688	12	0.1	2	1	2	41.51	2.175	12	1.9
1	1	0	14.02	6.317	6	59.5	3	1	1	34.36	2.610	12	10.2	1	4	1	41.75	2.164	12	6.7
2	0	0	16.20	5.471	6	3.7	1	3	1	34.36	2.610	12	19.8	4	1	1	41.75	2.164	12	4.4
1	0	1	19.15	4.634	12	100.0	0	0	2	35.09	2.557	2	42.0	3	3	0	42.95	2.106	6	21.6
1	2	0	21.49	4.136	6	0.6	3	2	0	35.77	2.510	6	1.3	3	0	2	43.20	2.094	12	13.6
2	1	0	21.49	4.136	6	6.1	2	3	0	35.77	2.510	6	1.7	2	4	0	43.78	2.068	6	0.2
1	1	1	22.36	3.975	12	1.0	1	0	2	36.06	2.490	12	1.2	5	0	1	45.06	2.012	12	4.9
2	0	1	23.81	3.736	12	2.7	4	0	1	37.27	2.412	12	34.0	2	2	2	45.64	1.988	12	0.5
3	0	0	24.40	3.647	6	55.1	1	4	0	37.67	2.388	6	3.4	5	1	0	46.19	1.965	6	0.5
1	2	1	27.74	3.216	12	56.9	4	1	0	37.67	2.388	6	0.7	3	1	2	46.43	1.956	12	0.4
2	1	1	27.74	3.216	12	61.4	1	1	2	37.95	2.371	12	1.1	1	3	2	46.43	1.956	12	0.5
2	2	0	28.25	3.159	6	0.3	2	0	2	38.87	2.317	12	2.6	2	4	1	47.42	1.917	12	1.3
3	1	0	29.43	3.035	6	0.9	2	3	1	40.01	2.254	12	10.4	4	2	1	47.42	1.917	12	1.1
1	3	0	29.43	3.035	6	3.2	3	2	1	40.01	2.254	12	10.4	4	0	2	48.74	1.868	12	13.1
3	0	1	30.09	2.970	12	4.3	5	0	0	41.25	2.188	6	1.7	1	5	1	49.70	1.835	12	0.5
4	0	0	32.74	2.736	6	42.6	1	2	2	41.51	2.175	12	1.5	5	1	1	49.70	1.835	12	0.3

Cancrinite CAN

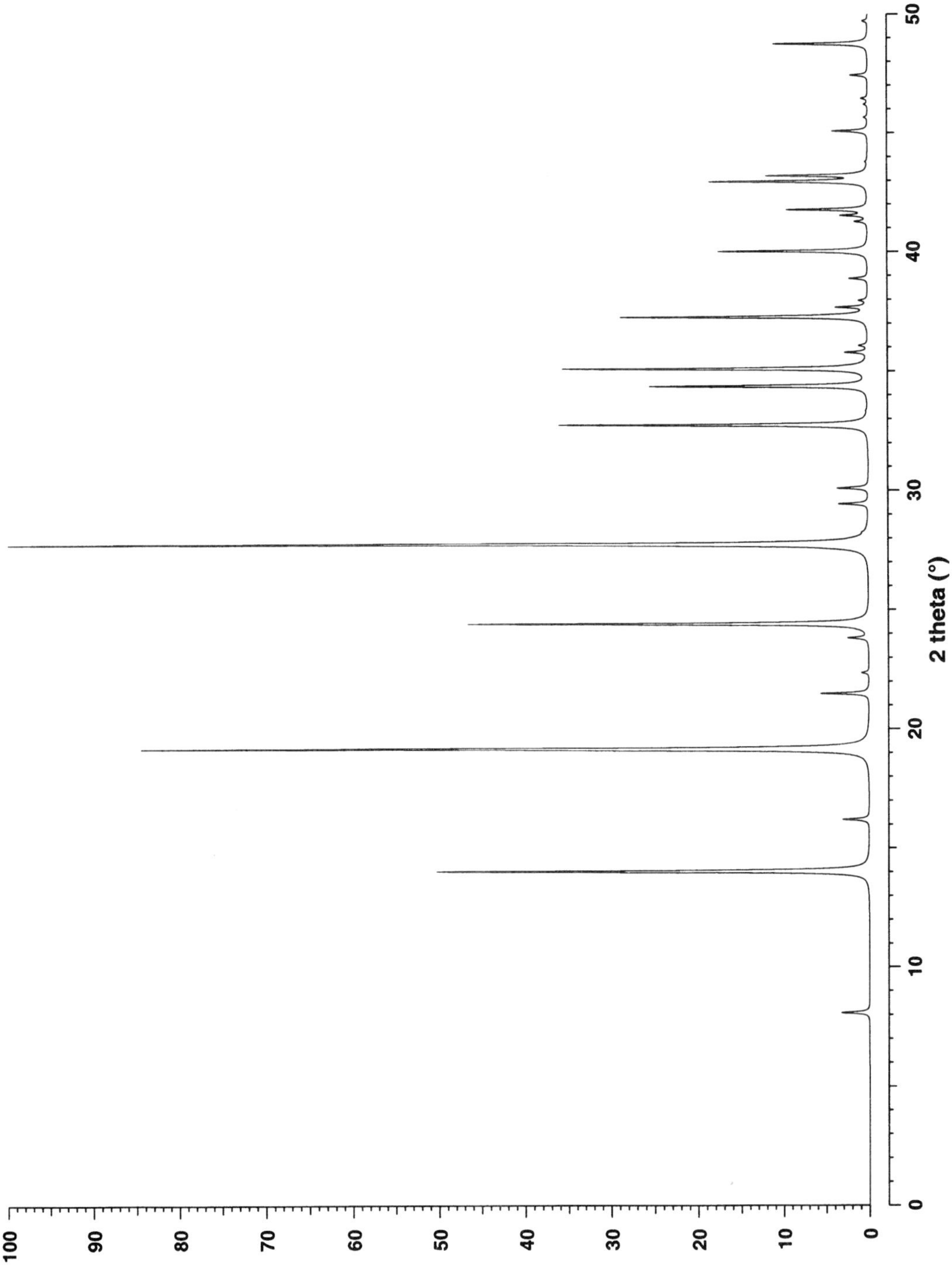

CAN Tiptopite

CHEMICAL COMPOSITION: $|K_2Li_{2.9}Na_{1.7}Ca_{0.7}(H_2O)_{1.3}(OH)_2\square_{0.7}|$ $[Be_6P_6O_{24}]$
$\square$ = cation vacancy
Tip Top Mine, Custer, S. Dakota, U.S.A.

REFINED COMPOSITION: $|Li_{2.9}Na_{1.7}K_2Ca_{0.7}O_6|$ $[Be_6P_6O_{24}]$

CRYSTAL DATA: $P6_3$ (No. 173)
a = 11.655 Å b = 11.655 Å c = 4.692 Å
α = 90° β = 90° γ = 120°
X-ray single crystal refinement, R_w = 0.048

REFERENCE: D. R. Peacor, R. C. Rouse and J. H. Ahn, *American Mineralogist* **72** 816–820 (1987).

h	k	l	2θ	d	M	I_{rel}
1	0	0	8.76	10.094	6	0.8
1	1	0	15.20	5.827	6	9.0
2	0	0	17.57	5.047	6	5.5
1	0	1	20.88	4.255	12	13.4
2	1	0	23.32	3.815	6	2.0
1	2	0	23.32	3.815	6	38.8
1	1	1	24.35	3.655	12	31.9
2	0	1	25.93	3.436	12	20.2
3	0	0	26.49	3.365	6	44.1
2	1	1	30.19	2.960	12	100.0
1	2	1	30.19	2.960	12	34.2
3	1	0	31.97	2.799	6	0.8
1	3	0	31.97	2.799	6	0.7
3	0	1	32.75	2.734	12	3.5
4	0	0	35.58	2.523	6	43.2
3	1	1	37.41	2.404	12	3.1
1	3	1	37.41	2.404	12	1.7
0	0	2	38.37	2.346	2	31.8
2	3	0	38.89	2.316	6	1.3
3	2	0	38.89	2.316	6	3.7
1	0	2	39.43	2.285	12	1.2
4	0	1	40.59	2.222	12	20.6
4	1	0	40.97	2.203	6	2.7
1	4	0	40.97	2.203	6	2.4
1	1	2	41.49	2.176	12	7.7
2	3	1	43.59	2.076	12	10.5
3	2	1	43.59	2.076	12	4.2
2	1	2	45.38	1.998	12	4.1
1	2	2	45.38	1.998	12	3.2
1	4	1	45.49	1.994	12	0.9
4	1	1	45.49	1.994	12	1.7
3	3	0	46.76	1.942	6	11.7
3	0	2	47.23	1.924	12	1.0
2	4	0	47.67	1.907	6	0.3
4	2	0	47.67	1.907	6	0.1
5	0	1	49.13	1.854	12	0.6
2	2	2	49.91	1.827	12	2.0

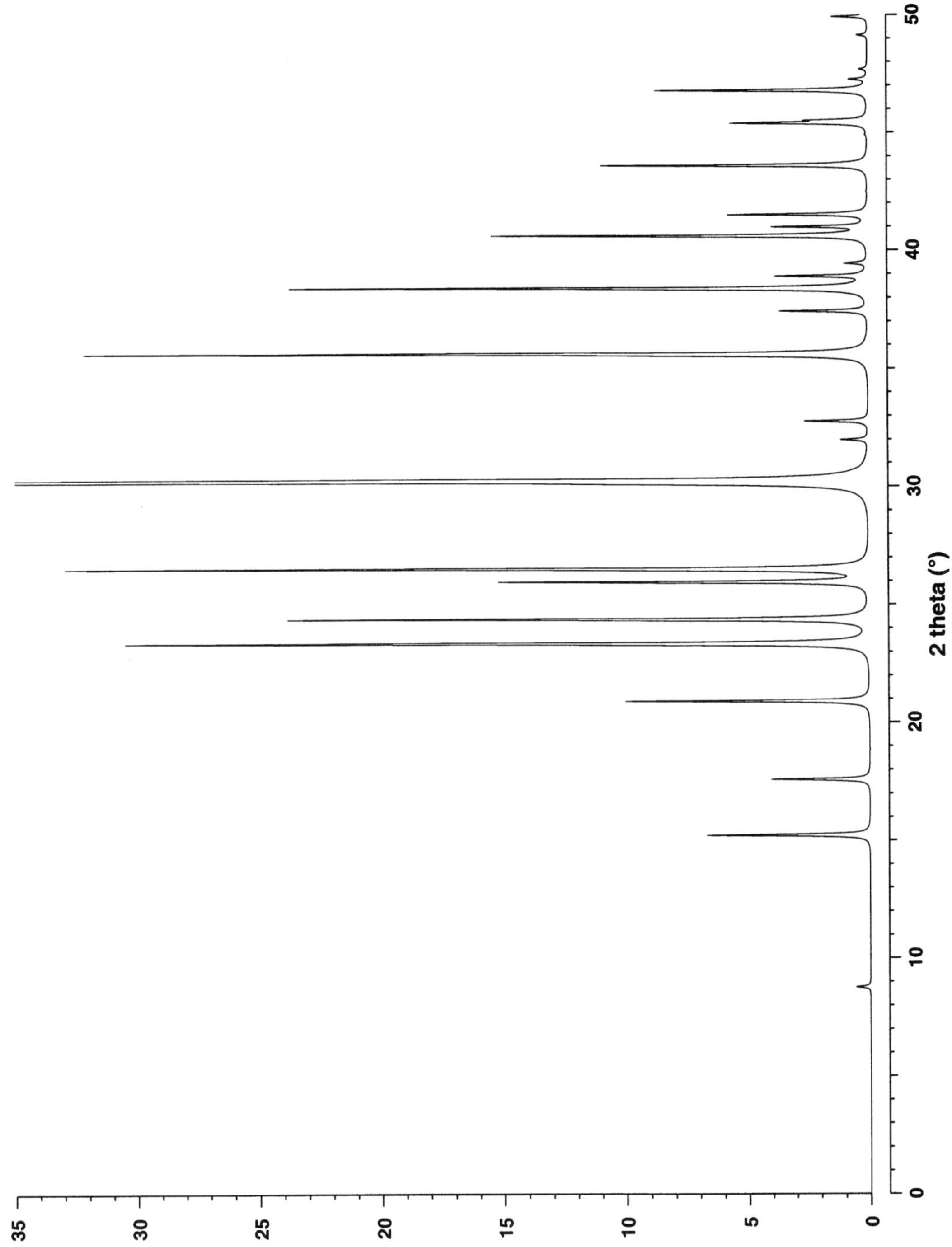
0
10
20
30
40
50
2 theta (°)
0
5
10
15
20
25
30
35

CAS Cs-Aluminosilicate

CHEMICAL COMPOSITION: $|Cs_4|\ [Si_{24}O_{48}]$

REFINED COMPOSITION: $|Cs_{3.26}|\ [Si_{24}O_{48}]$

CRYSTAL DATA: *Ama*2 (No. 40)
$a = 16.776$ Å $b = 13.828$ Å $c = 5.021$ Å
$\alpha = 90°$ $\beta = 90°$ $\gamma = 90°$
X-ray single crystal refinement, $R_w = 0.0549$

REFERENCE: T. Araki,
Z. Kristallogr. **152** 207–213 (1980).

h	k	l	2θ	d	M	I_{rel}
2	0	0	10.55	8.388	2	0.2
0	2	0	12.80	6.914	2	1.7
2	2	0	16.62	5.335	4	0.9
0	1	1	18.80	4.720	4	2.0
1	1	1	19.54	4.543	8	2.9
3	2	0	20.43	4.348	4	1.1
4	0	0	21.18	4.194	2	0.7
2	1	1	21.60	4.113	8	48.3
3	1	1	24.68	3.607	8	100.0
4	2	0	24.83	3.586	4	53.4
0	4	0	25.77	3.457	2	20.8
0	3	1	26.25	3.396	4	17.8
1	4	0	26.32	3.386	4	22.3
1	3	1	26.79	3.328	8	39.4
2	4	0	27.91	3.196	4	5.7
2	3	1	28.36	3.147	8	6.1
4	1	1	28.47	3.135	8	1.9
5	2	0	29.59	3.019	4	12.8
3	4	0	30.40	2.940	4	8.6
3	3	1	30.81	2.902	8	14.1
6	0	0	32.01	2.796	2	5.1
5	1	1	32.75	2.735	8	10.4

h	k	l	2θ	d	M	I_{rel}
4	4	0	33.59	2.668	4	2.1
4	3	1	33.97	2.639	8	3.9
6	2	0	34.60	2.592	4	4.5
0	0	2	35.77	2.510	2	12.5
0	5	1	37.11	2.422	4	3.4
5	4	0	37.35	2.408	4	1.8
6	1	1	37.38	2.406	8	6.7
2	0	2	37.39	2.405	4	1.7
1	5	1	37.51	2.398	8	1.8
5	3	1	37.69	2.387	8	0.1
0	2	2	38.14	2.360	4	8.4
1	2	2	38.53	2.337	8	1.8
2	5	1	38.69	2.327	8	7.8
1	6	0	39.47	2.283	4	2.2
2	2	2	39.68	2.272	8	1.3
7	2	0	39.81	2.264	4	1.6
3	5	1	40.58	2.223	8	1.0
2	6	0	40.59	2.222	4	0.2
3	2	2	41.54	2.174	8	1.7
6	4	0	41.54	2.174	4	0.4
6	3	1	41.85	2.158	8	7.2
4	0	2	41.94	2.154	4	0.2

h	k	l	2θ	d	M	I_{rel}
7	1	1	42.29	2.137	8	4.5
3	6	0	42.42	2.131	4	1.8
4	5	1	43.12	2.098	8	4.3
8	0	0	43.14	2.097	2	2.4
4	2	2	44.03	2.057	8	3.4
0	4	2	44.61	2.031	4	0.3
4	6	0	44.87	2.020	4	1.0
1	4	2	44.95	2.017	8	3.7
2	4	2	45.97	1.974	8	1.0
7	4	0	46.08	1.970	4	0.2
5	5	1	46.22	1.964	8	1.5
7	3	1	46.37	1.958	8	2.6
5	2	2	47.08	1.930	8	4.0
8	1	1	47.44	1.916	8	0.6
3	4	2	47.63	1.909	8	4.1
5	6	0	47.88	1.900	4	4.8
6	0	2	48.75	1.868	4	3.4
0	7	1	49.59	1.838	4	5.7
6	5	1	49.80	1.831	8	1.8
4	4	2	49.88	1.828	8	1.9
1	7	1	49.91	1.827	8	0.7

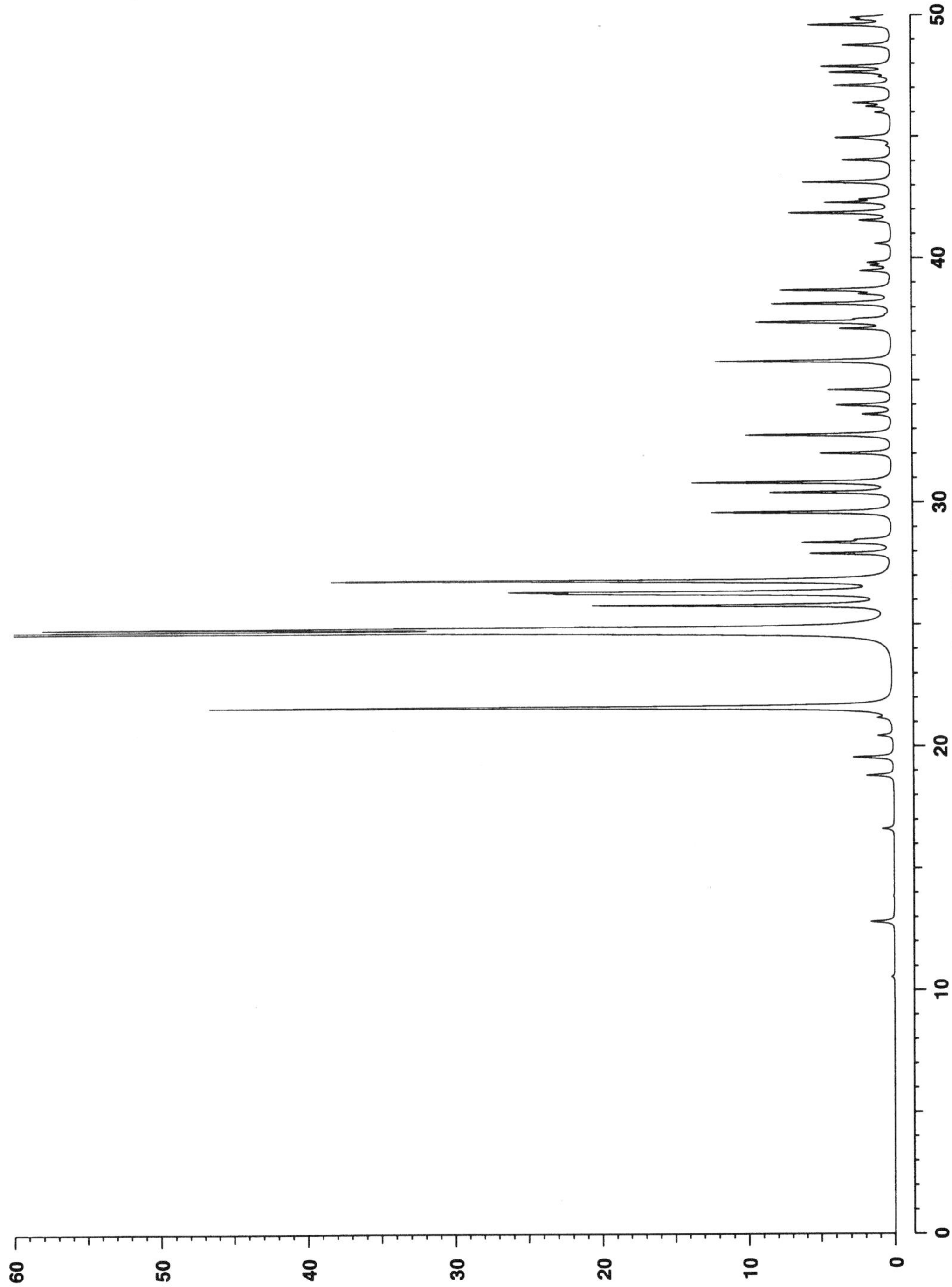
2 theta (°)
0
10
20
30
40
50
0
10
20
30
40
50
60

CFI CIT-5

CHEMICAL COMPOSITION: $[Si_{32}O_{64}]$

REFINED COMPOSITION: $[Si_{32}O_{64}]$

CRYSTAL DATA: *Imma* (No. 74)
$a = 13.695$ Å $b = 5.021$ Å $c = 25.497$ Å
$\alpha = 90.0°$ $\beta = 90.0°$ $\gamma = 90.0°$
X-ray Rietveld refinement, $R_p = 0.0875$, $R_{wp} = 0.1143$

REFERENCE: P. Wagner, M. Yoshikawa, M. Lavallo, K. Tsuji, M. Tsapatsis and M. E. Davis, *J. Chem. Soc., Chem. Commun.* 2179–2180 (1997).

h	k	l	2θ	d	M	I_{rel}	h	k	l	2θ	d	M	I_{rel}	h	k	l	2θ	d	M	I_{rel}
0	0	2	6.93	12.748	2	84.1	0	1	7	30.31	2.948	4	0.9	2	2	4	40.82	2.211	8	0.4
1	0	1	7.33	12.065	4	100.0	2	0	8	30.95	2.889	4	0.2	3	2	1	41.19	2.192	8	0.6
1	0	3	12.26	7.221	4	5.7	3	0	7	31.42	2.847	4	1.8	0	2	6	41.79	2.161	4	0.2
2	0	0	12.93	6.847	2	5.1	4	1	1	31.83	2.811	8	0.6	6	0	4	42.05	2.149	4	0.1
0	0	4	13.89	6.374	2	2.5	2	1	7	33.08	2.708	8	1.3	0	1	11	42.98	2.104	4	0.2
0	1	1	18.01	4.926	4	0.3	4	1	3	33.38	2.684	8	0.1	5	1	6	43.23	2.093	8	0.1
1	0	5	18.57	4.779	4	0.7	4	0	6	33.61	2.666	4	0.2	6	1	1	43.71	2.071	8	0.3
2	0	4	19.02	4.666	4	19.9	3	1	6	33.90	2.644	8	0.6	1	2	7	44.32	2.044	8	0.2
3	0	1	19.76	4.494	4	3.5	1	1	8	33.95	2.640	8	0.3	3	1	10	44.52	2.035	8	0.2
1	1	2	20.08	4.422	8	27.7	5	0	3	34.40	2.607	4	0.2	2	0	12	44.65	2.029	4	1.0
0	1	3	20.54	4.323	4	21.7	0	2	0	35.77	2.510	2	6.0	3	2	5	44.87	2.020	8	0.2
0	0	6	20.90	4.249	2	8.1	4	1	5	36.32	2.474	8	0.9	2	1	11	45.07	2.012	8	0.4
3	0	3	22.10	4.022	4	1.3	0	1	9	36.41	2.467	4	2.2	6	0	6	45.09	2.011	4	1.3
1	1	4	23.47	3.790	8	3.3	0	2	2	36.48	2.463	4	0.4	4	1	9	45.30	2.002	8	0.2
2	1	3	24.35	3.655	8	1.8	1	2	1	36.56	2.458	8	1.6	4	2	2	45.36	1.999	8	0.6
2	0	6	24.66	3.611	4	6.9	5	0	5	37.26	2.413	4	1.0	0	2	8	46.02	1.972	4	0.4
0	1	5	24.89	3.578	4	0.3	2	0	10	37.64	2.389	4	0.1	5	0	9	46.09	1.969	4	0.2
4	0	0	26.02	3.424	2	2.9	5	1	2	38.08	2.363	8	1.0	5	1	8	47.36	1.920	8	0.8
3	0	5	26.20	3.401	4	1.8	2	1	9	38.79	2.321	8	0.5	2	2	8	48.01	1.895	8	0.2
4	0	2	26.96	3.307	4	12.2	3	1	8	38.85	2.318	8	1.5	3	2	7	48.33	1.883	8	0.5
3	1	2	27.31	3.265	8	4.8	1	0	11	39.43	2.285	4	0.6	1	2	9	48.93	1.861	8	0.3
0	0	8	27.99	3.187	2	1.6	5	1	4	40.08	2.250	8	0.4	6	0	8	49.09	1.856	4	0.4
2	1	5	28.14	3.171	8	1.9	1	1	10	40.21	2.243	8	0.1	4	2	6	49.89	1.828	8	0.6
1	1	6	28.27	3.156	8	9.1	4	1	7	40.37	2.234	8	0.3	7	0	5	49.93	1.827	4	0.5

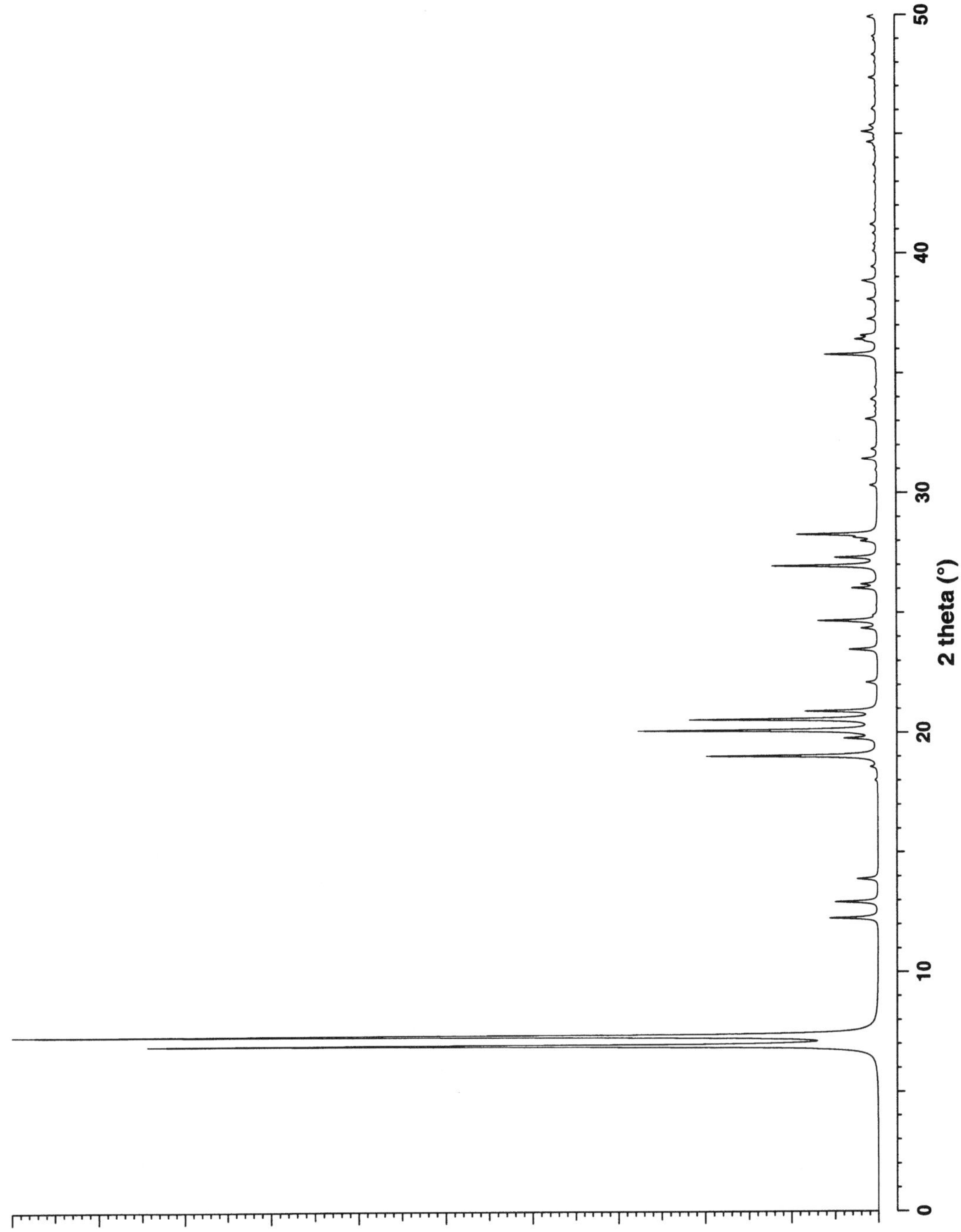
2 theta (°)
0
10
20
30
40
50
0
10
20
30
40
50
60
70
80
90
100

CGF Cobalt Gallium Phosphate-5

CHEMICAL COMPOSITION: $|(C_6H_{14}N_2)_8|$ $[(Co_{16}Ga_{20}P_{36}O_{144}]$
$C_6H_{12}N_2$ = 1,4-diazabicyclo[2.2.2]-octane (DABCO)

REFINED COMPOSITION: $|C_{48}N_{16}|$ $[Co_{16}Ga_{20}P_{36}O_{144}]$

CRYSTAL DATA: $I 1 2/a 1$ (No. 15) unique axis **b**, cell choice 3
a = 15.002 Å b = 17.688 Å c = 15.751 Å
α = 90° β = 97.24° γ = 90°
X-ray single crystal refinement, R = 0.0704, R_w = 0.0850

REFERENCE: A. M. Chippindale and A. R. Cowley, *Zeolites* **18** 176–181 (1997).

h	k	l	2θ	d	M	I_{rel}
0	2	0	10.00	8.844	2	100.0
0	0	2	11.33	7.813	2	20.7
2	0	0	11.89	7.441	2	2.7
-2	1	1	13.48	6.568	4	4.8
2	2	0	15.56	5.694	4	12.4
0	3	1	16.07	5.516	4	7.9
2	0	2	17.46	5.078	2	3.6
0	1	3	17.75	4.996	4	2.8
-2	2	2	18.37	4.828	4	12.0
3	1	0	18.58	4.776	4	0.7
-1	3	2	19.34	4.588	4	0.5
-2	3	1	19.60	4.529	4	13.2
-1	2	3	20.03	4.432	4	0.8
-2	1	3	20.17	4.402	4	8.7
1	3	2	20.22	4.392	4	0.9
2	3	1	20.46	4.340	4	15.7
-1	4	1	21.52	4.129	4	3.2
3	2	1	21.91	4.057	4	2.6
1	4	1	21.92	4.054	4	1.8
2	1	3	22.60	3.934	4	20.1
0	0	4	22.76	3.906	2	6.3
0	3	3	22.78	3.903	4	3.8
3	1	2	22.97	3.872	4	0.8
0	4	2	23.11	3.848	4	41.4
-1	1	4	23.35	3.809	4	4.2
-2	0	4	24.36	3.653	2	0.8
-4	1	1	24.41	3.647	4	0.8
-2	3	3	24.73	3.599	4	6.3
0	2	4	24.92	3.573	4	1.3
-4	0	2	25.18	3.537	2	36.0
-2	4	2	25.39	3.508	4	5.8
0	5	1	25.82	3.450	4	1.3
4	2	0	25.98	3.429	4	7.6
-2	2	4	26.39	3.377	4	1.6
-1	4	3	26.63	3.347	4	1.6
2	4	2	26.73	3.335	4	2.4
2	3	3	26.77	3.330	4	8.0
2	0	4	27.08	3.292	2	14.8
3	3	2	27.09	3.292	4	1.0
-3	4	1	27.14	3.286	4	0.5
-4	2	2	27.15	3.284	4	20.4
-1	3	4	27.41	3.253	4	0.7
4	0	2	27.83	3.206	2	1.8
3	4	1	28.09	3.176	4	1.0
-4	1	3	28.12	3.174	4	5.2
-2	5	1	28.20	3.164	4	1.0
-4	3	1	28.33	3.150	4	3.1
1	5	2	28.64	3.116	4	0.8
2	5	1	28.82	3.097	4	2.2
2	2	4	28.94	3.085	4	16.3
0	5	3	30.55	2.926	4	3.1
-3	4	3	30.83	2.900	4	1.5
3	1	4	31.29	2.859	4	0.8
1	6	1	31.60	2.831	4	0.8
-4	3	3	31.61	2.830	4	2.1
4	1	3	31.67	2.825	4	2.6
-2	5	3	32.06	2.792	4	10.1
0	3	5	32.42	2.761	4	4.3
0	6	2	32.46	2.758	4	2.7
2	6	0	32.67	2.741	4	10.3
-4	2	4	32.68	2.740	4	0.6
2	1	5	32.84	2.727	4	5.2
5	1	2	33.93	2.642	4	0.8
2	4	4	33.95	2.641	4	1.5
-1	5	4	34.22	2.621	4	1.0
3	3	4	34.50	2.600	4	0.6
-1	1	6	34.58	2.594	4	0.6
4	3	3	34.85	2.574	4	0.9
-4	5	1	34.97	2.566	4	3.2
-2	0	6	35.04	2.561	2	0.6
2	6	2	35.20	2.550	4	5.9
-4	1	5	35.47	2.531	4	1.4
2	3	5	35.93	2.500	4	2.6
0	2	6	35.95	2.498	4	0.7
-6	1	1	36.31	2.474	4	3.0
-2	2	6	36.53	2.460	4	2.4
-4	4	4	37.24	2.414	4	1.8
6	2	0	37.66	2.388	4	0.8
6	1	1	37.78	2.381	4	1.9
2	0	6	38.02	2.367	2	0.7
-6	2	2	38.04	2.366	4	1.7
-4	3	5	38.37	2.346	4	1.2
0	5	5	38.43	2.342	4	5.6
-3	6	3	38.49	2.339	4	1.1
4	6	0	38.98	2.311	4	1.1
-6	3	1	39.15	2.301	4	1.3
-2	6	4	39.27	2.294	4	1.2
2	2	6	39.41	2.286	4	1.8
6	0	2	39.48	2.283	2	2.6
0	7	3	39.64	2.273	4	1.5
-4	0	6	39.66	2.272	2	1.3
-4	6	2	39.81	2.264	4	1.9
6	3	1	40.53	2.225	4	2.8
4	5	3	40.55	2.225	4	0.5
6	2	2	40.83	2.210	4	1.2
-4	2	6	41.01	2.201	4	0.6
-2	1	7	41.04	2.199	4	0.5
-6	3	3	41.24	2.189	4	0.5
6	4	0	41.75	2.163	4	0.7
-6	2	4	41.87	2.157	4	1.7
3	7	2	42.41	2.131	4	0.5
0	8	2	42.49	2.127	4	0.9
2	4	6	43.36	2.087	4	2.0
1	7	4	43.49	2.081	4	0.6
-5	1	6	43.49	2.081	4	1.2
-2	3	7	43.63	2.075	4	0.6
-4	6	4	43.94	2.061	4	2.2
2	1	7	44.10	2.054	4	1.1
4	7	1	44.11	2.053	4	1.4
-5	4	5	44.31	2.044	4	0.8
4	0	6	44.93	2.017	2	0.7
6	3	3	45.12	2.009	4	1.4
-4	7	3	45.60	1.989	4	0.7
-6	4	4	45.65	1.987	4	0.8
-5	3	6	45.97	1.974	4	0.5
4	2	6	46.15	1.967	4	0.6
0	0	8	46.49	1.953	2	0.7
2	3	7	46.55	1.951	4	3.1
0	9	1	46.57	1.950	4	0.6
-2	7	5	46.88	1.938	4	0.9
-7	4	1	47.12	1.928	4	0.7
0	8	4	47.24	1.924	4	0.9
-6	0	6	47.32	1.921	2	1.0
-4	3	7	47.35	1.920	4	2.4
4	5	5	47.80	1.903	4	1.8
6	6	0	47.93	1.898	4	1.0
-1	9	2	47.93	1.898	4	0.6
4	7	3	48.03	1.894	4	0.5
-6	6	2	48.24	1.887	4	1.1
1	9	2	48.33	1.883	4	0.6
-2	5	7	48.47	1.878	4	1.5
3	5	6	48.83	1.865	4	0.7
0	9	3	49.57	1.839	4	1.2
5	1	6	49.60	1.838	4	0.5
6	5	3	49.85	1.829	4	1.5

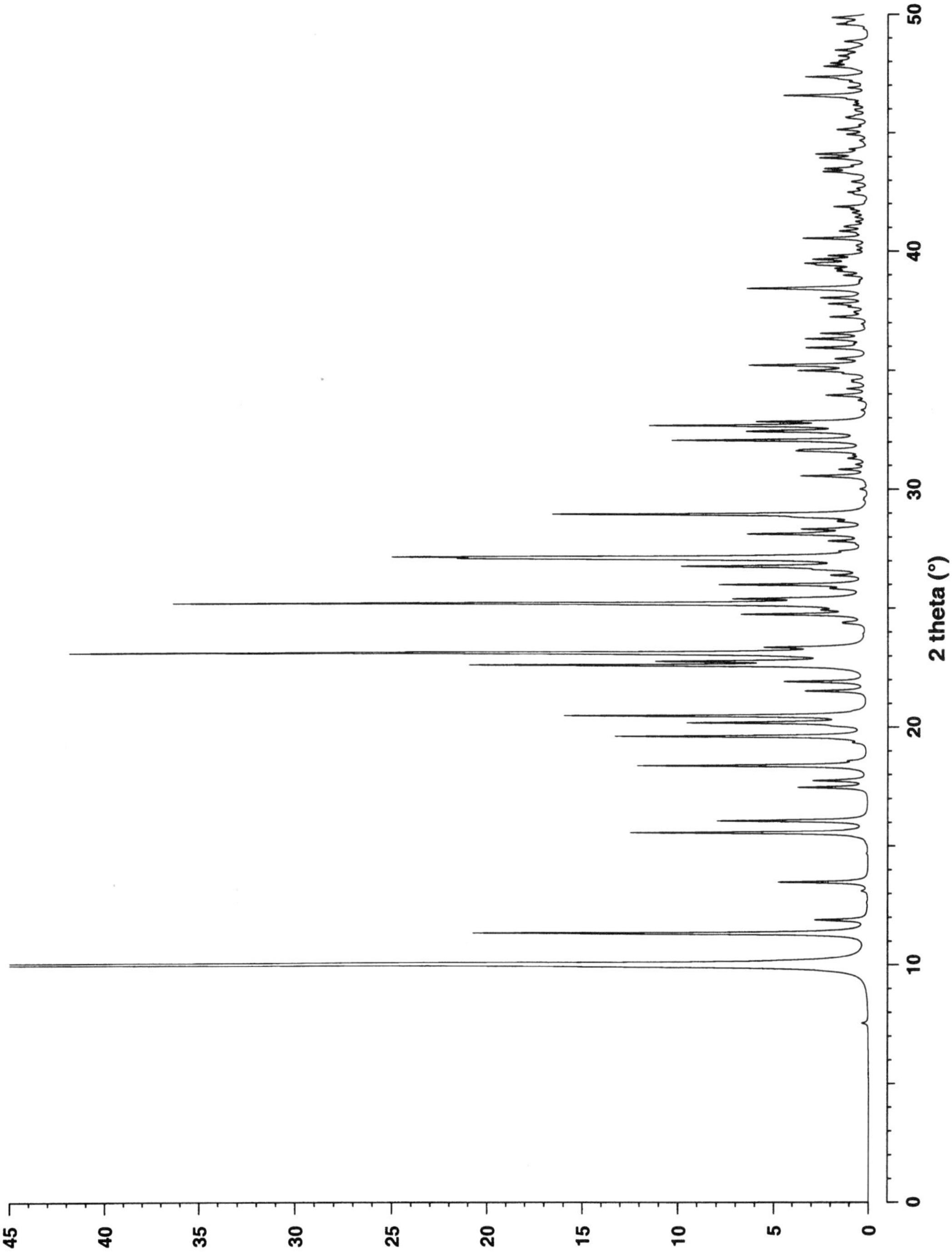
2 theta (°)
0
10
20
30
40
50
0
5
10
15
20
25
30
35
40
45

CGS Cobalt Gallium Phosphate-6

CHEMICAL COMPOSITION: $|(C_7H_{14}N)_4|$ $[Co_4Ga_{12}P_{16}O_{64}]$
$C_7H_{13}N$ = quinuclidine

REFINED COMPOSITION: $|C_{28}N_4|$ $[Co_4Ga_{12}P_{16}O_{64}]$

CRYSTAL DATA: $P1\,2_1/c\,1$ (No. 14) unique axis **b**, cell choice 1
$a = 14.365$ Å $b = 16.305$ Å $c = 8.734$ Å
$\alpha = 90°$ $\beta = 90.243°$ $\gamma = 90°$
X-ray single crystal refinement. $R = 0.0310$, $R_w = 0.0365$

REFERENCE: A. R. Cowley and A. M. Chippindale,
Microporous and Mesoporous Materials **28** 163–172 (1999).

h	k	l	2θ	d	M	I_{rel}
1	1	0	8.20	10.779	4	100.0
0	2	0	10.85	8.153	2	8.1
0	1	1	11.49	7.699	4	89.0
2	0	0	12.32	7.182	2	67.4
-1	1	1	13.03	6.796	4	22.9
1	1	1	13.07	6.775	4	29.9
0	2	1	14.86	5.960	4	10.2
2	2	0	16.45	5.389	4	71.6
-2	1	1	16.85	5.262	4	15.8
2	1	1	16.91	5.242	4	19.4
0	3	1	19.23	4.614	4	36.5
3	1	0	19.32	4.594	4	29.7
-2	2	1	19.33	4.593	4	47.0
2	2	1	19.38	4.580	4	31.7
-1	3	1	20.20	4.396	4	2.5
1	3	1	20.23	4.391	4	8.5
0	0	2	20.34	4.367	2	7.8
0	1	2	21.06	4.218	4	2.0
-1	0	2	21.24	4.183	2	3.0
1	0	2	21.29	4.173	2	1.9
3	2	0	21.52	4.129	4	2.9
0	4	0	21.80	4.076	2	2.4
-3	1	1	21.82	4.073	4	6.1
3	1	1	21.90	4.059	4	18.1
-1	1	2	21.94	4.052	4	14.9
1	4	0	22.67	3.921	4	7.4
2	3	1	22.93	3.878	4	21.6
0	2	2	23.10	3.849	4	16.8
-2	0	2	23.80	3.738	2	4.7
-3	2	1	23.80	3.738	4	18.3
3	2	1	23.87	3.727	4	7.8
-1	2	2	23.91	3.722	4	6.5
1	2	2	23.95	3.715	4	4.5
-2	1	2	24.43	3.644	4	1.4
2	1	2	24.52	3.631	4	9.7
4	0	0	24.79	3.591	2	1.6
-1	4	1	24.88	3.579	4	8.1
1	4	1	24.90	3.576	4	32.1
4	1	0	25.40	3.507	4	1.6
0	3	2	26.18	3.404	4	38.6
-2	2	2	26.22	3.398	4	1.7
-3	3	1	26.80	3.326	4	1.7
-2	4	1	27.13	3.287	4	1.7
4	2	0	27.13	3.286	4	18.2
2	4	1	27.17	3.282	4	4.8
-4	1	1	27.36	3.259	4	3.5
3	0	2	27.71	3.220	2	10.0
-3	1	2	28.13	3.172	4	3.5
-2	3	2	28.99	3.080	4	23.1
-4	2	1	28.99	3.080	4	19.7
2	3	2	29.06	3.072	4	13.9
4	2	1	29.07	3.072	4	15.1
0	5	1	29.23	3.055	4	5.1
2	5	0	30.10	2.969	4	3.0
-3	4	1	30.54	2.927	4	10.4
3	4	1	30.59	2.922	4	3.4
-1	4	2	30.62	2.919	4	5.3
1	4	2	30.66	2.916	4	8.8
0	1	3	31.21	2.866	4	6.4
-4	3	1	31.53	2.837	4	7.3
4	3	1	31.60	2.831	4	13.7
5	1	0	31.62	2.829	4	4.8
-1	1	3	31.81	2.813	4	3.2
-2	5	1	31.81	2.813	4	5.3
2	5	1	31.85	2.810	4	3.5
1	1	3	31.86	2.808	4	2.7
-4	0	2	32.20	2.780	2	2.8
2	4	2	32.57	2.750	4	4.3
0	2	3	32.66	2.742	4	6.7
-4	1	2	32.68	2.740	4	1.6
4	1	2	32.82	2.729	4	1.3
3	5	0	33.24	2.695	4	5.6
-5	1	1	33.24	2.695	4	3.6
4	4	0	33.25	2.695	4	1.7
1	2	3	33.29	2.691	4	7.4
1	6	0	33.56	2.670	4	1.5
2	1	3	33.72	2.658	4	2.3
-4	2	2	34.08	2.631	4	9.9
0	5	2	34.32	2.613	4	29.9
-5	2	1	34.62	2.591	4	5.1
4	4	1	34.87	2.573	4	1.4
0	3	3	34.96	2.566	4	2.3
-2	2	3	34.98	2.565	4	3.1
2	2	3	35.08	2.558	4	5.6
-1	6	1	35.14	2.554	4	5.9
1	6	1	35.15	2.553	4	3.0
5	3	0	35.34	2.540	4	4.0
-3	4	2	35.43	2.533	4	2.6
-1	3	3	35.51	2.528	4	1.9
-4	3	2	36.30	2.475	4	3.3
4	3	2	36.43	2.467	4	2.3
-2	5	2	36.56	2.457	4	4.2
2	5	2	36.63	2.453	4	1.4
5	0	2	37.55	2.395	2	3.5
5	1	2	37.96	2.370	4	1.7
6	1	0	37.99	2.369	4	2.4
-4	5	1	38.67	2.329	4	1.8
4	5	1	38.72	2.325	4	1.4
-5	2	2	39.05	2.307	4	8.3
1	7	0	39.18	2.299	4	2.8
5	2	2	39.20	2.298	4	10.1
4	4	2	39.35	2.290	4	4.0
-3	6	1	39.48	2.283	4	1.8
1	6	2	39.57	2.277	4	3.2
3	3	3	39.92	2.258	4	3.3
-2	4	3	40.03	2.252	4	1.3
-4	1	3	40.17	2.245	4	2.3
0	5	3	41.58	2.172	4	2.3
1	0	4	41.87	2.157	2	2.2
5	5	0	41.91	2.156	4	5.0
2	7	1	42.09	2.147	4	3.1
5	5	1	43.26	2.091	4	2.7
-1	2	4	43.33	2.088	4	1.4
-3	6	2	43.50	2.080	4	2.4
2	1	4	43.74	2.070	4	2.5
5	4	2	43.83	2.065	4	1.4
6	2	2	44.65	2.030	4	1.9
-2	2	4	44.73	2.026	4	2.1
0	6	3	45.67	1.987	4	2.6
0	8	1	45.71	1.985	4	4.8
-2	3	4	46.52	1.952	4	1.6
2	3	4	46.62	1.948	4	1.7
-7	2	1	46.78	1.942	4	2.1
7	2	1	46.86	1.939	4	1.4
5	6	1	47.23	1.924	4	1.6
7	3	0	47.35	1.920	4	3.3
-5	3	3	47.41	1.917	4	3.1
-4	7	1	47.66	1.908	4	2.2
4	7	1	47.71	1.906	4	1.3
-3	7	2	48.15	1.890	4	3.8
-6	4	2	48.72	1.869	4	1.8
-7	0	2	48.96	1.860	2	1.9
2	4	4	49.04	1.857	4	1.3
4	1	4	49.25	1.850	4	2.6
6	1	3	49.72	1.834	4	2.1

Cobalt Gallium Phosphate-6 CGS

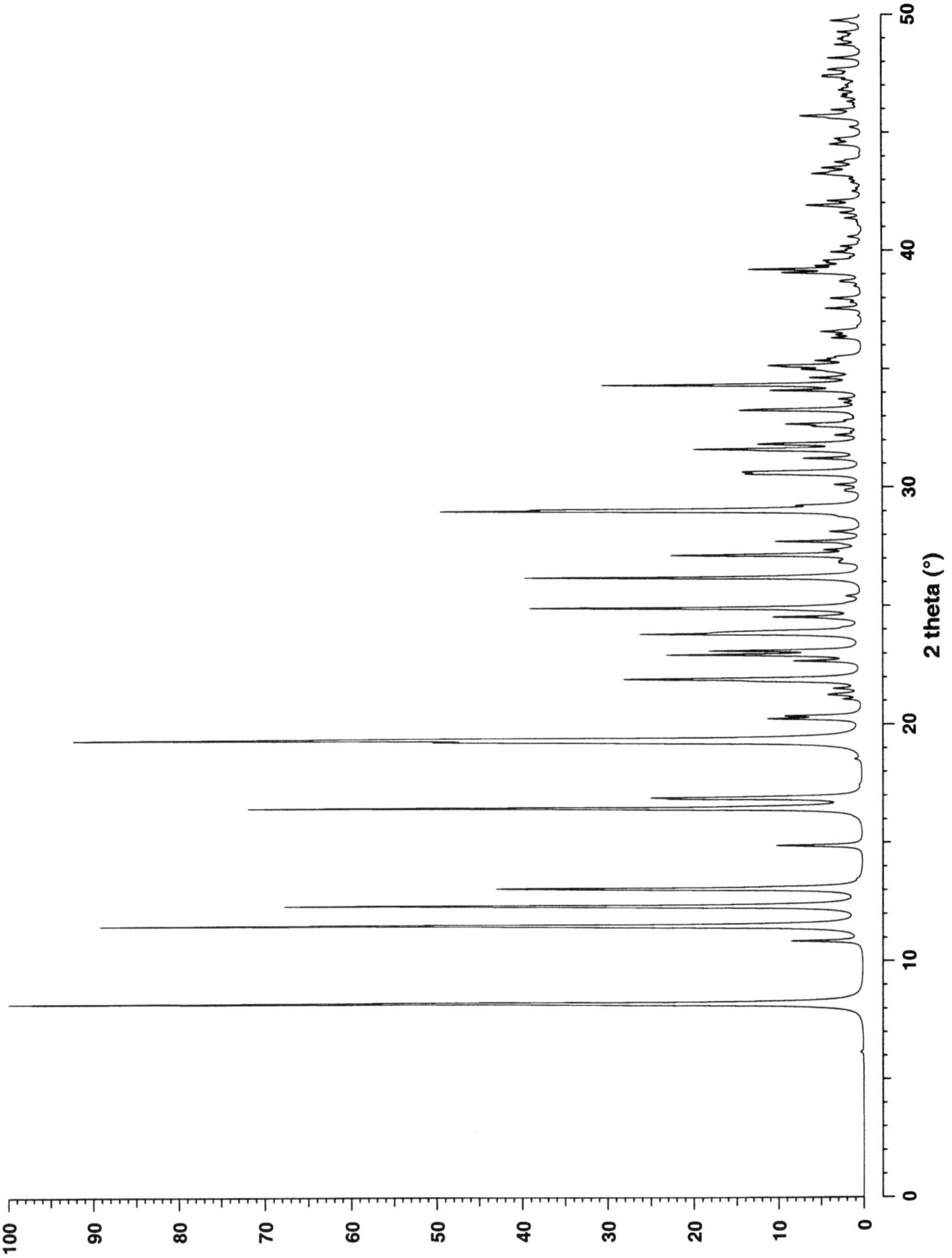

CHA Chabazite

CHEMICAL COMPOSITION: $|K_{3.2}Na_{0.75}(H_2O)_{8.4}|$ $[Si_{8.2}Al_{3.8}O_{24}]$
North East Azerbaijan, Iran

REFINED COMPOSITION: $|K_{4.16}(H_2O)_{7.23}|$ $[Si_{8.16}Al_{3.84}O_{24}]$

CRYSTAL DATA: $R\overline{3}m$ (No. 166) rhombohedral setting
$a = 9.459$ Å $b = 9.459$ Å $c = 9.459$ Å
$\alpha = 94.07°$ $\beta = 94.07°$ $\gamma = 94.07°$
X-ray single crystal refinement, $R = 0.05$

REFERENCE: M. Calligaris, G. Nardin and L. Randaccio, *Zeolites* **3** 205–208 (1983).

h	k	l	2θ	d	M	I_{rel}
1	0	0	9.40	9.408	6	100.0
1	0	-1	12.79	6.922	6	15.3
1	1	0	13.81	6.412	6	1.4
1	1	-1	15.90	5.575	6	6.9
1	1	1	17.53	5.059	2	12.6
2	0	0	18.87	4.704	6	7.3
2	0	-1	20.45	4.342	12	40.6
2	-1	-1	22.24	3.996	6	3.8
2	1	-1	22.86	3.890	12	26.7
2	1	1	24.61	3.617	6	23.3
2	0	-2	25.74	3.461	6	4.0
2	1	-2	27.46	3.248	12	3.0
2	2	0	27.83	3.206	6	7.2
3	0	0	28.46	3.136	6	2.4
2	2	-1	28.46	3.136	6	0.5
3	0	-1	29.32	3.046	12	1.2
2	2	1	30.38	2.942	6	4.3
3	-1	-1	30.40	2.940	6	63.6
3	1	0	30.73	2.909	12	38.2
3	1	-1	31.31	2.856	12	0.9
2	2	-2	32.11	2.788	6	3.2
3	1	1	33.08	2.708	6	5.0
3	0	-2	33.10	2.706	12	1.7
3	-1	-2	34.27	2.616	12	10.4
3	1	-2	34.69	2.586	12	4.3
3	2	0	35.60	2.522	12	14.4
3	2	1	37.87	2.376	12	0.9
2	2	-3	37.99	2.368	6	0.6
4	0	0	38.27	2.352	6	0.7
4	0	-1	38.75	2.324	12	0.1
3	2	-2	38.75	2.324	12	4.6
3	0	-3	39.04	2.307	6	6.9
4	0	-2	41.60	2.171	12	1.5
4	1	1	42.29	2.137	6	0.6
3	3	0	42.29	2.137	6	0.1
3	2	2	42.37	2.133	6	0.2
4	1	-2	43.09	2.099	12	5.9
3	2	-3	43.70	2.071	12	2.7
4	2	0	44.36	2.042	12	0.1
3	3	1	44.43	2.039	6	1.3
4	2	-1	44.45	2.038	12	1.0
4	2	1	46.43	1.956	12	1.7
4	0	-3	46.46	1.955	12	0.5
4	-1	-3	47.35	1.920	12	1.6
4	1	-3	47.67	1.908	12	0.4
5	0	0	48.37	1.882	6	5.4
3	3	2	48.59	1.874	6	3.1
5	0	-1	48.62	1.873	12	0.3
3	3	-3	49.01	1.858	6	0.3

Chabazite CHA

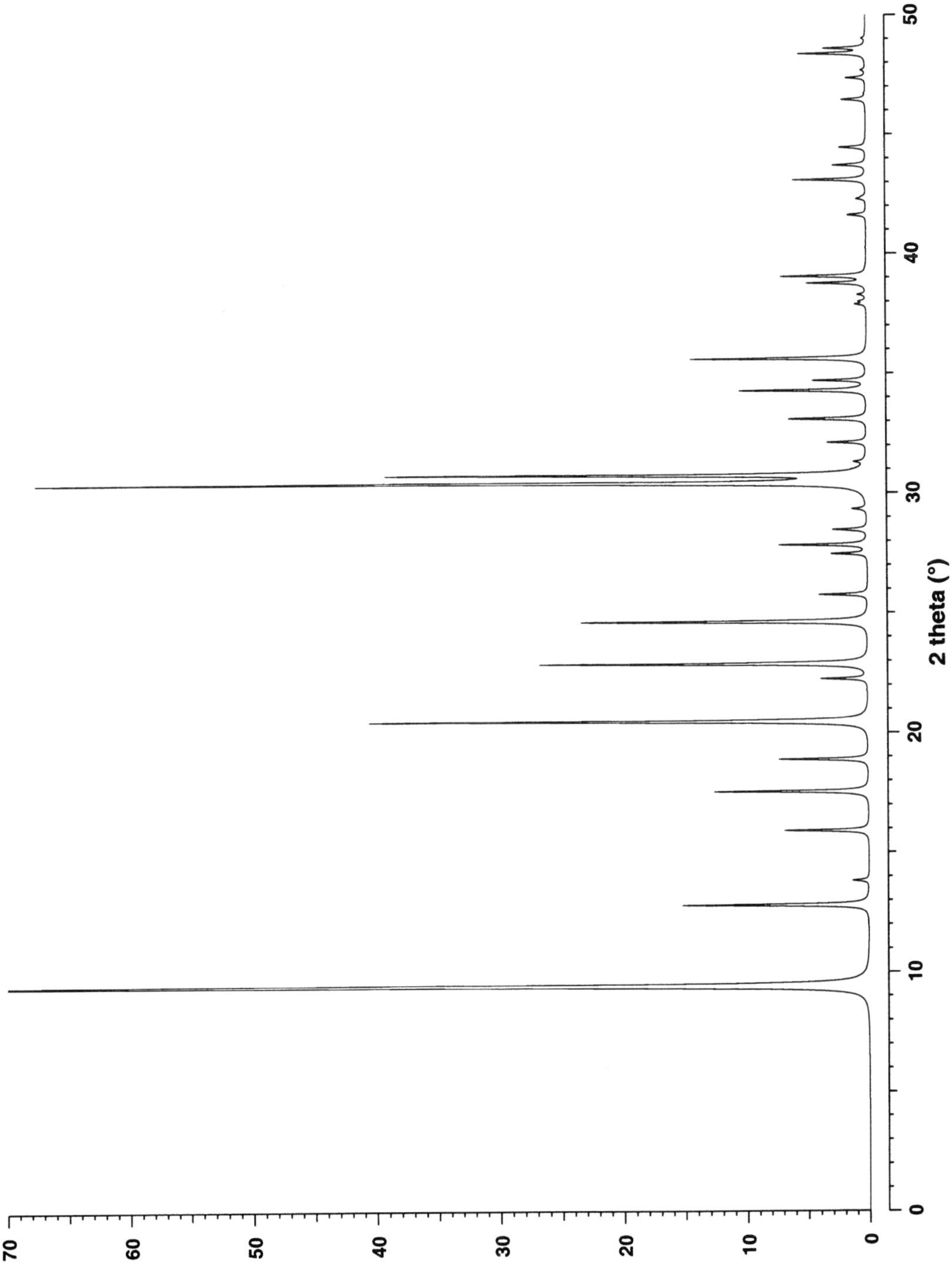

CHA Methylbutylamine SAPO-47

CHEMICAL COMPOSITION: $|(C_5H_{12}NH_2)_{1.4}(H_2O)_{2.5}|$ $[Al_{6.0}Si_{1.4}P_{4.6}O_{24}]$
$C_5H_{12}NH_2$ = methylbutylamine

REFINED COMPOSITION: $|(C_5N)_{2.06}|$ $[Al_{5.47}P_{5.37}O_{24}]$

CRYSTAL DATA: $R\bar{3}$ (No. 148) rhombohedral setting
$a = 9.3834$ Å $b = 9.3834$ Å $c = 9.3834$ Å
$\alpha = 94.085°$ $\beta = 94.085°$ $\gamma = 94.085°$
X-ray single crystal refinement, $R_w = 0.068$

REFERENCE: J. J. Pluth and J. V. Smith,
J. Phys. Chem. **93** 6516–6520 (1989).

h	k	l	2θ	d	M	I_{rel}	h	k	l	2θ	d	M	I_{rel}	h	k	l	2θ	d	M	I_{rel}
1	0	0	9.48	9.332	6	100.0	3	1	-1	31.57	2.834	6	1.5	4	0	-2	41.95	2.154	6	0.1
1	0	-1	12.89	6.867	6	34.9	3	-1	1	31.57	2.834	6	0.8	3	0	3	42.65	2.120	6	1.1
1	0	1	13.93	6.359	6	2.4	2	-2	2	32.37	2.766	6	1.6	4	1	1	42.65	2.120	6	0.9
1	-1	1	16.02	5.531	6	14.6	3	1	1	33.36	2.686	6	2.9	3	2	2	42.74	2.116	6	0.3
1	1	1	17.68	5.017	2	8.9	3	-2	0	33.37	2.685	6	0.4	3	-1	3	42.74	2.115	6	0.9
2	0	0	19.02	4.666	6	0.3	3	0	-2	33.37	2.685	6	0.8	4	-2	-1	42.75	2.115	6	0.3
2	0	-1	20.62	4.308	6	23.7	3	-1	-2	34.56	2.596	6	3.3	4	-1	-2	42.75	2.115	6	0.1
2	-1	0	20.62	4.308	6	45.3	3	-2	-1	34.56	2.596	6	2.1	4	1	-2	43.45	2.083	6	1.5
2	0	1	21.94	4.051	6	0.4	3	-2	1	34.98	2.565	6	0.3	4	-2	1	43.45	2.083	6	1.0
2	1	0	21.94	4.051	6	3.7	1	-2	3	34.98	2.565	6	0.9	3	-3	2	44.06	2.055	6	0.3
2	-1	-1	22.42	3.965	6	3.0	2	0	3	35.90	2.501	6	1.8	2	-3	3	44.06	2.055	6	0.1
1	-1	2	23.04	3.859	6	0.4	3	0	2	35.90	2.501	6	2.5	4	0	2	44.74	2.025	6	0.2
2	-1	1	23.04	3.859	6	7.0	3	2	1	38.20	2.356	6	0.2	3	1	3	44.82	2.022	6	0.1
2	1	1	24.82	3.587	6	26.9	3	1	2	38.20	2.356	6	0.1	3	-2	3	45.09	2.011	6	0.6
2	0	-2	25.95	3.434	6	9.7	2	-3	2	38.30	2.350	6	0.4	4	-2	-2	45.77	1.982	6	0.2
1	-2	2	27.68	3.222	6	0.8	4	0	0	38.59	2.333	6	1.0	4	1	2	46.84	1.940	6	0.2
2	-2	1	27.68	3.222	6	1.7	2	-2	3	39.07	2.305	6	0.2	4	0	-3	46.85	1.939	6	0.3
2	0	2	28.06	3.180	6	3.1	4	0	-1	39.07	2.305	6	0.2	4	-3	0	46.85	1.939	6	0.6
3	0	-1	29.56	3.021	6	1.7	3	-2	2	39.07	2.305	6	0.2	2	-2	4	47.09	1.930	6	0.1
3	-1	0	29.56	3.021	6	0.6	3	0	-3	39.36	2.289	6	0.7	4	-1	-3	47.75	1.905	6	2.2
2	1	2	30.64	2.918	6	7.9	4	-1	-1	39.74	2.268	6	2.7	4	-3	-1	47.75	1.905	6	1.2
3	-1	-1	30.65	2.917	6	24.9	4	-1	1	40.85	2.209	6	0.1	1	-3	4	48.07	1.893	6	0.1
3	1	0	30.99	2.885	6	5.3	4	1	-1	40.85	2.209	6	0.2	5	0	0	48.79	1.866	6	3.8
3	0	1	30.99	2.885	6	4.6	4	-2	0	41.95	2.154	6	0.2	3	2	3	49.02	1.858	6	1.5

Methylbutylamine SAPO-47   CHA

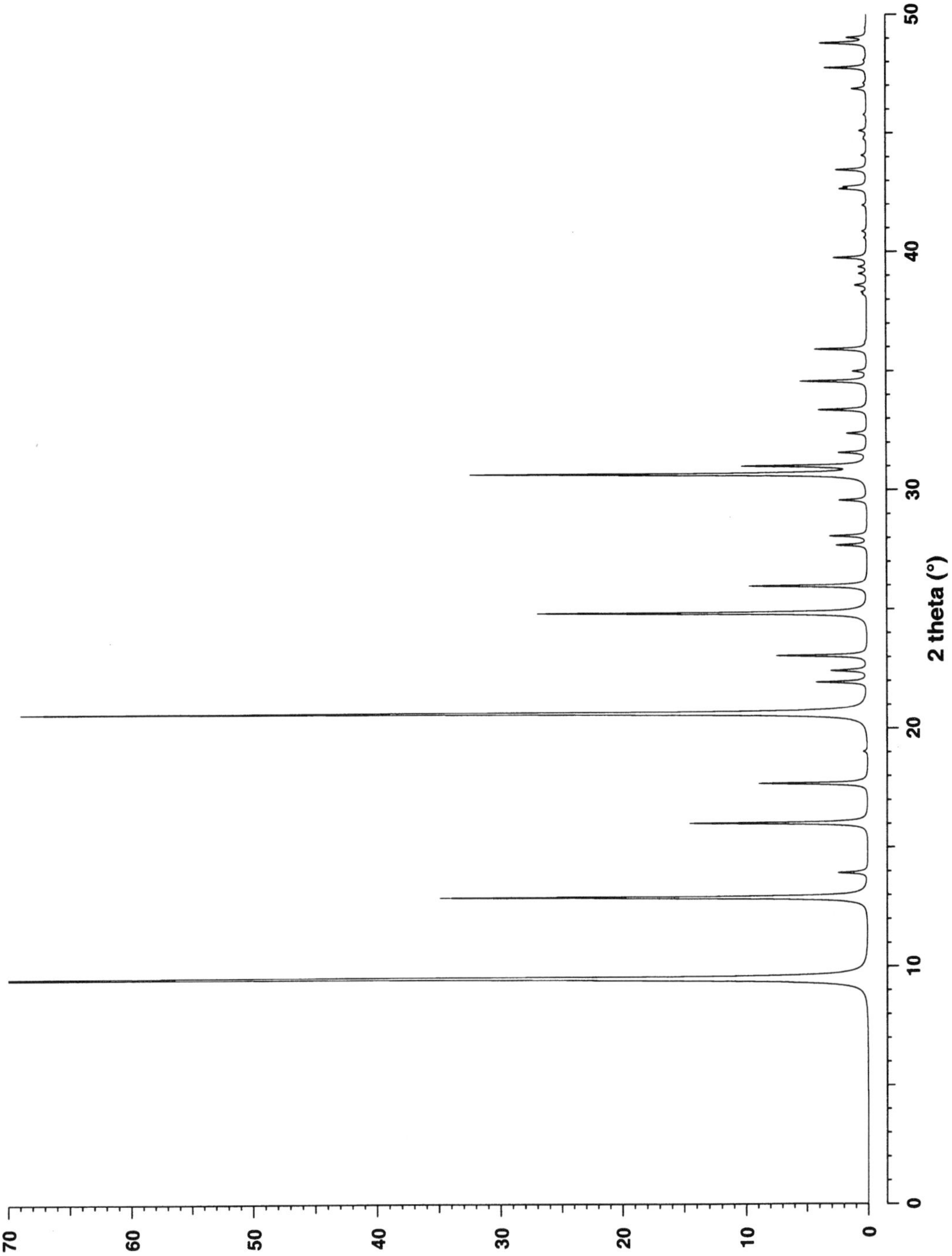

-CHI Chiavennite

CHEMICAL COMPOSITION: $|Ca_4Mn_4(H_2O)_8|\ [Si_{20}(BeOH)_8O_{52}]$
Val di San Giacomo, Chiavenna, Italy

REFINED COMPOSITION: $|Ca_4Mn_4(H_2O)_8|\ [Si_{20}(BeOH)_8O_{52}]$

CRYSTAL DATA: *Pnab* (No. 60) $\bar{c}$**ba** setting
$a = 8.729$ Å $b = 31.326$ Å $c = 4.903$ Å
$\alpha = 90°$ $\beta = 90°$ $\gamma = 90°$
X-ray single crystal refinement, $R = 0.041$

REFERENCE: V. Tazzoli, M. C. Domeneghetti, F. Mazzi and E. Cannillo
Eur. J. Mineral. **7** 1339–1344 (1995).

h	k	l	2θ	d	M	I_{rel}
0	2	0	5.64	15.663	2	100.0
0	4	0	11.30	7.832	2	20.6
1	2	0	11.61	7.625	4	6.7
1	4	0	15.20	5.829	4	24.3
0	1	1	18.31	4.844	4	4.8
1	6	0	19.81	4.481	4	2.2
0	3	1	20.01	4.438	4	2.7
2	0	0	20.35	4.365	2	3.3
1	1	1	20.97	4.236	8	1.9
2	2	0	21.13	4.204	4	0.1
1	2	1	21.55	4.124	8	20.9
1	3	1	22.47	3.956	8	20.7
0	8	0	22.71	3.916	2	34.8
0	5	1	23.03	3.861	4	0.6
2	4	0	23.33	3.812	4	20.4
1	4	1	23.71	3.752	8	1.1
1	8	0	24.92	3.573	4	0.7
1	5	1	25.22	3.531	8	0.8
2	6	0	26.62	3.349	4	1.0
1	6	1	26.96	3.308	8	0.2
0	7	1	26.97	3.305	4	4.1
2	0	1	27.36	3.260	4	30.6
2	1	1	27.51	3.242	8	5.4
2	2	1	27.95	3.192	8	10.6
0	10	0	28.49	3.133	2	1.0
2	3	1	28.69	3.112	8	8.8
2	4	1	29.68	3.010	8	3.2
1	10	0	30.31	2.948	4	7.7
2	8	0	30.67	2.915	4	0.6
2	5	1	30.92	2.892	8	43.9
3	2	0	31.27	2.861	4	2.5
0	9	1	31.52	2.838	4	8.5
2	6	1	32.38	2.765	8	1.3
3	4	0	32.84	2.728	4	0.2

h	k	l	2θ	d	M	I_{rel}
1	9	1	33.19	2.699	8	3.4
2	7	1	34.02	2.635	8	15.7
0	12	0	34.35	2.610	2	1.4
3	6	0	35.31	2.542	4	1.1
1	10	1	35.53	2.527	8	6.1
2	8	1	35.84	2.505	8	4.9
1	12	0	35.91	2.501	4	0.6
3	1	1	36.01	2.494	8	2.9
3	2	1	36.36	2.471	8	0.9
0	11	1	36.49	2.463	4	1.2
0	0	2	36.66	2.451	2	0.2
3	3	1	36.94	2.433	8	0.4
0	2	2	37.12	2.422	4	0.4
3	4	1	37.74	2.384	8	2.3
2	9	1	37.81	2.379	8	7.1
1	1	2	38.24	2.354	8	0.7
0	4	2	38.48	2.340	4	6.8
3	8	0	38.55	2.335	4	1.3
1	2	2	38.58	2.334	8	0.6
1	3	2	39.13	2.302	8	2.2
1	4	2	39.89	2.260	8	0.6
2	10	1	39.91	2.259	8	3.4
3	6	1	39.95	2.256	8	3.9
2	12	0	40.25	2.240	4	0.2
0	14	0	40.31	2.238	2	0.3
1	12	1	40.49	2.228	8	5.1
1	5	2	40.86	2.209	8	0.7
3	7	1	41.34	2.184	8	0.2
4	0	0	41.37	2.182	2	1.4
1	14	0	41.67	2.167	4	1.5
0	13	1	41.77	2.163	4	3.7
4	2	0	41.79	2.161	4	3.9
1	6	2	42.01	2.151	8	0.1
2	0	2	42.28	2.137	4	2.0

h	k	l	2θ	d	M	I_{rel}
3	8	1	42.89	2.108	8	3.4
4	4	0	43.03	2.102	4	0.5
2	3	2	43.20	2.094	8	1.1
0	8	2	43.56	2.078	4	1.8
2	4	2	43.91	2.062	8	3.5
3	9	1	44.60	2.032	8	1.0
2	5	2	44.80	2.023	8	0.8
1	8	2	44.84	2.021	8	0.8
4	6	0	45.02	2.013	4	1.1
4	0	1	45.49	1.994	4	0.5
2	14	0	45.56	1.991	4	0.1
4	1	1	45.59	1.990	8	1.8
1	14	1	45.77	1.982	8	0.3
2	6	2	45.87	1.978	8	6.5
4	2	1	45.88	1.978	8	0.7
4	3	1	46.36	1.958	8	1.3
0	16	0	46.38	1.958	2	0.7
3	10	1	46.45	1.955	8	2.6
1	9	2	46.49	1.953	8	0.3
2	13	1	46.88	1.938	8	4.6
4	4	1	47.03	1.932	8	3.3
0	10	2	47.07	1.931	4	4.8
2	7	2	47.12	1.929	8	1.0
0	15	1	47.31	1.921	4	0.5
4	8	0	47.71	1.906	4	0.3
1	10	2	48.28	1.885	8	0.4
3	11	1	48.42	1.880	8	0.1
1	15	1	48.51	1.876	8	1.9
2	8	2	48.52	1.876	8	0.5
3	1	2	48.65	1.871	8	0.3
4	6	1	48.90	1.863	8	0.9
2	14	1	49.40	1.845	8	0.8

Chiavennite -CHI

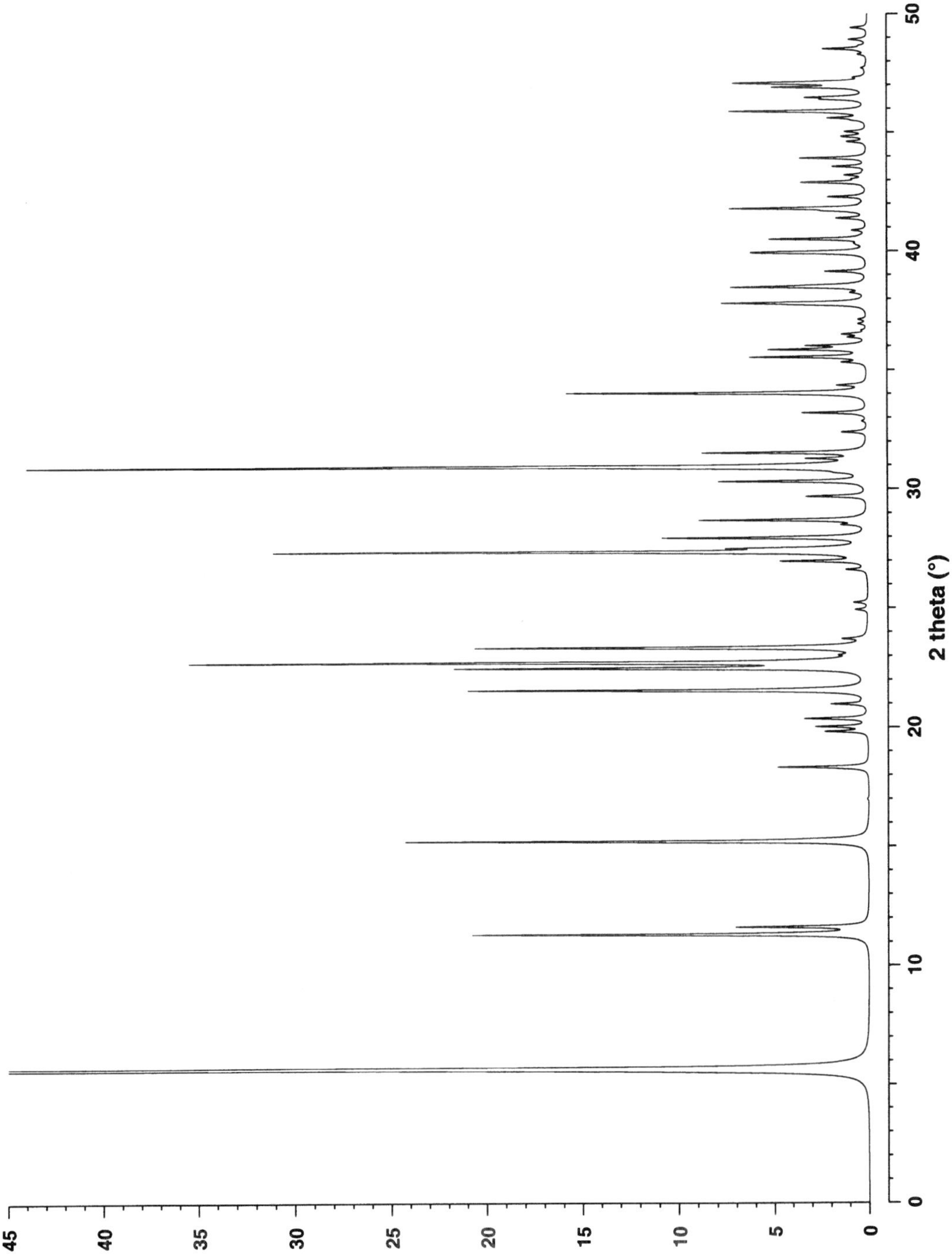

-CLO Cloverite

CHEMICAL COMPOSITION: $|(C_7H_{14}NF)_{192}|$ $[Ga_{768}P_{768}O_{2976}(OH)_{192}]$
$C_7H_{14}NF$ = quinuclidinium fluoride

REFINED COMPOSITION: $|F_{192}|$ $[Ga_{768}P_{768}O_{3168}]$

CRYSTAL DATA: $Fm\bar{3}c$ (No. 226)
$a = 51.7120$ Å $b = 51.7120$ Å $c = 51.7120$ Å
$\alpha = 90°$ $\beta = 90°$ $\gamma = 90°$
X-ray single crystal refinement, $R_w = 0.076$

REFERENCE: M. Estermann, L. B. McCusker, Ch. Baerlocher, A. Merrouche and H. Kessler, *Nature* **352** 320–322 (1991).

h	k	l	2θ	d	M	I_{rel}	h	k	l	2θ	d	M	I_{rel}	h	k	l	2θ	d	M	I_{rel}
2	0	0	3.42	25.856	6	100.0	10	10	2	24.59	3.621	24	0.7	16	10	4	33.42	2.681	48	0.4
2	2	0	4.83	18.283	12	15.7	12	8	0	24.83	3.586	24	0.8	18	6	4	33.60	2.667	48	0.1
2	2	2	5.92	14.928	8	11.8	14	4	0	25.07	3.552	24	0.3	18	8	0	34.15	2.625	24	0.2
4	0	0	6.84	12.928	6	8.6	12	8	2	25.07	3.552	48	0.2	16	10	6	34.33	2.612	48	0.1
4	2	0	7.65	11.563	24	1.2	12	6	6	25.31	3.519	24	1.7	14	14	0	34.33	2.612	12	0.3
4	2	2	8.38	10.556	24	5.2	12	8	4	25.78	3.455	48	0.1	18	8	2	34.33	2.612	48	0.2
4	4	0	9.67	9.141	12	18.4	14	4	4	26.02	3.425	24	0.8	15	11	7	34.47	2.602	48	0.6
6	0	0	10.26	8.619	6	15.3	10	8	8	26.02	3.425	24	0.3	15	13	3	34.83	2.576	48	0.2
6	2	0	10.82	8.176	24	0.4	14	6	2	26.48	3.366	48	0.4	16	12	2	34.87	2.573	48	0.1
6	2	2	11.35	7.796	24	1.2	10	10	6	26.48	3.366	24	0.2	14	12	8	34.87	2.573	48	1.6
4	4	4	11.86	7.464	8	13.7	12	10	2	27.16	3.284	48	0.2	14	14	4	35.05	2.560	24	0.2
6	4	0	12.34	7.171	24	3.0	14	8	0	27.82	3.207	24	0.3	16	12	4	35.40	2.535	48	0.4
6	4	2	12.81	6.910	48	1.4	10	10	8	28.04	3.183	24	0.5	19	7	3	35.53	2.526	48	0.6
8	0	0	13.70	6.464	6	1.0	14	8	2	28.04	3.183	48	0.2	16	10	8	35.58	2.523	48	0.6
8	2	0	14.12	6.271	24	2.9	16	4	0	28.47	3.136	24	0.4	18	10	0	35.75	2.511	24	0.1
6	4	4	14.12	6.271	24	1.5	12	8	8	28.47	3.136	24	0.9	14	12	10	36.44	2.465	48	0.2
8	2	2	14.53	6.094	24	0.6	16	4	2	28.68	3.113	48	0.2	16	14	2	37.12	2.422	48	0.3
6	6	0	14.53	6.094	12	0.4	14	8	4	28.68	3.113	48	2.5	16	10	10	37.12	2.422	24	0.2
8	4	4	16.80	5.278	24	0.2	12	10	6	28.89	3.090	48	0.2	18	10	6	37.29	2.411	48	0.3
8	6	0	17.15	5.171	24	0.3	12	12	0	29.31	3.047	12	0.4	20	8	0	37.46	2.401	24	0.2
8	6	2	17.49	5.071	48	0.7	16	6	0	29.52	3.026	24	0.3	14	12	12	38.29	2.351	24	0.2
6	6	6	17.82	4.976	8	0.4	14	8	6	29.72	3.006	48	0.2	22	0	0	38.29	2.351	6	0.5
10	4	0	18.48	4.801	24	1.3	14	10	0	29.72	3.006	24	0.3	18	12	4	38.29	2.351	48	0.1
10	4	2	18.80	4.721	48	0.1	17	3	1	29.88	2.991	48	0.2	22	2	0	38.45	2.341	24	0.6
8	8	0	19.42	4.571	12	0.6	15	9	1	30.28	2.951	48	0.1	22	4	2	39.11	2.303	48	0.1
8	8	2	19.72	4.501	24	0.4	16	6	4	30.33	2.947	48	0.2	20	10	2	39.11	2.303	48	0.2
12	0	0	20.61	4.309	6	0.2	14	10	4	30.53	2.928	48	0.3	14	14	12	40.38	2.234	24	0.3
8	8	4	20.61	4.309	24	3.7	15	9	3	30.68	2.914	48	0.1	24	0	0	41.93	2.155	6	0.1
10	6	4	21.18	4.194	48	1.2	17	5	3	31.08	2.877	48	0.5	20	14	4	43.28	2.090	48	0.1
11	5	3	21.39	4.154	48	0.5	12	12	6	31.13	2.873	24	0.2	18	16	8	44.46	2.038	48	0.2
9	7	5	21.39	4.154	48	0.1	16	8	2	31.13	2.873	48	0.6	18	18	0	44.60	2.031	12	0.1
12	4	0	21.74	4.088	24	0.4	18	4	0	31.91	2.804	24	0.4	16	16	12	44.89	2.019	24	0.1
8	8	6	22.01	4.038	24	0.7	14	12	0	31.91	2.804	24	0.2	20	16	2	45.04	2.013	48	0.2
10	8	0	22.01	4.038	24	1.3	12	10	10	32.10	2.788	24	0.1	20	14	8	45.04	2.013	48	0.2
10	8	2	22.28	3.990	48	0.3	15	11	1	32.25	2.776	48	0.4	26	2	0	45.75	1.983	24	0.1
11	7	1	22.48	3.955	48	0.3	12	12	8	32.48	2.756	24	0.4	26	4	2	46.32	1.960	48	0.2
12	4	4	22.81	3.898	24	0.2	15	9	7	32.63	2.745	48	0.1	16	16	14	46.74	1.943	24	0.1
11	7	3	23.01	3.865	48	0.4	15	11	3	32.63	2.745	48	0.4	26	6	0	46.88	1.938	24	0.1
13	3	1	23.01	3.865	48	0.3	16	8	6	32.67	2.741	48	0.1	20	16	8	47.16	1.927	48	0.2
12	6	0	23.07	3.854	24	0.2	18	4	4	32.67	2.741	24	0.4	18	16	12	47.30	1.922	48	0.2
10	8	4	23.07	3.854	48	0.1	18	6	0	32.86	2.725	24	0.5	20	14	12	47.85	1.901	48	0.2
12	6	2	23.33	3.812	48	0.2	16	10	2	32.86	2.725	48	1.0	22	14	8	47.99	1.896	48	0.1
14	0	0	24.09	3.694	6	0.6	13	11	9	33.37	2.685	48	0.1	22	16	4	48.40	1.881	48	0.3
10	10	0	24.34	3.657	12	0.4	19	3	1	33.37	2.685	48	0.4	22	14	10	49.21	1.852	48	0.2
14	2	0	24.34	3.657	24	0.1	15	11	5	33.37	2.685	48	0.2	28	0	0	49.34	1.847	6	0.2

Cloverite -CLO

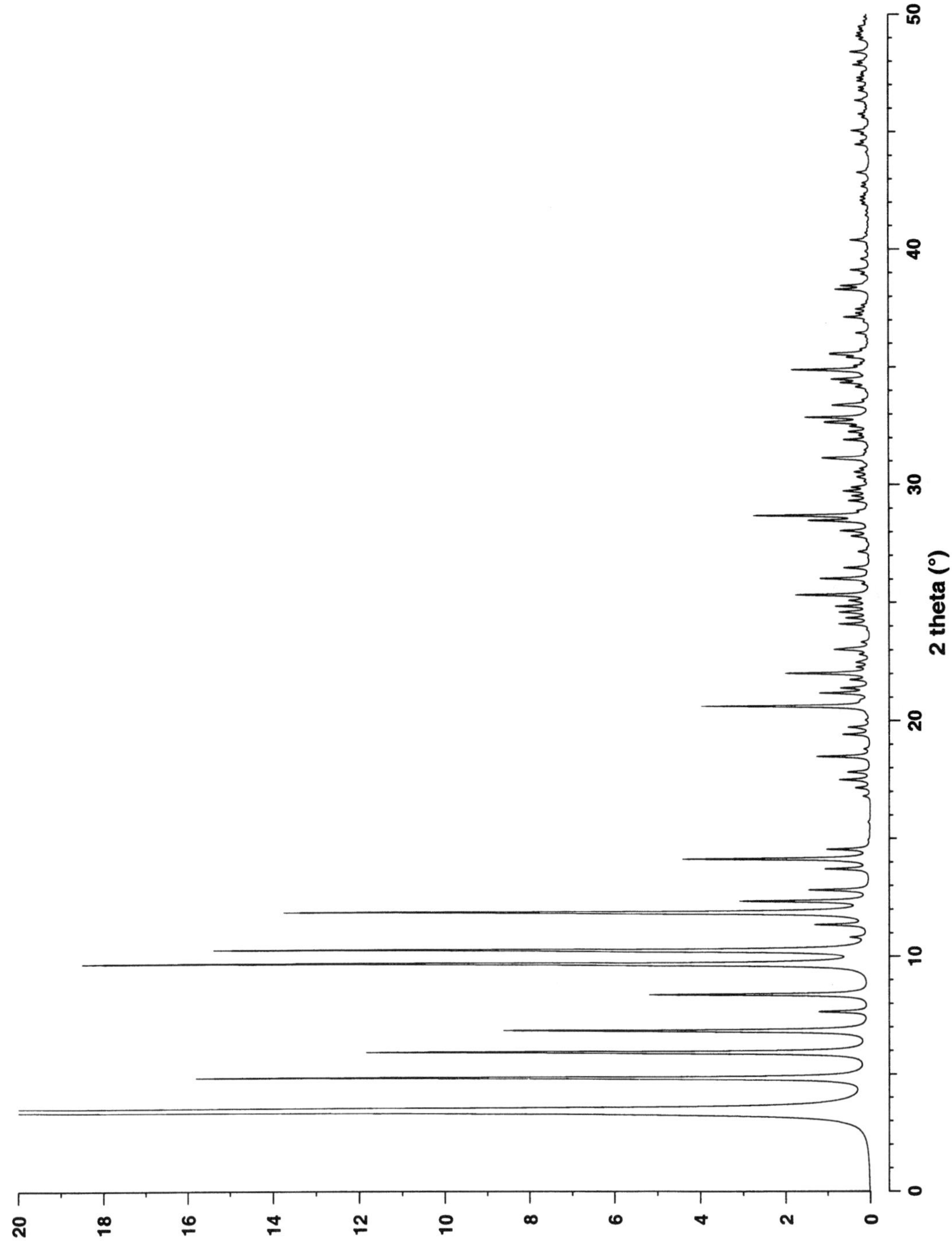

CON CIT-1

CHEMICAL COMPOSITION: $[Si_{56}O_{112}]$

REFINED COMPOSITION: $[Si_{56}O_{112}]$

CRYSTAL DATA: $C1\,2/m\,1$ (No. 12) unique axis **b**, cell choice 1
a = 22.6242 Å b = 13.3503 Å c = 12.3642 Å
α = 90° β = 68.913° γ = 90°
X-ray Rietveld refinement, R_{wp} = 0.136, R_F = 0.106

REFERENCE: R. F. Lobo and M. E. Davis,
J. Am. Chem. Soc. **117** 3766–3779 (1994).

h	k	l	2θ	d	M	I_{rel}
0	0	1	7.66	11.536	2	40.5
1	1	0	7.84	11.283	4	100.0
2	0	0	8.38	10.555	2	3.3
2	0	1	9.10	9.722	2	23.4
1	1	1	9.85	8.975	4	0.8
-1	1	1	11.98	7.387	4	1.3
-2	0	1	13.25	6.681	2	8.0
0	2	0	13.26	6.675	2	2.9
3	1	1	13.85	6.394	4	0.7
3	1	0	14.23	6.225	4	2.2
2	0	2	14.61	6.062	2	1.8
0	2	1	15.34	5.778	4	1.0
0	0	2	15.36	5.768	2	1.4
2	2	0	15.71	5.642	4	0.2
4	0	1	15.75	5.625	2	1.7
4	0	0	16.80	5.277	2	1.7
-3	1	1	18.22	4.869	4	0.3
4	0	2	18.25	4.861	2	0.7
-2	2	1	18.79	4.722	4	0.2
2	2	2	19.78	4.488	4	1.4
-2	0	2	20.02	4.435	2	0.1
0	2	2	20.35	4.364	4	0.4
1	3	0	20.39	4.354	4	10.8
4	2	1	20.65	4.302	4	0.2
5	1	1	20.73	4.285	4	0.2
4	2	0	21.46	4.140	4	2.7
2	0	3	21.56	4.121	2	0.6
5	1	0	22.08	4.025	4	8.1
5	1	2	22.17	4.009	4	0.9
4	2	2	22.63	3.929	4	0.1
1	1	3	22.95	3.874	4	2.0
3	1	3	22.99	3.868	4	3.5
0	0	3	23.13	3.845	2	5.0
4	0	3	23.21	3.833	2	7.2
3	3	0	23.66	3.761	4	0.2
-2	2	2	24.09	3.694	4	0.3
-3	1	2	24.14	3.686	4	1.7
6	0	2	24.44	3.642	2	2.0
-4	2	1	24.81	3.588	4	0.6
6	0	0	25.31	3.518	2	0.9
2	2	3	25.40	3.507	4	1.2
3	3	2	25.67	3.471	4	0.3
-5	1	1	25.82	3.450	4	0.9
5	1	3	25.98	3.430	4	0.2
-1	3	2	26.55	3.358	4	0.4
-4	0	2	26.68	3.341	2	2.0
0	4	0	26.71	3.338	2	6.6
6	0	3	27.52	3.241	2	0.2
0	4	1	27.83	3.206	4	0.5
6	2	2	27.91	3.197	4	1.1
2	4	1	28.27	3.157	4	0.9
7	1	1	28.56	3.125	4	0.6
6	2	0	28.68	3.112	4	1.4
2	0	4	29.01	3.078	2	1.7
5	3	0	29.16	3.063	4	0.9
5	3	2	29.23	3.056	4	1.1
4	0	4	29.47	3.031	2	0.3
-4	2	2	29.91	2.987	4	0.3
-2	4	1	29.93	2.986	4	1.2
7	1	0	30.39	2.941	4	0.1
-2	2	3	30.55	2.926	4	0.9
2	4	2	30.58	2.924	4	0.2
6	2	3	30.67	2.915	4	0.5
-3	3	2	30.78	2.905	4	1.1
-3	1	3	30.98	2.886	4	0.2
4	4	1	31.16	2.870	4	0.7
5	1	4	31.35	2.853	4	0.3
4	4	0	31.72	2.821	4	0.7
8	0	2	31.81	2.813	2	0.2
-5	3	1	32.13	2.785	4	0.4
-1	3	3	32.18	2.782	4	0.4
5	3	3	32.26	2.775	4	0.4
4	4	2	32.54	2.751	4	0.4
-4	0	3	33.41	2.682	2	1.0
-1	1	4	33.52	2.674	4	0.3
8	0	3	33.57	2.669	2	0.8
8	0	0	33.97	2.639	2	0.3
-7	1	1	33.99	2.637	4	0.3
-4	4	1	34.14	2.626	4	0.3
-1	5	1	35.10	2.557	4	0.1
7	1	4	35.11	2.556	4	0.6
0	4	3	35.62	2.521	4	0.8
4	4	3	35.67	2.517	4	0.7
7	3	0	35.97	2.496	4	0.1
3	5	0	35.97	2.496	4	0.5
4	0	5	36.46	2.464	2	0.3
8	2	0	36.62	2.454	4	0.4
6	4	0	37.13	2.421	4	1.3
3	5	2	37.38	2.406	4	0.3
-5	1	3	37.48	2.400	4	0.2
9	1	3	37.65	2.389	4	0.4
5	1	5	37.70	2.386	4	0.2
-1	5	2	38.01	2.367	4	0.2
-4	4	2	38.11	2.361	4	0.3
1	1	5	38.28	2.351	4	0.1
-3	1	4	38.35	2.347	4	0.2
-1	3	4	38.71	2.326	4	0.1
9	1	0	38.99	2.310	4	0.2
2	4	4	39.84	2.263	4	0.2
-6	4	1	39.89	2.260	4	0.2
7	3	4	40.12	2.248	4	0.1
1	5	3	40.47	2.229	4	0.2
0	6	0	40.54	2.225	2	0.7
6	2	5	40.59	2.223	4	0.2
-3	5	2	41.19	2.191	4	0.1
10	2	2	42.17	2.143	4	0.1
-8	0	2	42.49	2.127	2	0.3
4	0	6	43.94	2.061	2	1.3
8	2	5	43.96	2.060	4	0.1
4	6	0	44.17	2.050	4	0.6
7	5	2	44.27	2.046	4	0.2
11	1	2	44.59	2.032	4	0.1
9	1	5	44.60	2.032	4	0.2
-2	4	4	44.63	2.030	4	0.1
3	1	6	44.68	2.028	4	0.1
5	1	6	44.72	2.026	4	0.2
7	3	5	44.85	2.021	4	0.1
10	2	0	45.04	2.013	4	0.2
7	5	0	45.37	1.999	4	0.3
-1	3	5	45.85	1.979	4	0.1
-5	5	2	45.91	1.977	4	0.1
-3	1	5	46.10	1.969	4	0.3
7	1	6	46.48	1.954	4	0.1
10	0	5	46.72	1.944	2	0.2
-9	1	2	47.29	1.922	4	0.4
11	1	0	47.89	1.899	4	0.2
1	7	0	47.89	1.899	4	0.4
-8	0	3	48.35	1.883	2	0.1
-10	2	1	48.51	1.877	4	0.2
10	2	5	48.78	1.867	4	0.2
11	3	2	48.79	1.866	4	0.2
9	3	5	48.80	1.866	4	0.2
-4	6	2	49.20	1.852	4	0.1
3	7	0	49.52	1.841	4	0.1
6	6	3	49.70	1.834	4	0.2

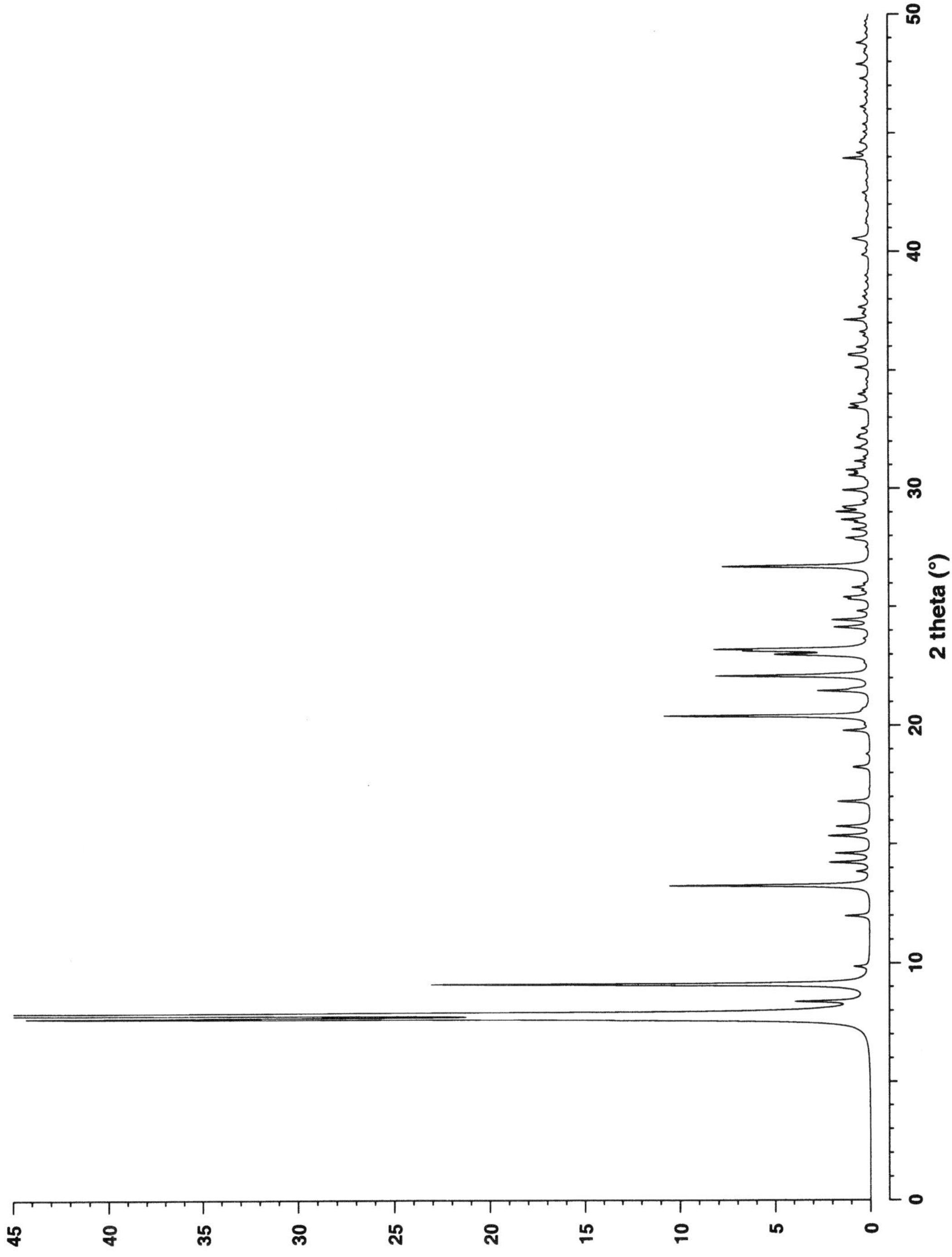

2 theta (°)
0
10
20
30
40
50
45
40
35
30
25
20
15
10
5
0

CZP Chiral Sodium Zincophosphate, $P6_122$

CHEMICAL COMPOSITION: $|Na_{12}(H_2O)_{12}|$ $[Zn_{12}P_{12}O_{48}]$

REFINED COMPOSITION: $|Na_{6.06}O_{9.12}|$ $[Zn_{12}P_{12}O_{48}]$

CRYSTAL DATA: $P6_122$ (No. 178)
$a = 10.4797$ Å $b = 10.4797$ Å $c = 15.089$ Å
$\alpha = 90.0°$ $\beta = 90.0°$ $\gamma = 120.0°$
X-ray single crystal refinement, $R = 0.0599$, $R_w = 0.0529$

REFERENCE: W. T. A. Harrison, T. E. Gier, G. D. Stucky, R. W. Broach and R. A. Bedard, *Chemistry of Materials* **8** 145–151 (1996).

h	*k*	*l*	*2θ*	*d*	*M*	I_{rel}	*h*	*k*	*l*	*2θ*	*d*	*M*	I_{rel}	*h*	*k*	*l*	*2θ*	*d*	*M*	I_{rel}
1	0	0	9.75	9.076	6	18.7	2	1	3	31.57	2.834	24	10.6	3	1	4	43.21	2.094	24	1.4
1	0	1	11.38	7.777	12	100.0	3	0	2	31.87	2.808	12	19.2	3	2	0	43.46	2.082	12	0.6
1	0	2	15.27	5.802	12	21.7	2	2	0	34.22	2.620	6	13.4	4	0	3	43.77	2.068	12	1.4
1	1	0	16.92	5.240	6	9.6	3	0	3	34.60	2.592	12	1.7	3	2	1	43.90	2.063	24	2.6
2	0	0	19.56	4.538	6	0.2	2	2	1	34.75	2.581	12	0.4	2	1	6	44.68	2.028	24	1.5
1	0	3	20.18	4.399	12	0.6	2	1	4	35.37	2.538	24	1.5	3	2	2	45.17	2.007	24	0.8
2	0	1	20.44	4.346	12	0.6	3	1	0	35.67	2.517	12	0.7	1	1	7	45.50	1.993	12	0.3
1	1	2	20.64	4.304	12	15.4	0	0	6	35.70	2.515	2	2.8	2	2	5	45.87	1.978	12	0.4
2	0	2	22.87	3.889	12	7.2	3	1	1	36.18	2.483	24	3.7	4	1	1	46.23	1.964	24	0.2
1	1	3	24.53	3.629	12	0.3	2	2	2	36.30	2.475	12	0.8	2	0	7	46.65	1.947	12	0.1
2	1	0	25.97	3.430	12	2.2	3	1	2	37.67	2.388	24	0.4	4	0	4	46.72	1.944	12	0.9
2	0	3	26.45	3.369	12	0.6	3	0	4	38.13	2.360	12	5.2	3	0	6	46.98	1.934	12	0.3
2	1	1	26.65	3.345	24	12.7	4	0	0	39.73	2.269	6	0.5	3	1	5	47.01	1.933	24	5.1
2	1	2	28.59	3.123	24	11.7	1	1	6	39.76	2.267	12	0.9	3	2	3	47.25	1.924	24	0.4
1	1	4	29.17	3.061	12	23.5	2	1	5	39.78	2.266	24	3.4	4	1	2	47.46	1.916	24	5.0
3	0	0	29.53	3.025	6	6.5	3	1	3	40.06	2.251	24	1.7	1	0	8	49.35	1.847	12	0.8
3	0	1	30.13	2.966	12	0.9	4	0	1	40.19	2.244	12	0.1	4	1	3	49.46	1.843	24	0.3
2	0	4	30.82	2.901	12	0.5	4	0	2	41.56	2.173	12	0.1	2	1	7	49.97	1.825	24	0.9
1	0	5	31.23	2.864	12	14.6	2	2	4	41.99	2.152	12	1.1							

Chiral Sodium Zincophosphate, $P6_122$ CZP

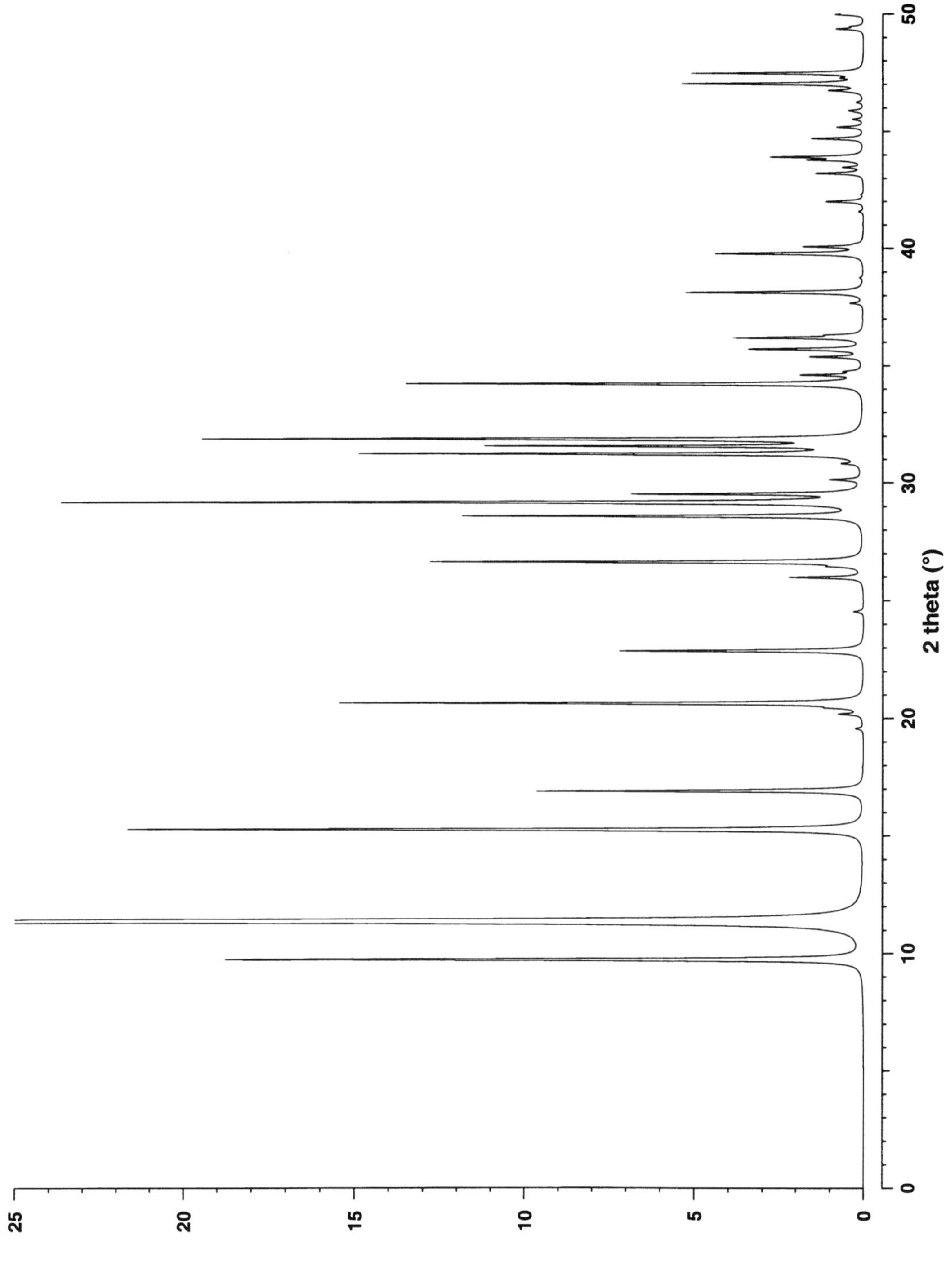

CZP Chiral Sodium Zincophosphate, $P6_522$

CHEMICAL COMPOSITION: $|Na_{12}(H_2O)_{12}|$ $[Zn_{12}P_{12}O_{48}]$

REFINED COMPOSITION: $|Na_{2.04}O_{4.08}|$ $[Zn_{12}P_{12}O_{48}]$

CRYSTAL DATA: $P6_522$ (No. 179)
a = 10.412 Å b = 10.412 Å c = 15.184 Å
α = 90.0° β = 90.0° γ = 120.0°
X-ray single crystal refinement, R = 0.0690, R_w = 0.0810

REFERENCE: W. T. A. Harrison, T. E. Gier, G. D. Stucky, R. W. Broach and R. A. Bedard, *Chemistry of Materials* **8** 145–151 (1996).

h	k	l	2θ	d	M	I_{rel}
1	0	0	9.81	9.017	6	10.9
1	0	1	11.41	7.753	12	100.0
1	0	2	15.26	5.808	12	23.8
1	1	0	17.03	5.206	6	16.4
1	0	3	20.12	4.414	12	1.5
2	0	1	20.55	4.322	12	0.6
1	1	2	20.69	4.294	12	11.3
2	0	2	22.94	3.877	12	6.6
1	1	3	24.53	3.629	12	0.1
2	1	0	26.15	3.408	12	1.6
2	1	1	26.81	3.325	24	12.8
2	1	2	28.71	3.109	24	9.6
1	1	4	29.11	3.067	12	18.8
3	0	0	29.72	3.006	6	6.7
3	0	1	30.31	2.948	12	0.6
2	0	4	30.79	2.904	12	0.4
1	0	5	31.07	2.878	12	12.8
2	1	3	31.65	2.827	24	4.1
3	0	2	32.03	2.795	12	14.0
2	2	0	34.45	2.603	6	9.5
3	0	3	34.71	2.584	12	1.7
2	2	1	34.97	2.566	12	0.3
2	1	4	35.39	2.536	24	1.4
0	0	6	35.47	2.531	2	2.0
3	1	0	35.91	2.501	12	0.5
3	1	1	36.41	2.468	24	2.4
2	2	2	36.49	2.462	12	0.5
1	0	6	36.89	2.437	12	0.2
3	1	2	37.88	2.375	24	0.3
3	0	4	38.19	2.356	12	3.8
2	2	3	38.91	2.315	12	0.1
1	1	6	39.60	2.276	12	1.4
2	1	5	39.75	2.267	24	3.6
4	0	0	39.99	2.254	6	0.4
3	1	3	40.22	2.242	24	0.7
4	0	1	40.45	2.230	12	0.1
4	0	2	41.80	2.161	12	0.3
2	2	4	42.09	2.147	12	0.9
3	1	4	43.32	2.088	24	1.4
3	2	0	43.76	2.069	12	0.2
4	0	3	43.97	2.059	12	1.4
3	2	1	44.18	2.050	24	2.3
2	1	6	44.60	2.032	24	0.9
1	1	7	45.29	2.002	12	0.3
3	2	2	45.44	1.996	24	0.8
2	2	5	45.92	1.976	12	0.2
4	0	4	46.87	1.938	12	0.8
3	0	6	46.93	1.936	12	0.6
3	1	5	47.07	1.931	24	3.5
3	2	3	47.48	1.915	24	0.3
4	1	2	47.75	1.905	24	4.1
1	0	8	49.05	1.857	12	0.6
4	1	3	49.71	1.834	24	0.5
2	1	7	49.83	1.830	24	0.5

Chiral Sodium Zincophosphate, $P6_522$ CZP

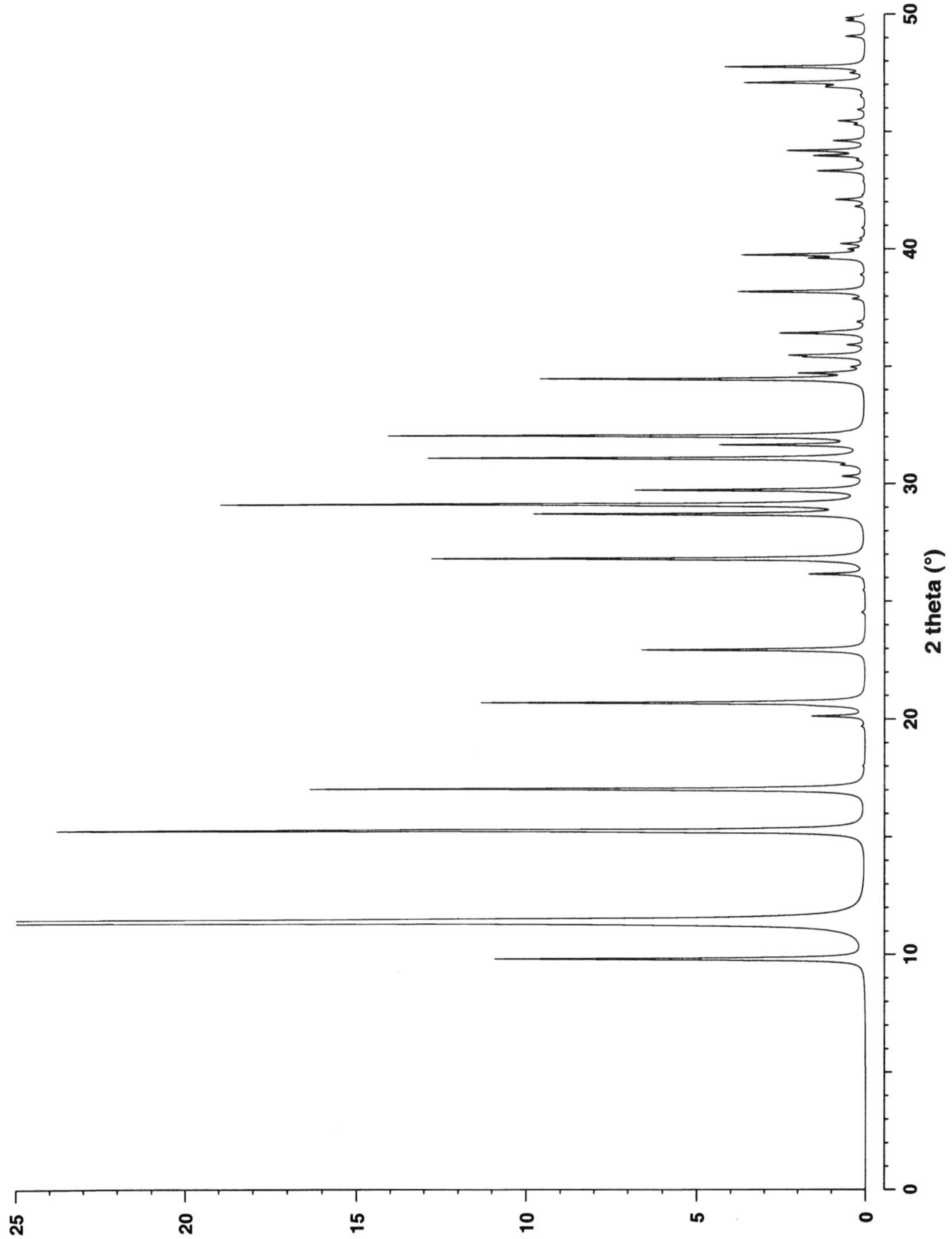

DAC Dachiardite

CHEMICAL COMPOSITION: $|Na_{0.42}K_{0.92}Cs_{0.11}Ca_{1.54}Sr_{0.12}Ba_{0.01}(H_2O)_{12.56}|$ $[Si_{18.96}Al_{4.86}Fe_{0.02}O_{48}]$
Elba, Italy

REFINED COMPOSITION: $|K_{0.62}Ca_{2.76}(H_2O)_{12}|$ $[Si_{19.2}Al_{4.8}O_{48}]$

CRYSTAL DATA: $C1\,2/m\,1$ (No. 12) unique axis **b**, cell choice 1
$a = 18.676$ Å $b = 7.518$ Å $c = 10.246$ Å
$\alpha = 90°$ $\beta = 107.87°$ $\gamma = 90°$
X-ray single crystal refinement, $R_w = 0.06$

REFERENCE: G. Vezzalini,
Z. Kristallogr. **166** 63–71 (1984).

h	k	l	2θ	d	M	I_{rel}
0	0	1	9.07	9.752	2	3.8
2	0	0	9.95	8.887	2	69.6
1	1	0	12.78	6.924	4	50.3
-1	1	1	14.78	5.994	4	25.9
1	1	1	16.56	5.352	4	9.0
-2	0	2	17.86	4.965	2	29.9
0	0	2	18.19	4.876	2	45.5
-3	1	1	19.05	4.659	4	2.0
3	1	0	19.07	4.654	4	2.3
-4	0	1	19.24	4.613	2	5.1
4	0	0	19.98	4.444	2	2.6
-1	1	2	20.99	4.232	4	3.4
-4	0	2	22.56	3.941	2	32.1
-3	1	2	23.02	3.863	4	8.7
3	1	1	23.08	3.854	4	1.7
2	0	2	23.35	3.810	2	38.7
1	1	2	23.54	3.780	4	12.6
0	2	0	23.67	3.759	2	7.8
4	0	1	24.43	3.644	2	8.6
0	2	1	25.39	3.507	4	9.0
2	2	0	25.73	3.462	4	100.0
-2	0	3	26.14	3.409	2	22.2
-2	2	1	26.27	3.393	4	9.2
-5	1	1	26.69	3.340	4	16.1
0	0	3	27.44	3.251	2	3.8
5	1	0	27.76	3.214	4	53.5
-4	0	3	28.62	3.119	2	4.6
-5	1	2	28.72	3.109	4	1.5
-1	1	3	28.91	3.088	4	6.7
-3	1	3	29.48	3.030	4	4.5
3	1	2	29.56	3.022	4	2.5
-2	2	2	29.81	2.997	4	1.3
0	2	2	30.02	2.977	4	1.8
-6	0	2	30.11	2.968	2	24.3
6	0	0	30.17	2.962	2	1.5
-4	2	1	30.68	2.914	4	0.2
4	0	2	31.11	2.874	2	10.7
4	2	0	31.16	2.870	4	20.4
5	1	1	31.64	2.828	4	0.3
1	1	3	31.78	2.816	4	1.3
2	0	3	32.09	2.789	2	0.2
-4	2	2	32.93	2.720	4	11.6
2	2	2	33.49	2.676	4	13.4
-6	0	3	34.12	2.627	2	1.8
4	2	1	34.27	2.616	4	1.0
-2	0	4	35.05	2.560	2	11.0
-2	2	3	35.55	2.525	4	4.9
-7	1	1	35.75	2.512	4	1.4
-4	0	4	36.18	2.483	2	2.7
1	3	0	36.20	2.481	4	1.2
-7	1	2	36.53	2.460	4	0.6
0	2	3	36.54	2.459	4	6.8
0	0	4	36.87	2.438	2	0.6
-1	3	1	37.00	2.430	4	1.3
-3	1	4	37.28	2.412	4	1.4
3	1	3	37.37	2.406	4	0.8
-4	2	3	37.47	2.400	4	4.8
5	1	2	37.55	2.395	4	3.8
-1	1	4	37.62	2.391	4	1.2
1	3	1	37.79	2.381	4	0.5
-6	2	2	38.65	2.329	4	0.2
6	2	0	38.70	2.327	4	0.7
4	0	3	39.03	2.308	2	4.7
-7	1	3	39.60	2.276	4	3.8
-5	1	4	39.73	2.269	4	0.3
-6	0	4	40.04	2.252	2	0.6
1	1	4	40.68	2.218	4	0.7
7	1	1	41.18	2.192	4	0.7
-3	3	2	41.22	2.190	4	0.8
3	3	1	41.26	2.188	4	0.3
2	0	4	41.30	2.186	2	1.7
1	3	2	41.53	2.174	4	1.1
-8	0	3	41.60	2.171	2	0.3
6	2	1	42.04	2.149	4	0.5
-2	2	4	42.73	2.116	4	0.2
-4	2	4	43.69	2.072	4	1.4
5	3	0	44.22	2.048	4	0.8
0	2	4	44.28	2.045	4	1.2
-2	0	5	44.39	2.041	2	4.6
8	0	1	44.51	2.035	2	0.2
-7	1	4	44.56	2.033	4	0.2
-4	0	5	44.64	2.030	2	0.1
5	1	3	44.82	2.022	4	0.8
-5	3	2	44.86	2.020	4	3.8
-1	3	3	45.00	2.014	4	0.4
-3	3	3	45.39	1.998	4	0.4
3	3	2	45.44	1.996	4	3.4
-9	1	2	45.57	1.990	4	1.0
-9	1	1	45.60	1.989	4	1.3
-3	1	5	45.91	1.977	4	0.2
3	1	4	46.00	1.973	4	1.5
-8	0	4	46.06	1.971	2	1.6
-8	2	2	46.17	1.966	4	2.6
7	1	2	46.71	1.945	4	1.5
-1	1	5	46.86	1.939	4	0.8
5	3	1	46.91	1.937	4	1.1
1	3	3	47.01	1.933	4	0.3
-6	2	4	47.04	1.932	4	1.9
6	2	2	47.16	1.927	4	1.9
-6	0	5	47.29	1.922	2	0.5
-5	1	5	47.34	1.920	4	0.1
-9	1	3	47.52	1.914	4	0.7
8	2	0	47.54	1.913	4	0.4
9	1	0	47.60	1.910	4	1.1
4	0	4	47.74	1.905	2	1.5
-5	3	3	48.14	1.890	4	0.2
2	2	4	48.15	1.890	4	2.5
0	4	0	48.43	1.880	2	11.9
-10	0	2	48.85	1.864	2	1.0
-10	0	1	49.20	1.852	2	1.8
0	4	1	49.38	1.846	4	0.4
2	4	0	49.57	1.839	4	0.4
-2	4	1	49.88	1.828	4	0.2
-7	3	1	49.96	1.826	4	2.8

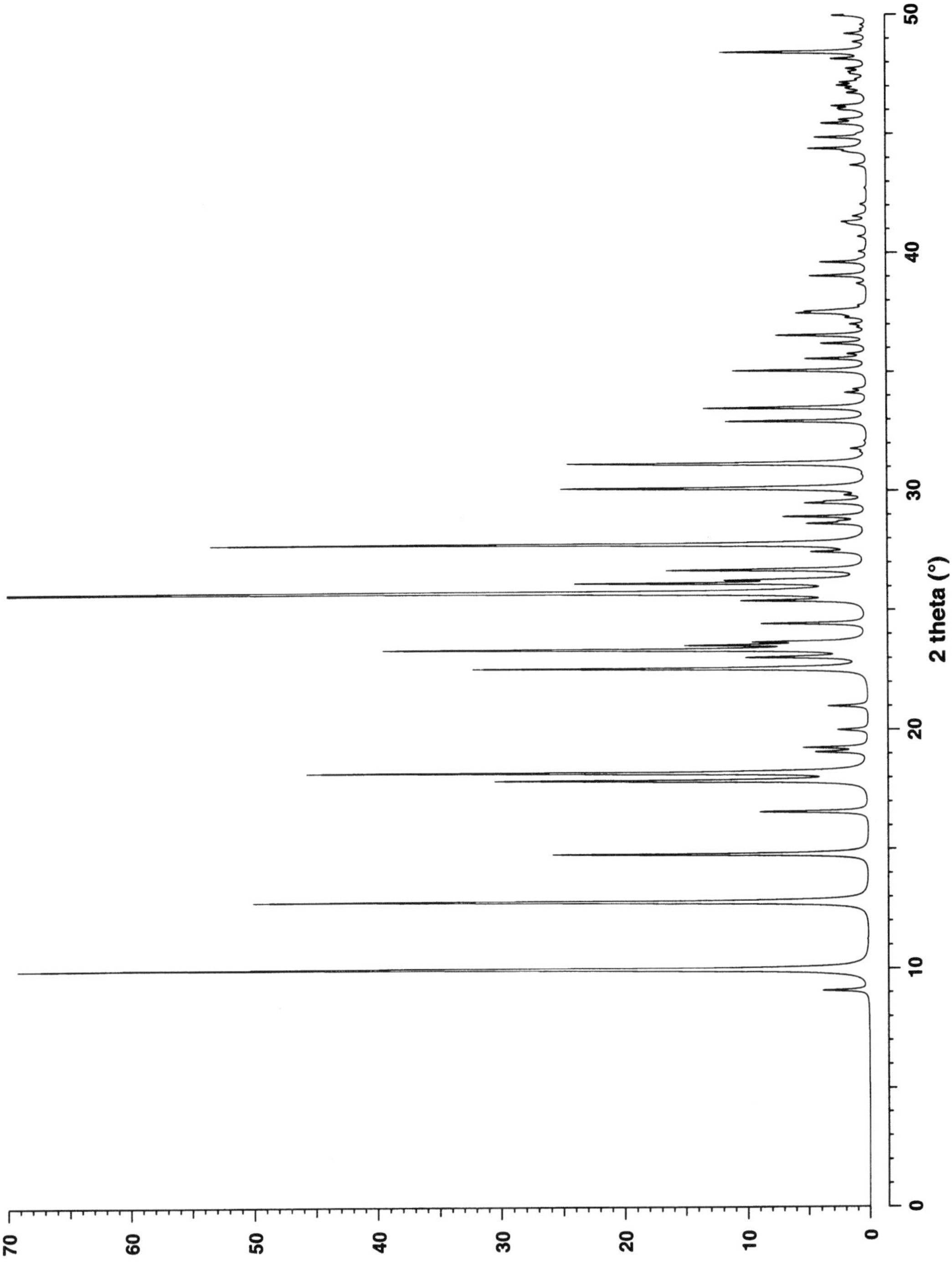
2 theta (°)
0
10
20
30
40
50
0
10
20
30
40
50
60
70

DDR N_2, 1-aminoadamantane Deca-Dodecasil 3R

CHEMICAL COMPOSITION: $|(C_{10}H_{17}N)_6(N_2)_9|$ $[Si_{120}O_{240}]$
$C_{10}H_{17}N$ = 1-aminoadamantane

REFINED COMPOSITION: $|C_{69}N_6|$ $[Si_{120}O_{240}]$

CRYSTAL DATA: $R\overline{3}m$ (No. 166) hexagonal setting
a = 13.860 Å b = 13.860 Å c = 40.891 Å
α = 90° β = 90° γ = 120°
X-ray single crystal refinement, R_w = 0.066

REFERENCE: H. Gies,
Z. Kristallogr. **175** 93–104 (1986).

h	k	l	2θ	d	M	I_{rel}
0	0	3	6.48	13.630	2	23.5
1	0	1	7.68	11.517	6	40.5
0	1	2	8.54	10.351	6	12.9
1	0	4	11.37	7.783	6	21.1
1	1	0	12.77	6.930	6	18.2
0	0	6	12.99	6.815	2	18.0
0	1	5	13.10	6.759	6	0.6
1	1	3	14.34	6.177	12	17.5
0	2	1	14.92	5.938	6	1.1
2	0	2	15.39	5.759	6	58.5
1	0	7	16.88	5.253	6	7.5
0	2	4	17.13	5.176	6	100.0
1	1	6	18.26	4.859	12	26.1
2	0	5	18.34	4.838	6	19.9
0	1	8	18.87	4.703	6	48.7
0	0	9	19.54	4.543	2	5.0
2	1	1	19.69	4.509	12	46.7
1	2	2	20.05	4.429	12	3.1
0	2	7	21.22	4.186	6	5.1
2	1	4	21.43	4.147	12	27.4
3	0	0	22.22	4.001	6	10.0
1	2	5	22.41	3.967	12	9.4
2	0	8	22.85	3.891	6	2.0
1	0	10	22.98	3.871	6	7.5
3	0	3	23.17	3.839	6	9.9
0	3	3	23.17	3.839	6	9.4
1	1	9	23.41	3.800	12	14.6
2	1	7	24.85	3.583	12	9.8
2	2	0	25.71	3.465	6	12.5
0	3	6	25.82	3.450	6	10.6
3	0	6	25.82	3.450	6	2.7
0	0	12	26.15	3.408	2	17.4
1	2	8	26.26	3.393	12	43.4
0	2	10	26.37	3.379	6	8.9
2	2	3	26.54	3.358	12	53.9
1	3	1	26.87	3.318	12	42.2
3	1	2	27.14	3.286	12	2.4
1	3	4	28.19	3.165	12	1.3
2	0	11	28.24	3.160	6	13.8
3	1	5	28.96	3.083	12	8.7
1	1	12	29.20	3.058	12	18.9
1	0	13	29.35	3.043	6	0.9
2	1	10	29.41	3.037	12	11.7
3	0	9	29.75	3.003	6	11.3
4	0	1	29.85	2.993	6	8.4
0	4	2	30.10	2.969	6	2.0
1	3	7	30.92	2.892	12	2.1
4	0	4	31.06	2.879	6	3.2
1	2	11	31.10	2.875	12	0.9
0	1	14	31.52	2.838	6	2.4
0	4	5	31.76	2.817	6	0.7
2	2	9	32.50	2.755	12	0.3
3	2	1	32.59	2.747	12	1.9
4	0	7	33.57	2.669	6	4.5
3	2	4	33.71	2.659	12	0.3
2	0	14	34.14	2.626	6	1.0
2	3	5	34.36	2.610	12	0.2
3	0	12	34.57	2.594	6	0.5
0	3	12	34.57	2.594	6	0.4
2	1	13	34.70	2.585	12	2.5
4	1	3	34.88	2.572	12	0.6
1	0	16	35.93	2.500	6	2.0
3	2	7	36.06	2.491	12	2.6
4	1	6	36.76	2.445	12	0.2
1	4	6	36.76	2.445	12	6.6
2	2	12	37.00	2.430	12	1.4
2	3	8	37.08	2.424	12	9.2
4	0	10	37.16	2.419	6	0.3
0	5	1	37.53	2.396	6	4.4
5	0	2	37.73	2.384	6	0.5
0	2	16	38.28	2.351	6	3.4
0	4	11	38.56	2.335	6	4.4
3	3	0	38.99	2.310	6	3.9
5	0	5	39.11	2.303	6	9.8
1	3	13	39.41	2.286	12	0.1
3	2	10	39.45	2.284	12	1.6
3	3	3	39.57	2.278	12	2.1
0	0	18	39.67	2.272	2	0.5
4	1	9	39.72	2.269	12	1.9
2	4	1	39.80	2.265	12	0.4
4	2	2	39.99	2.255	12	0.7
3	0	15	40.02	2.253	6	3.0
0	3	15	40.02	2.253	6	2.2
2	0	17	40.40	2.233	6	2.8
0	5	7	40.63	2.220	6	1.7
2	4	4	40.74	2.215	12	0.3
2	3	11	40.78	2.213	12	3.2
5	0	8	41.56	2.173	6	0.4
4	0	13	41.59	2.171	6	0.1
1	1	18	41.85	2.159	12	2.4
5	1	1	41.97	2.153	12	0.8
1	5	2	42.15	2.144	12	0.2
1	2	17	42.54	2.125	12	0.4
1	0	19	42.68	2.118	6	0.1
2	4	7	42.76	2.115	12	1.5
5	1	4	42.87	2.109	12	0.2
0	4	14	43.22	2.093	6	0.2
4	1	12	43.58	2.077	12	0.9
1	4	12	43.58	2.077	12	0.8
3	2	13	43.69	2.072	12	0.3
0	5	10	43.72	2.070	6	0.1
3	3	9	43.97	2.059	12	0.6
5	1	7	44.81	2.022	12	0.8
0	1	20	44.98	2.016	6	0.5
1	5	8	45.67	1.986	12	0.4
2	4	10	45.74	1.984	12	1.5
0	6	3	45.84	1.979	6	1.4
6	0	3	45.84	1.979	6	0.7
0	3	18	45.94	1.976	6	2.6
3	0	18	45.94	1.976	6	3.0
4	3	1	46.05	1.971	12	3.3
3	1	17	46.58	1.950	12	1.5
0	0	21	46.64	1.947	2	2.1
4	0	16	46.68	1.946	6	1.5
4	3	4	46.89	1.938	12	1.0
4	2	11	46.92	1.936	12	3.1
5	2	0	47.29	1.922	12	0.5
6	0	6	47.36	1.920	6	2.7
0	6	6	47.36	1.920	6	0.6
3	4	5	47.39	1.918	12	0.1
3	3	12	47.55	1.912	12	0.6
2	5	3	47.79	1.903	12	0.3
5	2	3	47.79	1.903	12	0.1
4	1	15	48.18	1.889	12	0.1
1	4	15	48.18	1.889	12	0.3
0	4	17	48.50	1.877	6	0.6
1	5	11	48.83	1.865	12	0.3
1	2	20	48.86	1.864	12	1.1
2	5	6	49.26	1.850	12	1.2
5	2	6	49.26	1.850	12	0.2
3	4	8	49.51	1.841	12	1.0
2	4	13	49.54	1.840	12	4.0
6	0	9	49.80	1.831	6	0.1
0	6	9	49.80	1.831	6	2.0
1	6	1	49.87	1.829	12	2.0

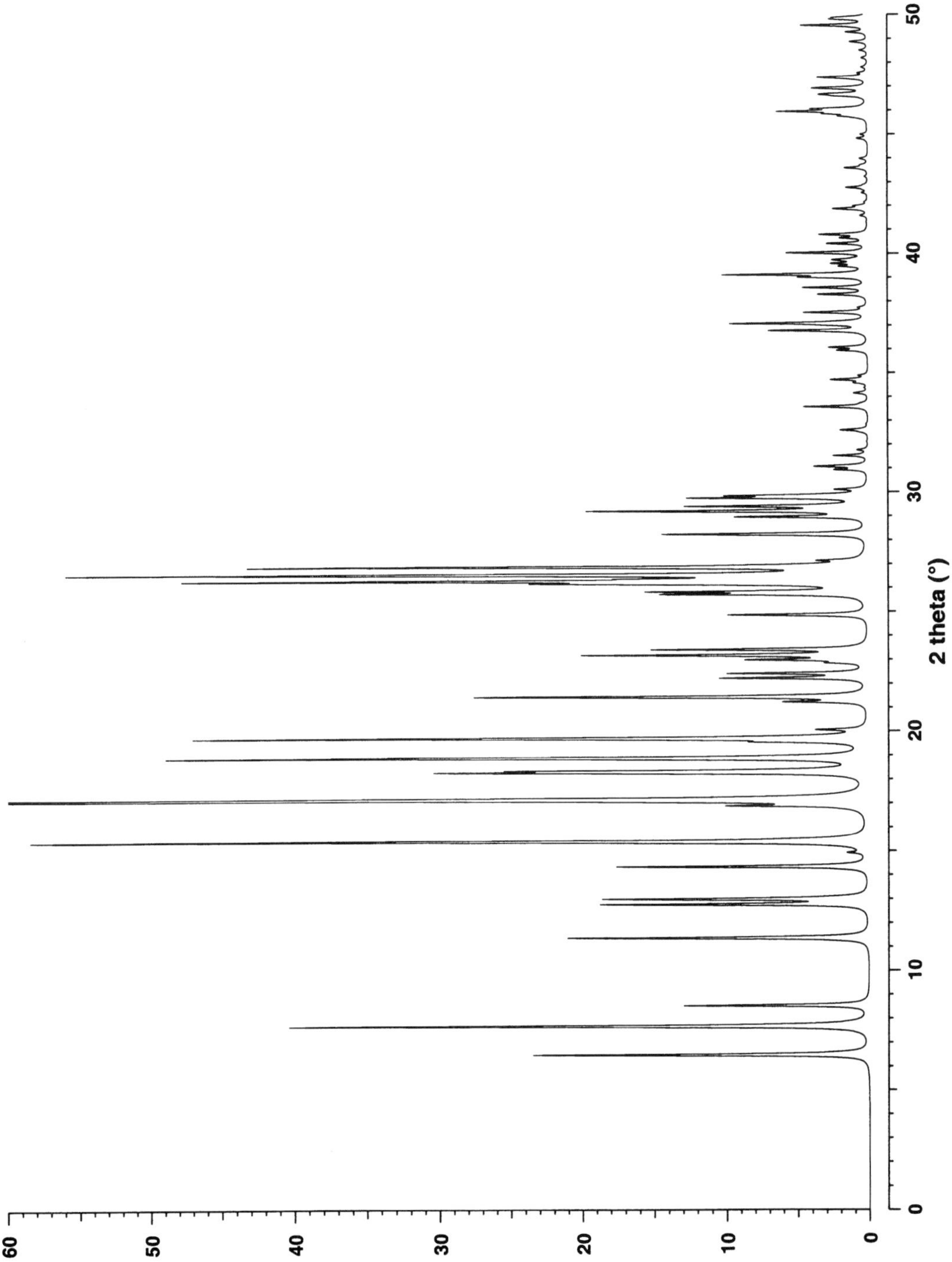
2 theta (°)
0
10
20
30
40
50
60

DFO Decamethonium DAF-1

CHEMICAL COMPOSITION: $|(C_{16}H_{38}N_2)_{7.92}(H_2O)_{40.26}|$ $[Mg_{14.52}Al_{51.48}P_{66}O_{264}]$
$C_{16}H_{38}N_2^{2+}$ = decamethonium ion

REFINED COMPOSITION: $|C_{23}|$ $[Si_{132}O_{264}]$

CRYSTAL DATA: $P6/mmm$ (No. 191)
a = 22.351 Å b = 22.351 Å c = 21.693 Å
α = 90° β = 90° γ = 120°
X-ray single crystal refinement, R_w = 0.122

REFERENCE: P. A. Wright, R. H. Jones, S. Natarajan, R. G. Bell, J. Chen, M. B. Hursthouse and J. M. Thomas, *J. Chem. Soc., Chem. Comm.* 633–635 (1993).

h	k	l	2θ	d	M	I_{rel}
0	0	1	4.07	21.693	2	6.3
1	0	0	4.56	19.357	6	39.6
1	0	1	6.12	14.443	12	18.5
1	1	0	7.91	11.175	6	100.0
0	0	2	8.15	10.846	2	39.2
1	1	1	8.90	9.935	12	4.2
2	0	0	9.14	9.678	6	0.2
1	0	2	9.35	9.462	12	0.2
2	0	1	10.01	8.839	12	20.1
1	1	2	11.37	7.783	12	1.9
2	1	0	12.10	7.316	12	4.5
2	0	2	12.26	7.221	12	2.6
1	0	3	13.07	6.774	12	2.2
3	0	0	13.72	6.452	6	0.7
3	0	1	14.32	6.184	12	1.6
2	1	2	14.60	6.065	24	0.4
2	0	3	15.30	5.793	12	5.0
2	2	0	15.86	5.588	6	0.8
3	0	2	15.98	5.545	12	0.3
0	0	4	16.34	5.423	2	0.9
2	2	1	16.38	5.411	12	4.0
3	1	0	16.51	5.369	12	0.6
1	0	4	16.98	5.222	12	2.8
3	1	1	17.01	5.211	24	0.8
2	1	3	17.24	5.143	24	0.3
1	1	4	18.18	4.879	12	0.2
4	0	0	18.33	4.839	6	1.5
2	0	4	18.76	4.731	12	0.5
4	0	1	18.79	4.723	12	0.2
2	2	3	20.08	4.421	12	0.2
2	1	4	20.38	4.357	24	8.1
3	2	1	20.41	4.350	24	1.2
0	0	5	20.47	4.339	2	0.6
1	0	5	20.98	4.234	12	0.4
4	1	0	21.03	4.224	12	12.3
4	1	1	21.43	4.146	24	1.8
4	0	3	22.10	4.022	12	1.8
2	0	5	22.46	3.959	12	6.3
4	1	2	22.59	3.936	24	2.0
2	2	4	22.85	3.892	12	0.3
5	0	0	22.97	3.871	6	3.4
3	1	4	23.31	3.815	24	2.1
2	1	5	23.84	3.732	24	5.6
3	3	0	23.89	3.725	6	3.5
3	3	1	24.24	3.671	12	0.2

h	k	l	2θ	d	M	I_{rel}
4	2	0	24.33	3.658	12	0.2
5	0	2	24.41	3.646	12	0.2
4	0	4	24.66	3.611	12	1.6
4	2	1	24.68	3.607	24	1.2
3	2	4	25.93	3.436	24	1.8
5	1	1	25.96	3.433	24	0.3
2	2	5	26.00	3.427	12	0.2
2	0	6	26.31	3.387	12	0.5
3	1	5	26.41	3.374	24	0.4
5	1	2	26.93	3.311	24	0.5
4	2	3	27.32	3.264	24	1.1
2	1	6	27.52	3.241	24	1.3
6	0	0	27.65	3.226	6	4.0
6	0	1	27.96	3.191	12	0.2
5	0	4	28.32	3.151	12	0.3
4	3	1	28.35	3.148	24	0.5
5	1	3	28.49	3.133	24	1.4
3	2	5	28.77	3.103	24	0.9
5	2	0	28.80	3.100	12	1.3
0	0	7	28.81	3.099	2	1.2
6	0	2	28.87	3.092	12	0.2
5	2	1	29.10	3.068	24	1.1
4	2	4	29.45	3.033	24	1.7
3	1	6	29.79	2.999	24	0.2
1	1	7	29.92	2.986	12	0.3
5	2	2	29.98	2.980	24	0.3
6	1	0	30.28	2.952	12	0.9
2	0	7	30.28	2.951	12	0.7
6	0	3	30.34	2.946	12	0.3
5	1	4	30.54	2.927	24	0.7
6	1	1	30.56	2.925	24	0.4
4	3	3	30.70	2.913	24	1.4
2	1	7	31.35	2.854	24	0.4
3	2	6	31.92	2.804	24	0.4
4	2	5	32.00	2.797	24	3.6
4	4	0	32.03	2.794	6	1.2
3	0	7	32.04	2.793	12	0.2
7	0	1	32.64	2.743	12	0.2
6	1	3	32.77	2.733	24	0.4
4	4	2	33.11	2.706	12	0.3
5	2	4	33.29	2.691	24	0.2
1	0	8	33.37	2.685	12	0.5
6	2	0	33.38	2.684	12	0.2
5	3	2	33.44	2.680	24	0.2
6	2	1	33.64	2.664	24	0.2

h	k	l	2θ	d	M	I_{rel}
2	0	8	34.34	2.611	12	0.6
6	2	2	34.42	2.606	24	0.2
6	1	4	34.60	2.593	24	0.3
5	3	3	34.73	2.583	24	0.8
7	0	3	34.73	2.583	12	1.0
4	2	6	34.89	2.571	24	0.5
4	3	5	34.97	2.566	24	0.5
7	1	0	35.00	2.564	12	0.2
7	1	1	35.25	2.546	24	0.2
3	2	7	35.32	2.541	24	0.2
4	1	7	35.94	2.499	24	0.3
7	0	4	36.47	2.463	12	0.6
6	1	5	36.83	2.441	24	0.8
6	3	0	36.86	2.439	12	0.5
8	0	0	37.16	2.420	6	0.2
7	1	3	37.21	2.416	24	0.3
6	2	4	37.38	2.406	24	1.1
8	0	1	37.40	2.405	12	0.2
7	2	0	38.05	2.365	12	0.4
7	0	5	38.61	2.332	12	0.2
3	2	8	38.91	2.314	24	0.3
6	3	3	38.98	2.311	24	0.2
5	5	0	40.35	2.235	6	0.4
7	0	6	41.09	2.196	12	0.2
3	3	8	41.18	2.192	12	0.3
7	3	0	41.46	2.178	12	0.2
9	0	0	42.01	2.151	6	0.5
5	1	8	42.27	2.138	24	0.3
6	4	3	42.59	2.123	24	0.3
2	0	10	42.71	2.117	12	0.2
8	0	5	42.79	2.113	12	0.4
8	2	1	43.03	2.102	24	0.3
9	1	0	44.66	2.029	12	0.2
7	3	4	44.85	2.021	24	0.6
7	4	0	45.17	2.007	12	0.6
6	1	8	45.42	1.997	24	0.5
1	0	11	46.27	1.962	12	0.2
9	1	3	46.48	1.954	24	0.2
7	3	5	46.67	1.946	24	0.3
6	2	8	47.67	1.908	24	0.5
2	1	11	47.76	1.904	24	0.7
3	1	11	49.22	1.851	24	0.2
8	3	4	49.32	1.848	24	0.3
9	2	3	49.42	1.844	24	0.3
8	4	0	49.86	1.829	12	0.2

Decamethonium DAF-1 DFO

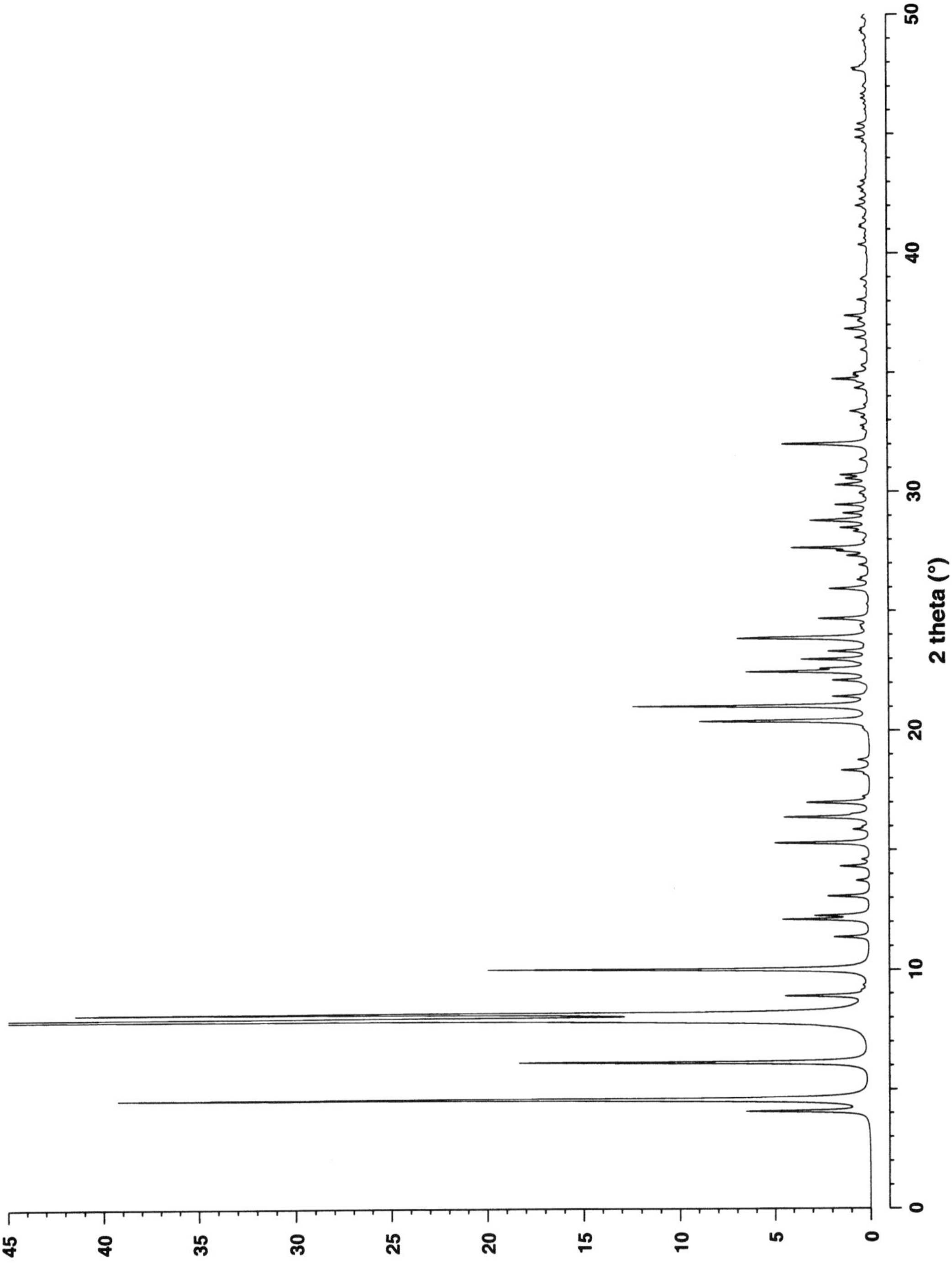

DFT Magnesium Phosphate UiO-20

CHEMICAL COMPOSITION: $|(NH_3(CH_2)_2NH_3)_{16}|$ $[Mg_{32}P_{32}O_{128}]$
$NH_2(CH_2)_2NH_2$ = ethylenediamine

REFINED COMPOSITION: $|C_{32}N_{32}|$ $[Mg_{32}P_{32}O_{128}]$

CRYSTAL DATA: $C1\,2/c\,1$ (No. 15) unique axis **b**, cell choice 1
$a = 20.9098$ Å $b = 17.8855$ Å $c = 14.7913$ Å
$\alpha = 90.0°$ $\beta = 134.842°$ $\gamma = 90.0°$
X-ray single crystal refinement, $R = 0.1007, R_w = 0.0994$

REFERENCE: K. O. Kongshaug, H. Fjellvåg and K. P. Lillerud, *Chemistry of Materials* **12** 1095–1099 (2000).

h	k	l	2θ	d	M	I_{rel}	h	k	l	2θ	d	M	I_{rel}	h	k	l	2θ	d	M	I_{rel}
-1	1	1	7.76	11.398	4	1.2	-7	1	5	33.02	2.713	4	0.8	-8	4	4	39.95	2.256	4	1.0
2	0	0	11.94	7.413	2	73.7	2	4	2	33.64	2.664	4	8.0	-5	3	6	40.02	2.253	4	1.7
-2	0	2	11.97	7.396	2	71.1	-2	4	4	33.67	2.662	4	13.6	-7	3	6	40.06	2.251	4	0.8
2	2	0	15.53	5.707	4	30.3	-6	4	2	33.69	2.660	4	12.5	0	8	0	40.34	2.236	2	12.3
-2	2	2	15.55	5.699	4	28.5	-6	4	4	33.72	2.658	4	6.5	2	6	2	40.69	2.217	4	11.5
0	0	2	16.91	5.244	2	19.5	0	0	4	34.20	2.622	2	13.8	-6	6	2	40.74	2.215	4	0.6
-4	0	2	16.96	5.227	2	21.7	-8	0	4	34.31	2.614	2	11.1	-6	6	4	40.76	2.214	4	10.7
0	2	2	19.62	4.524	4	1.2	0	6	2	34.61	2.591	4	6.9	2	4	3	40.80	2.212	4	0.8
-4	2	2	19.67	4.513	4	1.4	-4	6	2	34.64	2.589	4	6.6	-8	4	5	40.93	2.205	4	0.9
0	4	0	19.86	4.471	2	97.4	3	5	1	35.38	2.537	4	2.1	0	8	1	41.29	2.187	4	1.4
2	2	1	21.39	4.155	4	17.9	-5	5	1	35.41	2.535	4	1.5	-2	8	1	41.29	2.186	4	1.3
-2	2	3	21.42	4.149	4	29.7	-5	5	4	35.45	2.532	4	1.5	3	5	2	41.36	2.183	4	0.6
-4	2	1	21.42	4.149	4	27.2	-1	7	1	35.66	2.518	4	0.5	-7	5	5	41.47	2.177	4	0.8
-4	2	3	21.45	4.142	4	21.1	0	2	4	35.68	2.516	4	2.6	6	4	0	41.76	2.163	4	2.4
0	4	1	21.60	4.113	4	1.7	2	6	1	35.68	2.516	4	6.5	-6	4	6	41.84	2.159	4	3.5
-2	4	1	21.62	4.111	4	1.2	-4	6	1	35.70	2.515	4	0.9	-9	3	5	41.97	2.153	4	0.5
-5	1	3	22.25	3.996	4	0.5	-2	6	3	35.70	2.515	4	0.9	4	6	1	43.46	2.082	4	0.5
2	4	0	23.23	3.829	4	75.2	-4	6	3	35.72	2.513	4	7.3	4	4	2	43.57	2.077	4	3.5
-2	4	2	23.25	3.826	4	74.9	-8	2	4	35.79	2.509	4	2.6	-4	4	6	43.63	2.074	4	5.3
4	2	0	26.02	3.424	4	100.0	6	0	0	36.36	2.471	2	0.8	-8	4	2	43.63	2.074	4	4.8
-4	2	4	26.08	3.417	4	97.4	-6	0	6	36.45	2.465	2	0.9	-8	4	6	43.70	2.071	4	2.8
0	4	2	26.19	3.402	4	30.5	-2	2	5	36.76	2.445	4	0.5	0	8	2	44.03	2.057	4	3.0
-4	4	2	26.23	3.398	4	32.6	4	4	1	36.86	2.438	4	0.8	-2	0	6	44.05	2.056	2	0.8
1	3	2	26.32	3.387	4	1.2	-4	4	5	36.92	2.435	4	0.7	-4	8	2	44.05	2.056	4	2.9
-1	3	3	26.33	3.385	4	1.7	-6	4	5	36.95	2.433	4	0.9	-10	0	4	44.14	2.052	2	0.8
-5	3	2	26.37	3.380	4	1.3	5	1	1	37.21	2.416	4	1.2	0	2	5	44.36	2.042	4	0.6
-5	3	3	26.38	3.378	4	2.1	-7	1	1	37.25	2.414	4	2.2	6	2	1	44.36	2.042	4	3.8
2	0	2	26.87	3.318	2	83.4	-5	1	6	37.30	2.411	4	1.1	-6	2	7	44.46	2.038	4	0.6
-2	0	4	26.91	3.313	2	21.1	1	7	1	37.73	2.384	4	0.7	-8	2	7	44.50	2.036	4	3.1
-6	0	2	26.94	3.309	2	20.6	6	2	0	37.77	2.382	4	1.1	-2	2	6	45.26	2.003	4	4.0
-6	0	4	26.98	3.305	2	82.8	-6	2	6	37.86	2.377	4	1.5	-10	2	4	45.35	2.000	4	4.8
2	4	1	27.56	3.237	4	0.7	4	0	2	38.37	2.346	2	5.6	0	6	4	46.10	1.969	4	0.6
-4	4	3	27.61	3.231	4	0.7	-4	0	6	38.45	2.341	2	3.3	-8	6	4	46.19	1.965	4	0.6
-1	5	2	28.35	3.148	4	0.5	-8	0	2	38.45	2.341	2	3.5	1	5	4	46.71	1.945	4	0.9
2	2	2	28.70	3.111	4	0.5	-8	0	6	38.52	2.337	2	4.6	-9	5	5	46.84	1.940	4	0.8
-2	2	4	28.73	3.107	4	1.1	4	6	0	38.76	2.323	4	7.4	6	6	0	47.81	1.902	4	2.7
-6	2	2	28.76	3.104	4	0.9	-4	6	4	38.80	2.321	4	6.1	-6	6	6	47.88	1.900	4	2.7
3	3	1	28.98	3.081	4	0.7	-1	1	5	39.23	2.297	4	1.1	-10	4	5	48.03	1.894	4	0.5
-3	5	3	30.88	2.895	4	0.6	-9	1	5	39.35	2.290	4	1.9	8	0	0	49.16	1.853	2	14.1
4	2	1	32.36	2.766	4	0.7	-7	5	4	39.55	2.279	4	0.8	-8	0	8	49.28	1.849	2	16.6
2	6	0	32.37	2.766	4	18.4	4	2	2	39.72	2.269	4	4.3	4	6	2	49.44	1.843	4	3.0
-2	6	2	32.38	2.765	4	18.7	0	6	3	39.74	2.268	4	1.2	-4	6	6	49.50	1.841	4	0.6
3	1	2	32.89	2.723	4	0.5	-6	6	3	39.79	2.265	4	1.2	-8	6	2	49.50	1.841	4	1.0
-3	1	5	32.95	2.719	4	1.0	-8	2	6	39.87	2.261	4	4.2	-8	6	6	49.56	1.839	4	3.5
-7	1	2	32.96	2.717	4	1.0	5	3	1	39.94	2.257	4	1.4	-7	1	8	49.95	1.826	4	0.5

Magnesium Phosphate UiO-20 DFT

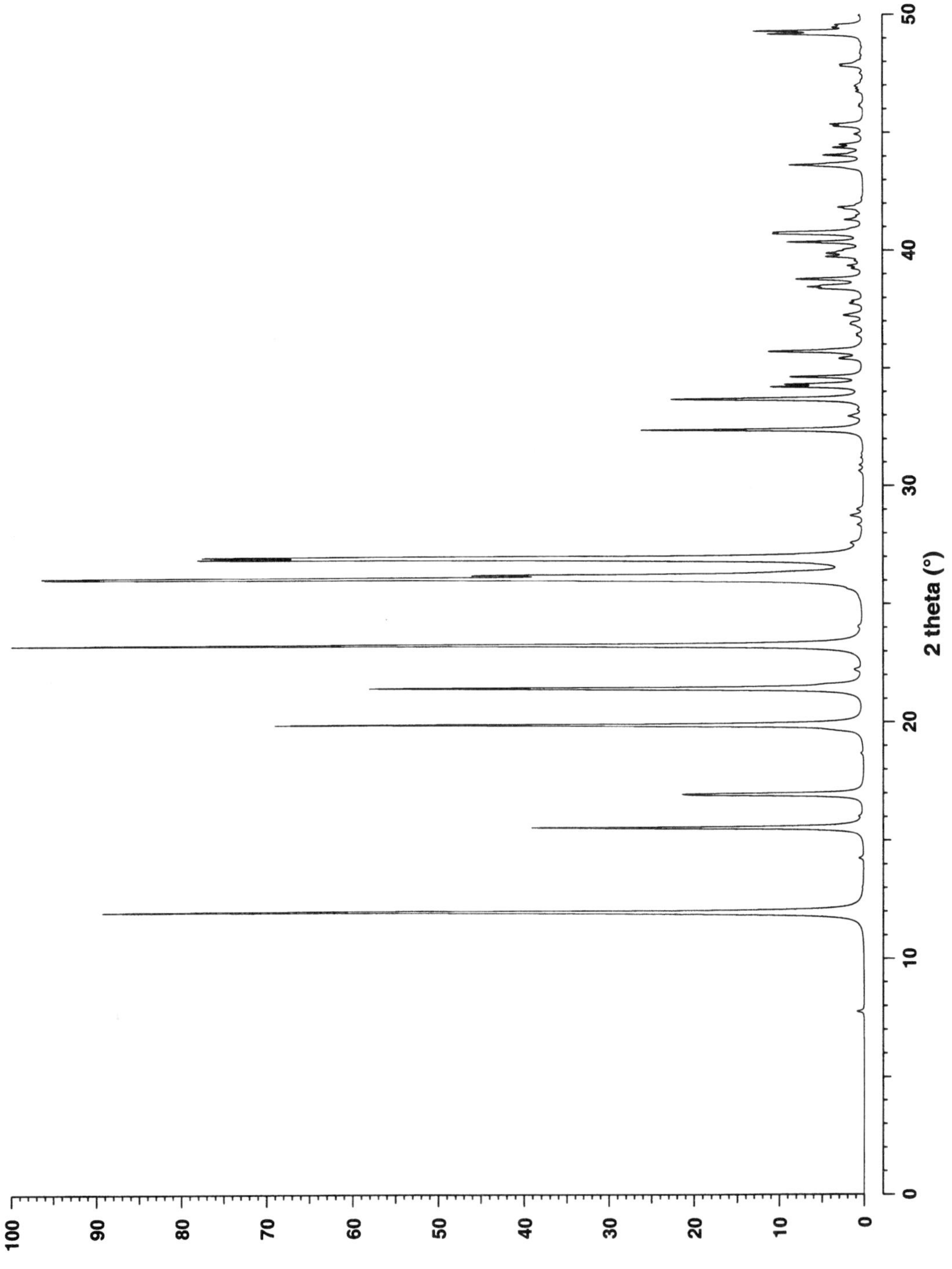

DOH N$_2$, Piperidine Dodecasil 1H

CHEMICAL COMPOSITION: $|(N_2)_3(N_2)_2C_5H_{10}NH|$ $[Si_{34}O_{68}]$
$C_5H_{10}NH$ = piperidine

REFINED COMPOSITION: $|C_{15.4}|$ $[Si_{34}O_{68}]$

CRYSTAL DATA: $P6/mmm$ (No. 191)
$a = 13.783$ Å $b = 13.783$ Å $c = 11.190$ Å
$\alpha = 90°$ $\beta = 90°$ $\gamma = 120°$
X-ray single crystal refinement, $R_w = 0.077$

REFERENCE: H. Gerke and H. Gies,
Z. Kristallogr. **166** 11–22 (1984).

h	k	l	2θ	d	M	I_{rel}
1	0	0	7.41	11.936	6	37.1
0	0	1	7.90	11.190	2	5.9
1	0	1	10.84	8.164	12	14.3
1	1	0	12.85	6.891	6	15.3
2	0	0	14.84	5.968	6	14.4
1	1	1	15.10	5.868	12	35.6
0	0	2	15.84	5.595	2	15.3
2	0	1	16.84	5.266	12	100.0
1	0	2	17.51	5.066	12	33.5
2	1	0	19.68	4.512	12	16.9
1	1	2	20.45	4.344	12	0.4
2	1	1	21.23	4.184	24	11.5
2	0	2	21.77	4.082	12	3.0
3	0	0	22.34	3.979	6	8.0
3	0	1	23.73	3.749	12	33.1
0	0	3	23.86	3.730	2	6.2
1	0	3	25.01	3.560	12	6.0
2	1	2	25.36	3.512	24	48.3
2	2	0	25.86	3.446	6	18.8
3	1	0	26.93	3.311	12	25.1
2	2	1	27.08	3.293	12	16.2
1	1	3	27.18	3.280	12	20.7
3	0	2	27.51	3.243	12	0.2
3	1	1	28.11	3.175	24	1.2
2	0	3	28.21	3.163	12	15.7
4	0	0	29.94	2.984	6	1.6
2	2	2	30.47	2.934	12	0.1
2	1	3	31.11	2.875	24	2.9
3	1	2	31.40	2.849	24	4.8
3	2	0	32.70	2.738	12	1.4
1	0	4	32.88	2.724	12	2.2
3	0	3	32.91	2.721	12	0.9
3	2	1	33.69	2.660	24	0.1
4	0	2	34.05	2.633	12	1.3
1	1	4	34.60	2.592	12	0.1
4	1	1	35.38	2.537	24	2.6
2	0	4	35.44	2.533	12	4.7
2	2	3	35.47	2.531	12	1.0
3	2	2	36.53	2.460	24	9.1
5	0	0	37.68	2.387	6	6.6
2	1	4	37.84	2.378	24	1.7
5	0	1	38.56	2.335	12	2.6
4	0	3	38.64	2.330	12	1.8
3	3	0	39.22	2.297	6	4.7
3	0	4	39.37	2.288	12	8.0
4	2	0	39.97	2.256	12	0.7
3	3	1	40.07	2.250	12	0.3
0	0	5	40.30	2.238	2	0.4
4	2	1	40.81	2.211	24	0.9
3	2	3	40.88	2.207	24	4.4
5	0	2	41.11	2.196	12	1.1
2	2	4	41.58	2.172	12	1.0
5	1	0	42.15	2.144	12	0.1
3	1	4	42.30	2.137	24	0.1
4	1	3	42.32	2.136	24	0.1
1	1	5	42.47	2.129	12	0.8
2	0	5	43.17	2.096	12	0.1
4	2	2	43.24	2.092	24	0.3
4	0	4	44.39	2.041	12	0.6
5	0	3	45.09	2.011	12	0.1
2	1	5	45.23	2.005	24	0.1
5	1	2	45.30	2.002	24	0.2
6	0	0	45.60	1.989	6	0.3
4	3	0	46.26	1.962	12	1.0
6	0	1	46.35	1.959	12	0.6
3	2	4	46.40	1.957	24	0.5
3	3	3	46.42	1.956	12	0.8
3	0	5	46.56	1.951	12	1.3
4	3	1	47.01	1.933	24	0.2
4	2	3	47.08	1.930	24	4.0
5	2	0	47.57	1.911	12	0.1
4	1	4	47.71	1.906	24	0.2
5	2	1	48.31	1.884	24	0.3
6	0	2	48.57	1.874	12	2.9
0	0	6	48.83	1.865	2	1.3
3	1	5	49.14	1.854	24	0.6
4	3	2	49.20	1.852	24	0.6

N_2, Piperidine Dodecasil 1H DOH

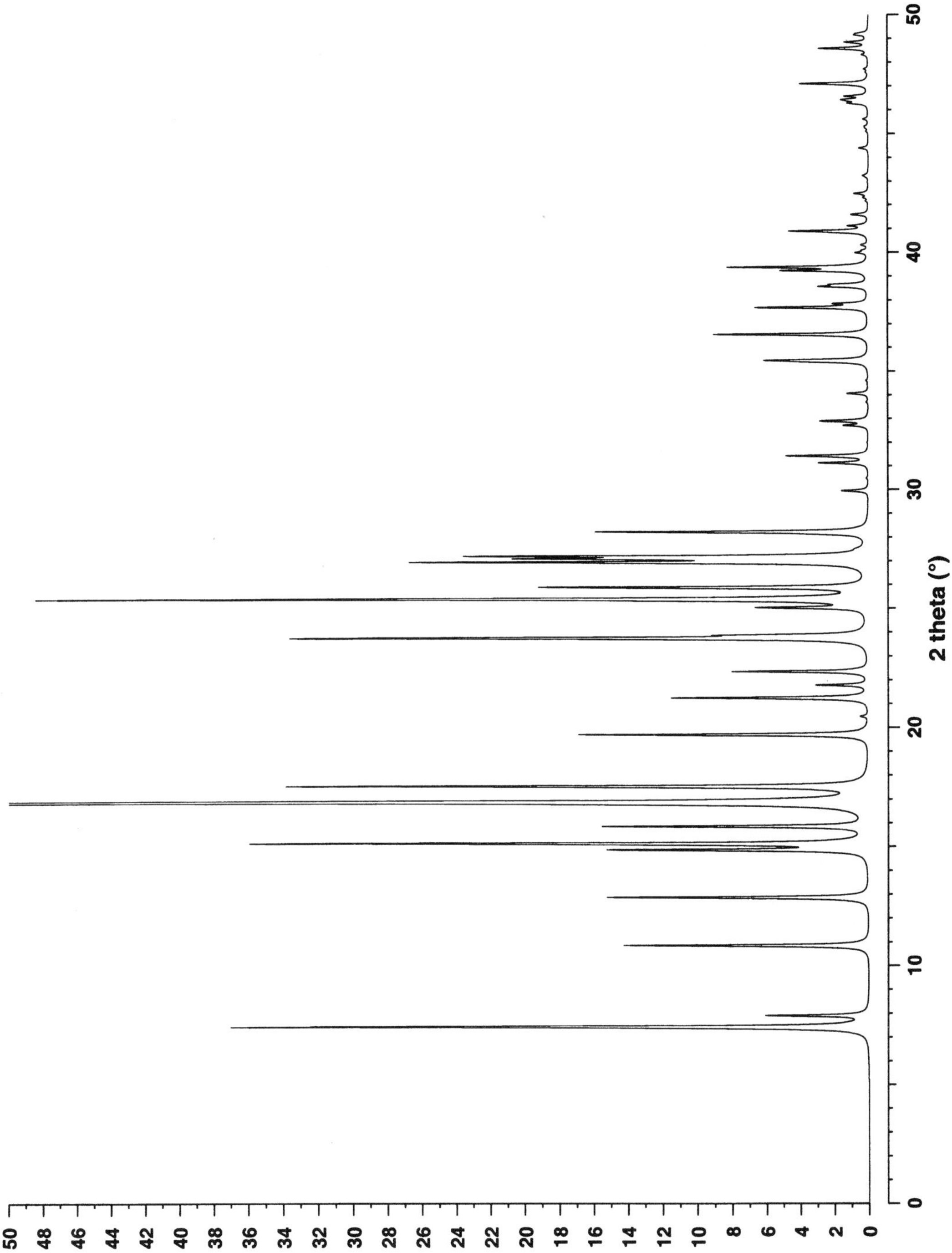

DON UTD-1

CHEMICAL COMPOSITION: $|((Cp^*)_2CoF_{0.075}OH_{0.25})_2|$ $[Si_{64}O_{128}]$
$(Cp^*)_2CoOH$ = bis(pentamethylcyclopentadienyl)cobalt(III) hydroxide

REFINED COMPOSITION: $|Co_2C_{35}|$ $[Si_{64}O_{128}]$

CRYSTAL DATA: *Pc* (No. 7) unique axis **b**, cell choice 1
$a = 14.9701$ Å $b = 8.4761$ Å $c = 30.0278$ Å
$\alpha = 90°$ $\beta = 102.65°$ $\gamma = 90°$
X-ray Rietveld refinement, $R_F = 0.041$, $R_{wp} = 0.134$

REFERENCE: T. Wessels, Ch. Baerlocher, L. B. McCusker and E. J. Creyghton, *J. Am. Chem. Soc.* **121** 6242–6247 (1999).

h	*k*	*l*	2*θ*	*d*	*M*	I_{rel}
0	0	2	6.03	14.649	2	62.1
1	0	0	6.05	14.607	2	65.5
-1	0	2	7.55	11.704	2	77.3
0	1	1	10.87	8.142	4	19.3
0	1	2	12.06	7.337	4	3.0
1	1	0	12.07	7.331	4	2.4
0	0	4	12.08	7.325	2	10.1
-1	1	1	12.12	7.303	4	12.5
2	0	0	12.12	7.303	2	9.2
-1	0	4	12.27	7.211	2	5.9
-2	0	2	12.30	7.196	2	4.3
1	1	1	12.77	6.935	4	8.1
-1	1	2	12.90	6.865	4	2.1
0	1	3	13.83	6.401	4	6.1
-1	1	3	14.29	6.199	4	5.6
1	0	4	14.67	6.039	2	22.3
2	0	2	14.69	6.030	2	22.5
-2	0	4	15.14	5.852	2	7.9
-2	1	1	15.80	5.609	4	3.7
1	1	3	15.89	5.576	4	3.2
2	1	1	16.79	5.279	4	2.2
-2	1	3	17.06	5.198	4	2.5
-1	0	6	17.85	4.970	2	10.6
-3	0	2	17.89	4.957	2	11.4
1	1	4	18.04	4.918	4	5.9
2	1	2	18.05	4.914	4	8.3
0	0	6	18.17	4.883	2	24.5
3	0	0	18.22	4.869	2	27.6
-1	1	5	18.31	4.844	4	2.1
-2	1	4	18.42	4.816	4	7.2
2	0	4	18.94	4.684	2	1.5
-2	0	6	19.53	4.545	2	23.5
-3	0	4	19.56	4.539	2	20.0
2	1	3	19.71	4.504	4	1.4
-2	1	5	20.16	4.404	4	1.3
1	0	6	20.40	4.353	2	3.5
1	1	5	20.42	4.350	4	1.8
3	0	2	20.44	4.344	2	2.3
-1	1	6	20.72	4.287	4	6.8
-3	1	2	20.76	4.279	4	9.3
0	2	0	20.96	4.238	2	100.0
0	1	6	21.00	4.231	4	32.0
3	1	0	21.04	4.222	4	32.2
0	2	2	21.83	4.071	4	22.4
3	1	1	21.83	4.071	4	1.2
1	2	0	21.84	4.070	4	19.6
-2	1	6	22.19	4.006	4	3.6
-3	1	4	22.22	4.001	4	2.4
-1	2	2	22.31	3.985	4	25.4
-3	0	6	22.79	3.901	2	8.9
1	1	6	22.97	3.872	4	16.8
3	1	2	23.00	3.866	4	14.0
1	2	2	23.03	3.861	4	5.4
-1	0	8	23.72	3.751	2	1.4
2	0	6	24.04	3.702	2	1.1
3	0	4	24.06	3.699	2	2.9
0	0	8	24.30	3.662	2	15.4
4	0	0	24.37	3.652	2	20.0
-2	0	8	24.69	3.605	2	14.2
-4	0	4	24.75	3.598	2	12.6
-3	1	6	25.13	3.544	4	17.4
1	2	4	25.68	3.469	4	2.3
2	2	2	25.69	3.467	4	2.4
1	0	8	26.35	3.382	2	12.2
4	0	2	26.41	3.374	2	8.9
-3	0	8	27.07	3.294	2	1.7
-1	2	6	27.66	3.225	4	2.2
-3	2	2	27.69	3.221	4	2.0
0	2	6	27.87	3.201	4	4.1
3	2	0	27.91	3.197	4	4.3
1	1	8	28.41	3.141	4	4.1
4	1	2	28.47	3.135	4	4.3
3	0	6	28.58	3.123	2	11.7
-2	2	6	28.80	3.100	4	5.0
-3	2	4	28.82	3.098	4	5.8
-1	0	10	29.76	3.002	2	11.8
-5	0	2	29.85	2.993	2	5.9
2	1	8	31.45	2.844	4	2.3
4	1	4	31.49	2.841	4	1.4
-3	0	10	32.02	2.795	2	1.9
-5	0	6	32.07	2.791	2	3.5
0	2	8	32.31	2.771	4	1.8
0	1	10	32.33	2.769	4	1.5
4	2	0	32.36	2.766	4	1.8
-4	1	8	32.37	2.766	4	2.8
5	1	0	32.42	2.762	4	1.1
1	0	10	32.46	2.759	2	3.9
5	0	2	32.54	2.752	2	7.2
-2	2	8	32.61	2.746	4	1.3
-4	2	4	32.65	2.743	4	1.6
-3	1	10	33.76	2.655	4	1.1
-5	1	6	33.82	2.651	4	1.3
1	2	8	33.91	2.643	4	2.5
4	2	2	33.96	2.640	4	2.8
-4	0	10	34.81	2.577	2	1.6
1	3	4	35.06	2.559	4	3.2
-2	3	4	35.27	2.544	4	4.9
5	0	4	35.49	2.530	2	1.7
3	2	6	35.71	2.514	4	6.0
-6	0	2	36.05	2.491	2	2.0
-2	0	12	36.14	2.485	2	3.1
-6	0	4	36.25	2.478	2	1.2
-1	3	6	36.58	2.456	4	2.6
-1	2	10	36.68	2.450	4	1.9
0	3	6	36.75	2.446	4	10.9
-5	2	2	36.76	2.445	4	1.8
3	3	0	36.78	2.444	4	8.2
-3	0	12	37.41	2.404	2	1.2
-3	3	4	37.49	2.399	4	2.7
-6	0	6	37.50	2.399	2	2.0
1	3	6	37.96	2.370	4	1.8
3	3	2	37.99	2.369	4	6.3
4	0	8	38.43	2.342	2	2.7
6	1	0	38.47	2.340	4	1.0
-3	2	10	38.58	2.333	4	1.6
-5	2	6	38.63	2.331	4	1.3
1	2	10	38.96	2.312	4	1.3
-6	0	8	39.72	2.269	2	1.7
3	3	4	40.16	2.245	4	2.0
-2	0	14	42.23	2.140	2	4.6
-7	0	4	42.36	2.134	2	2.9
0	4	0	42.67	2.119	2	9.8
-6	0	10	42.78	2.114	2	1.5
-5	1	12	44.13	2.052	4	1.1
-2	3	10	44.38	2.041	4	1.0
-4	3	8	44.58	2.033	4	1.3
7	1	0	44.73	2.026	4	1.6
3	0	12	45.03	2.013	2	3.7
6	0	6	45.10	2.010	2	2.3
-5	3	6	45.69	1.985	4	1.7
1	1	14	46.41	1.956	4	1.9
7	1	2	46.53	1.952	4	1.6
-2	2	14	47.61	1.910	4	1.0
-7	2	4	47.72	1.906	4	1.6
-6	1	12	47.85	1.901	4	1.2

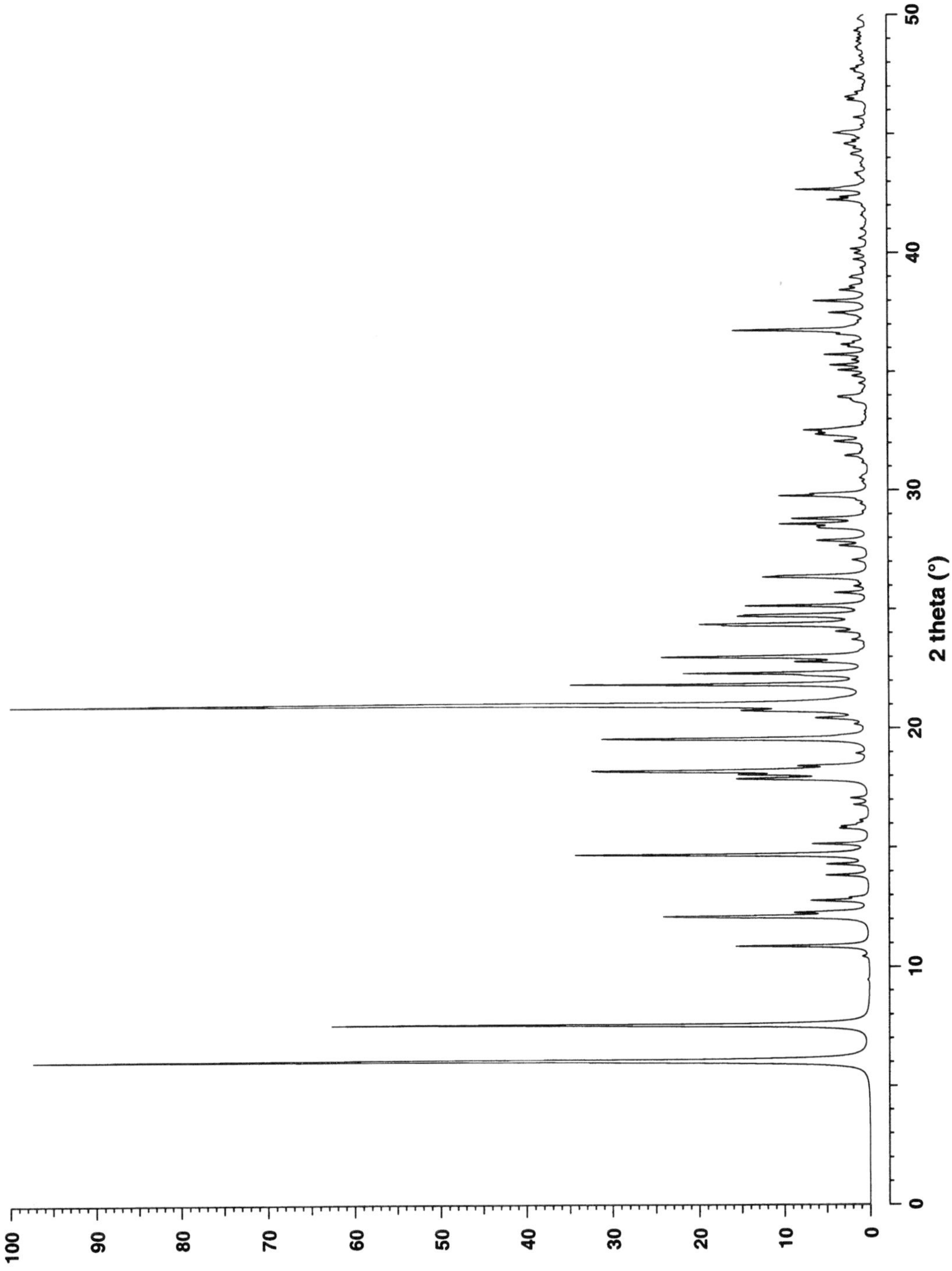
2 theta (°)
0
10
20
30
40
50
0
10
20
30
40
50
60
70
80
90
100

EAB Bellbergite

CHEMICAL COMPOSITION: $|Ba_{0.26}Na_{0.72}K_{1.33}Ca_{5.32}Sr_{2.36}(H_2O)_{30}|\ [Si_{18.36}Al_{17.55}O_{72}]$
Bellberg Volcano, Eifel, Germany

REFINED COMPOSITION: $|K_2Ca_6Sr_2(H_2O)_{30}|\ [Si_{18.45}Al_{17.55}O_{72}]$

CRYSTAL DATA: $P6_3/mmc$ (No. 194)
$a = 13.244$ Å $b = 13.244$ Å $c = 15.988$ Å
$\alpha = 90°$ $\beta = 90°$ $\gamma = 120°$
X-ray single crystal refinement, $R = 0.122$

REFERENCE: B. Rüdinger, E. Tillmanns and G. Hentschel, *Mineralogy and Petrology* **48** 147–152 (1993).

h	k	l	2θ	d	M	I_{rel}
1	0	0	7.71	11.470	6	16.4
1	0	1	9.49	9.320	12	42.6
0	0	2	11.07	7.994	2	5.0
1	1	0	13.37	6.622	6	0.6
1	0	2	13.50	6.558	12	100.0
2	0	0	15.45	5.735	6	2.5
2	0	1	16.42	5.398	12	13.6
1	1	2	17.39	5.100	12	1.0
1	0	3	18.36	4.833	12	10.5
2	0	2	19.05	4.660	12	8.1
2	1	0	20.49	4.335	12	0.3
2	1	1	21.23	4.184	24	38.6
0	0	4	22.24	3.997	2	8.0
3	0	0	23.27	3.823	6	78.1
2	1	2	23.34	3.811	24	23.9
1	0	4	23.57	3.774	12	59.8
3	0	2	25.83	3.449	12	5.1
1	1	4	26.04	3.422	12	0.4
2	1	3	26.50	3.363	24	0.5
2	2	0	26.93	3.311	6	1.1
2	0	4	27.19	3.279	12	1.0
3	1	1	28.61	3.120	24	17.5
3	0	3	28.74	3.107	12	1.0
1	0	5	28.99	3.080	12	4.5
2	2	2	29.19	3.059	12	0.4
3	1	2	30.24	2.956	24	13.4
2	1	4	30.42	2.939	24	21.1
4	0	0	31.19	2.867	6	0.6
4	0	1	31.70	2.822	12	35.0
2	0	5	32.05	2.793	12	6.5

h	k	l	2θ	d	M	I_{rel}
3	0	4	32.40	2.763	12	3.1
3	1	3	32.79	2.731	24	0.3
4	0	2	33.19	2.699	12	20.1
0	0	6	33.63	2.665	2	6.3
3	2	0	34.07	2.631	12	0.6
3	2	1	34.54	2.596	24	1.4
1	0	6	34.56	2.596	12	3.8
2	1	5	34.86	2.573	24	27.6
2	2	4	35.20	2.550	12	2.8
4	0	3	35.55	2.525	12	0.2
4	1	0	35.88	2.503	12	19.1
3	1	4	36.08	2.489	24	17.7
4	1	1	36.33	2.473	24	3.8
1	1	6	36.34	2.472	12	0.6
3	0	5	36.64	2.453	12	5.9
2	0	6	37.21	2.417	12	1.2
4	1	2	37.66	2.389	24	1.3
3	2	3	38.14	2.359	24	2.7
4	0	4	38.64	2.330	12	2.4
2	1	6	39.70	2.270	24	7.0
3	1	5	39.98	2.255	24	0.9
1	0	7	40.26	2.240	12	0.4
3	3	0	40.88	2.207	6	34.8
5	0	2	40.93	2.205	12	0.3
3	2	4	41.07	2.198	24	4.5
3	0	6	41.30	2.186	12	10.2
4	2	0	41.67	2.168	12	0.6
4	2	1	42.07	2.148	24	1.1
4	0	5	42.34	2.135	12	1.3
3	3	2	42.48	2.128	12	1.1

h	k	l	2θ	d	M	I_{rel}
2	0	7	42.61	2.122	12	0.1
4	1	4	42.62	2.121	24	4.2
5	0	3	42.92	2.107	12	0.3
5	1	0	43.95	2.060	12	0.7
5	1	1	44.33	2.043	24	0.3
3	1	6	44.34	2.043	24	0.4
3	2	5	44.59	2.032	24	3.3
2	1	7	44.85	2.021	24	0.5
4	2	3	45.16	2.008	24	0.5
0	0	8	45.38	1.998	2	7.4
5	1	2	45.47	1.995	24	1.5
5	0	4	45.59	1.990	12	4.7
1	0	8	46.10	1.969	12	1.1
3	0	7	46.30	1.961	12	0.5
4	0	6	46.52	1.952	12	3.2
3	3	4	47.03	1.932	12	2.1
1	1	8	47.52	1.913	12	0.2
6	0	0	47.57	1.912	6	0.2
4	2	4	47.73	1.905	24	1.7
6	0	1	47.93	1.898	12	0.2
4	3	1	48.62	1.873	24	1.5
3	2	6	48.63	1.872	24	2.6
6	0	2	48.99	1.859	12	0.5
5	2	0	49.64	1.837	12	0.4
4	3	2	49.68	1.835	24	5.7
5	1	4	49.80	1.831	24	0.2
5	2	1	49.99	1.825	24	0.6
4	1	6	49.99	1.824	24	13.2

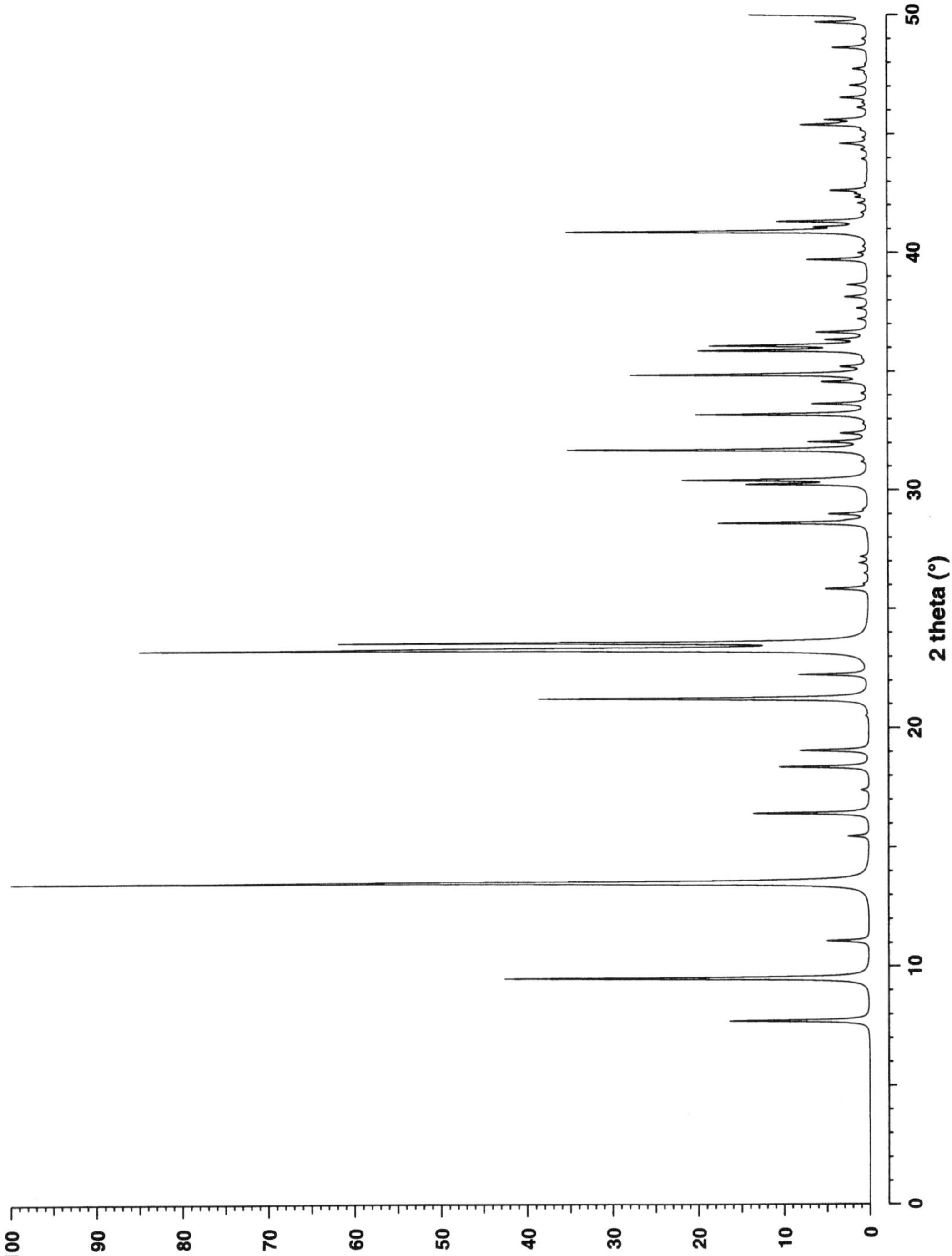

2 theta (°)
0
10
20
30
40
50
0
10
20
30
40
50
60
70
80
90
100

EAB (Na, Tetramethylammonium)-E

CHEMICAL COMPOSITION: $|Na_{8.1}((CH_3)_4N)_{2.2}(H_2O)_{25}|$ $[Si_{26.6}Al_{9.4}O_{72}]$
$(CH_3)_4N$ = tetramethylammonium

REFINED COMPOSITION: $|(H_2O)_{39.5}|$ $[Si_{26.6}Al_{9.4}O_{72}]$

CRYSTAL DATA: $P6_3/mmc$ (No. 194)
$a = 13.28$ Å $b = 13.28$ Å $c = 15.21$ Å
$\alpha = 90°$ $\beta = 90°$ $\gamma = 120°$
X-ray powder refinement, $R = 0.19$

REFERENCE: W. M. Meier and M. Groner, *J. Solid State Chem.* **37** 204–218 (1981).

h	k	l	2θ	d	M	I_{rel}
1	0	0	7.69	11.501	6	2.5
1	0	1	9.64	9.174	12	64.2
0	0	2	11.64	7.605	2	13.0
1	1	0	13.33	6.640	6	46.9
1	0	2	13.96	6.344	12	15.4
2	0	0	15.41	5.750	6	0.7
2	0	1	16.48	5.379	12	12.9
1	1	2	17.73	5.002	12	0.6
1	0	3	19.13	4.639	12	1.9
2	0	2	19.35	4.587	12	41.0
2	1	0	20.43	4.347	12	4.8
2	1	1	21.26	4.180	24	57.3
3	0	0	23.20	3.834	6	28.4
2	0	3	23.39	3.803	12	6.4
0	0	4	23.39	3.802	2	5.6
2	1	2	23.57	3.774	24	100.0
3	0	1	23.94	3.717	12	2.7
1	0	4	24.66	3.610	12	61.3
3	0	2	26.03	3.423	12	1.2
2	2	0	26.85	3.320	6	5.4
2	1	3	27.02	3.300	24	15.5
1	1	4	27.02	3.300	12	3.5
3	1	0	27.97	3.190	12	1.7
2	0	4	28.13	3.172	12	6.5
3	1	1	28.59	3.122	24	1.7
3	0	3	29.20	3.058	12	2.7
2	2	2	29.35	3.043	12	3.6
1	0	5	30.39	2.941	12	10.2
4	0	0	31.10	2.875	6	4.7
2	1	4	31.25	2.862	24	35.8
4	0	1	31.67	2.825	12	56.3
3	1	3	33.18	2.700	24	0.1
3	0	4	33.18	2.700	12	1.6
4	0	2	33.31	2.689	12	21.8
2	0	5	33.32	2.689	12	0.9
3	2	0	33.98	2.638	12	0.9
3	2	1	34.50	2.600	24	8.4
4	1	0	35.78	2.510	12	12.2
4	0	3	35.91	2.501	12	2.6
2	2	4	35.91	2.501	12	3.5
3	2	2	36.03	2.493	24	3.2
2	1	5	36.04	2.492	24	3.3
1	0	6	36.29	2.476	12	4.8
3	1	4	36.78	2.444	24	1.7
4	1	2	37.75	2.383	24	0.7
1	1	6	37.99	2.368	12	0.6
3	2	3	38.46	2.341	24	2.0
5	0	0	39.16	2.300	6	1.0
4	0	4	39.28	2.293	12	0.2
5	0	1	39.63	2.274	12	0.1
4	1	3	40.09	2.249	24	3.5
3	3	0	40.77	2.213	6	9.4
5	0	2	40.99	2.202	12	0.5
3	1	5	41.00	2.201	24	5.7
2	1	6	41.22	2.190	24	0.5
4	2	0	41.55	2.173	12	0.3
4	2	1	41.99	2.152	24	1.9
1	0	7	42.33	2.135	12	0.1
3	3	2	42.54	2.125	12	5.3
3	0	6	42.76	2.115	12	1.9
5	0	3	43.19	2.095	12	0.2
4	1	4	43.19	2.095	24	0.3
4	2	2	43.29	2.090	24	1.0
4	0	5	43.30	2.090	12	3.2
5	1	0	43.83	2.066	12	0.1
5	1	1	44.25	2.047	24	1.1
2	0	7	44.58	2.033	12	0.2
2	2	6	44.99	2.015	12	0.9
4	2	3	45.40	1.998	24	0.7
5	1	2	45.50	1.993	24	4.3
5	0	4	46.12	1.968	12	3.8
2	1	7	46.74	1.944	24	0.6
4	1	5	46.93	1.936	24	0.6
5	1	3	47.53	1.913	24	0.3
3	3	4	47.53	1.913	12	2.4
6	0	1	47.83	1.902	12	0.3
0	0	8	47.84	1.901	2	9.1
4	3	0	48.12	1.891	12	0.3
3	0	7	48.13	1.890	12	0.1
4	3	1	48.52	1.876	24	0.5
5	2	0	49.49	1.842	12	4.6
4	3	2	49.69	1.835	24	1.0
5	0	5	49.69	1.835	12	2.1
5	2	1	49.88	1.828	24	1.1
3	2	6	49.89	1.828	24	4.2
1	1	8	49.89	1.828	12	1.8

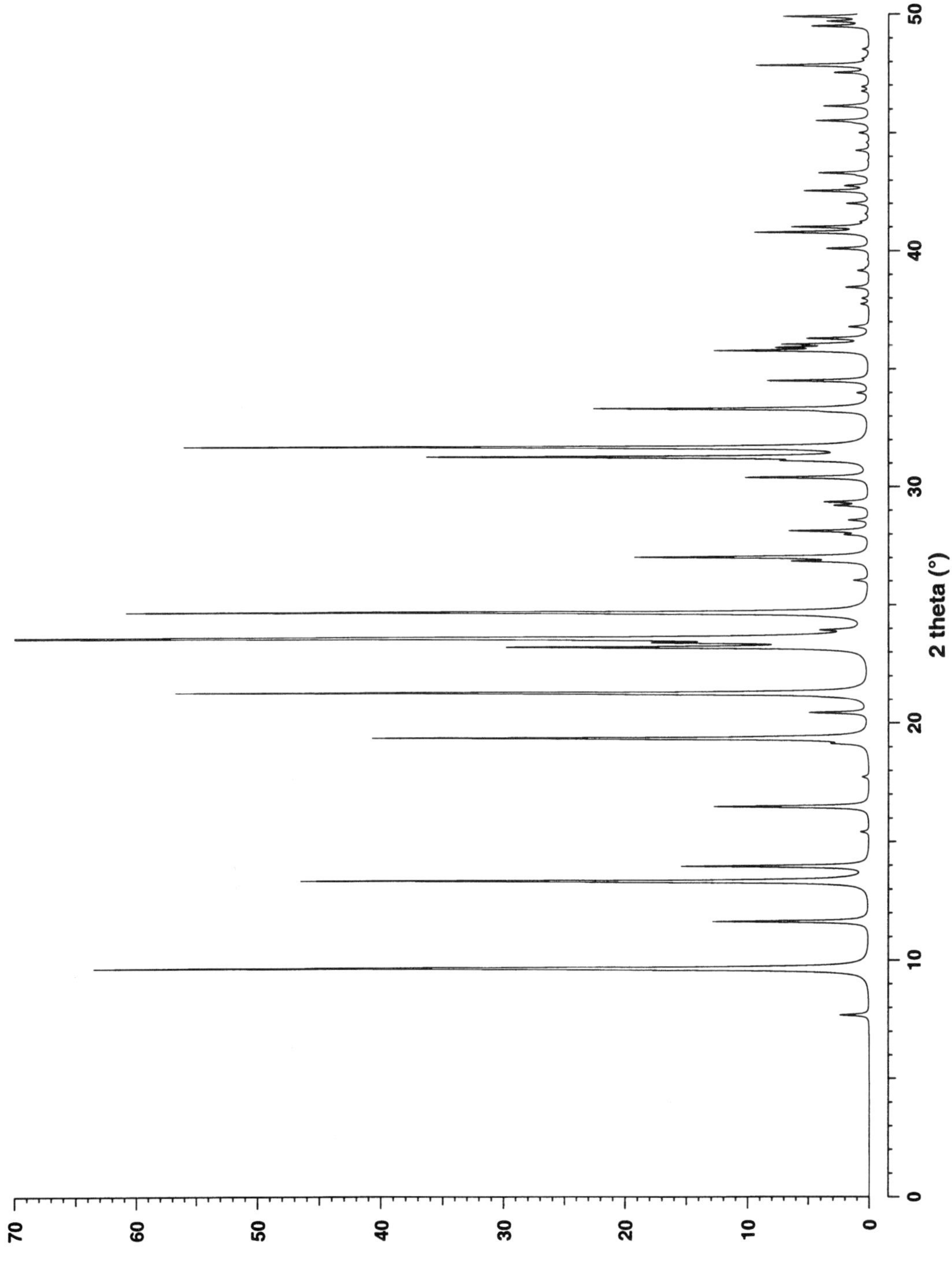
2 theta (°)
0
10
20
30
40
50
0
10
20
30
40
50
60
70

EDI Edingtonite

CHEMICAL COMPOSITION: $|Ba_2(H_2O)_7|$ $[Si_6Al_4O_{20}]$
Bathurst, New Brunswick, Canada

REFINED COMPOSITION: $|Ba_{1.95}(H_2O)_7|$ $[Si_6Al_4O_{20}]$

CRYSTAL DATA: $P2_12_12$ (No. 18)
$a = 9.537$ Å $b = 9.651$ Å $c = 6.509$ Å
$\alpha = 90°$ $\beta = 90°$ $\gamma = 90°$
Neutron single crystal refinement, $R_{wp} = 0.070$, $R_{F^2} = 0.055$

REFERENCE: Å. Kvick and J. V. Smith,
J. Chem. Phys. **79** 2356–2362 (1983).

h	k	l	2θ	d	M	I_{rel}
1	1	0	13.05	6.784	4	1.4
0	0	1	13.60	6.509	2	90.4
0	1	1	16.43	5.396	4	34.9
1	0	1	16.49	5.376	4	46.3
0	2	0	18.39	4.826	2	53.7
2	0	0	18.61	4.768	2	43.7
1	1	1	18.89	4.697	8	86.1
1	2	0	20.63	4.306	4	9.0
2	1	0	20.78	4.275	4	4.0
0	2	1	22.94	3.876	4	5.6
2	0	1	23.12	3.847	4	1.6
1	2	1	24.79	3.591	8	100.0
2	1	1	24.92	3.573	8	86.7
2	2	0	26.27	3.392	4	20.6
0	0	2	27.40	3.254	2	11.1
0	1	2	28.95	3.084	4	15.0
1	0	2	28.99	3.080	4	13.3
1	3	0	29.30	3.048	4	29.3
3	1	0	29.58	3.019	4	30.2
2	2	1	29.70	3.008	8	20.8
1	1	2	30.46	2.934	8	35.5
0	3	1	31.01	2.884	4	1.2
3	0	1	31.31	2.857	4	1.5
1	3	1	32.43	2.761	8	84.4
3	1	1	32.69	2.739	8	91.2
0	2	2	33.20	2.698	4	0.4
2	0	2	33.33	2.688	4	0.6
2	3	0	33.60	2.667	4	5.2
3	2	0	33.76	2.655	4	5.9
1	2	2	34.55	2.596	8	41.2
2	1	2	34.64	2.590	8	49.6
2	3	1	36.41	2.468	8	6.0
3	2	1	36.55	2.458	8	7.8
0	4	0	37.27	2.413	2	1.5
4	0	0	37.73	2.384	2	1.4
2	2	2	38.33	2.348	8	2.9
0	3	2	39.38	2.288	4	35.2
3	0	2	39.63	2.274	4	29.6
0	4	1	39.85	2.262	4	3.1
3	3	0	39.87	2.261	4	34.2
4	0	1	40.28	2.239	4	2.5
1	3	2	40.55	2.225	8	1.7
3	1	2	40.76	2.213	8	1.9
1	4	1	41.00	2.201	8	20.9
4	1	1	41.40	2.181	8	18.0
0	0	3	41.62	2.170	2	4.7
2	4	0	41.96	2.153	4	12.6
4	2	0	42.28	2.138	4	13.8
3	3	1	42.31	2.136	8	12.3
0	1	3	42.71	2.117	4	3.0
1	0	3	42.74	2.116	4	2.8
1	1	3	43.81	2.067	8	16.3
2	3	2	43.89	2.063	8	8.0
3	2	2	44.02	2.057	8	7.6
2	4	1	44.32	2.044	8	6.9
4	2	1	44.62	2.031	8	6.9
0	2	3	45.86	1.979	4	3.7
2	0	3	45.95	1.975	4	5.3
1	2	3	46.89	1.938	8	8.4
2	1	3	46.96	1.935	8	8.4
4	0	2	47.26	1.923	4	0.2
3	4	0	47.30	1.922	4	1.4
4	3	0	47.46	1.916	4	2.1
1	4	2	47.89	1.899	8	7.4
1	5	0	48.09	1.892	4	2.0
4	1	2	48.25	1.886	8	8.5
5	1	0	48.66	1.871	4	1.9
3	3	2	49.06	1.857	8	2.0
0	5	1	49.24	1.851	4	5.7
3	4	1	49.45	1.843	8	10.4
4	3	1	49.61	1.838	8	11.3
5	0	1	49.82	1.830	4	5.8
2	2	3	49.89	1.828	8	18.1

Edingtonite EDI

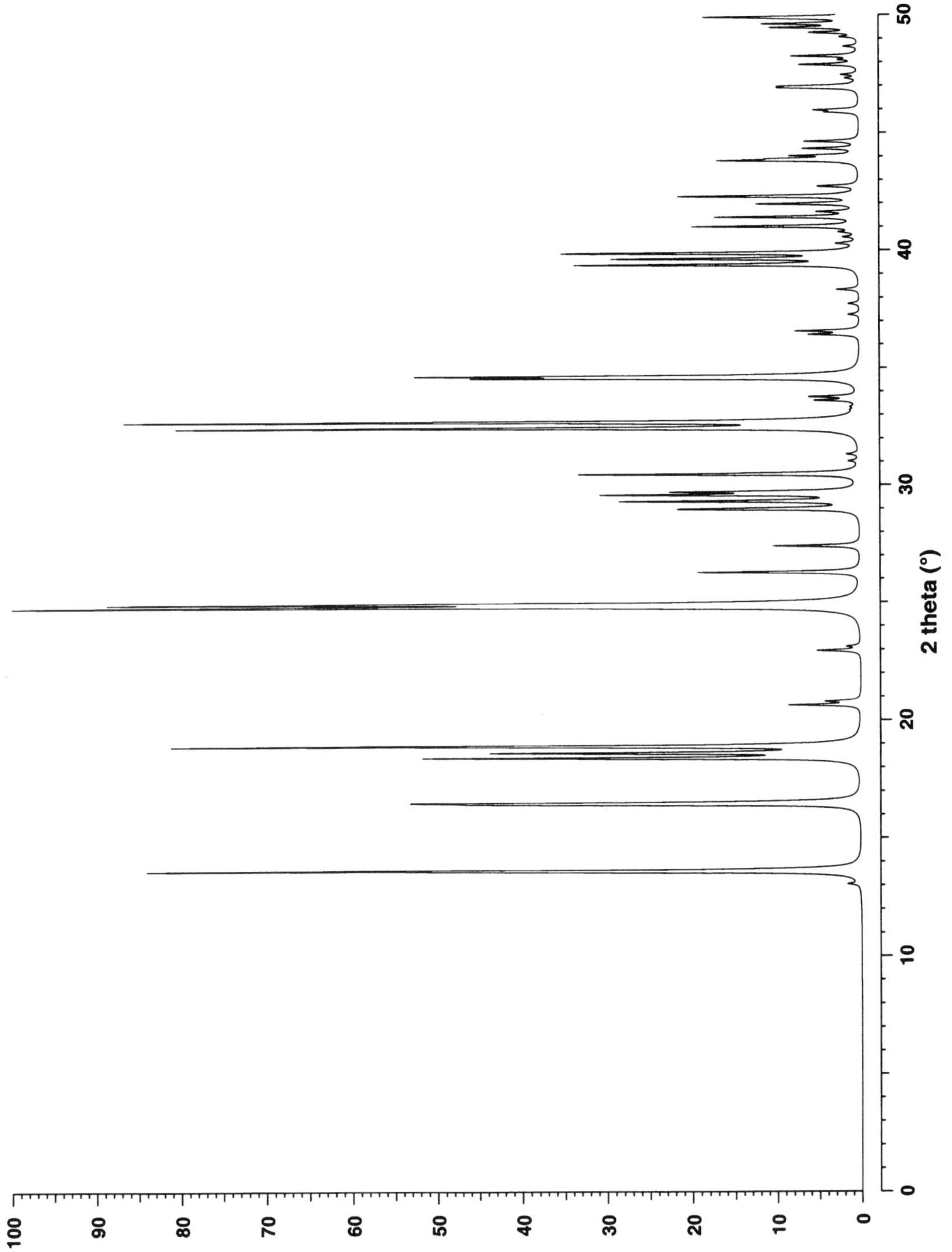

EMT

EMC-2, Calcined

CHEMICAL COMPOSITION: $|Na_{8.3}(H_2O)_6|$ $[Si_{76}Al_{20}O_{192}]$

REFINED COMPOSITION: $|Na_{8.3}(H_2O)_6|$ $[Si_{76.03}Al_{19.97}O_{192}]$

CRYSTAL DATA: $P6_3/mmc$ (No. 194)
$a = 17.3864$ Å $b = 17.3864$ Å $c = 28.3459$ Å
$\alpha = 90°$ $\beta = 90°$ $\gamma = 120°$
X-ray Rietveld refinement, $R_{exp} = 0.181$, $R_{wp} = 0.206$, $R_F = 0.074$

REFERENCE: Ch. Baerlocher, L. B. McCusker and R. Chiappetta, *Microporous Materials* **2** 269–280 (1994).

h	k	l	2θ	d	M	I_{rel}
1	0	0	5.87	15.057	6	100.0
0	0	2	6.24	14.173	2	41.5
1	0	1	6.65	13.297	12	29.6
1	0	2	8.57	10.320	12	0.2
1	1	0	10.18	8.693	6	9.9
1	0	3	11.05	8.003	12	12.7
2	0	0	11.75	7.529	6	2.7
1	1	2	11.94	7.410	12	6.1
2	0	1	12.16	7.276	12	1.5
0	0	4	12.49	7.086	2	0.3
2	0	2	13.32	6.649	12	0.7
1	0	4	13.81	6.412	12	2.1
2	0	3	15.05	5.888	12	0.6
2	1	0	15.57	5.691	12	7.0
2	1	1	15.88	5.580	24	3.8
1	0	5	16.71	5.306	12	1.2
2	1	2	16.79	5.281	24	0.8
2	0	4	17.18	5.160	12	6.1
3	0	1	17.95	4.942	12	0.1
2	1	3	18.20	4.875	24	2.5
3	0	2	18.76	4.731	12	1.4
0	0	6	18.78	4.724	2	0.2
2	0	5	19.60	4.529	12	0.4
1	0	6	19.69	4.508	12	1.0
2	1	4	20.01	4.437	24	0.7
3	0	3	20.03	4.432	12	0.1
2	2	0	20.43	4.347	6	3.8
1	1	6	21.41	4.151	12	0.3
3	1	1	21.51	4.131	24	0.2
2	1	5	22.13	4.016	24	1.0
2	0	6	22.21	4.002	12	1.5
1	0	7	22.74	3.910	12	2.0
3	1	3	23.29	3.820	24	4.4
4	0	0	23.63	3.764	6	2.1
4	0	1	23.85	3.732	12	0.7
4	0	2	24.47	3.638	12	0.4
2	1	6	24.49	3.635	24	0.1
3	1	4	24.75	3.598	24	2.1
2	0	7	24.97	3.566	12	0.3
3	2	0	25.79	3.454	12	0.7
1	0	8	25.83	3.449	12	0.7
3	0	6	25.90	3.440	12	0.2
3	1	5	26.51	3.362	24	1.4
3	2	2	26.56	3.356	24	0.3
4	1	0	27.14	3.286	12	1.7
1	1	8	27.18	3.281	12	2.1
3	2	3	27.49	3.244	24	0.9
3	0	7	28.32	3.152	12	0.3
3	1	6	28.53	3.129	24	0.2
3	2	4	28.75	3.105	24	1.6
5	0	0	29.66	3.011	6	0.1
5	0	1	29.84	2.995	12	0.1
5	0	2	30.34	2.946	12	0.2
2	0	9	30.77	2.906	12	0.3
3	3	0	30.86	2.898	6	1.3
3	0	8	30.89	2.895	12	1.4
5	0	3	31.17	2.869	12	3.4
3	3	2	31.51	2.839	12	0.2
0	0	10	31.56	2.835	2	1.6
4	2	1	31.60	2.831	24	0.2
4	2	2	32.08	2.790	24	0.5
3	2	6	32.10	2.788	24	0.3
1	0	10	32.13	2.786	12	0.3
5	0	4	32.30	2.772	12	1.5
2	2	8	32.60	2.746	12	1.0
4	2	3	32.87	2.725	24	0.3
5	1	0	33.13	2.704	12	0.3
3	1	8	33.16	2.702	24	0.6
1	1	10	33.24	2.695	12	0.2
5	0	5	33.70	2.659	12	0.7
5	1	2	33.74	2.656	24	0.2
4	2	4	33.95	2.641	24	1.2
3	2	7	34.12	2.628	24	0.6
5	1	3	34.50	2.600	24	1.9
4	1	7	35.17	2.551	24	0.2
1	0	11	35.34	2.540	12	0.4
6	0	0	35.78	2.510	6	0.1
6	0	3	37.06	2.425	12	0.2
6	0	4	38.04	2.366	12	0.4
0	0	12	38.10	2.362	2	0.4
6	0	5	39.26	2.295	12	0.7
4	2	8	40.66	2.219	24	0.1
4	4	0	41.55	2.173	6	0.5
5	1	8	42.03	2.150	24	0.3
6	1	5	42.47	2.128	24	0.1
7	0	2	42.51	2.127	12	0.1
4	0	11	42.51	2.126	12	0.6
4	3	7	42.82	2.112	24	0.4
5	3	3	43.13	2.097	24	0.1
2	0	13	43.19	2.094	12	0.1
6	2	2	43.82	2.066	24	0.3
6	0	8	44.23	2.048	12	0.5
6	2	3	44.43	2.039	24	0.3
2	0	14	46.44	1.955	12	0.2
3	1	13	47.01	1.933	24	0.2
7	1	4	47.35	1.920	24	0.2
6	3	0	47.95	1.897	12	0.3
5	4	3	48.17	1.889	24	0.2
1	0	15	48.55	1.875	12	0.1
5	1	11	48.82	1.866	24	0.3
5	4	4	48.96	1.860	24	0.1
5	2	10	49.64	1.837	24	0.1
7	2	1	49.66	1.836	24	0.1

EMC-2, Calcined **EMT**

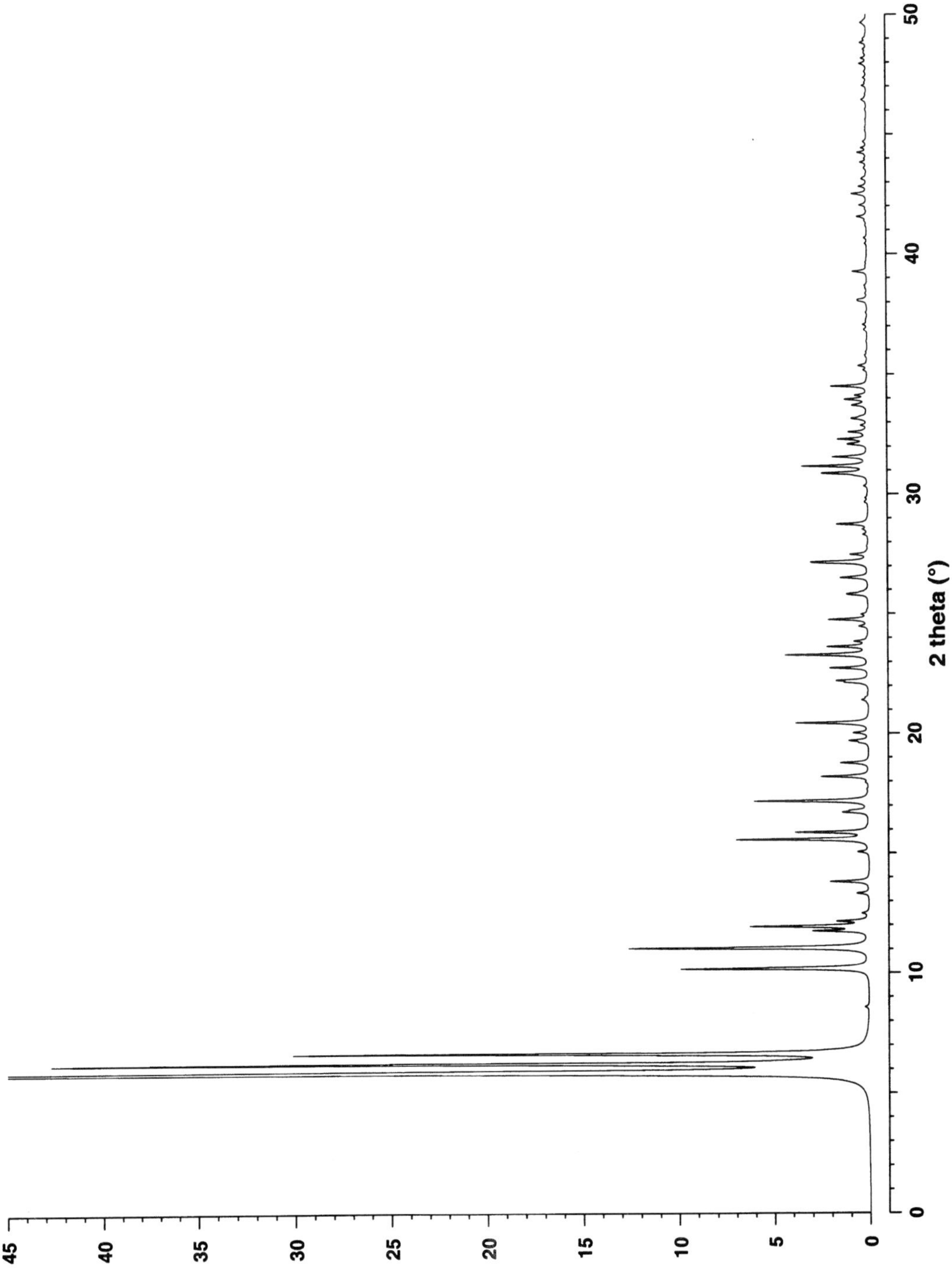

EMT EMC-2, Partially Dehydrated

CHEMICAL COMPOSITION: $|Na_{20}(C_{12}H_{24}O_6)_4(H_2O)_{22.6}|$ $[Si_{76}Al_{20}O_{192}]$
$C_{12}H_{24}O_6$ = 1,4,7,10,13,16-hexaoxacyclooctadecane (18-crown-6 ether)

REFINED COMPOSITION: $|Na_{20}C_{60}O_{24}(H_2O)_{22.6}|$ $[Si_{76.03}Al_{19.97}O_{192}]$

CRYSTAL DATA: $P6_3/mmc$ (No. 194)
a = 17.3741 Å b = 17.3741 Å c = 28.3646 Å
α = 90° β = 90° γ = 120°
X-ray Rietveld refinement, R_{exp} = 0.037, R_{wp} = 0.186, R_F = 0.068

REFERENCE: Ch. Baerlocher, L. B. McCusker and R. Chiappetta, *Microporous Materials* **2** 269–280 (1994).

h	k	l	2θ	d	M	I_{rel}
1	0	0	5.87	15.046	6	100.0
0	0	2	6.23	14.182	2	41.3
1	0	1	6.65	13.292	12	9.8
1	0	2	8.57	10.320	12	3.4
1	1	0	10.18	8.687	6	1.1
1	0	3	11.05	8.006	12	4.0
2	0	0	11.76	7.523	6	4.0
1	1	2	11.95	7.408	12	7.3
2	0	1	12.17	7.272	12	1.8
0	0	4	12.48	7.091	2	0.1
2	0	2	13.32	6.646	12	0.5
2	0	3	15.05	5.887	12	0.7
2	1	0	15.58	5.687	12	8.1
2	1	1	15.89	5.576	24	2.4
1	1	4	16.13	5.493	12	0.2
1	0	5	16.70	5.308	12	5.1
2	1	2	16.80	5.278	24	0.8
2	0	4	17.18	5.160	12	8.2
2	1	3	18.20	4.873	24	1.7
3	0	2	18.77	4.728	12	4.5
0	0	6	18.77	4.727	2	0.1
2	0	5	19.60	4.530	12	0.9
1	0	6	19.68	4.510	12	2.6
2	1	4	20.01	4.437	24	1.9
3	0	3	20.04	4.431	12	0.2
2	2	0	20.45	4.344	6	7.6
2	2	2	21.39	4.153	12	0.8
1	1	6	21.40	4.152	12	0.2
3	1	1	21.52	4.129	24	0.8
2	1	5	22.13	4.016	24	2.0
2	0	6	22.21	4.003	12	2.9
1	0	7	22.73	3.913	12	4.5
3	1	3	23.30	3.818	24	6.5
4	0	0	23.65	3.762	6	3.8
3	0	5	23.68	3.758	12	0.3
4	0	1	23.86	3.729	12	1.3
4	0	2	24.48	3.636	12	0.7
3	1	4	24.75	3.597	24	5.5
3	2	0	25.81	3.452	12	0.6
1	0	8	25.82	3.451	12	1.3
3	0	6	25.90	3.440	12	0.2
3	2	1	26.00	3.427	24	0.1
3	1	5	26.52	3.362	24	1.5
3	2	2	26.58	3.354	24	0.6
4	0	4	26.83	3.323	12	0.2
4	1	0	27.16	3.283	12	3.3
1	1	8	27.16	3.283	12	2.4
3	2	3	27.51	3.243	24	0.9
4	1	2	27.89	3.199	24	0.4
3	2	4	28.76	3.104	24	3.0
2	1	8	29.69	3.009	24	0.8
5	0	2	30.36	2.944	12	0.5
2	0	9	30.76	2.907	12	0.7
3	3	0	30.88	2.896	6	1.4
3	0	8	30.88	2.895	12	2.1
5	0	3	31.19	2.868	12	5.0
3	3	2	31.53	2.837	12	0.5
0	0	10	31.54	2.836	2	1.6
4	2	1	31.62	2.829	24	1.0
4	2	2	32.10	2.788	24	0.6
3	2	6	32.11	2.788	24	0.6
1	0	10	32.11	2.787	12	1.1
5	0	4	32.32	2.770	12	2.9
2	1	9	32.48	2.757	24	0.2
2	2	8	32.60	2.747	12	2.2
4	2	3	32.89	2.723	24	0.5
5	1	0	33.15	2.702	12	0.6
3	1	8	33.15	2.702	24	0.3
1	1	10	33.23	2.696	12	1.1
3	3	4	33.42	2.681	12	0.2
3	0	9	33.58	2.669	12	0.2
5	0	5	33.71	2.658	12	1.9
5	1	2	33.76	2.655	24	0.4
4	2	4	33.97	2.639	24	1.0
3	2	7	34.12	2.628	24	0.9
5	1	3	34.52	2.598	24	2.0
4	0	8	34.77	2.580	12	0.3
4	1	7	35.18	2.551	24	0.7
1	0	11	35.31	2.542	12	0.6
2	1	10	35.36	2.538	24	0.3
5	1	4	35.55	2.525	24	0.3
6	0	0	35.81	2.508	6	0.7
6	0	2	36.38	2.469	12	0.3
5	1	5	36.84	2.440	24	0.3
6	0	3	37.09	2.424	12	0.2
5	2	0	37.32	2.409	12	0.3
4	3	3	37.58	2.393	24	0.3
6	0	4	38.06	2.364	12	0.5
0	0	12	38.07	2.364	2	0.8
1	0	12	38.55	2.335	12	0.4
4	2	7	38.68	2.328	24	0.2
6	1	0	39.26	2.295	12	0.1
5	0	8	39.27	2.294	12	0.2
6	0	5	39.28	2.294	12	1.3
3	0	11	39.29	2.293	12	0.2
4	1	9	39.64	2.274	24	0.2
4	3	5	39.75	2.267	24	0.1
6	1	3	40.45	2.230	24	0.2
4	2	8	40.67	2.218	24	0.2
5	2	5	40.68	2.218	24	0.1
3	2	10	41.19	2.192	24	0.2
6	1	4	41.36	2.183	24	0.2
4	4	0	41.58	2.172	6	0.9
1	0	13	41.83	2.159	12	0.1
5	1	8	42.04	2.149	24	0.5
6	1	5	42.50	2.127	24	0.4
4	0	11	42.50	2.127	12	1.2
7	0	2	42.54	2.125	12	0.2
4	3	7	42.83	2.111	24	0.8
5	3	3	43.16	2.096	24	0.2
7	0	3	43.16	2.096	12	0.1
2	0	13	43.17	2.096	12	0.3
4	4	4	43.58	2.077	12	0.2
2	2	12	43.59	2.076	12	0.1
3	2	11	43.82	2.066	24	0.1
6	2	2	43.86	2.064	24	0.5
6	0	8	44.24	2.047	12	0.4
6	2	3	44.46	2.038	24	0.3
4	3	8	44.67	2.029	24	0.3
4	1	11	44.68	2.028	24	0.3
5	1	10	46.41	1.957	24	0.1
2	0	14	46.41	1.956	12	0.3
3	1	13	46.99	1.934	24	0.2
7	1	4	47.38	1.919	24	0.3
4	1	12	47.39	1.918	24	0.1
7	0	7	47.91	1.899	12	0.2
6	3	0	47.99	1.896	12	0.4
5	4	3	48.21	1.888	24	0.2
7	1	5	48.41	1.880	24	0.1
1	0	15	48.52	1.876	12	0.1
5	1	11	48.81	1.866	24	0.3
4	3	10	48.85	1.864	24	0.2
5	0	12	49.00	1.859	12	0.1
4	4	8	49.20	1.852	12	0.1
7	0	8	49.59	1.838	12	0.2

EMC-2, Partially Dehydrated — EMT

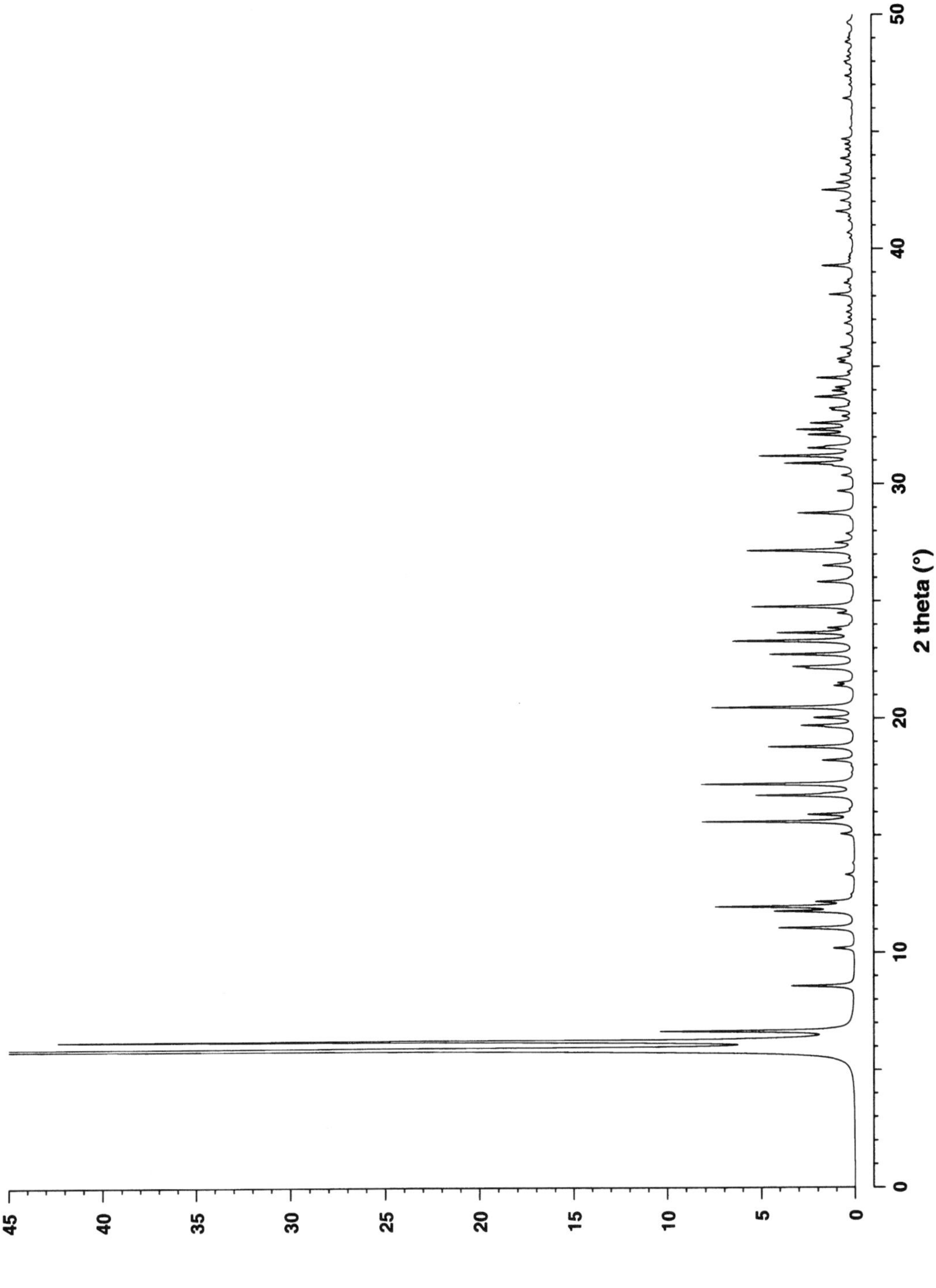

EPI

Epistilbite

CHEMICAL COMPOSITION: $|Na_{1.06}K_{0.1}Ca_{2.59}(H_2O)_{15.74}|\ [Si_{17.71}Al_{6.29}O_{48}]$
Teigarhorn, Iceland

REFINED COMPOSITION: $|Na_{1.04}Ca_{2.56}(H_2O)_{15.75}|\ [Si_{17.76}Al_{6.24}O_{48}]$

CRYSTAL DATA: $C1\,2/m\,1$ (No. 12) unique axis **b**, cell choice 1
$a = 9.08$ Å $b = 17.74$ Å $c = 10.25$ Å
$\alpha = 90°$ $\beta = 124.54°$ $\gamma = 90°$
X-ray single crystal refinement, $R_w = 0.16$

REFERENCE: A. J. Perrotta,
Mineral. Mag. **36** 480–490 (1967).

h	k	l	2θ	d	M	I_{rel}
0	2	0	9.97	8.870	2	53.4
0	0	1	10.48	8.443	2	17.1
-1	1	1	11.58	7.642	4	4.6
1	1	0	12.84	6.892	4	70.0
0	2	1	14.48	6.116	4	1.5
-1	1	2	18.02	4.924	4	73.6
-1	3	1	18.30	4.848	4	12.7
1	3	0	19.13	4.639	4	9.3
-2	0	1	19.78	4.489	2	15.4
0	4	0	20.02	4.435	2	4.5
1	1	1	20.45	4.342	4	37.7
-2	0	2	20.98	4.234	2	2.4
0	0	2	21.04	4.222	2	0.3
-2	2	1	22.19	4.006	4	12.6
0	4	1	22.65	3.926	4	13.3
-1	3	2	22.96	3.873	4	100.0
-2	2	2	23.28	3.821	4	11.0
0	2	2	23.34	3.812	4	32.9
2	0	0	23.79	3.740	2	14.8
1	3	1	24.94	3.570	4	1.3
2	2	0	25.85	3.446	4	83.0
-2	0	3	26.75	3.333	2	10.7
-1	5	1	27.25	3.272	4	16.9
-1	1	3	27.26	3.272	4	11.5
1	5	0	27.83	3.206	4	44.5
-2	4	1	28.29	3.155	4	0.3
-2	2	3	28.61	3.120	4	0.2
-2	4	2	29.16	3.062	4	6.6
0	4	2	29.20	3.058	4	0.7
-3	1	2	29.96	2.982	4	9.4
1	1	2	30.05	2.974	4	2.9
0	6	0	30.23	2.957	2	2.1
-1	5	2	30.65	2.917	4	50.1
-1	3	3	30.83	2.901	4	2.1
2	0	1	31.17	2.870	2	0.7
2	4	0	31.29	2.859	4	42.8
-3	1	1	31.53	2.837	4	3.6
0	0	3	31.79	2.814	2	1.1
0	6	1	32.07	2.791	4	0.7
-3	1	3	32.11	2.788	4	15.2
1	5	1	32.19	2.781	4	0.6
2	2	1	32.80	2.730	4	0.2
-3	3	2	33.26	2.693	4	12.6
1	3	2	33.34	2.687	4	15.2
0	2	3	33.40	2.683	4	6.3
-2	4	3	33.63	2.665	4	4.0
-2	0	4	35.02	2.562	2	12.6
-3	3	3	35.23	2.547	4	5.4
3	1	0	36.39	2.469	4	1.9
-2	2	4	36.50	2.462	4	7.8
-1	7	1	37.03	2.428	4	0.6
-1	5	3	37.03	2.428	4	11.8
-2	6	2	37.09	2.424	4	8.0
0	6	2	37.12	2.422	4	0.5
2	4	1	37.32	2.409	4	2.8
-3	1	4	37.41	2.404	4	1.5
1	7	0	37.47	2.400	4	0.3
-1	1	4	37.48	2.400	4	1.5
0	4	3	37.86	2.376	4	0.3
2	6	0	38.83	2.319	4	4.4
-3	5	2	39.13	2.302	4	0.7
3	3	0	39.21	2.297	4	0.2
-1	7	2	39.67	2.272	4	0.2
-4	0	3	39.99	2.254	2	3.9
-3	3	4	40.17	2.245	4	1.6
-4	0	2	40.17	2.245	2	0.2
-1	3	4	40.24	2.241	4	1.2
-3	5	1	40.38	2.233	4	0.7
1	1	3	40.46	2.230	4	6.7
-2	4	4	40.66	2.219	4	9.7
0	8	0	40.69	2.217	2	0.3
-2	6	3	40.80	2.212	4	4.5
-3	5	3	40.85	2.209	4	1.0
1	7	1	40.92	2.206	4	0.6
-4	2	3	41.32	2.185	4	2.9
-4	2	2	41.50	2.176	4	0.2
2	2	2	41.60	2.171	4	2.6
0	0	4	42.84	2.111	2	0.2
1	3	3	43.06	2.101	4	4.0
-4	0	1	43.22	2.093	2	3.2
2	6	1	43.97	2.059	4	0.7
-4	2	4	43.97	2.059	4	3.0
0	2	4	44.10	2.053	4	3.5
3	5	0	44.41	2.040	4	0.6
0	6	3	44.44	2.039	4	0.8
-4	2	1	44.47	2.037	4	0.6
-2	0	5	44.63	2.030	2	3.3
-3	1	5	44.90	2.019	4	4.2
-1	7	3	44.95	2.016	4	1.0
-3	5	4	45.27	2.003	4	1.4
-4	4	2	45.28	2.003	4	8.0
-1	5	4	45.34	2.000	4	0.7
2	4	2	45.37	1.999	4	3.8
-2	8	1	45.63	1.988	4	0.4
-2	2	5	45.85	1.979	4	0.5
3	3	1	46.05	1.971	4	1.3
-2	8	2	46.21	1.964	4	1.4
0	8	2	46.24	1.963	4	3.5
-3	7	2	46.76	1.943	4	3.4
1	7	2	46.82	1.940	4	5.3
-2	6	4	46.92	1.936	4	2.0
-3	3	5	47.30	1.922	4	0.1
-4	4	4	47.60	1.910	4	3.7
2	8	0	47.68	1.907	4	3.3
1	9	0	47.71	1.906	4	2.1
0	4	4	47.72	1.906	4	3.8
-3	7	1	47.85	1.901	4	0.5
-4	0	5	47.93	1.898	2	0.3
-3	7	3	48.26	1.886	4	0.1
4	0	0	48.70	1.870	2	16.8
-4	2	5	49.08	1.856	4	0.8
-2	4	5	49.37	1.846	4	0.4
-1	9	2	49.55	1.840	4	3.4

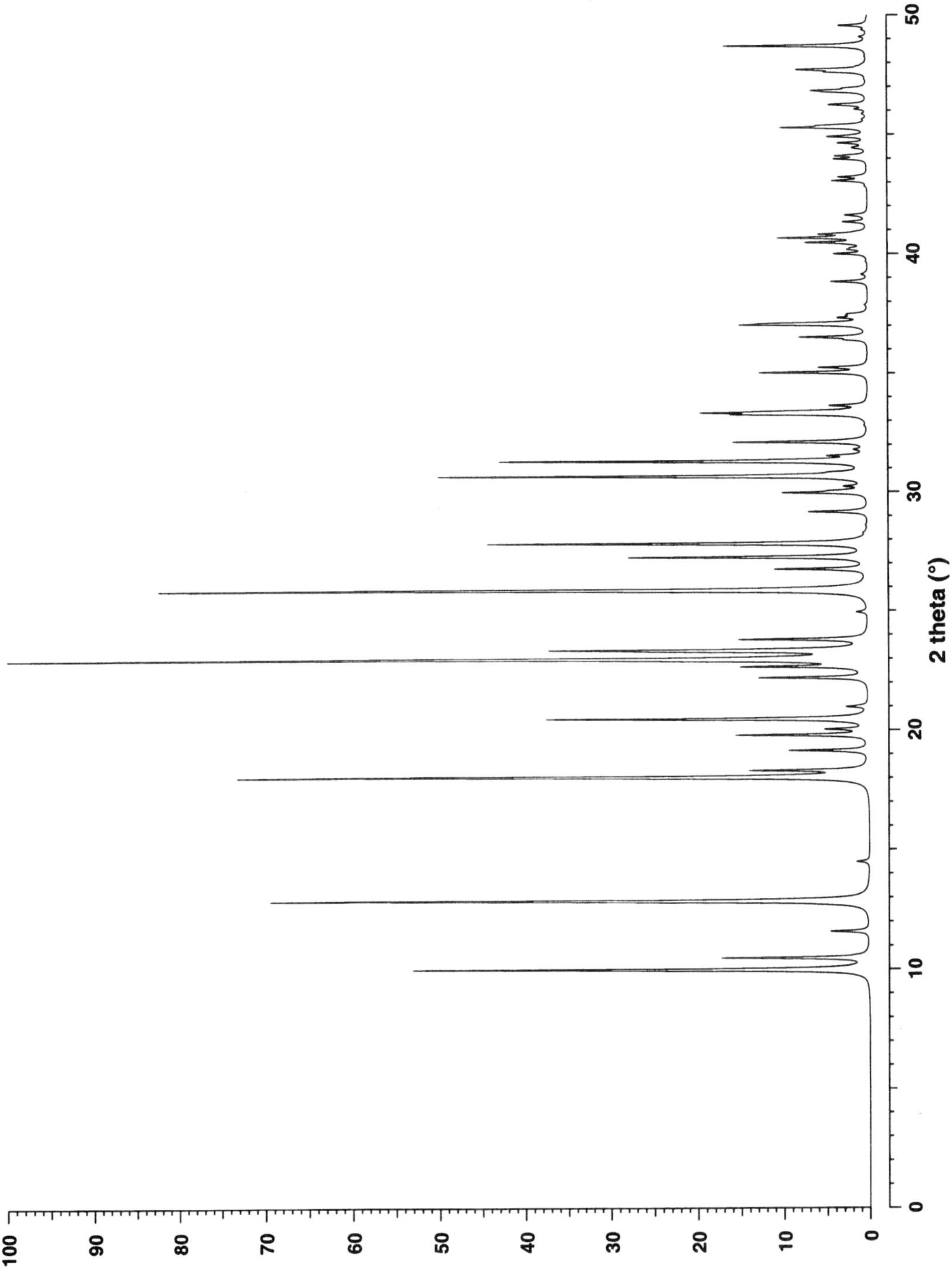
2 theta (°)
0
10
20
30
40
50
100
90
80
70
60
50
40
30
20
10
0

ERI Erionite

CHEMICAL COMPOSITION: $|K_{2.02}Na_{1.94}Ca_{1.25}Mg_{0.63}(H_2O)_{10.4}|$ $[Si_{26.16}Al_{9.31}Fe_{0.53}O_{72}]$
Wenatchee, Washington, U.S.A.

REFINED COMPOSITION: $|K_2Na_{1.86}Ca_{1.3}Mg_{0.7}(H_2O)_{6.12}|$ $[Si_{27}Al_9O_{72}]$

CRYSTAL DATA: $P6_3/mmc$ (No. 194)
$a = 13.27$ Å $b = 13.27$ Å $c = 15.05$ Å
$\alpha = 90°$ $\beta = 90°$ $\gamma = 120°$
X-ray single crystal refinement, $R = 0.15$

REFERENCE: J. A. Gard and J. M. Tait,
in *Proceedings of the Third International Conference on Molecular Sieves; Recent Progress Reports, Zurich*,
Ed. by J. B. Uytterhoeven (Leuven University Press) 94–99 (1973).

h	k	l	2θ	d	M	I_{rel}	h	k	l	2θ	d	M	I_{rel}	h	k	l	2θ	d	M	I_{rel}
1	0	0	7.69	11.492	6	100.0	3	1	2	30.46	2.935	24	0.9	5	0	2	41.06	2.198	12	0.4
1	0	1	9.68	9.134	12	23.6	1	0	5	30.70	2.912	12	0.5	3	3	2	42.61	2.122	12	1.5
0	0	2	11.76	7.525	2	13.3	4	0	0	31.13	2.873	6	8.2	1	0	7	42.79	2.113	12	0.2
1	1	0	13.34	6.635	6	24.9	2	1	4	31.46	2.844	24	18.0	3	0	6	43.11	2.098	12	0.9
1	0	2	14.07	6.295	12	6.5	4	0	1	31.71	2.822	12	8.1	5	0	3	43.30	2.090	12	0.3
2	0	0	15.42	5.746	6	4.6	3	1	3	33.30	2.690	24	0.8	4	2	2	43.36	2.087	24	1.1
2	0	1	16.51	5.368	12	4.3	4	0	2	33.38	2.684	12	4.2	5	1	1	44.29	2.045	24	0.3
1	0	3	19.30	4.598	12	5.3	2	0	5	33.61	2.666	12	0.9	5	1	2	45.57	1.991	24	1.1
2	0	2	19.44	4.567	12	7.9	3	2	1	34.54	2.597	24	0.2	3	1	6	46.04	1.971	24	0.2
2	1	0	20.45	4.344	12	11.2	0	0	6	35.80	2.508	2	0.9	5	0	4	46.29	1.961	12	0.7
2	1	1	21.29	4.173	24	7.1	4	1	0	35.81	2.508	12	5.1	2	1	7	47.17	1.927	24	0.2
3	0	0	23.22	3.831	6	9.0	4	0	3	36.02	2.493	12	0.3	4	1	5	47.17	1.927	24	0.4
2	0	3	23.54	3.779	12	1.4	2	2	4	36.09	2.488	12	0.8	5	1	3	47.64	1.909	24	0.2
0	0	4	23.65	3.762	2	2.2	3	2	2	36.10	2.488	24	2.4	3	3	4	47.70	1.907	12	0.2
2	1	2	23.65	3.762	24	15.1	2	1	5	36.31	2.474	24	1.3	4	0	6	48.16	1.890	12	0.9
1	0	4	24.90	3.576	12	13.3	4	1	1	36.32	2.474	24	0.2	4	3	0	48.16	1.889	12	0.1
3	0	2	26.10	3.414	12	0.6	1	1	6	38.36	2.346	12	0.3	0	0	8	48.38	1.881	2	2.9
2	2	0	26.87	3.317	6	4.0	2	0	6	39.19	2.299	12	0.1	4	2	4	48.39	1.881	24	0.2
2	1	3	27.15	3.284	24	3.7	5	0	0	39.19	2.298	6	0.3	4	3	1	48.57	1.875	24	0.3
1	1	4	27.25	3.273	12	1.0	4	0	4	39.46	2.283	12	0.3	1	0	8	49.07	1.857	12	0.9
3	1	0	27.99	3.187	12	0.8	5	0	1	39.67	2.272	12	0.4	5	2	0	49.53	1.840	12	1.4
2	0	4	28.35	3.148	12	6.5	4	1	3	40.20	2.243	24	0.4	4	3	2	49.76	1.832	24	0.2
3	1	1	28.63	3.118	24	0.7	3	3	0	40.80	2.212	6	1.7	5	0	5	49.92	1.827	12	0.6

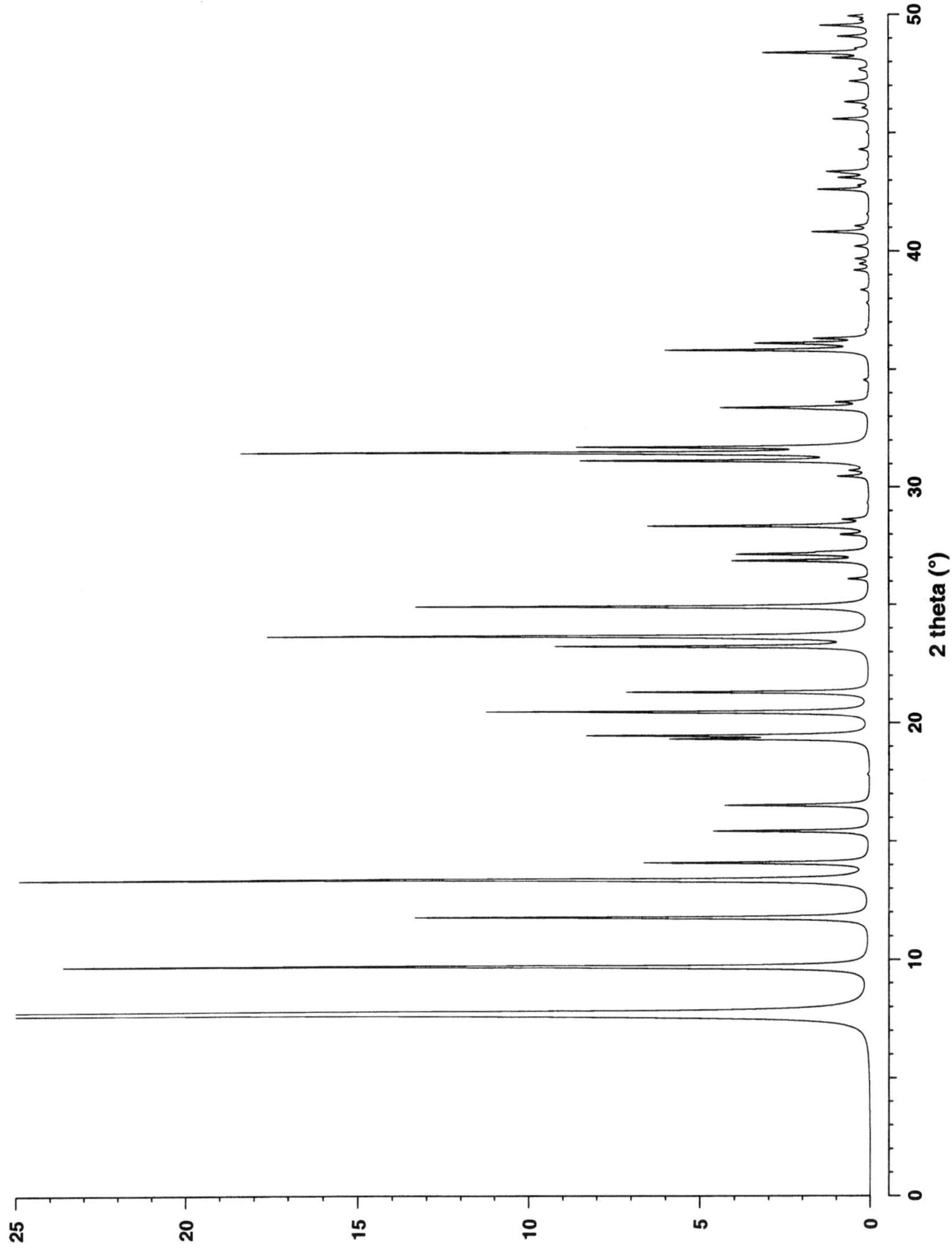
2 theta (°)
0
10
20
30
40
50
0
5
10
15
20
25

ERI Piperidine AlPO–17

CHEMICAL COMPOSITION: $|(C_5H_{10}NH)_4(H_2O)_4|$ $[Al_{18}P_{18}O_{72}]$
$C_5H_{10}NH$ = piperidine

REFINED COMPOSITION: $|C_{24}O_{4.14}|$ $[Al_{18}P_{18}O_{72}]$

CRYSTAL DATA: $P6_3/m$ (No. 176)
$a = 13.2371$ Å $b = 13.2371$ Å $c = 14.7708$ Å
$\alpha = 90°$ $\beta = 90°$ $\gamma = 120°$
X-ray single crystal refinement, $R_w = 0.081$

REFERENCE: J. J. Pluth, J. V. Smith and J. M. Bennett, *Acta Cryst.* **C42** 283–286 (1986).

h	k	l	2θ	d	M	I_{rel}
1	0	0	7.71	11.464	6	100.0
1	0	1	9.77	9.056	12	44.5
0	0	2	11.98	7.385	2	1.7
1	1	0	13.38	6.619	6	44.5
1	0	2	14.27	6.209	12	8.8
1	1	1	14.67	6.040	12	0.5
2	0	0	15.46	5.732	6	19.4
2	0	1	16.59	5.344	12	8.1
1	1	2	18.00	4.929	12	6.6
2	0	2	19.60	4.528	12	12.4
1	0	3	19.62	4.524	12	15.2
1	2	0	20.50	4.333	6	14.6
2	1	0	20.50	4.333	6	9.6
2	1	1	21.37	4.158	12	4.8
1	2	1	21.37	4.158	12	9.8
1	1	3	22.51	3.950	12	4.3
3	0	0	23.28	3.821	6	7.5
2	1	2	23.81	3.737	12	6.5
1	2	2	23.81	3.737	12	3.7
2	0	3	23.82	3.735	12	1.6
0	0	4	24.10	3.693	2	1.4
1	0	4	25.34	3.515	12	26.6
3	0	2	26.26	3.394	12	0.1
2	2	0	26.94	3.309	6	14.7
2	1	3	27.42	3.253	12	6.8
2	2	1	27.62	3.229	12	0.4
1	1	4	27.66	3.225	12	1.9
3	1	0	28.06	3.179	6	0.4
1	3	0	28.06	3.179	6	0.6
3	1	1	28.72	3.108	12	1.6
2	0	4	28.76	3.104	12	4.9
2	2	2	29.58	3.020	12	0.1
1	3	2	30.61	2.920	12	1.2
3	1	2	30.61	2.920	12	2.6
4	0	0	31.21	2.866	6	8.8
1	0	5	31.27	2.861	12	1.1
4	0	1	31.81	2.813	12	14.1
1	2	4	31.84	2.810	12	4.8
2	1	4	31.84	2.810	12	7.0
2	2	3	32.60	2.747	12	0.3
1	1	5	33.21	2.698	12	0.1
4	0	2	33.54	2.672	12	6.6
3	1	3	33.55	2.671	12	1.1
3	0	4	33.75	2.655	12	0.2
3	2	0	34.09	2.630	6	0.1
2	0	5	34.14	2.626	12	0.9
3	2	1	34.64	2.589	12	0.3
4	1	0	35.90	2.502	6	2.2
1	4	0	35.90	2.502	6	0.9
2	3	2	36.26	2.478	12	0.5
3	2	2	36.26	2.478	12	0.2
4	0	3	36.27	2.477	12	0.2
1	4	1	36.43	2.466	12	0.2
0	0	6	36.50	2.462	2	1.8
1	2	5	36.82	2.441	12	1.2
1	3	4	37.32	2.409	12	0.1
3	1	4	37.32	2.409	12	0.3
1	0	6	37.36	2.407	12	0.6
4	1	2	37.98	2.369	12	0.4
3	0	5	38.52	2.337	12	0.1
2	3	3	38.82	2.320	12	0.1
5	0	0	39.30	2.293	6	0.4
5	0	1	39.79	2.266	12	1.0
4	0	4	39.81	2.264	12	1.7
1	4	3	40.44	2.230	12	0.6
4	1	3	40.44	2.230	12	0.8
3	3	0	40.90	2.206	6	0.2
5	0	2	41.23	2.190	12	0.6
3	3	1	41.38	2.182	12	0.2
3	1	5	41.74	2.164	12	0.1
2	4	1	42.16	2.143	12	0.2
3	2	4	42.18	2.142	12	0.2
1	2	6	42.22	2.140	12	0.3
2	1	6	42.22	2.140	12	0.2
3	3	2	42.78	2.114	12	0.2
4	2	2	43.53	2.079	12	0.4
5	0	3	43.54	2.078	12	0.9
1	0	7	43.61	2.075	12	0.6
3	0	6	43.74	2.070	12	3.4
4	0	5	44.02	2.057	12	0.2
1	5	1	44.42	2.039	12	0.4
3	3	3	45.03	2.013	12	0.1
1	1	7	45.10	2.010	12	0.1
5	1	2	45.75	1.983	12	1.0
1	5	2	45.75	1.983	12	0.5
2	2	6	45.95	1.975	12	0.1
2	3	5	46.21	1.964	12	0.6
5	0	4	46.63	1.948	12	1.7
6	0	0	47.59	1.911	6	0.1
4	1	5	47.63	1.909	12	0.3
1	4	5	47.63	1.909	12	0.1
5	1	3	47.89	1.900	12	0.2
2	1	7	47.95	1.897	12	0.2
1	2	7	47.95	1.897	12	0.3
3	3	4	48.04	1.894	12	0.1
3	4	1	48.71	1.869	12	0.2
4	3	1	48.71	1.869	12	0.3
4	0	6	48.76	1.867	12	0.3
0	0	8	49.36	1.846	2	2.9
2	5	0	49.66	1.836	6	0.9
5	2	0	49.66	1.836	6	2.4
3	4	2	49.94	1.826	12	1.1
4	3	2	49.94	1.826	12	0.6

Piperidine AlPO–17

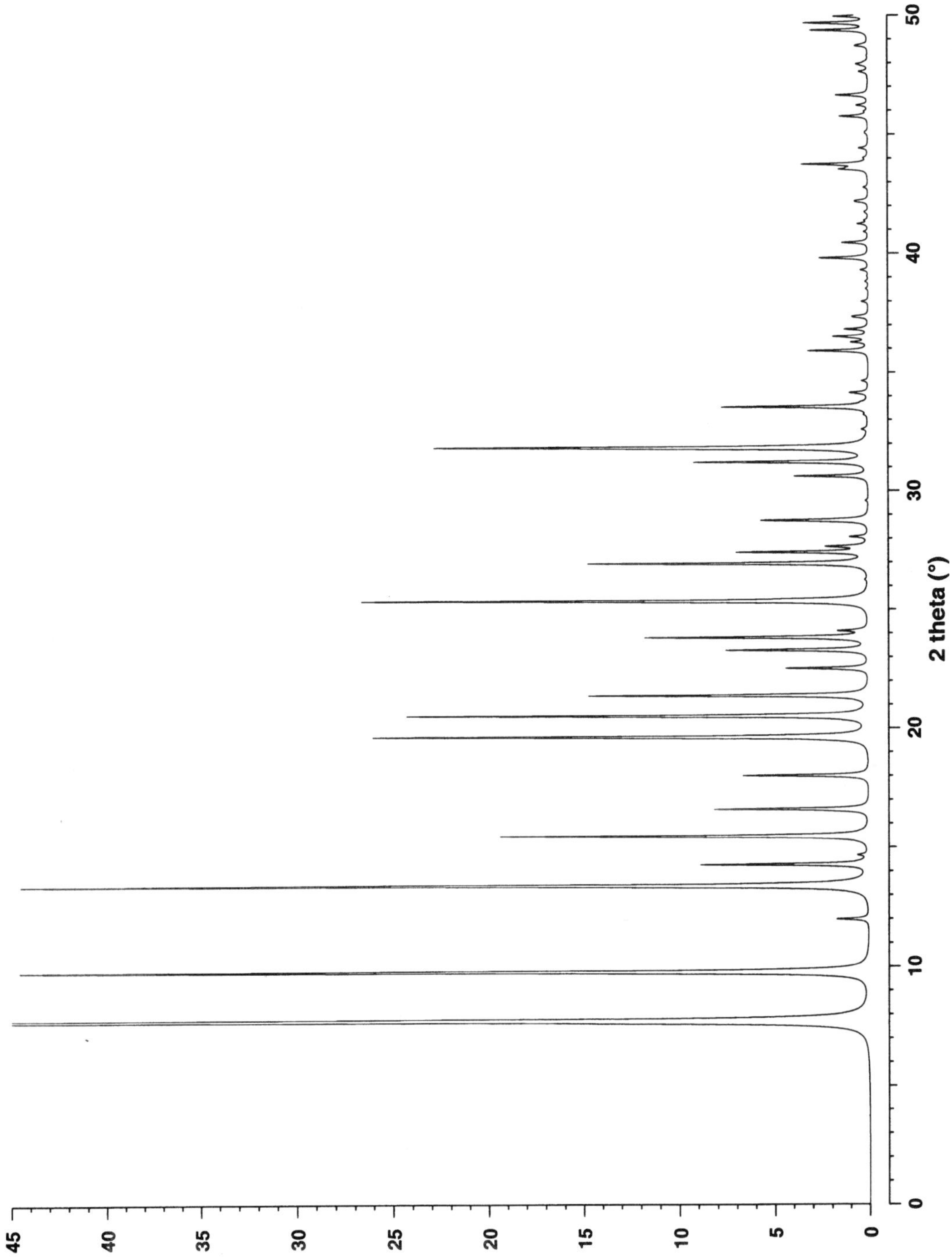

ESV ERS-7 Framework

CHEMICAL COMPOSITION: $[Al_{1.4}Si_{46.6}O_{96}]$

REFINED COMPOSITION: $[Si_{48}O_{96}]$

CRYSTAL DATA: *Pnma* (No. 62)
$a = 9.7998$ Å $b = 12.4116$ Å $c = 22.8606$ Å
$\alpha = 90.0°$ $\beta = 90.0°$ $\gamma = 90.0°$
X-ray Rietveld refinement, $R_p = 0.056$, $R_{wp} = 0.0611$

REFERENCE: R. Millini, G. Perego, L. Carluccio, G. Bellussi, D. E. Cox, B. J. Campbell and A. K. Cheetham, in *Proc. of the 12th International Zeolite Conference*, Ed. by M. M. J. Treacy, B. K. Marcus, J. B. Higgins and M. E. Bisher (Materials Research Society: Warrendale) vol. **I** 541–548 (1999).

h	*k*	*l*	2θ	*d*	*M*	I_{rel}	*h*	*k*	*l*	2θ	*d*	*M*	I_{rel}	*h*	*k*	*l*	2θ	*d*	*M*	I_{rel}
0	0	2	7.73	11.430	2	51.5	2	3	0	28.23	3.161	4	1.8	4	0	2	37.54	2.396	4	1.2
0	1	1	8.11	10.908	4	41.6	0	1	7	28.26	3.158	4	0.4	2	4	4	37.75	2.383	8	1.9
1	0	1	9.82	9.007	4	91.2	3	0	2	28.42	3.141	4	6.2	0	5	3	38.13	2.360	4	0.5
1	0	2	11.90	7.440	4	2.7	2	3	1	28.50	3.131	8	2.8	4	1	2	38.26	2.352	8	0.5
1	1	1	12.14	7.290	8	43.9	3	1	1	28.52	3.129	8	0.9	4	0	3	38.60	2.332	4	0.4
0	1	3	13.64	6.494	4	100.0	0	4	0	28.77	3.103	2	8.1	3	0	7	39.00	2.310	4	0.5
1	1	2	13.88	6.381	8	86.5	1	0	7	28.81	3.098	4	12.4	0	0	10	39.41	2.286	2	0.5
0	2	0	14.27	6.206	2	71.3	1	2	6	28.97	3.082	8	0.7	2	4	5	39.63	2.274	8	0.6
1	0	3	14.73	6.016	4	13.2	0	3	5	29.11	3.068	4	1.0	2	0	9	39.98	2.255	4	1.0
0	0	4	15.50	5.715	2	2.0	3	1	2	29.33	3.045	8	3.4	3	4	1	40.28	2.239	8	1.3
0	2	2	16.25	5.454	4	4.0	2	0	6	29.70	3.008	4	1.0	4	2	2	40.36	2.235	8	0.6
1	1	3	16.37	5.413	8	9.5	1	1	7	29.72	3.006	8	0.6	1	0	10	40.52	2.226	4	0.7
1	2	1	17.35	5.110	8	13.4	3	0	3	29.76	3.002	4	26.3	2	1	9	40.66	2.219	8	0.7
2	0	0	18.10	4.900	2	9.8	2	2	5	30.37	2.943	8	0.7	4	1	4	40.72	2.216	8	1.6
2	0	1	18.52	4.791	4	2.1	1	4	1	30.47	2.934	8	0.6	3	4	2	40.88	2.207	8	0.5
1	2	2	18.62	4.766	8	4.7	2	1	6	30.58	2.923	8	4.4	1	4	7	41.17	2.192	8	3.5
1	1	4	19.35	4.587	8	65.8	3	2	1	31.19	2.868	8	1.1	3	4	3	41.87	2.158	8	2.1
2	1	0	19.48	4.558	4	13.2	1	4	2	31.23	2.864	8	1.0	4	1	5	42.49	2.128	8	0.6
2	0	2	19.71	4.504	4	1.8	3	0	4	31.55	2.836	4	0.8	3	3	6	42.50	2.127	8	0.6
2	1	1	19.86	4.470	8	27.7	3	2	2	31.93	2.802	8	2.3	1	3	9	42.78	2.114	8	0.9
0	1	5	20.70	4.290	4	4.4	1	2	7	32.29	2.772	8	0.8	2	0	10	43.69	2.072	4	0.6
2	1	2	20.98	4.233	8	1.1	2	3	4	32.36	2.766	8	1.6	0	6	0	43.76	2.069	2	1.5
0	2	4	21.13	4.204	4	2.0	3	1	4	32.38	2.765	8	1.3	4	0	6	43.94	2.061	4	0.6
1	0	5	21.45	4.143	4	7.3	1	4	3	32.47	2.758	8	1.2	2	4	7	44.31	2.044	8	1.4
2	0	3	21.56	4.121	4	9.6	1	0	8	32.64	2.743	4	0.5	1	0	11	44.57	2.033	4	0.6
0	3	1	21.83	4.071	4	1.2	0	4	4	32.84	2.727	4	2.7	3	2	8	44.59	2.032	8	0.5
1	1	5	22.62	3.930	8	0.9	2	0	7	32.96	2.718	4	2.2	3	3	7	44.95	2.017	8	0.5
1	2	4	23.02	3.863	8	12.3	2	2	6	33.10	2.707	8	1.6	3	0	9	45.22	2.005	4	2.3
2	2	0	23.13	3.846	4	4.9	1	3	6	33.25	2.695	8	0.8	2	5	5	45.51	1.993	8	1.0
1	3	1	23.66	3.760	8	3.5	2	1	7	33.76	2.655	8	2.5	2	3	9	45.83	1.980	8	1.3
2	0	4	23.92	3.720	4	2.9	1	4	4	34.13	2.627	8	8.6	4	3	4	45.88	1.978	8	1.3
2	2	2	24.42	3.645	8	8.1	2	4	0	34.20	2.621	4	1.5	1	3	10	46.31	1.960	8	2.5
0	3	3	24.48	3.636	4	30.3	2	3	5	34.49	2.600	8	1.5	5	0	1	46.50	1.953	4	0.7
1	3	2	24.62	3.616	8	12.6	3	1	5	34.51	2.599	8	0.5	1	5	7	46.90	1.937	8	0.9
2	1	4	24.99	3.563	8	6.3	3	2	4	34.78	2.579	8	0.4	4	1	7	46.94	1.936	8	1.5
1	0	6	25.08	3.551	4	1.7	2	4	2	35.12	2.555	8	4.7	4	3	5	47.49	1.914	8	0.7
1	2	5	25.85	3.446	8	4.3	3	3	2	35.90	2.502	8	0.7	5	1	2	47.64	1.909	8	0.8
2	2	3	25.95	3.433	8	14.7	0	1	9	36.09	2.488	4	3.0	1	6	4	47.66	1.908	8	0.5
1	1	6	26.10	3.414	8	1.7	1	4	5	36.17	2.484	8	7.5	5	0	3	47.92	1.898	4	0.5
1	3	3	26.14	3.409	8	3.0	2	4	3	36.24	2.479	8	3.5	4	4	2	47.98	1.896	8	1.7
2	0	5	26.67	3.343	4	1.0	2	0	8	36.40	2.468	4	1.7	5	1	3	48.52	1.876	8	2.2
0	2	6	27.47	3.247	4	16.7	4	0	1	36.90	2.436	4	1.3	3	0	10	48.61	1.873	4	1.2
2	1	5	27.63	3.228	8	5.8	2	3	6	36.95	2.433	8	0.8	4	2	7	48.72	1.869	8	1.1
2	2	4	27.96	3.191	8	7.7	3	1	6	36.96	2.432	8	1.0	2	3	10	49.18	1.852	8	0.6
1	3	4	28.14	3.171	8	16.2	0	4	6	37.38	2.406	4	1.7	1	6	5	49.23	1.851	8	0.4

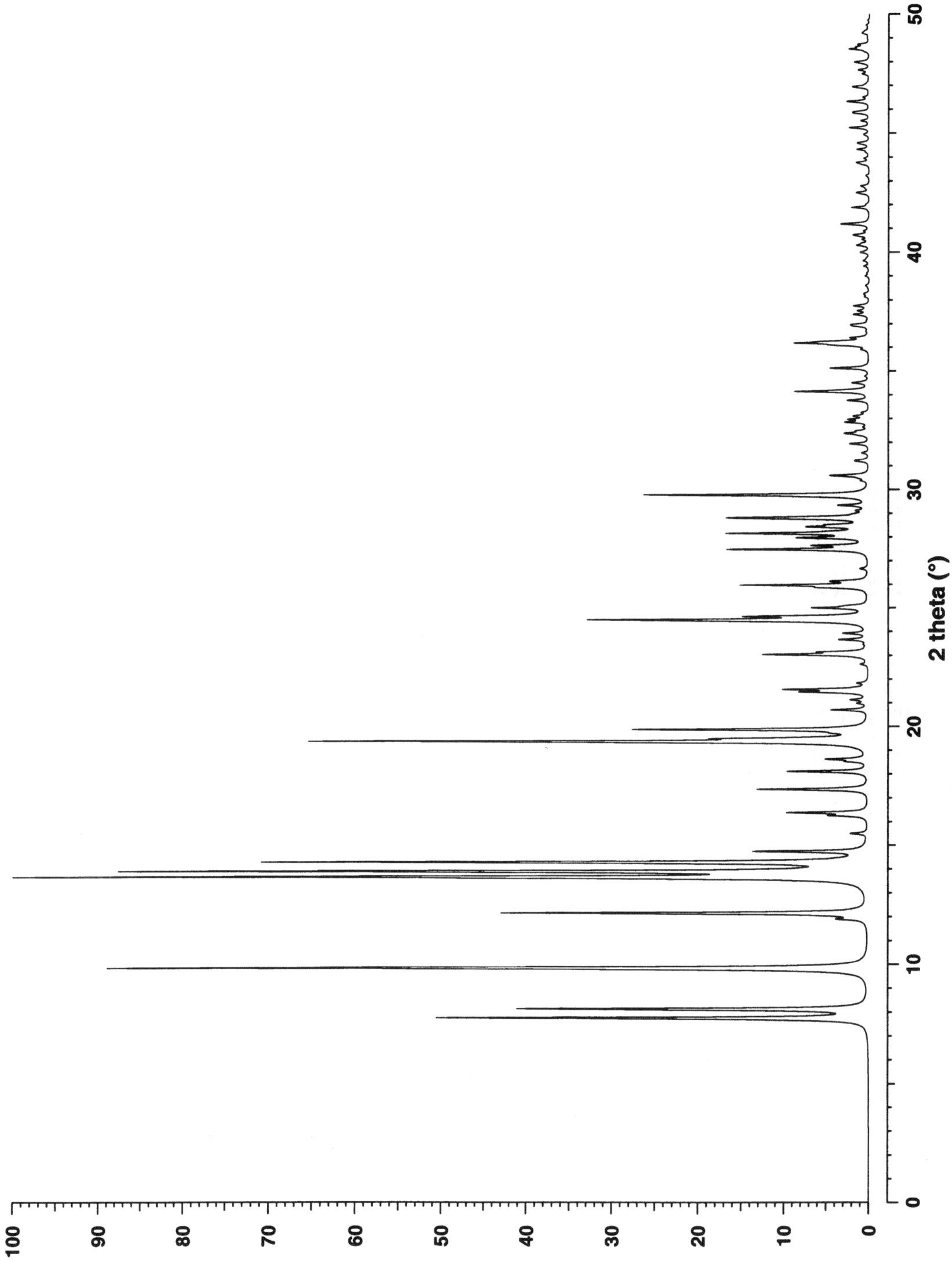
2 theta (°)
0
10
20
30
40
50
100
90
80
70
60
50
40
30
20
10
0

EUO EU-1, Calcined, Rehydrated

CHEMICAL COMPOSITION: $|(H_2O)_{36.16}|\ [Si_{112}O_{224}]$

REFINED COMPOSITION: $|O_{36.16}|\ [Si_{112}O_{224}]$

CRYSTAL DATA: *Cmme* (No. 67)
$a = 13.695$ Å $b = 22.326$ Å $c = 20.178$ Å
$\alpha = 90°$ $\beta = 90°$ $\gamma = 90°$
X-ray Rietveld limited refinement. No R-factor reported.

REFERENCE: N. A. Briscoe, D. W. Johnson, M. D. Shannon, G. T. Kokotailo and L. B. McCusker, *Zeolites* **8** 74–76 (1988).

h	k	l	2θ	d	M	I_{rel}	h	k	l	2θ	d	M	I_{rel}	h	k	l	2θ	d	M	I_{rel}
0	0	1	4.38	20.178	2	3.5	4	2	0	27.24	3.273	4	28.3	5	3	2	36.06	2.491	8	0.5
0	2	0	7.92	11.163	2	100.0	0	4	5	27.27	3.270	4	2.2	0	2	8	36.52	2.460	4	3.7
1	1	1	8.75	10.105	8	64.0	2	6	0	27.28	3.269	4	3.5	1	9	1	37.10	2.423	8	1.0
0	0	2	8.76	10.089	2	5.3	0	6	3	27.39	3.256	4	0.9	3	1	7	37.10	2.423	8	0.5
0	2	1	9.05	9.768	4	8.4	1	5	4	27.46	3.248	8	19.1	2	8	3	37.27	2.412	8	4.5
1	1	2	11.59	7.633	8	9.1	4	0	2	27.51	3.242	4	0.6	5	1	4	37.58	2.393	8	2.3
0	2	2	11.82	7.485	4	4.1	1	1	6	27.60	3.232	8	4.1	1	5	7	37.72	2.385	8	0.6
2	0	0	12.93	6.847	2	12.5	2	6	1	27.64	3.227	8	1.7	4	2	6	38.37	2.346	8	0.7
2	0	1	13.66	6.484	4	2.1	3	5	1	28.31	3.153	8	7.3	3	7	4	38.79	2.321	8	2.7
1	3	1	14.24	6.220	8	5.3	4	2	2	28.67	3.113	8	1.1	5	5	1	38.83	2.319	8	1.9
2	2	0	15.18	5.837	4	4.8	3	3	4	28.98	3.081	8	3.7	0	4	8	39.19	2.298	4	1.0
0	2	3	15.38	5.761	4	9.2	1	7	1	29.08	3.070	8	2.9	0	8	5	39.25	2.295	4	0.5
2	2	1	15.81	5.607	8	1.9	4	0	3	29.27	3.051	4	1.6	5	3	4	39.34	2.290	8	3.8
0	4	0	15.88	5.582	2	0.9	0	6	4	29.84	2.994	4	0.8	1	7	6	39.49	2.282	8	0.5
1	3	2	16.15	5.487	8	5.4	1	3	6	29.88	2.991	8	2.4	0	6	7	39.55	2.279	4	0.8
0	4	1	16.48	5.379	4	3.1	1	7	2	30.10	2.969	8	0.7	6	0	1	39.74	2.268	4	0.6
2	2	2	17.55	5.052	8	2.1	2	4	5	30.29	2.951	8	1.0	0	0	9	40.22	2.242	2	0.5
0	0	4	17.58	5.044	2	0.4	4	2	3	30.37	2.943	8	0.5	2	4	8	41.44	2.179	8	0.8
0	4	2	18.16	4.884	4	3.5	2	6	3	30.40	2.940	8	5.7	3	9	1	41.68	2.167	8	0.7
2	0	3	18.49	4.798	4	1.7	2	2	6	30.68	2.914	8	0.6	6	0	3	41.79	2.161	4	0.7
1	1	4	19.17	4.631	8	44.9	3	5	3	31.01	2.884	8	2.4	3	5	7	42.24	2.139	8	1.5
2	2	3	20.14	4.408	8	6.2	0	0	7	31.02	2.883	2	1.1	4	6	5	42.29	2.137	8	0.6
3	1	1	20.34	4.366	8	18.3	1	7	3	31.73	2.820	8	0.7	2	10	0	42.59	2.123	4	1.4
2	4	0	20.53	4.326	4	63.6	0	2	7	32.07	2.791	4	0.5	5	5	4	42.67	2.119	8	0.5
0	4	3	20.68	4.295	4	14.4	0	8	0	32.07	2.791	2	0.4	1	7	7	42.80	2.113	8	0.5
2	4	1	21.00	4.230	8	0.4	0	8	1	32.38	2.764	4	0.8	6	4	0	42.80	2.113	4	1.8
3	1	2	21.74	4.089	8	2.0	4	2	4	32.61	2.746	8	1.0	5	3	6	44.36	2.042	8	0.5
0	0	5	22.03	4.036	2	0.8	3	5	4	33.21	2.697	8	7.7	0	0	10	44.92	2.018	2	0.8
1	3	4	22.26	3.994	8	47.2	0	8	2	33.31	2.690	4	0.4	3	1	9	45.24	2.004	8	1.3
2	4	2	22.36	3.976	8	0.6	3	1	6	33.33	2.688	8	0.6	3	9	4	45.32	2.001	8	1.8
1	5	2	22.72	3.913	8	2.0	2	0	7	33.74	2.657	4	0.7	2	6	8	45.41	1.997	8	0.7
3	3	1	23.28	3.821	8	6.5	1	7	4	33.89	2.645	8	0.5	6	0	5	45.66	1.987	4	1.3
2	2	4	23.31	3.817	8	1.5	1	3	7	33.99	2.638	8	0.8	1	5	9	45.77	1.983	8	0.8
1	1	5	23.32	3.814	8	1.5	1	5	6	34.01	2.636	8	1.1	1	7	8	46.37	1.958	8	0.5
0	2	5	23.44	3.795	4	18.7	5	1	2	34.16	2.625	8	0.7	6	6	0	46.68	1.946	4	4.0
0	4	4	23.77	3.742	4	1.0	4	0	5	34.35	2.611	4	1.0	7	1	1	46.83	1.940	8	0.6
0	6	0	23.91	3.721	2	14.8	3	7	1	34.59	2.593	8	0.6	5	7	4	47.31	1.921	8	1.2
2	4	3	24.46	3.639	8	2.1	2	8	0	34.71	2.584	4	0.4	3	9	5	47.40	1.918	8	0.7
3	3	2	24.52	3.631	8	0.7	0	8	3	34.80	2.578	4	2.7	4	4	8	47.65	1.908	8	2.7
1	5	3	24.80	3.590	8	2.3	2	8	1	35.00	2.563	8	0.4	4	6	7	47.95	1.897	8	0.5
0	6	2	25.51	3.491	4	0.7	0	4	7	35.03	2.561	4	0.7	2	10	5	48.45	1.879	8	0.8
4	0	0	26.02	3.424	2	13.3	2	6	5	35.33	2.540	8	1.2	0	8	8	48.66	1.871	4	2.9
4	0	1	26.40	3.376	4	1.3	0	0	8	35.59	2.522	2	6.0	3	11	1	49.34	1.847	8	1.9
3	1	4	26.64	3.347	8	16.1	5	1	3	35.62	2.521	8	0.9	3	5	9	49.69	1.835	8	1.2
2	2	5	26.86	3.319	8	2.7	4	6	0	35.63	2.519	4	2.1	5	9	1	49.80	1.831	8	0.5

The space group symbol *Cmme* now replaces the older form, *Cmma*.

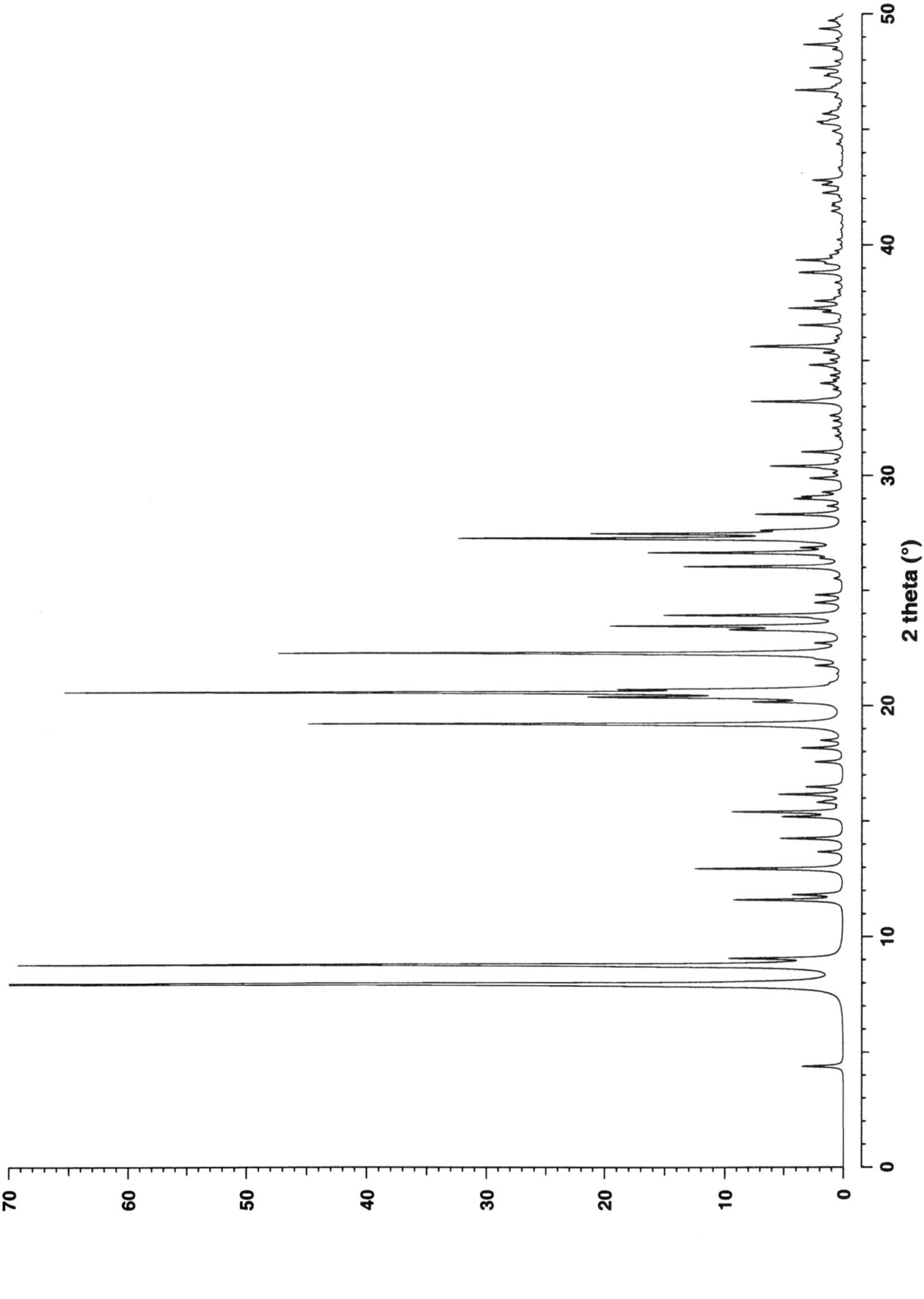
0
10
20
30
40
50
2 theta (°)
70
60
50
40
30
20
10
0

FAU Faujasite

CHEMICAL COMPOSITION: $|Na_{28.8}Ca_{14.4}(H_2O)_{263}|$ $[Si_{134.4}Al_{57.6}O_{384}]$
Kaiserstuhl, Germany

REFINED COMPOSITION: $|Na_{16}Ca_{16}(H_2O)_{42.7}|$ $[Si_{134.4}Al_{57.6}O_{384}]$

CRYSTAL DATA: $Fd\overline{3}m$ (No. 227) origin at centre ($\overline{3}m$)
$a = 24.74$ Å $b = 24.74$ Å $c = 24.74$ Å
$\alpha = 90°$ $\beta = 90°$ $\gamma = 90°$
X-ray single crystal refinement, $R = 0.13$

REFERENCE: W. H. Baur,
American Mineralogist **49** 697–704 (1964).

h	*k*	*l*	2*θ*	*d*	*M*	I_{rel}	*h*	*k*	*l*	2*θ*	*d*	*M*	I_{rel}	*h*	*k*	*l*	2*θ*	*d*	*M*	I_{rel}
1	1	1	6.19	14.284	8	100.0	7	3	1	27.70	3.221	48	0.5	8	6	2	37.06	2.426	48	0.3
2	2	0	10.11	8.747	12	1.4	8	0	0	28.87	3.092	6	0.2	10	2	0	37.06	2.426	24	0.3
3	1	1	11.86	7.459	24	2.0	7	3	3	29.55	3.022	24	0.5	6	6	6	37.79	2.381	8	0.6
4	0	0	14.32	6.185	6	0.5	8	2	2	30.66	2.916	24	1.1	7	7	5	40.43	2.231	24	0.2
3	3	1	15.61	5.676	24	4.5	6	6	0	30.66	2.916	12	0.7	8	8	0	41.29	2.187	12	0.5
4	2	2	17.56	5.050	24	0.3	5	5	5	31.31	2.857	8	2.0	11	3	1	41.79	2.162	48	0.2
5	1	1	18.64	4.761	24	2.7	7	5	1	31.31	2.857	48	0.3	9	7	1	41.79	2.162	48	0.3
4	4	0	20.30	4.373	12	2.5	8	4	0	32.37	2.766	24	1.1	10	6	0	42.62	2.121	24	0.2
4	4	2	21.55	4.123	24	0.4	9	1	1	32.98	2.716	24	0.5	11	3	3	43.11	2.098	24	0.2
6	2	0	22.73	3.912	24	0.2	7	5	3	32.98	2.716	48	0.5	9	7	3	43.11	2.098	48	0.2
5	3	3	23.58	3.773	24	5.6	8	4	2	33.19	2.699	48	0.1	8	8	4	43.92	2.062	24	0.5
4	4	4	24.93	3.571	8	0.2	6	6	4	33.99	2.637	24	1.3	12	4	2	47.04	1.932	48	0.3
7	1	1	25.72	3.464	24	0.1	9	3	1	34.58	2.593	48	1.0	10	8	2	47.64	1.909	48	0.1
5	5	1	25.72	3.464	24	0.5	8	4	4	35.55	2.525	24	0.1	13	3	1	49.28	1.849	48	0.1
6	4	2	26.97	3.306	48	2.4	7	5	5	36.12	2.486	24	0.1							

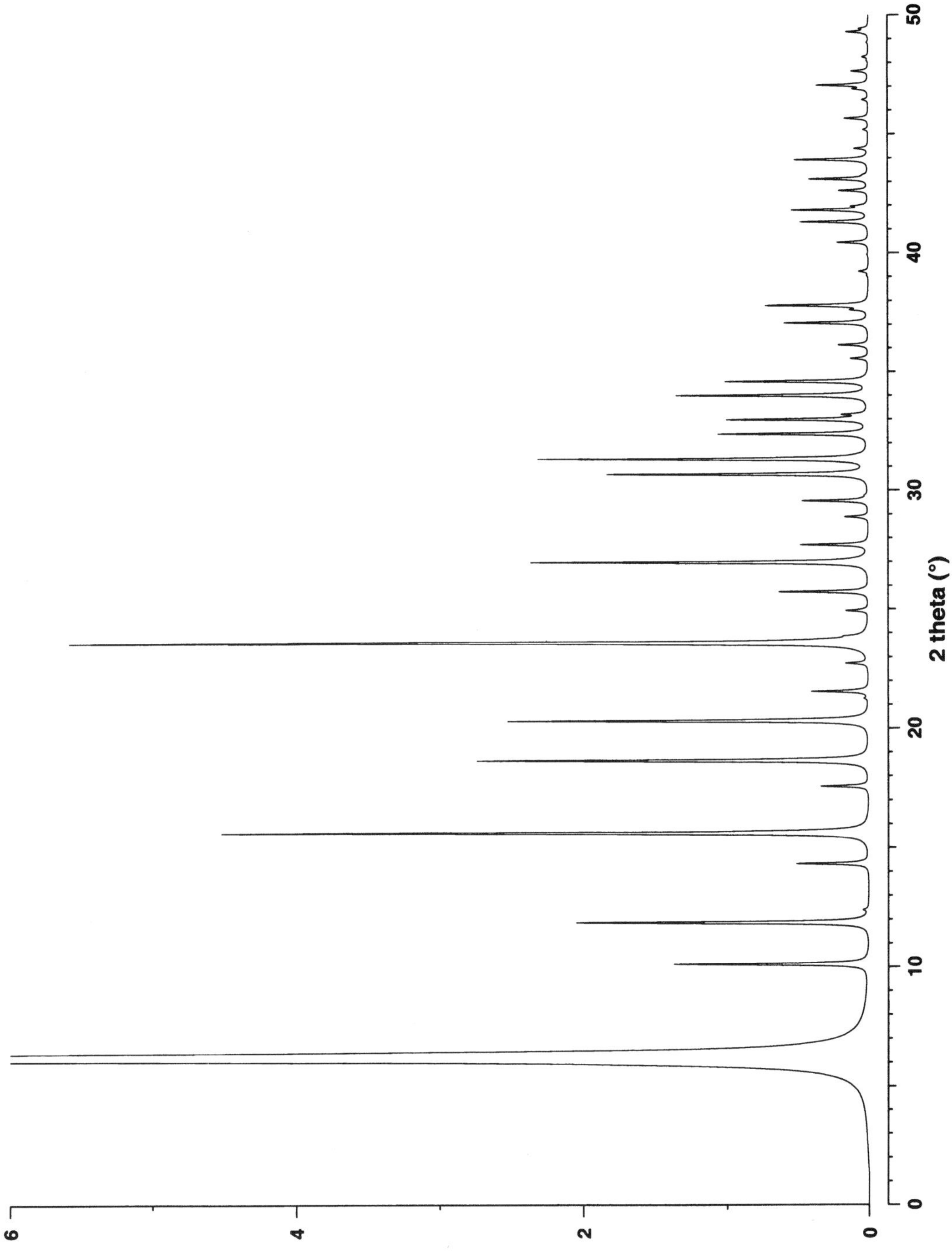
2 theta (°)
0
10
20
30
40
50
0
2
4
6

FAU

Na-X, Dehydrated

CHEMICAL COMPOSITION: $|Na_{88}|\ [Si_{104}Al_{88}O_{384}]$

REFINED COMPOSITION: $|Na_{92.9}|\ [Si_{103.68}Al_{88.32}O_{384}]$

CRYSTAL DATA: $Fd\bar{3}$ (No. 203) origin at centre ($\bar{3}$)
$a = 25.099$ Å $b = 25.099$ Å $c = 25.099$ Å
$\alpha = 90°$ $\beta = 90°$ $\gamma = 90°$
X-ray single crystal refinement, $R = 0.046$

REFERENCE: D. H. Olson,
Zeolites **15** 439–443 (1995).

h	k	l	2θ	d	M	I_{rel}	h	k	l	2θ	d	M	I_{rel}	h	k	l	2θ	d	M	I_{rel}
1	1	1	6.10	14.491	8	100.0	5	5	5	30.85	2.898	8	4.5	11	1	3	41.16	2.193	24	0.1
2	2	0	9.97	8.874	12	12.7	8	0	4	31.89	2.806	12	0.5	9	1	7	41.16	2.193	24	0.1
3	1	1	11.69	7.568	24	8.0	8	4	0	31.89	2.806	12	0.7	10	4	4	41.33	2.185	24	0.1
4	0	0	14.11	6.275	6	0.1	7	5	3	32.50	2.755	24	0.2	11	3	3	42.46	2.129	24	1.0
3	3	1	15.39	5.758	24	11.1	7	3	5	32.50	2.755	24	0.1	12	0	0	43.26	2.092	6	0.3
5	1	1	18.37	4.830	24	0.4	6	6	4	33.49	2.676	24	2.7	8	8	4	43.26	2.092	24	0.3
4	4	0	20.01	4.437	12	2.4	9	1	3	34.07	2.631	24	0.5	9	9	1	46.17	1.966	24	0.2
5	3	1	20.94	4.243	24	0.4	9	3	1	34.07	2.631	24	0.5	12	4	2	46.32	1.960	24	0.2
4	4	2	21.24	4.183	24	0.2	8	4	4	35.03	2.562	24	0.3	8	8	6	46.32	1.960	24	0.2
6	2	0	22.40	3.969	12	0.2	7	5	5	35.59	2.523	24	0.3	12	2	4	46.32	1.960	24	0.2
5	3	3	23.24	3.828	24	7.7	8	2	6	36.51	2.461	24	0.1	10	2	8	46.92	1.936	24	0.3
6	2	2	23.51	3.784	24	0.5	10	2	2	37.23	2.415	24	0.3	10	8	2	46.92	1.936	24	0.4
6	4	2	26.58	3.354	24	2.8	6	6	6	37.23	2.415	8	0.6	9	9	3	47.36	1.919	24	0.2
6	2	4	26.58	3.354	24	2.0	11	1	1	39.83	2.263	24	0.8	11	3	7	48.53	1.876	24	0.2
7	3	3	29.12	3.066	24	0.6	7	7	5	39.83	2.263	24	0.3	9	7	7	48.53	1.876	24	0.2
8	2	2	30.21	2.958	24	0.5	8	8	0	40.67	2.218	12	0.8	11	7	3	48.53	1.876	24	0.2
6	6	0	30.21	2.958	12	0.4	11	3	1	41.16	2.193	24	0.2	9	9	5	49.67	1.835	24	0.2

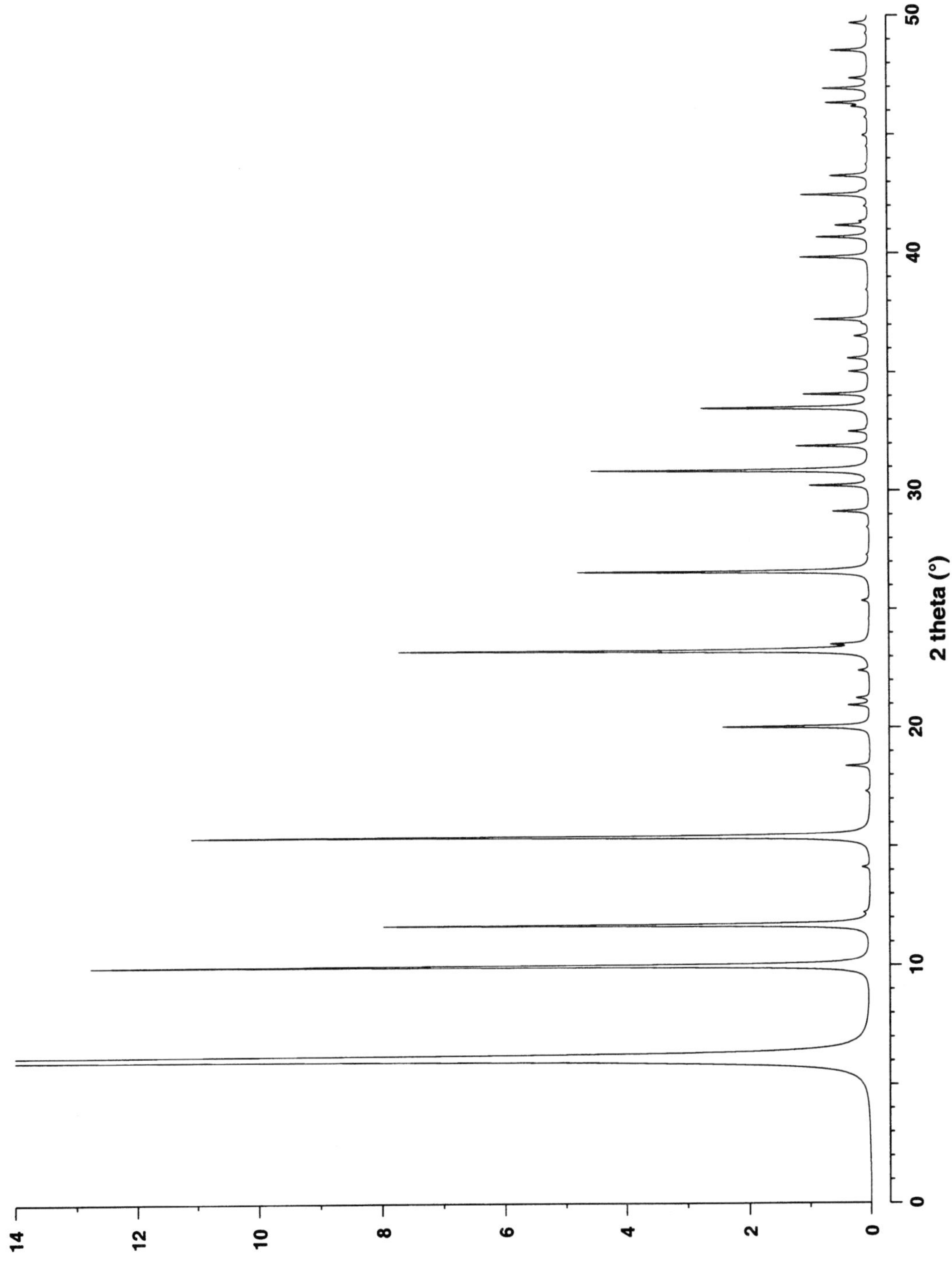
2 theta (°)
0
10
20
30
40
50
0
2
4
6
8
10
12
14

FAU

Na-X, Hydrated

CHEMICAL COMPOSITION: $|Na_{88}(H_2O)_{220}|$ $[Si_{104}Al_{88}O_{384}]$

REFINED COMPOSITION: $|Na_{40.32}(H_2O)_{171.84}|$ $[Si_{103.68}Al_{88.32}O_{384}]$

CRYSTAL DATA: $Fd\bar{3}$ (No. 203) origin at centre ($\bar{3}$)
$a = 25.028$ Å $b = 25.028$ Å $c = 25.028$ Å
$\alpha = 90°$ $\beta = 90°$ $\gamma = 90°$
X-ray single crystal refinement, $R = 0.09$

REFERENCE: D. H. Olson,
J. Phys. Chem. **74** 2758–2764 (1970).

h	k	l	2θ	d	M	I_{rel}
1	1	1	6.12	14.450	8	100.0
2	2	0	10.00	8.849	12	10.4
3	1	1	11.73	7.546	24	3.8
2	2	2	12.25	7.225	8	0.2
3	3	1	15.43	5.742	24	6.5
4	2	2	17.36	5.109	24	0.1
3	3	3	18.42	4.817	8	0.3
5	1	1	18.42	4.817	24	1.3
4	4	0	20.07	4.424	12	2.0
5	3	1	21.00	4.231	24	0.6
6	2	0	22.47	3.957	12	1.1
6	0	2	22.47	3.957	12	0.2
5	3	3	23.31	3.817	24	6.2
6	2	2	23.58	3.773	24	0.5
4	4	4	24.64	3.612	8	0.2
5	5	1	25.41	3.505	24	0.2
6	2	4	26.65	3.345	24	2.4
6	4	2	26.65	3.345	24	3.2
7	3	1	27.37	3.258	24	0.1
7	3	3	29.21	3.058	24	1.2
6	6	0	30.30	2.950	12	1.0
8	2	2	30.30	2.950	24	1.7
7	1	5	30.94	2.890	24	0.4
7	5	1	30.94	2.890	24	0.4
5	5	5	30.94	2.890	8	5.7
6	6	2	31.15	2.871	24	0.2
8	0	4	31.98	2.798	12	1.6
8	4	0	31.98	2.798	12	1.8
7	3	5	32.59	2.747	24	0.1
9	1	1	32.59	2.747	24	0.1
7	5	3	32.59	2.747	24	0.4
8	4	2	32.80	2.731	24	0.1
8	2	4	32.80	2.731	24	0.1
6	6	4	33.59	2.668	24	3.4
9	3	1	34.17	2.624	24	0.5
9	1	3	34.17	2.624	24	0.6
8	4	4	35.13	2.554	24	0.5
8	2	6	36.61	2.454	24	0.1
10	2	2	37.34	2.408	24	0.2
6	6	6	37.34	2.408	8	1.2
11	1	1	39.95	2.257	24	0.2
7	7	5	39.95	2.257	24	0.3
8	8	0	40.79	2.212	12	1.1
11	1	3	41.29	2.187	24	0.3
11	3	1	41.29	2.187	24	0.3
9	5	5	41.29	2.187	24	0.1
9	7	1	41.29	2.187	24	0.2
9	1	7	41.29	2.187	24	0.2
8	8	2	41.45	2.178	24	0.1
11	3	3	42.59	2.123	24	0.6
8	8	4	43.38	2.086	24	0.3
9	9	1	46.31	1.960	24	0.2
12	4	2	46.46	1.954	24	0.3
12	2	4	46.46	1.954	24	0.2
10	8	2	47.06	1.931	24	0.4
10	2	8	47.06	1.931	24	0.2
12	4	4	48.24	1.887	24	0.2
9	7	7	48.67	1.871	24	0.2
13	3	3	49.82	1.830	24	0.1
9	9	5	49.82	1.830	24	0.2

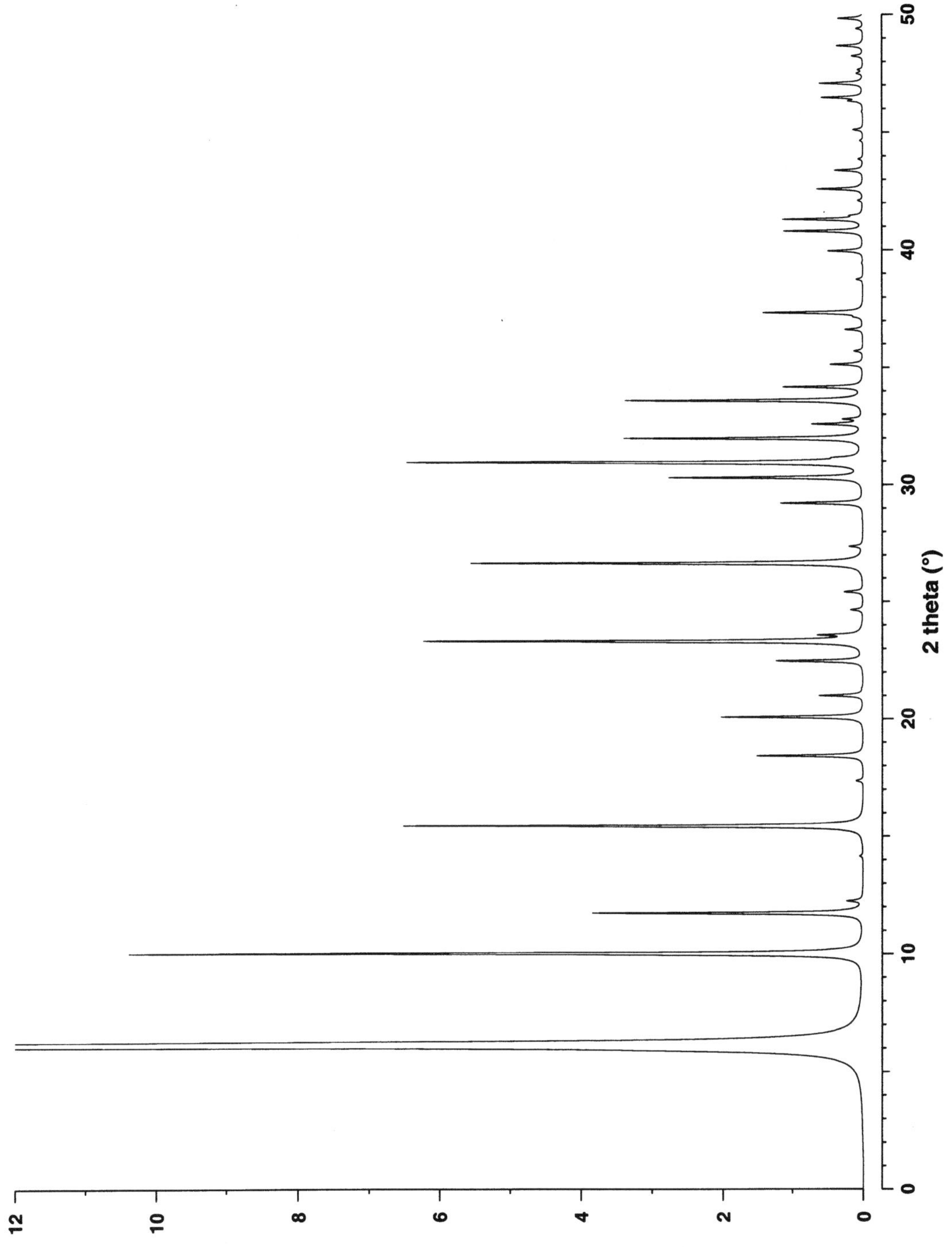
0
10
20
30
40
50
2 theta (°)
0
2
4
6
8
10
12

FAU Ultrastable Y, Dehydrated Dealuminated

CHEMICAL COMPOSITION: $|H_{18.9}|\ [Si_{173.1}Al_{18.9}O_{384}]$

REFINED COMPOSITION: $|Al_{5.6}O_{22.4}|\ [Si_{175.7}Al_{16.3}O_{384}]$

CRYSTAL DATA: $Fd\bar{3}m$ (No. 227) origin at centre ($\bar{3}m$)
$a = 24.188$ Å $b = 24.188$ Å $c = 24.188$ Å
$\alpha = 90°$ $\beta = 90°$ $\gamma = 90°$
Neutron Rietveld refinement, $R_{wp} = 0.1041$, $R_I = 0.0561$

REFERENCE: J. B. Parise, D. R. Corbin, L. Abrams and D. E. Cox, *Acta Cryst.* **C40** 1493–1497 (1984).

h	k	l	2θ	d	M	I_{rel}	h	k	l	2θ	d	M	I_{rel}	h	k	l	2θ	d	M	I_{rel}
1	1	1	6.33	13.965	8	100.0	5	5	1	26.31	3.387	24	0.5	8	6	2	37.93	2.372	48	0.2
2	2	0	10.34	8.552	12	8.7	6	4	2	27.60	3.232	48	4.1	9	5	1	38.50	2.338	48	0.1
3	1	1	12.14	7.293	24	5.8	7	3	3	30.24	2.955	24	0.9	6	6	6	38.69	2.327	8	0.4
4	0	0	14.65	6.047	6	0.8	8	2	2	31.38	2.851	24	0.5	8	6	4	40.15	2.246	48	0.2
3	3	1	15.97	5.549	24	9.0	6	6	0	31.38	2.851	12	1.1	11	1	1	41.40	2.181	24	0.2
4	2	2	17.97	4.937	24	0.7	7	5	1	32.04	2.793	48	0.4	8	8	0	42.27	2.138	12	0.4
5	1	1	19.07	4.655	24	4.4	5	5	5	32.04	2.793	8	2.5	9	7	3	44.14	2.052	48	0.2
3	3	3	19.07	4.655	8	0.3	8	4	0	33.13	2.704	24	0.8	11	3	3	44.14	2.052	24	0.2
4	4	0	20.77	4.276	12	3.8	7	5	3	33.76	2.655	48	0.6	10	6	2	44.31	2.044	48	0.1
5	3	1	21.74	4.089	48	0.5	8	4	2	33.97	2.639	48	0.2	8	8	4	44.97	2.016	24	0.5
4	4	2	22.05	4.031	24	0.2	6	6	4	34.79	2.578	24	1.4	12	0	0	44.97	2.016	6	0.2
6	2	0	23.26	3.824	24	1.3	9	3	1	35.40	2.536	48	0.7	8	8	6	48.18	1.889	24	0.2
5	3	3	24.13	3.689	24	7.1	8	4	4	36.39	2.469	24	0.2	10	8	2	48.80	1.866	48	0.5
4	4	4	25.51	3.491	8	0.3	7	5	5	36.98	2.431	24	0.2							

Ultrastable Y, Dehydrated Dealuminated FAU

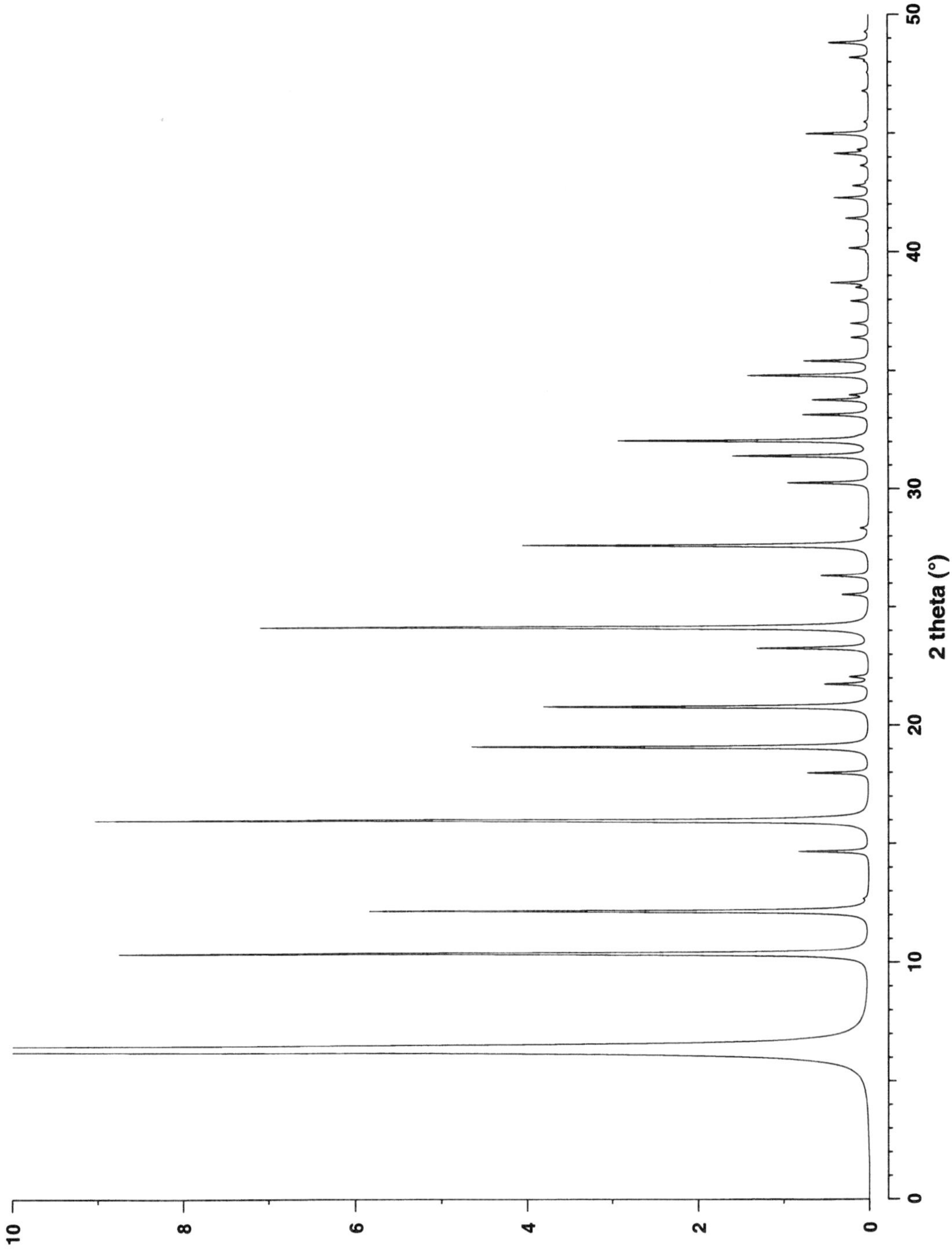

FAU

Na-Y, Siliceous

CHEMICAL COMPOSITION: $[Si_{192}O_{384}]$

REFINED COMPOSITION: $[Si_{188.16}O_{384}]$

CRYSTAL DATA: $Fd\overline{3}m$ (No. 227) origin at centre ($\overline{3}m$)
a = 24.2576 Å b = 24.2576 Å c = 24.2576 Å
$\alpha = 90°$ $\beta = 90°$ $\gamma = 90°$
Neutron Rietveld refinement, $R_{\text{exp}} = 0.022$, $R_{\text{wp}} = 0.031$

REFERENCE: J. J. Hriljac, M. M. Eddy, A. K. Cheetham, J. A. Donohue and G. J. Ray, *J. Solid State Chem.* **106** 66–72 (1993).

h	k	l	2θ	d	M	I_{rel}	h	k	l	2θ	d	M	I_{rel}	h	k	l	2θ	d	M	I_{rel}
1	1	1	6.31	14.005	8	100.0	7	3	1	28.26	3.158	48	0.2	8	6	4	40.03	2.252	48	0.2
2	2	0	10.31	8.576	12	18.6	7	3	3	30.16	2.964	24	1.5	11	1	1	41.28	2.187	24	0.4
3	1	1	12.10	7.314	24	12.5	8	2	2	31.29	2.859	24	0.7	7	7	5	41.28	2.187	24	0.2
4	0	0	14.61	6.064	6	0.3	6	6	0	31.29	2.859	12	1.9	8	8	0	42.14	2.144	12	0.6
3	3	1	15.92	5.565	24	16.0	7	5	1	31.95	2.801	48	0.5	9	7	1	42.66	2.119	48	0.1
4	2	2	17.91	4.952	24	0.1	5	5	5	31.95	2.801	8	4.4	10	6	0	43.51	2.080	24	0.1
5	1	1	19.01	4.668	24	4.3	8	4	0	33.03	2.712	24	1.2	9	7	3	44.01	2.058	48	0.1
3	3	3	19.01	4.668	8	0.2	7	5	3	33.66	2.663	48	0.5	11	3	3	44.01	2.058	24	0.5
4	4	0	20.71	4.288	12	6.4	8	4	2	33.87	2.647	48	0.1	10	6	2	44.18	2.050	48	0.2
5	3	1	21.67	4.100	48	0.3	6	6	4	34.69	2.586	24	2.4	12	0	0	44.84	2.021	6	0.3
4	4	2	21.98	4.043	24	0.1	9	3	1	35.29	2.543	48	1.0	8	8	4	44.84	2.021	24	0.7
6	2	0	23.19	3.835	24	1.3	8	4	4	36.28	2.476	24	0.3	9	7	5	46.61	1.948	48	0.1
5	3	3	24.06	3.699	24	7.6	7	5	5	36.87	2.438	24	0.2	12	4	2	48.03	1.894	48	0.1
4	4	4	25.44	3.501	8	0.3	8	6	2	37.82	2.379	48	0.2	8	8	6	48.03	1.894	24	0.2
5	5	1	26.24	3.397	24	1.2	9	5	1	38.38	2.345	48	0.1	10	8	2	48.65	1.872	48	1.1
6	4	2	27.52	3.242	48	4.6	6	6	6	38.57	2.334	8	1.1							

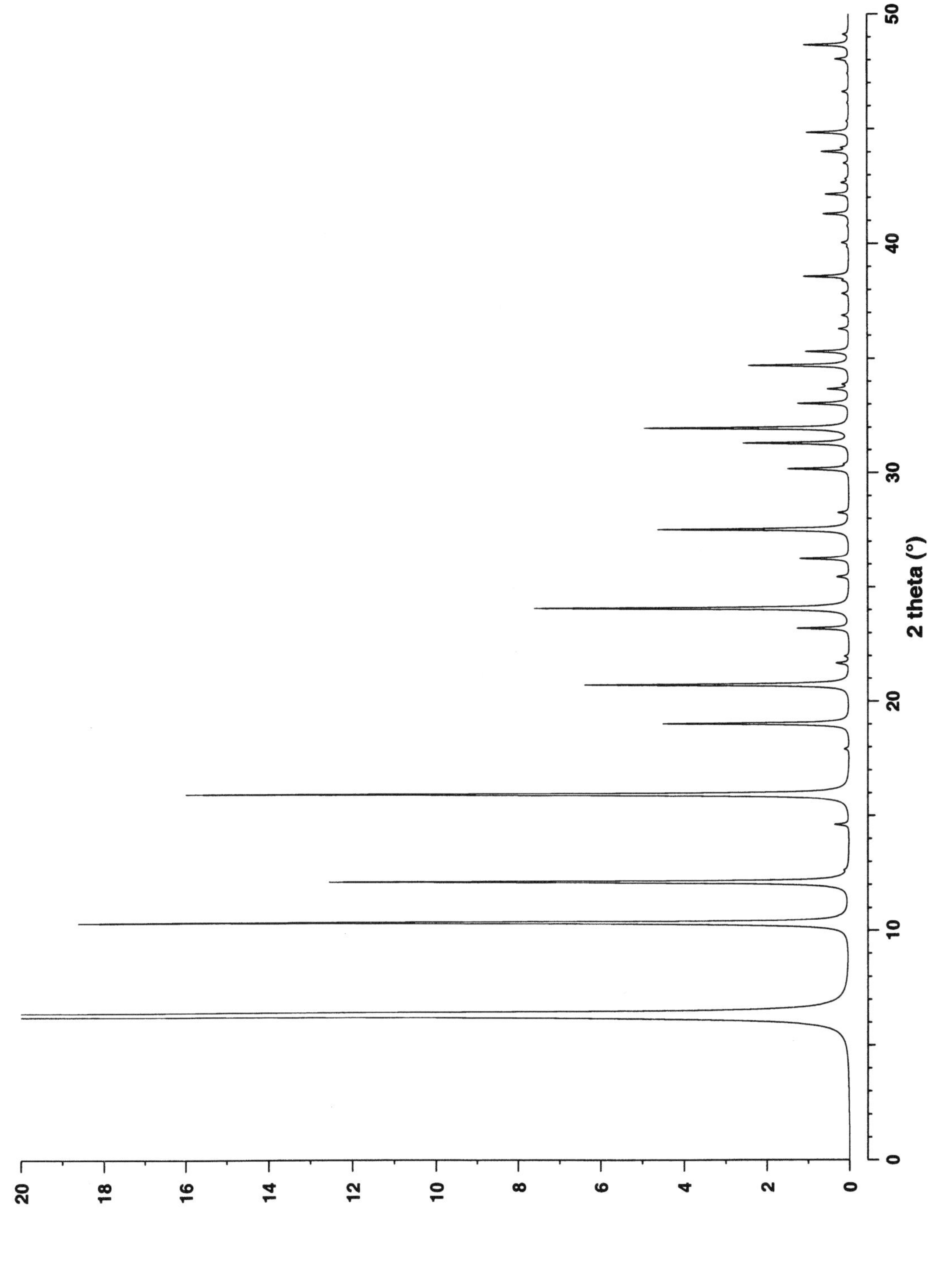
2 theta (°)
0
10
20
30
40
50
20
18
16
14
12
10
8
6
4
2
0

FAU Li-LSX

CHEMICAL COMPOSITION: $|Li_{96}|\ [Si_{96}Al_{96}O_{384}]$

REFINED COMPOSITION: $|Li_{87.36}|\ [Si_{96}Al_{96}O_{384}]$

CRYSTAL DATA: $Fd\bar{3}$ (No. 203) origin at centre ($\bar{3}$)
$a = 25.6957$ Å $b = 25.6957$ Å $c = 25.6957$ Å
$\alpha = 90.0°$ $\beta = 90.0°$ $\gamma = 90.0°$
X-ray and neutron Rietveld refinement, $R_p = 0.057$, $R_{wp} = 0.069$

REFERENCE: M. Feuerstein and R. F. Lobo,
Chemistry of Materials 2197–2204 (1998).

h	k	l	2θ	d	M	I_{rel}
1	1	1	5.96	14.835	8	100.0
2	2	0	9.74	9.085	12	9.6
3	1	1	11.42	7.748	24	9.9
3	3	1	15.03	5.895	24	12.3
4	2	2	16.90	5.245	24	0.1
5	1	1	17.94	4.945	24	1.8
4	4	0	19.54	4.542	12	3.0
5	3	1	20.45	4.343	24	0.2
6	0	2	21.88	4.063	12	0.1
6	2	0	21.88	4.063	12	0.4
5	3	3	22.69	3.919	24	5.7
4	4	4	23.99	3.709	8	0.5
5	5	1	24.74	3.598	24	0.3
6	4	2	25.95	3.434	24	2.5
6	2	4	25.95	3.434	24	1.4
5	5	3	26.65	3.345	24	0.1
7	3	3	28.43	3.139	24	0.6
8	2	2	29.50	3.028	24	0.4
6	6	0	29.50	3.028	12	1.1
7	1	5	30.12	2.967	24	0.1
5	5	5	30.12	2.967	8	3.6
8	0	4	31.13	2.873	12	0.3
8	4	0	31.13	2.873	12	0.3
7	3	5	31.72	2.820	24	0.2
7	5	3	31.72	2.820	24	0.4
6	6	4	32.69	2.739	24	2.1
9	1	3	33.26	2.694	24	0.2
9	3	1	33.26	2.694	24	0.1
8	4	4	34.19	2.623	24	0.2
7	7	1	34.74	2.583	24	0.2
10	2	2	36.33	2.473	24	0.2
6	6	6	36.33	2.473	8	2.1
8	6	4	37.70	2.386	24	0.2
7	7	5	38.87	2.317	24	0.8
11	1	1	38.87	2.317	24	0.2
9	5	5	40.17	2.245	24	0.1
8	6	6	40.96	2.203	24	0.1
11	3	3	41.43	2.179	24	0.3
9	7	3	41.43	2.179	24	0.2
9	3	7	41.43	2.179	24	0.2
8	8	4	42.20	2.141	24	0.6
12	4	2	45.19	2.006	24	0.2
12	2	4	45.19	2.006	24	0.1
10	2	8	45.77	1.982	24	0.1
10	8	2	45.77	1.982	24	0.3
9	7	7	47.33	1.921	24	0.2
9	9	5	48.44	1.879	24	0.6
8	8	8	49.13	1.854	8	0.1
11	5	7	49.54	1.840	24	0.1

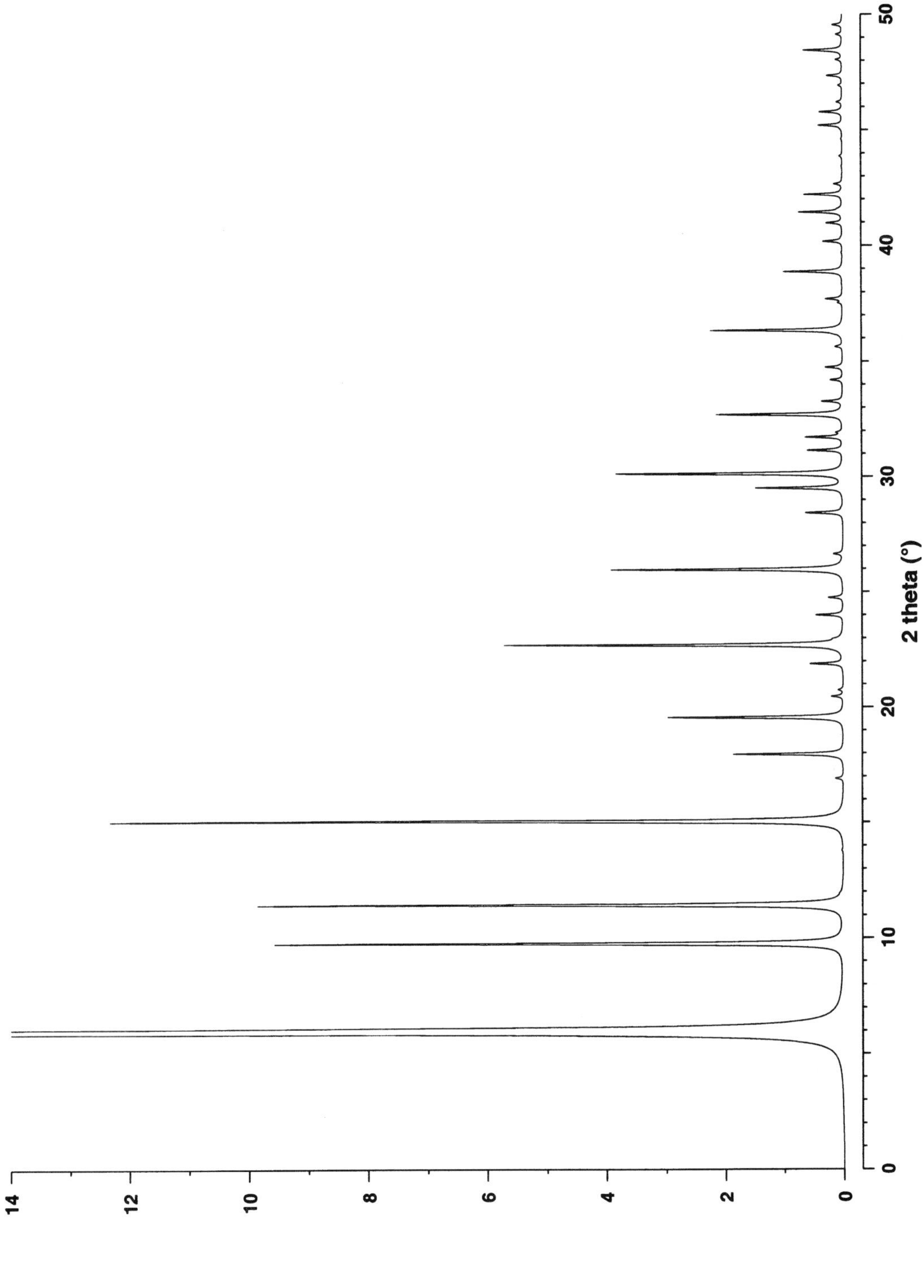
2 theta (°)
0
10
20
30
40
50
0
2
4
6
8
10
12
14

FER Ferrierite

CHEMICAL COMPOSITION: $|Na_{1.5}Mg_2(H_2O)_{18}|$ $[Si_{30.5}Al_{5.5}O_{72}]$
Kamloops Lake, British Columbia, Canada

REFINED COMPOSITION: $|Na_{0.96}Mg_2(H_2O)_{12}|$ $[Si_{30.24}Al_{5.76}O_{72}]$

CRYSTAL DATA: *Immm* (No. 71)
$a = 19.156$ Å $b = 14.127$ Å $c = 7.489$ Å
$\alpha = 90°$ $\beta = 90°$ $\gamma = 90°$
X-ray single crystal refinement, $R = 0.11$

REFERENCE: P. A. Vaughan,
Acta Cryst. **21** 983–990 (1966).

h	k	l	2θ	d	M	I_{rel}
1	1	0	7.78	11.370	4	58.8
2	0	0	9.23	9.578	2	100.0
0	2	0	12.53	7.063	2	9.2
1	0	1	12.69	6.975	4	13.7
0	1	1	13.38	6.617	4	7.9
3	1	0	15.23	5.819	4	26.1
2	2	0	15.59	5.685	4	2.5
2	1	1	16.28	5.444	8	0.2
1	2	1	17.87	4.963	8	5.5
3	0	1	18.26	4.859	4	0.5
4	0	0	18.53	4.789	2	2.2
1	3	0	19.41	4.573	4	1.8
3	2	1	22.21	4.003	8	26.2
0	3	1	22.30	3.986	4	7.7
4	2	0	22.43	3.964	4	21.1
4	1	1	22.92	3.879	8	15.0
3	3	0	23.47	3.790	4	16.8
0	0	2	23.76	3.744	2	3.6
5	1	0	24.07	3.698	4	6.7
1	1	2	25.04	3.557	8	15.7
0	4	0	25.22	3.532	2	29.2
2	0	2	25.54	3.487	4	35.6
5	0	1	26.13	3.411	4	6.1
2	4	0	26.91	3.314	4	6.8
0	2	2	26.95	3.308	4	0.6
6	0	0	27.95	3.193	2	1.2
1	4	1	28.32	3.151	8	1.6
3	1	2	28.34	3.149	8	13.5
2	2	2	28.54	3.127	8	1.8
5	2	1	29.07	3.071	8	9.2
4	3	1	29.15	3.064	8	0.1
5	3	0	30.07	2.972	4	7.6
4	0	2	30.30	2.950	4	2.3

h	k	l	2θ	d	M	I_{rel}
6	2	0	30.73	2.909	4	0.2
1	3	2	30.86	2.897	8	5.1
6	1	1	31.10	2.875	8	0.8
3	4	1	31.31	2.857	8	0.2
4	4	0	31.47	2.842	4	0.8
1	5	0	32.02	2.795	4	0.4
4	2	2	32.90	2.722	8	3.6
7	1	0	33.35	2.687	4	0.9
0	5	1	33.91	2.644	4	4.2
5	1	2	34.08	2.631	8	0.7
3	5	0	34.72	2.584	4	2.8
7	0	1	34.91	2.570	4	0.7
0	4	2	34.92	2.569	4	1.1
6	3	1	36.04	2.492	8	0.8
2	4	2	36.20	2.482	8	1.3
1	0	3	36.29	2.475	4	0.2
6	0	2	37.00	2.429	4	2.8
7	2	1	37.22	2.415	8	0.6
8	0	0	37.56	2.395	2	0.2
6	4	0	37.99	2.368	4	0.2
7	3	0	38.03	2.366	4	3.3
1	2	3	38.54	2.336	8	0.6
3	0	3	38.73	2.325	4	0.2
4	5	1	38.91	2.314	8	2.2
6	2	2	39.21	2.297	8	0.4
2	6	0	39.41	2.286	4	0.2
5	5	0	39.63	2.274	4	0.2
8	2	0	39.75	2.268	4	0.3
4	4	2	39.81	2.264	8	1.1
8	1	1	40.04	2.252	8	0.7
1	5	2	40.26	2.240	8	0.8
1	6	1	40.43	2.231	8	0.1
4	1	3	41.28	2.187	8	0.7

h	k	l	2θ	d	M	I_{rel}
7	1	2	41.36	2.183	8	0.2
2	3	3	42.04	2.149	8	1.9
3	5	2	42.51	2.127	8	0.4
3	6	1	42.67	2.119	8	0.6
4	6	0	42.80	2.113	4	2.5
9	1	0	42.97	2.105	4	0.5
9	0	1	44.24	2.047	4	1.4
6	5	1	44.49	2.036	8	0.7
1	4	3	44.70	2.027	8	0.2
8	0	2	44.93	2.017	4	0.5
1	7	0	45.18	2.007	4	0.2
5	2	3	45.21	2.005	8	3.1
6	4	2	45.30	2.002	8	1.8
7	3	2	45.34	2.000	8	0.3
0	6	2	45.51	1.993	4	0.2
8	4	0	45.78	1.982	4	1.2
9	2	1	46.16	1.966	8	1.5
0	7	1	46.61	1.949	4	1.7
6	1	3	46.63	1.948	8	0.6
5	5	2	46.74	1.944	8	0.2
3	4	3	46.78	1.942	8	0.2
8	2	2	46.83	1.940	8	0.1
9	3	0	46.84	1.940	4	2.0
5	6	1	46.89	1.938	8	0.2
3	7	0	47.23	1.924	4	2.3
10	0	0	47.46	1.916	2	0.3
0	0	4	48.63	1.872	2	5.7
1	1	4	49.33	1.847	8	0.4
4	6	2	49.53	1.840	8	0.3
2	0	4	49.61	1.837	4	0.3
2	5	3	49.65	1.836	8	0.4

Ferrierite FER

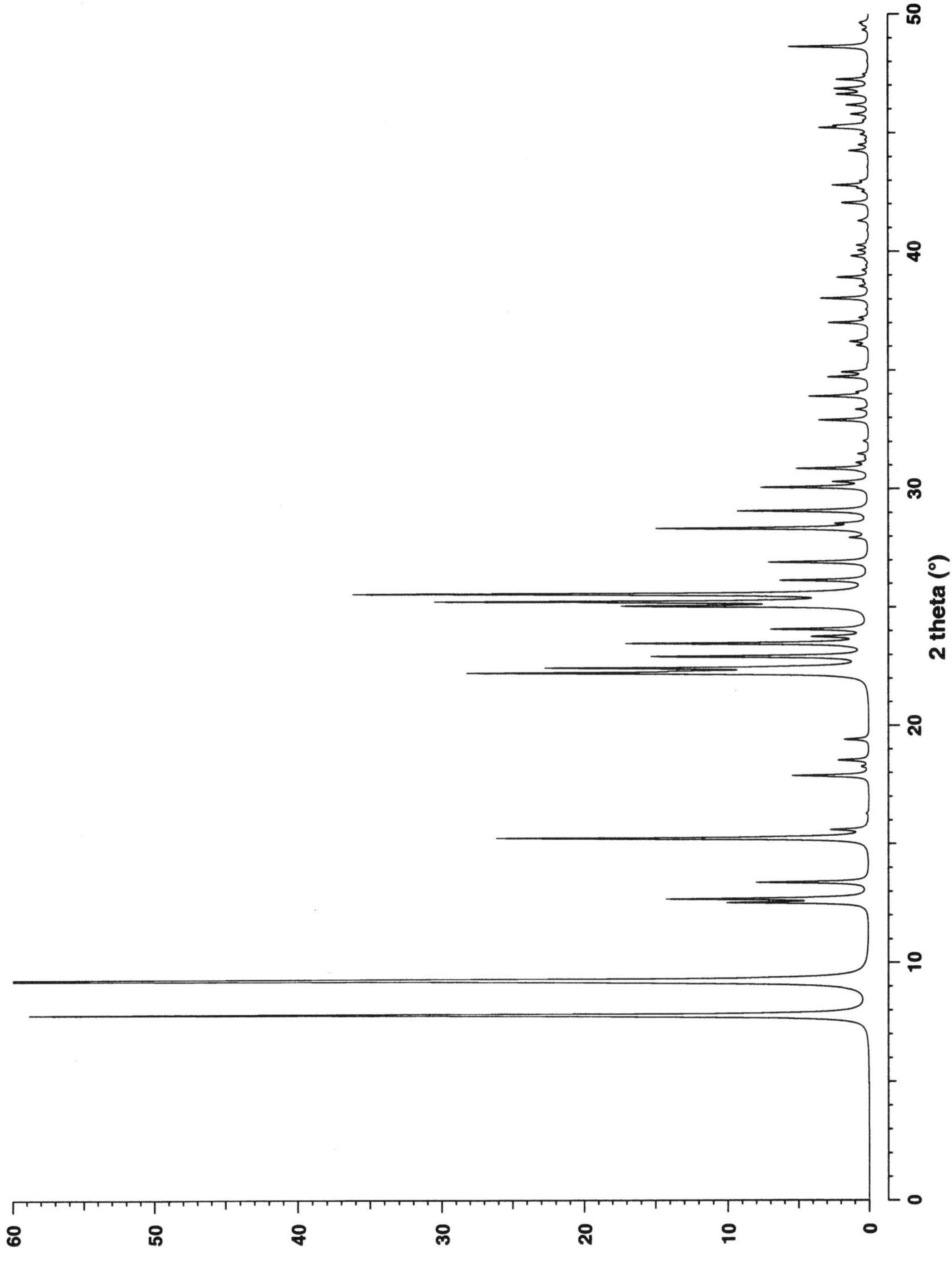

FER Ferrierite, Siliceous

CHEMICAL COMPOSITION: $[Si_{36}O_{72}]$
Dealuminated, synthetic powder.

REFINED COMPOSITION: $[Si_{36}O_{72}]$

CRYSTAL DATA: *Pnnm* (No. 58)
$a = 14.07025$ Å $b = 7.41971$ Å $c = 18.7202$ Å
$\alpha = 90°$ $\beta = 90°$ $\gamma = 90°$
Combined X-ray and neutron Rietveld refinement, $R_{wp} = 0.123$

REFERENCE: R. E. Morris, S. J. Weigel, N. J. Henson, L. M. Bull, M. T. Janicke, B. F. Chmelka and A. K. Cheetham, *J. Amer. Chem. Soc.* **116** 11849–11855 (1994).

h	k	l	2θ	d	M	I_{rel}	h	k	l	2θ	d	M	I_{rel}	h	k	l	2θ	d	M	I_{rel}
1	0	1	7.86	11.248	4	1.9	2	1	5	29.59	3.019	8	4.2	6	1	0	40.33	2.236	4	0.2
0	0	2	9.45	9.360	2	100.0	4	1	2	29.68	3.010	8	0.1	5	2	1	40.52	2.226	8	0.8
2	0	0	12.58	7.035	2	16.2	3	0	5	30.55	2.926	4	2.5	3	1	7	40.70	2.217	8	0.3
0	1	1	12.83	6.898	4	28.3	3	2	0	30.73	2.910	4	0.2	1	1	8	40.94	2.204	8	0.1
1	1	0	13.49	6.563	4	13.5	0	2	4	30.75	2.907	4	1.0	1	3	4	41.80	2.161	8	0.3
1	1	1	14.30	6.193	8	0.1	3	2	1	31.11	2.875	8	1.9	3	3	2	42.43	2.130	8	1.0
1	0	3	15.53	5.704	4	5.1	2	0	6	31.36	2.852	4	0.3	3	2	6	42.48	2.128	8	0.1
2	0	2	15.76	5.624	4	7.1	1	2	4	31.42	2.847	8	0.1	2	1	8	42.50	2.127	8	0.1
1	1	2	16.50	5.374	8	1.1	1	1	6	31.75	2.818	8	0.6	5	2	3	42.86	2.110	8	0.1
2	1	0	17.37	5.105	4	0.3	4	0	4	31.82	2.812	4	0.1	6	1	3	42.97	2.105	8	0.1
2	1	1	18.01	4.925	8	0.6	3	2	2	32.22	2.778	8	0.3	6	0	4	43.15	2.097	4	0.9
0	0	4	18.96	4.680	2	3.8	2	2	4	33.35	2.687	8	1.3	6	1	4	44.93	2.018	8	0.1
3	0	1	19.51	4.549	4	0.3	3	2	3	34.00	2.637	8	0.1	4	3	1	45.07	2.011	8	0.1
2	1	2	19.81	4.482	8	0.1	5	1	0	34.07	2.631	4	2.1	5	1	6	45.07	2.011	8	1.0
3	1	0	22.43	3.964	4	5.1	4	1	4	34.10	2.629	8	0.2	0	1	9	45.28	2.003	4	0.9
2	1	3	22.50	3.951	8	12.9	1	2	5	34.63	2.590	8	0.5	6	2	0	45.77	1.982	4	0.2
2	0	4	22.82	3.897	4	8.8	5	0	3	34.98	2.565	4	1.0	2	3	5	45.82	1.980	8	2.1
1	1	4	23.34	3.810	8	7.1	4	2	0	35.16	2.553	4	0.5	0	2	8	45.85	1.979	4	0.5
3	0	3	23.73	3.749	4	11.9	4	2	1	35.49	2.529	8	0.1	4	2	6	45.93	1.976	8	1.2
0	2	0	23.99	3.710	2	3.7	0	1	7	35.69	2.516	4	0.6	6	2	1	46.04	1.971	8	0.1
1	0	5	24.60	3.618	4	5.3	4	2	2	36.48	2.463	8	1.0	3	2	7	46.10	1.969	8	0.1
1	2	0	24.82	3.587	4	0.2	0	3	1	36.65	2.452	4	0.5	1	2	8	46.32	1.960	8	0.2
1	2	1	25.28	3.523	8	7.7	3	1	6	36.65	2.452	8	0.8	4	0	8	46.62	1.948	4	0.3
4	0	0	25.32	3.518	2	15.2	1	3	0	36.90	2.436	4	0.4	7	1	0	46.83	1.940	4	1.0
2	1	4	25.82	3.450	8	0.6	0	2	6	37.67	2.388	4	1.5	2	1	9	47.18	1.926	8	0.6
0	2	2	25.83	3.449	4	16.4	2	1	7	37.98	2.369	8	0.6	4	3	3	47.23	1.925	8	0.2
1	2	2	26.61	3.350	8	0.5	1	3	2	38.18	2.357	8	0.1	5	2	5	47.25	1.924	8	0.2
0	1	5	26.67	3.343	4	3.3	1	2	6	38.23	2.354	8	0.2	1	3	6	47.34	1.920	8	0.2
4	0	2	27.08	3.293	4	5.9	0	0	8	38.47	2.340	2	0.2	6	1	5	47.35	1.920	8	0.4
2	2	0	27.17	3.282	4	0.2	4	0	6	38.57	2.334	4	0.2	7	0	3	47.52	1.913	4	2.2
1	1	5	27.42	3.252	8	0.4	3	0	7	38.76	2.323	4	1.1	2	2	8	47.73	1.905	8	0.1
2	2	1	27.60	3.232	8	0.1	5	1	4	39.28	2.294	8	0.8	3	0	9	47.84	1.901	4	1.8
4	1	1	28.48	3.134	8	0.7	6	0	2	39.62	2.275	4	0.3	4	1	8	48.30	1.884	8	0.1
0	0	6	28.61	3.120	2	0.5	2	2	6	39.87	2.261	8	0.5	0	4	0	49.11	1.855	2	2.4
1	2	3	28.70	3.110	8	8.2	5	0	5	40.08	2.250	4	0.3	1	4	0	49.57	1.839	4	0.3
2	2	2	28.83	3.097	8	0.9	4	2	4	40.24	2.241	8	0.4	6	2	4	49.97	1.825	8	0.2

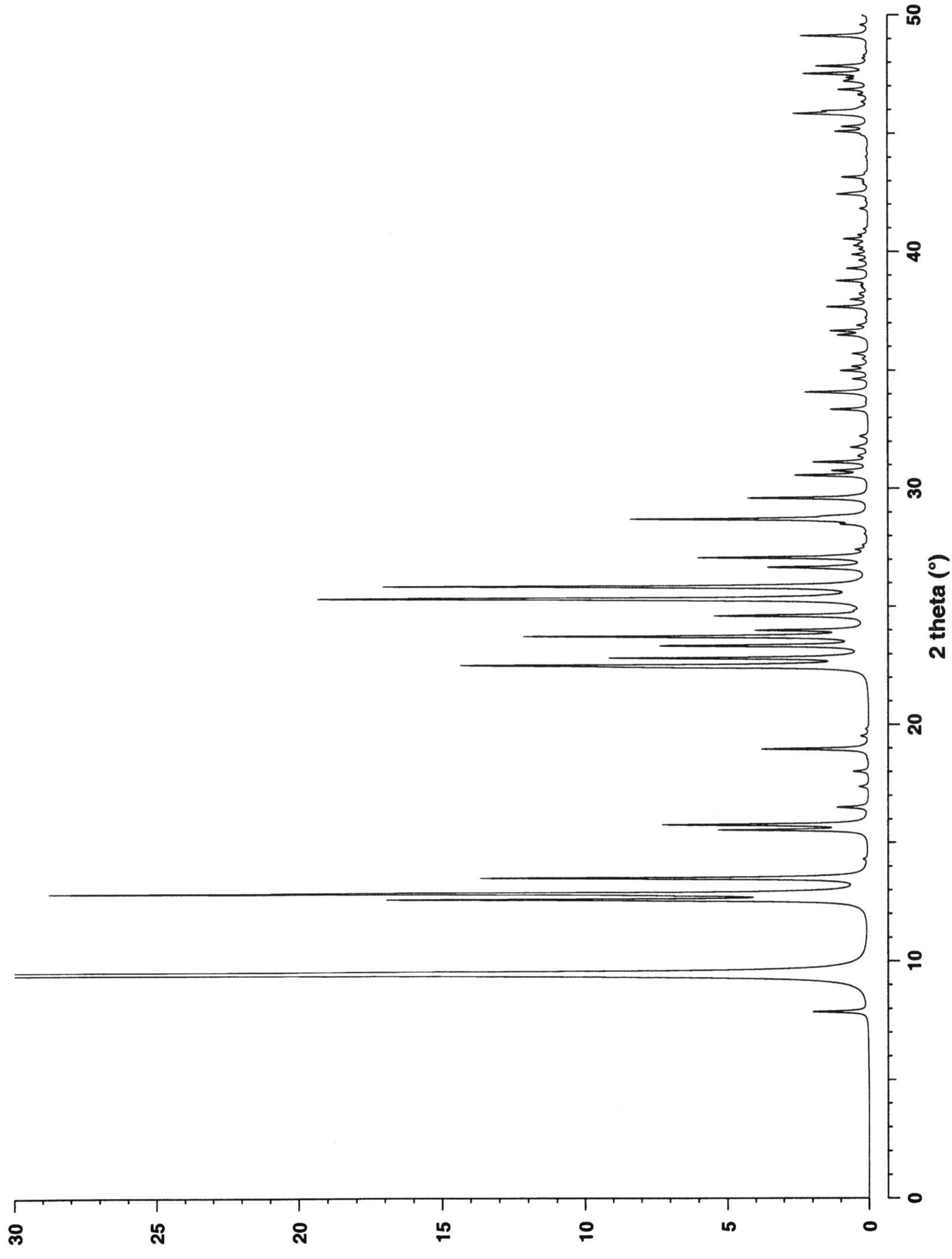
2 theta (°)
0
10
20
30
40
50
0
5
10
15
20
25
30

FRA Franzinite

CHEMICAL COMPOSITION: $|(Na,K)_{30}Ca_{10}(SO_4)_{10}(H_2O)_2|\ [Si_{30}Al_{30}O_{120}]$
Sacrofano, Latium, Italy

REFINED COMPOSITION: $|Ca_{30.72}S_{22}O_{70}(H_2O)_6|\ [Si_{30}Al_{30}O_{120}]$

CRYSTAL DATA: $P321$ (No. 150)
$a = 12.916$ Å $b = 12.916$ Å $c = 26.543$ Å
$\alpha = 90°$ $\beta = 90°$ $\gamma = 120°$
X-ray single crystal refinement, $R = 0.0596$

REFERENCE: P. Ballirano, E. Bonaccorsi, A. Maras and S. Merlino, *Canadian Mineralogist* **38** 657–668 (2000).

h	k	l	2θ	d	M	I_{rel}
0	0	1	3.33	26.543	2	85.0
0	0	2	6.66	13.271	2	4.8
1	0	0	7.90	11.186	6	0.8
1	0	1	8.58	10.308	6	9.8
0	0	3	10.00	8.848	2	2.3
1	0	2	10.34	8.553	6	13.1
0	1	2	10.34	8.553	6	8.6
0	1	3	12.76	6.939	6	0.4
1	0	3	12.76	6.939	6	35.7
1	1	0	13.71	6.458	6	37.5
1	1	1	14.11	6.275	12	15.0
1	1	2	15.26	5.807	12	4.1
1	0	4	15.53	5.707	6	4.2
0	0	5	16.70	5.309	2	0.9
1	1	3	17.00	5.216	12	2.2
2	0	2	17.20	5.154	6	1.6
0	2	2	17.20	5.154	6	1.5
1	0	5	18.50	4.796	6	0.7
0	1	5	18.50	4.796	6	2.6
2	0	3	18.77	4.727	6	0.5
0	2	3	18.77	4.727	6	26.0
2	0	4	20.77	4.276	6	1.4
0	2	4	20.77	4.276	6	15.2
2	1	1	21.28	4.175	12	1.6
1	2	1	21.28	4.175	12	0.6
1	0	6	21.60	4.114	6	5.5
0	1	6	21.60	4.114	6	16.7
1	1	5	21.67	4.101	12	2.2
1	2	2	22.07	4.028	12	0.7
2	1	2	22.07	4.028	12	4.0
2	1	3	23.32	3.815	12	85.9
1	2	3	23.32	3.815	12	3.1
3	0	0	23.86	3.729	6	100.0
0	3	1	24.10	3.692	6	1.1
1	1	6	24.39	3.650	12	0.4
0	1	7	24.79	3.591	6	55.3
1	0	7	24.79	3.591	6	1.0
3	0	2	24.80	3.590	6	1.6
2	1	4	24.97	3.566	12	69.1
1	2	4	24.97	3.566	12	9.0
0	2	6	25.67	3.470	6	0.5
2	0	6	25.67	3.470	6	2.8
0	3	3	25.93	3.436	6	1.9
1	2	5	26.96	3.307	12	10.5
2	1	5	26.96	3.307	12	12.3
2	2	0	27.62	3.229	6	10.5
2	2	1	27.83	3.205	12	0.6
1	0	8	28.05	3.181	6	0.9
0	1	8	28.05	3.181	6	0.6
2	0	7	28.44	3.139	6	6.6
1	2	6	29.22	3.056	12	6.9
2	1	6	29.22	3.056	12	1.6
0	3	5	29.27	3.051	6	0.6
2	2	3	29.45	3.033	12	0.5
3	1	2	29.57	3.021	12	1.8
1	3	2	29.57	3.021	12	1.5
1	1	8	30.28	2.951	12	0.5
0	0	9	30.31	2.949	2	2.4
1	3	3	30.53	2.928	12	2.9
3	1	3	30.53	2.928	12	1.9
2	2	4	30.79	2.903	12	0.5
0	2	8	31.35	2.854	6	0.4
0	1	9	31.37	2.852	6	1.8
4	0	0	32.00	2.796	6	2.6
4	0	1	32.19	2.781	6	0.5
0	4	1	32.19	2.781	6	0.7
2	2	5	32.45	2.759	12	1.4
0	4	2	32.73	2.736	6	2.6
1	1	9	33.40	2.683	12	1.9
3	1	5	33.45	2.678	12	0.7
4	0	3	33.61	2.666	6	13.7
0	4	3	33.61	2.666	6	1.8
3	0	7	33.71	2.659	6	0.6
0	0	10	33.77	2.654	2	12.9
2	0	9	34.38	2.609	6	0.7
4	0	4	34.81	2.577	6	3.1
0	4	4	34.81	2.577	6	1.1
2	3	1	35.13	2.554	12	1.0
1	3	6	35.34	2.540	12	1.1
3	1	6	35.34	2.540	12	5.0
0	3	8	36.24	2.479	6	0.5
3	0	8	36.24	2.479	6	1.8
0	4	5	36.31	2.474	6	1.4
4	0	5	36.31	2.474	6	0.8
3	2	3	36.46	2.465	12	5.1
1	1	10	36.60	2.455	12	4.6
4	1	0	36.82	2.441	12	7.3
1	4	1	36.98	2.431	12	0.8
1	2	9	37.17	2.419	12	0.6
3	1	7	37.45	2.401	12	7.6
3	2	4	37.58	2.393	12	1.1
4	0	6	38.07	2.364	6	0.6
0	4	6	38.07	2.364	6	3.5
3	0	9	38.94	2.313	6	2.9
1	4	4	39.33	2.291	12	0.5
1	3	8	39.78	2.266	12	0.5
0	4	7	40.06	2.251	6	10.9
4	0	7	40.06	2.251	6	0.6
2	2	9	41.47	2.178	12	2.2
3	0	10	41.77	2.162	6	9.7
0	3	10	41.77	2.162	6	2.6
3	3	0	41.97	2.153	6	17.8
2	3	7	42.54	2.125	12	1.3
4	2	0	42.78	2.114	12	0.5
2	4	1	42.92	2.107	12	0.8
2	2	10	44.17	2.050	12	0.5
2	3	8	44.64	2.030	12	1.1
0	4	9	44.65	2.029	6	1.4
0	3	11	44.73	2.026	6	0.8
5	1	0	45.13	2.009	12	0.8
5	1	1	45.27	2.003	12	0.8
5	0	6	45.43	1.996	6	0.6
5	1	2	45.67	1.986	12	0.8
2	4	5	46.22	1.964	12	0.4
2	1	12	46.32	1.960	12	0.6
5	1	3	46.34	1.959	12	1.6
3	2	9	46.93	1.936	12	1.4
2	3	9	46.93	1.936	12	0.6
5	0	7	47.17	1.927	6	2.0
4	0	10	47.21	1.925	6	0.7
0	4	10	47.21	1.925	6	1.1
5	1	4	47.27	1.923	12	1.8
4	2	6	47.68	1.907	12	0.8
4	1	9	48.41	1.880	12	2.2
1	4	9	48.41	1.880	12	0.8
1	5	5	48.45	1.879	12	0.7
5	1	5	48.45	1.879	12	0.9
1	0	14	48.71	1.869	6	2.3
6	0	0	48.85	1.864	6	0.9
4	2	7	49.36	1.846	12	1.0
1	2	13	49.58	1.839	12	1.0
2	1	13	49.58	1.839	12	1.3
1	5	6	49.85	1.829	12	1.4
0	4	11	49.92	1.827	6	0.7
4	0	11	49.92	1.827	6	0.6

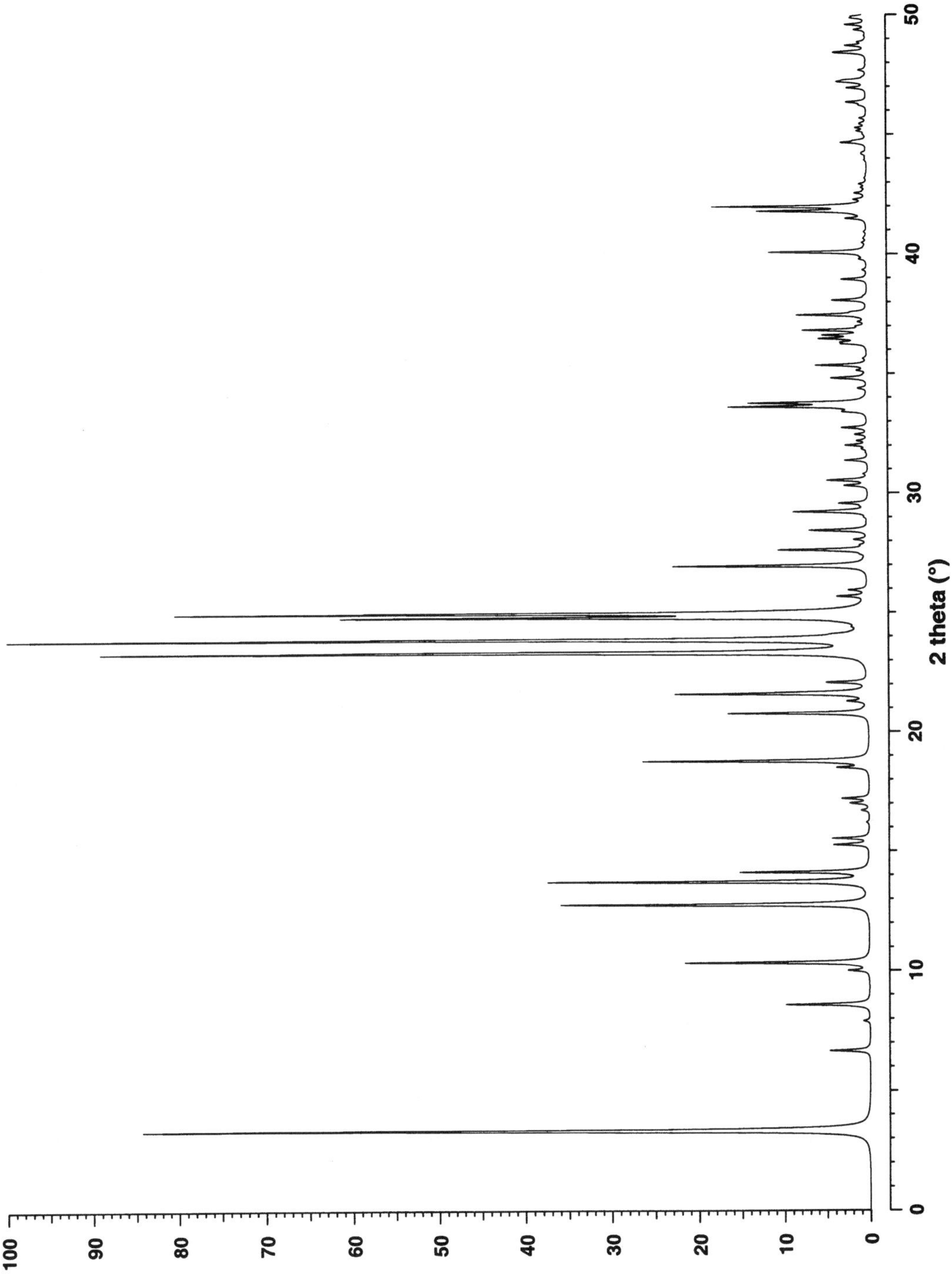
2 theta (°)
0
10
20
30
40
50
100
90
80
70
60
50
40
30
20
10
0

GIS Gismondine

CHEMICAL COMPOSITION: $|Ca_{3.68}(H_2O)_{17.2}|$ $[Si_{8.8}Al_{7.2}O_{32}]$
Hohenberg, Buhne, Westfalia

REFINED COMPOSITION: $|Ca_{3.68}(H_2O)_{17.28}|$ $[Si_{8.8}Al_{7.2}O_{32}]$

CRYSTAL DATA: $P11\,2_1/a$ (No. 14) unique axis **c**, cell choice 1
$a = 9.843$ Å $b = 10.023$ Å $c = 10.616$ Å
$\alpha = 90°$ $\beta = 90°$ $\gamma = 92.417°$
X-ray single crystal refinement, $R_w = 0.04$

REFERENCE: K. F. Fischer and V. Schramm,
In *Molecular Sieve Zeolites - I* (*ACS Adv. Chem. Ser. No. 101*),
Ed. by E. M. Flanigen and L. B. Sand
(American Chemical Society: Washington, DC) 250–258 (1971).

h	k	l	2θ	d	M	I_{rel}	h	k	l	2θ	d	M	I_{rel}	h	k	l	2θ	d	M	I_{rel}
0	1	0	8.83	10.014	2	5.6	1	2	3	32.50	2.755	4	2.8	2	3	3	42.00	2.151	4	4.5
0	1	1	12.15	7.285	4	69.0	2	1	3	32.64	2.743	4	6.7	2	4	1	42.05	2.149	4	0.9
1	0	1	12.27	7.214	4	23.5	1	-3	2	32.66	2.742	4	49.6	3	2	3	42.19	2.142	4	0.6
1	-1	1	14.91	5.941	4	8.5	2	-3	1	32.89	2.723	4	9.1	2	-2	4	42.37	2.133	4	5.8
1	1	1	15.36	5.770	4	21.1	3	-1	2	33.03	2.712	4	47.6	2	2	4	43.06	2.101	4	5.9
0	0	2	16.70	5.308	2	3.1	2	3	0	33.06	2.709	2	4.5	0	1	5	43.57	2.077	4	4.7
0	2	0	17.71	5.007	2	16.4	3	-2	1	33.12	2.705	4	6.2	1	0	5	43.61	2.075	4	0.3
2	0	0	18.04	4.917	2	76.7	1	3	2	33.30	2.690	4	58.4	3	0	4	43.89	2.063	4	1.1
0	1	2	18.92	4.690	4	0.7	3	1	2	33.66	2.662	4	23.3	0	4	3	44.32	2.044	4	2.1
1	0	2	19.00	4.671	4	14.9	0	0	4	33.77	2.654	2	39.1	1	-3	4	44.33	2.043	4	5.9
0	2	1	19.60	4.529	4	1.2	2	3	1	34.16	2.625	4	18.2	1	-1	5	44.50	2.036	4	0.4
2	-1	0	19.78	4.489	2	6.2	3	2	1	34.38	2.609	4	7.5	3	-1	4	44.61	2.031	4	8.4
2	0	1	19.90	4.462	4	10.9	1	0	4	35.02	2.562	4	4.2	2	4	2	44.69	2.028	4	1.4
2	1	0	20.45	4.342	2	3.8	2	-2	3	35.65	2.518	4	9.9	1	3	4	44.83	2.022	4	0.4
1	-1	2	20.82	4.266	4	100.0	0	4	0	35.87	2.504	2	2.5	4	-3	0	44.85	2.021	2	0.4
1	1	2	21.15	4.201	4	39.3	1	-1	4	36.09	2.489	4	2.9	1	-4	3	44.99	2.015	4	1.7
1	-2	1	21.28	4.174	4	28.7	2	-3	2	36.09	2.489	4	0.4	3	1	4	45.11	2.010	4	8.0
2	-1	1	21.49	4.135	4	5.4	1	1	4	36.28	2.476	4	1.0	0	5	0	45.28	2.003	2	1.7
1	2	1	21.92	4.055	4	24.0	3	-2	2	36.30	2.475	4	2.5	3	-4	1	45.46	1.995	4	0.6
2	1	1	22.12	4.019	4	6.9	2	2	3	36.44	2.465	4	4.7	4	-1	3	45.52	1.993	4	0.6
0	2	2	24.44	3.642	4	4.9	4	0	0	36.55	2.459	2	2.3	4	-3	1	45.70	1.985	4	2.7
2	0	2	24.68	3.607	4	3.7	0	3	3	37.02	2.428	4	1.9	3	-3	3	45.82	1.980	4	0.7
2	-2	0	24.84	3.585	2	4.6	3	0	3	37.39	2.405	4	9.3	0	5	1	46.12	1.968	4	6.2
1	-2	2	25.82	3.450	4	2.5	4	0	1	37.55	2.395	4	2.6	4	1	3	46.17	1.966	4	1.2
2	2	0	25.93	3.437	2	0.4	1	-4	1	37.66	2.388	4	5.6	5	0	1	46.98	1.934	4	8.1
2	-1	2	25.99	3.428	4	10.2	4	1	0	38.05	2.365	2	0.9	2	-3	4	47.04	1.932	4	0.3
2	-2	1	26.24	3.396	4	1.4	4	-1	1	38.28	2.351	4	2.8	1	-2	5	47.26	1.923	4	1.4
1	2	2	26.35	3.382	4	3.7	0	2	4	38.39	2.345	4	1.1	3	3	3	47.26	1.923	4	0.5
2	1	2	26.52	3.361	4	4.2	1	3	3	38.46	2.341	4	0.4	3	4	1	47.39	1.918	4	3.8
0	3	0	26.71	3.338	2	0.4	2	0	4	38.55	2.336	4	8.9	1	5	1	47.49	1.915	4	1.2
0	1	3	26.72	3.336	4	19.1	3	-3	1	38.62	2.331	4	1.4	5	-1	1	47.51	1.914	4	3.2
1	0	3	26.77	3.330	4	18.8	1	-2	4	39.32	2.291	4	2.5	1	2	5	47.58	1.911	4	1.0
2	2	1	27.28	3.270	4	1.0	1	2	4	39.69	2.271	4	4.9	2	-4	3	47.59	1.911	4	2.7
0	3	1	28.02	3.184	4	86.4	2	-4	0	39.70	2.270	2	0.7	4	3	1	47.63	1.909	4	2.3
3	0	1	28.50	3.132	4	69.1	0	4	2	39.81	2.264	4	2.8	2	1	5	47.68	1.907	4	0.4
1	-3	1	29.13	3.066	4	1.3	4	-2	0	40.17	2.245	2	6.5	3	-4	2	47.95	1.897	4	2.2
3	-1	1	29.54	3.024	4	5.0	4	0	2	40.43	2.231	4	1.1	5	1	1	48.30	1.884	4	3.1
1	3	1	29.84	2.994	4	13.3	2	-4	1	40.64	2.220	4	0.6	0	5	2	48.59	1.874	4	0.4
2	-2	2	30.08	2.971	4	0.6	4	-2	1	41.10	2.196	4	1.2	2	4	3	48.84	1.865	4	0.6
0	2	3	30.94	2.890	4	0.6	4	-1	2	41.11	2.195	4	2.7	1	-5	2	49.13	1.854	4	6.5
2	0	3	31.14	2.872	4	2.2	3	-2	3	41.13	2.194	4	0.4	2	-5	1	49.14	1.854	4	2.3
0	3	2	31.66	2.826	4	1.8	2	4	0	41.14	2.194	2	8.6	4	2	3	49.24	1.850	4	4.6
2	-3	0	31.76	2.818	2	4.8	1	4	2	41.25	2.188	4	1.1	1	5	2	49.90	1.828	4	0.4
3	0	2	32.09	2.789	4	2.5	4	2	0	41.60	2.171	2	1.5	2	-2	5	49.92	1.827	4	0.3
2	-1	3	32.21	2.779	4	3.6	4	1	2	41.82	2.160	4	0.9	5	-1	2	49.92	1.827	4	4.1

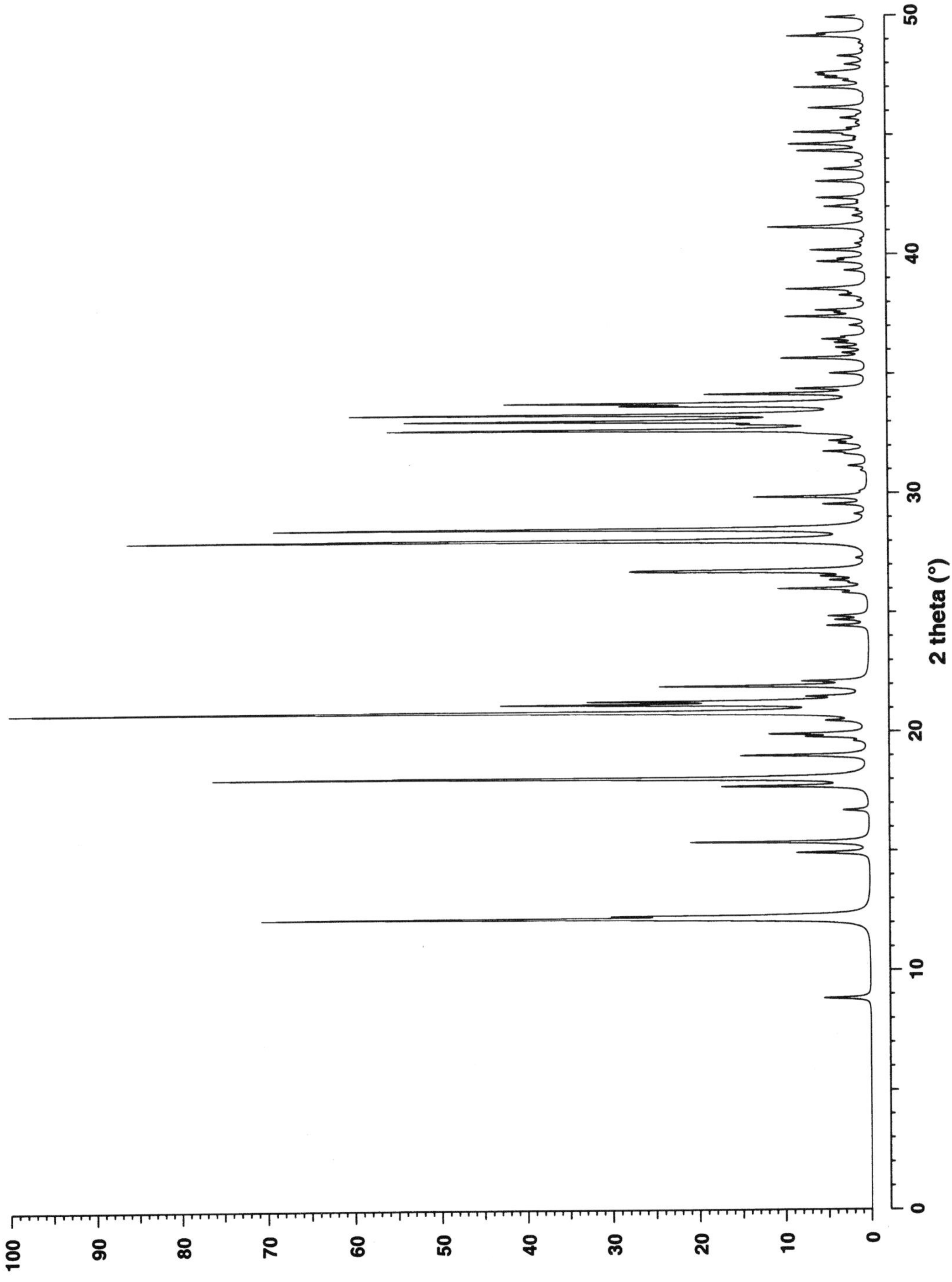
2 theta (°)
0
10
20
30
40
50
0
10
20
30
40
50
60
70
80
90
100

GIS Amicite

CHEMICAL COMPOSITION: $|Na_{3.6}K_{3.76}(H_2O)_{9.86}|$ $[Si_8Al_8O_{32}]$
Hoewenneg, Hegau, Germany

REFINED COMPOSITION: $|Na_{3.6}K_{3.76}(H_2O)_{9.86}|$ $[Si_8Al_8O_{32}]$

CRYSTAL DATA: $I121$ (No. 5) unique axis **b**, cell choice 3
$a = 10.226$ Å $b = 10.422$ Å $c = 9.884$ Å
$\alpha = 90°$ $\beta = 88.315°$ $\gamma = 90°$
X-ray single crystal refinement, $R_w = 0.033$

REFERENCE: A. Alberti and G. Vezzalini,
Acta Cryst. **B35** 2866–2869 (1979).

h	*k*	*l*	*2θ*	*d*	*M*	I_{rel}
1	1	0	12.13	7.298	4	100.0
1	0	1	12.27	7.211	2	0.8
0	1	1	12.34	7.170	4	21.7
-1	0	1	12.64	7.002	2	2.7
0	2	0	17.01	5.211	2	3.8
2	0	0	17.35	5.111	2	67.1
0	0	2	17.96	4.940	2	54.7
1	2	1	21.03	4.224	4	96.4
2	1	1	21.13	4.204	4	6.9
-1	2	1	21.25	4.180	4	53.0
1	1	2	21.51	4.131	4	3.6
-2	1	1	21.57	4.120	4	11.7
-1	1	2	21.94	4.051	4	8.8
2	2	0	24.39	3.649	4	8.7
2	0	2	24.69	3.605	2	0.3
0	2	2	24.83	3.585	4	5.7
-2	0	2	25.44	3.501	2	0.2
1	3	0	27.11	3.289	4	26.8
0	3	1	27.21	3.277	4	2.9
3	0	1	27.44	3.251	2	1.0
3	1	0	27.54	3.239	4	49.2
-3	0	1	27.95	3.192	2	0.2
1	0	3	28.22	3.162	2	0.2
0	1	3	28.42	3.140	4	89.4
-1	0	3	28.72	3.108	2	2.1
2	2	2	30.14	2.965	4	9.6
-2	2	2	30.77	2.906	4	0.4
2	3	1	32.30	2.771	4	7.0
3	2	1	32.46	2.758	4	36.6
1	3	2	32.56	2.750	4	1.7
-2	3	1	32.60	2.747	4	0.7
3	1	2	32.64	2.744	4	13.4
-1	3	2	32.85	2.726	4	11.5
-3	2	1	32.90	2.722	4	85.8
2	1	3	33.06	2.710	4	5.3
1	2	3	33.14	2.703	4	58.0
-3	1	2	33.51	2.674	4	21.2
-1	2	3	33.57	2.669	4	7.6
-2	1	3	33.92	2.643	4	2.9
0	4	0	34.42	2.605	2	39.7
0	0	4	36.37	2.470	2	7.2
1	4	1	36.67	2.450	4	1.6
3	3	0	36.95	2.433	4	0.4
4	1	1	37.10	2.424	4	10.8
3	0	3	37.42	2.404	2	0.2
-4	1	1	37.62	2.391	4	0.5
0	3	3	37.63	2.390	4	4.1
1	1	4	38.22	2.355	4	4.3
-3	0	3	38.57	2.334	2	0.6
-1	1	4	38.73	2.325	4	7.9
2	4	0	38.79	2.321	4	0.5
0	4	2	39.09	2.305	4	7.0
4	2	0	39.27	2.294	4	3.9
2	0	4	40.07	2.250	2	6.3
-4	0	2	40.21	2.243	2	5.6
0	2	4	40.41	2.232	4	1.6
3	3	2	41.01	2.201	4	1.4
2	3	3	41.36	2.183	4	1.9
3	2	3	41.37	2.183	4	0.5
-3	3	2	41.73	2.164	4	2.5
-2	3	3	42.08	2.147	4	0.3
-3	2	3	42.44	2.130	4	0.2
2	4	2	42.82	2.112	4	12.2
4	2	2	43.03	2.102	4	0.9
-2	4	2	43.29	2.090	4	11.4
2	2	4	43.83	2.066	4	0.2
-4	2	2	43.95	2.060	4	0.2
1	5	0	44.35	2.042	4	0.7
0	5	1	44.42	2.040	4	0.3
3	4	1	44.57	2.033	4	3.0
-2	2	4	44.74	2.026	4	2.0
4	3	1	44.76	2.025	4	5.3
-3	4	1	44.90	2.019	4	0.4
5	0	1	45.02	2.014	2	0.7
1	4	3	45.09	2.011	4	1.6
5	1	0	45.20	2.006	4	4.9
-4	3	1	45.21	2.006	4	0.8
-1	4	3	45.42	1.997	4	5.3
3	1	4	45.56	1.991	4	0.9
-5	0	1	45.58	1.990	2	2.3
-1	3	4	46.17	1.966	4	0.8
-4	1	3	46.44	1.955	4	0.2
0	1	5	46.79	1.941	4	4.9
-3	1	4	46.88	1.938	4	1.9
2	5	1	47.92	1.898	4	0.4
1	5	2	48.11	1.891	4	0.2
-1	5	2	48.32	1.884	4	1.2
5	2	1	48.46	1.878	4	5.3
5	1	2	48.48	1.878	4	3.6
-5	2	1	48.99	1.859	4	2.2
-5	1	2	49.53	1.840	4	1.4
2	1	5	49.76	1.832	4	0.4
1	2	5	49.92	1.827	4	3.9
4	4	0	49.99	1.824	4	14.0

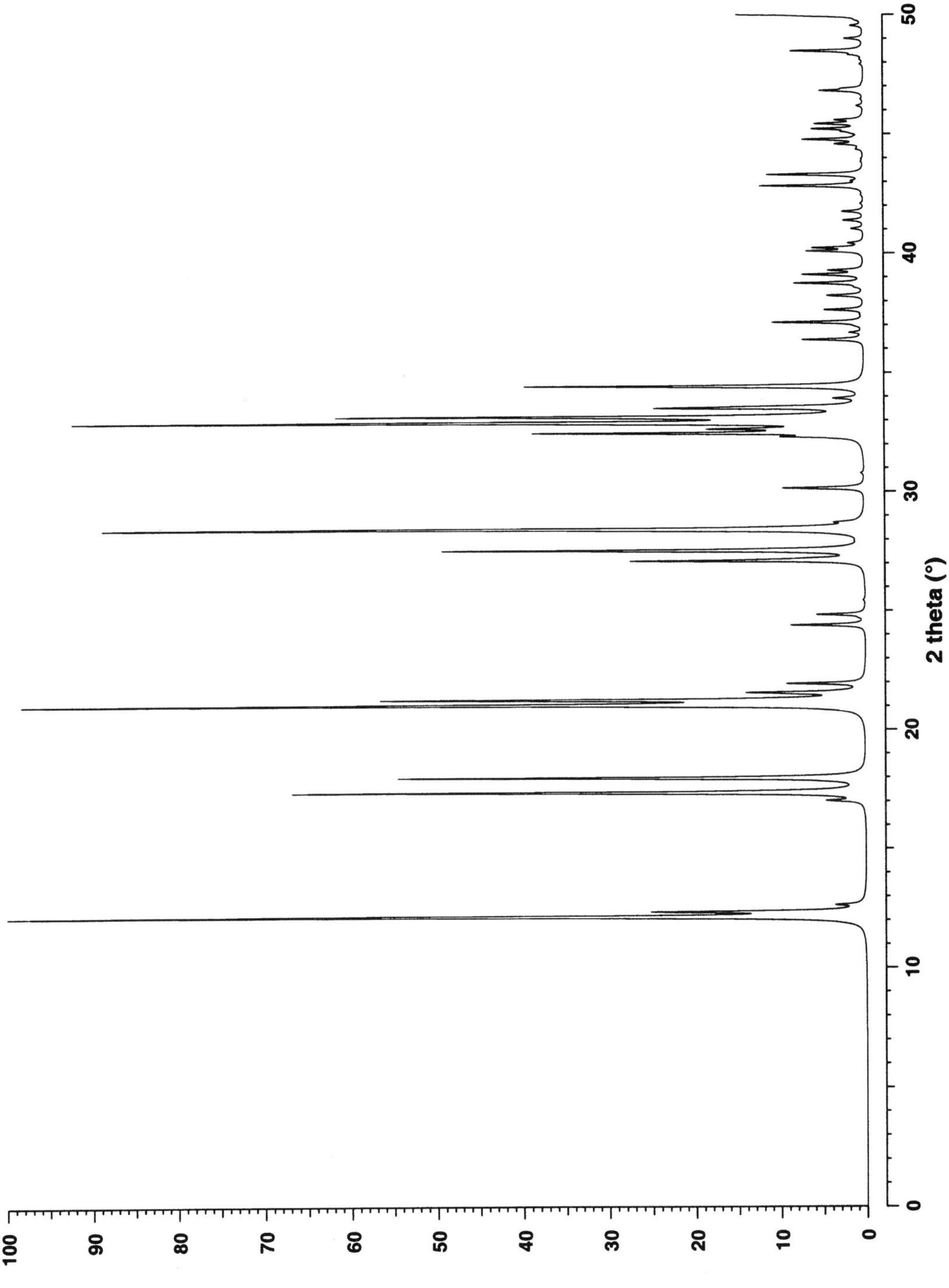

2 theta (°)
0
10
20
30
40
50
100
90
80
70
60
50
40
30
20
10
0

GIS Garronite

CHEMICAL COMPOSITION: $|Na_{0.8}Ca_{2.82}(H_2O)_{12.08}|$ $[Si_{10.4}Al_{5.6}O_{32}]$
Goble, Oregon, U.S.A.

REFINED COMPOSITION: $|Na_{0.8}Ca_{2.82}(H_2O)_{12.08}|$ $[Si_{10.4}Al_{5.6}O_{32}]$

CRYSTAL DATA: $I\bar{4}m2$ (No. 119)
$a = 9.9266$ Å $b = 9.9266$ Å $c = 10.3031$ Å
$\alpha = 90°$ $\beta = 90°$ $\gamma = 90°$
X-ray Rietveld refinement, $R_{exp} = 0.047$, $R_{wp} = 0.127$, $R_{F^2} = 0.111$

REFERENCE: G. Artioli,
American Mineralogist **77** 189–196 (1992).

h	k	l	2θ	d	M	I_{rel}
1	0	1	12.38	7.149	8	100.0
1	1	0	12.61	7.019	4	0.9
0	0	2	17.21	5.152	2	1.2
2	0	0	17.87	4.963	4	54.1
1	1	2	21.39	4.153	8	63.4
2	1	1	21.80	4.077	16	16.6
2	0	2	24.91	3.574	8	0.4
2	2	0	25.38	3.510	4	0.8
1	0	3	27.48	3.246	8	41.3
3	0	1	28.33	3.150	8	96.3
3	1	0	28.43	3.139	8	3.5
2	1	3	32.97	2.716	16	4.8
3	1	2	33.43	2.681	16	63.2
3	2	1	33.70	2.660	16	6.0
0	0	4	34.83	2.576	2	8.3
4	0	0	36.20	2.482	4	0.7
1	1	4	37.18	2.418	8	0.3
3	0	3	37.75	2.383	8	0.3
4	1	1	38.40	2.344	16	1.0
3	3	0	38.47	2.340	4	0.5
2	0	4	39.41	2.286	8	0.2
4	0	2	40.34	2.236	8	0.7
4	2	0	40.65	2.220	8	2.2
3	2	3	42.06	2.148	16	0.6
3	3	2	42.43	2.130	8	2.4
2	2	4	43.58	2.077	8	5.4
4	2	2	44.44	2.038	16	0.4
1	0	5	44.93	2.018	8	1.3
3	1	4	45.55	1.991	16	4.3
4	1	3	46.04	1.971	16	0.9
4	3	1	46.59	1.949	16	0.3
5	0	1	46.59	1.949	8	5.7
5	1	0	46.66	1.947	8	0.2
2	1	5	48.72	1.869	16	0.1

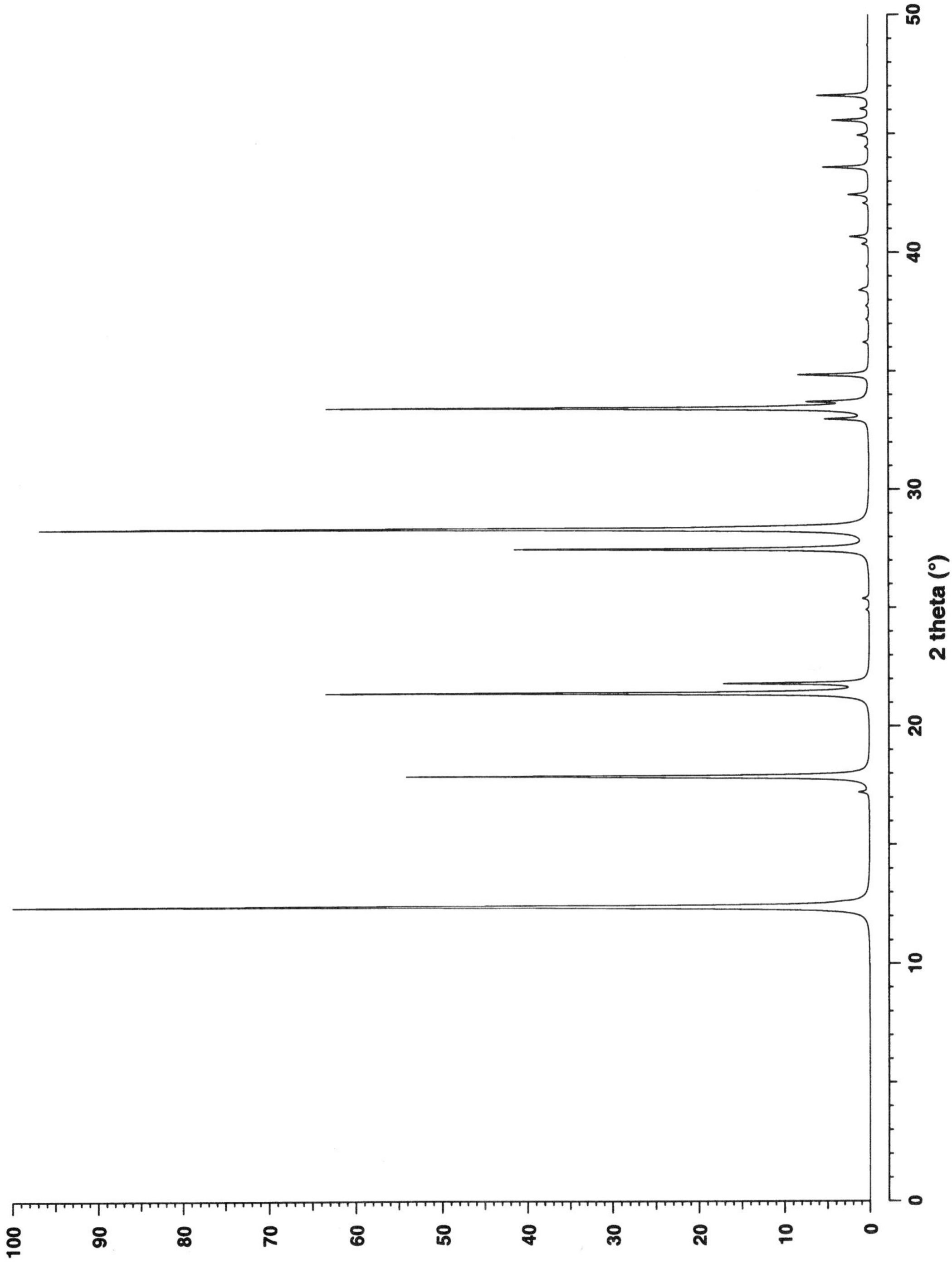
2 theta (°)
0
10
20
30
40
50
0
10
20
30
40
50
60
70
80
90
100

GIS Gobbinsite

CHEMICAL COMPOSITION: $|Ca_{0.6}Na_{2.6}K_{2.2}(H_2O)_{12}|$ $[Si_{10}Al_6O_{32}]$
Antrim District, Northern Ireland

REFINED COMPOSITION: $|Na_{2.6}K_{1.7}(H_2O)_{14.9}|$ $[Si_{10}Al_6O_{32}]$

CRYSTAL DATA: $Pmn2_1$ (No. 31)
$a = 10.108$ Å $b = 9.766$ Å $c = 10.171$ Å
$\alpha = 90°$ $\beta = 90°$ $\gamma = 90°$
X-ray Rietveld refinement, $R_{wp} = 0.136$, $R_F = 0.126$

REFERENCE: L. B. McCusker and Ch. Baerlocher,
Z. Kristallogr. **171** 281–289 (1985).

h	k	l	2θ	d	M	I_{rel}
0	1	0	9.05	9.766	2	2.4
1	0	1	12.35	7.170	4	51.0
0	1	1	12.57	7.044	4	72.1
1	1	0	12.60	7.023	4	6.3
1	1	1	15.33	5.779	8	8.8
0	0	2	17.44	5.085	2	5.8
2	0	0	17.55	5.054	2	29.7
0	2	0	18.17	4.883	2	21.1
0	1	2	19.68	4.511	4	1.1
2	1	0	19.78	4.489	4	0.3
0	2	1	20.17	4.402	4	6.9
1	2	0	20.20	4.397	4	4.4
1	1	2	21.57	4.119	8	100.0
2	1	1	21.64	4.106	8	14.4
1	2	1	22.02	4.036	8	14.0
2	0	2	24.84	3.585	4	1.2
0	2	2	25.29	3.522	4	4.4
2	2	0	25.36	3.512	4	0.9
1	2	2	26.80	3.326	8	17.8
2	2	1	26.86	3.319	8	4.7
0	3	0	27.40	3.255	2	1.1
1	0	3	27.75	3.214	4	54.0
0	1	3	27.85	3.203	4	4.4
3	0	1	27.89	3.198	4	82.8
3	1	0	28.01	3.185	4	1.7
0	3	1	28.79	3.100	4	84.1
1	3	0	28.81	3.099	4	6.1
1	1	3	29.25	3.053	8	3.1
3	1	1	29.38	3.040	8	6.7
1	3	1	30.15	2.964	8	7.5
2	2	2	30.95	2.890	8	3.4
0	2	3	32.14	2.785	4	2.0
3	2	0	32.28	2.773	4	0.6
0	3	2	32.66	2.742	4	6.6
2	3	0	32.72	2.737	4	2.5
2	1	3	33.11	2.705	8	0.8
3	1	2	33.19	2.699	8	42.5
1	2	3	33.37	2.685	8	7.7
3	2	1	33.49	2.676	8	8.8
1	3	2	33.88	2.646	8	21.3
2	3	1	33.92	2.643	8	2.9
0	0	4	35.30	2.543	2	18.7
4	0	0	35.52	2.527	2	1.8
0	1	4	36.51	2.461	4	0.1
2	2	3	36.85	2.439	8	2.3
3	2	2	36.92	2.435	8	3.3
2	3	2	37.31	2.410	8	0.7
1	1	4	37.62	2.391	8	0.7
4	1	1	37.82	2.379	8	5.0
0	4	1	37.90	2.374	4	3.9
0	3	3	38.33	2.348	4	2.6
3	3	0	38.45	2.341	4	0.2
3	1	3	38.79	2.321	8	0.4
1	4	1	38.97	2.311	8	0.8
3	3	1	39.50	2.281	8	1.1
2	0	4	39.68	2.271	4	2.0
4	0	2	39.83	2.263	4	1.1
0	2	4	39.98	2.255	4	3.3
4	2	0	40.18	2.244	4	0.5
2	1	4	40.78	2.212	8	3.6
4	1	2	40.94	2.205	8	0.4
1	2	4	41.00	2.201	8	1.0
0	4	2	41.01	2.201	4	0.6
2	4	0	41.06	2.198	4	4.4
4	2	1	41.19	2.192	8	0.3
1	4	2	42.01	2.151	8	2.3
2	4	1	42.05	2.149	8	4.6
3	2	3	42.09	2.147	8	0.6
2	3	3	42.45	2.130	8	2.7
3	3	2	42.51	2.127	8	1.5
2	2	4	43.96	2.060	8	4.2
4	2	2	44.11	2.053	8	1.5
2	4	2	44.92	2.018	8	0.5
0	3	4	45.25	2.004	4	1.9
1	0	5	45.48	1.994	4	1.1
0	1	5	45.55	1.991	4	1.3
3	1	4	45.65	1.987	8	1.5
4	1	3	45.73	1.984	8	3.2
5	0	1	45.76	1.983	4	2.7
0	4	3	45.80	1.981	4	1.5
1	3	4	46.18	1.966	8	7.0
1	1	5	46.48	1.954	8	0.2
0	5	0	46.49	1.953	2	0.2
1	4	3	46.72	1.944	8	1.9
5	1	1	46.75	1.943	8	0.4
3	4	1	46.81	1.941	8	2.3
3	3	3	47.18	1.926	8	1.8
0	5	1	47.39	1.918	4	4.4
1	5	1	48.29	1.885	8	0.3
0	2	5	48.48	1.878	4	1.1
3	2	4	48.58	1.874	8	0.1
4	2	3	48.65	1.871	8	1.1
5	2	0	48.75	1.868	4	0.1
2	3	4	48.89	1.863	8	0.8
4	3	2	49.02	1.858	8	1.5
2	1	5	49.17	1.853	8	0.2
5	1	2	49.40	1.845	8	3.1
3	4	2	49.46	1.843	8	2.5
5	2	1	49.62	1.837	8	0.3

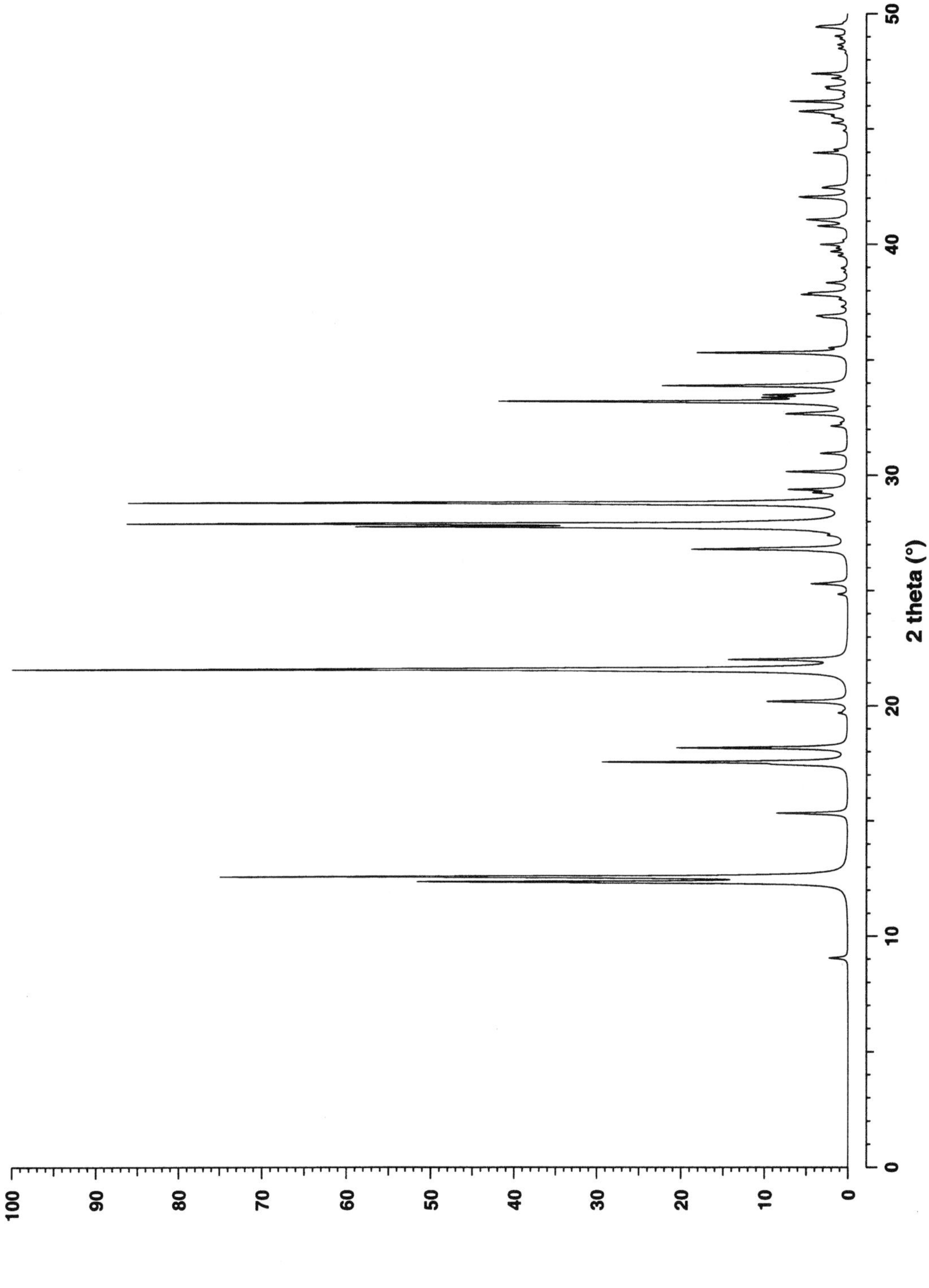
2 theta (°)
0
10
20
30
40
50
0
10
20
30
40
50
60
70
80
90
100

GIS Di-n-propylamine MAPO–43

CHEMICAL COMPOSITION: $|(C_6H_{16}N)_2|$ $[Al_6Mg_2P_8O_{32}]$
$C_6H_{16}N$ = di-n-propylamine

REFINED COMPOSITION: $|(C_6H_{16}N)_2|$ $[Al_8P_8O_{32}]$

CRYSTAL DATA: $I112/b$ (No. 15) unique axis **c**, cell choice 3
$a = 10.2192$ Å $b = 10.2198$ Å $c = 10.0126$ Å
$\alpha = 90°$ $\beta = 90°$ $\gamma = 90.987°$
X-ray single crystal refinement, $R_w = 0.049$

REFERENCE: J. J. Pluth, J. V. Smith and J. M. Bennett, *J. Am. Chem. Soc.* **111** 1692–1698 (1989).

h	k	l	2θ	d	M	I_{rel}
0	1	1	12.38	7.152	4	100.0
1	0	1	12.38	7.151	4	98.8
0	2	0	17.36	5.109	2	22.4
2	0	0	17.36	5.109	2	21.7
1	-2	1	21.25	4.181	4	2.8
2	-1	1	21.25	4.181	4	3.0
1	2	1	21.50	4.134	4	17.2
2	1	1	21.50	4.134	4	17.6
1	-1	2	21.53	4.127	4	81.9
1	1	2	21.66	4.104	4	26.4
2	-2	0	24.43	3.644	2	5.4
2	2	0	24.86	3.582	2	0.5
0	2	2	24.90	3.576	4	0.3
2	0	2	24.90	3.576	4	0.2
0	3	1	27.66	3.225	4	72.8
3	0	1	27.66	3.224	4	73.9
0	1	3	28.13	3.173	4	14.8
1	0	3	28.13	3.173	4	15.3
2	-3	1	32.59	2.747	4	0.6
3	-2	1	32.59	2.747	4	0.6
1	-3	2	32.87	2.725	4	13.8
3	-1	2	32.87	2.725	4	13.5
2	3	1	33.09	2.707	4	1.1
3	2	1	33.09	2.707	4	1.2
1	3	2	33.12	2.705	4	10.9
3	1	2	33.12	2.705	4	10.6
1	-2	3	33.16	2.702	4	0.9
2	-1	3	33.16	2.702	4	0.9
1	2	3	33.32	2.689	4	3.7
2	1	3	33.32	2.689	4	3.6
0	4	0	35.13	2.555	2	0.3
4	0	0	35.13	2.554	2	0.3
0	0	4	35.87	2.503	2	6.0
1	-4	1	37.23	2.415	4	1.0
4	-1	1	37.23	2.415	4	0.9
1	4	1	37.53	2.397	4	0.5
4	1	1	37.53	2.396	4	0.5
0	3	3	37.74	2.384	4	0.2
3	0	3	37.74	2.384	4	0.2
2	-4	0	39.15	2.301	2	0.3
4	-2	0	39.15	2.301	2	0.4
0	4	2	39.61	2.275	4	1.8
4	0	2	39.61	2.275	4	1.6
0	2	4	40.11	2.248	4	0.1
2	0	4	40.11	2.248	4	0.2
3	-3	2	41.31	2.186	4	3.3
3	3	2	41.92	2.155	4	3.9
2	-4	2	43.28	2.091	4	0.2
4	-2	2	43.28	2.090	4	0.2
2	4	2	43.80	2.067	4	0.9
4	2	2	43.80	2.067	4	0.8
2	-2	4	43.88	2.063	4	1.1
2	2	4	44.14	2.052	4	1.6
0	5	1	45.29	2.002	4	2.2
5	0	1	45.29	2.002	4	2.2
1	-4	3	45.46	1.995	4	4.5
4	-1	3	45.47	1.995	4	4.4
4	3	1	45.67	1.987	4	0.1
1	4	3	45.72	1.985	4	1.5
4	1	3	45.72	1.984	4	1.5
1	-3	4	45.76	1.983	4	0.8
3	-1	4	45.76	1.983	4	0.8
1	3	4	45.95	1.975	4	0.4
3	1	4	45.95	1.975	4	0.4
0	1	5	46.19	1.965	4	0.6
1	0	5	46.19	1.965	4	0.5
2	-5	1	48.55	1.875	4	0.2
5	-2	1	48.55	1.875	4	0.2
1	-5	2	48.81	1.866	4	1.2
5	-1	2	48.81	1.866	4	1.1
1	5	2	49.11	1.855	4	0.7
5	1	2	49.11	1.855	4	0.6
2	5	1	49.15	1.854	4	0.3
5	2	1	49.15	1.854	4	0.2
1	-2	5	49.65	1.836	4	0.1
1	2	5	49.77	1.832	4	0.1
2	1	5	49.77	1.832	4	0.1

Di-n-propylamine MAPO–43 GIS

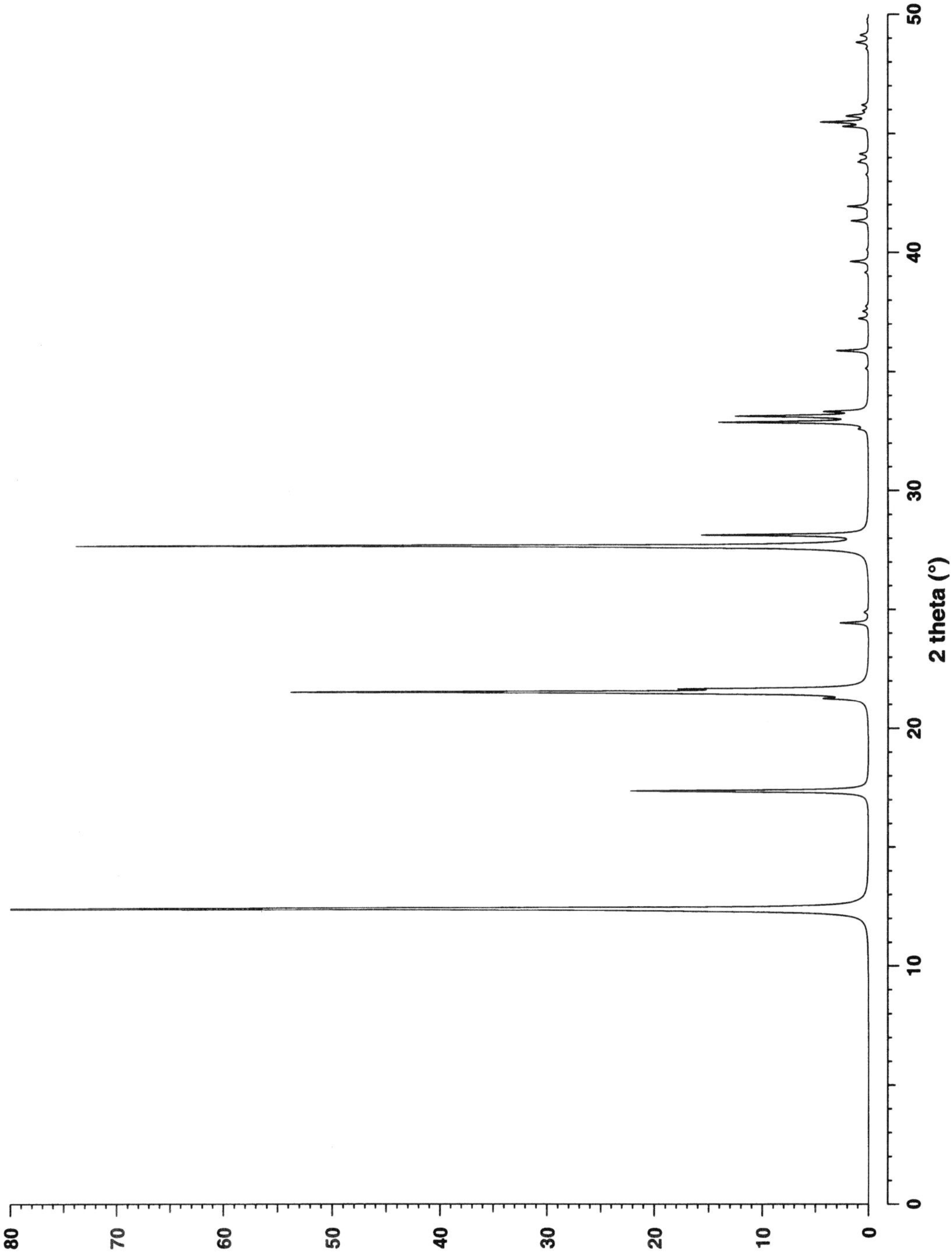

GIS Na-P1

CHEMICAL COMPOSITION: $|Na_6(H_2O)_{12}|\ [Si_{10}Al_6O_{32}]$

REFINED COMPOSITION: $|Na_{5.92}(H_2O)_{11.28}|\ [Si_{9.92}Al_{6.08}O_{32}]$

CRYSTAL DATA: $I\bar{4}$ (No. 82)
$a = 10.043$ Å $b = 10.043$ Å $c = 10.043$ Å
$\alpha = 90°$ $\beta = 90°$ $\gamma = 90°$
X-ray twinned crystal refinement, $R = 0.05$

REFERENCE: Ch. Baerlocher and W. M. Meier, *Z. Kristallogr.* **135** 339–354 (1972).

h	k	l	2θ	d	M	I_{rel}
1	0	1	12.46	7.101	8	92.1
1	1	0	12.46	7.101	4	1.0
2	0	0	17.66	5.022	4	60.6
0	0	2	17.66	5.022	2	2.7
2	1	1	21.67	4.100	8	13.3
1	1	2	21.67	4.100	8	66.1
1	2	1	21.67	4.100	8	7.0
2	0	2	25.08	3.551	8	0.8
2	2	0	25.08	3.551	4	1.1
3	1	0	28.10	3.176	4	1.6
3	0	1	28.10	3.176	8	100.0
1	0	3	28.10	3.176	8	34.9
2	2	2	30.84	2.899	8	9.0
2	1	3	33.38	2.684	8	3.5
3	1	2	33.38	2.684	8	45.0
3	2	1	33.38	2.684	8	1.2
2	3	1	33.38	2.684	8	11.5
1	3	2	33.38	2.684	8	12.7
1	2	3	33.38	2.684	8	7.5
0	0	4	35.76	2.511	2	6.8
3	0	3	38.01	2.367	8	0.4
4	1	1	38.01	2.367	8	0.6
1	1	4	38.01	2.367	8	1.0
3	3	0	38.01	2.367	4	0.3
1	4	1	38.01	2.367	8	12.0
4	0	2	40.15	2.246	8	1.4
2	0	4	40.15	2.246	8	0.9
4	2	0	40.15	2.246	4	2.3
3	2	3	42.20	2.141	8	1.0
2	3	3	42.20	2.141	8	0.3
3	3	2	42.20	2.141	8	1.0
4	2	2	44.18	2.050	8	6.5
2	4	2	44.18	2.050	8	0.5
2	2	4	44.18	2.050	8	0.4
1	4	3	46.08	1.970	8	0.2
3	1	4	46.08	1.970	8	0.1
5	0	1	46.08	1.970	8	3.3
5	1	0	46.08	1.970	4	2.2
1	3	4	46.08	1.970	8	11.1
4	3	1	46.08	1.970	8	0.9
1	0	5	46.08	1.970	8	1.6
4	1	3	46.08	1.970	8	2.0
3	4	1	46.08	1.970	8	1.8
2	5	1	49.72	1.834	8	0.2
5	1	2	49.72	1.834	8	2.9
2	1	5	49.72	1.834	8	0.7
1	5	2	49.72	1.834	8	0.3

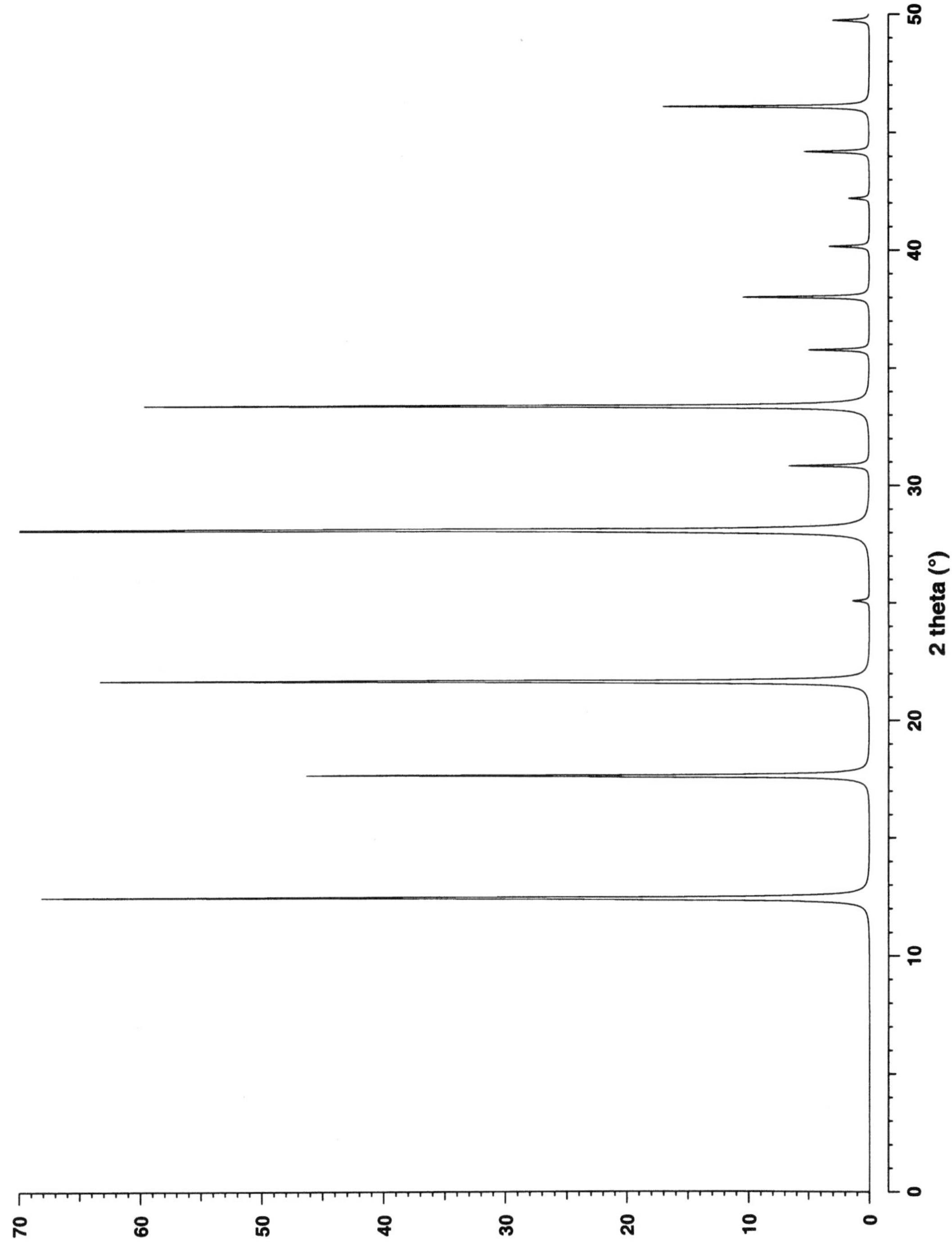
2 theta (°)
0
10
20
30
40
50
0
10
20
30
40
50
60
70

GME Gmelinite

CHEMICAL COMPOSITION: $|(Na_2, Ca)_4(H_2O)_{24}|$ $[Si_{16}Al_8O_{48}]$
Nova Scotia, Canada

REFINED COMPOSITION: $|Ca_4(H_2O)_{26.4}|$ $[Si_{16}Al_8O_{48}]$

CRYSTAL DATA: $P6_3/mmc$ (No. 194)
$a = 13.75$ Å $b = 13.75$ Å $c = 10.05$ Å
$\alpha = 90°$ $\beta = 90°$ $\gamma = 120°$
X-ray single crystal refinement, $R = 0.17$

REFERENCE: K. Fischer,
Neues Jahrbuch für Mineralogie Monatshefte **1** 1–13 (1966).

h	k	l	2θ	d	M	I_{rel}	h	k	l	2θ	d	M	I_{rel}	h	k	l	2θ	d	M	I_{rel}
1	0	0	7.42	11.908	6	100.0	2	0	3	30.62	2.920	12	7.2	5	0	2	41.98	2.152	12	0.3
1	0	1	11.52	7.680	12	18.8	4	0	1	31.34	2.854	12	7.4	5	1	0	42.26	2.139	12	0.1
1	1	0	12.88	6.875	6	2.1	2	2	2	31.53	2.837	12	0.2	3	0	4	42.58	2.123	12	0.4
2	0	0	14.88	5.954	6	0.9	3	2	0	32.78	2.732	12	0.1	3	2	3	42.71	2.117	24	0.2
2	0	1	17.31	5.122	12	12.3	2	1	3	33.34	2.687	24	12.4	5	1	1	43.25	2.092	24	0.6
0	0	2	17.65	5.025	2	25.0	3	2	1	34.01	2.636	24	1.0	3	3	2	43.40	2.085	12	3.4
1	0	2	19.17	4.630	12	5.0	4	1	0	34.52	2.599	12	4.1	4	2	2	44.09	2.054	24	0.2
2	1	0	19.72	4.501	12	6.4	4	0	2	35.03	2.561	12	1.1	4	1	3	44.11	2.053	24	1.7
2	1	1	21.63	4.108	24	23.5	4	1	1	35.69	2.516	24	0.5	1	0	5	45.78	1.982	12	0.4
1	1	2	21.91	4.057	12	0.2	3	2	2	37.47	2.400	24	1.2	5	1	2	46.13	1.968	24	0.2
3	0	0	22.40	3.969	6	1.0	1	1	4	38.13	2.360	12	0.3	5	0	3	46.80	1.941	12	1.0
3	0	1	24.11	3.692	12	0.4	3	1	3	38.27	2.352	24	0.6	4	0	4	47.34	1.920	12	0.3
2	2	0	25.92	3.437	6	2.5	5	0	1	38.86	2.317	12	0.7	5	2	0	47.69	1.907	12	0.7
2	1	2	26.59	3.353	24	0.6	4	1	2	39.02	2.308	24	0.4	2	0	5	47.76	1.904	12	0.5
3	1	0	27.00	3.303	12	0.9	3	3	0	39.31	2.292	6	1.0	5	2	1	48.60	1.873	24	0.1
1	0	3	27.66	3.225	12	8.7	4	2	0	40.07	2.250	12	0.2	4	2	3	48.75	1.868	24	0.7
4	0	0	30.02	2.977	6	13.5	4	2	1	41.10	2.196	24	0.8	2	1	5	49.67	1.835	24	1.7

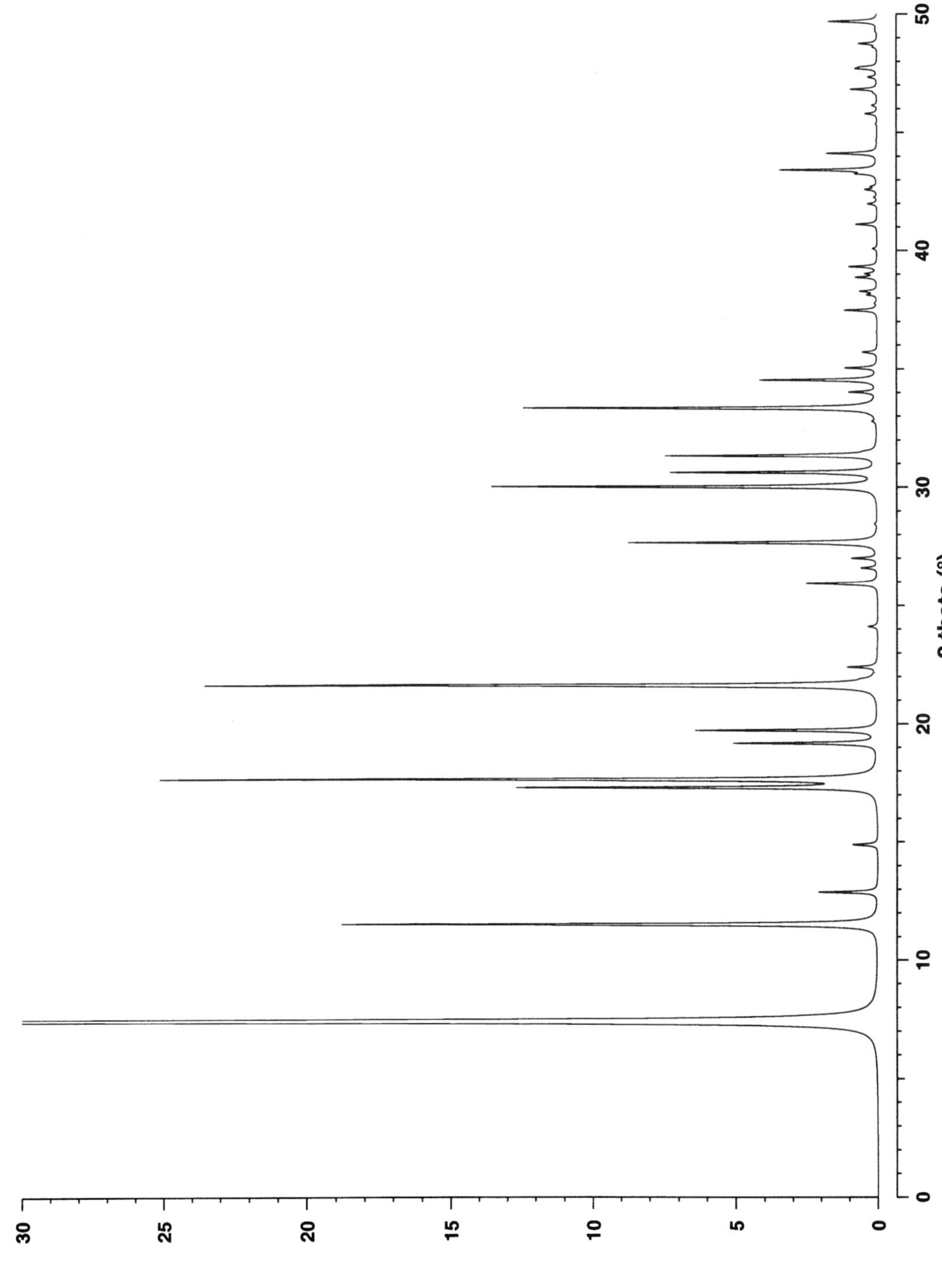
2 theta (°)
0
10
20
30
40
50
0
5
10
15
20
25
30

GON

GUS-1

CHEMICAL COMPOSITION: $[Si_{32}O_{64}]$

REFINED COMPOSITION: $[Si_{32}O_{64}]$

CRYSTAL DATA: $C222$ (No. 21)
$a = 16.4206$ Å $b = 20.0540$ Å $c = 5.0464$ Å
$\alpha = 90.0°$ $\beta = 90.0°$ $\gamma = 90.0°$
X-ray powder diffraction.

REFERENCE: J. Plevert, Y. Kubota, T. Honda, T. Okubo and Y. Sugi, *J. Chem. Soc., Chem. Commun.* 2363–2364 (2000).

h	k	l	2θ	d	M	I_{rel}
1	1	0	6.96	12.705	4	100.0
0	2	0	8.82	10.027	2	65.5
2	0	0	10.78	8.210	2	16.1
2	2	0	13.94	6.352	4	3.0
1	3	0	14.31	6.191	4	4.6
3	1	0	16.79	5.280	4	8.0
0	4	0	17.69	5.014	2	2.9
1	1	1	18.92	4.690	8	2.6
0	2	1	19.69	4.508	4	6.5
2	0	1	20.66	4.299	4	60.7
3	3	0	20.98	4.235	4	10.2
4	0	0	21.65	4.105	2	11.0
2	2	1	22.50	3.951	8	4.1
1	3	1	22.73	3.912	8	4.5
1	5	0	22.82	3.896	4	36.5
4	2	0	23.42	3.799	4	3.8
3	1	1	24.40	3.648	8	1.7
0	6	0	26.67	3.342	2	2.5
2	4	1	27.33	3.264	8	0.7
3	3	1	27.49	3.244	8	8.1
5	1	0	27.52	3.241	4	0.1
3	5	0	27.57	3.235	4	10.0
4	0	1	28.02	3.185	4	7.9
1	5	1	28.95	3.084	8	1.9
4	2	1	29.43	3.035	8	3.2
5	3	0	30.32	2.948	4	4.6

h	k	l	2θ	d	M	I_{rel}
1	7	0	31.70	2.822	4	1.2
6	0	0	32.72	2.737	2	0.2
5	1	1	32.84	2.727	8	1.7
3	5	1	32.88	2.724	8	5.5
4	4	1	33.33	2.688	8	1.3
6	2	0	33.95	2.640	4	1.8
2	6	1	33.97	2.639	8	1.6
4	6	0	34.61	2.592	4	0.2
5	3	1	35.26	2.545	8	1.1
5	5	0	35.32	2.541	4	0.1
3	7	0	35.36	2.538	4	0.2
0	0	2	35.58	2.523	2	8.0
0	8	0	35.82	2.507	2	0.6
1	1	2	36.30	2.475	8	1.8
1	7	1	36.48	2.463	8	2.0
0	2	2	36.73	2.447	4	0.8
2	0	2	37.28	2.412	4	0.4
6	0	1	37.38	2.406	4	2.3
2	8	0	37.51	2.397	4	0.3
2	2	2	38.39	2.345	8	0.3
6	2	1	38.48	2.339	8	1.2
1	3	2	38.53	2.337	8	0.1
4	6	1	39.07	2.306	8	0.1
3	1	2	39.58	2.277	8	0.4
5	5	1	39.71	2.270	8	1.4
3	7	1	39.75	2.268	8	0.4

h	k	l	2θ	d	M	I_{rel}
0	8	1	40.17	2.245	4	0.4
2	4	2	41.55	2.173	8	0.7
2	8	1	41.71	2.166	8	0.2
5	7	0	41.84	2.159	4	1.3
4	0	2	42.03	2.150	4	0.4
4	8	0	42.24	2.139	4	0.7
1	5	2	42.69	2.118	8	0.8
7	1	1	42.75	2.115	8	0.7
8	0	0	44.12	2.053	2	0.1
7	3	1	44.71	2.027	8	1.7
7	5	0	44.76	2.025	4	0.3
1	9	1	44.80	2.023	8	0.1
0	6	2	45.02	2.014	4	0.2
8	2	0	45.08	2.011	4	0.4
0	10	0	45.21	2.005	2	3.1
5	1	2	45.56	1.991	8	0.2
3	5	2	45.59	1.990	8	0.2
5	7	1	45.71	1.985	8	0.4
4	8	1	46.08	1.970	8	0.9
2	6	2	46.43	1.956	8	0.1
6	6	1	46.51	1.953	8	0.2
5	3	2	47.43	1.917	8	2.8
8	0	1	47.84	1.901	4	1.1
1	7	2	48.39	1.881	8	0.3
6	0	2	49.11	1.855	4	0.5
6	8	0	49.30	1.849	4	2.8

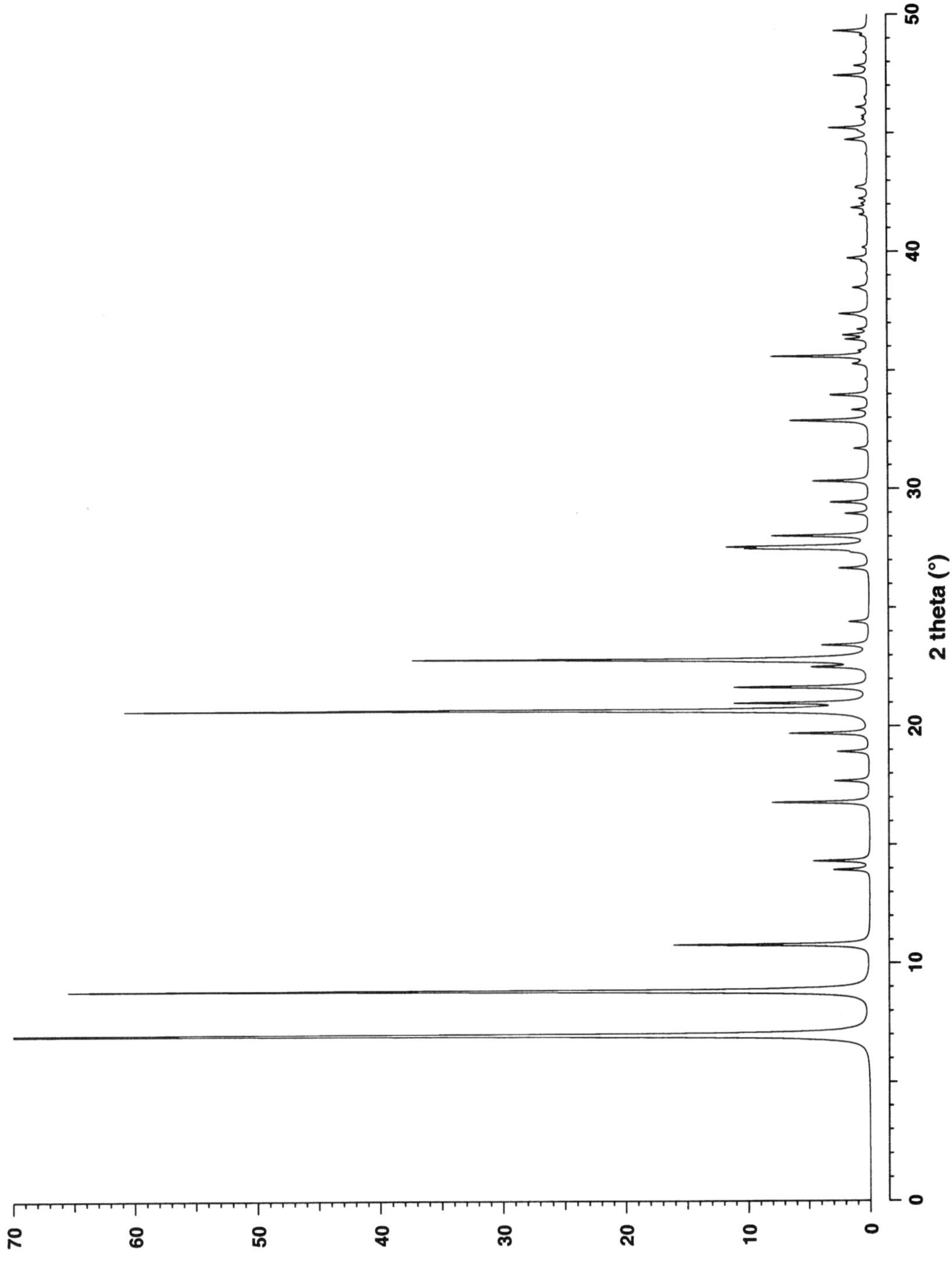
2 theta (°)
0
10
20
30
40
50
0
10
20
30
40
50
60
70

GOO Goosecreekite

CHEMICAL COMPOSITION: $|Ca_2(H_2O)_{10}|$ $[Si_{12}Al_4O_{32}]$
Loudon County, Virginia, U.S.A.

REFINED COMPOSITION: $|Ca_2(H_2O)_{10}|$ $[Si_{12}Al_4O_{32}]$

CRYSTAL DATA: $P12_11$ (No. 4) unique axis **b**
$a = 7.401$ Å $b = 17.439$ Å $c = 7.293$ Å
$\alpha = 90°$ $\beta = 105.44°$ $\gamma = 90°$
X-ray single crystal refinement, $R_w = 0.049$

REFERENCE: R. C. Rouse and D. R. Peacor,
American Mineralogist **71** 1494–1501 (1986).

h	*k*	*l*	2*θ*	*d*	*M*	I_{rel}
0	2	0	10.14	8.719	2	9.9
1	0	0	12.41	7.134	2	35.0
0	0	1	12.59	7.030	2	55.4
1	1	0	13.41	6.603	4	2.2
0	1	1	13.58	6.520	4	2.3
-1	0	1	15.16	5.845	2	5.2
-1	1	1	15.99	5.542	4	38.9
1	2	0	16.05	5.521	4	18.8
0	2	1	16.20	5.473	4	28.9
-1	2	1	18.27	4.855	4	81.0
1	3	0	19.70	4.506	4	64.6
0	3	1	19.82	4.480	4	100.0
1	0	1	19.95	4.450	2	12.8
0	4	0	20.37	4.360	2	0.4
1	1	1	20.60	4.312	4	8.2
-1	3	1	21.56	4.122	4	26.9
1	2	1	22.43	3.964	4	14.4
1	4	0	23.92	3.720	4	6.9
0	4	1	24.02	3.705	4	5.2
-2	0	1	24.80	3.590	2	5.3
2	0	0	24.96	3.567	2	9.2
-1	0	2	25.09	3.550	2	5.7
1	3	1	25.20	3.533	4	7.4
-2	1	1	25.33	3.516	4	11.1
0	0	2	25.34	3.515	2	6.1
-1	4	1	25.49	3.495	4	11.4
2	1	0	25.49	3.495	4	16.6
-1	1	2	25.61	3.478	4	22.1
0	1	2	25.86	3.446	4	8.7
-2	2	1	26.86	3.320	4	25.8
2	2	0	27.01	3.301	4	56.4
-1	2	2	27.12	3.288	4	13.3
0	2	2	27.36	3.260	4	41.7
1	5	0	28.49	3.133	4	19.3
0	5	1	28.57	3.124	4	18.9
1	4	1	28.66	3.114	4	11.3
-2	3	1	29.24	3.054	4	10.2
2	3	0	29.38	3.040	4	44.4
-1	3	2	29.48	3.030	4	6.6
0	3	2	29.70	3.008	4	45.2
-1	5	1	29.83	2.995	4	3.6
-2	0	2	30.59	2.923	2	7.5
0	6	0	30.76	2.906	2	14.1
-2	1	2	31.02	2.882	4	1.6
2	1	1	31.42	2.847	4	12.7
1	1	2	31.65	2.827	4	3.1
-2	4	1	32.30	2.771	4	10.0
-2	2	2	32.30	2.771	4	8.3
2	4	0	32.43	2.761	4	0.5
-1	4	2	32.53	2.753	4	15.0
1	5	1	32.62	2.745	4	2.9
2	2	1	32.68	2.740	4	9.0
1	2	2	32.91	2.722	4	2.9
1	6	0	33.29	2.692	4	18.1
0	6	1	33.36	2.686	4	8.7
-2	3	2	34.34	2.611	4	3.8
-1	6	1	34.46	2.603	4	8.9
2	3	1	34.70	2.585	4	8.9
1	3	2	34.91	2.570	4	7.5
-2	5	1	35.90	2.502	4	3.1
2	5	0	36.01	2.494	4	1.6
-1	5	2	36.10	2.488	4	3.5
0	5	2	36.28	2.476	4	1.9
-3	0	1	36.52	2.460	2	1.5
1	6	1	36.94	2.433	4	2.7
-2	4	2	37.03	2.428	4	9.4
-1	0	3	37.06	2.426	2	0.6
2	4	1	37.37	2.406	4	1.2
-1	1	3	37.43	2.403	4	1.4
1	4	2	37.56	2.394	4	2.4
3	0	0	37.83	2.378	2	1.8
-3	2	1	38.00	2.368	4	5.6
1	7	0	38.27	2.352	4	2.2
0	0	3	38.41	2.343	2	7.4
-1	2	3	38.52	2.337	4	14.3
0	1	3	38.77	2.322	4	0.6
3	2	0	39.27	2.294	4	3.1
-1	7	1	39.31	2.292	4	7.0
-3	3	1	39.79	2.266	4	3.3
0	2	3	39.83	2.263	4	3.5
-2	6	1	39.91	2.259	4	4.6
-3	1	2	40.06	2.251	4	0.7
-2	5	2	40.26	2.240	4	6.5
0	6	2	40.26	2.240	4	5.2
-1	3	3	40.28	2.239	4	3.3
-2	1	3	40.37	2.234	4	2.4
2	5	1	40.57	2.223	4	1.9
1	5	2	40.76	2.214	4	2.9
2	1	2	40.89	2.207	4	4.7
3	3	0	41.01	2.201	4	1.4
-3	2	2	41.09	2.197	4	1.6
-2	2	3	41.39	2.181	4	0.8
0	8	0	41.42	2.180	2	1.5
1	7	1	41.54	2.174	4	3.9
0	3	3	41.55	2.173	4	1.3
-3	4	1	42.18	2.143	4	5.4
-1	4	3	42.65	2.120	4	2.6
-3	3	2	42.76	2.114	4	1.4
-2	3	3	43.06	2.101	4	2.4
3	4	0	43.34	2.088	4	1.9
0	8	1	43.46	2.082	4	0.5
2	3	2	43.55	2.078	4	0.8
3	1	1	43.62	2.075	4	1.7
1	0	3	43.75	2.069	2	0.8
0	4	3	43.86	2.064	4	0.7
-2	6	2	43.93	2.061	4	2.0
2	6	1	44.23	2.048	4	4.9
2	7	0	44.35	2.042	4	1.4
1	6	2	44.40	2.040	4	1.2
-1	7	2	44.42	2.039	4	0.5
0	7	2	44.58	2.033	4	2.3
3	2	1	44.58	2.032	4	2.6
1	2	3	45.03	2.013	4	1.0
-3	5	1	45.10	2.010	4	1.3
-2	4	3	45.31	2.002	4	0.6
2	4	2	45.78	1.982	4	3.3
3	3	1	46.15	1.967	4	1.0
3	5	0	46.20	1.965	4	2.2
1	3	3	46.60	1.949	4	1.7
0	5	3	46.70	1.945	4	0.5
-3	1	3	46.92	1.936	4	3.0
-2	7	2	47.98	1.896	4	0.8
-2	5	3	48.08	1.892	4	1.4
3	4	1	48.29	1.885	4	3.2
2	5	2	48.53	1.876	4	1.5
0	9	1	48.75	1.868	4	1.7
-2	8	1	48.88	1.863	4	0.9
-1	6	3	48.90	1.862	4	1.1
-1	8	2	49.04	1.858	4	2.6
0	8	2	49.18	1.853	4	0.5
-4	0	1	49.25	1.850	2	10.4
-3	3	3	49.33	1.847	4	0.4
3	6	0	49.53	1.840	4	6.0
-4	1	1	49.55	1.840	4	0.5
-1	9	1	49.56	1.839	4	1.4

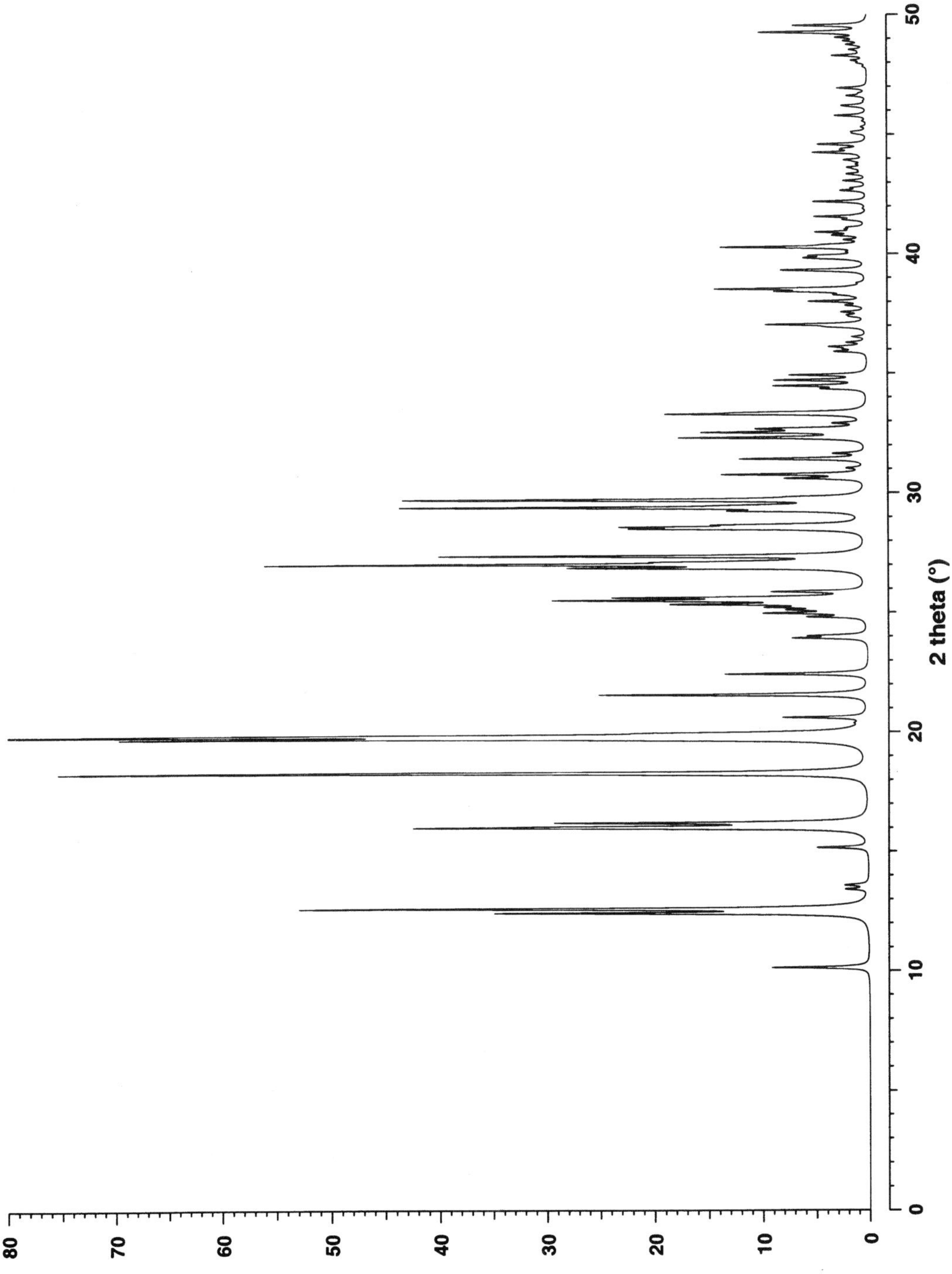
2 theta (°)
0
10
20
30
40
50
0
10
20
30
40
50
60
70
80

HEU Heulandite

CHEMICAL COMPOSITION: $|Na_{0.1}K_{8.57}Ba_{0.04}(H_2O)_{19.56}|$ $[Si_{26.83}Al_{9.31}O_{72}]$
Mossyrock Dam, Washington, U.S.A.

REFINED COMPOSITION: $|K_{8.48}(H_2O)_{18}|$ $[Si_{26.64}Al_{9.36}O_{72}]$

CRYSTAL DATA: $C1\,2/m\,1$ (No. 12) unique axis **b**, cell choice 1
$a = 17.767$ Å $b = 17.958$ Å $c = 7.431$ Å
$\alpha = 90°$ $\beta = 115.93°$ $\gamma = 90°$
X-ray single crystal refinement, $R_w = 0.07$

REFERENCE: E. Galli, G. Gottardi, H. Mayer, A. Preisinger and E. Passaglia, *Acta Cryst.* **B39** 189–197 (1983).

h	*k*	*l*	2θ	*d*	*M*	I_{rel}
1	1	0	7.41	11.937	4	3.3
0	2	0	9.85	8.979	2	100.0
2	0	0	11.07	7.989	2	64.0
-1	1	1	12.90	6.865	4	6.8
-2	0	1	13.03	6.792	2	17.8
0	0	1	13.25	6.683	2	2.9
2	2	0	14.84	5.969	4	12.9
1	3	0	15.81	5.606	4	2.2
-2	2	1	16.36	5.417	4	1.9
-3	1	1	16.86	5.258	4	11.3
1	1	1	17.19	5.157	4	32.0
3	1	0	17.37	5.106	4	4.3
-1	3	1	19.04	4.661	4	20.3
-4	0	1	20.31	4.373	2	8.2
2	0	1	20.72	4.286	2	1.1
-3	3	1	21.95	4.050	4	1.9
1	3	1	22.21	4.003	4	60.2
4	0	0	22.25	3.995	2	34.7
3	3	0	22.34	3.979	4	53.8
-4	2	1	22.62	3.931	4	21.9
2	4	0	22.72	3.914	4	29.4
2	2	1	22.99	3.868	4	3.1
-2	4	1	23.76	3.745	4	2.2
0	4	1	23.88	3.727	4	4.4
-2	0	2	23.96	3.715	2	5.9
-3	1	2	24.99	3.563	4	29.4
-1	1	2	25.22	3.531	4	0.9
1	5	0	25.42	3.504	4	0.9
-5	1	1	25.58	3.482	4	4.5
-2	2	2	25.96	3.433	4	50.8
3	1	1	26.03	3.424	4	6.8
-4	0	2	26.24	3.396	2	18.5
0	0	2	26.68	3.341	2	19.0
-4	2	2	28.09	3.176	4	45.1
-4	4	1	28.49	3.132	4	34.1
0	2	2	28.50	3.132	4	0.8
-3	3	2	28.73	3.107	4	1.0
2	4	1	28.80	3.100	4	2.4
-1	3	2	28.93	3.086	4	22.8
-5	1	2	29.45	3.033	4	5.3
3	3	1	29.64	3.013	4	1.6
-3	5	1	29.71	3.007	4	15.0
0	6	0	29.85	2.993	2	4.1
1	5	1	29.90	2.988	4	41.6
4	4	0	29.94	2.984	4	2.2

h	*k*	*l*	2θ	*d*	*M*	I_{rel}
3	5	0	30.01	2.978	4	27.8
1	1	2	30.03	2.975	4	12.7
-6	0	1	30.21	2.958	2	4.5
4	0	1	30.69	2.913	2	2.3
-2	4	2	31.25	2.862	4	0.5
5	3	0	31.74	2.819	4	43.6
-6	2	1	31.85	2.810	4	14.0
4	2	1	32.30	2.771	4	0.7
-6	0	2	32.57	2.749	2	1.5
-2	6	1	32.70	2.739	4	10.6
-5	3	2	32.72	2.737	4	1.6
0	6	1	32.79	2.732	4	12.8
-4	4	2	33.07	2.708	4	0.5
1	3	2	33.25	2.694	4	7.7
2	0	2	33.28	2.692	2	0.5
0	4	2	33.43	2.681	4	3.2
6	0	0	33.65	2.663	2	1.3
2	2	2	34.79	2.579	4	9.4
-3	5	2	35.12	2.555	4	1.4
6	2	0	35.15	2.553	4	4.8
-1	5	2	35.29	2.543	4	2.8
1	7	0	35.44	2.533	4	1.7
-5	5	1	35.56	2.525	4	6.7
3	5	1	35.89	2.502	4	2.5
-7	1	1	35.93	2.499	4	2.1
-6	4	1	36.37	2.470	4	3.2
-4	6	1	36.37	2.470	4	3.9
5	1	1	36.42	2.467	4	2.4
2	6	1	36.62	2.454	4	11.3
-3	1	3	36.63	2.453	4	2.3
4	4	1	36.78	2.444	4	10.4
-2	0	3	36.84	2.439	2	3.9
-7	1	2	37.06	2.426	4	5.6
4	6	0	37.55	2.395	4	1.3
3	1	2	37.86	2.377	4	0.9
-4	2	3	37.99	2.368	4	1.5
-5	1	3	38.15	2.359	4	1.1
-2	2	3	38.23	2.354	4	1.7
-6	4	2	38.39	2.345	4	0.7
-3	7	1	38.73	2.325	4	0.5
1	7	1	38.89	2.316	4	0.4
2	4	2	39.01	2.309	4	1.3
5	3	1	39.18	2.299	4	1.3
-3	3	3	39.37	2.288	4	1.1
-4	6	2	40.16	2.245	4	1.0

h	*k*	*l*	2θ	*d*	*M*	I_{rel}
0	8	0	40.17	2.245	2	0.9
0	0	3	40.49	2.228	2	1.4
-6	2	3	41.12	2.195	4	2.0
-1	3	3	41.25	2.189	4	2.6
2	8	0	41.80	2.161	4	0.4
-2	4	3	42.16	2.143	4	1.9
4	0	2	42.17	2.143	2	0.8
7	3	0	42.38	2.133	4	2.4
-2	8	1	42.41	2.131	4	0.8
-7	1	3	42.85	2.110	4	0.6
6	2	1	42.89	2.108	4	5.9
-6	6	1	42.99	2.104	4	2.6
-3	7	2	43.16	2.096	4	2.2
4	6	1	43.34	2.088	4	1.3
-5	7	1	43.52	2.079	4	0.7
1	1	3	43.70	2.071	4	2.6
3	7	1	43.81	2.067	4	3.9
5	5	1	44.26	2.046	4	1.4
-7	5	2	44.80	2.023	4	5.2
-6	4	3	44.84	2.021	4	4.7
3	5	2	45.48	1.994	4	0.7
6	6	0	45.60	1.990	4	0.5
1	9	0	45.83	1.980	4	1.1
-8	4	1	46.03	1.972	4	4.4
-5	7	2	46.06	1.971	4	0.4
-1	5	3	46.14	1.967	4	4.3
-8	4	2	46.18	1.966	4	1.1
6	4	1	46.50	1.953	4	2.6
-9	1	2	46.53	1.952	4	7.1
8	2	0	46.58	1.950	4	0.5
2	0	3	46.89	1.938	2	0.6
-8	2	3	47.05	1.932	4	1.2
-1	9	1	47.16	1.927	4	1.3
-2	8	2	47.31	1.921	4	2.1
5	1	2	47.46	1.916	4	0.8
7	1	1	47.67	1.908	4	0.5
2	2	3	48.03	1.894	4	0.5
-4	8	2	48.62	1.873	4	1.1
3	9	0	48.73	1.869	4	2.1
-9	3	2	48.81	1.866	4	0.5
-4	0	4	49.05	1.857	2	1.2
-5	1	4	49.55	1.840	4	2.0
5	3	2	49.71	1.834	4	0.7
-3	1	4	49.80	1.831	4	5.0
-7	5	3	49.86	1.829	4	2.0

Heulandite HEU

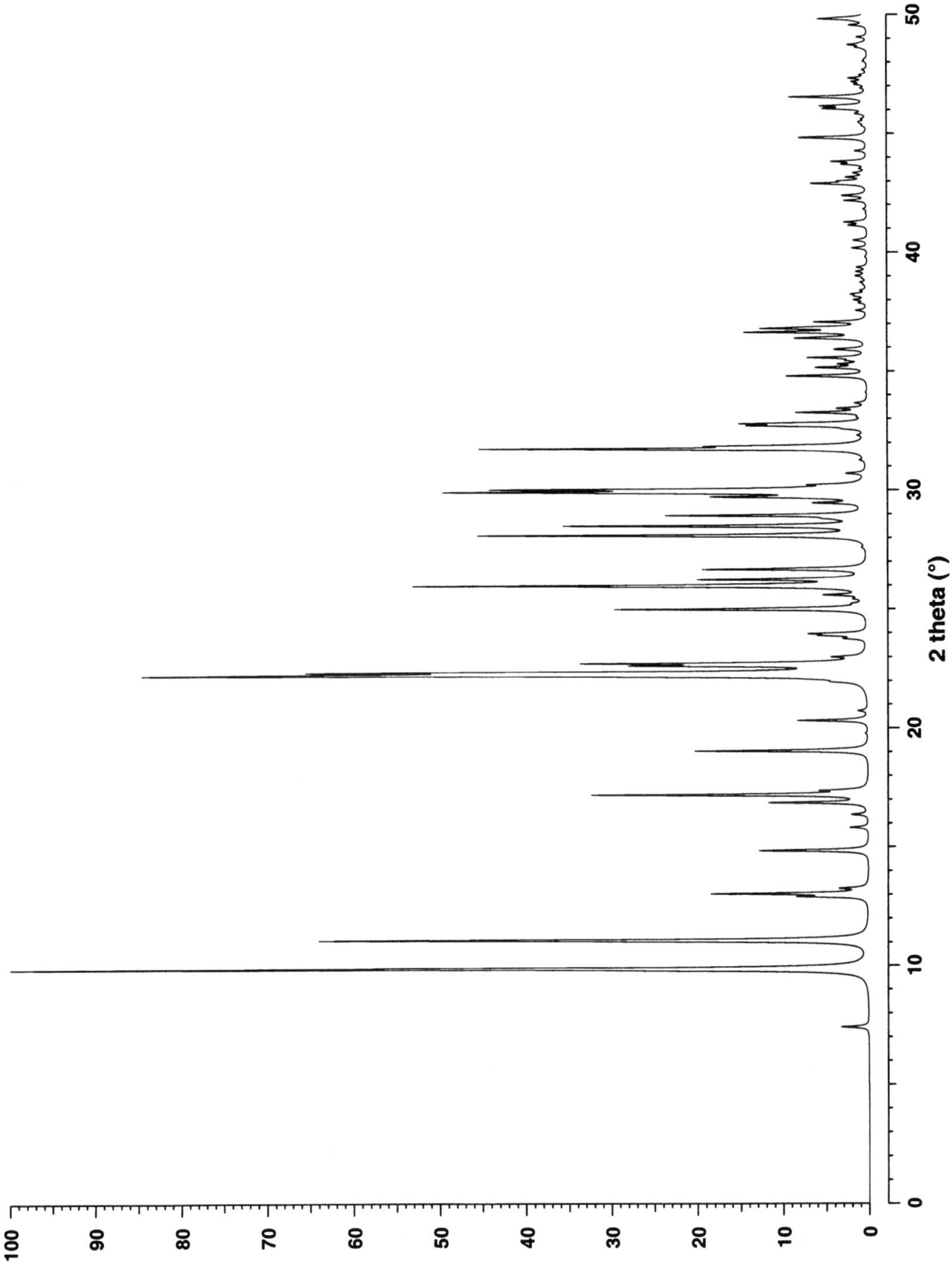

HEU Clinoptilolite

CHEMICAL COMPOSITION: $|Na_{1.84}K_{1.76}Mg_{0.2}Ca_{1.24}(H_2O)_{21.36}|$ $[Si_{29.84}Al_{6.16}O_{72}]$
Agoura, California, U.S.A.

REFINED COMPOSITION: $|Na_{1.84}K_{1.76}Mg_{0.2}Ca_{1.24}(H_2O)_{21.36}|$ $[Si_{29.84}Al_{6.16}O_{72}]$

CRYSTAL DATA: $C1\,2/m\,1$ (No. 12) unique axis **b**, cell choice 1
a = 17.662 Å b = 17.911 Å c = 7.407 Å
α = 90° β = 116.40° γ = 90°
X-ray single crystal refinement, R_w = 0.088

REFERENCE: K. Koyama and Y. Takeuchi,
Z. Kristallogr. **145** 216–239 (1977).

h	k	l	2θ	d	M	I_{rel}
1	1	0	7.46	11.857	4	4.7
0	2	0	9.88	8.955	2	100.0
2	0	0	11.19	7.910	2	40.0
-1	1	1	12.94	6.842	4	1.4
-2	0	1	13.06	6.780	2	14.5
0	0	1	13.35	6.635	2	7.0
2	2	0	14.94	5.929	4	6.3
1	3	0	15.87	5.586	4	1.9
-2	2	1	16.40	5.405	4	0.6
0	2	1	16.63	5.331	4	1.6
-3	1	1	16.91	5.243	4	15.8
1	1	1	17.36	5.110	4	24.8
3	1	0	17.53	5.059	4	6.0
-1	3	1	19.10	4.648	4	15.6
-4	0	1	20.40	4.354	2	5.6
1	3	1	22.36	3.976	4	49.2
4	0	0	22.48	3.955	2	18.1
3	3	0	22.49	3.952	4	37.2
-4	2	1	22.71	3.916	4	21.0
2	4	0	22.82	3.897	4	24.4
2	2	1	23.21	3.832	4	4.9
-2	4	1	23.81	3.736	4	3.1
0	4	1	23.98	3.712	4	1.5
-2	0	2	24.04	3.702	2	6.8
4	2	0	24.61	3.618	4	0.3
-3	1	2	25.05	3.555	4	14.7
-1	1	2	25.35	3.513	4	0.6
1	5	0	25.49	3.494	4	1.0
-5	1	1	25.72	3.463	4	5.0
-2	2	2	26.04	3.421	4	32.1
-4	0	2	26.29	3.390	2	13.6
3	1	1	26.32	3.386	4	2.8
0	0	2	26.88	3.317	2	9.6
-4	2	2	28.15	3.170	4	27.2
-4	4	1	28.60	3.121	4	13.8
0	2	2	28.70	3.111	4	0.5
-3	3	2	28.80	3.100	4	2.7
2	4	1	29.00	3.078	4	1.0
-1	3	2	29.07	3.072	4	14.0
-5	3	1	29.39	3.039	4	0.7
-5	1	2	29.51	3.027	4	2.8
-3	5	1	29.79	2.999	4	9.5
3	3	1	29.92	2.986	4	1.3
0	6	0	29.93	2.985	2	2.0
1	5	1	30.05	2.973	4	26.4
4	4	0	30.15	2.964	4	0.4
3	5	0	30.16	2.963	4	17.8
1	1	2	30.30	2.950	4	11.2
-6	0	1	30.41	2.940	2	2.7
4	0	1	31.05	2.880	2	1.9
5	3	0	32.01	2.796	4	23.1
-6	2	1	32.04	2.793	4	8.7
4	2	1	32.66	2.742	4	0.6
-6	0	2	32.66	2.742	2	0.8
-2	6	1	32.78	2.732	4	9.7
-5	3	2	32.79	2.731	4	0.4
0	6	1	32.90	2.722	4	4.9
1	3	2	33.51	2.674	4	4.0
2	0	2	33.62	2.666	2	1.0
0	4	2	33.62	2.665	4	1.2
6	0	0	34.00	2.637	2	0.3
2	2	2	35.13	2.555	4	4.1
-3	5	2	35.21	2.549	4	0.7
-1	5	2	35.44	2.533	4	2.3
6	2	0	35.49	2.529	4	2.9
1	7	0	35.54	2.526	4	1.8
-5	5	1	35.71	2.514	4	3.2
3	5	1	36.16	2.484	4	2.0
-7	1	1	36.19	2.482	4	0.9
-4	6	1	36.49	2.462	4	1.3
-6	4	1	36.57	2.457	4	2.4
-4	0	3	36.69	2.449	2	0.4
2	6	1	36.82	2.441	4	6.2
5	1	1	36.85	2.439	4	1.3
-2	0	3	37.02	2.428	2	3.6
4	4	1	37.11	2.422	4	5.7
-7	1	2	37.20	2.417	4	4.1
4	6	0	37.76	2.383	4	1.5
-4	2	3	38.09	2.362	4	2.0
-5	1	3	38.22	2.355	4	0.6
3	1	2	38.27	2.352	4	1.1
-2	2	3	38.41	2.344	4	1.9
-6	4	2	38.50	2.338	4	0.4
-2	6	2	38.75	2.324	4	0.5
-3	7	1	38.84	2.319	4	0.8
-7	3	1	38.97	2.311	4	0.2
2	4	2	39.34	2.290	4	0.6
5	3	1	39.59	2.276	4	0.4
6	4	0	39.67	2.272	4	1.2
-6	0	3	39.89	2.260	2	0.7
-7	3	2	39.92	2.258	4	0.4
-4	6	2	40.25	2.240	4	0.3
0	0	3	40.80	2.212	2	1.1
3	3	2	40.93	2.205	4	0.3
-6	2	3	41.20	2.191	4	2.1
-1	3	3	41.49	2.176	4	1.5
2	8	0	41.94	2.154	4	0.4
6	0	1	42.13	2.145	2	0.3
-2	4	3	42.34	2.134	4	2.1
-2	8	1	42.52	2.126	4	0.5
4	0	2	42.66	2.120	2	0.8
7	3	0	42.78	2.114	4	1.9
-7	1	3	42.94	2.106	4	0.4
-6	6	1	43.19	2.095	4	1.3
-3	7	2	43.27	2.091	4	1.4
6	2	1	43.38	2.086	4	3.4
4	6	1	43.67	2.073	4	2.3
3	7	1	44.07	2.055	4	1.5
1	1	3	44.08	2.054	4	0.9
5	5	1	44.66	2.029	4	1.0
-6	4	3	44.93	2.018	4	2.7
-7	5	2	44.96	2.016	4	2.3
-7	3	3	45.38	1.998	4	0.4
8	0	0	45.89	1.978	2	0.4
6	6	0	45.92	1.976	4	0.8
1	9	0	45.96	1.975	4	0.9
-5	7	2	46.18	1.966	4	0.6
-8	4	1	46.34	1.959	4	2.6
-8	4	2	46.38	1.958	4	1.2
-1	5	3	46.38	1.958	4	4.5
-9	1	2	46.77	1.942	4	2.6
6	4	1	46.97	1.934	4	2.2
-8	2	3	47.16	1.927	4	0.8
-1	9	1	47.30	1.922	4	0.6
2	0	3	47.35	1.920	2	0.4
-2	8	2	47.46	1.916	4	1.3
5	1	2	48.02	1.895	4	0.8
-2	6	3	48.32	1.884	4	0.3
2	2	3	48.49	1.877	4	0.6
-4	8	2	48.74	1.868	4	0.5
3	9	0	48.92	1.862	4	1.2
-9	3	2	49.06	1.857	4	0.6
-4	0	4	49.22	1.851	2	0.7
-5	1	4	49.69	1.835	4	1.0
-7	5	3	49.97	1.825	4	0.8

Clinoptilolite HEU

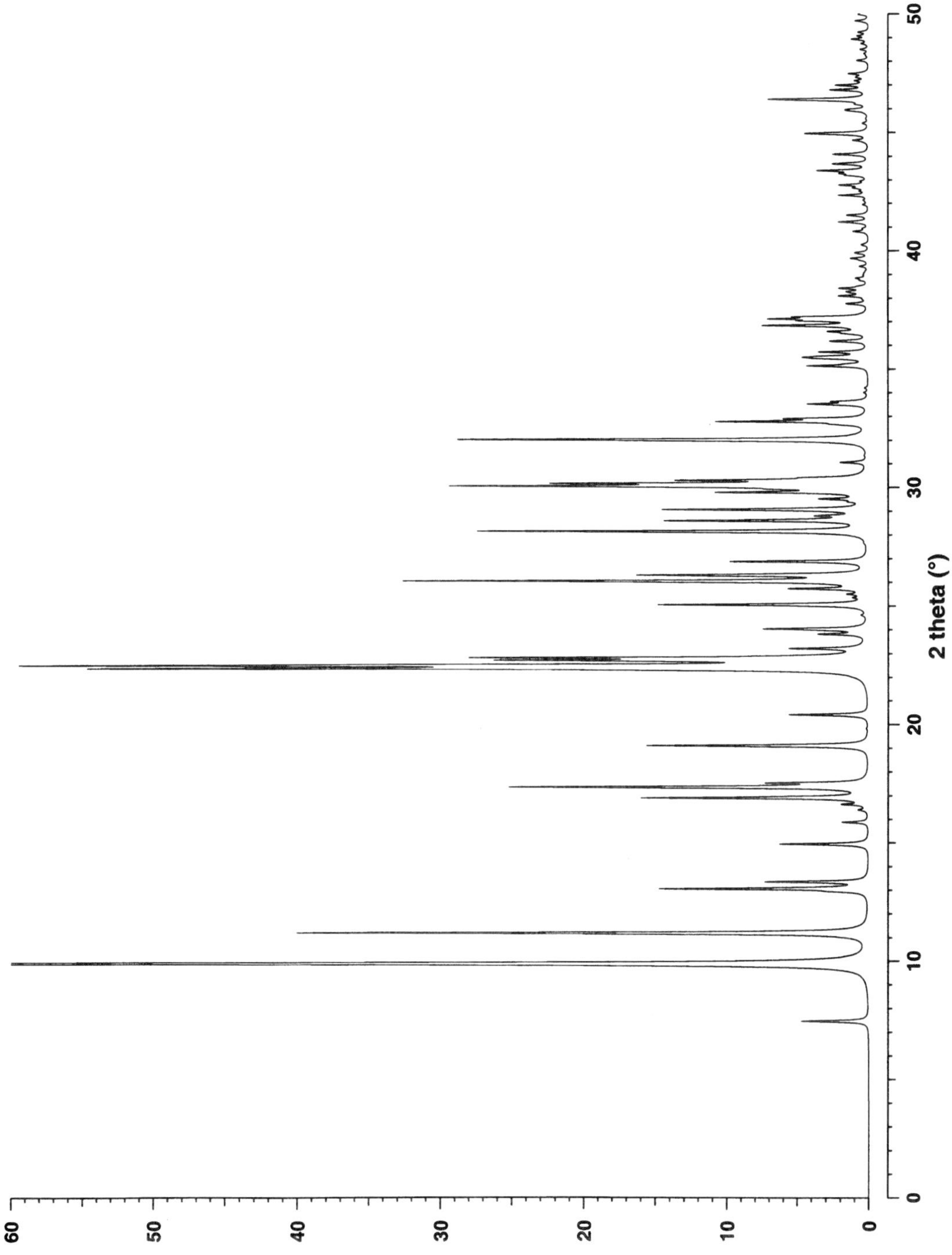

IFR ITQ-4, Calcined

CHEMICAL COMPOSITION: $[Si_{32}O_{64}]$

REFINED COMPOSITION: $[Si_{32}O_{64}]$

CRYSTAL DATA: $I\,1\,2/m\,1$ (No. 12) unique axis **b**, cell choice 3
a = 18.65243 Å b = 13.49597 Å c = 7.63109 Å
α = 90.0° β = 101.9781° γ = 90.0°
X-ray Rietveld refinement, R_{wp} = 0.0767, R_p = 0.0558, R_b = 0.0644

REFERENCE: P. A. Barrett, M. A. Camblor, A. Corma, R. H. Jones and L. A. Villaescusa, *Chemistry of Materials* **9** 1713–1715 (1997).

h	k	l	2θ	d	M	I_{rel}
1	1	0	8.15	10.850	4	100.0
2	0	0	9.69	9.123	2	9.4
-1	0	1	11.84	7.474	2	1.0
0	2	0	13.12	6.748	2	2.0
0	1	1	13.56	6.532	4	3.0
1	0	1	13.72	6.455	2	4.4
-2	1	1	15.18	5.837	4	5.8
3	1	0	15.98	5.545	4	0.5
2	2	0	16.34	5.425	4	0.5
-3	0	1	16.78	5.283	2	0.5
2	1	1	18.08	4.906	4	0.4
1	2	1	19.02	4.665	4	2.7
4	0	0	19.46	4.562	2	0.8
1	3	0	20.33	4.368	4	8.8
3	0	1	20.66	4.299	2	6.7
-3	2	1	21.36	4.160	4	2.3
-4	1	1	21.64	4.106	4	6.7
4	2	0	23.54	3.779	4	1.3
-2	0	2	23.81	3.737	2	3.0
0	0	2	23.84	3.732	2	3.8
-1	1	2	24.24	3.672	4	0.5
3	2	1	24.55	3.625	4	0.2
3	3	0	24.61	3.617	4	1.1
-5	0	1	24.83	3.585	2	2.2
5	1	0	25.28	3.523	4	1.6
2	3	1	26.05	3.421	4	0.1
-3	1	2	26.13	3.410	4	0.8
1	1	2	26.19	3.403	4	5.3
0	4	0	26.42	3.374	2	5.6
-2	2	2	27.28	3.269	4	0.9
0	2	2	27.30	3.266	4	0.3
-4	0	2	27.56	3.237	2	0.6
-5	2	1	28.18	3.166	4	1.5
-4	3	1	28.68	3.112	4	0.6
-1	4	1	29.04	3.075	4	0.3

h	k	l	2θ	d	M	I_{rel}
6	0	0	29.37	3.041	2	1.7
5	0	1	29.39	3.039	2	1.4
1	4	1	29.88	2.990	4	0.3
-6	1	1	30.08	2.971	4	0.2
2	2	2	30.70	2.912	4	0.4
-1	3	2	30.73	2.910	4	0.3
-5	1	2	31.23	2.864	4	0.2
3	1	2	31.32	2.856	4	0.2
-3	4	1	31.46	2.843	4	1.2
5	3	0	31.57	2.834	4	0.7
4	3	1	31.97	2.799	4	0.3
-3	3	2	32.27	2.774	4	0.4
6	2	0	32.29	2.773	4	0.3
5	2	1	32.31	2.771	4	0.3
1	3	2	32.31	2.770	4	0.1
4	4	0	33.02	2.713	4	0.9
1	5	0	33.56	2.670	4	0.1
3	4	1	33.77	2.654	4	0.7
4	0	2	34.05	2.633	2	0.3
6	1	1	34.73	2.583	4	0.5
7	1	0	35.06	2.559	4	0.4
0	5	1	35.36	2.538	4	1.3
-1	0	3	35.38	2.537	2	1.1
0	4	2	35.88	2.503	4	0.7
-2	1	3	36.01	2.494	4	0.1
-3	0	3	36.05	2.491	2	0.6
-2	5	1	36.05	2.491	4	0.6
3	5	0	36.42	2.467	4	0.1
-6	2	2	36.53	2.460	4	0.5
-5	4	1	36.57	2.457	4	1.4
-5	3	2	36.59	2.456	4	0.4
-4	1	3	38.01	2.367	4	0.2
-7	1	2	38.38	2.345	4	0.3
5	1	2	38.49	2.339	4	0.3
-4	4	2	38.54	2.336	4	0.1

h	k	l	2θ	d	M	I_{rel}
1	2	3	39.87	2.261	4	0.1
0	6	0	40.09	2.249	2	0.4
7	2	1	41.21	2.190	4	0.1
5	5	0	41.62	2.170	4	0.3
8	2	0	41.81	2.161	4	0.5
4	5	1	41.93	2.154	4	0.1
3	0	3	41.99	2.152	2	0.3
-3	5	2	42.17	2.143	4	0.2
1	5	2	42.21	2.141	4	0.3
-6	1	3	42.42	2.131	4	0.3
1	6	1	42.56	2.124	4	0.2
-3	6	1	43.74	2.070	4	0.1
-9	0	1	43.77	2.068	2	0.3
6	2	2	44.23	2.048	4	0.1
8	1	1	44.38	2.041	4	0.3
-7	0	3	44.83	2.022	2	0.4
-6	5	1	44.87	2.020	4	0.5
4	1	3	45.50	1.994	4	0.2
-6	3	3	46.69	1.946	4	0.2
-9	1	2	46.83	1.940	4	0.2
1	7	0	47.42	1.917	4	0.2
7	4	1	47.63	1.909	4	0.2
-2	0	4	47.67	1.908	2	0.4
8	4	0	48.16	1.890	4	0.1
5	0	3	48.36	1.882	2	0.2
-1	1	4	48.46	1.878	4	0.2
8	3	1	48.51	1.877	4	0.2
0	7	1	48.78	1.867	4	0.2
-2	7	1	49.32	1.848	4	0.4
-10	1	1	49.36	1.846	4	0.1
4	3	3	49.55	1.839	4	0.1
0	5	3	49.84	1.830	4	0.1
10	0	0	49.98	1.825	2	0.1

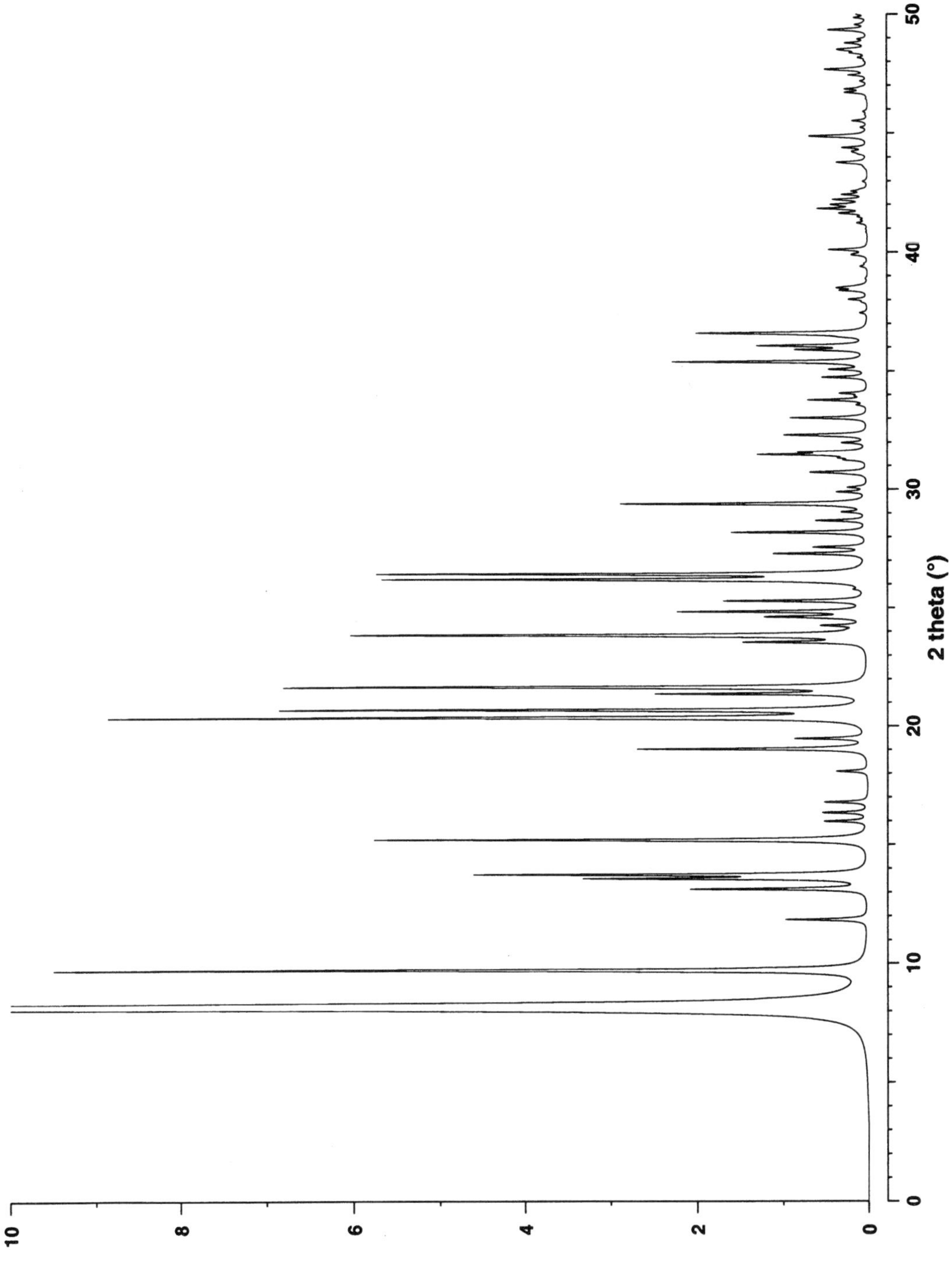

2 theta (°)
0
10
20
30
40
50
0
2
4
6
8
10

ISV ITQ–7, Siliceous, Calcined

CHEMICAL COMPOSITION: $[Si_{64}O_{128}]$

REFINED COMPOSITION: $[Si_{64}O_{128}]$

CRYSTAL DATA: $P\,4_2/m\,m\,c$ (No. 131)
a = 12.8528 Å b = 12.8528 Å c = 25.2136 Å
α = 90.0° β = 90.0° γ = 90.0°
X-ray Rietveld refinement, R_p = 0.0827, R_{wp} = 0.1076, R_b = 0.0.0626

REFERENCE: L. A. Villaescusa, P. A. Barrett and M. A. Camblor, *Angew. Chem., Int. ed.* **38** 1997–2000 (1999).

h	k	l	2θ	d	M	I_{rel}
1	0	0	6.88	12.853	4	52.2
0	0	2	7.01	12.607	2	100.0
1	0	1	7.72	11.451	8	65.5
1	1	0	9.73	9.088	4	60.5
1	0	2	9.83	9.000	8	41.4
1	0	3	12.58	7.034	8	0.7
2	0	0	13.78	6.426	4	6.3
0	0	4	14.05	6.303	2	1.1
2	0	1	14.22	6.227	8	2.3
2	1	0	15.42	5.748	8	4.3
2	0	2	15.48	5.725	8	0.3
1	0	4	15.66	5.659	8	2.7
2	1	1	15.81	5.604	16	0.5
2	1	2	16.95	5.230	16	2.8
1	1	4	17.12	5.180	8	2.0
2	1	3	18.70	4.744	16	0.9
1	0	5	18.90	4.694	8	3.4
2	2	0	19.53	4.544	4	5.8
2	0	4	19.73	4.500	8	7.2
3	0	0	20.73	4.284	4	0.9
2	2	2	20.78	4.275	8	1.0
2	1	4	20.92	4.247	16	0.6
3	0	1	21.03	4.224	8	1.3
0	0	6	21.14	4.202	2	2.2
3	1	0	21.87	4.064	8	21.3
3	0	2	21.91	4.056	8	8.4
3	1	1	22.15	4.013	16	4.3
1	0	6	22.26	3.994	8	21.5
2	0	5	22.41	3.967	8	0.5
3	1	2	22.99	3.868	16	1.7
3	0	3	23.30	3.817	8	0.2
1	1	6	23.32	3.814	8	3.4
2	1	5	23.47	3.791	16	0.2
2	2	4	24.14	3.686	8	2.9
3	1	3	24.33	3.659	16	0.5
3	2	0	24.98	3.565	8	3.0
3	2	1	25.23	3.530	16	0.6
2	0	6	25.32	3.517	8	2.9
1	0	7	25.68	3.468	8	1.9
3	2	2	25.97	3.430	16	1.6
3	1	4	26.09	3.416	16	1.3
2	1	6	26.27	3.392	16	2.1
3	2	3	27.17	3.282	16	0.4
3	0	5	27.31	3.265	8	0.9
4	0	0	27.76	3.213	4	8.1

h	k	l	2θ	d	M	I_{rel}
4	0	1	27.99	3.187	8	0.2
3	1	5	28.20	3.164	16	0.2
0	0	8	28.32	3.152	2	4.2
2	0	7	28.40	3.142	8	0.3
4	1	0	28.64	3.117	8	1.6
4	0	2	28.67	3.114	8	0.4
3	2	4	28.77	3.103	16	1.2
2	2	6	28.94	3.085	8	2.1
1	0	8	29.17	3.061	8	0.3
3	3	0	29.48	3.029	4	2.5
4	1	2	29.52	3.026	16	1.6
4	0	3	29.77	3.001	8	0.9
3	0	6	29.78	3.000	8	2.9
1	1	8	30.01	2.978	8	0.8
3	3	2	30.34	2.946	8	0.5
3	1	6	30.60	2.921	16	0.2
3	2	5	30.71	2.911	16	1.1
4	2	0	31.12	2.874	8	3.6
4	0	4	31.24	2.863	8	3.3
4	2	1	31.33	2.855	16	1.5
2	0	8	31.62	2.830	8	0.4
4	2	2	31.94	2.802	16	0.6
4	1	4	32.03	2.794	16	1.3
2	1	8	32.40	2.764	16	0.2
3	3	4	32.80	2.730	8	0.4
3	2	6	32.95	2.718	16	2.8
4	0	5	33.06	2.710	8	0.3
3	1	7	33.23	2.696	16	0.2
4	2	4	34.29	2.615	16	0.3
5	0	0	34.90	2.571	4	0.3
4	3	0	34.90	2.571	8	0.2
4	3	1	35.09	2.557	16	0.8
5	0	1	35.09	2.557	8	0.3
4	0	6	35.16	2.553	8	0.7
3	0	8	35.35	2.539	8	0.2
0	0	10	35.61	2.521	2	0.5
5	1	0	35.62	2.521	8	1.0
5	0	2	35.64	2.519	8	1.5
4	3	2	35.64	2.519	16	1.3
2	1	9	35.65	2.518	16	0.3
4	1	6	35.87	2.504	16	2.8
4	2	5	35.97	2.497	16	0.3
3	1	8	36.06	2.491	16	1.3
1	0	10	36.31	2.474	8	0.8
5	2	0	37.69	2.387	8	0.2

h	k	l	2θ	d	M	I_{rel}
4	3	4	37.79	2.380	16	0.5
3	2	8	38.11	2.361	16	0.3
2	0	10	38.35	2.347	8	0.2
5	2	2	38.38	2.345	16	0.2
3	0	9	38.39	2.345	8	0.2
5	1	4	38.46	2.340	16	0.8
2	1	10	39.01	2.309	16	0.3
4	4	0	39.67	2.272	4	1.0
1	0	11	39.95	2.257	8	0.2
4	0	8	40.07	2.250	8	1.0
4	2	7	40.14	2.247	16	0.1
5	3	0	40.94	2.204	8	0.2
5	3	1	41.11	2.196	16	0.2
5	2	5	41.87	2.157	16	0.2
6	0	0	42.19	2.142	4	1.4
4	4	4	42.28	2.137	8	0.3
6	0	1	42.34	2.134	8	0.5
4	2	8	42.57	2.124	16	0.3
6	0	2	42.82	2.112	8	0.3
4	0	9	42.82	2.112	8	0.2
0	0	12	43.05	2.101	2	2.6
5	3	4	43.49	2.081	16	2.1
5	2	6	43.61	2.075	16	1.1
1	0	12	43.65	2.074	8	0.1
3	2	10	43.99	2.058	16	0.8
1	1	12	44.24	2.047	8	0.4
3	0	11	44.84	2.021	8	0.3
5	3	5	44.88	2.020	16	0.1
5	4	0	45.17	2.007	8	0.3
5	0	8	45.54	1.992	8	0.2
5	4	2	45.77	1.982	16	0.4
5	1	8	46.11	1.969	16	0.4
4	1	10	46.31	1.960	16	0.5
5	4	3	46.51	1.952	16	0.2
5	3	6	46.52	1.952	16	0.3
3	3	10	46.88	1.938	8	0.5
3	2	11	47.14	1.928	16	0.3
1	0	13	47.40	1.918	8	0.2
2	2	12	47.68	1.907	8	0.2
4	2	10	48.00	1.895	16	0.4
6	2	5	48.28	1.885	16	0.2
5	3	7	48.41	1.880	16	0.2
5	1	9	48.59	1.874	16	0.2
3	1	12	48.79	1.866	16	0.3
5	4	5	48.83	1.865	16	0.2

ITQ-7, Siliceous, Calcined ISV

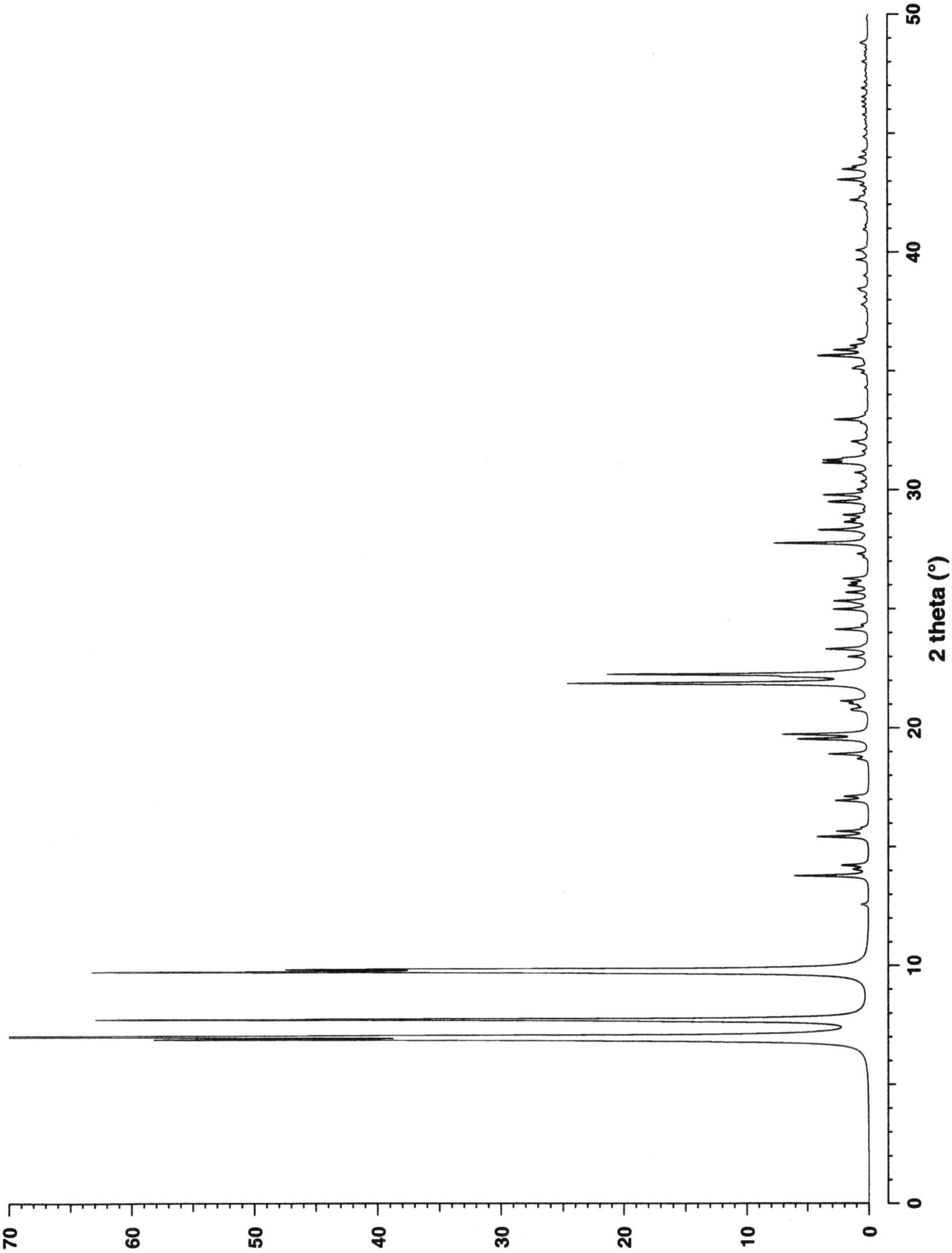

ITE ITQ–3, Calcined

CHEMICAL COMPOSITION: $[Si_{64}O_{128}]$

REFINED COMPOSITION: $[Si_{64}O_{128}]$

CRYSTAL DATA: *Cmcm* (No. 63)
$a = 20.622$ Å $b = 9.7242$ Å $c = 19.623$ Å
$\alpha = 90.0°$ $\beta = 90.0°$ $\gamma = 90.0°$
X-ray Rietveld refinement, $R_{wp} = 0.086$, $R_p = 0.062$

REFERENCE: M.A. Camblor, A. Corma, P. Lightfoot, L.A. Villaescusa and P.A. Wright, *Angew. Chem., Int. ed.* **36** 2659–2661 (1997).

h	k	l	2θ	d	M	I_{rel}
2	0	0	8.58	10.311	2	100.0
0	0	2	9.01	9.811	2	87.8
1	1	0	10.06	8.795	4	53.6
1	1	1	11.02	8.026	8	5.9
2	0	2	12.45	7.108	4	4.3
1	1	2	13.52	6.549	8	0.2
3	1	0	15.79	5.613	4	2.2
1	1	3	16.89	5.249	8	5.0
4	0	0	17.20	5.155	2	1.7
0	0	4	18.08	4.906	2	0.4
3	1	2	18.21	4.872	8	40.7
0	2	1	18.80	4.719	4	4.6
4	0	2	19.45	4.564	4	8.0
2	0	4	20.04	4.430	4	9.6
2	2	0	20.19	4.398	4	5.9
0	2	2	20.38	4.357	4	6.0
2	2	1	20.70	4.291	8	0.5
1	1	4	20.73	4.284	8	3.8
3	1	3	20.85	4.260	8	2.3
2	2	2	22.15	4.013	8	7.0
0	2	3	22.79	3.902	4	1.3
5	1	1	23.87	3.728	8	1.1
3	1	4	24.09	3.694	8	8.0
2	2	3	24.39	3.650	8	6.9
1	1	5	24.84	3.584	8	2.7
4	0	4	25.06	3.554	4	14.2
5	1	2	25.15	3.541	8	8.1
4	2	0	25.18	3.537	4	4.0
4	2	1	25.59	3.481	8	1.6
0	2	4	25.80	3.453	4	0.1
6	0	0	25.92	3.437	2	0.2
4	2	2	26.79	3.328	8	0.3
5	1	3	27.15	3.284	8	2.7
2	2	4	27.23	3.275	8	2.2
0	0	6	27.27	3.270	2	1.9
6	0	2	27.50	3.244	4	0.9
1	3	0	27.86	3.202	4	1.6
1	3	1	28.24	3.160	8	3.4
2	0	6	28.63	3.117	4	1.8
4	2	3	28.69	3.111	8	0.8
1	1	6	29.13	3.065	8	0.3
1	3	2	29.34	3.044	8	0.5
5	1	4	29.75	3.003	8	0.8
3	3	0	30.49	2.932	4	4.4
2	2	5	30.53	2.928	8	3.3
3	3	1	30.84	2.900	8	2.0
1	3	3	31.10	2.876	8	2.9
4	2	4	31.17	2.869	8	3.0
3	1	6	31.66	2.826	8	1.6
7	1	0	31.74	2.819	4	2.4
6	0	4	31.79	2.815	4	1.7
3	3	2	31.86	2.809	8	1.8
6	2	0	31.89	2.807	4	1.0
4	0	6	32.42	2.762	4	1.7
5	1	5	32.82	2.729	8	0.2
0	2	6	33.01	2.714	4	1.5
7	1	2	33.06	2.710	8	1.8
6	2	2	33.20	2.698	8	1.7
1	3	4	33.42	2.681	8	0.3
1	1	7	33.55	2.671	8	1.2
4	2	5	34.12	2.627	8	0.1
2	2	6	34.17	2.624	8	2.8
7	1	3	34.64	2.589	8	0.5
6	2	3	34.78	2.579	8	0.3
5	3	0	35.21	2.549	4	0.2
3	3	4	35.68	2.517	8	1.0
3	1	7	35.80	2.508	8	0.4
8	0	2	36.02	2.493	4	0.2
1	3	5	36.20	2.481	8	0.4
5	1	6	36.25	2.478	8	2.2
5	3	2	36.42	2.467	8	0.3
7	1	4	36.77	2.444	8	0.1
6	2	4	36.90	2.436	8	1.0
0	4	1	37.27	2.413	4	1.7
2	0	8	37.70	2.386	4	0.3
6	0	6	37.98	2.369	4	0.4
2	2	7	38.07	2.364	8	0.1
1	1	8	38.09	2.363	8	0.7
3	3	5	38.32	2.349	8	0.2
7	1	5	39.35	2.290	8	1.6
1	3	6	39.38	2.288	8	0.4
8	0	4	39.49	2.282	4	0.1
0	4	3	39.55	2.279	4	0.7
8	2	0	39.57	2.277	4	0.4
5	3	4	39.86	2.262	8	0.3
9	1	0	40.44	2.230	4	0.1
8	2	2	40.67	2.218	8	0.4
4	0	8	40.74	2.215	4	0.6
4	4	0	41.05	2.199	4	0.3
4	4	1	41.32	2.185	8	0.2
3	3	6	41.36	2.183	8	0.2
7	3	0	41.42	2.180	4	0.6
9	1	2	41.52	2.175	8	0.3
2	2	8	42.18	2.142	8	0.2
5	3	5	42.28	2.137	8	0.2
7	1	6	42.32	2.135	8	2.0
2	4	4	42.41	2.131	8	0.4
6	2	6	42.44	2.130	8	0.2
1	1	9	42.73	2.116	8	0.5
9	1	3	42.84	2.111	8	0.4
4	4	3	43.42	2.084	8	0.3
7	3	3	43.77	2.068	8	0.6
8	2	4	43.82	2.066	8	0.5
5	1	8	43.94	2.060	8	0.1
3	1	9	44.58	2.032	8	0.6
9	1	4	44.63	2.030	8	0.3
4	2	8	44.97	2.016	8	1.2
5	3	6	45.10	2.010	8	0.1
4	4	4	45.19	2.007	8	0.1
7	1	7	45.63	1.988	8	0.9
6	2	7	45.74	1.983	8	0.2
8	2	5	46.08	1.970	8	0.5
0	0	10	46.26	1.962	2	0.2
1	3	8	46.64	1.947	8	0.1
6	4	2	46.69	1.945	8	0.6
9	1	5	46.85	1.939	8	0.6
1	5	0	46.92	1.936	4	0.7
1	5	1	47.17	1.927	8	0.4
4	4	5	47.39	1.918	8	0.4
2	4	6	47.42	1.917	8	1.4
1	1	10	47.47	1.915	8	1.0
7	3	5	47.72	1.906	8	0.1
1	5	2	47.89	1.900	8	0.5
6	4	3	47.89	1.899	8	1.4
10	2	0	47.91	1.898	4	0.3
5	1	9	48.12	1.891	8	0.1
9	3	0	48.66	1.871	4	0.7
8	2	6	48.72	1.869	8	0.2
3	5	1	48.89	1.863	8	0.4
3	1	10	49.19	1.852	8	0.6
9	1	6	49.46	1.843	8	0.1
11	1	0	49.52	1.841	4	0.8
6	4	4	49.54	1.840	8	0.3
4	0	10	49.71	1.834	4	0.1
4	4	6	49.98	1.825	8	0.3

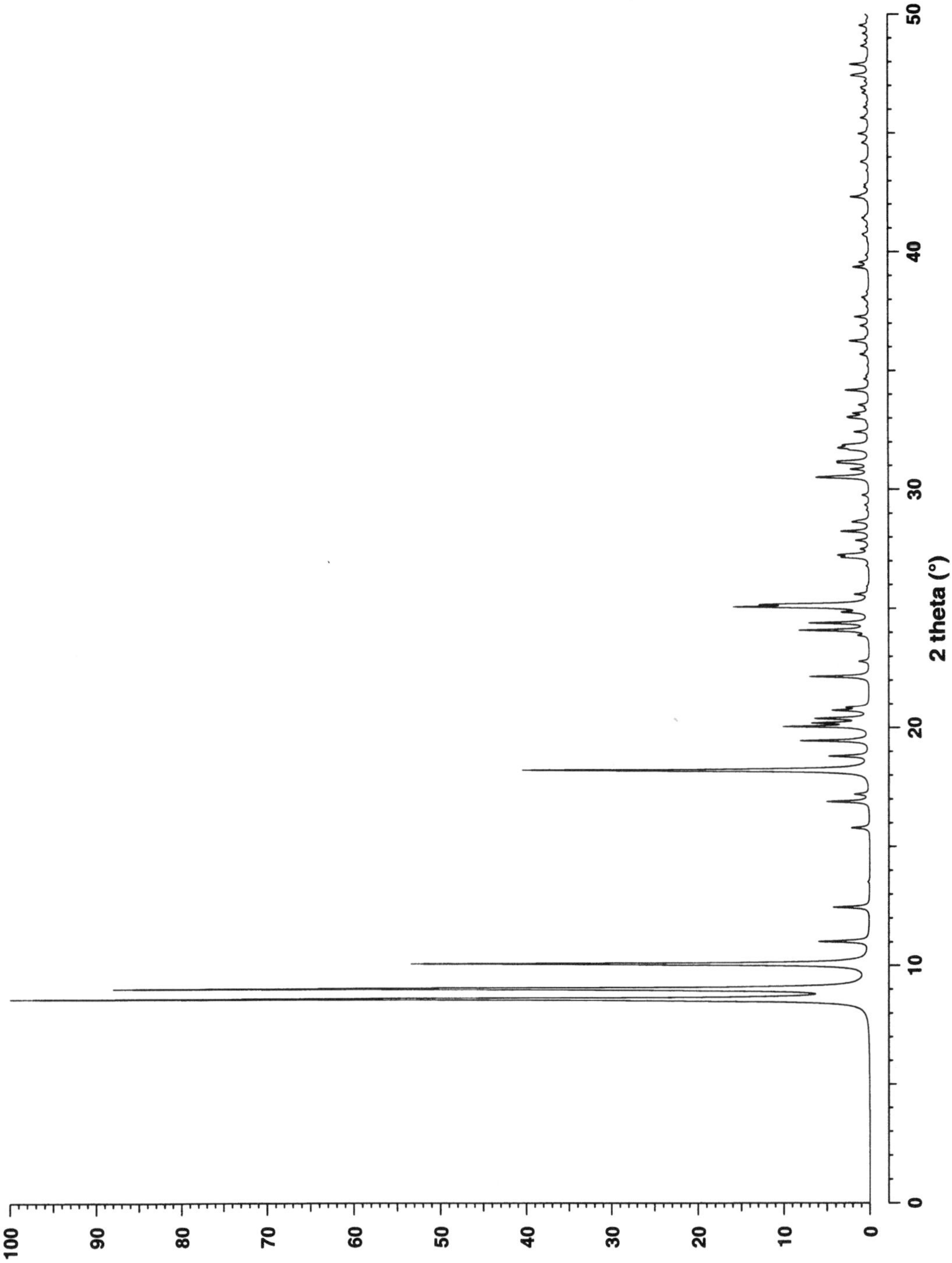
2 theta (°)
0
10
20
30
40
50
100
90
80
70
60
50
40
30
20
10
0

JBW Nepheline Hydrate

CHEMICAL COMPOSITION: $|Na_3H_2O|\ [Si_3Al_3O_{12}]$

REFINED COMPOSITION: $|Na_3H_2O|\ [Si_3Al_3O_{12}]$

CRYSTAL DATA: $Pmc2_1$ (No. 26)
$a = 7.503$ Å $b = 8.233$ Å $c = 5.230$ Å
$\alpha = 90°$ $\beta = 90°$ $\gamma = 90°$
X-ray single crystal refinement, $R = 0.039$

REFERENCE: A. Rheinhardt, E. Hellner and H. Ahsbahs, *Fortsch. der Mineralogie* **60** 175–176 (1982).

h	*k*	*l*	2θ	*d*	*M*	I_{rel}	*h*	*k*	*l*	2θ	*d*	*M*	I_{rel}	*h*	*k*	*l*	2θ	*d*	*M*	I_{rel}
0	1	0	10.75	8.233	2	100.0	0	0	2	34.29	2.615	2	23.4	2	0	2	42.12	2.145	4	1.1
1	0	0	11.79	7.503	2	41.0	1	3	0	34.81	2.577	4	0.1	3	2	0	42.28	2.137	4	0.2
1	1	0	15.98	5.546	4	7.4	3	0	0	35.91	2.501	2	0.3	1	2	2	42.70	2.118	8	2.0
0	1	1	20.11	4.415	4	63.3	0	1	2	36.04	2.492	4	1.4	2	1	2	43.60	2.076	8	5.2
0	2	0	21.59	4.116	2	5.0	1	0	2	36.38	2.469	4	0.2	0	4	0	43.99	2.058	2	4.4
1	1	1	23.38	3.805	8	3.9	2	2	1	36.68	2.450	8	39.1	2	3	1	44.42	2.040	8	0.2
1	2	0	24.67	3.609	4	0.2	0	3	1	36.99	2.430	4	15.3	3	2	1	45.86	1.979	8	4.3
2	1	0	26.10	3.414	4	81.7	3	1	0	37.59	2.393	4	1.7	0	4	1	47.47	1.915	4	0.8
0	2	1	27.57	3.235	4	14.4	1	1	2	38.04	2.365	8	1.4	2	2	2	47.81	1.902	8	0.7
1	2	1	30.08	2.970	8	30.8	1	3	1	38.96	2.312	8	3.2	0	3	2	48.06	1.893	4	6.9
2	1	1	31.29	2.859	8	13.9	2	3	0	40.74	2.215	4	0.3	4	0	0	48.53	1.876	2	10.4
2	2	0	32.28	2.773	4	1.1	0	2	2	40.88	2.207	4	3.3	1	4	1	49.09	1.856	8	0.3
0	3	0	32.63	2.744	2	6.2	3	1	1	41.50	2.176	8	2.2	4	1	0	49.86	1.829	4	0.2

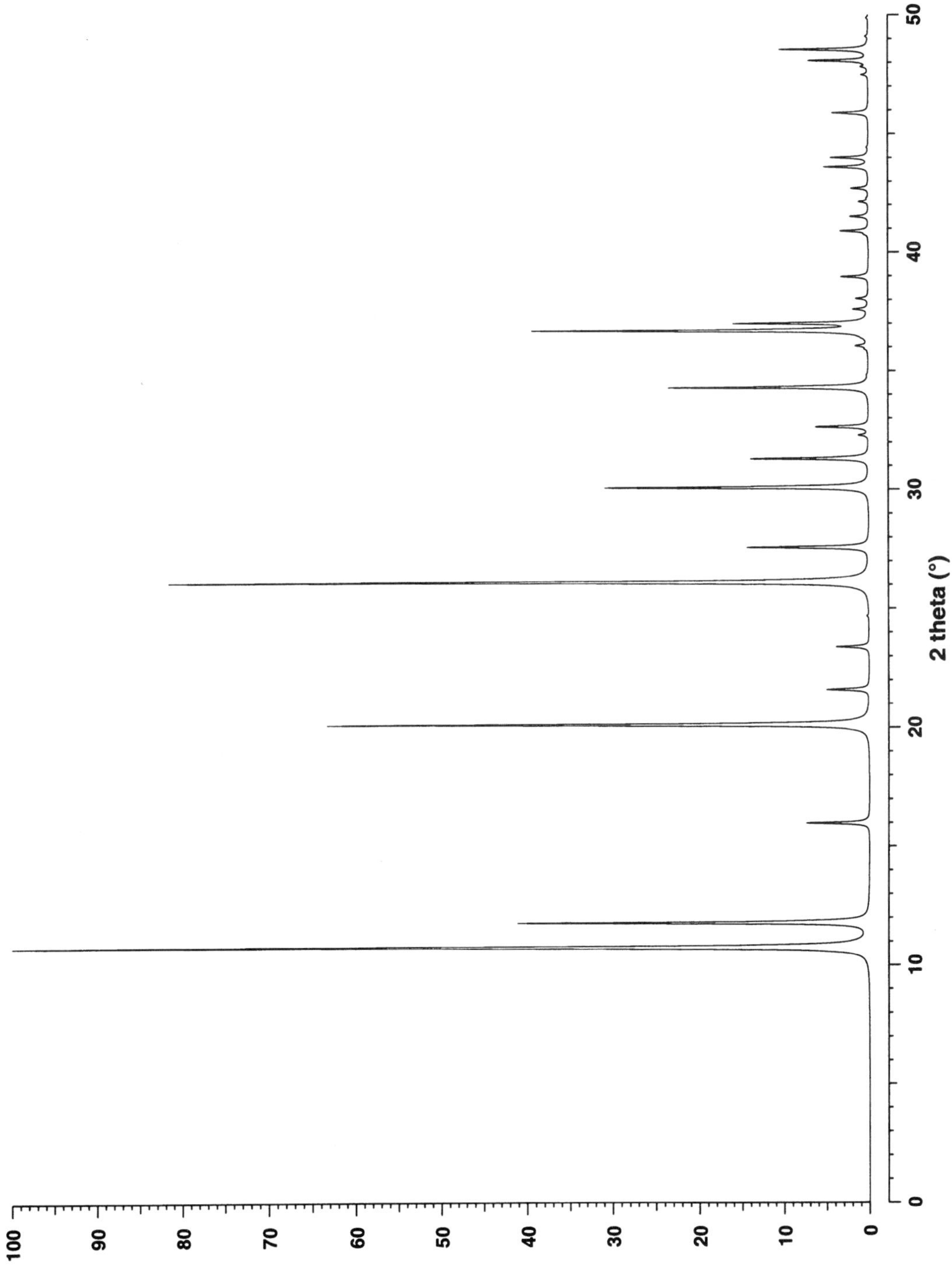
2 theta (°)
0
10
20
30
40
50
0
10
20
30
40
50
60
70
80
90
100

KFI (Cs,K) ZK5, Dehydrated

CHEMICAL COMPOSITION: $|Cs_{9.7}K_{13}|$ $[Si_{73.2}Al_{22.8}O_{192}]$

REFINED COMPOSITION: $|Cs_{9.72}K_{12.88}|$ $[Si_{72.96}Al_{23.04}O_{192}]$

CRYSTAL DATA: $Im\bar{3}m$ (No. 229)
a = 18.671 Å b = 18.671 Å c = 18.671 Å
α = 90° β = 90° γ = 90°
X-ray Rietveld refinement, R_{wp} = 0.09

REFERENCE: J. B. Parise, R. D. Shannon, E. Prince and D. E. Cox, *Z. Kristallogr.* **165** 175–190 (1983).

h	k	l	2θ	d	M	I_{rel}
1	1	0	6.69	13.202	12	60.6
2	0	0	9.47	9.335	6	78.9
2	1	1	11.61	7.622	24	16.8
2	2	0	13.41	6.601	12	28.8
3	1	0	15.00	5.904	24	100.0
2	2	2	16.45	5.390	8	0.2
3	2	1	17.77	4.990	48	6.3
4	0	0	19.01	4.668	6	1.3
4	1	1	20.18	4.401	24	20.6
3	3	0	20.18	4.401	12	0.5
4	2	0	21.28	4.175	24	27.2
3	3	2	22.33	3.981	24	1.0
4	2	2	23.34	3.811	24	29.5
4	3	1	24.31	3.662	48	1.9
5	1	0	24.31	3.662	24	4.9
5	2	1	26.14	3.409	48	45.1
4	4	0	27.01	3.301	12	1.5
4	3	3	27.86	3.202	24	15.5
5	3	0	27.86	3.202	24	23.1
6	0	0	28.69	3.112	6	4.1
4	4	2	28.69	3.112	24	0.1
6	1	1	29.49	3.029	24	4.4
5	3	2	29.49	3.029	48	7.2
6	2	0	30.27	2.952	24	18.9
5	4	1	31.04	2.881	48	4.1
6	2	2	31.79	2.815	24	30.9
4	4	4	33.24	2.695	8	0.9
7	1	0	33.95	2.640	24	0.2
5	5	0	33.95	2.640	12	6.2
5	4	3	33.95	2.640	48	8.0
6	4	0	34.64	2.589	24	0.3
6	3	3	35.32	2.541	24	24.1
5	5	2	35.32	2.541	24	10.1
6	4	2	36.00	2.495	48	0.2
7	3	0	36.65	2.452	24	0.1
6	5	1	37.94	2.371	48	0.1
7	3	2	37.94	2.371	48	3.0
7	4	1	39.20	2.298	48	0.4
8	1	1	39.20	2.298	24	6.2
5	5	4	39.20	2.298	24	3.6
8	2	0	39.81	2.264	24	0.3
6	4	4	39.81	2.264	24	1.8
6	5	3	40.42	2.232	48	0.6
8	2	2	41.02	2.200	24	2.5
6	6	0	41.02	2.200	12	9.2
7	4	3	41.61	2.170	48	1.8
7	5	0	41.61	2.170	24	0.2
6	6	2	42.19	2.142	24	0.7
7	5	2	42.77	2.114	48	1.1
8	4	0	43.34	2.087	24	0.1
8	3	3	43.91	2.062	24	0.3
9	1	0	43.91	2.062	24	2.6
7	6	1	45.03	2.013	48	0.1
9	2	1	45.03	2.013	48	0.6
6	5	5	45.03	2.013	24	0.3
8	5	1	46.12	1.968	48	3.3
7	5	4	46.12	1.968	48	0.2
9	3	0	46.12	1.968	24	1.1
7	6	3	47.19	1.926	48	8.9
8	4	4	47.73	1.906	24	0.5
8	5	3	48.25	1.886	48	5.2
7	7	0	48.25	1.886	12	2.8
10	0	0	48.77	1.867	6	4.5
8	6	0	48.77	1.867	24	0.2
10	1	1	49.29	1.849	24	0.2
7	7	2	49.29	1.849	24	0.5
10	2	0	49.80	1.831	24	0.8
8	6	2	49.80	1.831	48	5.8

(Cs,K) ZK5, Dehydrated

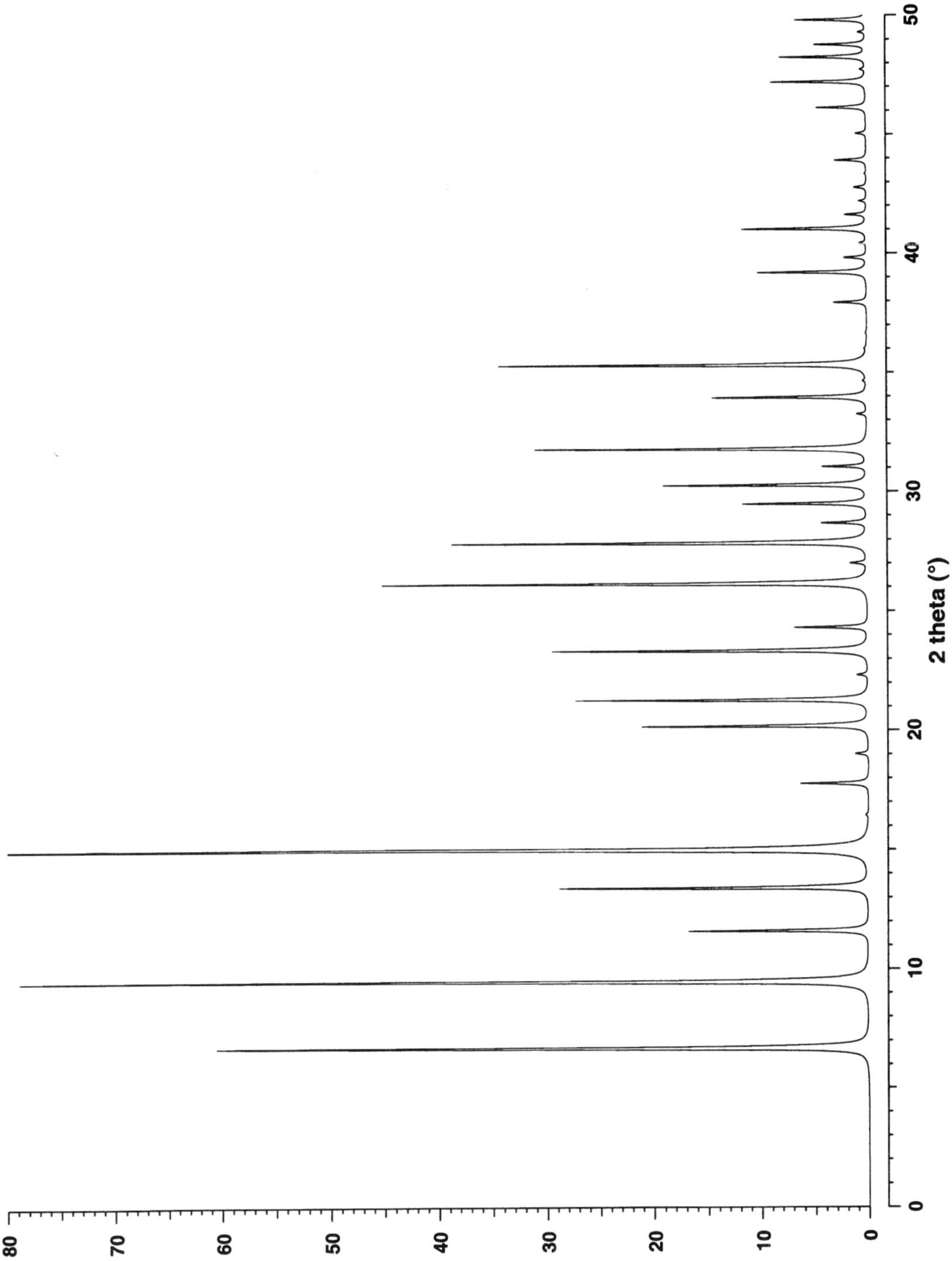

LAU Laumontite

CHEMICAL COMPOSITION: $|Ca_4(H_2O)_{18}|$ $[Si_{16}Al_8O_{48}]$
Nasik, India

REFINED COMPOSITION: $|Ca_4(H_2O)_{18}|$ $[Si_{16}Al_8O_{48}]$

CRYSTAL DATA: $C1\,2/m\,1$ (No. 12) unique axis **b**, cell choice 1
$a = 14.8538$ Å $b = 13.1695$ Å $c = 7.5421$ Å
$\alpha = 90°$ $\beta = 110.323°$ $\gamma = 90°$
X-ray Rietveld refinement, $R_{wp} = 0.115$, $R_p = 0.090$, $R_{F^2} = 0.046$

REFERENCE: G. Artioli and K. Ståhl,
Zeolites **13** 249–255 (1993).

h	k	l	2θ	d	M	I_{rel}	h	k	l	2θ	d	M	I_{rel}	h	k	l	2θ	d	M	I_{rel}
1	1	0	9.24	9.570	4	100.0	-4	2	2	32.16	2.783	4	15.2	-1	3	3	41.97	2.153	4	2.8
0	0	1	12.52	7.073	2	1.7	4	0	1	32.36	2.766	2	0.3	4	0	2	42.28	2.138	2	0.2
2	0	0	12.71	6.965	2	44.3	5	1	0	32.86	2.726	4	0.5	2	4	2	42.50	2.127	4	0.2
0	2	0	13.45	6.585	2	0.2	-3	3	2	33.03	2.712	4	0.8	4	4	1	42.69	2.118	4	3.3
-1	1	1	13.68	6.474	4	4.1	3	3	1	33.46	2.678	4	0.6	0	6	1	43.15	2.096	4	0.6
-2	0	1	14.42	6.142	2	7.4	-5	1	2	34.03	2.635	4	0.2	2	6	0	43.22	2.093	4	0.2
1	1	1	17.28	5.132	4	14.4	2	4	1	34.38	2.609	4	14.9	-3	5	2	43.22	2.093	4	0.8
0	2	1	18.41	4.819	4	0.1	1	5	0	34.66	2.588	4	0.3	5	3	1	43.99	2.059	4	0.6
2	2	0	18.54	4.785	4	11.4	1	3	2	34.88	2.572	4	4.9	-6	0	3	44.24	2.047	2	0.5
-2	2	1	19.77	4.492	4	9.6	2	2	2	34.96	2.566	4	4.0	-7	1	2	44.34	2.043	4	1.6
-3	1	1	20.05	4.428	4	0.7	4	2	1	35.19	2.551	4	0.2	4	2	2	44.56	2.033	4	3.5
3	1	0	20.28	4.379	4	1.3	-2	0	3	35.72	2.514	2	4.5	2	0	3	44.65	2.029	2	2.4
2	0	1	20.78	4.275	2	10.7	-1	5	1	36.18	2.483	4	2.4	1	5	2	44.70	2.027	4	2.0
1	3	0	21.22	4.187	4	34.4	-2	4	2	36.45	2.465	4	4.4	6	0	1	45.14	2.009	2	1.0
-1	3	1	23.54	3.780	4	1.6	-5	3	1	36.55	2.458	4	3.1	-2	4	3	45.39	1.998	4	0.6
-2	0	2	23.94	3.717	2	0.8	-4	4	1	36.67	2.451	4	14.3	-5	3	3	45.81	1.981	4	0.6
-4	0	1	24.25	3.670	2	29.6	-3	1	3	36.91	2.435	4	1.2	-6	4	1	45.87	1.978	4	0.6
-1	1	2	24.67	3.609	4	0.4	-1	1	3	37.03	2.428	4	0.2	-5	5	1	46.08	1.970	4	5.6
2	2	1	24.83	3.586	4	15.4	0	4	2	37.32	2.410	4	0.3	1	3	3	46.12	1.968	4	0.8
0	0	2	25.18	3.536	2	52.2	3	1	2	37.56	2.394	4	0.2	7	1	0	46.13	1.968	4	0.2
1	3	1	25.83	3.449	4	8.4	4	4	0	37.60	2.392	4	4.5	-6	2	3	46.44	1.955	4	0.2
-3	1	2	26.72	3.337	4	26.3	1	5	1	37.78	2.381	4	2.5	2	6	1	46.51	1.953	4	1.0
0	4	0	27.08	3.292	2	20.6	-4	0	3	37.94	2.372	2	2.1	2	2	3	46.85	1.939	4	1.7
3	1	1	27.24	3.274	4	2.6	0	0	3	38.17	2.358	2	0.3	6	2	1	47.31	1.921	4	0.6
-2	2	2	27.56	3.237	4	1.5	5	3	0	38.26	2.352	4	0.4	5	5	0	47.51	1.914	4	0.4
-3	3	1	27.80	3.209	4	26.6	-2	2	3	38.32	2.349	4	3.8	-7	3	1	47.70	1.907	4	0.9
-4	2	1	27.83	3.206	4	1.0	6	0	0	38.79	2.322	2	0.2	-2	6	2	48.14	1.890	4	1.8
3	3	0	27.97	3.190	4	5.7	-5	3	2	39.29	2.293	4	1.2	-5	5	2	48.37	1.882	4	0.6
0	2	2	28.65	3.115	4	2.0	3	5	0	39.33	2.291	4	2.1	-2	0	4	48.52	1.876	2	0.7
1	1	2	28.93	3.086	4	26.0	-4	4	2	40.15	2.246	4	0.8	5	1	2	48.61	1.873	4	1.1
4	2	0	29.01	3.078	4	11.4	-4	2	3	40.42	2.231	4	0.3	-7	3	2	48.67	1.871	4	0.5
-4	0	2	29.08	3.071	2	2.4	-6	2	2	40.75	2.214	4	6.5	-3	1	4	48.82	1.865	4	4.5
2	4	0	30.02	2.977	4	2.4	0	6	0	41.12	2.195	2	7.3	0	6	2	48.83	1.865	4	4.8
-2	4	1	30.81	2.902	4	0.4	6	2	0	41.23	2.189	4	8.6	-4	0	4	49.01	1.859	2	0.6
-5	1	1	30.89	2.895	4	22.4	1	1	3	41.58	2.172	4	0.2	-7	1	3	49.07	1.856	4	3.8
-1	3	2	31.36	2.852	4	0.2	-3	3	3	41.86	2.158	4	4.0	-8	0	1	49.37	1.846	2	0.5
2	0	2	32.12	2.787	2	4.6	-1	5	2	41.90	2.156	4	4.8	3	1	3	49.55	1.839	4	1.9

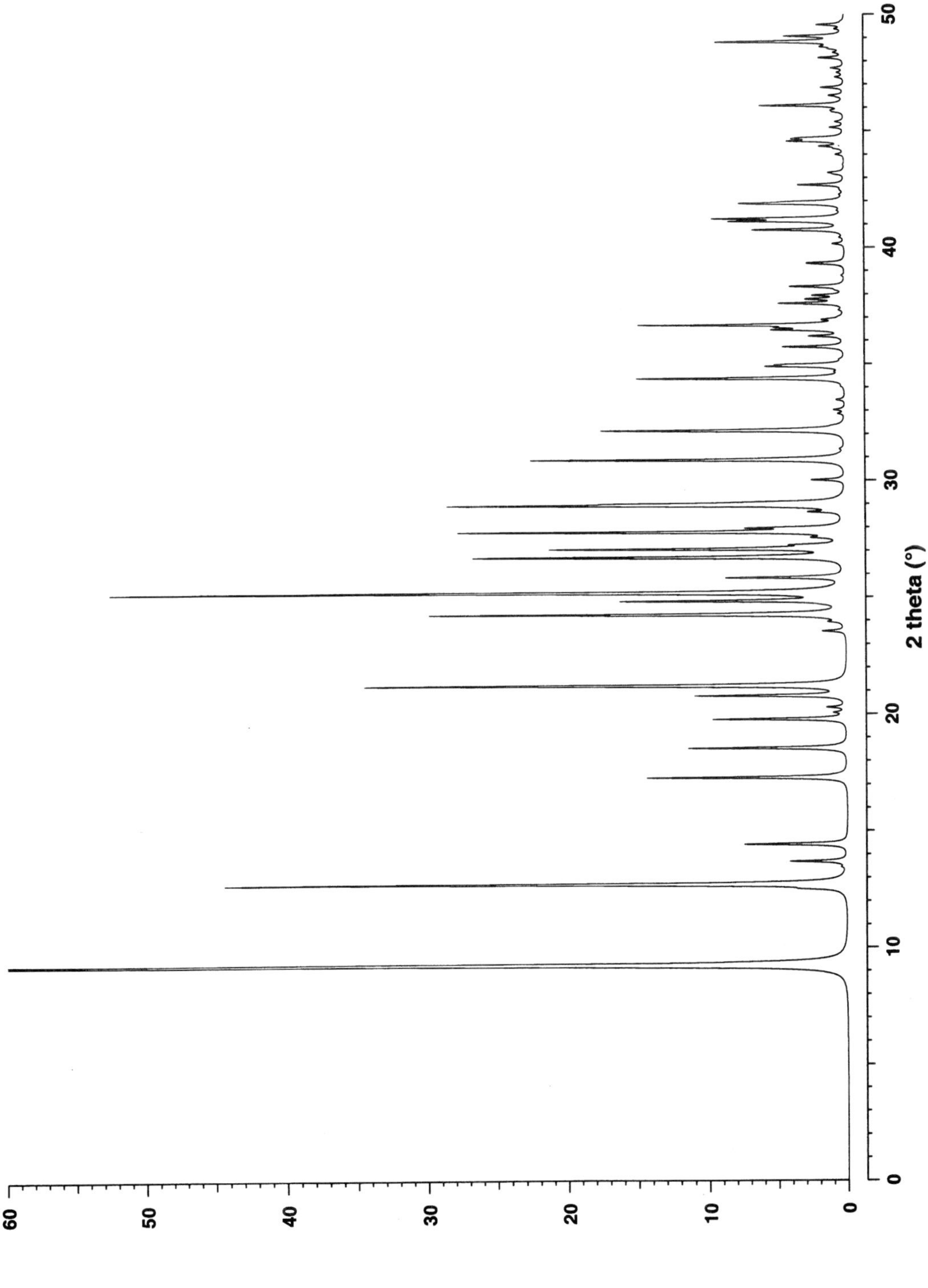
0
10
20
30
40
50
2 theta (°)
0
10
20
30
40
50
60

LAU Leonhardite

CHEMICAL COMPOSITION: $|K_{0.1}Na_{0.30}Ca_{3.60}(H_2O)_{25.2}|\ [Si_{16.40}Al_{7.60}O_{48}]$
Teigahorn, Berufjördur, Iceland

REFINED COMPOSITION: $|Ca_4H_{25.184}O_{13.36}|\ [Si_{16.4}Al_{7.6}O_{48}]$

CRYSTAL DATA: $C1\,2/m\,1$ (No. 12) unique axis **b**, cell choice 1
$a = 14.690$ Å $b = 13.061$ Å $c = 7.574$ Å
$\alpha = 90°$ $\beta = 112.01°$ $\gamma = 90°$
Neutron single crystal refinement, $R_{F^2} = 0.084$; At T = 15K.

REFERENCE: G. Artioli, J. V. Smith and Å. Kvick, *Zeolites* **9** 377–391 (1989).

h	k	l	2θ	d	M	I_{rel}
1	1	0	9.38	9.427	4	100.0
2	0	0	13.00	6.810	2	53.7
0	2	0	13.56	6.530	2	1.4
-1	1	1	13.63	6.496	4	1.2
-2	0	1	14.33	6.182	2	17.7
1	1	1	17.60	5.040	4	15.5
0	2	1	18.55	4.782	4	0.7
2	2	0	18.83	4.713	4	11.4
-2	2	1	19.78	4.489	4	30.3
-3	1	1	20.08	4.422	4	1.7
3	1	0	20.71	4.288	4	1.8
2	0	1	21.31	4.170	2	25.3
1	3	0	21.43	4.147	4	57.3
-1	3	1	23.64	3.764	4	5.3
-2	0	2	23.76	3.744	2	0.9
-4	0	1	24.41	3.647	2	35.6
-1	1	2	24.66	3.611	4	1.3
2	2	1	25.34	3.514	4	20.4
0	0	2	25.37	3.511	2	77.0
4	0	0	26.17	3.405	2	1.3
1	3	1	26.17	3.405	4	15.2
-3	1	2	26.51	3.362	4	28.2
0	4	0	27.31	3.265	2	29.7
-2	2	2	27.46	3.248	4	0.9
-3	3	1	27.94	3.194	4	16.2
3	1	1	27.96	3.192	4	1.3
3	3	0	28.40	3.142	4	12.7
0	2	2	28.87	3.092	4	2.6
-4	0	2	28.89	3.091	2	3.9
1	1	2	29.36	3.042	4	16.9
4	2	0	29.59	3.019	4	16.5
2	4	0	30.36	2.944	4	1.5
-5	1	1	31.21	2.866	4	20.0
-1	3	2	31.45	2.844	4	0.8
-4	2	2	32.04	2.794	4	18.4
2	0	2	32.79	2.731	2	0.8
-3	3	2	32.96	2.718	4	0.6
4	0	1	33.27	2.693	2	0.2
5	1	0	33.61	2.667	4	0.2
-5	1	2	33.95	2.641	4	3.5
3	3	1	34.15	2.625	4	3.0
2	4	1	34.90	2.571	4	15.7
1	5	0	34.97	2.565	4	0.6
1	3	2	35.33	2.540	4	5.7
-2	0	3	35.58	2.523	2	6.2
2	2	2	35.63	2.520	4	10.7
4	2	1	36.08	2.490	4	0.5
-1	5	1	36.43	2.466	4	2.6
-2	4	2	36.51	2.461	4	5.6
-3	1	3	36.65	2.452	4	0.6
-5	3	1	36.91	2.435	4	2.7
-4	4	1	36.95	2.433	4	19.6
-1	1	3	37.10	2.423	4	0.2
-4	0	3	37.60	2.392	2	1.9
0	4	2	37.62	2.391	4	1.4
4	4	0	38.19	2.357	4	5.0
1	5	1	38.19	2.357	4	4.7
-2	2	3	38.24	2.354	4	3.9
3	1	2	38.46	2.341	4	0.5
5	3	0	39.00	2.309	4	0.8
-5	3	2	39.30	2.292	4	0.5
6	0	0	39.71	2.270	2	1.6
3	5	0	39.81	2.264	4	6.0
-4	2	3	40.14	2.246	4	0.3
-4	4	2	40.17	2.245	4	0.5
-6	2	2	40.82	2.211	4	11.0
-5	1	3	40.92	2.205	4	0.1
0	2	3	40.96	2.203	4	0.6
0	6	0	41.48	2.177	2	5.9
-3	3	3	41.71	2.165	4	8.3
-1	3	3	42.11	2.146	4	0.5
1	1	3	42.12	2.145	4	1.2
-1	5	2	42.13	2.145	4	3.5
6	2	0	42.15	2.144	4	10.1
2	4	2	43.18	2.095	4	1.1
-3	5	2	43.31	2.089	4	3.0
3	3	2	43.34	2.088	4	1.7
4	0	2	43.40	2.085	2	2.0
0	6	1	43.53	2.079	4	0.2
4	4	1	43.56	2.078	4	0.4
2	6	0	43.65	2.073	4	1.1
-6	0	3	43.94	2.061	2	2.8
-2	6	1	44.10	2.053	4	0.3
3	5	1	44.27	2.046	4	0.1
-7	1	2	44.52	2.035	4	2.4
5	3	1	45.06	2.012	4	0.1
1	5	2	45.22	2.005	4	0.8
-2	4	3	45.43	1.997	4	0.6
2	0	3	45.45	1.996	2	4.1
-5	3	3	45.58	1.990	4	0.3
4	2	2	45.68	1.986	4	3.8
-6	2	3	46.20	1.965	4	0.7
-6	4	1	46.40	1.957	4	3.3
6	0	1	46.42	1.956	2	3.6
-5	5	1	46.52	1.952	4	7.4
1	3	3	46.69	1.946	4	3.5
-4	4	3	47.09	1.930	4	0.1
2	6	1	47.09	1.930	4	0.8
2	2	3	47.65	1.908	4	3.3
0	4	3	47.81	1.902	4	0.1
5	5	0	48.27	1.885	4	1.6
-7	3	1	48.36	1.882	4	0.9
-2	6	2	48.37	1.882	4	3.2
-5	5	2	48.52	1.876	4	0.5
-3	1	4	48.59	1.874	4	6.0
6	2	1	48.59	1.874	4	0.3
-4	0	4	48.64	1.872	2	0.5
-4	6	1	48.72	1.869	4	0.6
6	4	0	48.86	1.864	4	0.6
-7	1	3	48.87	1.863	4	3.4
-7	3	2	48.91	1.862	4	0.2
0	6	2	49.25	1.850	4	11.5
1	7	0	49.29	1.849	4	0.4
5	1	2	49.96	1.825	4	2.5

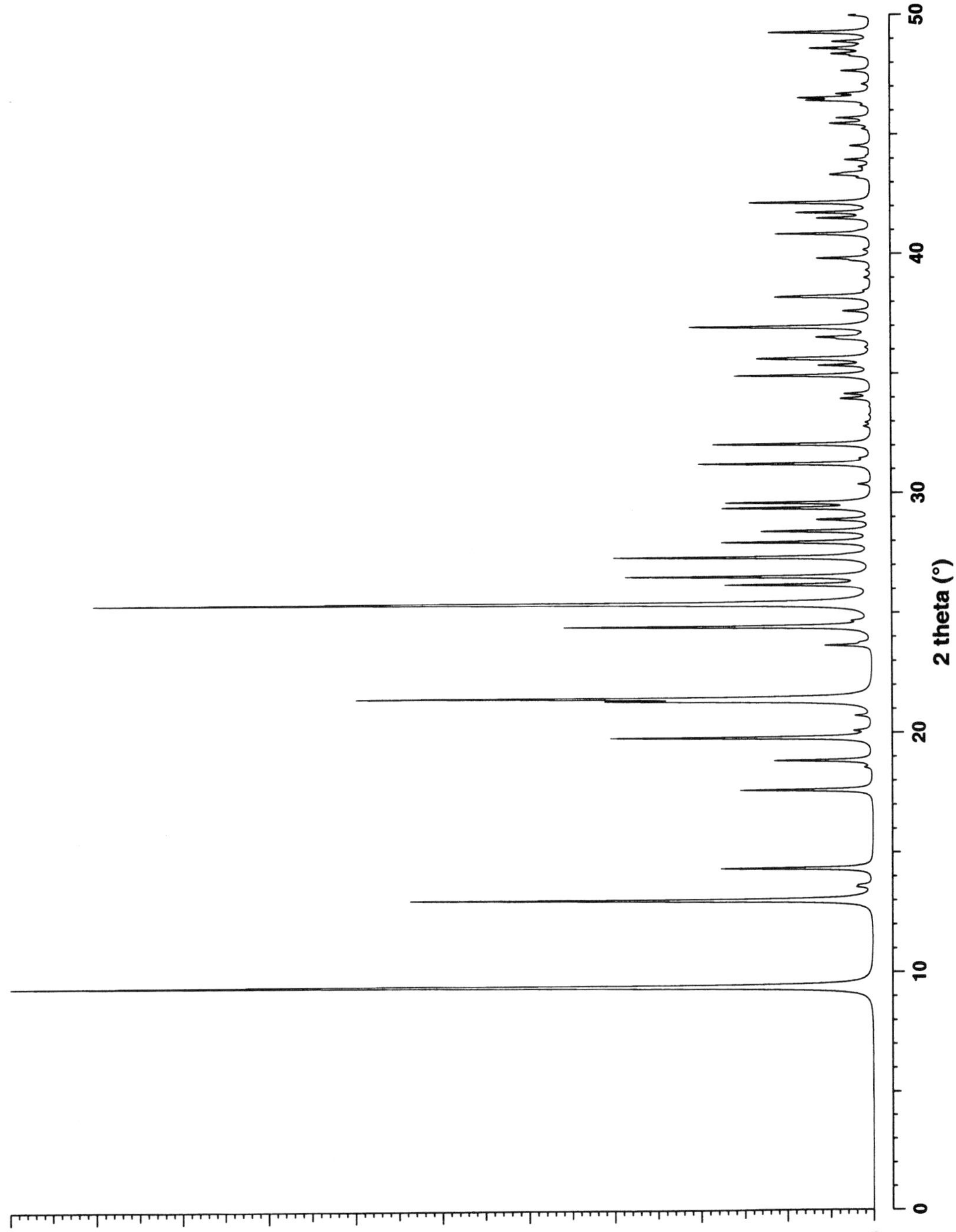
2 theta (°)
0
10
20
30
40
50
0
10
20
30
40
50
60
70
80
90
100

LAU — Cobalt Gallophosphate

CHEMICAL COMPOSITION: $|(C_5H_5NH)_8|$ $[Co_8Ga_{16}P_{24}O_{96}]$
C_5H_5NH = pyridinium cation

REFINED COMPOSITION: $|(C_5H_6N)_8|$ $[Co_8Ga_{16}P_{24}O_{96}]$

CRYSTAL DATA: $C1\,2/c\,1$ (No. 15) unique axis **b**, cell choice 1
a = 15.058 Å b = 13.197 Å c = 15.273 Å
α = 90° β = 112.20° γ = 90°

REFERENCE: A. M. Chippindale and R. I. Walton, *J. Chem. Soc., Chem. Commun.* 2453–2454 (1994).

h	k	l	2θ	d	M	I_{rel}
1	1	0	9.23	9.584	4	100.0
1	1	1	12.43	7.121	4	0.7
2	0	0	12.70	6.971	2	5.6
0	2	0	13.42	6.599	2	4.5
-1	1	2	13.49	6.562	4	5.1
-2	0	2	14.07	6.293	2	17.2
1	1	2	17.41	5.093	4	15.1
2	2	0	18.51	4.792	4	0.7
-2	2	2	19.49	4.554	4	26.6
-3	1	2	19.66	4.515	4	5.1
3	1	0	20.26	4.383	4	3.9
2	0	2	21.01	4.229	2	12.6
2	2	1	21.06	4.218	4	3.0
1	3	0	21.18	4.195	4	45.4
-1	3	1	21.40	4.153	4	2.4
-3	1	3	22.23	3.998	4	2.2
-2	2	3	22.77	3.906	4	4.4
1	3	1	22.79	3.903	4	1.2
1	1	3	23.09	3.852	4	0.7
0	2	3	23.19	3.836	4	1.4
3	1	1	23.29	3.820	4	3.8
-2	0	4	23.53	3.782	2	0.9
-4	0	2	23.82	3.736	2	29.2
-1	1	4	24.47	3.638	4	0.4
2	2	2	25.01	3.560	4	16.1
0	0	4	25.19	3.535	2	62.9
1	3	2	25.90	3.440	4	1.8
-3	1	4	26.14	3.409	4	10.5
0	4	0	27.03	3.299	2	23.7
-2	2	4	27.18	3.281	4	2.8
-4	2	1	27.48	3.245	4	0.4
-3	3	2	27.49	3.245	4	4.0
0	4	1	27.77	3.213	4	3.1
3	3	0	27.93	3.195	4	9.6
-4	0	4	28.36	3.147	2	1.6
0	2	4	28.65	3.116	4	1.7
4	2	0	28.97	3.082	4	7.0
1	1	4	29.10	3.068	4	7.8
-3	3	3	29.42	3.036	4	0.4
0	4	2	29.88	2.990	4	0.5
2	4	0	29.96	2.982	4	2.6
3	3	1	30.24	2.956	4	0.9
-5	1	2	30.45	2.936	4	28.1
-5	1	1	31.00	2.884	4	0.5
-1	3	4	31.18	2.869	4	1.4
-4	2	4	31.50	2.840	4	4.6
2	0	4	32.44	2.760	2	1.0
2	4	2	34.48	2.601	4	15.5
0	2	5	34.50	2.599	4	0.7
1	3	4	35.00	2.564	4	9.2
2	2	4	35.25	2.546	4	0.4
-2	0	6	35.29	2.543	2	6.0
1	5	1	35.63	2.520	4	0.4
5	1	1	35.75	2.512	4	0.9
-6	0	2	35.83	2.506	2	1.6
-1	5	2	36.04	2.492	4	1.8
-2	4	4	36.13	2.486	4	0.8
-5	3	2	36.15	2.485	4	5.8
-1	3	5	36.23	2.479	4	1.8
-3	1	6	36.28	2.476	4	1.4
-4	4	2	36.33	2.473	4	8.2
-5	1	5	36.33	2.473	4	1.1
-4	4	1	36.37	2.470	4	1.2
-5	3	1	36.63	2.453	4	0.5
-4	0	6	37.11	2.423	2	4.1
0	4	4	37.28	2.412	4	0.5
-6	0	4	37.43	2.403	2	0.4
1	5	2	37.78	2.381	4	5.1
-2	2	6	37.91	2.373	4	2.4
3	1	4	37.97	2.370	4	1.0
-5	3	4	38.58	2.333	4	0.4
-3	5	2	38.93	2.313	4	0.8
5	1	2	39.56	2.278	4	2.0
-4	4	4	39.58	2.277	4	2.1
-6	2	4	39.93	2.258	4	1.9
4	2	3	39.94	2.257	4	0.7
-5	1	6	40.27	2.239	4	3.7
3	5	1	41.00	2.201	4	1.0
0	6	0	41.03	2.199	2	1.0
6	2	0	41.19	2.192	4	1.3
-3	3	6	41.27	2.187	4	1.7
0	6	1	41.55	2.173	4	0.4
-1	5	4	41.72	2.165	4	1.4
-1	3	6	41.77	2.162	4	0.9
1	1	6	41.78	2.162	4	0.4
0	4	5	42.08	2.147	4	1.3
-4	4	5	42.65	2.120	4	0.4
4	0	4	42.76	2.114	2	2.4
-3	5	4	42.78	2.114	4	5.5
3	3	4	42.80	2.113	4	1.6
-7	1	2	42.82	2.112	4	0.8
-2	6	1	42.86	2.110	4	0.7
2	6	0	43.13	2.098	4	0.6
-1	1	7	43.31	2.089	4	0.4
-2	6	2	43.59	2.076	4	1.5
3	5	2	43.68	2.072	4	1.2
3	1	5	43.85	2.065	4	0.4
-2	2	7	43.88	2.063	4	0.6
-7	1	1	43.99	2.058	4	0.5
6	2	1	44.00	2.058	4	0.6
-5	3	6	44.90	2.019	4	0.7
-2	4	6	45.01	2.014	4	0.9
2	0	6	45.02	2.014	2	1.6
4	2	4	45.02	2.014	4	1.5
-7	1	5	45.22	2.005	4	0.9
6	0	2	45.46	1.995	2	2.8
-5	5	2	45.71	1.985	4	1.8
7	1	0	46.09	1.969	4	1.2
-5	5	1	46.10	1.969	4	0.5
-6	4	1	46.21	1.965	4	0.8
-4	4	6	46.50	1.953	4	1.2
2	6	2	46.54	1.951	4	1.7
-6	4	4	46.77	1.942	4	0.8
4	4	3	46.78	1.942	4	0.9
2	2	6	47.19	1.926	4	1.9
-7	3	2	47.24	1.924	4	0.4
5	5	0	47.43	1.917	4	0.4
-7	3	4	47.82	1.902	4	1.4
-2	6	4	47.84	1.901	4	1.2
-7	1	6	47.86	1.900	4	0.4
6	4	0	47.88	1.900	4	2.9
-4	6	1	48.03	1.894	4	0.8
-2	0	8	48.09	1.892	2	0.6
-3	1	8	48.16	1.889	4	2.5
1	1	7	48.40	1.881	4	1.0
5	3	3	48.40	1.880	4	0.6
0	6	4	48.76	1.868	4	1.0
-1	7	1	48.85	1.864	4	0.7
-8	0	2	48.87	1.864	2	0.5
4	6	0	48.97	1.860	4	0.4
5	1	4	49.15	1.854	4	0.4
-5	3	7	49.16	1.853	4	0.5
2	6	3	49.53	1.840	4	0.6
1	7	1	49.54	1.840	4	0.4
-1	7	2	49.85	1.829	4	1.0

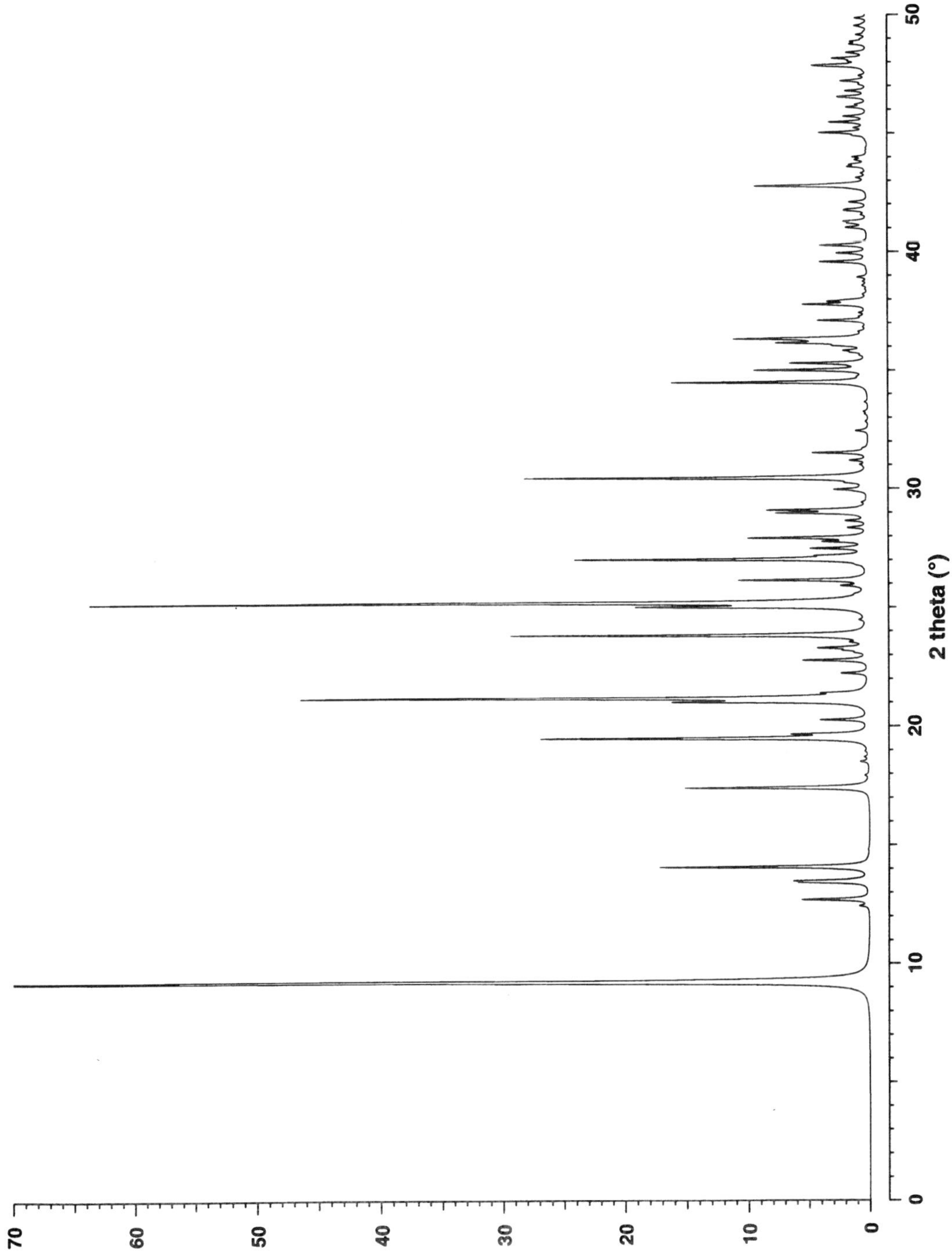
0
10
20
30
40
50
60
70
2 theta (°)

LEV Levyne

CHEMICAL COMPOSITION: $|Ca_{8.19}Na_{1.95}K_{0.6}(H_2O)_{50}|\ [Si_{35}Al_{19}O_{108}]$
Nurri, Nuoro, Sardinia, Italy

REFINED COMPOSITION: $|Ca_{7.8}Na_{2.16}K_{0.72}(H_2O)_{46.08}|\ [Si_{35.1}Al_{18.9}O_{108}]$

CRYSTAL DATA: $R\overline{3}m$ (No. 166) hexagonal setting
a = 13.338 Å b = 13.338 Å c = 23.014 Å
α = 90° β = 90° γ = 120°
X-ray single crystal refinement, R = 0.07

REFERENCE: S. Merlino, E. Galli and A. Alberti,
Tschermaks Mineral. Petrogr. Mitt. **22** 117–129 (1975).

h	k	l	2θ	d	M	I_{rel}	h	k	l	2θ	d	M	I_{rel}	h	k	l	2θ	d	M	I_{rel}
1	0	1	8.56	10.324	6	73.2	0	4	2	31.95	2.801	6	68.3	0	3	9	42.43	2.130	6	3.2
0	1	2	10.85	8.152	6	83.0	1	3	4	31.97	2.799	12	0.8	3	0	9	42.43	2.130	6	0.9
0	0	3	11.53	7.671	2	41.1	3	0	6	32.96	2.717	6	2.5	1	4	6	42.93	2.107	12	0.9
1	1	0	13.28	6.669	6	13.2	0	3	6	32.96	2.717	6	2.6	4	1	6	42.93	2.107	12	1.1
0	2	1	15.82	5.602	6	0.2	3	1	5	34.10	2.629	12	7.6	5	1	1	43.81	2.066	12	1.5
2	0	2	17.18	5.162	6	36.4	2	1	7	34.14	2.626	12	33.9	5	0	5	43.85	2.065	6	2.0
1	1	3	17.62	5.033	12	5.1	2	3	2	34.74	2.582	12	5.3	3	2	7	43.88	2.063	12	2.5
2	1	1	20.71	4.289	12	4.0	0	0	9	35.09	2.557	2	1.4	1	5	2	44.37	2.042	12	0.6
0	1	5	20.77	4.276	6	41.6	4	1	0	35.62	2.521	12	17.0	2	4	4	44.38	2.041	12	1.6
1	2	2	21.77	4.082	12	100.0	2	2	6	35.68	2.517	12	1.0	0	4	8	44.45	2.038	6	0.2
0	2	4	21.80	4.076	6	6.3	0	4	5	36.74	2.446	6	3.2	2	1	10	44.50	2.036	12	0.9
3	0	0	23.10	3.850	6	23.1	3	2	4	37.36	2.407	12	4.0	4	2	5	46.01	1.972	12	0.9
0	0	6	23.19	3.836	2	8.9	1	2	8	37.44	2.402	12	5.0	2	0	11	46.15	1.967	6	1.4
2	0	5	24.73	3.600	6	6.9	1	4	3	37.56	2.395	12	1.7	5	1	4	46.53	1.952	12	3.6
2	1	4	25.61	3.478	12	20.7	4	1	3	37.56	2.395	12	7.0	2	3	8	46.60	1.949	12	2.5
3	0	3	25.89	3.441	6	4.9	0	5	1	39.19	2.299	6	0.6	6	0	0	47.21	1.925	6	0.3
0	3	3	25.89	3.441	6	1.9	2	3	5	39.23	2.297	12	1.8	3	3	6	47.26	1.923	12	0.2
2	2	0	26.73	3.334	6	11.4	1	3	7	39.26	2.294	12	4.5	0	0	12	47.40	1.918	2	3.3
1	1	6	26.81	3.325	12	3.3	1	0	10	39.94	2.257	6	2.0	0	5	7	48.14	1.890	6	4.4
1	3	1	28.12	3.173	12	0.2	3	3	0	40.58	2.223	6	8.4	1	2	11	48.23	1.887	12	0.3
1	2	5	28.17	3.168	12	26.8	2	4	1	41.55	2.173	12	1.2	3	4	2	48.59	1.874	12	1.2
1	0	7	28.22	3.162	6	23.4	4	0	7	41.62	2.170	6	0.3	1	3	10	48.72	1.869	12	2.5
3	1	2	28.93	3.086	12	15.6	4	2	2	42.13	2.145	12	1.1	0	6	3	48.77	1.867	6	0.1
2	2	3	29.20	3.058	12	0.1	0	5	4	42.15	2.144	6	2.1	6	0	3	48.77	1.867	6	0.3
4	0	1	31.22	2.865	6	10.4	0	2	10	42.27	2.138	6	3.5	5	2	0	49.26	1.850	12	2.2
0	2	7	31.31	2.857	6	11.1	3	3	3	42.33	2.135	12	3.8	1	1	12	49.45	1.843	12	2.4

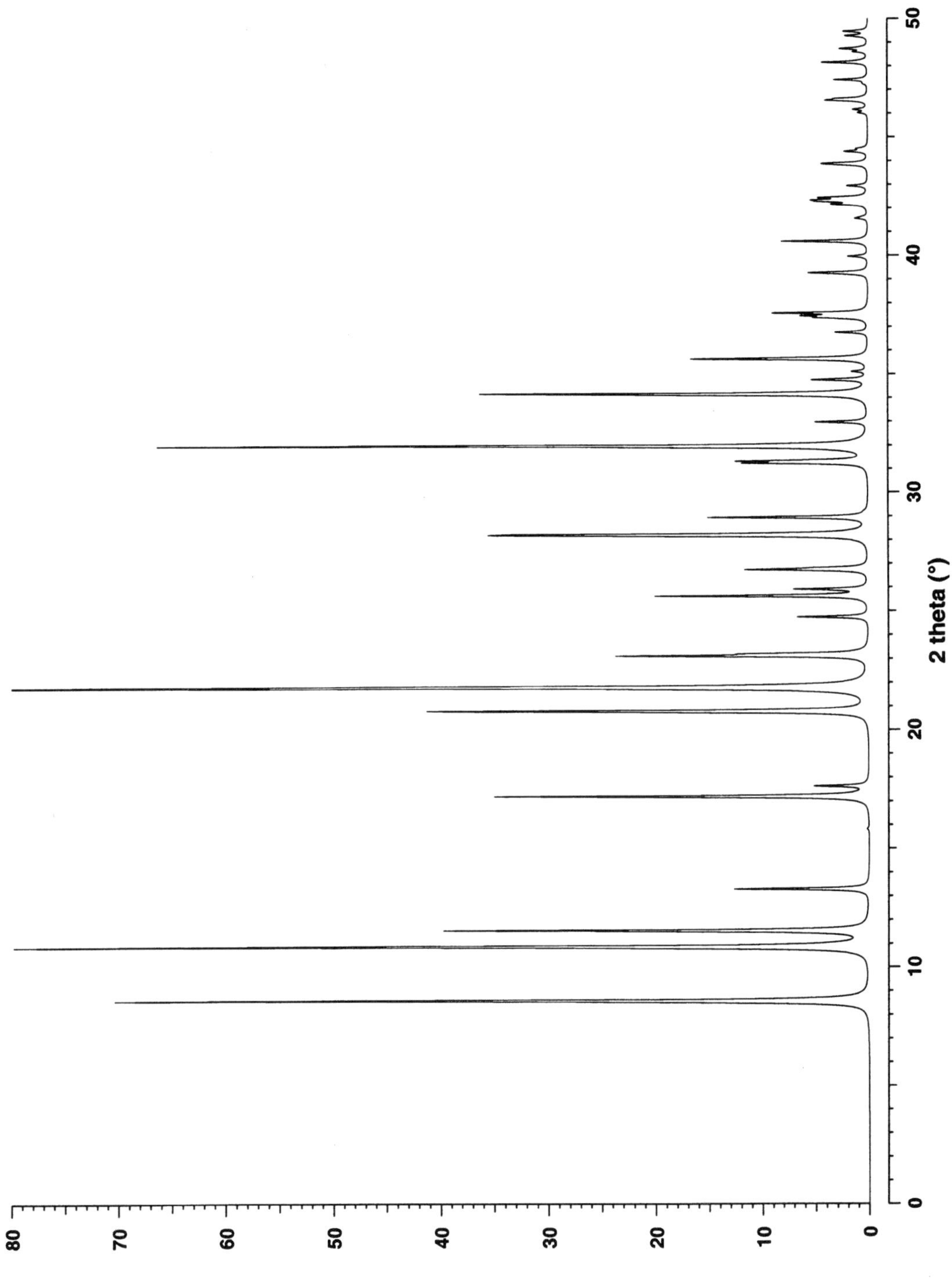
2 theta (°)
0
10
20
30
40
50
0
10
20
30
40
50
60
70
80

LEV 1-aminoadamantane NU-3

CHEMICAL COMPOSITION: $|(C_{10}H_{15}NH_2)_6|$ $[Si_{54}O_{108}]$
$C_{10}H_{15}NH_2$ = 1-aminoadamantane

REFINED COMPOSITION: $|(C_{10}N)_6|$ $[Si_{54}O_{108}]$

CRYSTAL DATA: $R\bar{3}m$ (No. 166) hexagonal setting
$a = 13.2251$ Å $b = 13.2251$ Å $c = 22.2916$ Å
$\alpha = 90°$ $\beta = 90°$ $\gamma = 120°$
X-ray Rietveld refinement, $R_{wp} = 0.104$, $R_F = 0.060$

REFERENCE: L. B. McCusker,
Mat. Sci. Forum Vol. **133-136** 423–434 (1993).
And L. B. McCusker, Private communication (1995).

h	*k*	*l*	2θ	*d*	*M*	I_{rel}
1	0	1	8.68	10.187	6	17.6
0	1	2	11.08	7.988	6	54.6
0	0	3	11.91	7.431	2	0.5
1	1	0	13.39	6.613	6	27.1
0	2	1	15.98	5.547	6	4.9
2	0	2	17.41	5.094	6	66.4
1	0	4	17.70	5.011	6	7.1
1	1	3	17.96	4.940	12	13.6
2	1	1	20.90	4.250	12	14.5
0	1	5	21.39	4.155	6	55.2
1	2	2	22.03	4.035	12	100.0
0	2	4	22.26	3.994	6	0.4
3	0	0	23.30	3.818	6	12.3
0	0	6	23.95	3.715	2	4.6
2	0	5	25.32	3.518	6	5.5
2	1	4	26.06	3.419	12	1.2
0	3	3	26.24	3.396	6	0.7
2	2	0	26.97	3.306	6	22.9
1	1	6	27.54	3.239	12	1.5
1	3	1	28.38	3.145	12	3.8
1	2	5	28.74	3.106	12	26.0
1	0	7	29.10	3.068	6	29.5
3	1	2	29.23	3.055	12	9.1
4	0	1	31.50	2.840	6	1.1
0	2	7	32.16	2.783	6	4.7
0	4	2	32.28	2.773	6	49.9
1	3	4	32.44	2.760	12	1.8
0	1	8	33.09	2.707	6	1.0
0	3	6	33.66	2.663	6	0.5
3	2	1	34.37	2.609	12	0.3
3	1	5	34.67	2.587	12	1.4
2	1	7	34.98	2.565	12	8.2
2	3	2	35.09	2.557	12	1.5
4	0	4	35.24	2.547	6	2.2
2	0	8	35.84	2.506	6	0.1
4	1	0	35.93	2.499	12	3.6
0	0	9	36.27	2.477	2	0.1
2	2	6	36.37	2.470	12	0.7
3	2	4	37.85	2.377	12	0.5
4	1	3	37.98	2.369	12	0.7
1	2	8	38.42	2.343	12	0.3
1	1	9	38.82	2.319	12	0.2
0	5	1	39.55	2.279	6	4.5
3	3	0	40.94	2.204	6	1.1
1	0	10	41.26	2.188	6	0.6
4	0	7	42.45	2.129	6	3.1
4	2	2	42.55	2.125	12	0.1
0	5	4	42.68	2.119	6	3.3
3	3	3	42.79	2.113	12	2.8
0	3	9	43.56	2.078	6	3.5
3	0	9	43.56	2.078	6	1.2
0	2	10	43.57	2.077	6	0.7
1	4	6	43.65	2.074	12	0.2
5	1	1	44.22	2.048	12	0.1
5	0	5	44.46	2.037	6	0.8
3	2	7	44.71	2.027	12	0.7
1	5	2	44.80	2.023	12	0.4
2	4	4	44.93	2.018	12	0.6
0	4	8	45.42	1.997	6	1.0
2	2	9	45.77	1.982	12	0.3
2	1	10	45.78	1.982	12	0.5
4	2	5	46.65	1.947	12	0.5
5	1	4	47.09	1.930	12	0.8
2	3	8	47.56	1.912	12	0.2
2	0	11	47.60	1.910	6	0.4
6	0	0	47.64	1.909	6	1.5
3	3	6	47.99	1.896	12	0.5
1	5	5	48.75	1.868	12	1.1
0	5	7	48.98	1.860	6	8.4
0	0	12	49.04	1.858	2	6.6
3	4	2	49.07	1.857	12	2.8
0	6	3	49.29	1.849	6	0.5
5	2	0	49.71	1.834	12	6.5

1-aminoadamantane NU-3 LEV

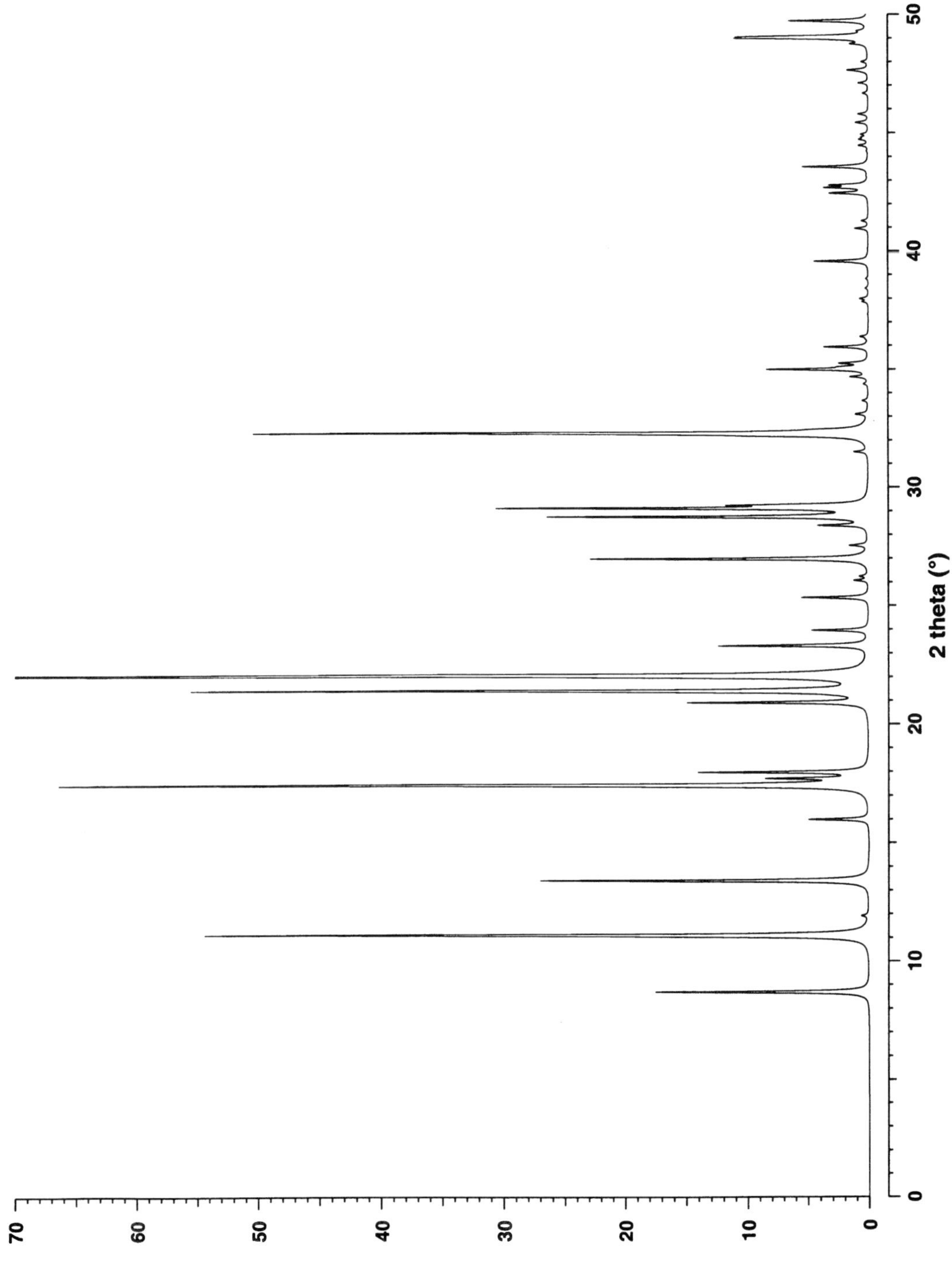

LIO Liottite

CHEMICAL COMPOSITION: $|Na_{9.96}(K,Ca)_{8.04}Ca_6(SO_4)_5Cl_{3.5}F_{0.5}|$ $[Si_{18}Al_{18}O_{72}]$
Pitigliano, Tuscany, Italy

REFINED COMPOSITION: $|K_{13.788}Ca_6(SO_4)_5Cl_{3.5}F_{0.5}|$ $[Si_{18}Al_{18}O_{72}]$

CRYSTAL DATA: $P\overline{6}$ (No. 174)
$a = 12.8701$ Å $b = 12.8701$ Å $c = 16.096$ Å
$\alpha = 90°$ $\beta = 90°$ $\gamma = 120°$
X-ray single crystal refinement, $R_p = 0.0365$, $R_{wp} = 0.0829$

REFERENCE: P. Ballirano, S. Merlino and E. Bonaccorsi,
Canadian Mineralogist **34** 1021–1030 (1996).

h	k	l	2θ	d	M	I_{rel}
0	0	1	5.49	16.096	2	0.8
1	0	0	7.93	11.146	6	5.5
1	0	1	9.65	9.163	12	4.8
1	0	2	13.57	6.525	12	5.7
1	1	0	13.76	6.435	6	0.4
1	1	1	14.83	5.975	12	0.7
2	0	0	15.90	5.573	6	1.2
0	0	3	16.52	5.365	2	4.9
2	0	1	16.84	5.266	12	1.2
1	1	2	17.65	5.026	12	0.3
1	0	3	18.35	4.834	12	48.7
2	0	2	19.37	4.582	12	0.5
2	1	0	21.09	4.213	6	0.7
1	2	0	21.09	4.213	6	3.1
1	1	3	21.56	4.121	12	3.5
2	1	1	21.81	4.075	12	1.8
1	2	1	21.81	4.075	12	1.6
2	0	3	23.01	3.865	12	0.3
1	0	4	23.50	3.785	12	23.5
1	2	2	23.84	3.732	12	3.1
2	1	2	23.84	3.732	12	2.9
3	0	0	23.95	3.715	6	100.0
3	0	1	24.59	3.620	12	1.0
3	0	2	26.42	3.373	12	3.3
2	1	3	26.91	3.313	12	58.6
1	2	3	26.91	3.313	12	58.7
2	0	4	27.34	3.262	12	2.6
0	0	5	27.71	3.219	2	0.2
2	2	0	27.73	3.218	6	4.0
1	0	5	28.87	3.093	12	2.5
1	3	0	28.88	3.091	6	0.2
3	1	0	28.88	3.091	6	0.6
3	0	3	29.24	3.054	12	2.8
1	3	1	29.42	3.036	12	3.6
3	1	1	29.42	3.036	12	3.4
2	2	2	29.91	2.988	12	0.1
3	1	2	30.99	2.886	12	4.4
1	3	2	30.99	2.886	12	4.5
2	0	5	32.11	2.788	12	0.1
4	0	0	32.12	2.786	6	14.6
2	2	3	32.45	2.759	12	6.0
4	0	1	32.61	2.746	12	9.8
3	0	4	32.81	2.730	12	3.6
0	0	6	33.40	2.683	2	35.5
3	1	3	33.45	2.679	12	8.6
1	3	3	33.45	2.679	12	6.1
4	0	2	34.05	2.633	12	9.2
1	0	6	34.38	2.608	12	3.0
2	1	5	35.08	2.558	12	0.2
1	2	5	35.08	2.558	12	0.1
2	3	0	35.09	2.557	6	0.2
3	2	0	35.09	2.557	6	0.3
1	1	6	36.28	2.476	12	0.3
4	0	3	36.33	2.473	12	22.8
3	1	4	36.66	2.451	12	3.7
1	3	4	36.66	2.451	12	3.7
3	0	5	36.95	2.433	12	2.1
4	1	0	36.96	2.432	6	3.0
1	4	0	36.96	2.432	6	2.6
2	0	6	37.20	2.417	12	2.2
1	4	1	37.39	2.405	12	0.2
4	1	1	37.39	2.405	12	0.2
4	1	2	38.67	2.328	12	0.3
1	4	2	38.67	2.328	12	0.3
0	0	7	39.18	2.299	2	0.2
4	0	4	39.33	2.291	12	5.0
2	2	5	39.60	2.276	12	0.2
1	2	6	39.84	2.263	12	0.5
2	1	6	39.84	2.263	12	0.5
1	0	7	40.04	2.252	12	0.1
1	3	5	40.45	2.230	12	1.2
3	1	5	40.45	2.230	12	1.1
5	0	0	40.46	2.229	6	1.7
1	4	3	40.73	2.215	12	0.1
5	0	1	40.87	2.208	12	0.4
3	0	6	41.52	2.175	12	3.8
3	2	4	41.86	2.158	12	0.3
2	3	4	41.86	2.158	12	0.3
5	0	2	42.06	2.148	12	0.1
3	3	0	42.13	2.145	6	31.1
2	0	7	42.53	2.126	12	0.4
4	0	5	42.93	2.107	12	1.7
4	2	0	42.94	2.106	6	2.2
2	4	0	42.94	2.106	6	2.3
2	2	6	43.94	2.060	12	9.1
1	3	6	44.73	2.026	12	0.2
1	2	7	44.91	2.018	12	1.0
2	1	7	44.91	2.018	12	1.0
0	0	8	45.06	2.012	2	0.4
3	2	5	45.29	2.002	12	0.6
2	3	5	45.29	2.002	12	0.6
5	1	0	45.30	2.002	6	1.2
1	5	0	45.30	2.002	6	1.2
3	3	3	45.54	1.992	12	1.1
1	0	8	45.83	1.980	12	0.4
4	2	3	46.30	1.961	12	1.8
2	4	3	46.30	1.961	12	2.2
5	0	4	46.57	1.950	12	0.4
4	1	5	46.81	1.941	12	0.2
1	4	5	46.81	1.941	12	0.3
4	0	6	47.02	1.933	12	14.6
1	1	8	47.34	1.920	12	0.4
2	0	8	48.08	1.892	12	0.1
5	1	3	48.54	1.876	12	3.3
1	5	3	48.54	1.876	12	3.3
6	0	0	49.04	1.858	6	1.8
2	3	6	49.23	1.851	12	0.2
6	0	1	49.38	1.845	12	0.6
1	3	7	49.40	1.845	12	0.3
3	1	7	49.40	1.845	12	0.3
5	0	5	49.75	1.833	12	0.2
3	4	0	49.76	1.832	6	0.6

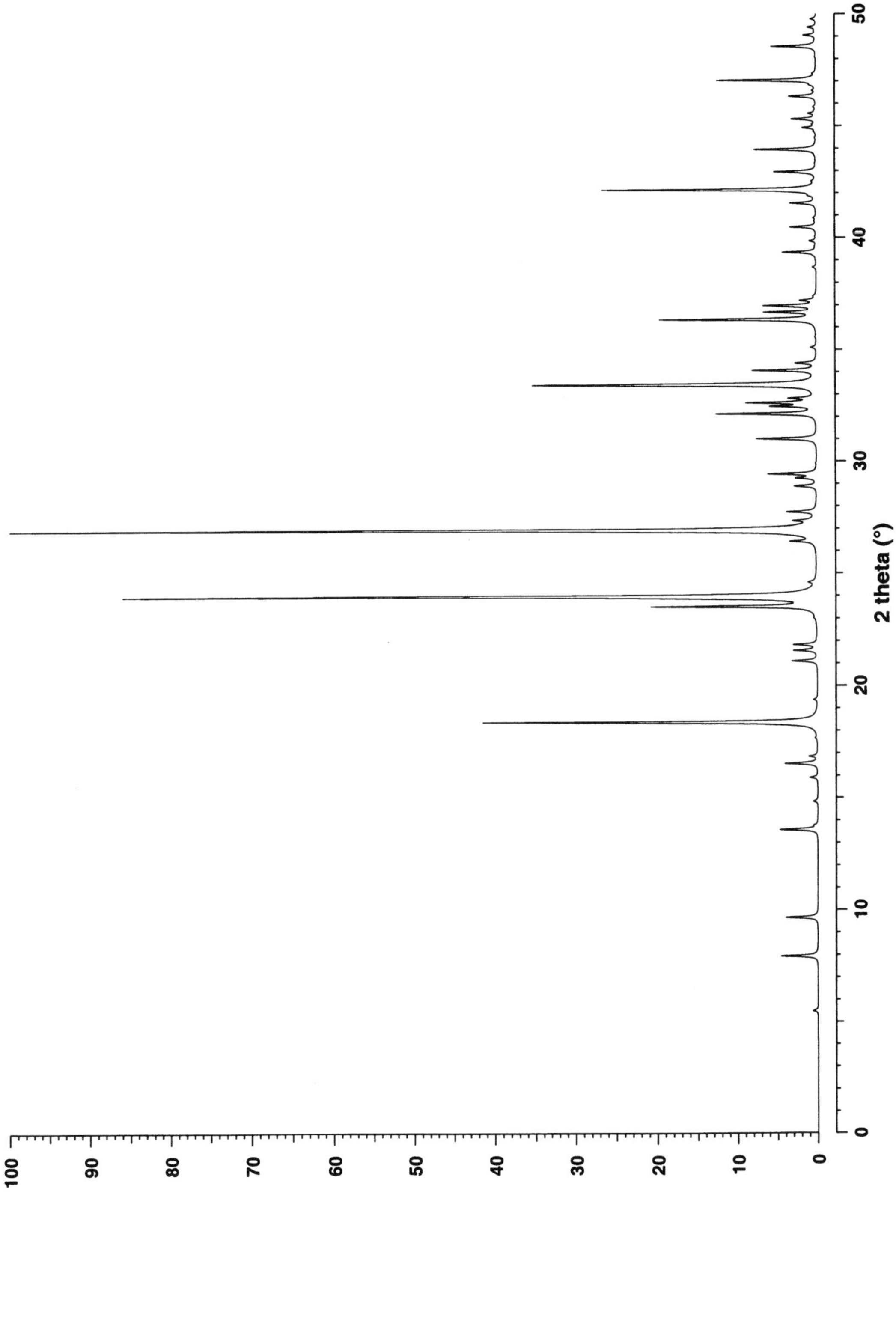
2 theta (°)
0
10
20
30
40
50
0
10
20
30
40
50
60
70
80
90
100

LOS Losod

CHEMICAL COMPOSITION: $|Na_{12}(H_2O)_{18}|\ [Si_{12}Al_{12}O_{48}]$
Synthetic material

REFINED COMPOSITION: $|Na_{12.5}(H_2O)_{14.32}|\ [Si_{12}Al_{12}O_{48}]$

CRYSTAL DATA: $P6_3mc$ (No. 186)
a = 12.906 Å b = 12.906 Å c = 10.541 Å
α = 90° β = 90° γ = 120°
X-ray Rietveld refinement, R_{wp} = 0.154, R_F = 0.084, R_{exp} = 0.136

REFERENCE: P. A. Schicker,
Ph.D. Thesis No. 8494, ETH Zurich (1988).

h	k	l	2θ	d	M	I_{rel}
1	0	0	7.91	11.177	6	10.8
1	0	1	11.54	7.669	12	42.3
1	1	0	13.72	6.453	6	100.0
2	0	0	15.86	5.588	6	3.5
0	0	2	16.82	5.270	2	0.1
2	0	1	17.96	4.937	12	5.4
1	0	2	18.61	4.767	12	71.2
2	1	0	21.03	4.224	12	6.7
1	1	2	21.77	4.082	12	2.7
2	1	1	22.68	3.921	24	52.1
2	0	2	23.20	3.834	12	8.2
3	0	0	23.88	3.726	6	64.7
3	0	1	25.35	3.513	12	5.0
1	0	3	26.59	3.352	12	9.8
2	1	2	27.05	3.296	24	81.4
2	2	0	27.65	3.226	6	3.9
3	1	0	28.80	3.100	12	6.8
3	0	2	29.36	3.042	12	0.7
2	0	3	30.04	2.975	12	0.9
3	1	1	30.05	2.974	24	15.5
4	0	0	32.03	2.794	6	11.6
2	2	2	32.54	2.752	12	1.9
2	1	3	33.16	2.701	24	13.7
4	0	1	33.17	2.701	12	47.3
3	1	2	33.54	2.672	24	25.9
0	0	4	34.02	2.635	2	34.6
1	0	4	34.98	2.565	12	0.9
3	2	0	34.99	2.564	12	0.2
3	0	3	35.11	2.556	12	4.4
3	2	1	36.05	2.492	24	1.1
4	0	2	36.39	2.469	12	19.2
1	1	4	36.84	2.440	12	2.1
4	1	0	36.85	2.439	12	1.7
2	0	4	37.74	2.384	12	0.5
3	1	3	38.74	2.325	24	0.2
3	2	2	39.06	2.306	24	6.0
2	1	4	40.34	2.236	24	1.0
5	0	0	40.35	2.235	6	0.1
4	1	2	40.76	2.213	24	0.7
4	0	3	41.28	2.187	12	6.7
5	0	1	41.28	2.187	12	1.0
3	0	4	41.99	2.151	12	13.1
3	3	0	42.00	2.151	6	12.8
1	0	5	43.69	2.072	12	0.3
3	2	3	43.70	2.071	24	2.9
4	2	1	43.71	2.071	24	0.7
5	0	2	44.00	2.058	12	5.3
2	2	4	44.38	2.041	12	0.1
3	1	4	45.16	2.008	24	0.8
3	3	2	45.55	1.992	12	0.9
2	0	5	46.01	1.973	12	0.3
5	1	1	46.02	1.972	24	3.3
4	2	2	46.31	1.961	24	1.4
4	0	4	47.42	1.917	12	3.3
2	1	5	48.24	1.886	24	4.5
5	0	3	48.25	1.886	12	0.1
5	1	2	48.53	1.876	24	3.4
3	2	4	49.60	1.838	24	0.2
4	3	0	49.61	1.837	12	0.3
3	0	5	49.69	1.835	12	1.1
6	0	1	49.70	1.834	12	1.4

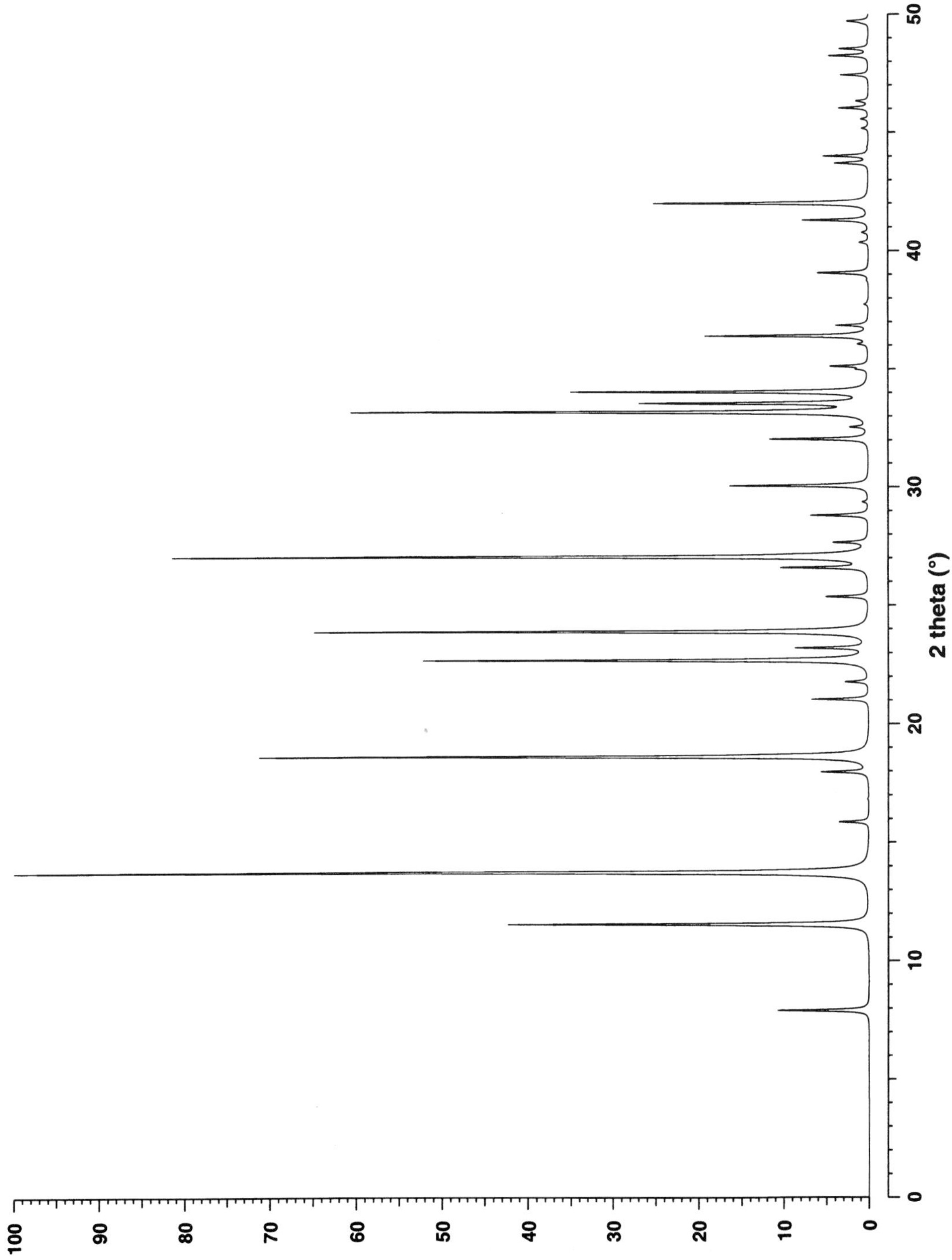
2 theta (°)
0
10
20
30
40
50
0
10
20
30
40
50
60
70
80
90
100

LOV Lovdarite

CHEMICAL COMPOSITION: $|Na_{12}K_4(H_2O)_{18}|$ $[Si_{28}Be_8O_{72}]$
Mt. Karnasurt, Lovezero Pluton, Russia

REFINED COMPOSITION: $|Na_{12}K_4(H_2O)_{18}|$ $[Si_{28}Be_8O_{72}]$

CRYSTAL DATA: *Pma*2 (No. 28)
$a = 39.576$ Å $b = 6.9308$ Å $c = 7.1526$ Å
$\alpha = 90°$ $\beta = 90°$ $\gamma = 90°$
X-ray single crystal refinement, $R_w = 0.09$

REFERENCE: S. Merlino,
Eur. J. Mineral. **2** 809–817 (1990).

h	*k*	*l*	2θ	*d*	*M*	I_{rel}
4	0	0	8.94	9.894	2	33.1
0	0	1	12.37	7.153	2	19.3
2	0	1	13.16	6.727	4	58.9
6	0	0	13.42	6.596	2	3.3
2	1	0	13.54	6.541	4	50.8
3	1	0	14.44	6.136	4	10.0
4	0	1	15.29	5.797	4	60.8
4	1	0	15.61	5.677	4	3.3
0	1	1	17.82	4.977	4	100.0
8	0	0	17.93	4.947	2	10.6
1	1	1	17.96	4.938	8	6.1
6	0	1	18.30	4.849	4	4.2
6	1	0	18.57	4.778	4	3.0
4	1	1	19.97	4.446	8	5.3
7	1	0	20.27	4.381	4	2.7
8	0	1	21.84	4.069	4	4.5
7	1	1	23.82	3.736	8	2.4
9	1	0	23.97	3.713	4	4.5
2	0	2	25.31	3.519	4	21.2
8	1	1	25.38	3.509	8	10.9
0	2	0	25.71	3.465	2	5.6
10	0	1	25.73	3.463	4	25.4
1	2	0	25.81	3.452	4	2.6
10	1	0	25.92	3.437	4	16.3
2	2	0	26.10	3.413	4	14.4
4	0	2	26.50	3.363	4	15.6
3	2	0	26.59	3.352	4	4.0
12	0	0	27.04	3.298	2	82.3
9	1	1	27.06	3.295	8	4.7
4	2	0	27.27	3.271	4	5.1
11	1	0	27.94	3.193	4	44.4
0	1	2	28.08	3.178	4	29.2
5	2	0	28.11	3.174	4	3.1
6	0	2	28.39	3.144	4	49.7
2	1	2	28.44	3.138	8	78.9
0	2	1	28.62	3.119	4	26.4
1	2	1	28.71	3.109	8	16.0
10	1	1	28.82	3.098	8	7.4
2	2	1	28.98	3.081	8	57.0
6	2	0	29.11	3.068	4	34.7
3	2	1	29.43	3.035	8	19.1
4	1	2	29.52	3.026	8	10.9
12	0	1	29.83	2.995	4	9.5
4	2	1	30.04	2.974	8	6.5
7	2	0	30.25	2.955	4	15.3
5	1	2	30.30	2.949	8	3.7
11	1	1	30.66	2.916	8	4.7
5	2	1	30.82	2.902	8	21.3
6	1	2	31.24	2.863	8	10.6
14	0	0	31.65	2.827	2	1.9
6	2	1	31.74	2.819	8	2.9
12	1	1	32.57	2.749	8	27.9
7	2	1	32.80	2.731	8	3.1
8	1	2	33.51	2.674	8	32.4
10	0	2	33.78	2.653	4	3.0
8	2	1	33.98	2.638	8	5.0
14	0	1	34.10	2.629	4	7.1
10	2	0	34.40	2.607	4	7.8
9	2	1	35.28	2.544	8	9.2
0	2	2	36.09	2.489	4	12.9
10	1	2	36.25	2.478	8	17.6
16	0	0	36.32	2.474	2	7.3
2	2	2	36.38	2.469	8	2.2
15	1	0	36.44	2.466	4	7.0
14	1	1	36.55	2.458	8	5.2
10	2	1	36.69	2.450	8	5.3
3	2	2	36.75	2.446	8	2.2
12	0	2	37.08	2.424	4	2.8
12	2	0	37.65	2.389	4	2.2
0	0	3	37.73	2.384	2	2.3
11	1	2	37.77	2.382	8	5.0
5	2	2	37.90	2.374	8	4.1
2	0	3	38.01	2.367	4	5.1
11	2	1	38.19	2.357	8	2.5
16	0	1	38.51	2.338	4	4.9
15	1	1	38.62	2.331	8	2.5
4	0	3	38.85	2.318	4	1.8
2	3	0	39.26	2.295	4	5.1
12	1	2	39.37	2.288	8	23.1
13	2	0	39.40	2.287	4	6.9
3	3	0	39.60	2.276	4	7.4
12	2	1	39.78	2.266	8	3.6
0	1	3	39.99	2.255	4	2.6
1	1	3	40.06	2.251	8	2.0
4	3	0	40.08	2.250	4	5.7
8	2	2	40.58	2.223	8	2.8
3	1	3	40.59	2.222	8	2.0
14	0	2	40.68	2.218	4	2.2
16	1	1	40.73	2.215	8	1.7
17	1	0	40.89	2.207	4	4.0
13	1	2	41.06	2.198	8	3.3
0	3	1	41.06	2.198	4	3.1
4	1	3	41.06	2.198	8	8.9
14	2	0	41.21	2.190	4	2.1
13	2	1	41.45	2.178	8	1.9
5	1	3	41.65	2.168	8	1.9
4	3	1	42.10	2.146	8	6.0
6	1	3	42.37	2.133	8	4.2
5	3	1	42.68	2.118	8	4.2
14	1	2	42.81	2.112	8	4.6
15	2	0	43.09	2.099	4	2.9
14	2	1	43.19	2.094	8	4.0
7	1	3	43.20	2.094	8	3.3
6	3	1	43.38	2.086	8	3.1
7	3	1	44.20	2.049	8	3.0
11	2	2	44.25	2.047	8	2.6
9	3	0	44.29	2.045	4	1.8
10	0	3	44.36	2.042	4	2.5
16	0	2	44.54	2.034	4	1.5
15	1	2	44.64	2.030	8	8.2
15	2	1	45.00	2.014	8	3.4
18	1	1	45.08	2.011	8	3.4
8	3	1	45.13	2.009	8	3.8
20	0	0	45.86	1.979	2	6.1
10	1	3	46.35	1.959	8	3.1
0	3	2	46.81	1.941	4	2.5
16	2	1	46.88	1.938	8	5.1
4	2	3	47.17	1.927	8	11.7
3	3	2	47.35	1.920	8	2.4
4	3	2	47.76	1.904	8	3.7
20	1	0	47.80	1.903	4	1.6
6	2	3	48.35	1.883	8	3.2
17	1	2	48.47	1.878	8	8.2
11	3	1	48.53	1.876	8	1.8
18	0	2	48.61	1.873	4	2.4
14	2	2	48.75	1.868	8	2.3
17	2	1	48.82	1.866	8	1.9
6	3	2	48.92	1.862	8	2.5
12	1	3	48.94	1.861	8	4.8
18	2	0	49.07	1.857	4	2.4
7	2	3	49.10	1.855	8	1.6
13	3	0	49.53	1.840	4	1.5
20	1	1	49.57	1.839	8	3.6
12	3	1	49.85	1.829	8	2.8
8	2	3	49.96	1.826	8	2.5

Lovdarite **LOV**

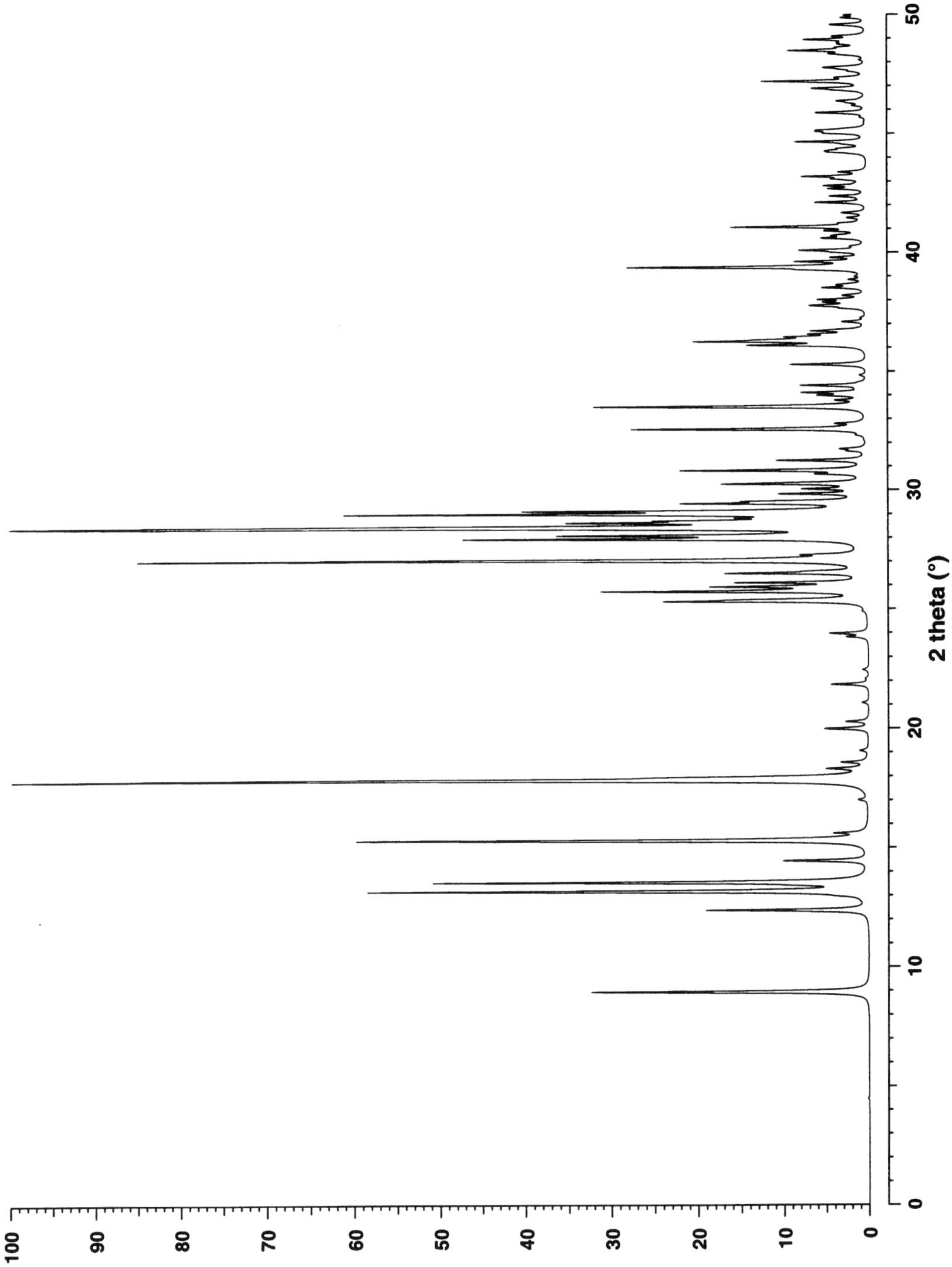

LTA Linde Type A, Hydrated

CHEMICAL COMPOSITION: $|Na_{96}(H_2O)_{216}|$ $[Si_{96}Al_{96}O_{384}]$

REFINED COMPOSITION: $|Na_{64}(H_2O)_{326.71}|$ $[Si_{96}Al_{96}O_{384}]$

CRYSTAL DATA: $Fm\bar{3}c$ (No. 226)
$a = 24.61$ Å $b = 24.61$ Å $c = 24.61$ Å
$\alpha = 90°$ $\beta = 90°$ $\gamma = 90°$
X-ray single crystal refinement, $R_w = 0.04$

REFERENCE: V. Gramlich and W. M. Meier,
Z. Kristallogr. **133** 134–149 (1971).

h	k	l	2θ	d	M	I_{rel}
2	0	0	7.18	12.305	6	100.0
2	2	0	10.17	8.701	12	51.3
2	2	2	12.46	7.104	8	31.8
4	0	0	14.40	6.153	6	0.5
4	2	0	16.11	5.503	24	20.3
4	2	2	17.65	5.023	24	2.4
4	4	0	20.41	4.350	12	3.6
5	3	1	21.36	4.160	48	2.1
6	0	0	21.67	4.102	6	10.6
4	4	2	21.67	4.102	24	22.8
6	2	0	22.85	3.891	24	1.2
6	2	2	23.99	3.710	24	44.3
4	4	4	25.07	3.552	8	0.7
6	4	0	26.11	3.413	24	10.1
6	4	2	27.11	3.289	48	41.0
8	0	0	29.03	3.076	6	0.4
6	4	4	29.94	2.984	24	19.7
8	2	0	29.94	2.984	24	19.9
8	2	2	30.83	2.900	24	5.4
6	6	0	30.83	2.900	12	2.3
6	6	2	31.70	2.823	24	0.2
8	4	0	32.54	2.751	24	9.3
8	4	2	33.37	2.685	48	3.0
6	6	4	34.18	2.623	24	27.1
9	3	1	34.77	2.580	48	0.1
8	4	4	35.75	2.512	24	4.7
10	0	0	36.51	2.461	6	4.1
8	6	0	36.51	2.461	24	0.4
10	2	0	37.26	2.413	24	0.1
10	2	2	38.00	2.368	24	1.6
6	6	6	38.00	2.368	8	1.4
8	6	4	39.43	2.285	48	1.0
10	4	0	39.43	2.285	24	0.3
10	4	2	40.14	2.247	48	2.2
8	8	0	41.51	2.175	12	4.3
10	4	4	42.19	2.142	24	3.4
8	6	6	42.85	2.110	24	2.3
10	6	0	42.85	2.110	24	1.2
10	6	2	43.51	2.080	48	1.8
8	8	4	44.16	2.051	24	0.9
12	0	0	44.16	2.051	6	7.8
12	2	0	44.80	2.023	24	0.5
12	2	2	45.44	1.996	24	0.1
12	4	0	46.69	1.946	24	0.2
10	8	0	47.30	1.922	24	2.3
8	8	6	47.30	1.922	24	4.2
12	4	2	47.30	1.922	48	0.3
10	8	2	47.91	1.899	48	5.1
10	6	6	48.51	1.876	24	0.3
12	4	4	49.11	1.855	24	0.9
10	8	4	49.70	1.834	48	1.8
12	6	0	49.70	1.834	24	0.9

Linde Type A, Hydrated LTA

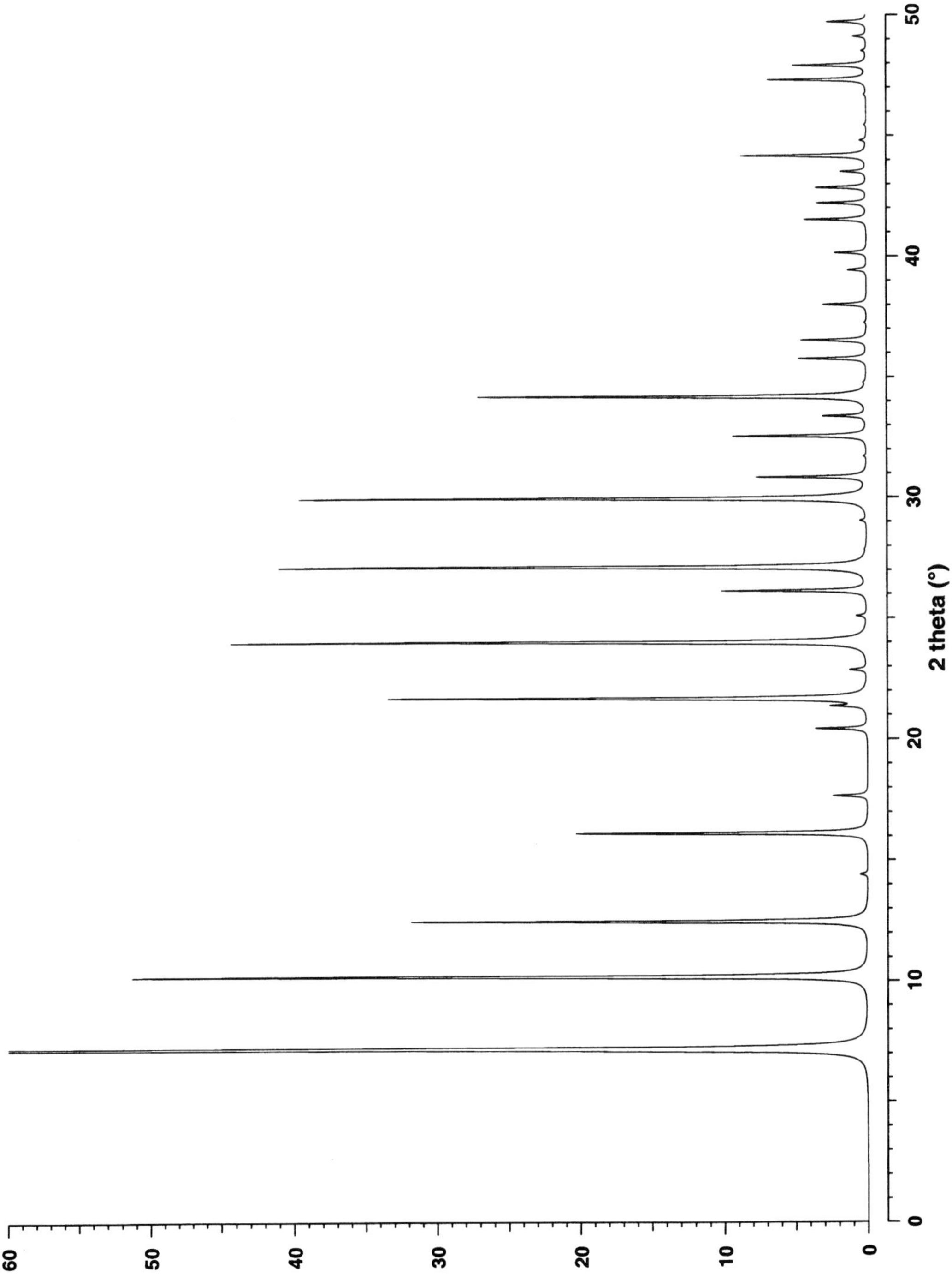

LTA Linde Type A, Dehydrated

CHEMICAL COMPOSITION: $|Na_{91.7}|$ $[Si_{96}Al_{96}O_{384}]$

REFINED COMPOSITION: $|Na_{91.78}|$ $[Si_{96}Al_{96}O_{384}]$

CRYSTAL DATA: $Fm\overline{3}c$ (No. 226)
a = 24.555 Å b = 24.555 Å c = 24.555 Å
α = 90° β = 90° γ = 90°
X-ray single crystal refinement, R_w = 0.023

REFERENCE: J. J. Pluth and J. V. Smith,
J. Am. Chem. Soc. **102** 4704–4708 (1980).

h	k	l	2θ	d	M	I_{rel}	h	k	l	2θ	d	M	I_{rel}	h	k	l	2θ	d	M	I_{rel}
2	0	0	7.20	12.278	6	100.0	6	4	4	30.01	2.978	24	4.6	8	8	0	41.61	2.170	12	1.3
2	2	0	10.19	8.682	12	49.6	8	2	0	30.01	2.978	24	6.4	10	4	4	42.29	2.137	24	1.3
2	2	2	12.49	7.088	8	22.4	6	6	0	30.90	2.894	12	0.6	8	6	6	42.95	2.106	24	1.0
4	2	0	16.14	5.491	24	20.1	8	2	2	30.90	2.894	24	1.2	10	6	0	42.95	2.106	24	0.5
4	2	2	17.69	5.012	24	1.9	8	4	0	32.62	2.745	24	1.4	10	6	2	43.61	2.075	48	0.9
4	4	0	20.46	4.341	12	3.0	8	4	2	33.45	2.679	48	1.0	12	0	0	44.26	2.046	6	2.2
5	3	1	21.41	4.151	48	0.8	6	6	4	34.26	2.618	24	8.4	8	8	4	44.26	2.046	24	0.5
6	0	0	21.72	4.093	6	1.3	8	4	4	35.83	2.506	24	0.8	12	2	0	44.91	2.018	24	0.1
4	4	2	21.72	4.093	24	8.4	10	0	0	36.59	2.455	6	3.0	12	4	0	46.80	1.941	24	0.2
6	2	0	22.91	3.882	24	0.1	6	6	6	38.08	2.363	8	1.3	8	8	6	47.41	1.917	24	1.2
6	2	2	24.04	3.702	24	13.7	10	2	2	38.08	2.363	24	0.2	10	8	0	47.41	1.917	24	0.6
4	4	4	25.13	3.544	8	0.3	8	6	4	39.53	2.280	48	0.2	10	8	2	48.02	1.894	48	1.8
6	4	0	26.17	3.405	24	0.6	10	4	0	39.53	2.280	24	0.1	10	8	4	49.82	1.830	48	0.4
6	4	2	27.18	3.281	48	9.0	10	4	2	40.23	2.242	48	0.5							

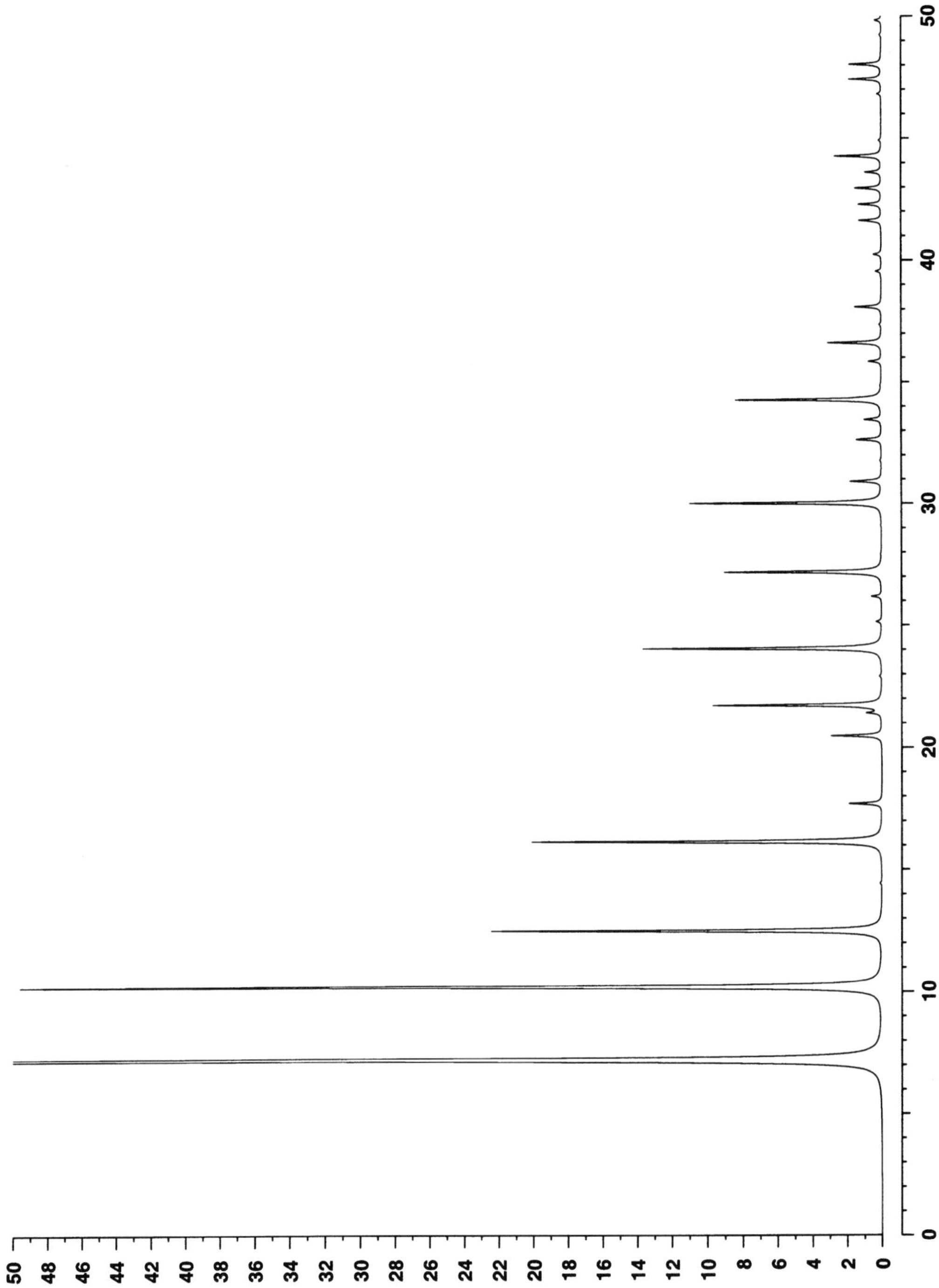
2 theta (°)

LTL Perlialite

CHEMICAL COMPOSITION: $|K_{12}(H_2O)_{20}|$ $[Si_{24}Al_{12}O_{72}]$
Murun Massif, Russia

REFINED COMPOSITION: $|K_{7.56}Tl_{3.8}(H_2O)_{22.46}|$ $[Si_{24}Al_{12}O_{72}]$

CRYSTAL DATA: $P6/mmm$ (No. 191)
$a = 18.5432$ Å $b = 18.5432$ Å $c = 7.5310$ Å
$\alpha = 90°$ $\beta = 90°$ $\gamma = 120°$
X-ray powder refinement, $R_{wp} = 0.20$, $R_p = 0.17$

REFERENCE: G. Artioli and Å. Kvick,
Eur. J. Mineral. **2** 749–759 (1990).

h	k	l	2θ	d	M	I_{rel}
1	0	0	5.50	16.059	6	100.0
1	1	0	9.54	9.272	6	8.8
2	0	0	11.02	8.029	6	2.5
1	0	1	12.98	6.818	12	1.6
2	1	0	14.59	6.070	12	6.3
1	1	1	15.16	5.846	12	0.3
2	0	1	16.14	5.493	12	4.6
3	0	0	16.56	5.353	6	18.9
2	1	1	18.78	4.726	24	1.0
2	2	0	19.14	4.636	6	19.0
3	1	0	19.93	4.454	12	3.0
3	0	1	20.35	4.363	12	17.0
4	0	0	22.14	4.015	6	3.5
2	2	1	22.52	3.948	12	51.0
3	1	1	23.20	3.834	24	12.3
0	0	2	23.63	3.765	2	5.1
3	2	0	24.16	3.684	12	3.1
1	0	2	24.28	3.666	12	5.4
4	0	1	25.14	3.543	12	0.7
4	1	0	25.42	3.504	12	2.8
1	1	2	25.53	3.489	12	25.7
3	2	1	26.94	3.309	24	3.2
5	0	0	27.78	3.212	6	18.7
2	1	2	27.88	3.200	24	6.3
4	1	1	28.08	3.177	24	2.6
3	3	0	28.89	3.091	6	0.1
3	0	2	28.99	3.080	12	34.8
4	2	0	29.43	3.035	12	7.0
2	2	2	30.59	2.923	12	6.7
5	1	0	31.00	2.884	12	3.0
3	3	1	31.28	2.859	12	15.8
4	2	1	31.79	2.815	24	6.8
4	0	2	32.60	2.746	12	6.1
5	1	1	33.26	2.693	24	3.6
6	0	0	33.48	2.676	6	27.1
4	3	0	33.96	2.640	12	0.1
3	2	2	34.04	2.633	24	12.2
5	2	0	34.89	2.571	12	0.1
4	1	2	34.98	2.565	24	6.5
6	0	1	35.60	2.522	12	7.2
0	0	3	35.77	2.510	2	0.4
4	3	1	36.05	2.491	24	11.7
1	0	3	36.22	2.480	12	0.7
5	0	2	36.78	2.444	12	3.1
5	2	1	36.94	2.434	24	5.3
1	1	3	37.10	2.423	12	0.4
2	0	3	37.54	2.396	12	0.8
3	3	2	37.65	2.389	12	6.1
4	2	2	38.08	2.363	24	0.5
6	1	1	38.66	2.329	24	0.4
2	1	3	38.82	2.320	24	0.5
4	4	0	38.85	2.318	6	0.8
7	0	0	39.27	2.294	6	2.8
5	3	0	39.27	2.294	12	2.1
5	1	2	39.35	2.290	24	0.7
3	0	3	39.65	2.273	12	6.4
6	2	0	40.51	2.227	12	0.2
4	4	1	40.73	2.215	12	8.7
2	2	3	40.88	2.207	12	7.7
7	0	1	41.13	2.195	12	2.0
3	1	3	41.28	2.187	24	0.4
6	0	2	41.39	2.182	12	4.7
4	3	2	41.79	2.162	24	1.5
6	2	1	42.32	2.136	24	1.9
5	2	2	42.57	2.124	24	1.3
5	4	0	44.04	2.056	12	0.6
6	1	2	44.11	2.053	24	0.4
7	1	1	44.25	2.047	24	2.0
4	1	3	44.39	2.041	24	0.2
6	3	0	44.79	2.023	12	0.4
8	0	0	45.17	2.007	6	0.3
5	0	3	45.88	1.978	12	0.4
5	3	2	46.34	1.959	24	2.2
7	0	2	46.34	1.959	12	0.9
6	3	1	46.47	1.954	24	4.8
3	3	3	46.61	1.949	12	2.6
8	0	1	46.84	1.940	12	3.6
4	2	3	46.97	1.934	24	1.1
7	2	1	47.91	1.899	24	0.7
5	1	3	48.05	1.894	24	0.2
0	0	4	48.34	1.883	2	5.3
1	0	4	48.69	1.870	12	0.6
5	5	0	49.13	1.854	6	0.3
7	1	2	49.20	1.852	24	1.4
1	1	4	49.39	1.845	12	1.2
6	4	0	49.48	1.842	12	0.5

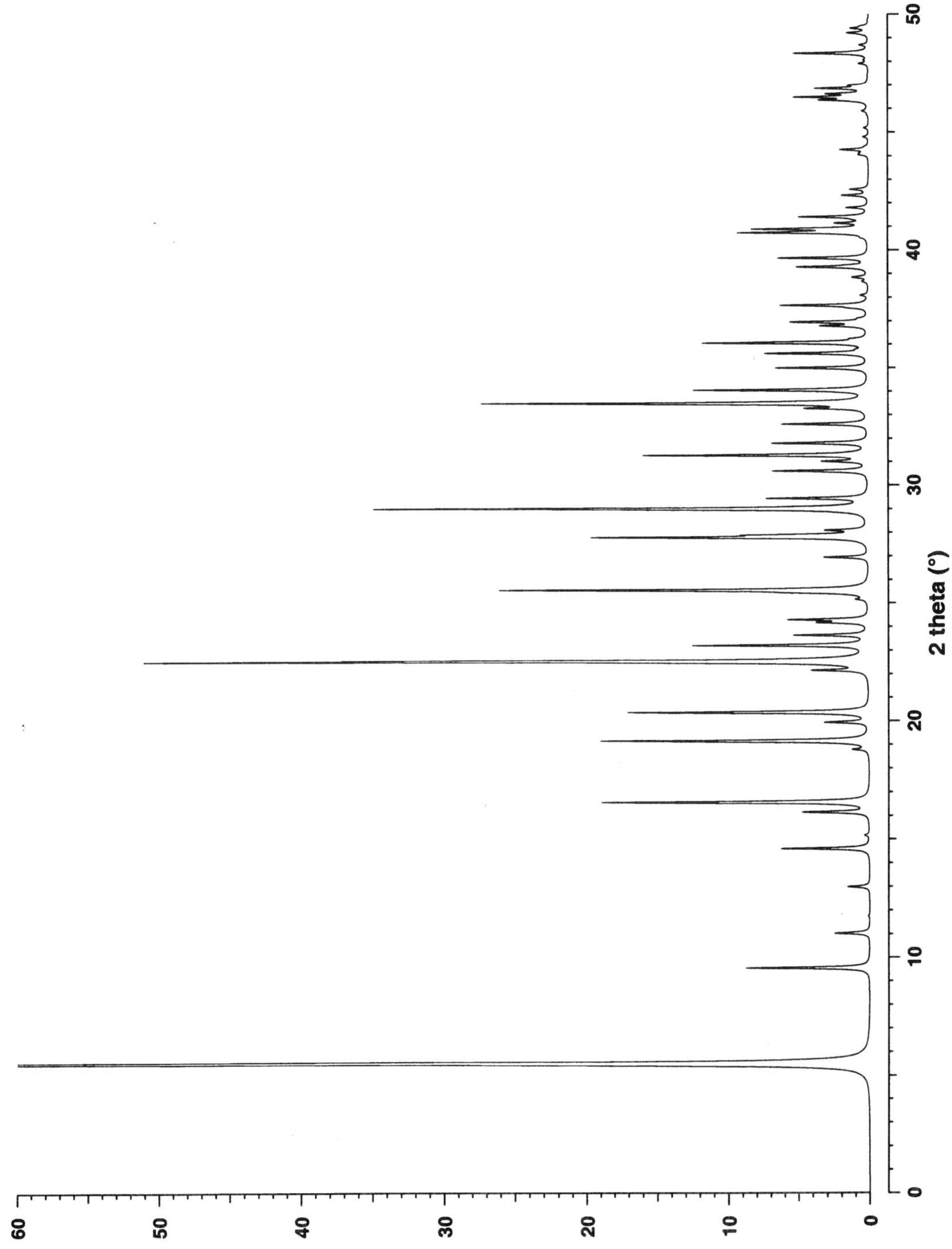
2 theta (°)
0
10
20
30
40
50
60

LTL Linde Type L

CHEMICAL COMPOSITION: $|Na_3K_6(H_2O)_{21}|$ $[Si_{27}Al_9O_{72}]$
Synthetic material

REFINED COMPOSITION: $|Na_5K_{4.7}(H_2O)_{20.8}|$ $[Si_{27}Al_9O_{72}]$

CRYSTAL DATA: $P6/mmm$ (No. 191)
$a = 18.40$ Å $b = 18.40$ Å $c = 7.52$ Å
$\alpha = 90°$ $\beta = 90°$ $\gamma = 120°$
X-ray powder refinement, $R_F = 0.13$

REFERENCE: R. M. Barrer and H. Villiger,
Z. Kristallogr. **128** 352–270 (1969).

h	k	l	2θ	d	M	I_{rel}	h	k	l	2θ	d	M	I_{rel}	h	k	l	2θ	d	M	I_{rel}
1	0	0	5.55	15.935	6	100.0	5	0	1	30.46	2.934	12	0.8	3	1	3	41.41	2.180	24	0.1
1	1	0	9.61	9.200	6	2.2	2	2	2	30.71	2.911	12	25.2	5	3	1	41.44	2.179	24	0.4
2	0	0	11.10	7.967	6	1.9	3	1	2	31.23	2.864	24	2.3	6	0	2	41.63	2.169	12	0.4
0	0	1	11.77	7.520	2	24.3	5	1	0	31.25	2.862	12	0.2	4	0	3	42.61	2.122	12	0.8
1	0	1	13.02	6.801	12	1.7	3	3	1	31.50	2.840	12	1.6	6	2	1	42.64	2.120	24	0.2
2	1	0	14.71	6.023	12	17.5	4	2	1	32.01	2.796	24	2.9	7	1	0	42.85	2.111	12	0.1
1	1	1	15.22	5.822	12	10.6	5	1	1	33.50	2.675	24	2.8	3	2	3	43.79	2.067	24	0.4
2	0	1	16.21	5.469	12	0.2	6	0	0	33.75	2.656	6	11.3	6	1	2	44.39	2.041	24	1.1
3	0	0	16.69	5.312	6	0.7	3	2	2	34.21	2.621	24	8.3	4	1	3	44.56	2.033	24	0.9
2	1	1	18.88	4.701	24	1.0	4	1	2	35.15	2.553	24	0.1	7	1	1	44.59	2.032	24	0.5
2	2	0	19.30	4.600	6	18.3	5	2	0	35.17	2.552	12	3.1	6	3	0	45.16	2.008	12	0.9
3	1	0	20.09	4.420	12	3.5	0	0	3	35.82	2.507	2	0.8	8	0	0	45.54	1.992	6	0.4
3	0	1	20.47	4.339	12	11.6	6	0	1	35.86	2.504	12	4.6	5	0	3	46.07	1.970	12	0.4
4	0	0	22.32	3.984	6	1.5	1	0	3	36.28	2.476	12	0.1	5	4	1	46.10	1.969	24	0.9
2	2	1	22.66	3.924	12	22.7	4	3	1	36.31	2.474	24	4.4	4	4	2	46.27	1.962	12	0.2
3	1	1	23.35	3.810	24	0.3	5	0	2	36.97	2.431	12	0.1	5	3	2	46.64	1.947	24	0.7
1	0	2	24.32	3.660	12	16.1	6	1	0	36.99	2.430	12	5.5	7	0	2	46.64	1.947	12	0.2
3	2	0	24.35	3.656	12	0.5	1	1	3	37.17	2.419	12	0.3	7	2	0	46.66	1.947	12	0.6
4	0	1	25.30	3.520	12	0.4	5	2	1	37.21	2.416	24	1.8	3	3	3	46.81	1.941	12	0.5
1	1	2	25.59	3.481	12	11.5	3	3	2	37.86	2.376	12	0.3	8	0	1	47.20	1.925	12	0.4
4	1	0	25.62	3.477	12	5.3	4	2	2	38.29	2.350	24	0.4	6	2	2	47.74	1.905	24	0.7
2	0	2	26.21	3.400	12	1.3	6	1	1	38.95	2.312	24	0.2	5	1	3	48.26	1.886	24	0.8
3	2	1	27.12	3.288	24	12.1	4	4	0	39.17	2.300	6	3.7	7	2	1	48.29	1.885	24	0.3
2	1	2	27.97	3.189	24	5.7	5	1	2	39.57	2.277	24	0.5	0	0	4	48.42	1.880	2	7.2
5	0	0	28.00	3.187	6	20.9	7	0	0	39.59	2.276	6	0.2	1	0	4	48.77	1.867	12	0.4
4	1	1	28.28	3.156	24	3.2	5	3	0	39.59	2.276	12	0.4	1	1	4	49.48	1.842	12	0.1
3	0	2	29.10	3.069	12	15.4	6	2	0	40.84	2.210	12	0.4	7	1	2	49.52	1.840	24	0.2
3	3	0	29.12	3.067	6	3.0	2	2	3	41.00	2.201	12	2.0	5	5	0	49.54	1.840	6	1.7
4	2	0	29.66	3.011	12	4.8	4	4	1	41.04	2.199	12	6.7	2	0	4	49.84	1.830	12	0.1

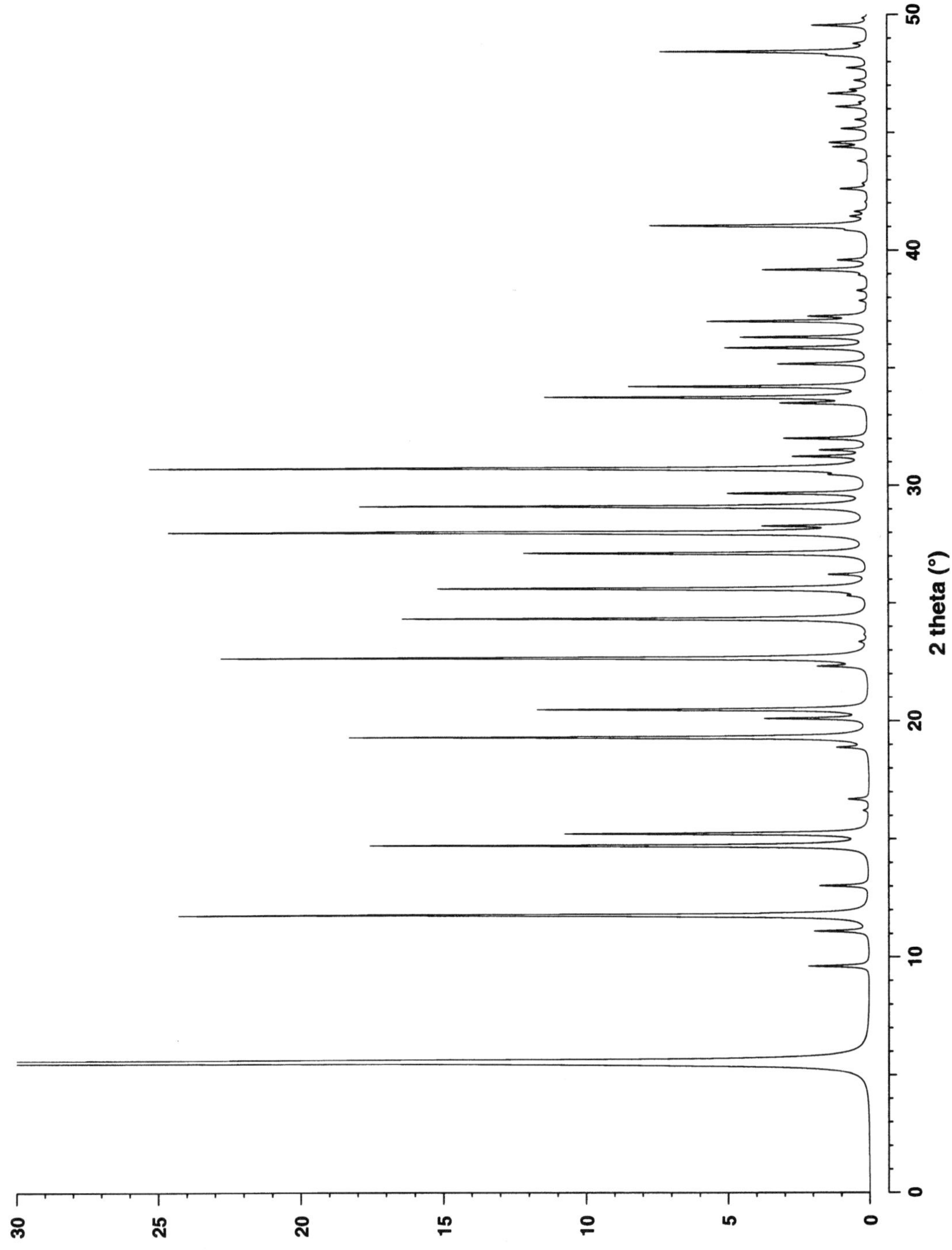
2 theta (°)
0
10
20
30
40
50
0
5
10
15
20
25
30

LTN NaZ-21

CHEMICAL COMPOSITION: $|Na_{384}(H_2O)_{394}|\ [Si_{384}Al_{384}O_{1536}]$

REFINED COMPOSITION: $|Na_{426.6}(H_2O)_{393.6}|\ [Si_{384}Al_{384}O_{1536}]$

CRYSTAL DATA: $Fd\bar{3}$ (No. 203) origin at centre ($\bar{3}$)
$a = 36.95$ Å $b = 36.95$ Å $c = 36.95$ Å
$\alpha = 90°$ $\beta = 90°$ $\gamma = 90°$
X-ray single crystal refinement, $R = 0.051$

REFERENCE: Y. F. Shepelev, Y. I. Smolin, I. K. Butikova and V. I. Tarasov, *Doklady Akad. Nauk SSSR* **272** 1133–1137 (1975).

h	k	l	2θ	d	M	I_{rel}	h	k	l	2θ	d	M	I_{rel}	h	k	l	2θ	d	M	I_{rel}
1	1	1	4.14	21.333	8	34.7	9	5	3	25.86	3.446	24	2.3	14	0	2	34.32	2.613	12	2.0
2	2	0	6.77	13.064	12	100.0	8	4	6	25.97	3.431	24	1.5	13	5	3	34.59	2.593	24	0.7
3	1	1	7.94	11.141	24	54.6	8	6	4	25.97	3.431	24	1.0	13	3	5	34.59	2.593	24	0.7
2	2	2	8.29	10.667	8	7.5	10	2	4	26.42	3.373	24	5.0	14	2	2	34.67	2.587	24	0.6
4	0	0	9.57	9.238	6	24.0	10	4	2	26.42	3.373	24	2.3	9	9	7	35.28	2.544	24	3.9
3	3	1	10.44	8.477	24	4.8	8	8	0	27.31	3.266	12	0.8	11	9	3	35.28	2.544	24	0.6
5	1	1	12.45	7.111	24	10.4	11	3	1	27.63	3.228	24	2.1	12	6	6	35.71	2.514	24	1.9
3	3	3	12.45	7.111	8	0.8	11	1	3	27.63	3.228	24	1.1	12	8	4	36.39	2.469	24	0.8
4	4	0	13.56	6.532	12	58.6	9	1	7	27.63	3.228	24	2.2	15	1	1	36.64	2.452	24	8.2
5	3	1	14.18	6.246	24	14.3	9	7	1	27.63	3.228	24	3.6	11	5	9	36.64	2.452	24	3.3
5	1	3	14.18	6.246	24	13.5	8	8	2	27.74	3.216	24	1.1	11	9	5	36.64	2.452	24	2.1
4	4	2	14.38	6.158	24	0.8	10	4	4	27.74	3.216	24	2.0	10	8	8	36.73	2.447	24	5.1
6	0	2	15.16	5.842	12	0.9	8	6	6	28.16	3.168	24	0.9	15	3	1	37.31	2.410	24	0.6
6	2	0	15.16	5.842	12	4.4	10	0	6	28.16	3.168	12	0.7	9	9	9	37.96	2.370	8	3.1
5	3	3	15.73	5.635	24	6.5	9	7	3	28.48	3.134	24	3.9	11	7	9	38.60	2.332	24	0.6
6	2	2	15.91	5.570	24	2.8	11	3	3	28.48	3.134	24	6.6	16	0	0	39.00	2.309	6	2.0
4	4	4	16.62	5.333	8	2.9	9	3	7	28.48	3.134	24	1.5	15	5	3	39.24	2.296	24	1.5
5	5	1	17.14	5.174	24	12.6	11	5	1	29.30	3.048	24	1.3	15	3	5	39.24	2.296	24	1.3
6	2	4	17.96	4.938	24	2.1	11	1	5	29.30	3.048	24	2.0	15	5	5	40.48	2.228	24	1.8
7	1	3	18.44	4.810	24	6.2	7	7	7	29.30	3.048	8	0.8	16	2	4	40.56	2.224	24	0.9
7	3	1	18.44	4.810	24	4.8	10	4	6	29.81	2.997	24	11.7	16	4	2	40.56	2.224	24	0.8
7	3	3	19.67	4.514	24	4.6	10	6	4	29.81	2.997	24	7.0	15	3	7	41.09	2.196	24	0.7
8	2	2	20.39	4.355	24	3.9	12	2	2	29.81	2.997	24	7.7	12	12	0	41.47	2.177	12	9.0
6	6	0	20.39	4.355	12	3.6	11	5	3	30.11	2.968	24	0.8	17	1	1	41.70	2.166	24	0.9
5	5	5	20.82	4.267	8	1.0	12	0	4	30.60	2.921	12	2.7	11	11	7	41.70	2.166	24	2.1
7	1	5	20.82	4.267	24	1.0	12	4	0	30.60	2.921	12	2.2	12	12	2	41.77	2.162	24	1.3
6	6	2	20.96	4.238	24	1.7	12	2	4	30.99	2.885	24	3.2	13	7	9	42.29	2.137	24	0.7
8	4	0	21.51	4.131	12	1.3	12	4	2	30.99	2.885	24	1.9	17	3	3	42.88	2.109	24	2.2
8	0	4	21.51	4.131	12	6.7	10	2	8	31.38	2.851	24	1.0	12	10	8	42.96	2.105	24	0.6
7	5	3	21.91	4.056	24	14.1	11	1	7	31.66	2.826	24	1.6	16	0	8	43.83	2.066	12	0.7
7	3	5	21.91	4.056	24	7.7	11	5	5	31.66	2.826	24	5.7	17	5	3	44.04	2.056	24	1.3
9	1	1	21.91	4.056	24	6.7	11	7	1	31.66	2.826	24	3.4	15	7	7	44.04	2.056	24	0.6
8	2	4	22.05	4.032	24	1.0	13	1	1	31.66	2.826	24	0.9	17	3	5	44.04	2.056	24	2.1
6	6	4	22.57	3.939	24	5.5	12	4	4	32.14	2.785	24	6.8	11	11	9	44.04	2.056	24	1.0
8	4	4	23.59	3.771	24	17.9	11	3	7	32.42	2.762	24	0.6	16	6	6	44.40	2.040	24	0.8
7	7	1	23.96	3.714	24	1.7	13	3	1	32.42	2.762	24	10.6	13	9	9	44.61	2.031	24	0.9
9	3	3	23.96	3.714	24	24.0	9	7	7	32.42	2.762	24	9.7	13	13	1	45.18	2.007	24	0.6
7	5	5	23.96	3.714	24	3.1	13	1	3	32.42	2.762	24	8.9	19	1	1	46.84	1.939	24	0.8
8	6	2	24.57	3.623	24	1.2	10	8	4	32.51	2.754	24	0.9	16	10	4	47.46	1.916	24	0.7
9	5	1	24.93	3.572	24	2.1	12	2	6	32.88	2.724	24	2.3	16	4	10	47.46	1.916	24	0.6
9	1	5	24.93	3.572	24	1.2	12	6	2	32.88	2.724	24	2.2	14	14	0	48.80	1.866	12	2.7
7	7	3	24.93	3.572	24	1.9	9	9	5	33.15	2.702	24	6.1	18	6	6	49.06	1.857	24	0.9
6	6	6	25.04	3.556	8	0.8	8	8	8	33.61	2.667	8	12.2	16	0	12	49.32	1.847	12	0.8
10	2	2	25.04	3.556	24	1.6	10	8	6	34.32	2.613	24	2.3	16	12	0	49.32	1.847	12	0.6
9	3	5	25.86	3.446	24	2.9	10	6	8	34.32	2.613	24	1.8	20	0	0	49.32	1.847	6	0.6

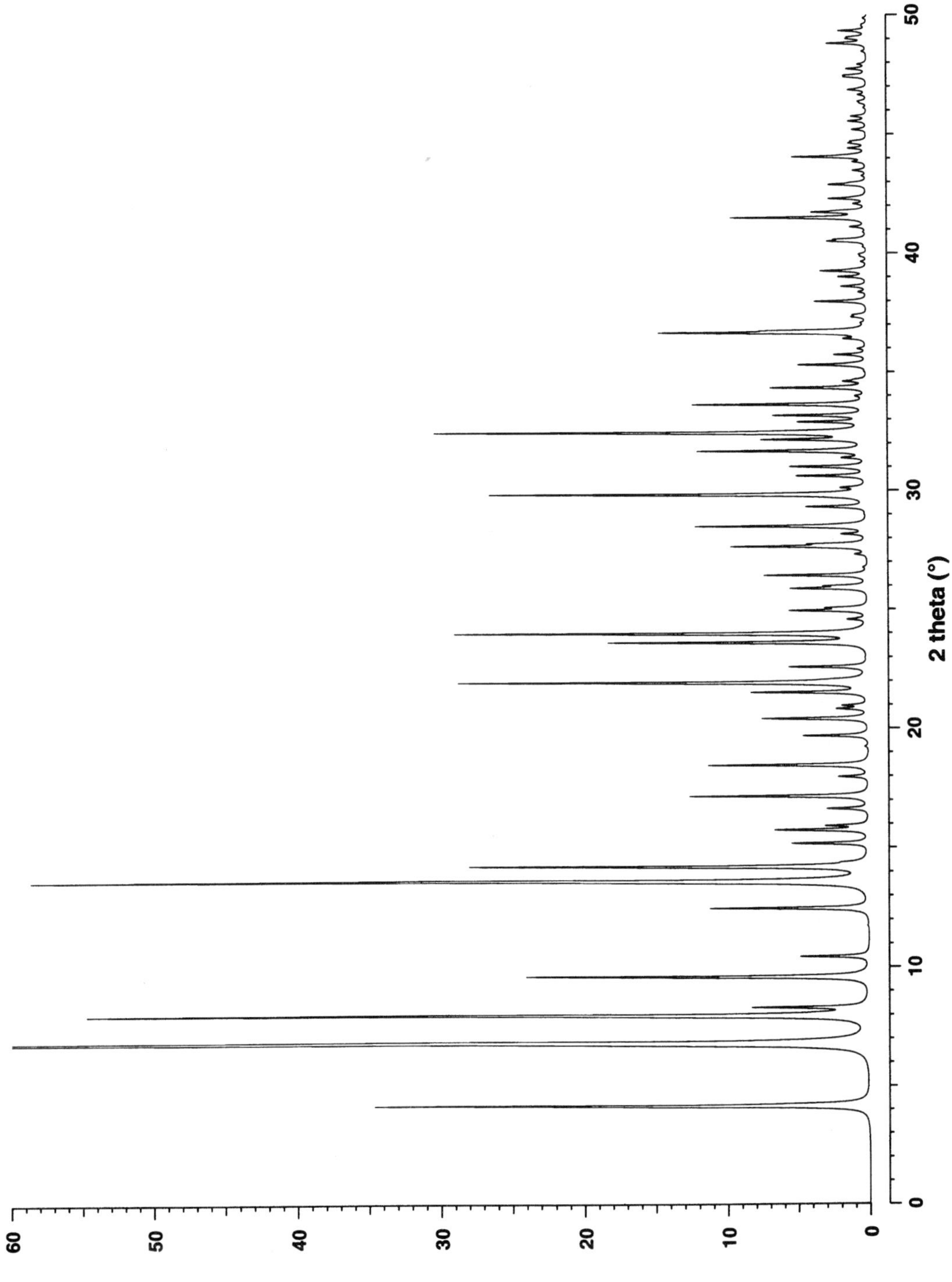
2 theta (°)
0
10
20
30
40
50
0
10
20
30
40
50
60

MAZ Mazzite

CHEMICAL COMPOSITION: $|Na_{0.8}K_{2.5}Ca_{1.4}Mg_{2.1}(H_2O)_{28}|$ $[Si_{26.5}Al_{9.9}O_{72}]$
Mont Semiol, Loire, France

REFINED COMPOSITION: $|Na_{0.3}K_{2.52}Ca_{1.06}Mg_2(H_2O)_{31.48}|$ $[Si_{26.28}Al_{9.72}O_{72}]$

CRYSTAL DATA: $P6_3/mmc$ (No. 194)
a = 18.392 Å b = 18.392 Å c = 7.646 Å
α = 90° β = 90° γ = 120°
X-ray single crystal refinement, R_w = 0.05

REFERENCE: E. Galli,
Rend. Soc. Ital. Mineral. Petrol. **31** 599–612 (1975).

h	k	l	2θ	d	M	I_{rel}	h	k	l	2θ	d	M	I_{rel}	h	k	l	2θ	d	M	I_{rel}
1	0	0	5.55	15.928	6	98.1	3	1	2	30.93	2.891	24	0.3	6	0	2	41.41	2.181	12	0.6
1	1	0	9.62	9.196	6	92.2	5	1	0	31.27	2.861	12	3.7	4	0	3	42.09	2.147	12	8.9
2	0	0	11.11	7.964	6	39.7	4	2	1	31.95	2.801	24	2.0	6	2	1	42.60	2.122	24	9.8
1	0	1	12.84	6.893	12	25.3	5	1	1	33.44	2.679	24	13.9	5	2	2	42.61	2.122	24	0.8
2	1	0	14.71	6.020	12	45.0	6	0	0	33.76	2.655	6	4.6	3	2	3	43.28	2.090	24	3.6
2	0	1	16.07	5.516	12	5.1	3	2	2	33.94	2.642	24	16.3	6	1	2	44.17	2.050	24	6.8
3	0	0	16.70	5.309	6	16.4	4	3	0	34.24	2.619	12	0.7	7	1	1	44.55	2.034	24	4.0
2	1	1	18.76	4.730	24	50.4	4	1	2	34.89	2.572	24	0.2	6	3	0	45.18	2.007	12	4.1
2	2	0	19.30	4.598	6	1.3	5	2	0	35.19	2.551	12	18.0	8	0	0	45.56	1.991	6	3.6
3	1	0	20.10	4.418	12	9.1	1	0	3	35.68	2.517	12	1.4	5	0	3	45.58	1.990	12	8.4
3	0	1	20.36	4.361	12	1.4	6	0	1	35.81	2.508	12	3.2	5	4	1	46.06	1.970	24	2.1
4	0	0	22.33	3.982	6	11.6	5	0	2	36.72	2.447	12	3.1	4	4	2	46.07	1.970	12	3.7
3	1	1	23.25	3.825	24	97.3	6	1	0	37.01	2.429	12	1.0	5	3	2	46.44	1.955	24	1.2
0	0	2	23.27	3.823	2	14.3	2	0	3	37.03	2.427	12	3.5	7	0	2	46.44	1.955	12	0.4
1	0	2	23.94	3.717	12	36.2	5	2	1	37.16	2.419	24	3.7	7	2	0	46.68	1.946	12	0.5
3	2	0	24.36	3.654	12	31.9	3	3	2	37.61	2.392	12	7.4	4	2	3	46.70	1.945	24	1.4
4	0	1	25.22	3.532	12	0.2	4	2	2	38.05	2.365	24	3.1	6	3	1	46.80	1.941	24	3.1
1	1	2	25.23	3.530	12	100.0	2	1	3	38.35	2.347	24	0.5	8	0	1	47.17	1.927	12	1.1
4	1	0	25.63	3.476	12	5.2	6	1	1	38.90	2.315	24	0.4	6	2	2	47.54	1.913	24	0.5
2	0	2	25.85	3.446	12	8.7	4	4	0	39.18	2.299	6	12.7	0	0	4	47.57	1.911	2	19.6
3	2	1	27.04	3.297	24	2.1	3	0	3	39.21	2.298	12	0.1	5	1	3	47.79	1.903	24	5.6
5	0	0	28.01	3.186	6	83.7	5	1	2	39.34	2.290	24	0.5	1	0	4	47.93	1.898	12	0.8
4	1	1	28.20	3.164	24	0.4	7	0	0	39.61	2.275	6	0.3	7	2	1	48.26	1.886	24	2.9
3	0	2	28.78	3.102	12	35.4	5	3	0	39.61	2.275	12	0.5	1	1	4	48.65	1.871	12	0.2
3	3	0	29.13	3.065	6	25.4	6	2	0	40.85	2.209	12	3.4	2	0	4	49.01	1.859	12	2.8
4	2	0	29.68	3.010	12	31.6	3	1	3	40.88	2.208	24	1.2	5	5	0	49.56	1.839	6	10.5
5	0	1	30.40	2.941	12	54.5	7	0	1	41.40	2.181	12	0.7	6	0	3	49.58	1.839	12	0.6
2	2	2	30.41	2.940	12	48.2	5	3	1	41.40	2.181	24	0.3	6	4	0	49.91	1.827	12	0.3

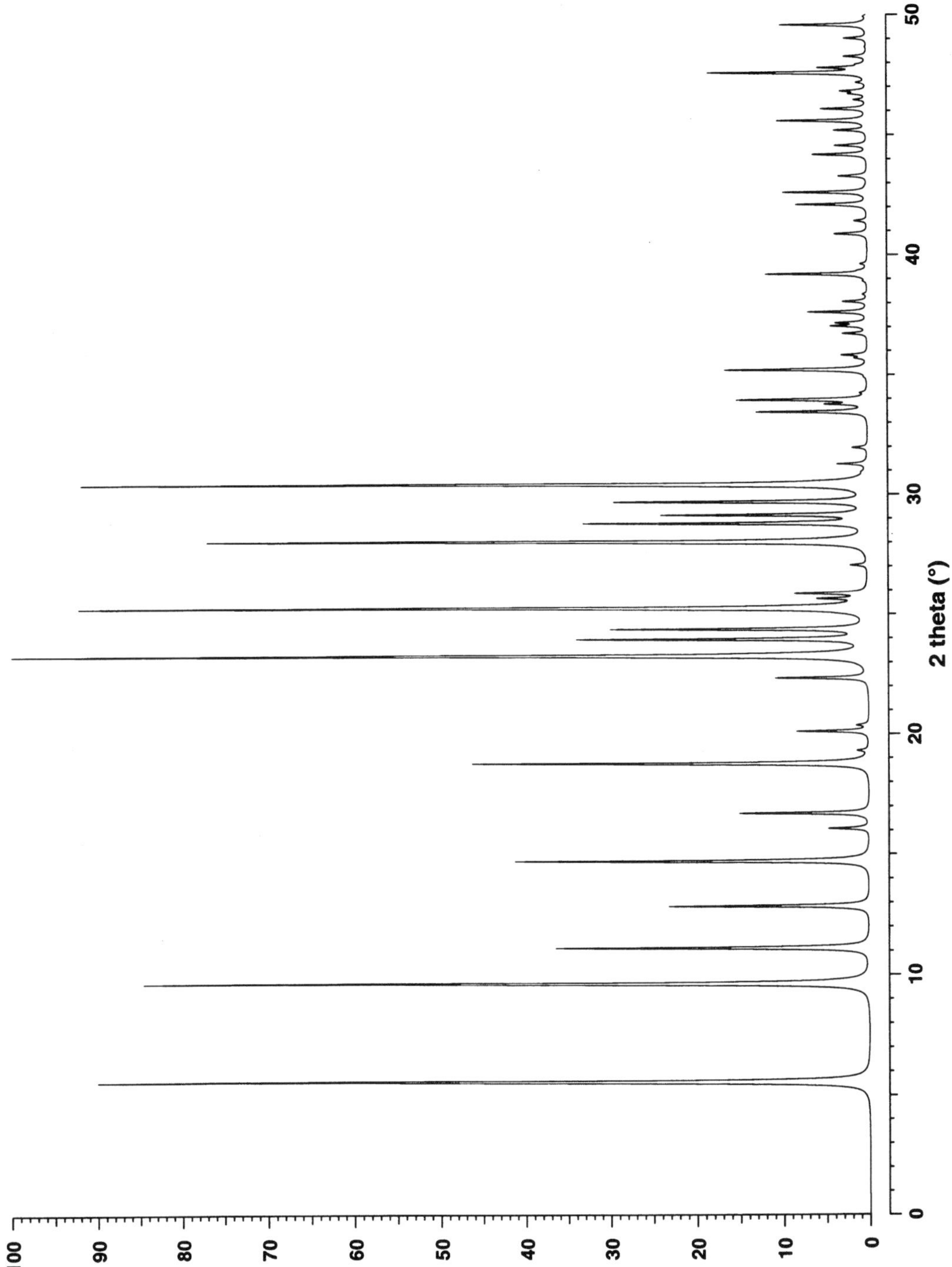
2 theta (°)

MEI ZSM-18, SiO_2 Framework

CHEMICAL COMPOSITION: $[Si_{34}O_{68}]$

REFINED COMPOSITION: $[Si_{34}O_{68}]$

CRYSTAL DATA: $P6_3/m$ (No. 176)
$a = 13.175$ Å $b = 13.175$ Å $c = 15.848$ Å
$\alpha = 90°$ $\beta = 90°$ $\gamma = 120°$
DLS refinement

REFERENCE: S. L. Lawton and W. J. Rohrbaugh, *Science* **247** 1319–1322 (1990).

h	k	l	2θ	d	M	I_{rel}
1	0	0	7.75	11.410	6	100.0
1	0	1	9.55	9.260	12	5.2
0	0	2	11.17	7.924	2	17.2
1	1	0	13.44	6.587	6	4.5
1	0	2	13.60	6.508	12	1.7
1	1	1	14.56	6.083	12	8.0
2	0	0	15.53	5.705	6	0.3
2	0	1	16.51	5.368	12	2.1
1	1	2	17.51	5.066	12	5.9
1	0	3	18.51	4.794	12	2.6
2	0	2	19.17	4.630	12	1.6
2	1	0	20.59	4.313	6	2.8
1	2	0	20.59	4.313	6	0.5
1	2	1	21.35	4.161	12	7.5
1	1	3	21.56	4.121	12	6.9
0	0	4	22.44	3.962	2	1.5
2	0	3	22.94	3.876	12	1.2
3	0	0	23.39	3.803	6	0.1
1	2	2	23.49	3.788	12	4.8
1	0	4	23.77	3.743	12	0.4
3	0	2	25.99	3.429	12	0.7
1	2	3	26.68	3.341	12	3.6
2	1	3	26.68	3.341	12	0.2
2	2	0	27.07	3.294	6	2.4
2	2	1	27.66	3.225	12	2.6
3	1	1	28.77	3.103	12	2.2
3	0	3	28.93	3.087	12	0.3
1	0	5	29.24	3.054	12	2.2
2	2	2	29.37	3.041	12	1.2
1	3	2	30.41	2.939	12	0.2
3	1	2	30.41	2.939	12	0.9
1	2	4	30.64	2.918	12	1.7
1	1	5	31.32	2.856	12	2.0
4	0	0	31.36	2.852	6	0.7
2	2	3	32.02	2.795	12	0.8
2	0	5	32.31	2.771	12	0.7
3	0	4	32.64	2.744	12	0.8
1	3	3	32.99	2.715	12	0.6
3	1	3	32.99	2.715	12	0.1
4	0	2	33.38	2.684	12	0.7
2	3	0	34.26	2.618	6	0.3
2	3	1	34.73	2.583	12	0.4
3	2	1	34.73	2.583	12	0.3
2	1	5	35.14	2.554	12	0.7
1	2	5	35.14	2.554	12	0.3
2	2	4	35.44	2.533	12	0.3
4	0	3	35.77	2.510	12	0.1
4	1	0	36.07	2.490	6	0.2
1	4	0	36.07	2.490	6	1.2
2	3	2	36.14	2.486	12	0.3
3	2	2	36.14	2.486	12	0.5
3	1	4	36.33	2.473	12	0.2
1	4	1	36.53	2.460	12	0.2
4	1	1	36.53	2.460	12	0.1
3	0	5	36.92	2.435	12	0.3
2	0	6	37.52	2.397	12	0.1
4	1	2	37.88	2.375	12	0.3
1	4	2	37.88	2.375	12	0.2
3	2	3	38.38	2.345	12	0.3
2	3	3	38.38	2.345	12	0.5
2	2	5	39.45	2.284	12	0.1
1	2	6	40.03	2.252	12	0.4
1	4	3	40.03	2.252	12	0.3
3	2	4	41.34	2.184	12	0.1
2	3	4	41.34	2.184	12	0.1
3	0	6	41.63	2.169	12	0.1
2	4	0	41.90	2.156	6	0.4
4	2	0	41.90	2.156	6	0.3
1	1	7	42.21	2.141	12	0.5
4	2	1	42.30	2.137	12	0.3
4	0	5	42.64	2.120	12	0.2
3	3	2	42.73	2.116	12	0.2
5	0	3	43.18	2.095	12	0.2
2	2	6	43.94	2.061	12	0.3
5	1	0	44.19	2.049	6	0.3
1	5	0	44.19	2.049	6	0.2
1	3	6	44.69	2.028	12	0.3
3	3	3	44.69	2.028	12	0.4
1	2	7	45.23	2.005	12	0.2
2	4	3	45.43	1.996	12	0.1
5	1	2	45.73	1.984	12	0.1
1	4	5	46.37	1.958	12	0.2
6	0	0	47.83	1.902	6	0.8
1	1	8	47.95	1.897	12	0.6
4	2	4	48.04	1.894	12	0.3
2	4	4	48.04	1.894	12	0.1
2	0	8	48.65	1.871	12	0.4
4	3	1	48.89	1.863	12	0.5
3	2	6	48.99	1.859	12	0.1
2	3	6	48.99	1.859	12	0.3
4	3	2	49.96	1.825	12	0.1

ZSM-18, SiO_2 Framework MEI

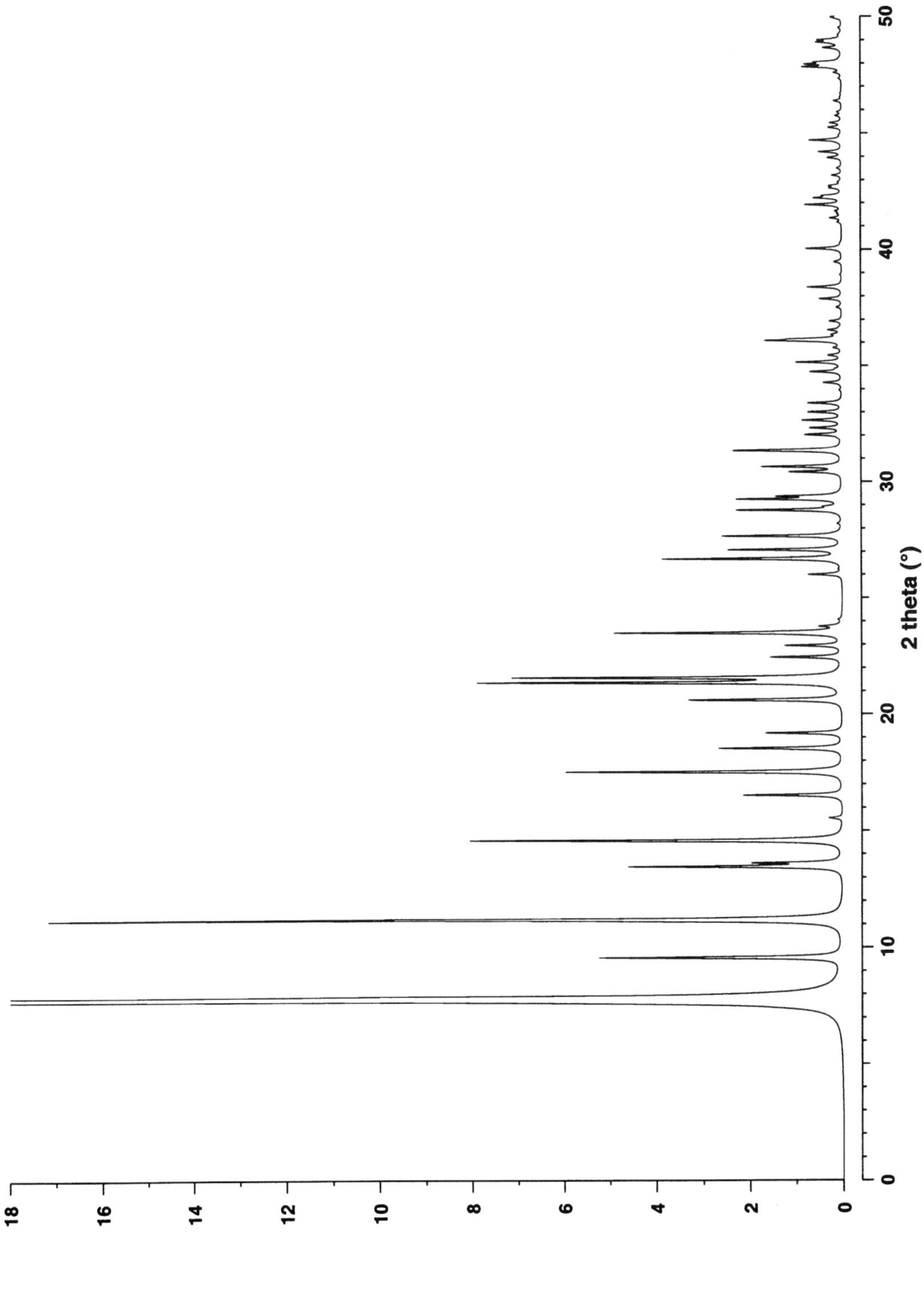

MEL ZSM-11, Calcined

CHEMICAL COMPOSITION: $[Si_{96}O_{192}]$

REFINED COMPOSITION: $[Si_{96}O_{192}]$

CRYSTAL DATA: $I\bar{4}m2$ (No. 119)
$a = 20.067$ Å $b = 20.067$ Å $c = 13.411$ Å
$\alpha = 90°$ $\beta = 90°$ $\gamma = 90°$
X-ray Rietveld refinement, $R_p = 0.13$

REFERENCE: C. A. Fyfe, H. Gies, G. T. Kokotailo, C. Pasztor, H. Strobl and D. E. Cox, *J. Am. Chem. Soc.* **111** 2470–2474 (1989).

h	k	l	2θ	d	M	I_{rel}
1	1	0	6.23	14.190	4	2.4
1	0	1	7.93	11.150	8	100.0
2	0	0	8.81	10.033	4	53.6
2	1	1	11.87	7.458	16	3.9
2	2	0	12.48	7.095	4	0.4
0	0	2	13.20	6.706	2	9.2
1	1	2	14.61	6.063	8	0.7
3	0	1	14.80	5.986	8	15.5
2	0	2	15.90	5.575	8	8.0
3	2	1	17.25	5.140	16	3.6
4	0	0	17.68	5.017	4	7.1
3	3	0	18.76	4.730	4	1.6
3	1	2	19.26	4.609	16	3.7
4	1	1	19.40	4.575	16	0.3
4	2	0	19.79	4.487	8	0.2
1	0	3	20.35	4.363	8	3.2
4	0	2	22.13	4.017	8	0.2
2	1	3	22.22	4.001	16	0.6
5	1	0	22.59	3.935	8	1.2
3	3	2	23.01	3.865	8	1.2
5	0	1	23.13	3.845	8	52.5
4	3	1	23.13	3.845	16	0.3
4	2	2	23.86	3.729	16	0.1
3	0	3	23.94	3.717	8	22.7
5	2	1	24.80	3.590	16	5.1
4	4	0	25.10	3.547	4	0.5
3	2	3	25.56	3.485	16	2.6
5	3	0	25.89	3.441	8	0.1
5	1	2	26.26	3.394	16	2.1
0	0	4	26.59	3.353	2	2.4
6	0	0	26.65	3.345	4	2.7
1	1	4	27.33	3.263	8	0.5
6	1	1	27.85	3.203	16	0.4
2	0	4	28.06	3.180	8	2.0
5	3	2	29.17	3.062	16	7.6

h	k	l	2θ	d	M	I_{rel}
5	4	1	29.26	3.052	16	0.4
5	0	3	29.92	2.986	8	9.6
4	3	3	29.92	2.986	16	0.3
3	1	4	30.15	2.964	16	0.7
6	3	1	30.62	2.920	16	0.2
6	2	2	31.18	2.868	16	0.1
5	2	3	31.25	2.862	16	2.2
7	1	0	31.52	2.838	8	0.1
5	5	0	31.52	2.838	4	0.6
7	0	1	31.92	2.803	8	0.2
4	0	4	32.11	2.788	8	0.4
3	3	4	32.74	2.735	8	0.2
1	0	5	33.71	2.659	8	0.5
6	1	3	33.77	2.654	16	0.2
7	3	0	34.02	2.635	8	0.4
7	1	2	34.31	2.613	16	0.3
5	5	2	34.31	2.613	8	2.3
2	1	5	34.91	2.570	16	0.1
5	4	3	34.96	2.566	16	0.3
5	1	4	35.16	2.552	16	1.4
6	5	1	35.58	2.523	16	0.4
8	0	0	35.80	2.508	4	1.4
3	0	5	36.08	2.490	8	1.6
6	3	3	36.13	2.486	16	0.6
7	3	2	36.64	2.452	16	0.5
8	1	1	36.72	2.447	16	0.3
8	2	0	36.94	2.433	8	0.4
3	2	5	37.21	2.416	16	0.2
7	0	3	37.26	2.413	8	1.3
5	3	4	37.45	2.402	16	1.9
6	0	4	38.00	2.368	8	0.2
7	5	0	38.59	2.333	8	0.2
8	3	1	38.93	2.313	16	0.8
8	2	2	39.39	2.288	16	0.2
0	0	6	40.35	2.235	2	0.2

h	k	l	2θ	d	M	I_{rel}
5	0	5	40.45	2.230	8	0.1
6	5	3	40.49	2.228	16	0.3
7	5	2	40.96	2.203	16	1.0
9	0	1	41.03	2.199	8	0.4
2	0	6	41.38	2.182	8	0.3
5	2	5	41.48	2.177	16	0.2
5	5	4	41.70	2.166	8	0.2
8	4	2	42.49	2.128	16	0.2
3	1	6	42.90	2.108	16	0.4
8	5	1	43.06	2.101	16	0.2
6	1	5	43.48	2.081	16	0.1
8	3	3	43.53	2.079	16	1.3
7	3	4	43.69	2.072	16	0.3
4	0	6	44.37	2.042	8	0.1
3	3	6	44.85	2.021	8	0.4
9	4	1	45.00	2.014	16	0.2
8	0	4	45.14	2.008	8	3.6
10	0	0	45.18	2.007	4	4.7
8	6	0	45.18	2.007	8	0.2
10	1	1	45.95	1.975	16	0.4
8	2	4	46.09	1.969	16	0.9
10	2	0	46.13	1.968	8	0.4
7	0	5	46.36	1.959	8	0.2
10	0	2	47.28	1.922	8	0.3
8	5	3	47.33	1.921	16	1.2
7	5	4	47.48	1.915	16	0.2
10	3	1	47.80	1.903	16	0.2
10	2	2	48.20	1.888	16	0.2
5	3	6	48.57	1.875	16	2.4
8	7	1	48.71	1.869	16	0.5
8	4	4	48.84	1.865	16	0.3
3	0	7	49.49	1.842	8	0.7
8	1	5	49.99	1.824	16	0.2

ZSM-11, Calcined

MEL

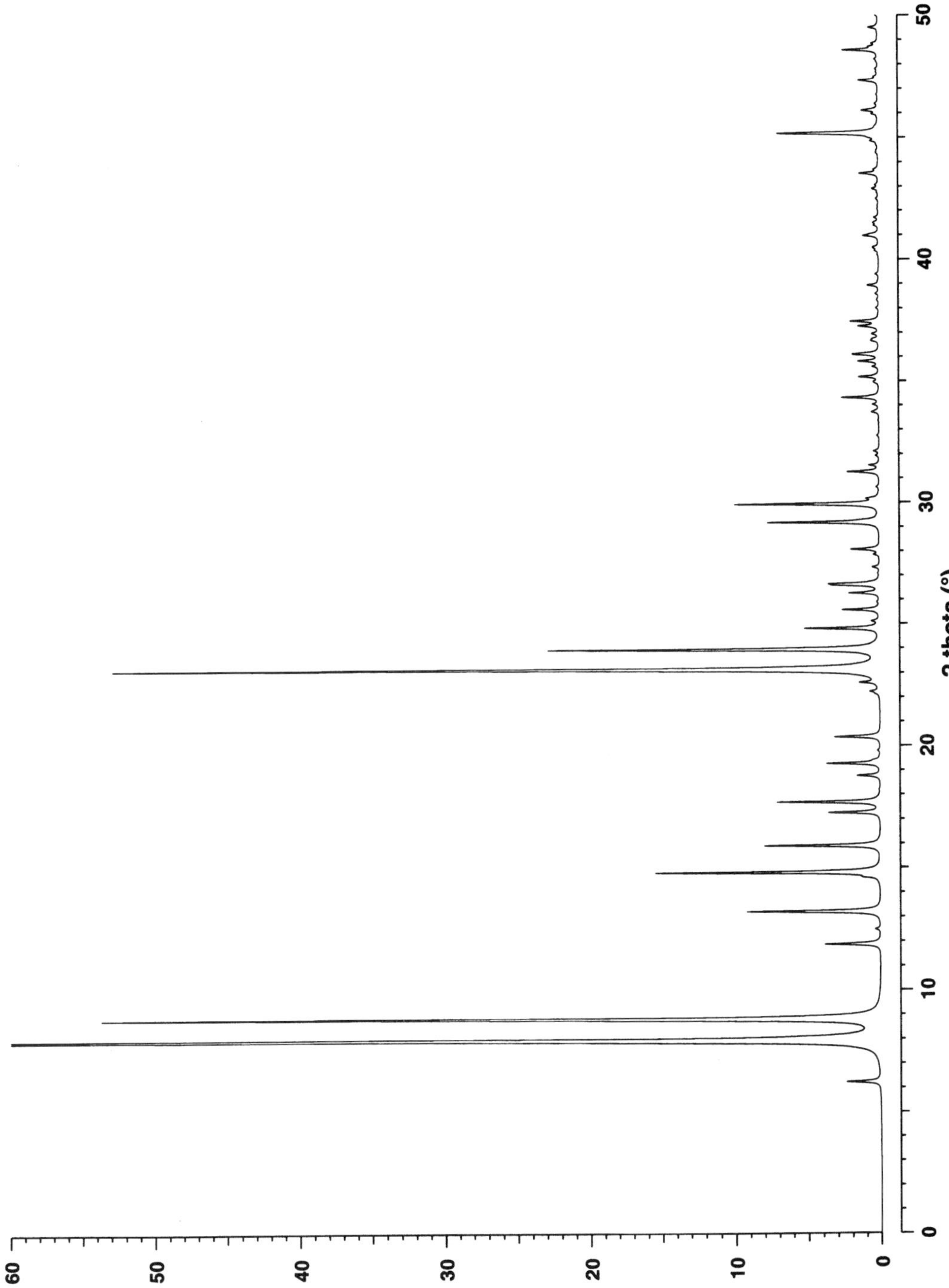

MEP Melanophlogite

CHEMICAL COMPOSITION: $|(CH_4, N_2)_2(CO_2, N_2)_6|$ $[Si_{46}O_{92}]$
Mt. Hamilton, California, U.S.A.

REFINED COMPOSITION: $|C_8|$ $[Si_{46}O_{92}]$

CRYSTAL DATA: $Pm\bar{3}n$ (No. 223)
$a = 13.436$ Å $b = 13.436$ Å $c = 13.436$ Å
$\alpha = 90°$ $\beta = 90°$ $\gamma = 90°$
X-ray single crystal refinement, $R_w = 0.04$

REFERENCE: H. Gies,
Z. Kristallogr. **164** 247–257 (1983).

h	k	l	2θ	d	M	I_{rel}
1	1	0	9.31	9.501	12	9.3
2	0	0	13.18	6.718	6	15.2
2	1	0	14.74	6.009	24	100.0
2	1	1	16.16	5.485	24	75.2
3	1	0	20.91	4.249	24	2.7
2	2	2	22.93	3.879	8	34.7
3	2	0	23.88	3.726	24	21.4
3	2	1	24.79	3.591	48	88.1
4	0	0	26.54	3.359	6	2.2
4	1	0	27.37	3.259	24	24.2
4	1	1	28.18	3.167	24	1.0
3	3	0	28.18	3.167	12	15.3
4	2	0	29.74	3.004	24	0.5
4	2	1	30.49	2.932	48	12.7
3	3	2	31.22	2.865	24	4.6
4	2	2	32.65	2.743	24	2.0
4	3	0	33.34	2.687	24	2.3
5	1	0	34.02	2.635	24	1.1
5	2	0	36.00	2.495	24	0.2
4	3	2	36.00	2.495	48	9.6
5	2	1	36.63	2.453	48	0.2
4	4	0	37.88	2.375	12	1.5
5	3	0	39.09	2.304	24	7.6
4	3	3	39.09	2.304	24	3.1
5	3	1	39.69	2.271	48	2.2
4	4	2	40.27	2.239	24	1.0
6	0	0	40.27	2.239	6	3.7
6	1	0	40.85	2.209	24	1.0
6	1	1	41.43	2.180	24	0.2
5	3	2	41.43	2.180	48	4.3
5	4	0	43.11	2.098	24	0.3
5	4	1	43.66	2.073	48	3.7
6	3	0	45.27	2.003	24	0.3
5	4	2	45.27	2.003	48	2.2
6	3	1	45.80	1.981	48	2.8
4	4	4	46.85	1.939	8	2.9
6	3	2	47.36	1.919	48	1.0
7	1	0	47.87	1.900	24	0.2
5	5	0	47.87	1.900	12	1.9
7	2	0	49.38	1.846	24	0.8
6	4	1	49.38	1.846	48	0.3
5	5	2	49.87	1.828	24	0.1
7	2	1	49.87	1.828	48	0.3
6	3	3	49.87	1.828	24	2.3

Melanophlogite MEP

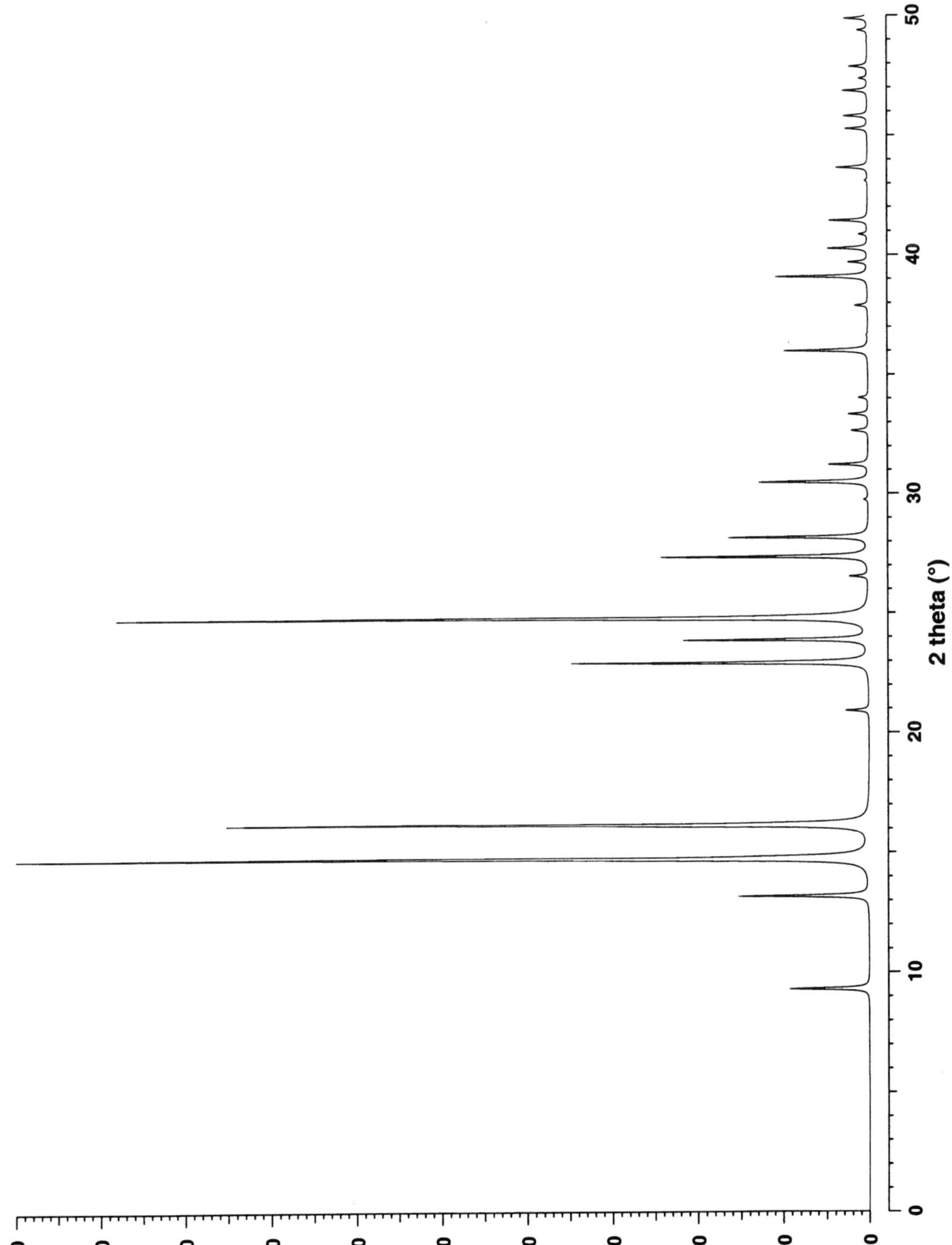

MER Merlinoite

CHEMICAL COMPOSITION: $|Na_{0.55}K_{4.21}Ca_{1.49}Ba_{0.43}(H_2O)_{22.74}|$ $[Si_{22.69}Al_{9.31}O_{64}]$
Cupaello, Italy

REFINED COMPOSITION: $|Na_{0.68}K_{5.28}Ca_{1.844}Ba_{0.444}(H_2O)_{19.44}|$ $[Si_{22.72}Al_{9.28}O_{64}]$

CRYSTAL DATA: *Immm* (No. 71)
$a = 14.116$ Å $b = 14.229$ Å $c = 9.946$ Å
$\alpha = 90°$ $\beta = 90°$ $\gamma = 90°$
X-ray single crystal refinement, $R = 0.09$

REFERENCE: E. Galli, G. Gottardi and D. Pongiluppi,
Neues Jahrb. Miner. Monatsh. **1** 1–9 (1979).

h	k	l	2θ	d	M	I_{rel}	h	k	l	2θ	d	M	I_{rel}	h	k	l	2θ	d	M	I_{rel}
1	1	0	8.82	10.021	4	12.4	4	3	1	32.89	2.723	8	6.5	2	5	3	43.84	2.065	8	1.8
0	1	1	10.85	8.152	4	26.1	0	3	3	32.96	2.717	4	26.1	6	3	1	43.91	2.062	8	4.4
1	0	1	10.88	8.131	4	10.7	5	0	1	32.98	2.716	4	32.3	5	2	3	44.01	2.058	8	1.1
0	2	0	12.44	7.114	2	88.3	3	0	3	33.05	2.710	4	4.7	2	6	2	44.21	2.048	8	2.7
2	0	0	12.54	7.058	2	73.5	2	4	2	33.47	2.677	8	5.3	0	4	4	44.45	2.038	4	0.7
1	2	1	16.56	5.354	8	80.9	4	2	2	33.59	2.668	8	16.4	6	2	2	44.46	2.038	8	1.3
2	1	1	16.61	5.336	8	61.2	2	5	1	35.18	2.551	8	27.4	4	0	4	44.58	2.033	4	0.9
2	2	0	17.70	5.011	4	44.3	5	2	1	35.38	2.537	8	12.1	1	7	0	45.06	2.012	4	0.6
0	0	2	17.84	4.973	2	46.8	2	3	3	35.40	2.536	8	0.3	5	5	0	45.24	2.004	4	6.6
1	3	0	19.75	4.496	4	1.4	3	2	3	35.44	2.533	8	14.8	7	1	0	45.42	1.997	4	2.8
3	1	0	19.87	4.467	4	34.5	4	4	0	35.84	2.505	4	8.2	3	3	4	45.47	1.995	8	0.3
1	1	2	19.93	4.455	8	0.2	0	0	4	36.12	2.487	2	2.9	0	7	1	45.55	1.992	4	0.2
0	3	1	20.75	4.281	4	41.2	3	5	0	36.91	2.435	4	0.7	7	0	1	45.92	1.976	4	7.1
3	0	1	20.88	4.253	4	1.1	1	5	2	36.95	2.433	8	10.2	0	1	5	46.07	1.970	4	4.2
0	2	2	21.80	4.076	4	12.4	5	3	0	37.06	2.426	4	4.9	1	0	5	46.08	1.970	4	0.9
2	0	2	21.86	4.065	4	21.7	5	1	2	37.16	2.419	8	1.0	4	6	0	46.11	1.968	4	0.3
2	3	1	24.32	3.660	8	10.5	1	1	4	37.26	2.413	8	1.3	6	4	0	46.26	1.962	4	1.5
3	2	1	24.38	3.651	8	32.4	1	4	3	37.63	2.390	8	10.4	2	4	4	46.37	1.958	8	4.2
0	4	0	25.03	3.557	2	1.6	4	1	3	37.76	2.382	8	2.5	4	2	4	46.46	1.954	8	0.9
2	2	2	25.23	3.530	8	0.9	0	6	0	37.94	2.372	2	0.2	2	7	1	47.43	1.917	8	1.7
4	0	0	25.24	3.529	2	6.1	6	0	0	38.25	2.353	2	1.8	1	6	3	47.58	1.911	8	2.9
3	3	0	26.69	3.340	4	1.5	0	2	4	38.35	2.347	4	1.0	7	2	1	47.76	1.904	8	0.1
3	1	2	26.83	3.323	8	0.6	2	0	4	38.38	2.345	4	0.2	6	1	3	47.84	1.901	8	0.6
1	4	1	27.37	3.259	8	67.2	1	6	1	39.58	2.277	8	2.2	1	2	5	47.92	1.898	8	1.6
4	1	1	27.54	3.239	8	33.6	4	4	2	40.31	2.237	8	0.6	2	1	5	47.94	1.898	8	2.4
0	1	3	27.63	3.229	4	48.4	2	2	4	40.50	2.227	8	0.1	3	7	0	48.80	1.866	4	0.3
1	0	3	27.64	3.228	4	57.7	3	5	2	41.28	2.187	8	0.2	1	7	2	48.83	1.865	8	1.9
2	4	0	28.09	3.177	4	100.0	5	3	2	41.41	2.180	8	12.9	5	5	2	49.00	1.859	8	0.7
4	2	0	28.23	3.161	4	76.4	1	3	4	41.50	2.176	8	4.2	1	5	4	49.08	1.856	8	4.2
1	2	3	30.41	2.939	8	63.1	3	1	4	41.57	2.173	8	2.5	7	1	2	49.17	1.853	8	3.2
2	1	3	30.44	2.936	8	75.5	4	5	1	41.77	2.162	8	0.6	5	1	4	49.25	1.850	8	1.2
0	4	2	30.91	2.893	4	0.5	0	5	3	41.83	2.159	4	0.3	4	5	3	49.48	1.842	8	7.7
4	0	2	31.07	2.878	4	0.2	3	4	3	41.91	2.156	8	4.8	5	4	3	49.55	1.840	8	0.4
1	5	0	32.08	2.790	4	0.1	4	3	3	41.96	2.153	8	0.5	0	3	5	49.70	1.834	4	3.0
3	3	2	32.28	2.773	8	3.6	5	0	3	42.03	2.149	4	1.5	3	0	5	49.76	1.832	4	0.1
5	1	0	32.33	2.769	4	19.8	0	6	2	42.22	2.141	4	1.8	4	6	2	49.82	1.830	8	0.7
0	5	1	32.73	2.736	4	8.4	6	0	2	42.51	2.127	4	0.2	6	4	2	49.96	1.825	8	1.0
3	4	1	32.82	2.729	8	38.2	3	6	1	43.70	2.071	8	0.7							

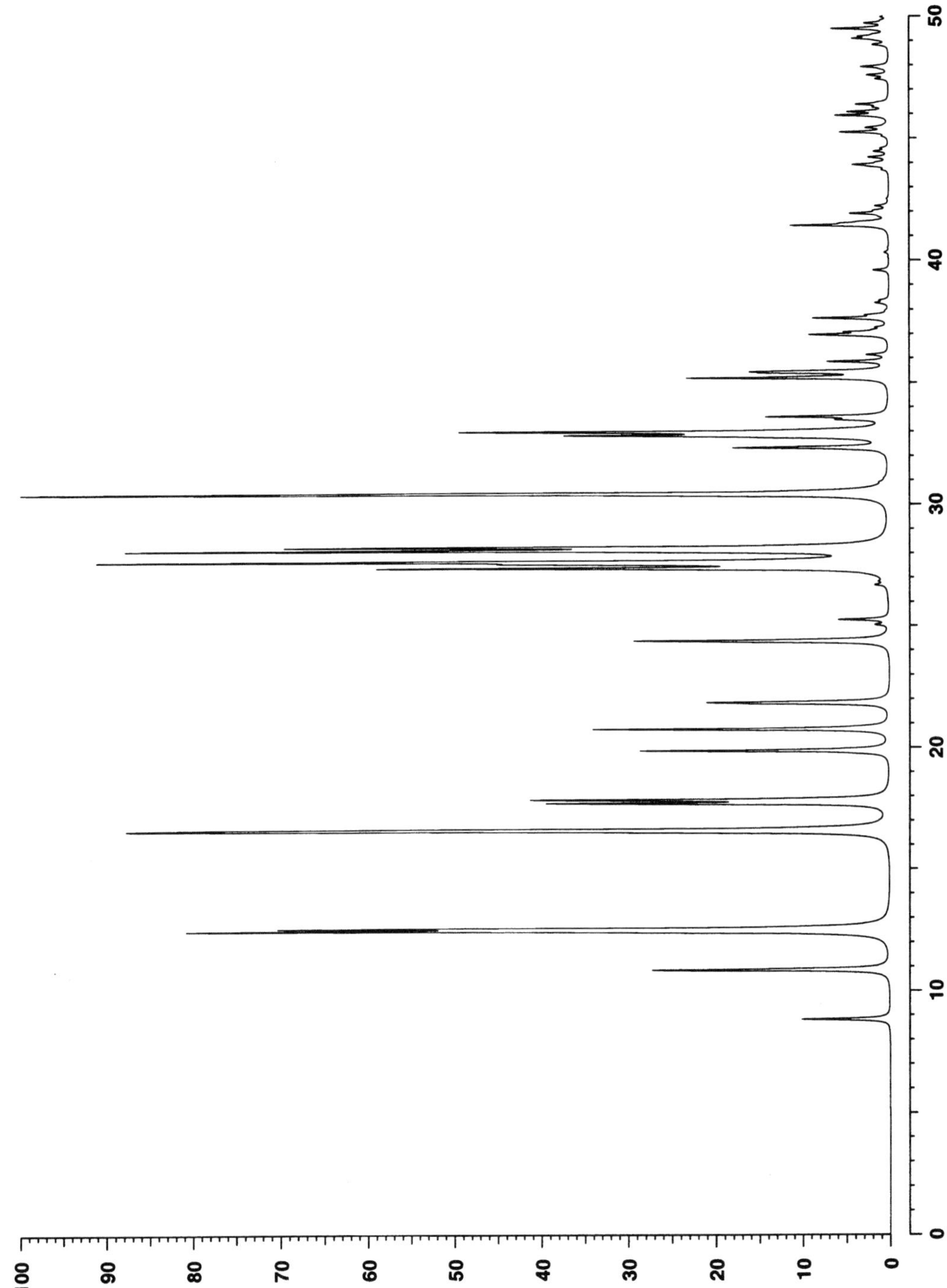
2 theta (°)
0
10
20
30
40
50
0
10
20
30
40
50
60
70
80
90
100

MER Barium Chloroaluminosilicate

CHEMICAL COMPOSITION: $|Ba_4(H_2O)_{12}(BaCl_2)_4|$ $[Si_{21.33}Al_{10.67}O_{64}]$

REFINED COMPOSITION: $|(Ba_3Cl_2O)_4|$ $[Si_{21.34}Al_{10.66}O_{64}]$

CRYSTAL DATA: $I\,4/m\,m\,m$ (No. 139)
$a = 14.194$ Å $b = 14.194$ Å $c = 9.234$ Å
$\alpha = 90°$ $\beta = 90°$ $\gamma = 90°$
X-ray single crystal refinement, $R = 0.14$

REFERENCE: L. P. Solov'eva, S. V. Borisov and V. V. Bakakin, *Sov. Phys. - Crystallogr.* **16** 1035–1038 (1972).

h	k	l	2θ	d	M	I_{rel}
1	1	0	8.81	10.037	4	0.6
1	0	1	11.43	7.740	8	25.9
2	0	0	12.47	7.097	4	0.8
2	1	1	16.95	5.231	16	1.5
2	2	0	17.67	5.018	4	25.8
0	0	2	19.22	4.617	2	0.3
3	1	0	19.78	4.489	8	55.4
3	0	1	21.10	4.211	8	100.0
1	1	2	21.18	4.194	8	36.2
2	0	2	22.98	3.870	8	0.4
3	2	1	24.58	3.621	16	0.2
4	0	0	25.09	3.549	4	5.3
2	2	2	26.23	3.398	8	39.3
3	3	0	26.64	3.346	4	36.3
4	1	1	27.65	3.226	16	35.5
3	1	2	27.72	3.218	16	1.8
4	2	0	28.11	3.174	8	29.6
1	0	3	29.70	3.008	8	21.0
4	0	2	31.80	2.814	8	3.9
5	1	0	32.15	2.784	8	10.9
2	1	3	32.32	2.770	16	79.3

h	k	l	2θ	d	M	I_{rel}
4	3	1	33.01	2.713	16	36.4
5	0	1	33.01	2.713	8	7.7
3	3	2	33.07	2.709	8	23.3
4	2	2	34.28	2.615	16	44.2
3	0	3	34.77	2.580	8	1.2
5	2	1	35.42	2.535	16	37.0
4	4	0	35.79	2.509	4	0.4
5	3	0	36.93	2.434	8	3.8
3	2	3	37.07	2.425	16	0.2
5	1	2	37.73	2.384	16	33.6
6	0	0	38.04	2.366	4	8.3
0	0	4	39.02	2.309	2	6.3
4	1	3	39.26	2.295	16	4.1
6	1	1	39.85	2.262	16	3.3
1	1	4	40.08	2.250	8	1.4
6	2	0	40.18	2.244	8	1.4
4	4	2	40.93	2.205	8	1.2
2	0	4	41.12	2.195	8	3.9
5	4	1	41.91	2.155	16	1.0
5	3	2	41.96	2.153	16	0.3
6	0	2	42.96	2.105	8	3.8

h	k	l	2θ	d	M	I_{rel}
2	2	4	43.13	2.097	8	2.1
5	0	3	43.36	2.087	8	3.3
4	3	3	43.36	2.087	16	4.1
6	3	1	43.90	2.062	16	23.4
3	1	4	44.11	2.053	16	4.4
6	2	2	44.91	2.018	16	0.2
7	1	0	45.17	2.007	8	1.7
5	2	3	45.29	2.002	16	6.3
7	0	1	45.81	1.981	8	1.4
6	4	0	46.11	1.968	8	1.6
4	0	4	46.95	1.935	8	1.2
7	2	1	47.67	1.908	16	0.3
3	3	4	47.87	1.900	8	3.9
4	2	4	48.78	1.867	16	2.9
7	3	0	48.87	1.864	8	0.8
6	1	3	48.98	1.860	16	1.3
5	5	2	49.51	1.841	8	3.8
7	1	2	49.51	1.841	16	0.2
1	0	5	49.79	1.831	8	1.0

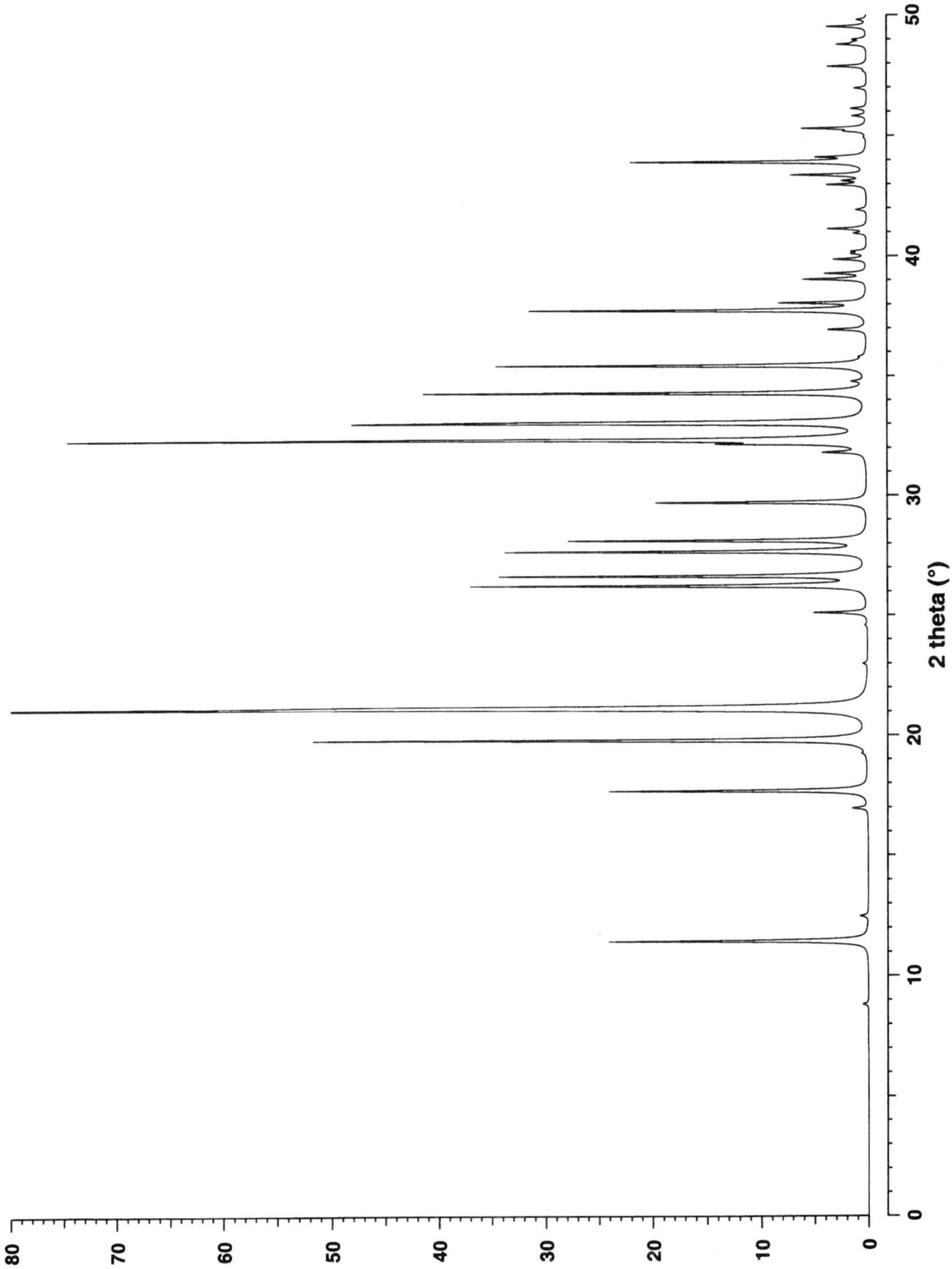
2 theta (°)
0
10
20
30
40
50
80
70
60
50
40
30
20
10
0

MFI Tetrapropylammonium ZSM-5

CHEMICAL COMPOSITION: $|((C_3H_7)_4NOH)_4|$ $[Si_{95.7}Al_{0.3}O_{192}]$
$(C_3H_7)_4NOH$ = tetrapropylammonium hydroxide

REFINED COMPOSITION: $|(C_{12}N)_4|$ $[Si_{96}O_{192}]$

CRYSTAL DATA: *Pnma* (No. 62)
$a = 20.022$ Å $b = 19.899$ Å $c = 13.383$ Å
$\alpha = 90°$ $\beta = 90°$ $\gamma = 90°$
X-ray single crystal refinement, $R_w = 0.044$

REFERENCE: H. van Koningsveld, H. van Bekkum and J. C. Jansen, *Acta Cryst.* **B43** 127–132 (1987).

h	k	l	2θ	d	M	I_{rel}
1	0	1	7.95	11.126	4	68.4
0	1	1	7.96	11.105	4	50.3
2	0	0	8.83	10.011	2	31.3
0	2	0	8.89	9.950	2	36.8
1	1	1	9.11	9.711	8	25.4
2	1	0	9.89	8.943	4	4.7
2	0	1	11.04	8.016	4	1.0
2	1	1	11.90	7.436	8	2.8
1	2	1	11.93	7.417	8	10.4
2	2	0	12.54	7.057	4	5.4
0	0	2	13.23	6.692	2	6.8
1	0	2	13.95	6.346	4	17.4
1	1	2	14.65	6.046	8	8.6
3	0	1	14.83	5.973	4	10.9
0	3	1	14.91	5.943	4	4.1
3	1	1	15.49	5.720	8	0.8
1	3	1	15.55	5.697	8	10.0
2	0	2	15.93	5.563	4	6.2
0	2	2	15.96	5.553	4	5.6
2	3	0	16.03	5.529	4	3.5
2	1	2	16.55	5.358	8	1.8
1	2	2	16.57	5.351	8	1.5
3	2	1	17.32	5.121	8	3.5
2	3	1	17.35	5.110	8	0.5
4	0	0	17.72	5.005	2	2.6
0	4	0	17.83	4.975	2	5.6
3	1	2	19.31	4.598	8	8.4
1	3	2	19.36	4.586	8	2.1
4	1	1	19.45	4.563	8	0.8
4	2	0	19.86	4.472	4	0.5
3	3	1	20.00	4.438	8	2.6
1	0	3	20.40	4.354	4	4.7
0	1	3	20.40	4.353	4	7.1
2	3	2	20.84	4.262	8	3.2
1	1	3	20.88	4.254	8	3.9
4	2	1	20.95	4.241	8	8.9
2	0	3	21.81	4.075	4	3.0
4	3	0	22.25	3.995	4	5.6
0	4	2	22.27	3.992	4	0.7
1	2	3	22.29	3.989	8	1.8
4	1	2	22.63	3.929	8	0.7
1	4	2	22.71	3.915	8	0.5
3	3	2	23.11	3.849	8	4.0
5	0	1	23.18	3.836	4	100.0
4	3	1	23.23	3.829	8	0.8
0	5	1	23.32	3.815	4	78.7
1	5	1	23.74	3.747	8	31.0
3	0	3	23.99	3.709	4	47.7
0	3	3	24.04	3.702	4	4.0
2	5	0	24.06	3.698	4	1.8
1	3	3	24.45	3.640	8	39.7
5	2	1	24.87	3.579	8	1.2
3	2	3	25.63	3.475	8	3.0
2	3	3	25.66	3.472	8	0.6
4	3	2	25.97	3.430	8	13.6
5	1	2	26.32	3.386	8	1.3
1	5	2	26.43	3.372	8	1.6
0	0	4	26.64	3.346	2	1.2
6	0	0	26.71	3.337	2	2.7
4	0	3	26.77	3.330	4	5.9
0	6	0	26.88	3.316	2	2.9
1	0	4	27.02	3.300	4	9.6
1	1	4	27.40	3.256	8	0.7
6	0	1	27.55	3.238	4	0.7
3	3	3	27.55	3.237	8	1.6
2	5	2	27.56	3.237	8	2.1
1	6	1	28.07	3.178	8	0.7
2	0	4	28.12	3.173	4	1.3
4	2	3	28.26	3.158	8	0.7
1	2	4	28.50	3.132	8	2.4
4	5	0	28.66	3.115	4	0.9
3	5	2	29.34	3.044	8	11.9
5	4	1	29.40	3.038	8	0.8
4	5	1	29.44	3.034	8	0.6
6	3	0	29.97	2.981	4	1.2
5	0	3	29.99	2.980	4	7.4
3	4	3	30.05	2.973	8	1.1
0	5	3	30.09	2.970	4	9.1
3	1	4	30.22	2.958	8	0.9
1	5	3	30.43	2.938	8	7.3
5	2	3	31.33	2.855	8	1.8
4	0	4	32.18	2.782	4	0.5
0	4	4	32.24	2.776	4	0.6
6	3	2	32.89	2.723	8	4.5
6	0	3	33.54	2.672	4	1.1
1	0	5	33.78	2.653	4	0.5
5	5	2	34.48	2.601	8	6.4
2	0	5	34.69	2.586	4	0.9
7	3	1	34.81	2.577	8	1.8
4	3	4	34.98	2.565	8	1.7
6	5	0	35.09	2.557	4	1.0
5	1	4	35.24	2.546	8	1.4
1	5	4	35.33	2.540	8	1.0
5	6	1	35.79	2.509	8	0.5
8	0	0	35.88	2.503	2	2.6
0	8	0	36.11	2.487	2	1.9
3	0	5	36.16	2.484	4	1.9
0	3	5	36.19	2.482	4	2.1
2	5	4	36.20	2.481	8	1.0
3	6	3	36.34	2.472	8	0.6
4	7	0	36.34	2.472	4	0.9
8	1	1	36.81	2.442	8	0.9
8	2	0	37.04	2.427	4	1.0
2	8	0	37.25	2.414	4	0.5
7	0	3	37.35	2.408	4	2.7
5	3	4	37.56	2.394	8	0.8
3	5	4	37.62	2.391	8	2.3
6	5	2	37.66	2.389	8	2.1
1	7	3	37.79	2.380	8	1.6
8	0	2	38.40	2.344	4	0.9
4	7	2	38.84	2.319	8	1.6
0	0	6	40.44	2.230	2	0.6
4	8	1	41.08	2.197	8	1.3
3	5	5	42.91	2.107	8	0.6
3	1	6	43.00	2.104	8	0.8
6	0	5	43.33	2.088	4	0.5
8	3	3	43.66	2.073	8	1.5
8	0	4	45.25	2.004	4	5.7
10	0	0	45.29	2.002	2	5.8
0	8	4	45.44	1.996	4	2.6
4	8	3	45.51	1.993	8	3.2
0	10	0	45.59	1.990	2	7.5
1	8	4	45.67	1.986	8	2.4
1	10	1	46.35	1.959	8	0.6
2	8	4	46.38	1.958	8	0.7
4	3	6	46.64	1.948	8	2.9
8	5	3	47.51	1.914	8	1.6
3	8	4	47.54	1.912	8	1.2
0	10	2	47.68	1.907	4	0.6
9	3	3	47.69	1.907	8	1.9
5	3	6	48.70	1.870	8	1.8
3	5	6	48.75	1.868	8	3.4
3	0	7	49.60	1.838	4	0.7
0	3	7	49.62	1.837	4	0.7
8	0	5	49.88	1.828	4	0.6

Tetrapropylammonium ZSM-5 MFI

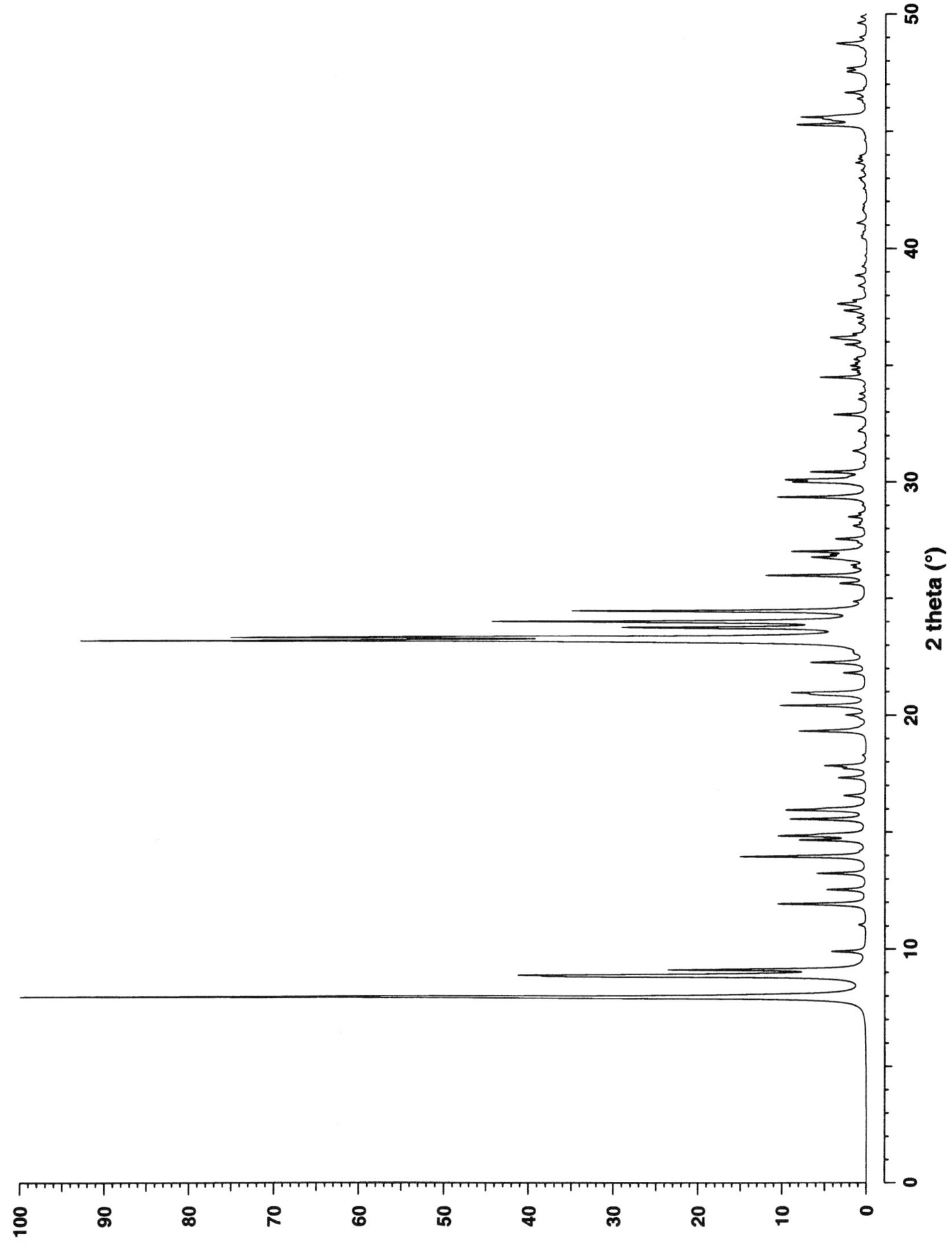

MFI ZSM-5, Calcined

CHEMICAL COMPOSITION: $|H_{0.32}|$ $[Si_{95.68}Al_{0.32}O_{192}]$

REFINED COMPOSITION: $[Si_{96}O_{192}]$

CRYSTAL DATA: $P1\,2_1/n\,1$ (No. 14) unique axis **b**, cell choice 2
a = 19.879 Å b = 20.107 Å c = 13.369 Å
α = 90° β = 90.67° γ = 90°
X-ray single crystal refinement, R_w = 0.045

REFERENCE: H. van Koningsveld, J. C. Jansen and H. van Bekkum, *Zeolites* **10** 235–242 (1990).

h	k	l	2θ	d	M	I_{rel}	h	k	l	2θ	d	M	I_{rel}	h	k	l	2θ	d	M	I_{rel}
-1	0	1	7.93	11.153	2	37.1	-5	0	1	23.27	3.823	2	16.1	5	4	2	31.78	2.816	4	0.4
0	1	1	7.94	11.132	4	100.0	5	0	1	23.42	3.798	2	18.2	-7	0	1	32.15	2.784	2	0.7
1	0	1	8.01	11.033	2	31.9	-5	1	1	23.69	3.756	4	8.1	-3	6	2	32.74	2.735	4	1.6
0	2	0	8.80	10.054	2	47.8	5	1	1	23.84	3.732	4	10.4	3	6	2	32.87	2.724	4	1.2
2	0	0	8.90	9.939	2	51.6	-3	0	3	23.93	3.718	2	1.4	3	3	4	32.99	2.715	4	0.4
-1	1	1	9.07	9.753	4	13.9	0	3	3	23.98	3.711	4	23.0	0	6	3	33.46	2.678	4	1.3
1	1	1	9.14	9.673	4	14.1	5	2	0	24.07	3.697	4	0.9	-1	0	5	33.77	2.654	2	0.9
1	2	0	9.86	8.971	4	1.4	3	0	3	24.20	3.678	2	2.1	0	7	2	33.97	2.639	4	0.4
0	2	1	11.01	8.035	4	1.2	-3	1	3	24.35	3.656	4	7.6	-5	5	2	34.34	2.612	4	1.6
-2	1	1	11.88	7.450	4	1.1	3	1	3	24.61	3.618	4	8.2	5	5	2	34.55	2.596	4	1.1
2	1	1	11.99	7.378	4	0.6	-2	5	1	24.77	3.595	4	1.0	-3	7	1	34.66	2.588	4	0.9
2	2	0	12.52	7.068	4	0.5	2	5	1	24.82	3.587	4	0.5	-5	1	4	35.16	2.552	4	0.6
0	0	2	13.25	6.684	2	7.3	-2	3	3	25.54	3.488	4	1.5	-1	5	4	35.17	2.552	4	0.9
0	1	2	13.96	6.343	4	11.0	-3	2	3	25.54	3.487	4	0.4	0	8	0	35.72	2.513	2	1.6
-1	1	2	14.61	6.062	4	0.7	2	3	3	25.71	3.465	4	0.9	8	0	0	36.15	2.485	2	1.1
0	3	1	14.79	5.991	4	13.7	3	2	3	25.79	3.454	4	1.0	0	3	5	36.17	2.483	4	1.4
-3	0	1	14.85	5.964	2	2.7	-3	4	2	25.85	3.446	4	3.1	7	4	0	36.34	2.472	4	0.7
3	0	1	14.99	5.909	2	4.9	3	4	2	26.02	3.425	4	4.1	-3	1	5	36.36	2.471	4	0.6
-3	1	1	15.50	5.718	4	5.0	-1	5	2	26.22	3.399	4	1.2	3	0	5	36.38	2.469	2	0.4
3	1	1	15.63	5.669	4	5.0	-5	1	2	26.32	3.385	4	0.4	2	8	0	36.89	2.437	4	0.4
-2	0	2	15.89	5.577	2	2.0	5	1	2	26.59	3.352	4	0.7	0	7	3	37.24	2.414	4	1.3
0	2	2	15.92	5.566	4	6.1	0	6	0	26.60	3.351	2	2.8	-5	3	4	37.44	2.402	4	0.5
3	2	0	16.02	5.532	4	1.3	6	0	0	26.91	3.313	2	2.7	-5	6	2	37.50	2.399	4	0.8
2	0	2	16.07	5.517	2	1.4	0	1	4	27.05	3.297	4	5.1	-7	1	3	37.63	2.390	4	0.9
-2	1	2	16.50	5.374	4	1.5	-1	1	4	27.37	3.258	4	0.6	3	5	4	37.65	2.389	4	0.7
2	1	2	16.66	5.320	4	2.0	0	6	1	27.44	3.251	4	1.1	5	3	4	37.83	2.378	4	0.9
-2	3	1	17.24	5.143	4	0.6	-5	2	2	27.44	3.250	4	0.8	-7	4	2	38.70	2.327	4	0.4
2	3	1	17.32	5.119	4	1.1	5	2	2	27.70	3.220	4	0.7	8	4	1	41.15	2.193	4	0.4
0	4	0	17.64	5.027	2	3.6	0	2	4	28.14	3.171	4	0.6	0	7	4	41.45	2.178	4	0.7
4	0	0	17.85	4.969	2	5.4	-2	1	4	28.42	3.140	4	1.1	0	10	0	45.09	2.011	2	3.2
1	4	0	18.20	4.873	4	0.7	2	1	4	28.62	3.118	4	0.4	0	8	4	45.14	2.009	4	3.4
0	4	1	18.86	4.705	4	0.6	-5	3	2	29.22	3.057	4	3.0	-8	0	4	45.22	2.005	2	0.9
-1	3	2	19.24	4.613	4	2.2	5	3	2	29.46	3.032	4	3.3	-8	4	3	45.32	2.001	4	1.1
1	3	2	19.31	4.596	4	1.5	3	6	0	29.88	2.990	4	0.7	10	0	0	45.64	1.988	2	4.6
-1	0	3	20.37	4.359	2	2.2	0	5	3	29.93	2.985	4	6.6	8	4	3	45.73	1.984	4	0.8
0	1	3	20.41	4.350	4	1.9	-5	0	3	29.94	2.984	2	4.0	8	0	4	45.76	1.983	2	1.4
1	0	3	20.48	4.337	2	1.3	-1	3	4	30.17	2.962	4	0.8	-3	4	6	46.50	1.953	4	1.4
-3	2	2	20.74	4.282	4	0.8	1	6	2	30.19	2.960	4	0.4	8	2	4	46.69	1.945	4	0.5
-2	4	1	20.85	4.259	4	2.6	-5	1	3	30.28	2.952	4	2.6	3	4	6	46.80	1.941	4	0.8
2	4	1	20.92	4.246	4	3.4	5	0	3	30.30	2.949	2	1.9	-5	8	3	47.28	1.922	4	0.8
3	2	2	20.94	4.242	4	0.7	5	1	3	30.64	2.918	4	0.9	-3	9	3	47.48	1.915	4	0.6
0	2	3	21.82	4.074	4	2.1	-2	5	3	31.21	2.865	4	1.0	5	8	3	47.53	1.913	4	0.6
3	4	0	22.20	4.005	4	2.8	-5	2	3	31.27	2.861	4	0.4	3	9	3	47.62	1.909	4	0.8
0	5	1	23.10	3.851	4	44.7	2	5	3	31.35	2.853	4	0.6	-5	3	6	48.54	1.875	4	1.4
3	3	2	23.19	3.836	4	1.0	-5	4	2	31.55	2.836	4	0.6	5	3	6	49.02	1.858	4	1.0

ZSM-5, Calcined MFI

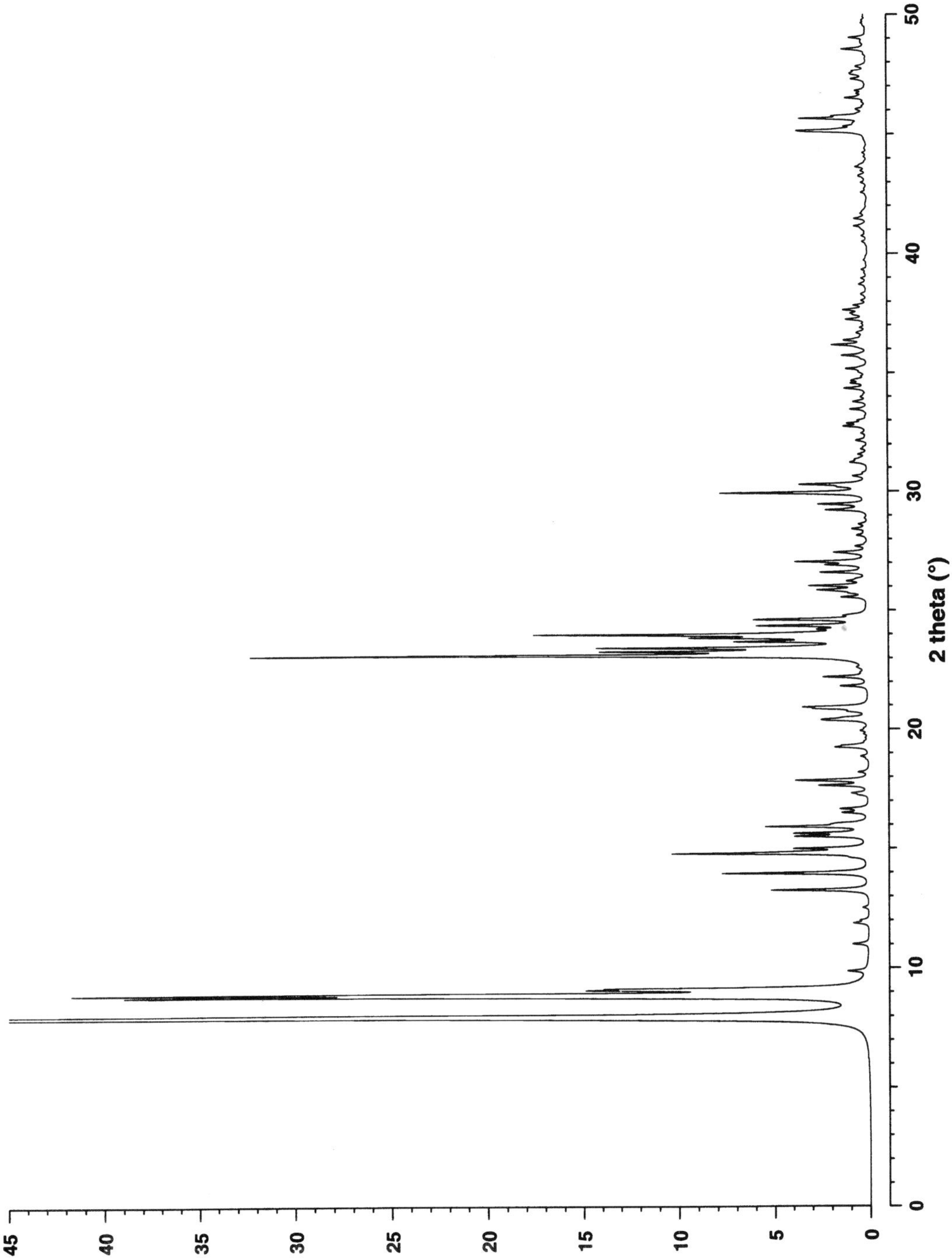

MFS — ZSM-57, SiO_2 Framework

CHEMICAL COMPOSITION: $[Si_{36}O_{72}]$

REFINED COMPOSITION: $[Si_{36}O_{72}]$

CRYSTAL DATA: *Imm*2 (No. 44)
$a = 7.4510$ Å $b = 14.1711$ Å $c = 18.767$ Å
$\alpha = 90°$ $\beta = 90°$ $\gamma = 90°$
DLS refinement.

REFERENCE: J. L. Schlenker, J. B. Higgins and E. W. Valyocsik, *Zeolites* **10** 293–296 (1990).

h	k	l	2θ	d	M	I_{rel}
0	1	1	7.82	11.309	4	100.0
0	0	2	9.42	9.384	2	57.8
0	2	0	12.49	7.086	2	18.3
1	0	1	12.78	6.925	4	37.0
1	1	0	13.43	6.595	4	0.8
0	1	3	15.48	5.723	4	8.7
0	2	2	15.67	5.655	4	11.8
1	1	2	16.43	5.396	8	16.5
1	2	1	17.91	4.953	8	1.4
1	0	3	18.52	4.791	4	11.2
0	0	4	18.91	4.692	2	4.1
0	3	1	19.38	4.581	4	0.5
1	3	0	22.28	3.990	4	1.6
1	2	3	22.40	3.969	8	3.6
0	2	4	22.73	3.912	4	3.8
1	1	4	23.27	3.823	8	11.1
0	3	3	23.60	3.770	4	31.2
2	0	0	23.88	3.726	2	4.7
1	3	2	24.24	3.671	8	0.8
0	1	5	24.53	3.628	4	4.1
0	4	0	25.14	3.543	2	17.5
2	1	1	25.17	3.538	8	2.6
2	0	2	25.73	3.463	4	26.4
1	0	5	26.59	3.352	4	3.3
0	4	2	26.90	3.314	4	3.6
2	2	0	27.04	3.297	4	0.2
1	4	1	28.30	3.154	8	1.4
0	0	6	28.54	3.128	2	2.3
2	1	3	28.59	3.122	8	17.6
2	2	2	28.69	3.111	8	2.5
1	3	4	29.39	3.039	8	3.8
1	2	5	29.48	3.030	8	5.9
0	3	5	30.42	2.939	4	3.9
2	3	1	30.94	2.890	8	0.7
0	2	6	31.26	2.861	4	0.7
1	4	3	31.40	2.849	8	0.8
0	4	4	31.65	2.827	4	0.3
1	1	6	31.66	2.826	8	0.3
0	5	1	31.93	2.802	4	2.7
2	3	3	33.83	2.650	8	1.0
1	5	0	33.84	2.649	4	0.8
2	1	5	34.50	2.599	8	0.3
0	5	3	34.75	2.582	4	0.8
2	4	0	34.95	2.567	4	0.3
1	5	2	35.20	2.549	8	2.1
1	0	7	35.59	2.523	4	0.4
2	4	2	36.28	2.476	8	0.9
3	0	1	36.49	2.462	4	1.3
1	3	6	36.50	2.462	8	1.4
2	0	6	37.55	2.395	4	1.2
1	2	7	37.86	2.377	8	1.6
0	0	8	38.37	2.346	2	0.3
0	4	6	38.39	2.345	4	0.2
3	2	1	38.71	2.326	8	0.2
2	3	5	39.04	2.307	8	3.9
1	5	4	39.05	2.307	8	1.7
0	6	2	39.34	2.290	4	0.1
2	2	6	39.72	2.269	8	0.4
0	5	5	39.86	2.262	4	0.2
2	5	1	40.27	2.240	8	0.4
1	6	1	40.35	2.235	8	0.2
0	2	8	40.51	2.227	4	0.4
3	3	0	41.06	2.198	4	0.2
3	2	3	41.13	2.195	8	0.2
3	1	4	41.63	2.169	8	1.5
3	3	2	42.22	2.140	8	0.5
2	5	3	42.61	2.122	8	0.5
3	0	5	43.70	2.071	4	0.5
0	1	9	43.89	2.063	4	0.2
1	4	7	44.07	2.055	8	0.6
1	3	8	44.82	2.022	8	0.2
1	5	6	44.84	2.021	8	1.5
0	7	1	45.04	2.013	4	0.1
1	0	9	45.15	2.008	4	0.3
3	2	5	45.63	1.988	8	2.2
2	0	8	45.70	1.985	4	1.6
2	4	6	45.72	1.984	8	1.7
0	4	8	46.42	1.956	4	1.9
1	7	0	46.48	1.954	4	0.3
2	6	2	46.54	1.951	8	0.2
0	5	7	46.63	1.948	4	0.1
3	4	3	46.98	1.934	8	0.3
2	5	5	47.00	1.933	8	0.2
1	2	9	47.03	1.932	8	1.2
1	6	5	47.07	1.931	8	0.2
3	1	6	47.16	1.927	8	1.2
0	7	3	47.19	1.926	4	5.8
1	7	2	47.54	1.913	8	0.4
0	3	9	47.67	1.908	4	2.2
3	5	0	48.75	1.868	4	0.1
4	0	0	48.89	1.863	2	3.9
4	1	1	49.60	1.838	8	0.5
3	5	2	49.77	1.832	8	0.3
4	0	2	49.91	1.827	4	0.4

ZSM-57, SiO_2 Framework **MFS**

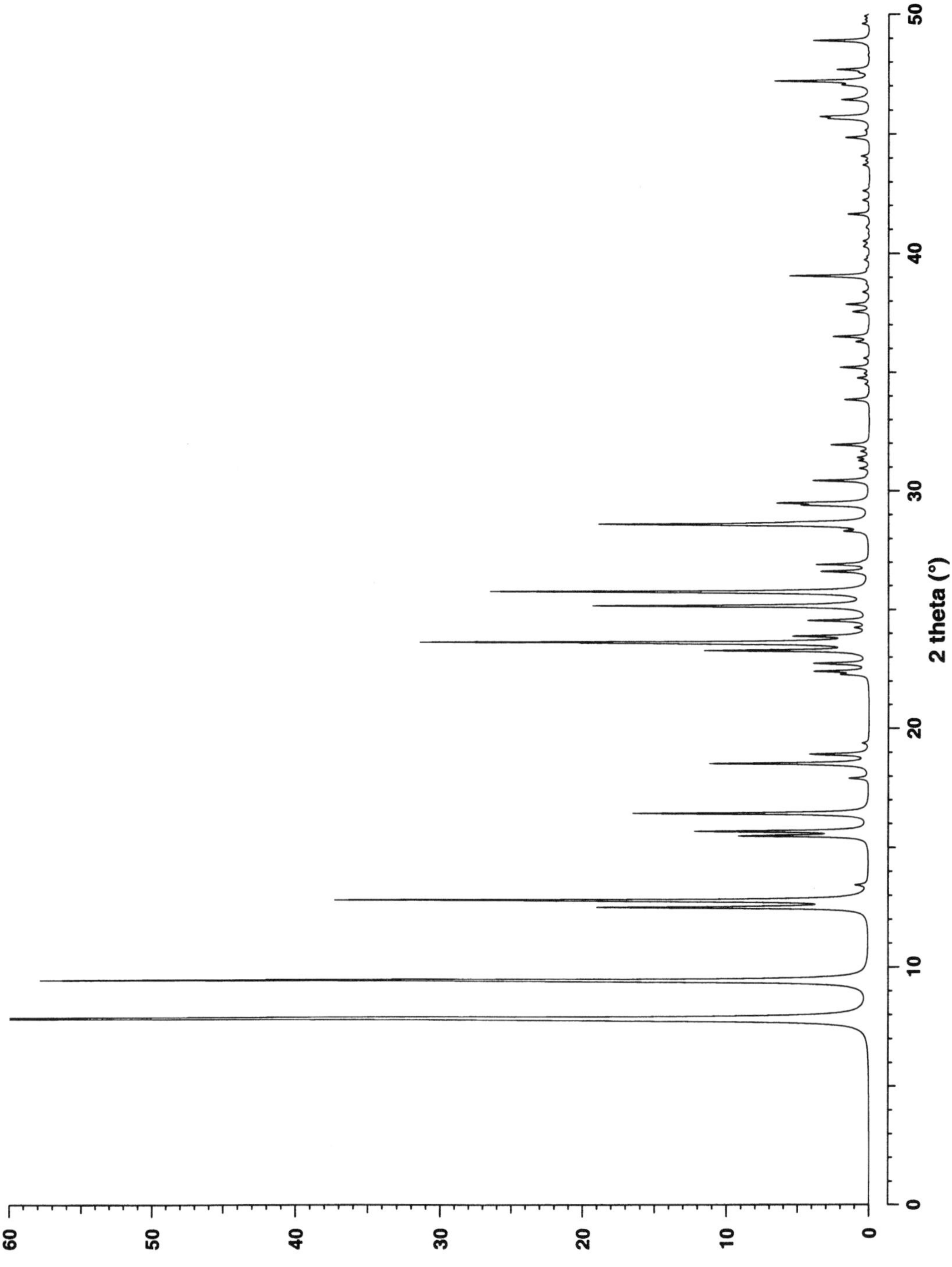

MON Montesommaite

CHEMICAL COMPOSITION: $|Na_{0.1}K_{4.3}(H_2O)_5|$ $[Si_{11.3}Al_{4.7}O_{32}]$
Pollena, Mt. Somma-Vesuvius, Italy

REFINED COMPOSITION: $|K_{4.24}(H_2O)_4|$ $[Si_{11.3}Al_{4.7}O_{32}]$

CRYSTAL DATA: $I\,4_1/amd$ (No. 141) origin at centre $(2/m)$
$a = 7.141$ Å $b = 7.141$ Å $c = 17.307$ Å
$\alpha = 90°$ $\beta = 90°$ $\gamma = 90°$
X-ray powder refinement of idealized substructure, $R = 0.10$

REFERENCE: R. C. Rouse, P. J. Dunn, J. D. Grice, J. L. Schlenker and J. B. Higgins, *American Mineralogist* **75** 1415–1420 (1990).

h	k	l	2θ	d	M	I_{rel}
1	0	1	13.41	6.601	8	81.3
1	1	2	20.36	4.361	8	13.5
0	0	4	20.53	4.327	2	43.9
2	0	0	24.94	3.570	4	0.5
2	0	2	27.01	3.301	8	100.0
2	1	1	28.42	3.141	16	57.0
1	0	5	28.66	3.115	8	37.4
2	1	3	32.03	2.794	16	20.7
2	2	0	35.56	2.525	4	13.5
1	1	6	35.85	2.505	8	1.4
3	0	1	38.16	2.358	8	0.3
2	1	5	38.35	2.347	16	17.5
1	0	7	38.53	2.336	8	4.8
2	0	6	40.19	2.244	8	2.3
3	1	2	41.32	2.185	16	1.2
2	2	4	41.41	2.181	8	11.1
0	0	8	41.75	2.163	2	8.8
3	1	4	45.30	2.002	16	4.3
3	0	5	46.29	1.961	8	0.1
2	1	7	46.45	1.955	16	0.2
3	2	3	48.60	1.873	16	3.8
1	0	9	49.06	1.857	8	3.7
2	0	8	49.25	1.850	8	0.3

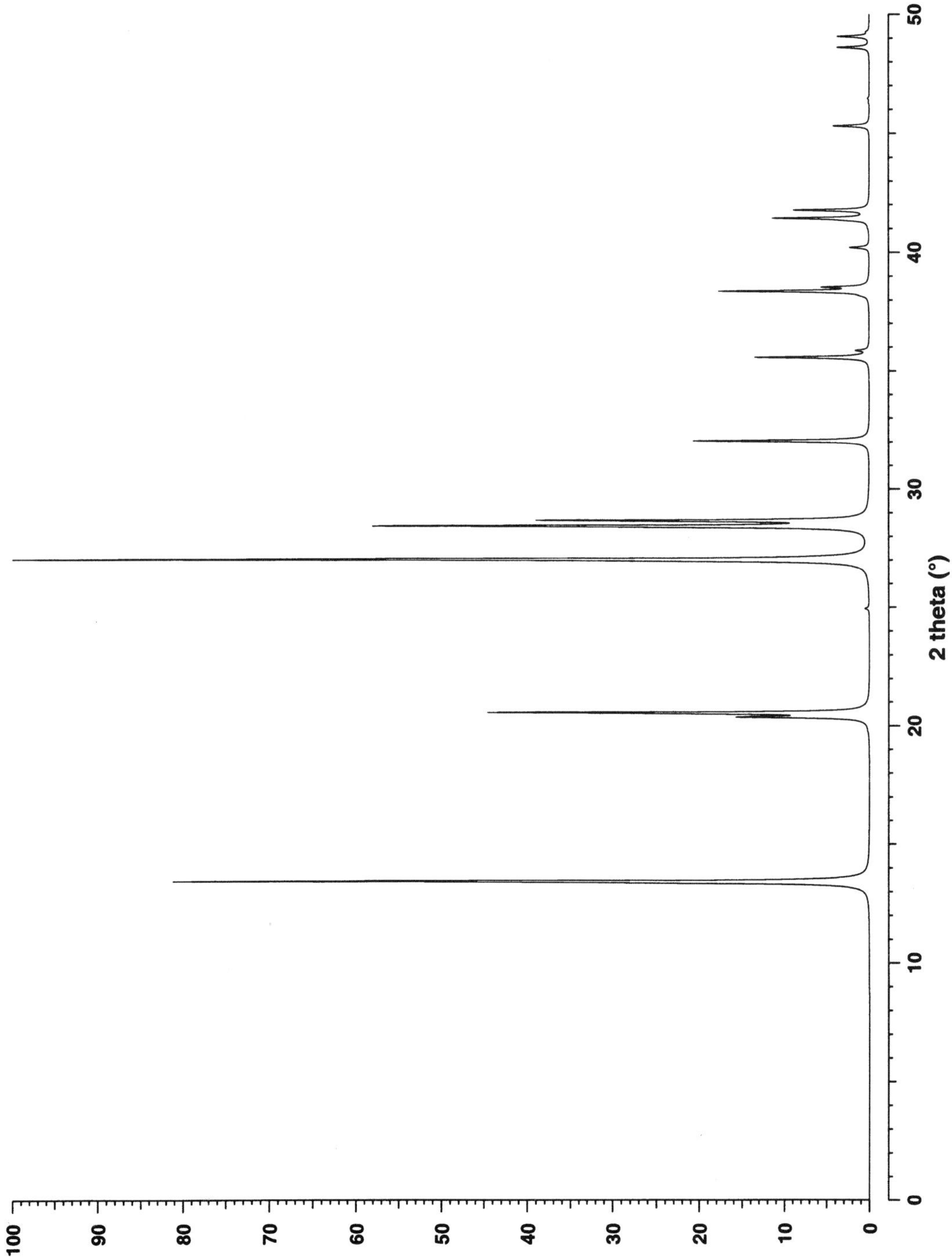
2 theta (°)
0
10
20
30
40
50
100
90
80
70
60
50
40
30
20
10
0

MOR Mordenite

CHEMICAL COMPOSITION: $|Na_8(H_2O)_{24}|$ $[Si_{40}Al_8O_{96}]$
Challis, Idaho, U.S.A.

REFINED COMPOSITION: $|Na_8(H_2O)_{25}|$ $[Si_{40}Al_8O_{96}]$

CRYSTAL DATA: *Cmcm* (No. 63)
$a = 18.11$ Å $b = 20.53$ Å $c = 7.528$ Å
$\alpha = 90°$ $\beta = 90°$ $\gamma = 90°$
X-ray single crystal refinement, $R = 0.07$

REFERENCE: V. Gramlich,
PhD dissertation, ETH, Zurich, (1971).
And V. Gramlich, Private communication.

h	k	l	2θ	d	M	I_{rel}
1	1	0	6.51	13.581	4	100.0
0	2	0	8.61	10.265	2	13.1
2	0	0	9.77	9.055	2	56.4
2	2	0	13.04	6.791	4	0.3
1	1	1	13.45	6.584	8	40.4
1	3	0	13.83	6.402	4	29.1
0	2	1	14.59	6.071	4	13.3
3	1	0	15.30	5.791	4	9.5
0	4	0	17.28	5.133	2	0.6
2	2	1	17.59	5.042	8	2.3
1	3	1	18.19	4.877	8	4.4
3	1	1	19.34	4.590	8	3.2
4	0	0	19.61	4.528	2	1.4
3	3	0	19.61	4.527	4	22.7
2	4	0	19.88	4.465	4	0.6
0	4	1	20.95	4.241	4	1.4
4	2	0	21.45	4.142	4	5.7
1	5	0	22.20	4.004	4	46.1
3	3	1	22.92	3.880	8	1.0
2	4	1	23.16	3.840	8	16.9
0	0	2	23.64	3.764	2	6.9
4	2	1	24.53	3.629	8	4.5
1	1	2	24.54	3.627	8	0.3
5	1	0	24.96	3.567	4	2.3
1	5	1	25.19	3.535	8	0.3
0	2	2	25.20	3.534	4	4.3
2	0	2	25.63	3.476	4	75.7
0	6	0	26.04	3.422	2	5.0
3	5	0	26.25	3.395	4	43.5
2	2	2	27.09	3.292	8	4.7
1	3	2	27.49	3.245	8	12.6
5	1	1	27.67	3.223	8	46.1
5	3	0	27.87	3.201	4	28.8
2	6	0	27.87	3.201	4	0.1
3	1	2	28.28	3.156	8	4.0
0	6	1	28.66	3.115	4	2.7
4	4	1	28.85	3.095	8	3.3
3	5	1	28.85	3.095	8	0.5
0	4	2	29.43	3.035	4	0.2
5	3	1	30.34	2.946	8	0.3
2	6	1	30.34	2.946	8	5.5
6	2	0	30.88	2.896	4	2.0
4	0	2	30.89	2.894	4	12.9
3	3	2	30.89	2.894	8	8.8
2	4	2	31.08	2.878	8	0.2
1	5	2	32.65	2.743	8	3.2
5	5	0	32.98	2.716	4	0.6
6	2	1	33.15	2.703	8	2.8
1	7	1	33.15	2.702	8	6.6
3	7	0	33.98	2.638	4	1.6
5	1	2	34.65	2.589	8	0.6
7	1	0	34.95	2.567	4	0.2
0	8	0	34.96	2.566	2	5.4
5	5	1	35.12	2.555	8	0.5
4	4	2	35.61	2.521	8	0.6
3	5	2	35.61	2.521	8	15.5
2	8	0	36.39	2.469	4	1.3
1	1	3	36.41	2.468	8	1.2
6	4	1	36.54	2.459	8	5.3
5	3	2	36.86	2.439	8	1.2
2	6	2	36.86	2.438	8	1.8
0	2	3	36.87	2.438	4	0.2
7	1	1	37.00	2.429	8	0.4
0	8	1	37.01	2.429	4	0.5
7	3	0	37.15	2.420	4	0.1
6	0	2	38.22	2.355	4	0.1
2	2	3	38.24	2.354	8	0.4
1	3	3	38.53	2.336	8	0.7
7	3	1	39.10	2.304	8	0.2
6	2	2	39.25	2.295	8	0.4
5	7	0	39.54	2.279	4	1.1
8	0	0	39.82	2.264	2	0.1
0	4	3	39.99	2.254	4	0.1
4	8	0	40.40	2.233	4	2.2
8	2	0	40.82	2.211	4	0.4
7	5	0	41.24	2.189	4	0.3
2	4	3	41.27	2.188	8	2.2
6	6	1	41.67	2.168	8	1.4
3	7	2	41.81	2.160	8	1.8
4	2	3	42.10	2.146	8	0.5
6	4	2	42.22	2.140	8	0.2
3	9	0	42.36	2.134	4	0.3
0	8	2	42.64	2.120	4	0.5
2	8	2	43.85	2.064	8	0.5
0	10	0	44.11	2.053	2	5.3
3	9	1	44.11	2.053	8	0.8
5	1	3	44.13	2.052	8	0.6
7	3	2	44.51	2.036	8	3.3
0	6	3	44.79	2.024	4	0.3
4	4	3	44.92	2.018	8	4.9
9	1	0	45.28	2.003	4	0.5
8	4	1	45.41	1.997	8	2.9
5	3	3	45.95	1.975	8	0.8
2	6	3	45.95	1.975	8	0.7
6	8	0	46.44	1.955	4	7.5
5	7	2	46.58	1.950	8	0.6
9	1	1	46.95	1.935	8	1.8
2	10	1	46.96	1.935	8	0.2
9	3	0	47.07	1.930	4	0.9
5	9	0	47.08	1.930	4	0.2
4	8	2	47.34	1.920	8	5.2
6	2	3	47.97	1.896	8	0.6
1	7	3	47.98	1.896	8	1.1
7	5	2	48.08	1.892	8	0.7
0	0	4	48.36	1.882	2	9.4
7	7	1	48.45	1.879	8	0.5
4	10	0	48.70	1.870	4	0.1
5	9	1	48.70	1.870	8	2.5
1	1	4	48.85	1.864	8	1.2
1	11	0	49.07	1.857	4	0.5
3	9	2	49.07	1.856	8	0.5
4	6	3	49.33	1.847	8	0.4
5	5	3	49.45	1.843	8	0.1
2	0	4	49.46	1.843	4	0.3
8	6	1	49.79	1.831	8	0.1

Mordenite MOR

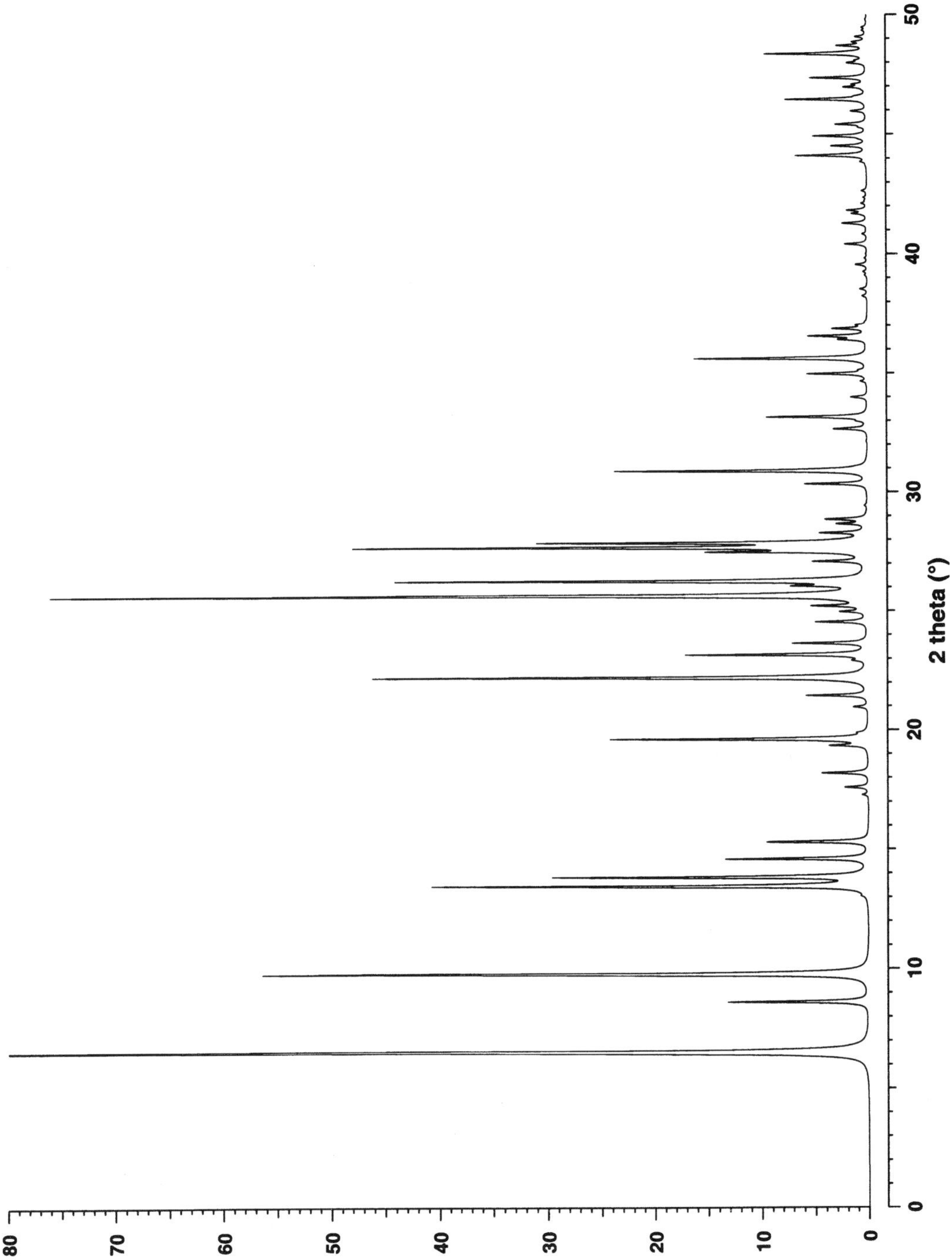

MOR Maricopaite

CHEMICAL COMPOSITION: $|Pb_7Ca_2(H_2O)_{28}|\ [Si_{36}Al_{12}(O,OH)_{100}]$
Moon Anchor Mine, Tonopah, AZ, U.S.A.

REFINED COMPOSITION: $|Pb_{10.68}(H_2O)_{26.5}O_4|\ [Si_{36.38}Al_{11.02}O_{100}]$

CRYSTAL DATA: $Cm2m$ (No. 38) **bca** setting
$a = 19.432$ Å $b = 19.702$ Å $c = 7.538$ Å
$\alpha = 90°$ $\beta = 90°$ $\gamma = 90°$
X-ray single crystal refinement, $R_w = 0.051$

REFERENCE: R. C. Rouse and D. R. Peacor,
American Mineralogist **79** 175–184 (1994).

h	k	l	2θ	d	M	I_{rel}
1	1	0	6.39	13.835	4	100.0
0	2	0	8.98	9.851	2	4.0
2	0	0	9.10	9.716	2	0.3
0	0	1	11.74	7.538	2	1.0
2	2	0	12.80	6.917	4	2.3
1	1	1	13.38	6.619	8	1.8
1	3	0	14.24	6.222	4	0.9
3	1	0	14.39	6.153	4	1.9
0	2	1	14.80	5.986	4	1.7
2	2	1	17.40	5.097	8	0.4
1	3	1	18.49	4.798	8	4.6
3	1	1	18.61	4.767	8	0.6
3	3	0	19.25	4.612	4	0.5
2	4	0	20.21	4.393	4	0.6
4	2	0	20.38	4.357	4	4.3
0	4	1	21.55	4.123	4	1.9
4	0	1	21.76	4.083	4	0.3
3	3	1	22.60	3.934	8	0.6
1	5	0	23.03	3.862	4	1.9
5	1	0	23.33	3.813	4	0.7
2	4	1	23.44	3.796	8	0.1
4	2	1	23.58	3.772	8	0.2
0	0	2	23.60	3.769	2	0.1
1	1	2	24.48	3.636	8	0.9
0	2	2	25.30	3.520	4	1.0
2	0	2	25.35	3.514	4	2.0
4	4	0	25.76	3.459	4	1.5
1	5	1	25.92	3.437	8	0.6
5	1	1	26.19	3.402	8	0.5
3	5	0	26.48	3.366	4	3.3
5	3	0	26.65	3.345	4	3.3
2	2	2	26.94	3.310	8	1.9
0	6	0	27.16	3.284	2	4.1
1	3	2	27.67	3.224	8	4.2
3	1	2	27.76	3.214	8	5.8
4	4	1	28.39	3.144	8	0.2
2	6	0	28.70	3.111	4	0.8
6	2	0	29.02	3.077	4	0.4
5	3	1	29.21	3.057	8	2.9
0	6	1	29.67	3.010	4	1.3
0	4	2	29.85	2.993	4	2.1
4	0	2	30.01	2.978	4	5.6
6	0	1	30.03	2.976	4	0.2
2	6	1	31.10	2.876	8	0.2
2	4	2	31.27	2.861	8	0.8
4	2	2	31.38	2.850	8	3.6
6	2	1	31.40	2.849	8	2.0
1	7	0	32.13	2.786	4	0.6
5	5	0	32.35	2.767	4	0.3
7	1	0	32.57	2.749	4	0.4
4	6	0	32.92	2.720	4	0.2
6	4	0	33.10	2.706	4	1.1
1	5	2	33.21	2.697	8	1.1
5	1	2	33.43	2.680	8	3.5
1	7	1	34.32	2.613	8	0.3
5	5	1	34.53	2.598	8	0.5
4	6	1	35.07	2.559	8	0.9
7	3	0	35.09	2.557	4	0.8
4	4	2	35.22	2.548	8	0.1
6	4	1	35.24	2.547	8	0.8
0	0	3	35.73	2.513	2	0.1
3	5	2	35.76	2.511	8	0.2
5	3	2	35.90	2.502	8	0.3
0	6	2	36.28	2.476	4	0.2
1	1	3	36.34	2.472	8	0.3
0	8	0	36.48	2.463	2	0.4
6	0	2	36.58	2.456	4	2.5
0	2	3	36.92	2.435	4	0.3
8	0	0	37.01	2.429	2	0.2
7	3	1	37.13	2.421	8	0.7
2	6	2	37.49	2.399	8	0.7
2	8	0	37.68	2.387	4	0.3
6	2	2	37.74	2.383	8	0.5
8	2	0	38.16	2.358	4	0.6
0	8	1	38.45	2.341	4	0.2
1	3	3	38.64	2.330	8	0.3
6	6	0	39.06	2.306	4	0.4
2	8	1	39.60	2.276	8	0.1
8	2	1	40.06	2.251	8	0.2
5	5	2	40.44	2.230	8	0.8
7	1	2	40.62	2.221	8	1.7
3	3	3	40.90	2.206	8	0.2
4	6	2	40.91	2.206	8	0.9
6	6	1	40.93	2.205	8	0.3
6	4	2	41.06	2.198	8	0.2
4	8	0	41.09	2.197	4	0.3
5	7	1	41.38	2.182	8	0.6
8	4	0	41.45	2.179	4	0.1
4	2	3	41.48	2.177	8	0.4
1	9	0	41.51	2.175	4	0.2
9	1	0	42.10	2.146	4	0.7
3	7	2	42.44	2.130	8	0.3
5	1	3	43.11	2.098	8	0.4
1	9	1	43.29	2.090	8	0.3
0	8	2	43.92	2.062	4	0.2
4	4	3	44.57	2.033	8	0.4
2	8	2	44.95	2.017	8	0.2
3	5	3	45.02	2.014	8	0.1
8	2	2	45.36	1.999	8	0.3
0	6	3	45.45	1.995	4	0.3
9	3	1	45.85	1.979	8	0.7
6	6	2	46.15	1.967	8	0.2
2	6	3	46.46	1.955	8	0.1
8	6	0	46.50	1.953	4	0.3
5	7	2	46.56	1.951	8	0.5
7	5	2	46.72	1.944	8	0.1
10	0	0	46.75	1.943	2	0.5
2	10	0	47.06	1.931	4	0.1
7	7	1	47.56	1.912	8	0.1
5	9	0	47.68	1.907	4	1.0
10	2	0	47.70	1.906	4	0.1
4	8	2	47.93	1.898	8	0.3
6	8	1	47.95	1.897	8	0.8
8	4	2	48.25	1.886	8	0.4
0	0	4	48.29	1.885	2	1.6
1	9	2	48.31	1.884	8	1.1
1	1	4	48.77	1.867	8	1.1
1	7	3	48.81	1.866	8	0.4
5	5	3	48.97	1.860	8	0.1
0	2	4	49.23	1.851	4	0.2
5	9	1	49.28	1.849	8	0.2
10	2	1	49.30	1.848	8	0.3
4	6	3	49.37	1.846	8	0.1
6	4	3	49.50	1.841	8	0.3
4	10	0	49.95	1.826	4	0.3

Maricopaite MOR

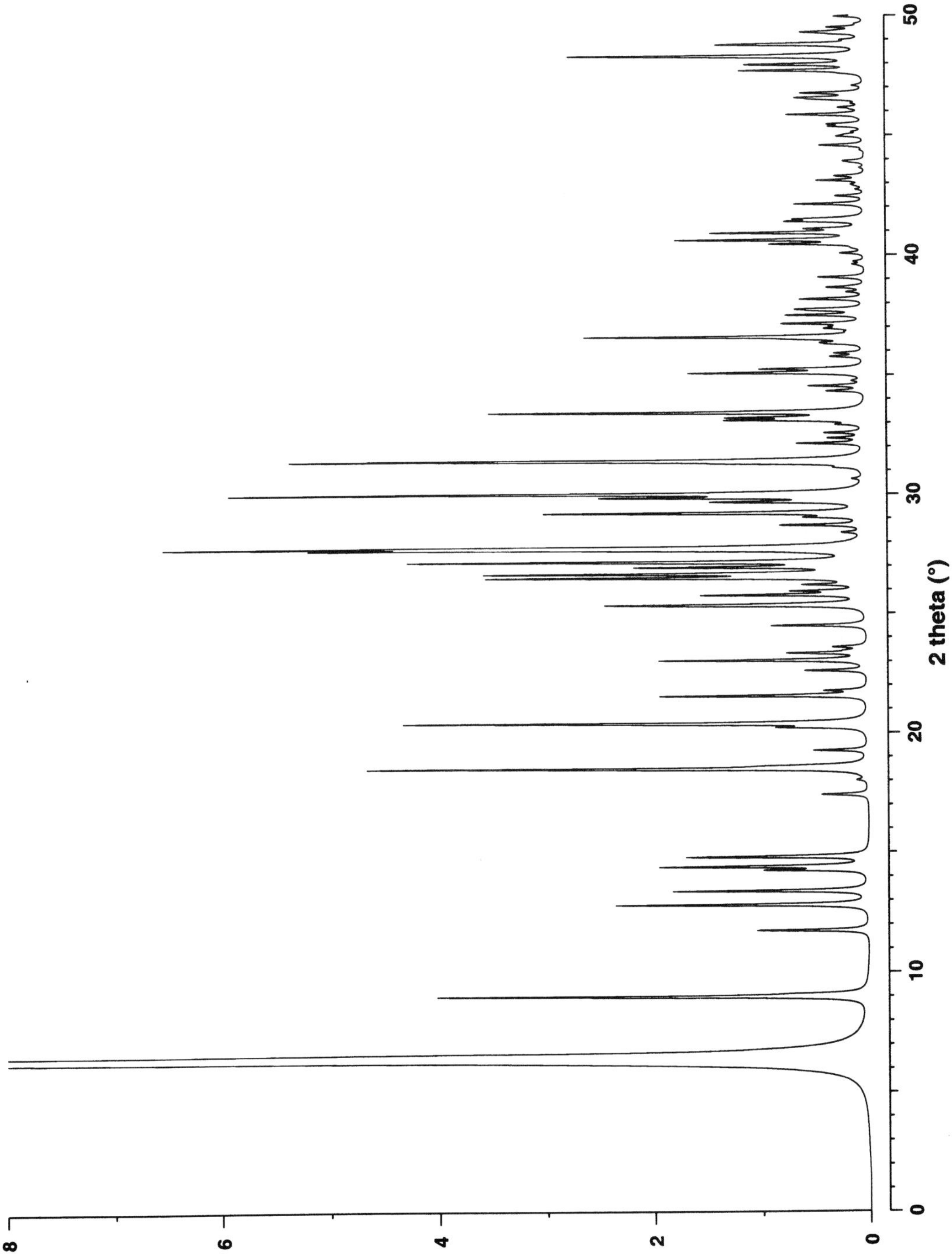

MSO MCM-61

CHEMICAL COMPOSITION: $|(C_{12}H_{24}O_6)_{36}|$ $[Si_{33.48}Al_{2.52}O_{78}]$
$C_{12}H_{24}O_6$ = 1,4,7,10,13,16-hexaoxacyclooctadecane (18-crown-6 ether)

REFINED COMPOSITION: $|K_{9.38}C_{12}H_{48}|$ $[Si_{36}O_{78}]$

CRYSTAL DATA: $R\bar{3}m$ (No. 166) rhombohedral setting
a = 11.841 Å b = 11.841 Å c = 11.841 Å
α = 93.29° β = 93.29° γ = 93.29°
X-ray Rietveld refinement, R_p = 0.0533, R_{wp} = 0.0704

REFERENCE: D. F. Shantz, A. Burton and R. F. Lobo,
Microporous and Mesoporous Materials **31** 61–73 (1999).

h	k	l	2θ	d	M	I_{rel}
1	0	0	7.49	11.800	6	100.0
1	0	-1	10.27	8.610	6	31.6
1	1	0	10.92	8.101	6	0.8
1	1	-1	12.73	6.955	6	12.2
1	1	1	13.77	6.432	2	13.9
2	0	0	15.02	5.900	6	0.1
2	0	-1	16.38	5.410	12	2.2
2	1	0	17.21	5.153	12	3.6
2	1	-1	18.23	4.867	12	0.6
2	1	1	19.34	4.590	6	10.9
2	0	-2	20.63	4.305	6	0.3
2	2	0	21.94	4.050	6	1.3
2	1	-2	21.98	4.044	12	0.2
3	0	0	22.61	3.933	6	1.1
2	2	-1	22.61	3.933	6	0.4
2	2	1	23.82	3.736	6	1.1
3	-1	-1	24.32	3.660	6	0.8
3	1	-1	24.89	3.578	12	0.7
2	2	-2	25.62	3.478	6	0.4
3	1	1	26.00	3.427	6	0.3
3	0	-2	26.46	3.369	12	0.7
3	-1	-2	27.41	3.254	12	0.2
3	1	-2	27.66	3.224	12	0.4
3	2	0	28.02	3.184	12	0.7
4	0	0	30.30	2.950	6	0.5
2	2	-3	30.32	2.948	6	0.2
3	2	-2	30.79	2.904	12	0.7
4	1	0	31.71	2.822	12	0.3
4	1	-1	32.07	2.791	12	0.1
3	2	2	33.05	2.711	6	0.2
3	3	1	34.69	2.586	6	0.2
3	2	-3	34.76	2.581	12	0.1
4	2	0	34.82	2.576	12	0.1
4	2	1	36.28	2.476	12	0.3
3	3	2	37.78	2.381	6	0.1
5	-1	-1	38.84	2.318	6	0.2
3	3	-3	38.84	2.318	6	0.2
4	3	0	39.28	2.294	12	0.1
5	0	-2	40.29	2.239	12	0.2
4	3	1	40.69	2.218	12	0.2
5	1	1	40.71	2.217	6	0.2
5	-1	-2	40.76	2.213	12	0.2
3	3	3	42.15	2.144	2	0.1
5	2	1	43.43	2.084	12	0.1
4	3	2	43.50	2.080	12	0.3
4	3	-3	44.02	2.057	12	0.2
4	4	0	44.75	2.025	6	0.1
4	4	1	46.11	1.969	6	0.2
6	0	0	46.16	1.967	6	0.2
6	1	-1	47.42	1.917	12	0.1
4	3	3	47.51	1.914	6	0.3
6	0	-2	47.86	1.901	12	0.1

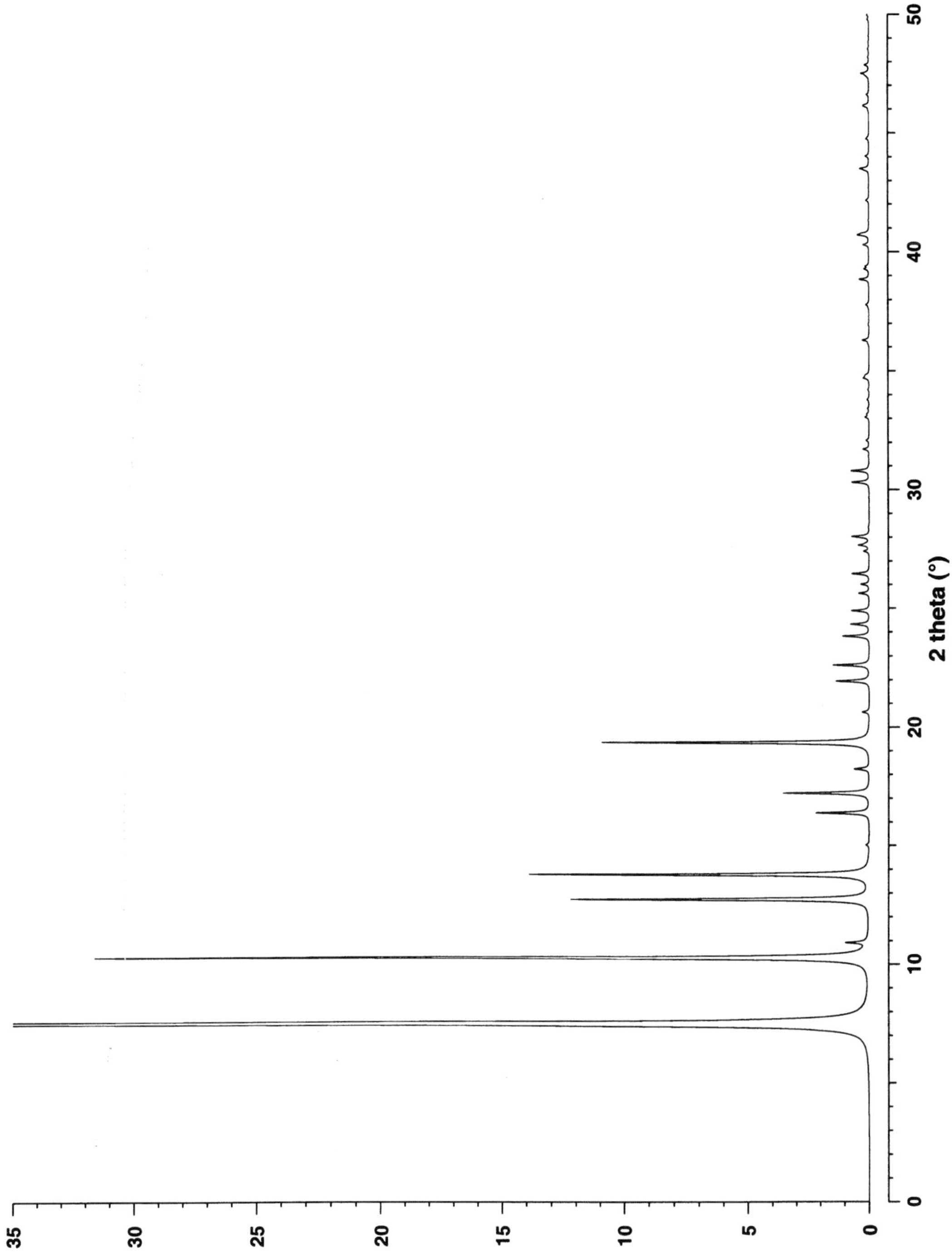
2 theta (°)
0
10
20
30
40
50
35
30
25
20
15
10
5
0

MTF MCM-35

CHEMICAL COMPOSITION: $[Si_{44}O_{88}]$

REFINED COMPOSITION: $[Si_{44}O_{88}]$

CRYSTAL DATA: $C1\,2/m\,1$ (No. 12) unique axis **b**, cell choice 1
$a = 9.5000$ Å $b = 30.7096$ Å $c = 7.3133$ Å
$\alpha = 90.0°$ $\beta = 91.7113°$ $\gamma = 90.0°$
X-ray synchrotron Rietveld refinement, $R_p = 0.0729$, $R_{wp} = 0.0916$, $R_b = 0.0302$

REFERENCE: P. A. Barrett, M.-J. Diaz-Cabanas and M. A. Camblor, *Chemistry of Materials* **11** 2919–2927 (1999).

h	k	l	2θ	d	M	I_{rel}	h	k	l	2θ	d	M	I_{rel}	h	k	l	2θ	d	M	I_{rel}
0	2	0	5.76	15.355	2	100.0	-3	1	1	30.58	2.923	4	0.2	4	0	1	40.29	2.238	2	0.2
1	1	0	9.75	9.072	4	48.1	-2	2	2	30.98	2.887	4	0.2	-3	9	1	40.53	2.226	4	1.2
0	4	0	11.53	7.677	2	4.9	3	1	1	31.26	2.861	4	0.5	-1	5	3	40.69	2.217	4	0.4
0	0	1	12.11	7.310	2	3.8	2	0	2	31.33	2.855	2	1.6	3	5	2	41.00	2.201	4	0.3
1	3	0	12.72	6.962	4	5.0	0	10	1	31.60	2.831	4	0.7	-1	13	1	41.18	2.192	4	0.4
0	2	1	13.41	6.600	4	10.9	-3	3	1	31.70	2.823	4	3.4	-4	4	1	41.34	2.184	4	0.7
-1	1	1	15.35	5.773	4	13.3	2	2	2	31.88	2.807	4	0.5	-1	11	2	41.63	2.169	4	0.2
1	5	0	17.19	5.157	4	1.2	-2	8	1	32.18	2.782	4	0.2	-2	12	1	41.79	2.161	4	0.2
0	6	0	17.33	5.118	2	0.9	3	3	1	32.36	2.767	4	0.6	4	6	0	41.95	2.154	4	0.5
-1	3	1	17.40	5.097	4	0.3	2	8	1	32.61	2.746	4	0.1	1	11	2	41.99	2.152	4	0.4
1	3	1	17.78	4.987	4	8.9	-2	4	2	32.62	2.745	4	0.9	4	4	1	42.05	2.149	4	0.3
2	0	0	18.69	4.748	2	1.5	-1	7	2	33.06	2.710	4	1.2	2	12	1	42.14	2.144	4	0.4
-1	5	1	20.92	4.247	4	1.3	1	11	0	33.45	2.678	4	1.1	2	2	3	42.62	2.121	4	0.5
0	6	1	21.19	4.193	4	2.4	2	4	2	33.48	2.676	4	0.4	-2	4	3	42.86	2.110	4	0.5
1	5	1	21.24	4.182	4	24.5	1	7	2	33.48	2.676	4	0.3	-1	7	3	43.29	2.090	4	0.3
2	4	0	22.01	4.038	4	13.6	-3	5	1	33.83	2.649	4	0.2	-4	6	1	43.48	2.081	4	1.4
-2	0	1	22.02	4.037	2	0.9	0	8	2	33.86	2.647	4	0.4	1	7	3	43.80	2.067	4	0.6
1	7	0	22.32	3.983	4	10.5	3	5	1	34.46	2.603	4	1.0	2	4	3	43.88	2.063	4	0.3
2	0	1	22.63	3.929	2	0.7	3	7	0	34.95	2.567	4	0.8	-4	0	2	44.90	2.019	2	0.3
-2	2	1	22.77	3.904	4	23.0	0	12	0	35.06	2.559	2	0.1	-2	6	3	44.94	2.017	4	0.1
0	8	0	23.17	3.839	2	0.2	-1	11	1	35.60	2.522	4	0.3	3	11	1	45.29	2.002	4	6.1
2	2	1	23.37	3.806	4	0.3	1	11	1	35.80	2.508	4	0.6	-4	2	2	45.31	2.001	4	0.3
0	0	2	24.35	3.655	2	0.8	-2	10	1	36.77	2.444	4	0.5	1	15	0	45.31	2.001	4	0.2
-2	4	1	24.92	3.573	4	0.5	-3	7	1	36.83	2.440	4	0.3	2	14	0	45.55	1.991	4	0.1
0	2	2	25.04	3.556	4	13.5	0	0	3	36.89	2.437	2	0.4	-3	9	2	45.86	1.979	4	0.2
-1	7	1	25.33	3.516	4	3.4	-1	9	2	37.08	2.425	4	0.1	2	6	3	45.93	1.976	4	0.3
2	4	1	25.47	3.497	4	6.8	-3	1	2	37.13	2.421	4	0.3	-4	8	1	46.35	1.959	4	0.9
2	6	0	25.59	3.481	4	16.0	2	10	1	37.16	2.419	4	1.1	-1	9	3	46.56	1.951	4	0.4
1	7	1	25.60	3.479	4	17.4	0	12	1	37.22	2.415	4	0.6	1	13	2	46.94	1.936	4	0.7
-1	1	2	26.02	3.424	4	5.0	0	2	3	37.37	2.407	4	0.6	1	9	3	47.04	1.932	4	2.1
0	8	1	26.22	3.399	4	3.2	3	7	1	37.41	2.404	4	1.4	-2	12	2	47.09	1.930	4	0.3
1	1	2	26.55	3.357	4	3.1	1	9	2	37.47	2.400	4	1.3	-2	14	1	47.15	1.927	4	0.3
0	4	2	27.02	3.300	4	8.2	-1	1	3	37.96	2.370	4	1.0	-3	3	3	47.23	1.924	4	0.6
-1	3	2	27.31	3.266	4	1.0	3	1	2	38.28	2.351	4	0.4	-2	8	3	47.73	1.905	4	0.5
1	9	0	27.78	3.211	4	0.6	0	10	2	38.28	2.351	4	0.7	5	1	0	47.99	1.896	4	0.3
1	3	2	27.81	3.207	4	1.1	4	2	0	38.37	2.346	4	0.9	0	14	2	48.39	1.881	4	0.3
-2	6	1	28.15	3.170	4	5.2	3	9	0	38.81	2.321	4	0.1	-4	6	2	48.48	1.878	4	0.1
3	1	0	28.34	3.149	4	0.3	-1	3	3	38.89	2.316	4	0.2	3	3	3	48.65	1.871	4	0.4
2	6	1	28.64	3.116	4	0.5	3	3	2	39.20	2.298	4	0.1	2	8	3	48.68	1.871	4	0.7
0	10	0	29.08	3.071	2	0.7	1	13	0	39.30	2.292	4	0.6	5	3	0	48.77	1.867	4	0.2
3	3	0	29.54	3.024	4	4.4	2	8	2	39.33	2.291	4	0.2	-3	5	3	48.79	1.867	4	0.3
0	6	2	30.04	2.974	4	1.0	1	3	3	39.45	2.284	4	0.3	0	16	1	49.07	1.856	4	0.3
1	5	2	30.20	2.959	4	2.7	-4	0	1	39.56	2.278	2	0.2	4	6	2	49.72	1.834	4	0.4
-1	9	1	30.29	2.951	4	1.0	-3	5	2	39.91	2.259	4	0.8	-4	10	1	49.84	1.830	4	0.2
-2	0	2	30.41	2.939	2	1.4	-4	2	1	40.01	2.253	4	0.2	3	13	1	49.98	1.825	4	0.6

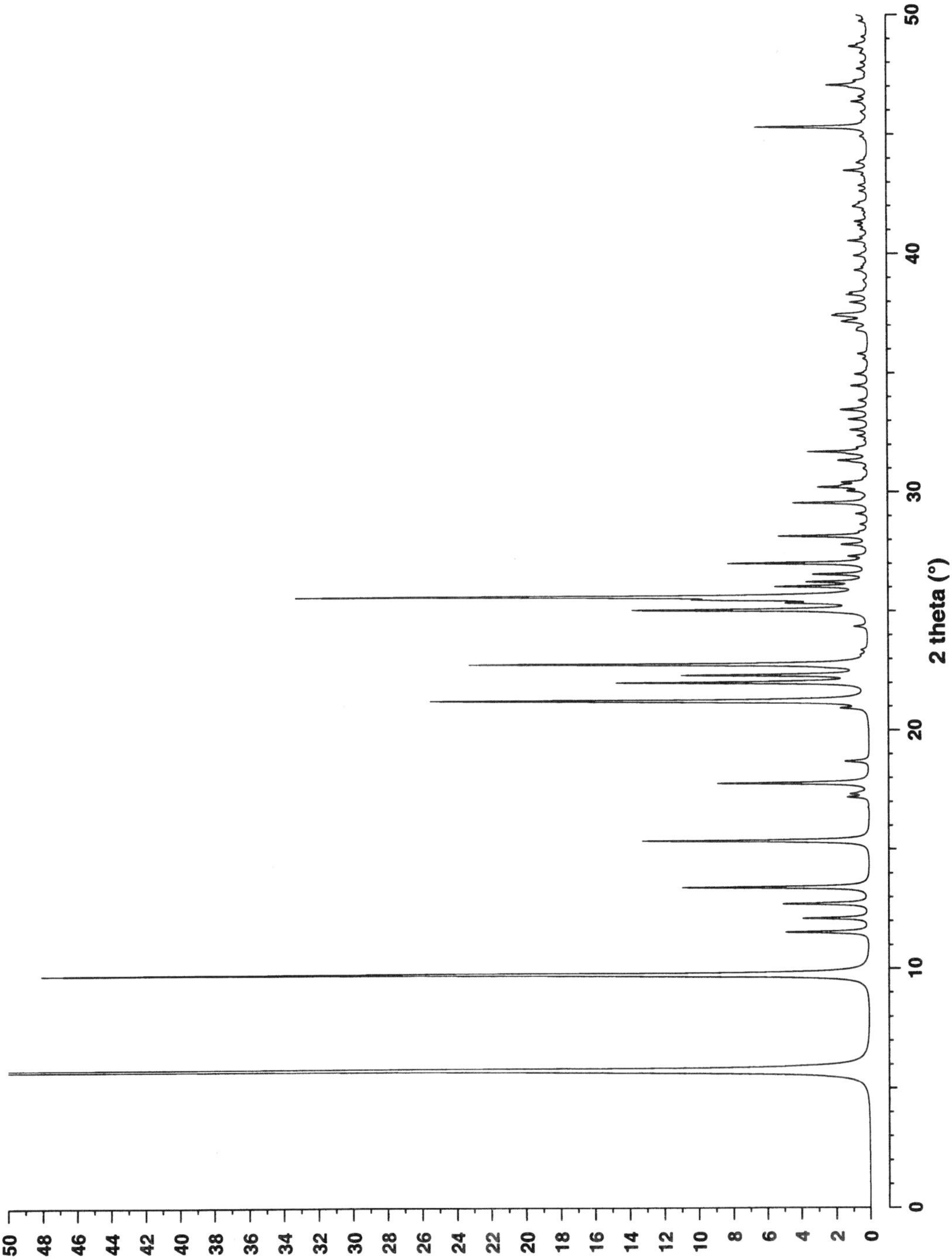
2 theta (°)
0
10
20
30
40
50
0
2
4
6
8
10
12
14
16
18
20
22
24
26
28
30
32
34
36
38
40
42
44
46
48
50

MTN Dodecasil 3C

CHEMICAL COMPOSITION: $|((CH_3)_3N, CO_2)_y(N_2, Ar, CH_4)_x|$ $[Si_{136}O_{272}]$

REFINED COMPOSITION: $|N_{24}|$ $[Si_{136}O_{272}]$

CRYSTAL DATA: $Fd\overline{3}$ (No. 203) origin at centre ($\overline{3}$)
$a = 19.402$ Å $b = 19.402$ Å $c = 19.402$ Å
$\alpha = 90°$ $\beta = 90°$ $\gamma = 90°$
X-ray single crystal refinement, $R_w = 0.06$

REFERENCE: H. Gies,
Z. Kristallogr. **167** 73–82 (1984).

h	*k*	*l*	2*θ*	*d*	*M*	I_{rel}
1	1	1	7.89	11.202	8	43.4
2	2	0	12.91	6.860	12	55.2
3	1	1	15.14	5.850	24	100.0
2	2	2	15.82	5.601	8	52.1
4	0	0	18.29	4.851	6	27.2
3	3	1	19.95	4.451	24	16.3
4	2	2	22.45	3.960	24	29.2
5	1	1	23.83	3.734	24	67.1
3	3	3	23.83	3.734	8	21.7
4	4	0	25.98	3.430	12	35.3
5	1	3	27.19	3.280	24	40.9
5	3	1	27.19	3.280	24	40.5
4	4	2	27.58	3.234	24	6.0
6	2	0	29.11	3.068	12	5.9
6	0	2	29.11	3.068	12	6.0
5	3	3	30.21	2.959	24	7.0
6	2	2	30.56	2.925	24	0.1
5	5	1	32.97	2.717	24	1.5
7	1	1	32.97	2.717	24	0.2
7	1	3	35.54	2.526	24	2.4
5	5	3	35.54	2.526	24	2.6
7	3	1	35.54	2.526	24	2.4
8	0	0	37.07	2.425	6	0.8
7	3	3	37.96	2.370	24	11.0
6	4	4	38.25	2.353	24	0.6
8	2	2	39.41	2.287	24	13.0
6	6	0	39.41	2.287	12	9.4
5	5	5	40.25	2.240	8	3.4
7	1	5	40.25	2.240	24	0.4
7	5	1	40.25	2.240	24	0.4
6	6	2	40.53	2.226	24	0.9
8	0	4	41.63	2.169	12	1.0
8	4	0	41.63	2.169	12	1.0
7	5	3	42.44	2.130	24	0.2
7	3	5	42.44	2.130	24	0.2
8	4	4	45.82	1.980	24	2.3
7	7	1	46.57	1.950	24	2.5
9	3	3	46.57	1.950	24	1.9
7	5	5	46.57	1.950	24	0.1
8	2	6	47.81	1.903	24	0.6
8	6	2	47.81	1.903	24	0.6
9	5	1	48.54	1.876	24	0.1
9	1	5	48.54	1.876	24	0.1
7	7	3	48.54	1.876	24	0.2
6	6	6	48.78	1.867	8	2.9
10	2	2	48.78	1.867	24	5.9

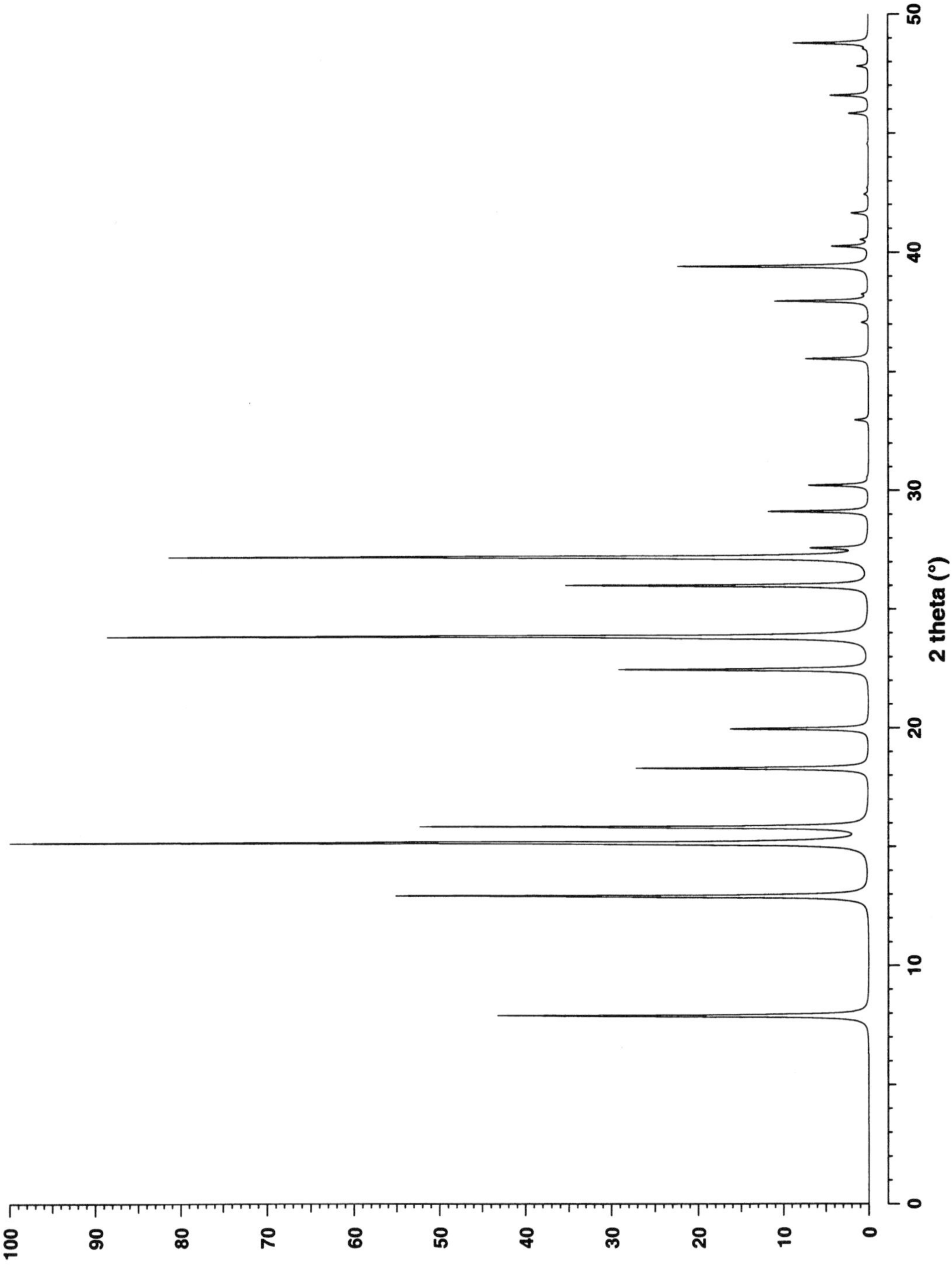
2 theta (°)
0
10
20
30
40
50
0
10
20
30
40
50
60
70
80
90
100

MTT Ammonium Fluoride ZSM-23

CHEMICAL COMPOSITION: $|(NH_4F)_{1.72}|\ [Si_{24}O_{48}]$

REFINED COMPOSITION: $|(NH_4F)_{1.72}|\ [Si_{24}O_{48}]$

CRYSTAL DATA: $P12_11$ (No. 4) unique axis **b**
$a = 11.129$ Å $b = 5.025$ Å $c = 21.519$ Å
$\alpha = 90°$ $\beta = 89.85°$ $\gamma = 90°$
X-ray Rietveld refinement, $R_{exp} = 0.033$, $R_{wp} = 0.085$, $R_F = 0.089$

REFERENCE: B. Marler, C. Deroche, H. Gies, C. A. Fyfe, H. Grondey, G. T. Kokotailo, Y. Feng, S. Ernst, J. Weitkamp and D. E. Cox, *J. Appl. Cryst.* **26** 636–644 (1993).

h	k	l	2θ	d	M	I_{rel}	h	k	l	2θ	d	M	I_{rel}	h	k	l	2θ	d	M	I_{rel}
1	0	0	7.94	11.129	2	81.9	1	1	5	28.44	3.138	4	4.8	4	1	3	39.05	2.307	4	2.0
0	0	2	8.22	10.759	2	100.0	-1	1	5	28.47	3.135	4	5.0	-4	1	3	39.11	2.303	4	1.8
1	0	1	8.94	9.896	2	21.8	2	1	4	29.11	3.067	4	1.2	4	1	4	40.65	2.220	4	0.6
-1	0	1	8.96	9.875	2	19.0	-2	1	4	29.16	3.062	4	2.0	2	1	8	41.35	2.183	4	0.5
1	0	2	11.42	7.746	2	33.2	2	0	6	29.60	3.018	2	2.3	0	0	10	41.98	2.152	2	0.8
-1	0	2	11.45	7.725	2	26.0	-2	0	6	29.67	3.011	2	2.4	-1	1	9	42.68	2.119	4	0.6
1	0	3	14.67	6.036	2	2.7	3	1	0	29.94	2.984	4	0.6	-4	1	5	42.72	2.117	4	0.5
-1	0	3	14.71	6.022	2	4.9	-1	0	7	30.18	2.961	2	0.5	3	2	1	43.71	2.071	4	1.1
2	0	0	15.93	5.564	2	8.8	3	1	1	30.22	2.957	4	3.3	-3	2	1	43.72	2.070	4	1.9
0	0	4	16.48	5.380	2	9.4	-3	1	2	31.11	2.874	4	1.6	0	2	6	44.00	2.058	4	3.2
0	1	1	18.13	4.893	4	2.7	1	1	6	31.67	2.825	4	3.3	3	2	2	44.34	2.043	4	0.8
1	0	4	18.30	4.848	2	3.7	-1	1	6	31.70	2.822	4	1.3	-3	2	2	44.36	2.042	4	1.0
-1	0	4	18.34	4.839	2	3.4	2	1	5	31.72	2.821	4	0.9	5	1	0	44.52	2.035	4	3.0
1	1	1	19.82	4.480	4	34.1	-2	1	5	31.78	2.816	4	1.9	2	2	5	44.81	2.022	4	1.5
-1	1	1	19.82	4.478	4	30.2	3	0	5	31.80	2.814	2	5.9	-2	2	5	44.85	2.021	4	1.2
2	0	3	20.17	4.402	2	5.2	-3	0	5	31.89	2.806	2	4.5	4	1	6	44.96	2.016	4	3.2
-2	0	3	20.22	4.391	2	4.9	4	0	0	32.17	2.782	2	2.1	2	1	9	45.00	2.014	4	0.6
1	1	2	21.07	4.216	4	30.6	3	1	3	32.47	2.758	4	0.9	-4	1	6	45.06	2.012	4	0.6
-1	1	2	21.09	4.212	4	27.0	-3	1	3	32.52	2.753	4	1.0	-2	1	9	45.08	2.011	4	0.8
0	1	3	21.59	4.116	4	8.4	4	0	2	33.24	2.695	2	0.5	2	0	10	45.13	2.009	2	1.4
1	0	5	22.12	4.018	2	0.7	0	0	8	33.31	2.690	2	0.6	-2	0	10	45.22	2.005	2	1.7
-1	0	5	22.16	4.011	2	1.4	1	0	8	34.28	2.616	2	0.8	-3	1	8	45.44	1.996	4	0.5
2	0	4	22.96	3.873	2	17.1	-1	0	8	34.32	2.613	2	0.6	5	0	5	45.85	1.979	2	1.0
-2	0	4	23.02	3.863	2	18.4	3	1	4	34.33	2.612	4	1.7	0	1	10	45.87	1.978	4	0.8
1	1	3	23.03	3.862	4	24.7	-3	1	4	34.39	2.607	4	1.0	-5	0	5	45.95	1.975	2	0.9
-1	1	3	23.05	3.858	4	25.2	-2	1	6	34.73	2.583	4	1.4	1	1	10	46.61	1.948	4	0.5
2	1	0	23.86	3.729	4	2.2	-1	1	7	35.18	2.551	4	0.7	-1	1	10	46.65	1.947	4	1.0
3	0	0	23.99	3.710	2	25.6	0	2	0	35.74	2.513	2	24.9	3	2	4	46.80	1.941	4	1.3
2	1	1	24.21	3.676	4	21.0	-4	0	4	36.39	2.469	2	0.5	-3	2	4	46.85	1.939	4	1.1
-2	1	1	24.23	3.674	4	25.8	3	1	5	36.60	2.455	4	2.8	4	1	7	47.60	1.910	4	0.9
3	0	1	24.34	3.657	2	4.2	1	2	0	36.67	2.451	4	1.1	5	1	4	47.74	1.905	4	0.6
-3	0	1	24.36	3.654	2	4.2	-3	1	5	36.68	2.450	4	5.0	-5	1	4	47.82	1.902	4	0.5
0	0	6	24.82	3.586	2	44.1	0	2	2	36.73	2.447	4	2.0	5	0	6	48.05	1.893	2	2.3
2	1	2	25.26	3.526	4	5.9	-1	2	1	36.92	2.435	4	0.5	-5	0	6	48.17	1.889	2	2.4
-2	1	2	25.29	3.522	4	7.0	4	1	0	36.93	2.434	4	6.0	3	2	5	48.58	1.874	4	1.8
3	0	2	25.38	3.510	2	11.6	4	1	1	37.16	2.419	4	2.1	-3	2	5	48.64	1.872	4	1.2
-3	0	2	25.42	3.504	2	12.3	-4	1	1	37.18	2.418	4	0.9	3	1	9	48.75	1.868	4	0.5
1	1	4	25.53	3.489	4	3.7	1	2	2	37.64	2.390	4	1.1	4	2	0	48.84	1.865	4	0.7
-1	1	4	25.56	3.485	4	7.7	4	1	2	37.88	2.375	4	1.3	-3	1	9	48.86	1.864	4	0.6
1	0	6	26.08	3.416	2	4.6	2	1	7	37.90	2.374	4	2.6	2	0	11	49.34	1.847	2	3.5
-1	0	6	26.12	3.411	2	4.7	-4	1	2	37.92	2.373	4	2.0	-2	0	11	49.42	1.844	2	3.1
2	0	5	26.14	3.409	2	18.1	0	1	8	37.94	2.371	4	0.6	0	2	8	49.65	1.836	4	1.1
-2	0	5	26.21	3.400	2	14.1	-2	1	7	37.96	2.370	4	2.5	-2	2	7	49.67	1.835	4	1.0
2	1	3	26.93	3.311	4	1.6	1	1	8	38.81	2.320	4	4.8	6	0	2	49.87	1.829	2	0.9
0	1	5	27.28	3.269	4	3.7	-1	1	8	38.84	2.318	4	2.4	-6	0	2	49.91	1.827	2	0.9

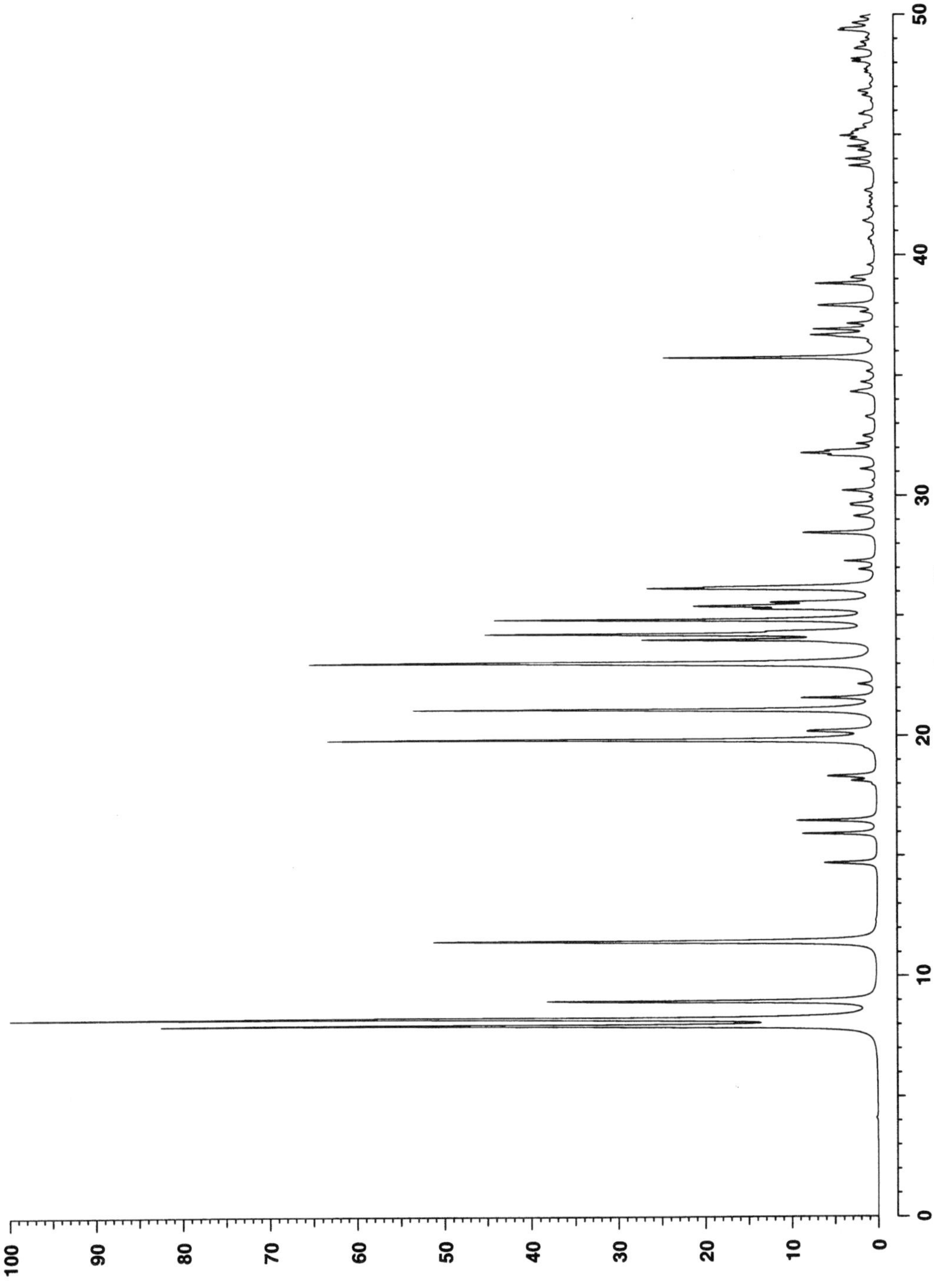
2 theta (°)
0
10
20
30
40
50
100
90
80
70
60
50
40
30
20
10
0

MTW ZSM-12, Calcined

CHEMICAL COMPOSITION: $[Si_{56}O_{112}]$

REFINED COMPOSITION: $[Si_{56}O_{112}]$

CRYSTAL DATA: $C1\,2/c\,1$ (No. 15) unique axis **b**, cell choice 1
a = 24.8633 Å b = 5.01238 Å c = 24.3275 Å
$\alpha = 90°$ $\beta = 107.7215°$ $\gamma = 90°$
X-ray Rietveld refinement, $R_{exp} = 0.058$, $R_{wp} = 0.181$, $R_I = 0.069$

REFERENCE: C. A. Fyfe, H. Gies, G. T. Kokotailo, B. Marler and D. E. Cox, *J. Phys. Chem.* **94** 3718–3721 (1990).

h	k	l	2θ	d	M	I_{rel}
2	0	0	7.47	11.842	2	78.3
0	0	2	7.63	11.587	2	100.0
-2	0	2	8.91	9.929	2	63.6
2	0	2	12.21	7.251	2	5.0
-2	0	4	14.84	5.971	2	7.7
4	0	0	14.96	5.921	2	0.4
0	0	4	15.29	5.793	2	8.3
1	1	0	18.09	4.904	4	0.2
1	1	1	18.73	4.737	4	1.0
4	0	2	18.79	4.722	2	8.8
2	0	4	18.99	4.673	2	3.8
-1	1	2	19.20	4.622	4	7.0
1	1	2	20.10	4.417	4	4.4
-1	1	3	20.84	4.263	4	0.4
3	1	0	20.99	4.232	4	75.7
-3	1	2	21.14	4.202	4	1.1
-6	0	2	21.46	4.141	2	2.9
-2	0	6	21.93	4.054	2	1.4
3	1	1	21.96	4.048	4	2.3
1	1	3	22.07	4.027	4	0.7
-3	1	3	22.25	3.996	4	0.4
6	0	0	22.52	3.947	2	7.0
-1	1	4	23.01	3.865	4	0.5
0	0	6	23.03	3.862	2	18.8
-6	0	4	23.08	3.853	2	9.7
-4	0	6	23.36	3.809	2	22.6
3	1	2	23.53	3.781	4	1.2
-3	1	4	23.93	3.718	4	2.2
1	1	4	24.51	3.632	4	0.5
4	0	4	24.55	3.626	2	2.0
-5	1	1	25.30	3.520	4	6.0
-5	1	2	25.30	3.520	4	1.1
3	1	3	25.60	3.480	4	1.3
5	1	0	25.88	3.443	4	10.2
-5	1	3	25.90	3.440	4	2.6
6	0	2	25.97	3.431	2	0.5
-3	1	5	26.10	3.415	4	1.1
2	0	6	26.36	3.381	2	11.9
-6	0	6	26.94	3.310	2	8.1
5	1	1	27.01	3.301	4	0.2
-5	1	4	27.04	3.297	4	5.7
3	1	4	28.07	3.179	4	6.0
-1	1	6	28.50	3.132	4	1.0
5	1	2	28.64	3.117	4	0.7
-3	1	6	28.64	3.117	4	0.3

h	k	l	2θ	d	M	I_{rel}
-8	0	2	28.76	3.104	2	0.4
-8	0	4	29.41	3.037	2	2.1
-2	0	8	29.43	3.035	2	4.1
-4	0	8	29.93	2.985	2	0.4
1	1	6	30.35	2.945	4	0.2
5	1	3	30.68	2.914	4	0.3
3	1	5	30.85	2.898	4	0.4
-7	1	2	30.86	2.898	4	1.4
6	0	4	31.06	2.880	2	3.9
-7	1	3	31.06	2.879	4	2.7
-1	1	7	31.64	2.828	4	0.6
7	1	0	31.91	2.804	4	0.7
-6	0	8	32.29	2.772	2	0.4
-7	1	5	32.89	2.723	4	1.8
7	1	1	33.13	2.704	4	1.9
8	0	2	33.44	2.679	2	0.6
3	1	6	33.88	2.646	4	3.1
-7	1	6	34.46	2.603	4	1.5
7	1	2	34.76	2.581	4	2.7
-1	1	8	34.97	2.566	4	0.9
5	1	5	35.77	2.511	4	0.6
-5	1	8	35.83	2.506	4	2.1
0	2	0	35.83	2.506	2	9.8
-10	0	2	36.33	2.473	2	0.3
-7	1	7	36.39	2.469	4	1.0
-2	2	1	36.61	2.455	4	0.6
2	2	0	36.65	2.452	4	0.3
0	2	2	36.69	2.450	4	1.4
-2	2	2	36.99	2.430	4	1.0
1	1	8	37.03	2.428	4	0.7
-4	0	10	37.11	2.423	2	1.8
3	1	7	37.11	2.423	4	1.5
-9	1	3	37.18	2.418	4	0.7
-9	1	4	37.52	2.397	4	0.6
-2	2	3	37.79	2.381	4	1.0
-10	0	6	38.05	2.365	2	0.4
-9	1	5	38.28	2.351	4	0.2
-1	1	9	38.45	2.341	4	0.6
4	0	8	38.53	2.336	2	0.6
9	1	0	38.64	2.330	4	1.2
-7	1	8	38.64	2.330	4	0.2
5	1	6	38.70	2.327	4	1.2
-4	2	1	38.75	2.324	4	0.2
-5	1	9	38.76	2.323	4	1.6
4	2	0	39.03	2.308	4	0.3

h	k	l	2θ	d	M	I_{rel}
7	1	4	39.06	2.306	4	0.2
2	2	3	39.24	2.296	4	0.5
-9	1	6	39.42	2.286	4	0.5
9	1	1	39.92	2.258	4	2.1
3	1	8	40.51	2.227	4	0.3
-9	1	7	40.92	2.206	4	0.3
-10	0	8	41.24	2.189	2	0.9
-3	1	10	41.25	2.188	4	0.3
-8	0	10	41.54	2.174	2	0.4
7	1	5	41.64	2.169	4	0.3
5	1	7	41.83	2.159	4	0.3
2	0	10	41.94	2.154	2	0.6
-1	1	10	42.06	2.148	4	0.2
-6	2	2	42.14	2.144	4	0.3
4	2	3	42.17	2.143	4	0.3
-6	2	3	42.41	2.131	4	0.3
6	2	0	42.74	2.116	4	0.7
0	2	6	43.02	2.102	4	0.7
-6	2	4	43.05	2.101	4	0.3
-4	2	6	43.21	2.094	4	0.4
9	1	3	43.49	2.081	4	0.3
8	0	6	43.83	2.066	2	0.4
-11	1	3	43.96	2.060	4	0.3
-11	1	4	44.04	2.056	4	1.0
-11	1	5	44.48	2.037	4	0.3
-4	0	12	44.71	2.027	2	5.0
-12	0	6	44.76	2.025	2	0.2
-11	1	1	44.88	2.020	4	0.7
-10	0	10	45.69	1.986	2	0.4
9	1	4	45.72	1.984	4	1.3
11	1	0	45.87	1.978	4	0.7
6	2	3	46.30	1.961	4	0.6
-11	1	7	46.42	1.956	4	0.5
-8	2	1	46.88	1.938	4	0.8
-8	2	4	47.01	1.933	4	1.9
-2	2	8	47.02	1.932	4	0.6
-12	0	8	47.17	1.927	2	0.2
11	1	1	47.19	1.926	4	0.3
-9	1	10	47.28	1.922	4	0.9
8	2	0	47.53	1.913	4	0.5
3	1	10	47.72	1.906	4	1.1
6	2	4	48.13	1.890	4	1.3
-1	1	12	49.63	1.837	4	1.0
-11	1	9	49.64	1.836	4	0.5
-9	1	11	49.93	1.827	4	0.4

ZSM-12, Calcined **MTW**

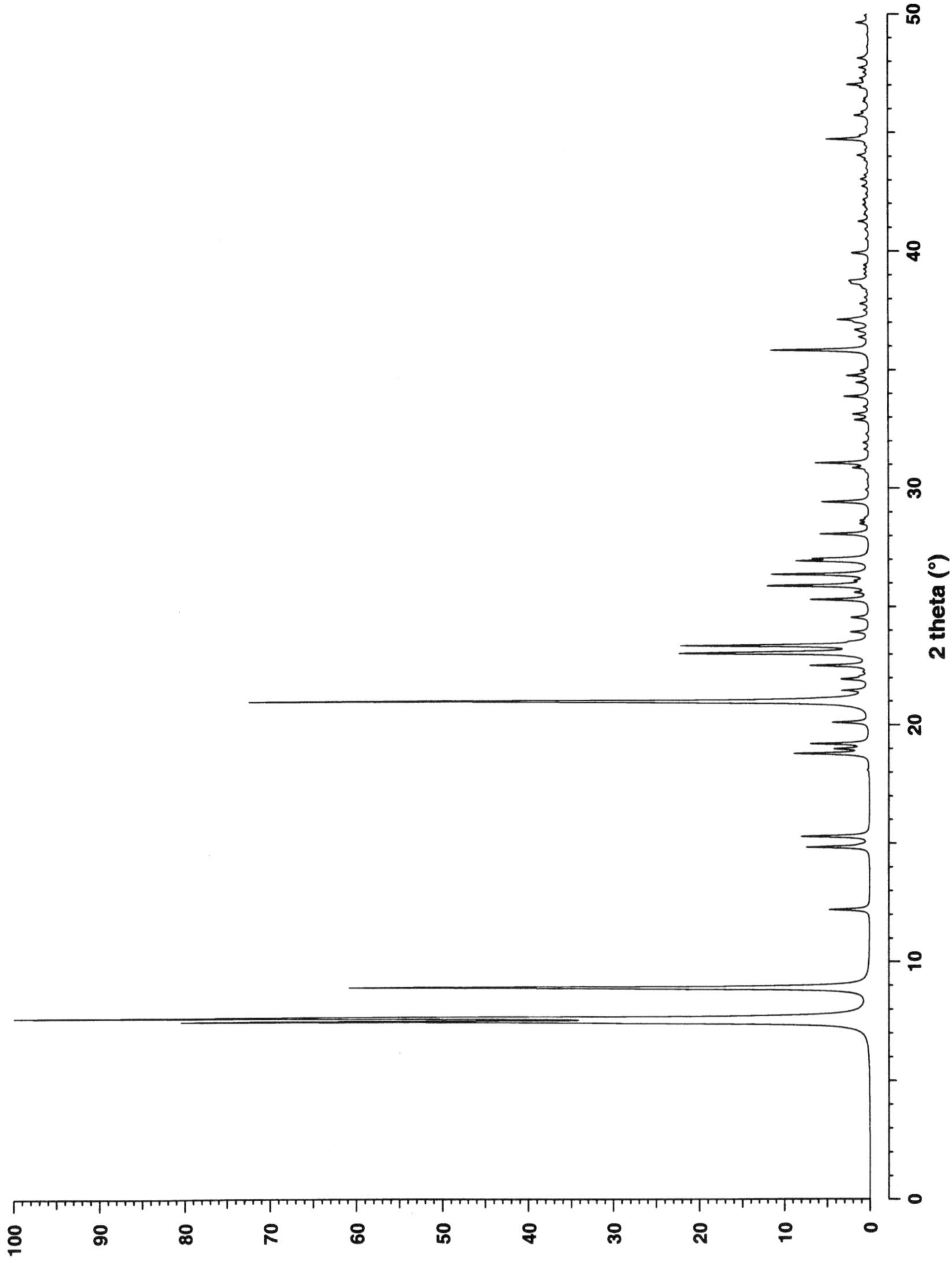

MWW ITQ-1, Calcined (Siliceous MCM-22)

CHEMICAL COMPOSITION: $[Si_{72}O_{144}]$

REFINED COMPOSITION: $[Si_{72}O_{144}]$

CRYSTAL DATA: $P6/mmm$ (No. 191)
a = 14.2081 Å b = 14.2081 Å c = 24.945 Å
α = 90° β = 90° γ = 120°
X-ray Rietveld refinement. R_{exp} = 0.103, R_{wp} = 0.159, R_F = 0.065

REFERENCE: M. A. Camblor, A. Corma, M.-J. Diaz-Cabanas and Ch. Baerlocher, *J. Phys. Chem. B* **102** 44–51 (1998).

h	k	l	2θ	d	M	I_{rel}	h	k	l	2θ	d	M	I_{rel}	h	k	l	2θ	d	M	I_{rel}
0	0	1	3.54	24.945	2	23.9	2	2	3	27.30	3.266	12	1.7	5	1	0	40.83	2.210	12	0.4
0	0	2	7.09	12.472	2	74.3	1	1	7	28.01	3.185	12	15.1	5	0	5	40.89	2.207	12	0.3
1	0	0	7.18	12.305	6	90.6	3	0	5	28.16	3.168	12	0.2	4	0	8	41.22	2.190	12	0.4
1	0	1	8.01	11.035	12	100.0	0	0	8	28.63	3.118	2	0.8	4	2	4	41.44	2.179	24	0.1
1	0	2	10.10	8.759	12	69.6	2	1	6	28.80	3.100	24	5.7	1	1	11	41.81	2.160	12	1.9
0	0	3	10.64	8.315	2	2.6	2	0	7	28.95	3.084	12	4.2	3	0	10	42.41	2.131	12	0.4
1	1	0	12.46	7.104	6	0.3	4	0	1	29.25	3.053	12	1.8	4	2	5	42.91	2.108	24	1.5
1	0	3	12.85	6.889	12	2.4	1	0	8	29.55	3.023	12	0.4	3	2	8	43.23	2.093	24	0.6
1	1	1	12.96	6.832	12	8.3	3	1	4	29.84	2.994	24	2.4	5	1	4	43.44	2.083	24	0.9
0	0	4	14.20	6.236	2	28.6	3	0	6	30.62	2.920	12	0.3	6	0	1	44.32	2.044	12	0.3
1	1	2	14.35	6.173	12	6.0	4	0	3	31.00	2.885	12	0.3	2	1	11	44.44	2.038	24	1.2
2	0	0	14.40	6.152	6	15.8	1	1	8	31.33	2.855	12	0.1	4	1	8	44.53	2.035	24	0.3
2	0	1	14.83	5.973	12	7.0	3	2	0	31.70	2.823	12	2.4	4	2	6	44.65	2.029	24	0.1
1	0	4	15.93	5.563	12	5.3	3	1	5	31.77	2.817	24	0.4	5	0	7	44.75	2.025	12	0.7
2	0	2	16.06	5.518	12	13.4	3	2	1	31.90	2.805	24	0.5	6	0	2	44.79	2.024	12	0.1
1	1	3	16.41	5.401	12	3.0	0	0	9	32.30	2.772	2	0.1	4	3	0	44.80	2.023	12	2.3
0	0	5	17.78	4.989	2	2.8	4	0	4	32.45	2.759	12	3.7	5	1	5	44.86	2.021	24	0.1
2	0	3	17.93	4.946	12	1.2	1	0	9	33.13	2.704	12	0.7	4	3	1	44.96	2.016	24	0.3
2	1	0	19.08	4.651	12	1.8	3	0	7	33.31	2.690	12	0.2	3	1	10	45.01	2.014	24	0.5
1	0	5	19.20	4.623	12	2.4	3	2	3	33.52	2.673	24	6.3	4	3	2	45.42	1.997	24	0.5
2	1	1	19.41	4.572	24	1.4	3	1	6	33.99	2.638	24	1.5	3	2	9	45.88	1.978	24	0.4
2	0	4	20.28	4.380	12	6.2	2	1	8	34.63	2.590	24	4.0	5	2	0	46.07	1.970	12	0.4
2	1	2	20.38	4.358	24	4.4	1	1	9	34.74	2.582	12	0.2	4	3	3	46.18	1.966	24	0.6
0	0	6	21.37	4.157	2	3.6	3	2	4	34.89	2.572	24	0.9	5	2	1	46.22	1.964	24	1.7
3	0	0	21.67	4.102	6	6.4	4	1	3	35.12	2.555	24	1.3	5	1	6	46.54	1.951	24	0.2
1	1	5	21.77	4.083	12	3.1	2	0	9	35.52	2.527	12	1.8	6	0	4	46.62	1.948	12	1.0
2	1	3	21.90	4.059	24	0.7	2	2	7	35.69	2.516	12	0.7	4	2	7	46.64	1.947	24	2.1
3	0	1	21.96	4.047	12	9.9	0	0	10	36.00	2.494	2	0.4	5	2	2	46.67	1.946	24	1.0
1	0	6	22.57	3.939	12	5.7	3	0	8	36.19	2.482	12	0.4	4	0	10	46.89	1.938	12	0.2
3	0	2	22.82	3.896	12	20.2	4	0	6	36.33	2.473	12	0.2	4	1	9	47.12	1.929	24	1.1
2	0	5	22.95	3.875	12	0.9	4	1	4	36.43	2.466	24	0.2	6	0	5	47.96	1.897	12	0.5
2	1	4	23.87	3.728	24	16.4	3	1	7	36.45	2.465	24	0.3	1	0	13	47.98	1.896	12	0.1
3	0	3	24.20	3.678	12	0.2	5	0	0	36.51	2.461	6	0.9	5	1	7	48.47	1.878	24	0.2
1	1	6	24.81	3.588	12	1.0	3	2	5	36.57	2.457	24	0.2	6	1	0	48.51	1.876	12	1.7
0	0	7	24.99	3.564	2	4.3	5	0	1	36.69	2.449	12	0.5	4	3	5	48.56	1.875	24	0.8
2	2	0	25.07	3.552	6	5.7	5	0	2	37.24	2.414	12	2.1	4	2	8	48.86	1.864	24	0.1
2	2	1	25.33	3.517	12	3.0	3	3	0	38.00	2.368	6	2.9	6	1	2	49.10	1.856	24	1.5
2	0	6	25.86	3.445	12	0.2	3	2	6	38.55	2.335	24	0.5	1	1	13	49.18	1.852	12	0.5
3	0	4	26.00	3.427	12	4.4	4	0	7	38.67	2.329	12	0.3	5	0	9	49.53	1.840	12	0.4
1	0	7	26.03	3.423	12	2.4	4	2	0	38.72	2.325	12	0.5	6	0	6	49.56	1.839	12	0.7
2	2	2	26.08	3.416	12	0.6	2	0	10	38.96	2.312	12	0.2	5	2	5	49.75	1.833	24	0.2
3	1	0	26.11	3.413	12	30.4	3	1	8	39.13	2.302	24	0.1	2	0	13	49.78	1.832	12	0.9
2	1	5	26.20	3.402	24	2.8	3	0	9	39.23	2.296	12	1.1	6	1	3	49.82	1.830	24	0.2
3	1	1	26.36	3.381	24	1.6	0	0	11	39.75	2.268	2	0.4	4	1	10	49.90	1.828	24	0.1
3	1	2	27.09	3.292	24	8.8	3	2	7	40.78	2.213	24	1.1	4	0	11	49.96	1.825	12	0.2

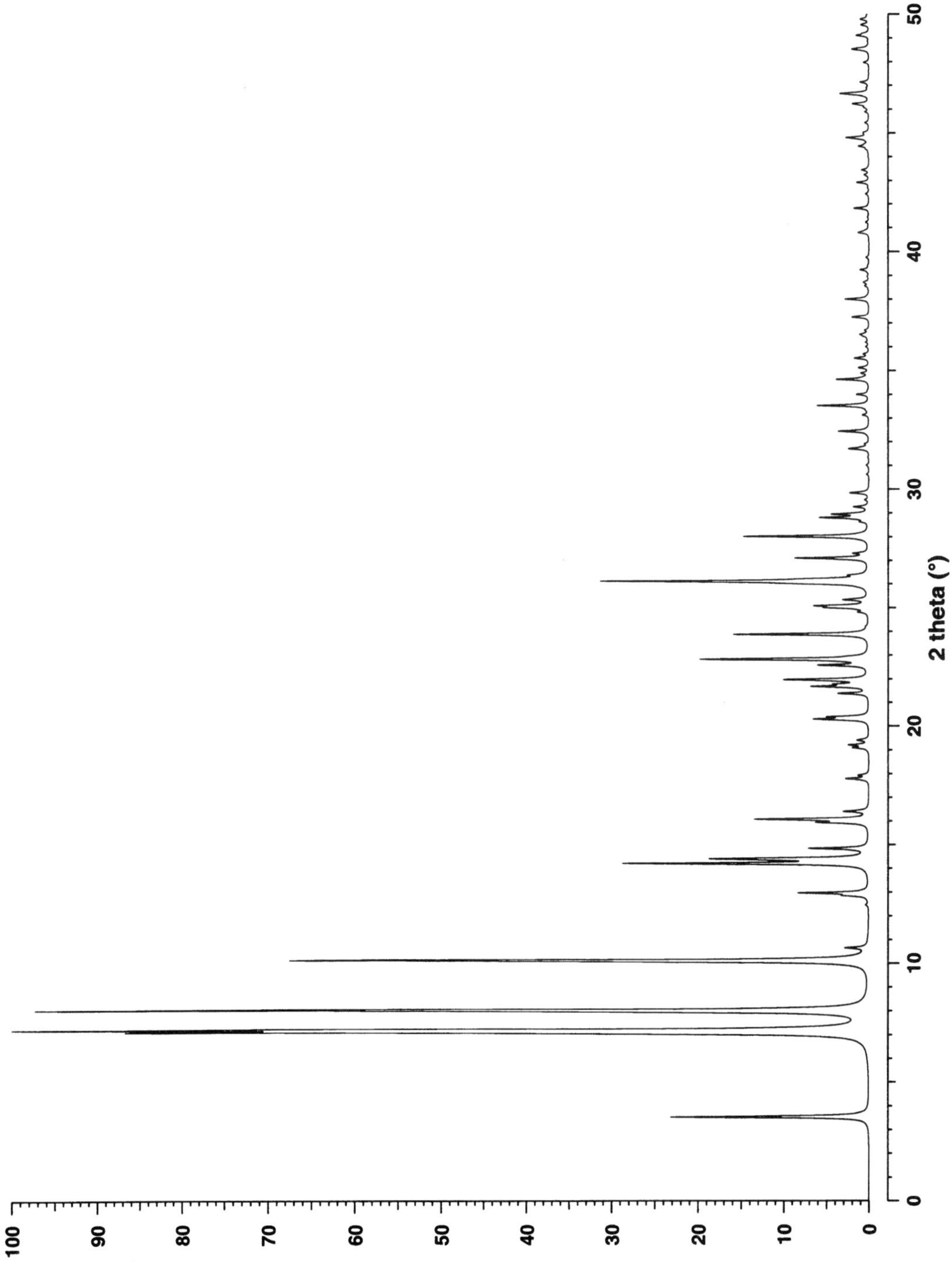
2 theta (°)
0
10
20
30
40
50
0
10
20
30
40
50
60
70
80
90
100

NAT Natrolite

CHEMICAL COMPOSITION: $|Na_{16}(H_2O)_{16}|\ [Si_{24}Al_{16}O_{80}]$
Aussig, Bohemia

REFINED COMPOSITION: $|Na_{16}(H_2O)_{16}|\ [Si_{24}Al_{16}O_{80}]$

CRYSTAL DATA: $Fdd2$ (No. 43)
$a = 18.30$ Å $b = 18.63$ Å $c = 6.60$ Å
$\alpha = 90°$ $\beta = 90°$ $\gamma = 90°$
X-ray single crystal refinement, $R_{wp} = 0.08$

REFERENCE: W. M. Meier,
Z. Kristallogr. **113** 430–444 (1960).

h	k	l	2θ	d	M	I_{rel}
2	2	0	13.56	6.528	4	59.6
1	1	1	15.04	5.890	8	100.0
0	4	0	19.05	4.657	2	24.9
4	0	0	19.40	4.575	2	16.0
1	3	1	20.22	4.391	8	40.5
3	1	1	20.39	4.356	8	47.2
2	4	0	21.41	4.151	4	22.5
4	2	0	21.64	4.106	4	13.6
3	3	1	24.50	3.633	8	1.6
4	4	0	27.32	3.264	4	3.2
1	5	1	27.93	3.195	8	32.0
5	1	1	28.29	3.155	8	37.3
0	2	2	28.70	3.111	4	11.5
2	0	2	28.76	3.104	4	13.8
2	2	2	30.35	2.945	8	33.4
2	6	0	30.40	2.940	4	2.6
6	2	0	30.85	2.899	4	3.8
3	5	1	31.22	2.865	8	61.7
5	3	1	31.44	2.845	8	54.9
2	4	2	34.73	2.583	8	7.5
4	2	2	34.88	2.572	8	12.2
4	6	0	34.92	2.569	4	0.7
6	4	0	35.17	2.552	4	1.3
1	7	1	36.74	2.446	8	15.7
7	1	1	37.31	2.410	8	15.3
0	8	0	38.66	2.329	2	3.0
4	4	2	38.81	2.321	8	5.5
3	7	1	39.38	2.288	8	1.3
8	0	0	39.39	2.287	2	2.1
7	3	1	39.83	2.263	8	0.6
0	6	2	39.86	2.261	4	5.2
2	8	0	39.95	2.257	4	1.1
6	0	2	40.26	2.240	4	3.0
8	2	0	40.61	2.221	4	1.7
2	6	2	41.12	2.195	8	12.5
6	2	2	41.46	2.178	8	10.3
6	6	0	41.50	2.176	4	4.3
1	1	3	41.63	2.169	8	0.6
4	8	0	43.61	2.075	4	0.6
1	3	3	43.94	2.061	8	2.1
3	1	3	44.02	2.057	8	1.9
8	4	0	44.11	2.053	4	1.6
5	7	1	44.26	2.046	8	0.2
7	5	1	44.50	2.036	8	0.2
4	6	2	44.70	2.027	8	0.4
6	4	2	44.90	2.019	8	0.6
1	9	1	46.23	1.964	8	1.5
3	3	3	46.24	1.963	8	3.2
9	1	1	47.01	1.933	8	1.1
1	5	3	48.30	1.884	8	4.3
3	9	1	48.44	1.879	8	0.8
5	1	3	48.53	1.876	8	6.0
2	8	2	48.89	1.863	8	0.8
9	3	1	49.12	1.855	8	0.6
8	2	2	49.46	1.843	8	1.7
2	10	0	49.96	1.826	4	6.1

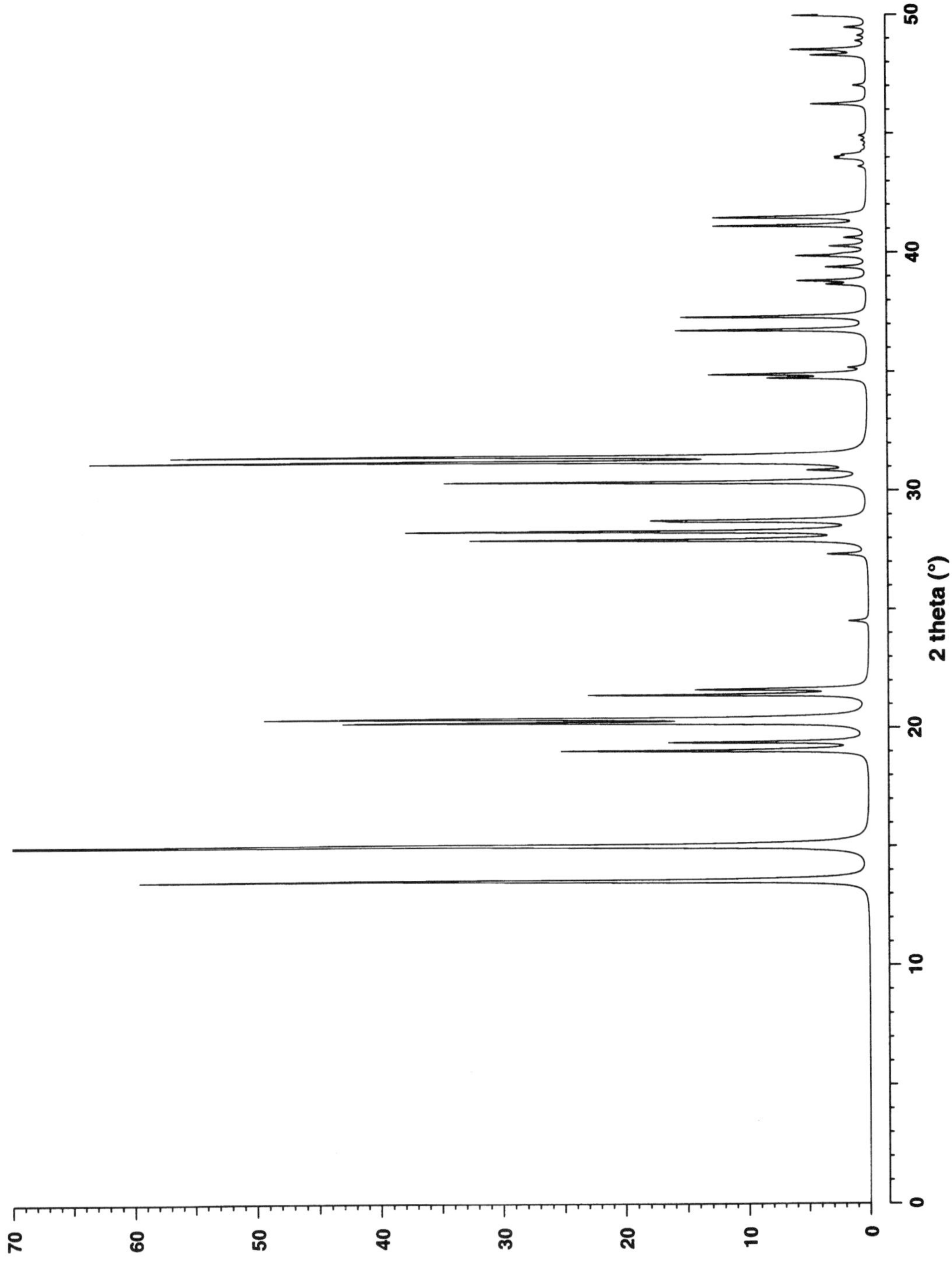
2 theta (°)
0
10
20
30
40
50
0
10
20
30
40
50
60
70

NAT Gonnardite

CHEMICAL COMPOSITION: $|Na_{6.42}K_{0.01}Ca_{1.50}(H_2O)_{12.37}|$ $[Si_{10.73}Al_{9.22}O_{40}]$
Tvedalen, Langesund District, Norway

REFINED COMPOSITION: $|Na_{6.528}Ca_{1.472}(H_2O)_{12}H_8|$ $[Si_{10.25}Al_{9.75}O_{40}]$

CRYSTAL DATA: $I\bar{4}2d$ (No. 122)
$a = 13.21$ Å $b = 13.21$ Å $c = 6.622$ Å
$\alpha = 90°$ $\beta = 90°$ $\gamma = 90°$
X-ray single crystal refinement, $R = 0.03$

REFERENCE: F. Mazzi, A. O. Larsen, G. Gottardi and E. Galli, *Neues Jarhb. Mineral. Monatsh.* 219–228 (1986).

h	k	l	2θ	d	M	I_{rel}	h	k	l	2θ	d	M	I_{rel}	h	k	l	2θ	d	M	I_{rel}
2	0	0	13.41	6.605	4	23.6	3	1	2	34.57	2.595	16	15.2	6	2	0	43.32	2.089	8	1.4
1	0	1	14.96	5.920	8	62.1	5	1	0	34.62	2.591	8	4.8	2	1	3	43.78	2.068	16	3.1
2	2	0	19.00	4.670	4	31.9	4	3	1	36.62	2.454	16	21.9	6	1	1	43.87	2.064	16	0.3
2	1	1	20.14	4.408	16	49.8	4	0	2	38.50	2.338	8	4.6	5	1	2	44.40	2.040	16	0.3
3	1	0	21.27	4.177	8	17.5	4	4	0	38.55	2.335	4	5.1	3	0	3	45.99	1.973	8	3.0
4	0	0	27.00	3.303	4	2.1	5	2	1	39.16	2.300	16	3.0	5	4	1	46.08	1.970	16	2.7
3	2	1	27.83	3.206	16	46.6	3	3	2	39.74	2.268	8	8.7	3	2	3	48.12	1.891	16	7.4
1	1	2	28.60	3.121	8	17.2	5	3	0	39.79	2.265	8	1.0	6	3	1	48.21	1.888	16	1.5
2	0	2	30.19	2.960	8	33.4	4	2	2	40.94	2.204	16	19.7	5	3	2	48.70	1.870	16	2.1
4	2	0	30.26	2.954	8	4.4	6	0	0	40.99	2.202	4	6.0	6	4	0	49.77	1.832	8	9.8
4	1	1	31.01	2.884	16	100.0	1	0	3	41.47	2.177	8	0.4							

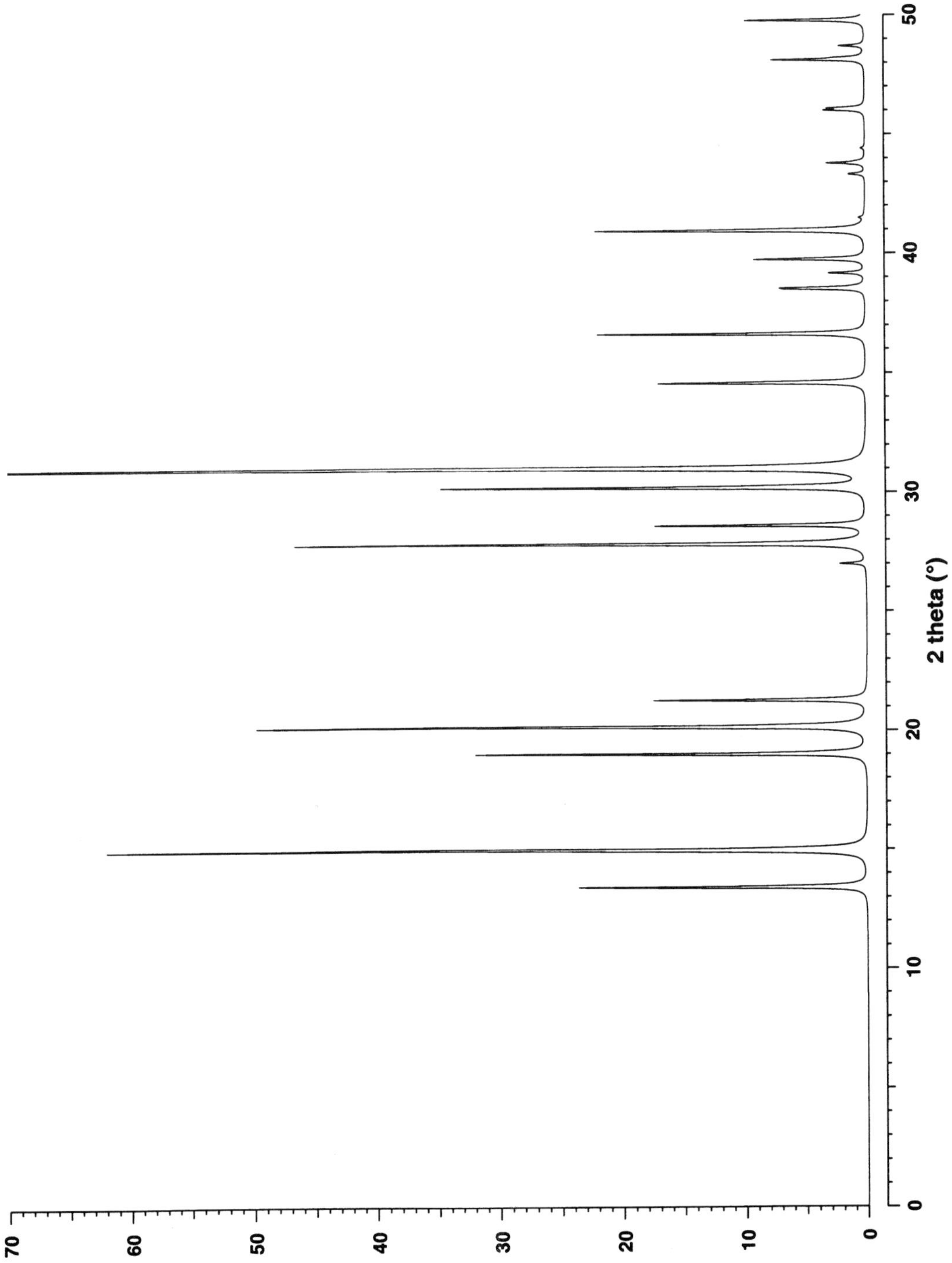
0
10
20
30
40
50
2 theta (°)
70
60
50
40
30
20
10
0

NAT Mesolite

CHEMICAL COMPOSITION: $|Na_{16}Ca_{16}(H_2O)_{64}|$ $[Si_{72}Al_{48}O_{240}]$
Poona, India

REFINED COMPOSITION: $|Na_{16}Ca_{16}(H_2O)_{64}|$ $[Si_{72}Al_{48}O_{240}]$

CRYSTAL DATA: *Fdd*2 (No. 43)
$a = 18.4049$ Å $b = 56.655$ Å $c = 6.5443$ Å
$\alpha = 90°$ $\beta = 90°$ $\gamma = 90°$
X-ray single crystal refinement, $R_w = 0.052$

REFERENCE: G. Artioli, J. V. Smith and J. J. Pluth,
Acta Cryst. **C42** 937–942 (1986).

h	k	l	2θ	d	M	I_{rel}
0	4	0	6.24	14.164	2	2.2
2	2	0	10.11	8.752	4	0.8
2	4	0	11.47	7.717	4	0.9
0	8	0	12.50	7.082	2	0.5
2	6	0	13.44	6.590	4	65.9
1	1	1	14.45	6.130	8	13.1
1	3	1	15.11	5.862	8	100.0
1	5	1	16.37	5.416	8	14.4
1	7	1	18.09	4.905	8	2.9
0	12	0	18.80	4.721	2	40.0
4	0	0	19.29	4.601	2	26.7
3	1	1	19.90	4.462	8	8.7
1	9	1	20.16	4.405	8	42.3
4	4	0	20.29	4.376	4	0.5
3	3	1	20.39	4.355	8	50.1
2	12	0	21.15	4.201	4	20.4
3	5	1	21.34	4.163	8	3.7
4	6	0	21.48	4.136	4	11.5
1	11	1	22.49	3.953	8	4.7
3	7	1	22.70	3.917	8	6.7
4	8	0	23.05	3.858	4	2.4
2	14	0	24.02	3.704	4	1.1
3	9	1	24.40	3.648	8	1.4
4	10	0	24.93	3.572	4	0.7
3	11	1	26.38	3.378	8	5.2
2	16	0	26.98	3.305	4	1.5
4	12	0	27.06	3.295	4	2.0
0	2	2	27.44	3.251	4	1.6
1	15	1	27.70	3.221	8	28.3
5	1	1	27.85	3.203	8	1.2
5	3	1	28.21	3.163	8	38.4
3	13	1	28.59	3.122	8	1.6
0	6	2	28.88	3.092	4	14.1
5	5	1	28.92	3.087	8	0.4
2	0	2	28.96	3.083	4	13.9
2	2	2	29.14	3.065	8	2.1
2	4	2	29.65	3.013	8	2.1
6	4	0	29.80	2.998	4	1.3
5	7	1	29.96	2.983	8	5.5
2	18	0	30.00	2.978	4	4.2
1	17	1	30.49	2.932	8	2.5
2	6	2	30.50	2.931	8	51.0
6	6	0	30.64	2.917	4	5.6
3	15	1	30.98	2.887	8	99.2
5	9	1	31.29	2.858	8	79.3
0	10	2	31.57	2.834	4	1.4
2	8	2	31.65	2.827	8	2.0
6	8	0	31.79	2.815	4	1.2
5	11	1	32.89	2.723	8	2.9
2	20	0	33.09	2.707	4	0.7
6	10	0	33.21	2.697	4	0.6
3	17	1	33.52	2.673	8	1.3
4	4	2	34.22	2.621	8	0.9
4	18	0	34.52	2.598	4	1.0
5	13	1	34.72	2.584	8	1.5
2	12	2	34.75	2.581	8	5.9
6	12	0	34.88	2.572	4	1.9
4	6	2	34.96	2.566	8	7.9
0	14	2	35.27	2.544	4	0.8
4	8	2	35.99	2.496	8	0.7
3	19	1	36.20	2.482	8	1.0
1	21	1	36.35	2.472	8	13.7
2	14	2	36.64	2.452	8	1.0
6	14	0	36.76	2.445	4	1.3
7	1	1	36.88	2.437	8	0.5
7	3	1	37.16	2.420	8	15.3
4	10	2	37.27	2.413	8	1.0
4	20	0	37.27	2.412	4	0.5
0	24	0	38.12	2.361	2	2.2
7	7	1	38.54	2.336	8	1.1
2	16	2	38.72	2.325	8	1.0
4	12	2	38.78	2.322	8	1.9
5	17	1	38.97	2.311	8	2.3
3	21	1	38.98	2.311	8	2.4
8	0	0	39.16	2.301	2	1.5
2	24	0	39.41	2.287	4	1.9
7	9	1	39.62	2.275	8	1.4
0	18	2	39.73	2.268	4	7.3
6	0	2	40.30	2.238	4	3.9
8	6	0	40.35	2.235	4	2.8
6	2	2	40.43	2.231	8	1.6
4	14	2	40.51	2.227	8	1.4
6	4	2	40.82	2.210	8	0.7
7	11	1	40.93	2.205	8	0.9
2	18	2	40.98	2.202	8	15.0
6	18	0	41.09	2.197	4	9.6
8	8	0	41.26	2.188	4	0.7
6	6	2	41.47	2.178	8	10.9
1	1	3	41.72	2.165	8	1.2
3	23	1	41.86	2.158	8	1.2
1	3	3	41.98	2.152	8	0.8
8	10	0	42.40	2.132	4	0.7
7	13	1	42.46	2.129	8	1.8
1	5	3	42.48	2.128	8	1.2
1	25	1	42.50	2.127	8	0.4
2	26	0	42.64	2.120	4	1.2
4	24	0	43.07	2.100	4	0.8
1	7	3	43.23	2.093	8	1.9
6	10	2	43.48	2.081	8	0.4
6	20	0	43.48	2.081	4	1.6
8	12	0	43.77	2.068	4	2.0
3	1	3	44.09	2.054	8	1.2
7	15	1	44.19	2.049	8	0.6
1	9	3	44.22	2.048	8	2.4
3	3	3	44.33	2.043	8	2.7
4	18	2	44.53	2.035	8	0.4
0	22	2	44.78	2.024	4	0.4
3	5	3	44.81	2.022	8	1.8
6	12	2	44.82	2.022	8	0.5
3	25	1	44.83	2.022	8	0.6
8	14	0	45.34	2.000	4	1.5
1	11	3	45.42	1.997	8	1.2
3	7	3	45.53	1.992	8	1.3
1	27	1	45.67	1.986	8	2.7
2	22	2	45.91	1.976	8	0.7
4	26	0	46.09	1.969	4	0.5
6	14	2	46.36	1.958	8	1.4
3	9	3	46.48	1.954	8	5.8
5	23	1	46.48	1.954	8	0.8
4	20	2	46.79	1.942	8	1.3
9	3	1	46.79	1.942	8	2.1
1	13	3	46.83	1.940	8	0.9
3	11	3	47.64	1.909	8	1.8
3	27	1	47.88	1.900	8	1.8
1	15	3	48.44	1.879	8	2.2
2	24	2	48.57	1.874	8	1.0
5	3	3	48.76	1.867	8	4.5
8	4	2	48.81	1.866	8	1.6
9	9	1	48.85	1.864	8	1.7
8	18	0	49.05	1.857	4	0.9
5	5	3	49.21	1.851	8	1.1
5	25	1	49.22	1.851	8	0.6
2	30	0	49.25	1.850	4	5.5
8	6	2	49.38	1.846	8	2.5
5	7	3	49.88	1.828	8	2.4

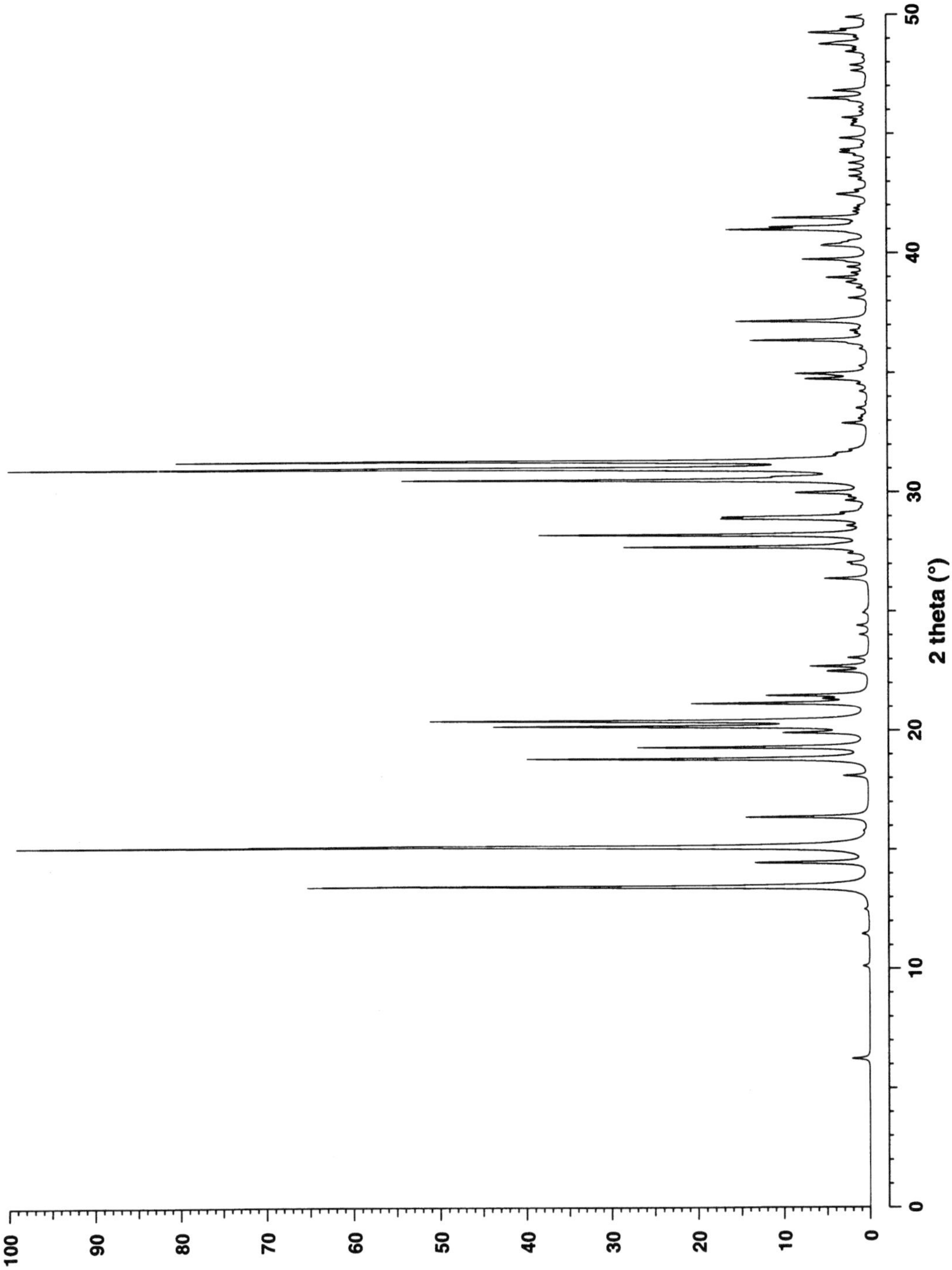
2 theta (°)
0
10
20
30
40
50
100
90
80
70
60
50
40
30
20
10
0

NAT Scolecite

CHEMICAL COMPOSITION: $|Ca_4(H_2O)_{12}|\ [Si_{12}Al_8O_{40}]$
Bombay, India

REFINED COMPOSITION: $|Ca_4(H_2O)_{12}|\ [Si_{12}Al_8O_{40}]$

CRYSTAL DATA: $C1c1$ (No. 9) unique axis **b**, cell choice 1
$a = 6.5222$ Å $b = 18.9678$ Å $c = 9.8398$ Å
$\alpha = 90°$ $\beta = 109.972°$ $\gamma = 90°$
X-ray single crystal refinement, $R_w = 0.031$

REFERENCE: J. V. Smith, J. J. Pluth, G. Artioli and F. K. Ross, in *Proceedings of the 6th International Zeolite Conference, Reno 1983*, Ed. by D. H. Olson and A. Bisio (Butterworth: Guildford) 842–850 (1984).

h	k	l	2θ	d	M	I_{rel}
0	2	1	13.37	6.621	4	49.3
-1	1	1	15.10	5.869	4	100.0
1	1	0	15.19	5.833	4	13.1
0	4	0	18.71	4.742	2	37.5
0	0	2	19.19	4.624	2	35.1
-1	3	1	20.11	4.416	4	26.1
1	3	0	20.18	4.401	4	24.4
-1	1	2	20.26	4.383	4	42.7
1	1	1	20.47	4.338	4	10.3
0	4	1	21.05	4.220	4	12.3
0	2	2	21.38	4.156	4	7.8
-1	3	2	24.26	3.669	4	4.6
1	3	1	24.44	3.642	4	10.6
0	4	2	26.93	3.311	4	2.2
-1	5	1	27.60	3.232	4	10.4
1	5	0	27.65	3.226	4	9.0
-1	1	3	28.00	3.187	4	9.6
0	6	0	28.23	3.161	2	0.2
1	1	2	28.26	3.158	4	15.2
-2	0	2	28.93	3.086	2	6.1
-2	2	1	28.95	3.084	4	9.3
2	0	0	29.13	3.065	2	7.3
0	6	1	29.87	2.991	4	6.3
-2	2	2	30.46	2.934	4	31.5
0	2	3	30.49	2.932	4	11.5
2	2	0	30.65	2.916	4	13.1
-1	5	2	30.81	2.902	4	33.9
1	5	1	30.96	2.889	4	54.7
-1	3	3	31.07	2.878	4	26.2
1	3	2	31.31	2.857	4	52.9
-2	4	1	33.35	2.687	4	1.0
0	6	2	34.36	2.610	4	1.6
-2	4	2	34.68	2.586	4	3.6

h	k	l	2θ	d	M	I_{rel}
0	4	3	34.71	2.585	4	2.0
-2	2	3	34.80	2.578	4	2.8
2	4	0	34.85	2.574	4	1.2
2	2	1	35.14	2.554	4	2.0
-1	7	1	36.20	2.481	4	3.0
1	7	0	36.25	2.478	4	5.2
-1	5	3	36.52	2.460	4	0.6
1	5	2	36.72	2.447	4	3.3
-1	1	4	36.87	2.438	4	6.3
1	1	3	37.15	2.420	4	9.5
0	8	0	37.95	2.371	2	0.9
-2	4	3	38.60	2.332	4	2.8
-1	7	2	38.78	2.322	4	4.1
1	7	1	38.90	2.315	4	1.3
2	4	1	38.92	2.314	4	1.0
0	0	4	38.95	2.312	2	2.3
0	8	1	39.22	2.297	4	4.0
-1	3	4	39.32	2.291	4	0.9
1	3	3	39.59	2.276	4	0.8
-2	6	1	39.71	2.270	4	8.0
-2	0	4	40.03	2.252	2	3.4
0	2	4	40.14	2.246	4	4.7
2	0	2	40.48	2.228	2	0.5
-2	6	2	40.87	2.208	4	9.7
0	6	3	40.89	2.207	4	9.9
2	6	0	41.01	2.201	4	3.0
-2	2	4	41.19	2.191	4	4.7
2	2	2	41.64	2.169	4	8.0
-3	1	2	42.07	2.148	4	1.2
-3	1	1	42.18	2.143	4	3.0
0	8	2	42.86	2.110	4	1.3
0	4	4	43.55	2.078	4	3.3
1	7	2	43.76	2.069	4	0.9

h	k	l	2θ	d	M	I_{rel}
-1	5	4	43.89	2.063	4	1.0
1	5	3	44.13	2.052	4	1.5
-3	3	2	44.28	2.046	4	2.4
-3	1	3	44.29	2.045	4	0.1
-2	6	3	44.32	2.044	4	0.5
-3	3	1	44.38	2.041	4	2.7
-2	4	4	44.53	2.034	4	2.6
3	1	0	44.60	2.032	4	5.0
2	6	1	44.60	2.032	4	1.7
2	4	2	44.95	2.017	4	0.2
-1	9	1	45.48	1.994	4	2.2
1	9	0	45.51	1.993	4	0.6
-3	3	3	46.42	1.956	4	8.4
-1	1	5	46.43	1.956	4	1.3
3	3	0	46.72	1.944	4	0.7
1	1	4	46.73	1.944	4	0.3
-2	8	1	47.41	1.918	4	1.2
-1	9	2	47.64	1.909	4	0.5
1	9	1	47.74	1.905	4	3.9
0	10	0	47.96	1.897	2	1.4
-2	8	2	48.41	1.880	4	1.5
0	8	3	48.43	1.879	4	0.9
-1	3	5	48.48	1.878	4	0.8
2	8	0	48.54	1.875	4	0.1
-3	5	1	48.56	1.875	4	1.7
-3	1	4	48.58	1.874	4	1.7
1	3	4	48.78	1.867	4	3.0
0	6	4	48.80	1.866	4	2.6
0	10	1	49.02	1.858	4	3.5
3	1	1	49.07	1.857	4	3.2
2	2	3	49.51	1.841	4	4.3
-2	6	4	49.70	1.834	4	0.5

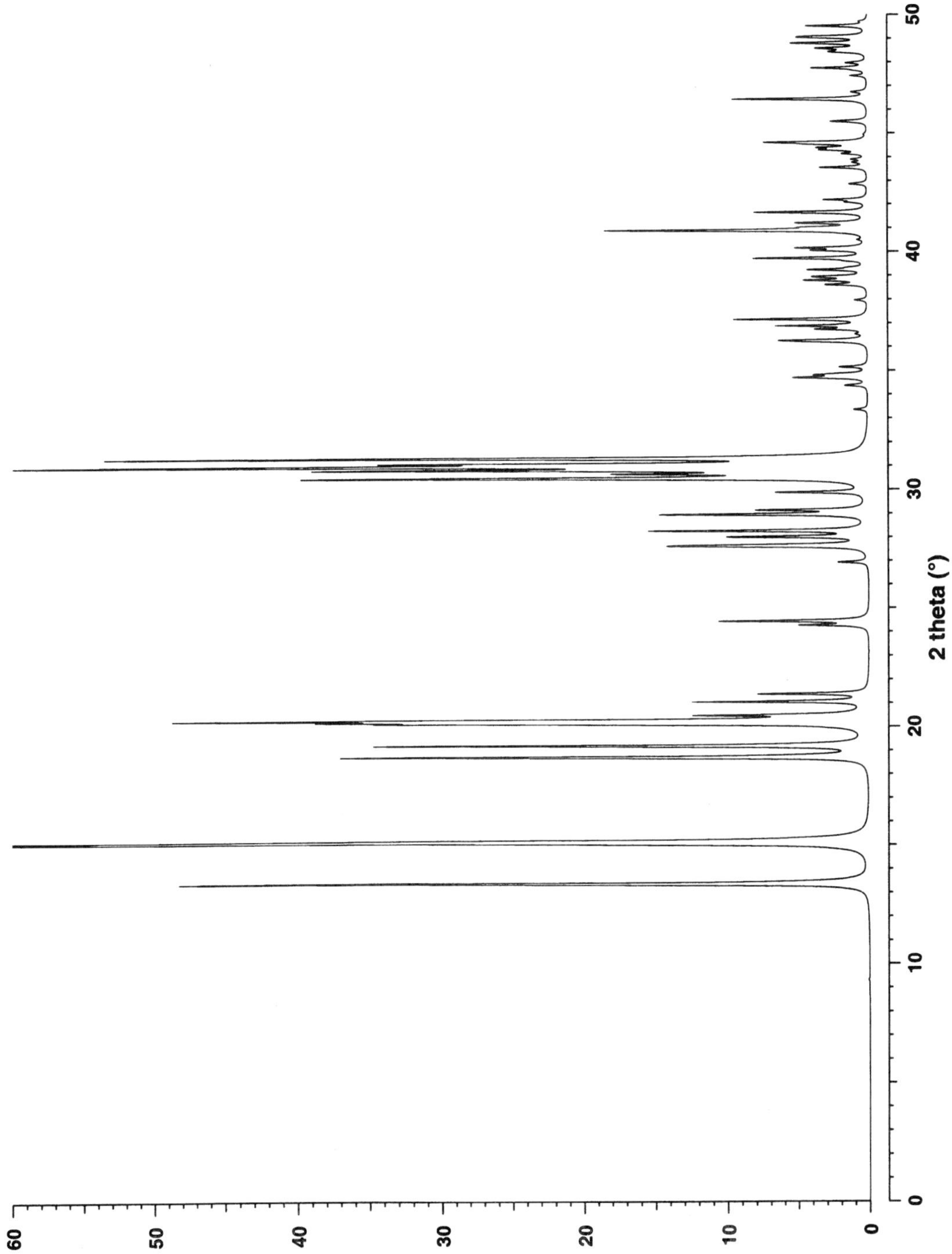
2 theta (°)
0
10
20
30
40
50
60

NES NU-87

CHEMICAL COMPOSITION: $|(H_2O)_{20}|$ $[Si_{68}O_{136}]$

REFINED COMPOSITION: $|(H_2O)_{20}|$ $[Si_{68}O_{136}]$

CRYSTAL DATA: $P1\,2_1/c\,1$ (No. 14) unique axis **b**, cell choice 1
a = 14.324 Å b = 22.376 Å c = 25.092 Å
α = 90° β = 151.515° γ = 90°
X-ray Rietveld refinement, R_{exp} = 0.075, R_{wp} = 0.14, R_I = 0.065

REFERENCE: M. D. Shannon, J. L. Casci, P. A. Cox and S. J. Andrews, *Nature* **353** 417–420 (1991).

h	k	l	2θ	d	M	I_{rel}	h	k	l	2θ	d	M	I_{rel}	h	k	l	2θ	d	M	I_{rel}
-1	0	2	7.05	12.546	2	10.9	-4	2	6	26.15	3.407	4	1.1	-1	0	6	33.99	2.637	2	0.7
0	2	0	7.90	11.188	2	100.0	-2	2	6	26.29	3.389	4	1.0	-3	7	4	34.07	2.632	4	0.5
-1	1	1	8.34	10.597	4	12.9	2	1	0	26.39	3.377	4	3.1	-2	8	4	35.13	2.555	4	1.2
0	1	1	8.38	10.553	4	19.6	-3	4	6	26.61	3.349	4	1.3	2	6	0	35.64	2.519	4	0.7
-1	2	2	10.59	8.350	4	9.3	-4	1	5	26.65	3.345	4	6.0	-5	0	10	35.79	2.509	2	3.9
0	2	1	10.83	8.173	4	0.5	-1	1	5	26.82	3.324	4	1.9	-4	6	5	35.83	2.506	4	0.7
1	0	0	12.96	6.832	2	5.7	-1	6	3	26.99	3.303	4	0.5	-5	2	10	36.71	2.448	4	2.2
-2	1	3	12.99	6.814	4	2.4	-3	0	2	27.01	3.301	2	0.9	0	9	1	36.93	2.434	4	0.4
1	1	0	13.55	6.534	4	0.5	1	0	2	27.10	3.290	2	1.1	-3	8	5	37.37	2.406	4	0.5
-1	3	1	13.97	6.339	4	0.3	1	6	0	27.24	3.273	4	3.5	-4	6	8	37.47	2.400	4	0.7
0	3	1	13.99	6.330	4	1.4	2	2	0	27.30	3.267	4	13.2	-3	8	4	37.56	2.395	4	0.9
-2	0	4	14.12	6.273	2	0.6	0	4	3	27.46	3.248	4	0.5	-5	1	5	37.59	2.393	4	0.8
-2	2	4	16.20	5.472	4	3.7	-3	5	5	27.46	3.248	4	4.4	-1	8	4	37.62	2.391	4	1.0
-2	3	3	17.17	5.163	4	1.0	-2	5	5	27.52	3.241	4	5.3	-4	7	6	37.74	2.383	4	0.4
1	3	0	17.60	5.038	4	0.4	-1	2	5	27.71	3.220	4	1.9	0	1	5	37.80	2.380	4	0.6
-3	0	4	19.15	4.635	2	0.4	-2	6	4	27.83	3.206	4	0.8	-3	8	6	38.73	2.325	4	0.6
-3	1	5	19.22	4.619	4	8.3	-2	6	2	28.15	3.170	4	0.7	-3	5	1	38.76	2.323	4	0.5
-2	1	5	19.29	4.600	4	11.1	-2	5	1	28.16	3.168	4	1.8	2	5	1	38.80	2.321	4	0.3
-2	1	1	20.19	4.399	4	1.9	-3	2	2	28.18	3.166	4	0.4	-5	3	5	39.34	2.290	4	0.8
1	1	1	20.23	4.389	4	4.1	0	6	2	28.20	3.165	4	0.5	-5	4	5	40.82	2.211	4	0.4
-3	2	5	20.42	4.349	4	1.6	1	5	1	28.20	3.165	4	1.1	-5	5	10	41.25	2.189	4	0.7
-2	2	5	20.49	4.334	4	0.4	1	2	2	28.27	3.157	4	0.5	-4	8	5	41.91	2.156	4	0.6
1	4	0	20.52	4.328	4	26.6	-4	0	8	28.46	3.136	2	1.3	1	10	0	42.51	2.126	4	0.4
-1	2	4	20.86	4.259	4	1.6	-1	7	1	28.90	3.089	4	1.0	0	5	5	42.85	2.111	4	0.4
-2	4	4	21.28	4.175	4	6.7	0	7	1	28.91	3.088	4	1.3	3	4	0	42.88	2.109	4	0.6
-3	3	5	22.29	3.989	4	10.7	-4	3	5	28.98	3.081	4	1.0	-6	5	9	42.98	2.105	4	0.5
-2	3	5	22.35	3.977	4	12.5	-1	3	5	29.14	3.065	4	0.9	-3	5	9	43.18	2.095	4	0.5
-1	3	4	22.69	3.919	4	0.5	-4	4	6	29.64	3.014	4	0.4	-6	0	12	43.27	2.091	2	1.1
-3	2	6	22.70	3.917	4	13.1	-2	4	6	29.77	3.001	4	0.4	-7	1	11	44.49	2.036	4	0.4
-2	3	1	23.14	3.844	4	0.7	-3	5	3	29.92	2.987	4	0.7	3	5	0	44.65	2.030	4	0.8
1	3	1	23.17	3.838	4	1.4	0	5	3	30.01	2.978	4	0.5	-4	1	11	44.73	2.026	4	0.4
-2	5	3	23.45	3.794	4	1.2	-3	6	5	30.55	2.926	4	1.8	-4	9	5	45.26	2.003	4	0.6
-1	5	3	23.49	3.788	4	0.4	-3	6	4	30.77	2.906	4	0.9	-2	10	1	45.29	2.002	4	0.4
1	5	0	23.77	3.743	4	2.9	-4	2	4	30.78	2.905	4	0.5	-1	9	5	45.37	1.999	4	0.5
0	6	0	23.86	3.729	2	9.2	-1	6	4	30.85	2.898	4	0.9	0	0	6	45.48	1.995	2	0.6
-3	4	5	24.68	3.608	4	1.6	-4	4	5	30.89	2.895	4	0.6	3	6	0	46.74	1.944	4	1.0
-2	4	5	24.74	3.599	4	0.4	-4	3	8	30.93	2.891	4	0.5	0	7	5	47.45	1.916	4	0.4
-3	4	4	24.95	3.569	4	1.4	-1	4	5	31.04	2.881	4	0.4	-7	4	10	47.69	1.907	4	0.7
-1	6	1	25.00	3.562	4	0.8	1	6	1	31.22	2.865	4	1.2	-3	4	10	47.97	1.897	4	1.0
-1	4	4	25.04	3.556	4	0.4	-3	4	2	31.47	2.843	4	0.4	-5	8	10	48.75	1.868	4	1.2
-2	4	1	25.45	3.500	4	0.6	-4	4	8	32.73	2.736	4	0.3	-7	5	11	48.97	1.860	4	0.4
1	4	1	25.49	3.495	4	0.5	-1	8	2	32.81	2.730	4	0.8	3	7	0	49.12	1.855	4	0.4
-4	1	7	25.96	3.432	4	0.9	0	6	3	32.88	2.724	4	0.3	-4	5	11	49.20	1.852	4	0.5
-3	1	7	26.05	3.421	4	1.7	-1	5	5	33.33	2.688	4	0.7	1	11	1	49.20	1.852	4	0.4
2	0	0	26.09	3.416	2	7.4	-5	0	6	33.77	2.654	2	0.5	-3	5	10	49.59	1.838	4	0.3

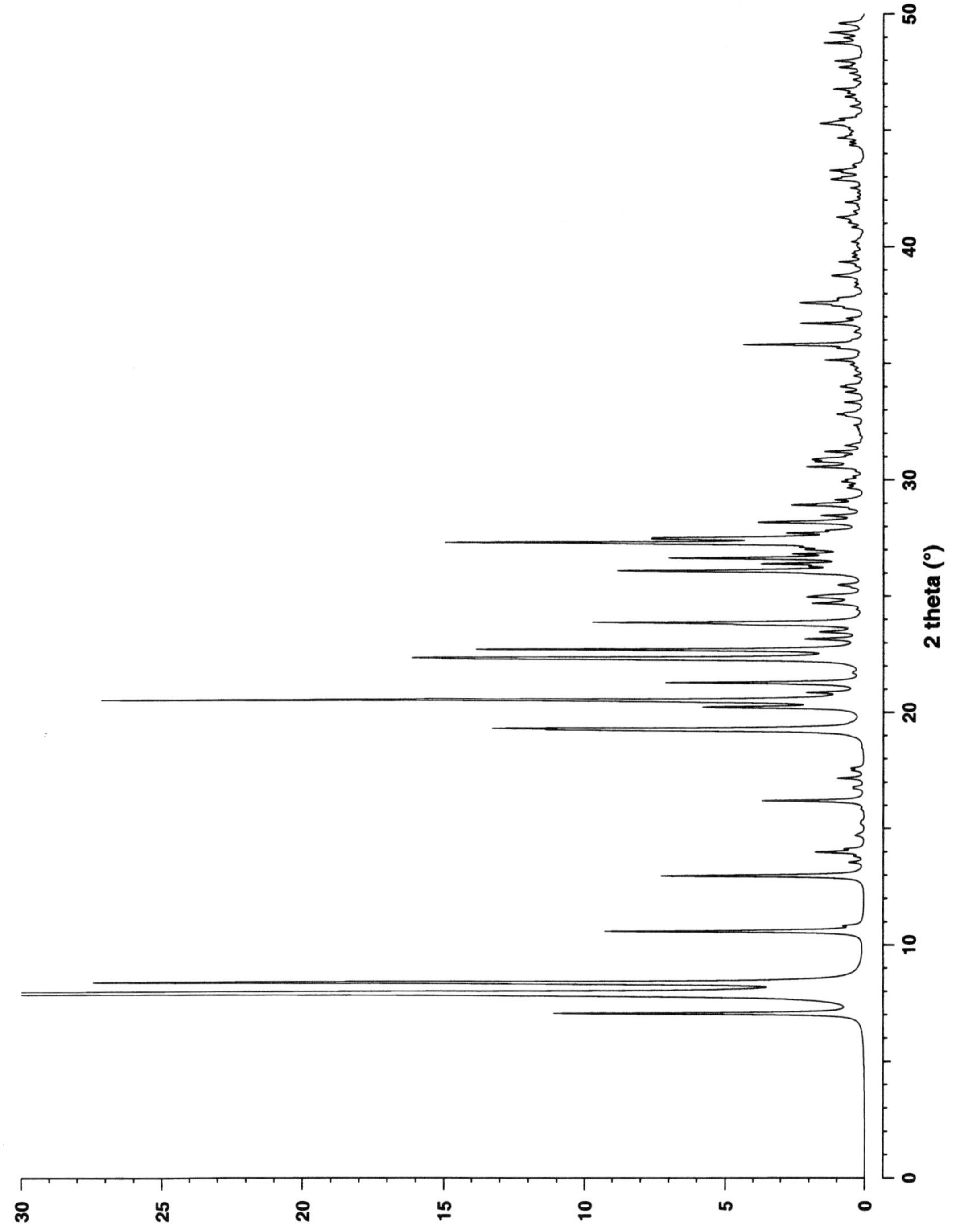
2 theta (°)
0
10
20
30
40
50
0
5
10
15
20
25
30

NES Gottardiite

CHEMICAL COMPOSITION: $|Na_{2.5}K_{0.2}Mg_{3.1}Ca_{4.9}(H_2O)_{93}|$ $[Si_{117.2}Al_{18.8}O_{272}]$
Mt. Adamson, Northern Victoria Land, Antarctica

REFINED COMPOSITION: $|Na_{66.8}|^{\dagger}$ $[Si_{115.6}Al_{20.4}O_{272}]$

CRYSTAL DATA: *Cmce* (No. 64)
a = 13.698 Å b = 25.213 Å c = 22.660 Å
α = 90.0° β = 90.0° γ = 90.0°
X-ray single crystal refinement, R_F = 0.1195, R_w = 0.0968

REFERENCE: A. Alberti, G. Vezzalini, E. Galli and S. Quartieri, *Eur. J. Mineral.* **8** 69–75 (1996).

h	k	l	2θ	d	M	I_{rel}
0	2	0	7.01	12.606	2	1.1
0	0	2	7.80	11.330	2	100.0
1	1	1	8.32	10.630	8	42.9
0	2	2	10.50	8.427	4	4.0
1	1	2	10.72	8.250	8	0.5
2	0	0	12.93	6.849	2	3.6
1	3	1	12.96	6.830	8	2.5
1	1	3	13.84	6.398	8	3.1
0	4	0	14.05	6.303	2	0.6
2	2	0	14.72	6.018	4	0.5
0	4	2	16.09	5.508	4	4.1
2	2	2	16.68	5.315	8	2.9
1	3	3	17.06	5.198	8	1.3
1	5	1	19.16	4.632	8	17.9
3	1	1	20.15	4.407	8	11.4
1	5	2	20.34	4.367	8	1.4
2	0	4	20.34	4.365	4	36.1
2	4	2	20.69	4.292	8	3.3
0	4	4	21.08	4.213	4	9.4
0	6	0	21.14	4.202	2	0.3
2	2	4	21.54	4.125	8	1.2
1	5	3	22.17	4.010	8	30.2
0	6	2	22.57	3.940	4	19.9
3	1	3	23.03	3.861	8	3.7
1	3	5	23.22	3.830	8	2.6
0	0	6	23.56	3.777	2	10.3
1	5	4	24.51	3.632	8	3.6
2	2	5	24.59	3.620	8	0.2
0	2	6	24.61	3.618	4	1.5
2	4	4	24.81	3.589	8	1.2
2	6	0	24.86	3.582	4	1.3
1	7	1	25.88	3.443	8	5.2
4	0	0	26.02	3.424	2	14.1
2	6	2	26.09	3.415	8	2.7
0	6	4	26.41	3.375	4	2.4
3	5	1	26.63	3.347	8	8.2
2	0	6	26.96	3.307	4	7.8
4	2	0	26.98	3.305	4	1.6
4	0	2	27.20	3.278	4	22.6
1	5	5	27.25	3.273	8	13.9
3	5	2	27.50	3.243	8	2.2
2	4	5	27.52	3.241	8	0.2
0	4	6	27.53	3.240	4	0.3
2	6	3	27.56	3.236	8	0.6
2	2	6	27.89	3.199	8	2.3
3	1	5	27.96	3.191	8	9.0
4	2	2	28.13	3.173	8	0.6
1	7	3	28.21	3.163	8	0.4
0	8	0	28.32	3.152	2	1.6
1	1	7	28.55	3.126	8	2.2
3	5	3	28.91	3.089	8	4.5
2	6	4	29.50	3.027	8	1.9
3	3	5	29.74	3.004	8	1.8
1	5	6	30.28	2.952	8	4.4
2	4	6	30.52	2.929	8	3.9
4	4	2	30.74	2.908	8	1.0
3	5	4	30.77	2.906	8	3.7
4	2	4	31.34	2.855	8	1.2
2	8	1	31.50	2.840	8	0.3
2	6	5	31.84	2.810	8	0.3
3	7	1	31.89	2.806	8	2.6
2	8	2	32.25	2.776	8	0.7
0	2	8	32.39	2.764	4	2.9
1	7	5	32.42	2.762	8	0.3
1	9	1	32.87	2.725	8	0.3
3	5	5	33.03	2.712	8	1.3
4	6	0	33.76	2.655	4	1.5
3	1	7	34.14	2.626	8	0.6
2	0	8	34.26	2.617	4	0.3
2	6	6	34.51	2.599	8	1.1
4	6	2	34.71	2.585	8	0.4
0	4	8	34.72	2.584	4	3.1
1	9	3	34.78	2.580	8	0.6
4	0	6	35.38	2.537	4	1.1
0	10	0	35.61	2.521	2	5.9
3	5	6	35.62	2.521	8	0.5
0	10	1	35.83	2.506	4	1.6
4	2	6	36.11	2.487	8	0.5
0	10	2	36.51	2.461	4	3.1
1	5	8	36.99	2.430	8	0.4
0	8	6	37.15	2.420	4	0.2
2	4	8	37.19	2.417	8	2.8
5	5	1	37.57	2.394	8	1.8
0	10	3	37.61	2.392	4	0.2
3	9	1	37.89	2.375	8	1.0
5	5	2	38.22	2.355	8	0.8
2	10	1	38.24	2.353	8	0.6
0	6	8	38.32	2.349	4	1.6
1	9	5	38.34	2.348	8	0.6
3	5	7	38.48	2.339	8	0.7
5	1	5	38.57	2.334	8	1.6
5	5	3	39.28	2.294	8	1.8
6	0	0	39.47	2.283	2	0.3
4	8	2	39.67	2.272	8	0.3
1	5	9	40.59	2.223	8	0.4
0	10	5	40.96	2.203	4	0.3
3	1	9	41.09	2.196	8	1.2
3	5	8	41.57	2.172	8	0.3
4	6	6	41.58	2.172	8	0.7
1	11	3	41.70	2.166	8	0.2
2	0	10	42.00	2.151	4	1.6
6	4	0	42.09	2.147	4	0.8
5	5	5	42.52	2.126	8	0.3
2	2	10	42.63	2.121	8	0.2
6	0	4	42.70	2.118	4	1.5
3	9	5	42.80	2.113	8	0.4
0	8	8	42.93	2.107	4	0.2
0	12	0	43.05	2.101	2	1.2
3	11	1	44.40	2.040	8	0.5
2	4	10	44.50	2.036	8	0.3
5	5	6	44.64	2.030	8	1.0
5	3	7	44.65	2.029	8	0.4
4	10	1	44.82	2.022	8	0.3
3	5	9	44.87	2.020	8	0.4
2	12	0	45.14	2.009	4	0.4
6	6	0	45.20	2.006	4	1.2
0	10	7	45.60	1.989	4	0.3
1	3	11	45.83	1.980	8	0.3
4	8	6	45.92	1.976	8	0.5
6	0	6	46.48	1.954	4	1.8
5	5	7	47.04	1.932	8	1.1
3	9	7	47.30	1.922	8	0.3
2	6	10	47.48	1.915	8	0.5
1	13	1	47.52	1.913	8	0.6
4	10	4	47.57	1.911	8	2.1
7	1	3	48.17	1.889	8	0.2
0	10	8	48.33	1.883	4	1.7
3	5	10	48.34	1.883	8	0.3
3	1	11	48.62	1.873	8	1.6
3	11	5	48.79	1.867	8	2.3
1	9	9	49.10	1.855	8	0.3
1	11	7	49.16	1.853	8	0.6
7	3	3	49.31	1.848	8	0.2
5	5	8	49.70	1.834	8	1.3
3	3	11	49.75	1.833	8	0.5

The space group symbol *Cmce* now replaces the older form, *Cmca*.

†Na atoms represent unresolved cations and water in the channels.

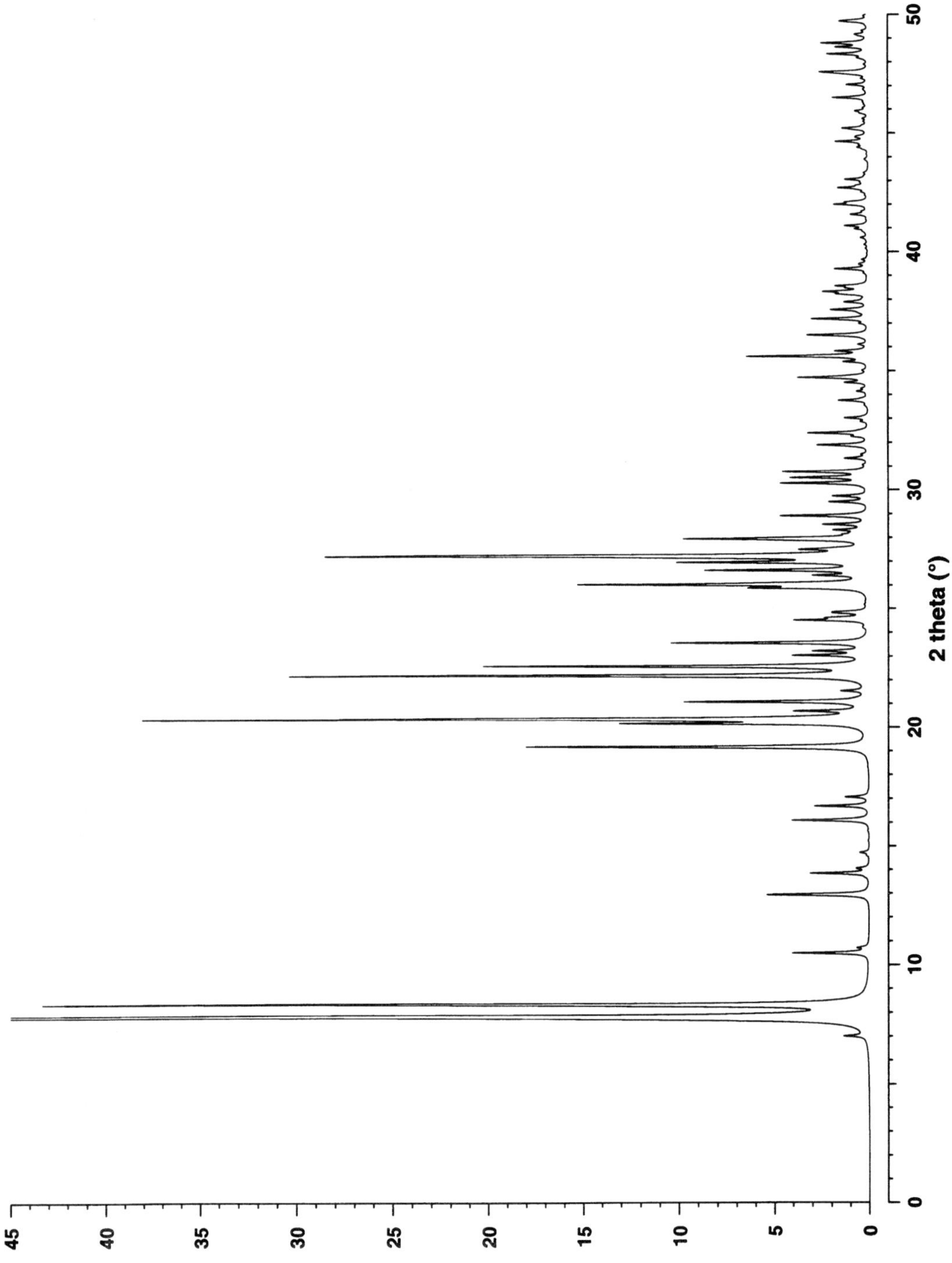
2 theta (°)

NON 2-aminopentane Nonasil

CHEMICAL COMPOSITION: $|(C_5H_{11}NH_2)_4|$ $[Si_{88}O_{176}]$
$C_5H_{11}NH_2$ = 2-aminopentane

REFINED COMPOSITION: $|C_{36}|$ $[Si_{88}O_{176}]$

CRYSTAL DATA: *Fmmm* (No. 69)
$a = 22.232$ Å $b = 15.058$ Å $c = 13.627$ Å
$\alpha = 90°$ $\beta = 90°$ $\gamma = 90°$
X-ray single crystal refinement, $R_w = 0.125$

REFERENCE: B. Marler, N. Dehnbostel, H. H. Eulert, H. Gies and F. Liebau, *J. Inclusion Phenomena* **4** 339–349 (1986).

h	*k*	*l*	2*θ*	*d*	*M*	I_{rel}
2	0	0	7.95	11.116	2	24.5
1	1	1	9.61	9.198	8	52.7
0	2	0	11.75	7.529	2	5.4
0	0	2	12.99	6.814	2	4.3
2	2	0	14.21	6.234	4	2.9
3	1	1	14.82	5.976	8	1.6
2	0	2	15.25	5.809	4	2.8
1	3	1	19.26	4.608	8	61.7
2	2	2	19.30	4.599	8	14.4
4	2	0	19.85	4.472	4	31.2
4	0	2	20.62	4.307	4	100.0
1	1	3	20.81	4.268	8	30.3
5	1	1	21.84	4.070	8	0.2
3	3	1	22.36	3.975	8	47.3
0	4	0	23.63	3.764	2	4.7
3	1	3	23.72	3.751	8	10.4
4	2	2	23.80	3.738	8	2.0
6	0	0	24.02	3.705	2	21.8
2	4	0	24.97	3.566	4	26.1
0	0	4	26.16	3.407	2	5.4
1	3	3	26.77	3.330	8	24.4
6	2	0	26.82	3.325	4	3.5
2	0	4	27.38	3.257	4	26.4
6	0	2	27.40	3.255	4	0.4
5	3	1	27.59	3.233	8	14.7
4	4	0	28.64	3.117	4	1.6
5	1	3	28.71	3.109	8	3.2
0	2	4	28.76	3.104	4	0.5
3	3	3	29.12	3.066	8	1.9
7	1	1	29.48	3.030	8	2.1
2	2	4	29.89	2.989	8	4.5
6	2	2	29.90	2.988	8	1.9
1	5	1	30.67	2.915	8	1.1
4	4	2	31.56	2.834	8	4.2
8	0	0	32.21	2.779	2	1.7
3	5	1	32.76	2.733	8	3.2
5	3	3	33.37	2.685	8	11.0
6	4	0	33.95	2.641	4	0.4
7	3	1	34.05	2.633	8	0.2
8	2	0	34.40	2.607	4	0.5
8	0	2	34.87	2.573	4	1.0
7	1	3	34.98	2.565	8	2.0
0	4	4	35.54	2.526	4	2.1
3	1	5	35.60	2.522	8	0.7
0	6	0	35.78	2.510	2	11.4
6	0	4	35.80	2.508	4	1.3
6	4	2	36.49	2.462	8	2.3
5	5	1	36.64	2.453	8	2.6
2	6	0	36.71	2.448	4	3.5
8	2	2	36.92	2.435	8	4.2
9	1	1	37.48	2.400	8	1.4
1	3	5	37.78	2.381	8	2.6
6	2	4	37.81	2.379	8	0.2
0	6	2	38.22	2.355	4	0.2
7	3	3	38.98	2.311	8	3.1
2	6	2	39.10	2.304	8	0.8
4	4	4	39.17	2.300	8	1.5
5	1	5	39.23	2.296	8	3.5
4	6	0	39.39	2.287	4	1.8
3	3	5	39.54	2.279	8	4.0
0	0	6	39.68	2.271	2	0.8
8	4	0	40.34	2.236	4	1.6
2	0	6	40.54	2.225	4	0.3
5	5	3	41.30	2.186	8	1.7
0	2	6	41.53	2.174	4	1.3
4	6	2	41.65	2.168	8	0.2
9	1	3	42.07	2.148	8	0.7
2	2	6	42.35	2.134	8	0.1
10	2	0	42.39	2.132	4	0.8
8	4	2	42.56	2.124	8	0.8
1	7	1	42.75	2.115	8	0.5
10	0	2	42.78	2.114	4	2.3
5	3	5	42.89	2.109	8	0.2
4	0	6	43.02	2.102	4	0.7
6	4	4	43.35	2.087	8	0.2
6	6	0	43.55	2.078	4	0.2
8	2	4	43.72	2.070	8	0.4
7	1	5	44.20	2.049	8	0.1
3	7	1	44.35	2.043	8	0.3
1	5	5	45.05	2.012	8	0.5
9	3	3	45.54	1.992	8	1.8
2	6	4	45.63	1.988	8	0.4
6	6	2	45.64	1.988	8	0.4
7	5	3	46.09	1.969	8	0.4
3	5	5	46.58	1.950	8	0.4
0	4	6	46.71	1.945	4	1.2
1	7	3	46.91	1.937	8	2.4
6	0	6	46.92	1.936	4	2.5
1	1	7	47.26	1.923	8	1.2
5	7	1	47.42	1.917	8	0.7
2	4	6	47.46	1.916	8	0.1
10	4	0	47.49	1.914	4	0.1
7	3	5	47.55	1.912	8	0.6
4	6	4	47.90	1.899	8	2.5
9	5	1	48.11	1.891	8	0.6
0	8	0	48.35	1.882	2	0.6
8	6	0	48.90	1.863	4	1.2
2	8	0	49.09	1.856	4	0.2
10	4	2	49.45	1.843	8	1.1
4	4	6	49.67	1.836	8	0.2
11	1	3	49.75	1.833	8	1.3

2-aminopentane Nonasil **NON**

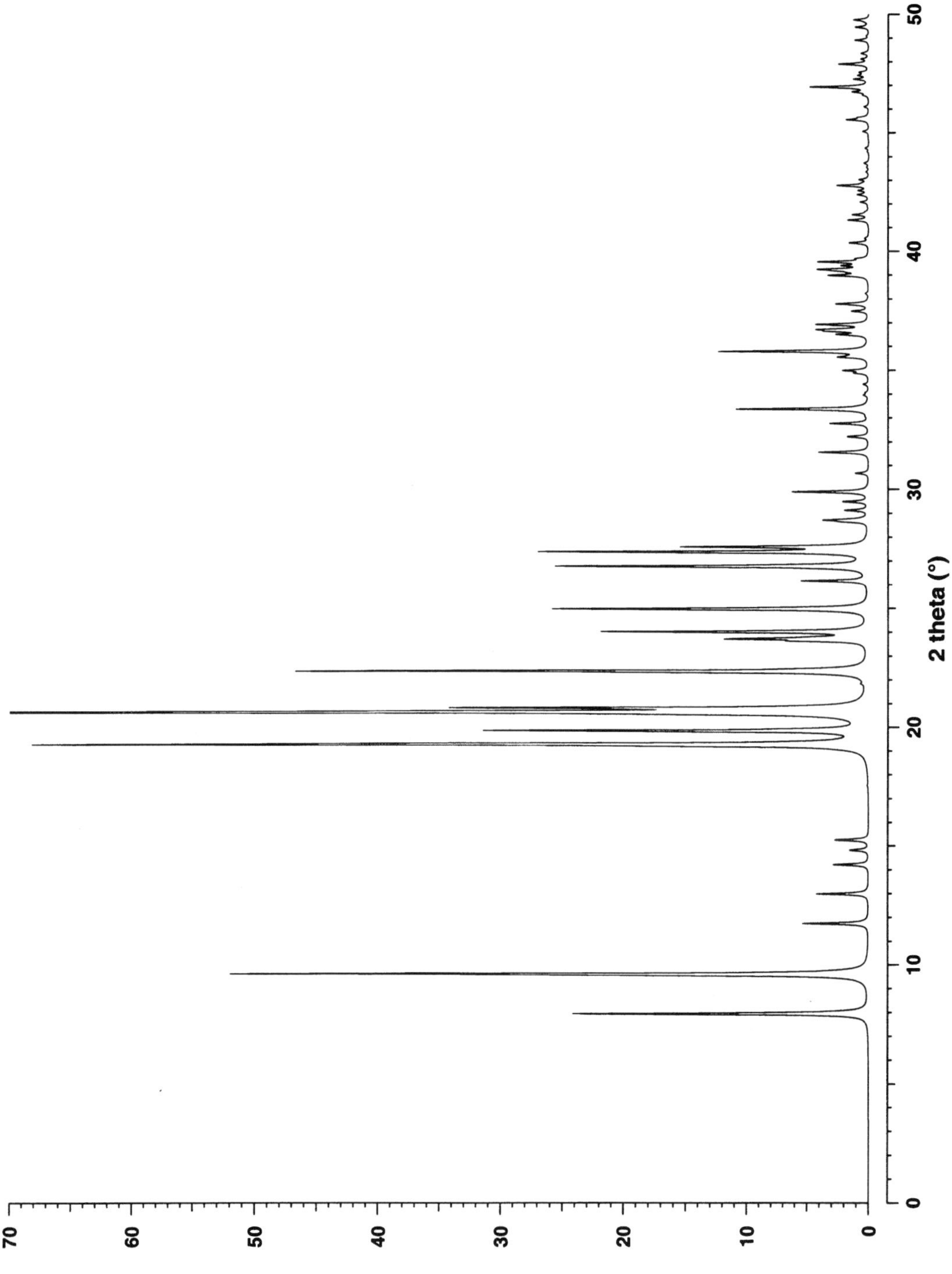

OFF Offretite

CHEMICAL COMPOSITION: $|Na_{0.1}K_{1.09}Mg_{1.11}Ca_{1.07}Sr_{0.01}Ba_{0.01}(H_2O)_{17.63}|$ $[Al_{5.52}Fe^{3+}_{0.01}Si_{12.49}O_{36}]$
Poia Creek, Adamello, Italy

REFINED COMPOSITION: $|KMgCa_{1.32}O_{16.04}|$ $[Si_{12.6}Al_{5.4}O_{36}]$

CRYSTAL DATA: $P\overline{6}m2$ (No. 187)
$a = 13.331$ Å $b = 13.291$ Å $c = 7.593$ Å
$\alpha = 90°$ $\beta = 90°$ $\gamma = 120°$
X-ray single crystal refinement, $R = 0.058$, $R_w = 0.038$

REFERENCE: A. Alberti, G. Cruciani, E. Galli and G. Vezzalini, *Zeolites* **17** 457–461 (1996).

h	k	l	2θ	d	M	I_{rel}
1	0	0	7.66	11.545	6	100.0
0	0	1	11.65	7.593	2	13.8
1	1	0	13.30	6.655	6	12.1
1	0	1	13.96	6.344	12	9.7
2	0	0	15.35	5.772	6	19.3
2	0	1	19.31	4.595	12	23.1
2	1	0	20.37	4.360	12	31.9
3	0	0	23.11	3.848	6	20.0
0	0	2	23.43	3.796	2	3.0
2	1	1	23.53	3.781	24	57.1
1	0	2	24.68	3.607	12	37.0
3	0	1	25.96	3.433	12	3.1
2	2	0	26.79	3.328	6	7.6
1	1	2	27.04	3.298	12	2.0
3	1	0	27.88	3.200	12	3.5
2	0	2	28.13	3.172	12	8.1
2	2	1	29.30	3.048	12	0.6
3	1	1	30.31	2.949	24	7.4
4	0	0	30.98	2.886	6	25.3
2	1	2	31.24	2.863	24	46.9
3	0	2	33.15	2.703	12	1.3
4	0	1	33.21	2.698	12	8.8
3	2	0	33.88	2.646	12	1.5
0	0	3	35.47	2.531	2	1.3
4	1	0	35.65	2.518	12	15.5
2	2	2	35.88	2.502	12	0.5
3	2	1	35.95	2.498	24	4.9
1	0	3	36.34	2.472	12	0.4
3	1	2	36.73	2.447	24	1.1
4	1	1	37.63	2.390	24	0.1
2	0	3	38.85	2.318	12	1.5
5	0	0	39.01	2.309	6	1.2
4	0	2	39.21	2.298	12	0.2
3	3	0	40.67	2.218	6	7.6
5	0	1	40.85	2.209	12	0.9
2	1	3	41.24	2.189	24	1.0
4	2	0	41.42	2.180	12	0.8
3	2	2	41.61	2.171	24	0.1
3	3	1	42.45	2.129	12	3.4
3	0	3	42.76	2.115	12	4.8
4	1	2	43.10	2.099	24	1.0
4	2	1	43.18	2.095	24	3.4
5	1	0	43.67	2.073	12	0.4
2	2	3	45.00	2.015	12	0.1
5	1	1	45.35	2.000	24	3.9
3	1	3	45.70	1.985	24	1.0
5	0	2	46.00	1.973	12	3.0
3	3	2	47.46	1.915	12	0.4
4	0	3	47.80	1.903	12	3.0
0	0	4	47.92	1.898	2	7.7
4	2	2	48.13	1.890	24	0.3
1	0	4	48.61	1.873	12	2.5
6	0	1	48.83	1.865	12	0.3
5	2	0	49.33	1.847	12	1.2
4	3	1	49.56	1.839	24	0.6

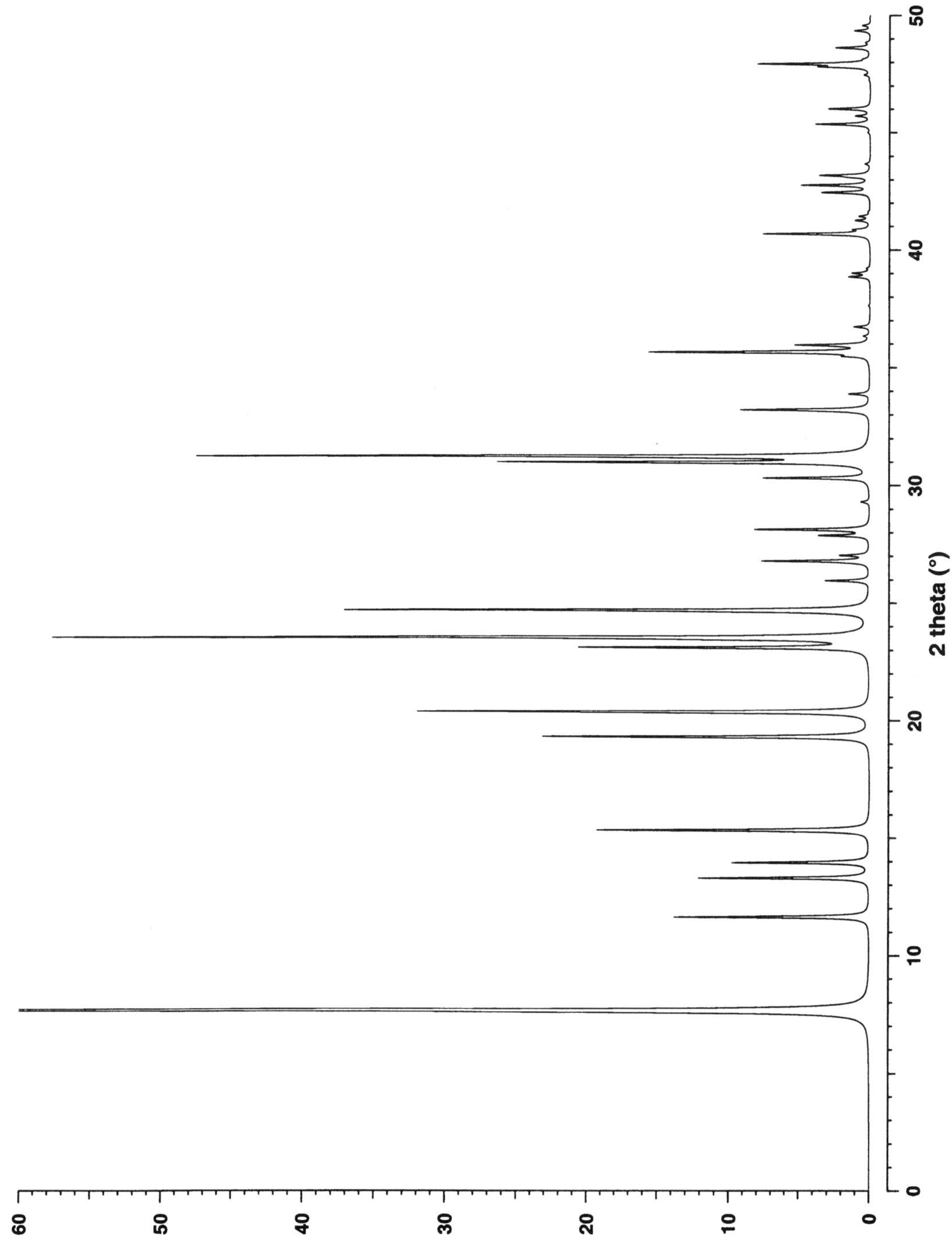
2 theta (°)
0
10
20
30
40
50
0
10
20
30
40
50
60

OSI UiO-6, Calcined

CHEMICAL COMPOSITION: $[Al_{16}P_{16}O_{64}]$

REFINED COMPOSITION: $[Al_{16}P_{16}O_{64}]$

CRYSTAL DATA: *Imm*2 (No. 44)
$a = 18.3549$ Å $b = 18.3206$ Å $c = 5.0530$ Å
$\alpha = 90.0°$ $\beta = 90.0°$ $\gamma = 90.0°$
X-ray Rietveld refinement, $R_p = 0.044$, $R_{wp} = 0.059$

REFERENCE: D. E. Akporiaye, H. Fjellvåg, E. N. Halvorsen, T. Haug, A. Karlsson, and K. P Lillerud, *J. Chem. Soc., Chem. Commun.* 1553–1554 (1996).

h	k	l	2θ	d	M	I_{rel}
1	1	0	6.82	12.967	4	100.0
2	0	0	9.64	9.177	2	16.9
0	2	0	9.66	9.160	2	30.5
2	2	0	13.66	6.483	4	1.2
3	1	0	15.27	5.803	4	1.6
1	3	0	15.29	5.795	4	2.4
1	0	1	18.21	4.872	4	1.1
0	1	1	18.21	4.871	4	0.2
4	0	0	19.34	4.589	2	3.3
0	4	0	19.38	4.580	2	5.9
3	3	0	20.55	4.322	4	43.0
2	1	1	20.64	4.303	8	32.1
1	2	1	20.65	4.301	8	27.6
2	4	0	21.68	4.098	4	2.2
3	0	1	22.82	3.896	4	5.9
0	3	1	22.84	3.893	4	0.3
1	5	0	24.78	3.593	4	0.7
3	2	1	24.83	3.585	8	2.3
2	3	1	24.84	3.584	8	2.3
4	1	1	26.69	3.340	8	7.0
1	4	1	26.71	3.337	8	13.0
4	4	0	27.51	3.242	4	7.4
6	0	0	29.19	3.059	2	1.8
0	6	0	29.25	3.053	2	1.6
5	0	1	30.09	2.970	4	0.5
4	3	1	30.10	2.969	8	0.9
3	4	1	30.11	2.968	8	1.0
0	5	1	30.13	2.966	4	0.5
6	2	0	30.81	2.902	4	5.0
2	6	0	30.86	2.897	4	3.7
5	2	1	31.67	2.825	8	1.6
2	5	1	31.70	2.823	8	0.6
7	1	0	34.55	2.596	4	0.9
5	5	0	34.59	2.593	4	0.3
1	7	0	34.62	2.591	4	1.1
6	1	1	34.62	2.591	8	0.6
1	6	1	34.67	2.587	8	0.1
6	4	0	35.28	2.544	4	0.4
4	6	0	35.31	2.542	4	0.6
0	0	2	35.53	2.526	2	8.3
5	4	1	36.04	2.492	8	3.1
4	5	1	36.05	2.491	8	4.2
1	1	2	36.22	2.480	8	1.1
2	0	2	36.90	2.436	4	0.1
0	2	2	36.91	2.436	4	0.5
7	3	0	37.32	2.409	4	2.4
3	7	0	37.37	2.406	4	1.5
6	3	1	37.39	2.405	8	0.2
3	6	1	37.42	2.403	8	0.4
2	2	2	38.23	2.354	8	0.2
7	0	1	38.69	2.327	4	0.9
0	7	1	38.75	2.324	4	0.6
3	1	2	38.88	2.316	8	0.2
8	0	0	39.27	2.294	2	0.1
7	2	1	39.97	2.256	8	0.8
2	7	1	40.02	2.253	8	0.1
8	2	0	40.53	2.226	4	0.2
2	8	0	40.60	2.222	4	0.5
4	0	2	40.77	2.213	4	0.7
0	4	2	40.79	2.212	4	0.5
3	3	2	41.39	2.181	8	0.5
6	6	0	41.80	2.161	4	0.4
4	2	2	42.00	2.151	8	0.2
7	5	0	42.39	2.132	4	0.2
6	5	1	42.45	2.130	8	0.2
5	6	1	42.46	2.129	8	0.7
8	1	1	43.60	2.076	8	0.2
7	4	1	43.62	2.075	8	0.1
4	7	1	43.66	2.073	8	1.3
1	8	1	43.67	2.073	8	0.2
1	5	2	43.80	2.067	8	0.2
4	8	0	44.20	2.049	4	0.6
1	9	0	44.79	2.023	4	0.3
4	4	2	45.52	1.993	8	1.0
8	3	1	45.91	1.977	8	0.2
3	8	1	45.97	1.974	8	0.4
5	3	2	46.07	1.970	8	0.5
3	5	2	46.09	1.969	8	0.8
6	0	2	46.62	1.948	4	0.4
0	6	2	46.66	1.947	4	0.4
3	9	0	47.05	1.932	4	0.2
6	2	2	47.73	1.905	8	1.2
2	6	2	47.76	1.904	8	0.7
9	0	1	48.11	1.891	4	1.2
0	9	1	48.19	1.888	4	1.1
7	7	0	49.19	1.852	4	0.5
9	2	1	49.19	1.852	8	0.6
7	6	1	49.22	1.851	8	1.2
6	7	1	49.24	1.851	8	0.5
2	9	1	49.27	1.849	8	0.2
10	0	0	49.67	1.835	2	0.1
6	8	0	49.73	1.833	4	0.2
0	10	0	49.77	1.832	2	0.8

UiO-6, Calcined

OSI

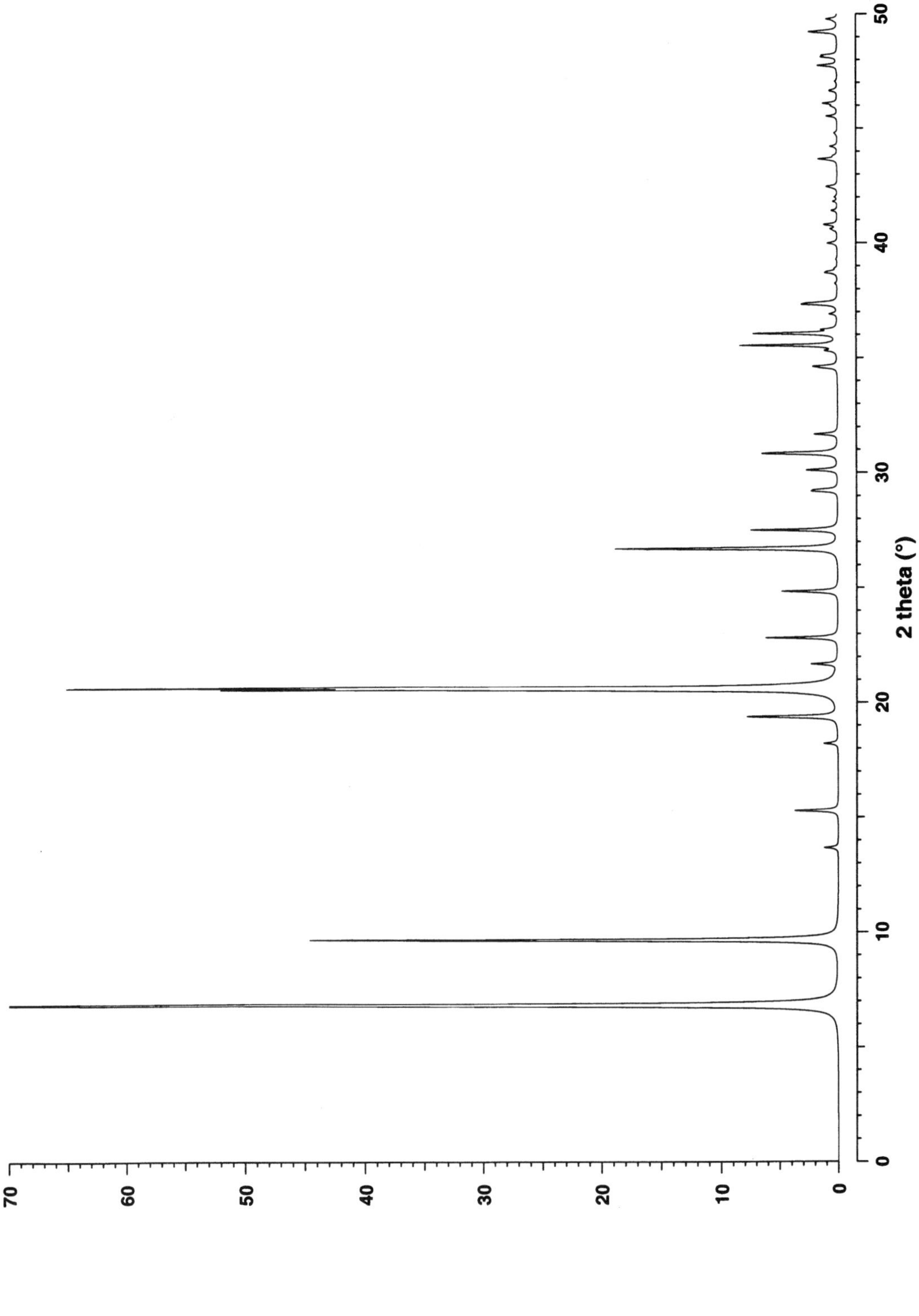

OSO

OSB-1

CHEMICAL COMPOSITION: $|K_6(H_2O)_9|\ [Be_3Si_6O_{18}]$

REFINED COMPOSITION: $|K_{5.37}(H_2O)_9|\ [Be_{3.51}Si_{5.52}O_{18}]$

CRYSTAL DATA: $P3_2$ (No. 145)
$a = 10.0928$ Å $b = 10.0928$ Å $c = 7.6264$ Å
$\alpha = 90°$ $\beta = 90°$ $\gamma = 120°$
X-ray single crystal refinement, $R = 0.0596$

REFERENCE: K. O. Kongshaug, H. Fjellvåg, K. P. Lillerud, T. E. Gier, G. D. Stucky and A. K. Cheetham, Private communication (2000).

h	k	l	2θ	d	M	I_{rel}
0	1	0	10.12	8.741	6	100.0
1	0	1	15.42	5.746	6	15.5
0	1	1	15.42	5.746	6	13.9
1	1	0	17.57	5.046	6	24.9
0	2	0	20.32	4.370	6	20.1
2	-1	1	21.11	4.208	6	1.1
1	1	1	21.11	4.208	6	0.4
0	2	1	23.46	3.792	6	3.3
2	0	1	23.46	3.792	6	4.0
0	1	2	25.48	3.495	6	1.0
1	0	2	25.48	3.495	6	0.8
1	2	0	26.99	3.304	6	13.0
2	1	0	26.99	3.304	6	12.8
1	1	2	29.36	3.042	6	50.3
2	-1	2	29.36	3.042	6	43.7
3	-2	1	29.46	3.031	6	2.2
3	-1	1	29.46	3.031	6	2.1
1	2	1	29.46	3.031	6	2.2
2	1	1	29.46	3.031	6	2.7
0	3	0	30.69	2.914	6	0.4
2	0	2	31.13	2.873	6	8.5
0	2	2	31.13	2.873	6	5.8
3	0	1	32.91	2.722	6	4.7
0	3	1	32.91	2.722	6	8.0
0	0	3	35.31	2.542	2	6.3
2	2	0	35.58	2.523	6	19.3
3	-2	2	35.97	2.497	6	7.0
1	2	2	35.97	2.497	6	7.7
3	-1	2	35.97	2.497	6	8.2
2	1	2	35.97	2.497	6	5.7
1	0	3	36.82	2.441	6	0.2
1	3	0	37.08	2.424	6	8.2
3	1	0	37.08	2.424	6	6.2
2	2	1	37.55	2.395	6	0.7
4	-2	1	37.55	2.395	6	0.4
3	0	2	38.90	2.315	6	0.2
0	3	2	38.90	2.315	6	0.2
4	-3	1	38.98	2.310	6	4.5
1	3	1	38.98	2.310	6	3.3
4	-1	1	38.98	2.310	6	2.7
3	1	1	38.98	2.310	6	5.2
2	-1	3	39.70	2.270	6	0.3
2	0	3	41.08	2.197	6	1.9
0	2	3	41.08	2.197	6	1.6
4	-2	2	42.98	2.104	6	0.2
2	2	2	42.98	2.104	6	0.1
0	4	1	43.06	2.101	6	1.8
4	0	1	43.06	2.101	6	2.1
4	-3	2	44.27	2.046	6	6.2
4	-1	2	44.27	2.046	6	4.8
3	1	2	44.27	2.046	6	5.8
1	3	2	44.27	2.046	6	4.6
3	-2	3	44.99	2.015	6	0.9
1	2	3	44.99	2.015	6	0.8
3	-1	3	44.99	2.015	6	1.3
2	1	3	44.99	2.015	6	1.2
2	3	0	45.22	2.005	6	0.2
3	2	0	45.22	2.005	6	0.1
2	3	1	46.85	1.939	6	0.3
5	-3	1	46.85	1.939	6	0.2
3	2	1	46.85	1.939	6	0.1
5	-2	1	46.85	1.939	6	0.2
3	0	3	47.46	1.915	6	0.3
0	3	3	47.46	1.915	6	0.3
4	1	0	47.68	1.907	6	1.0
1	4	0	47.68	1.907	6	1.0
0	4	2	47.98	1.896	6	0.1
4	0	2	47.98	1.896	6	0.2
0	1	4	48.89	1.863	6	0.1
1	0	4	48.89	1.863	6	0.2
5	-4	1	49.24	1.850	6	0.2
4	1	1	49.24	1.850	6	0.3
5	-1	1	49.24	1.850	6	0.2

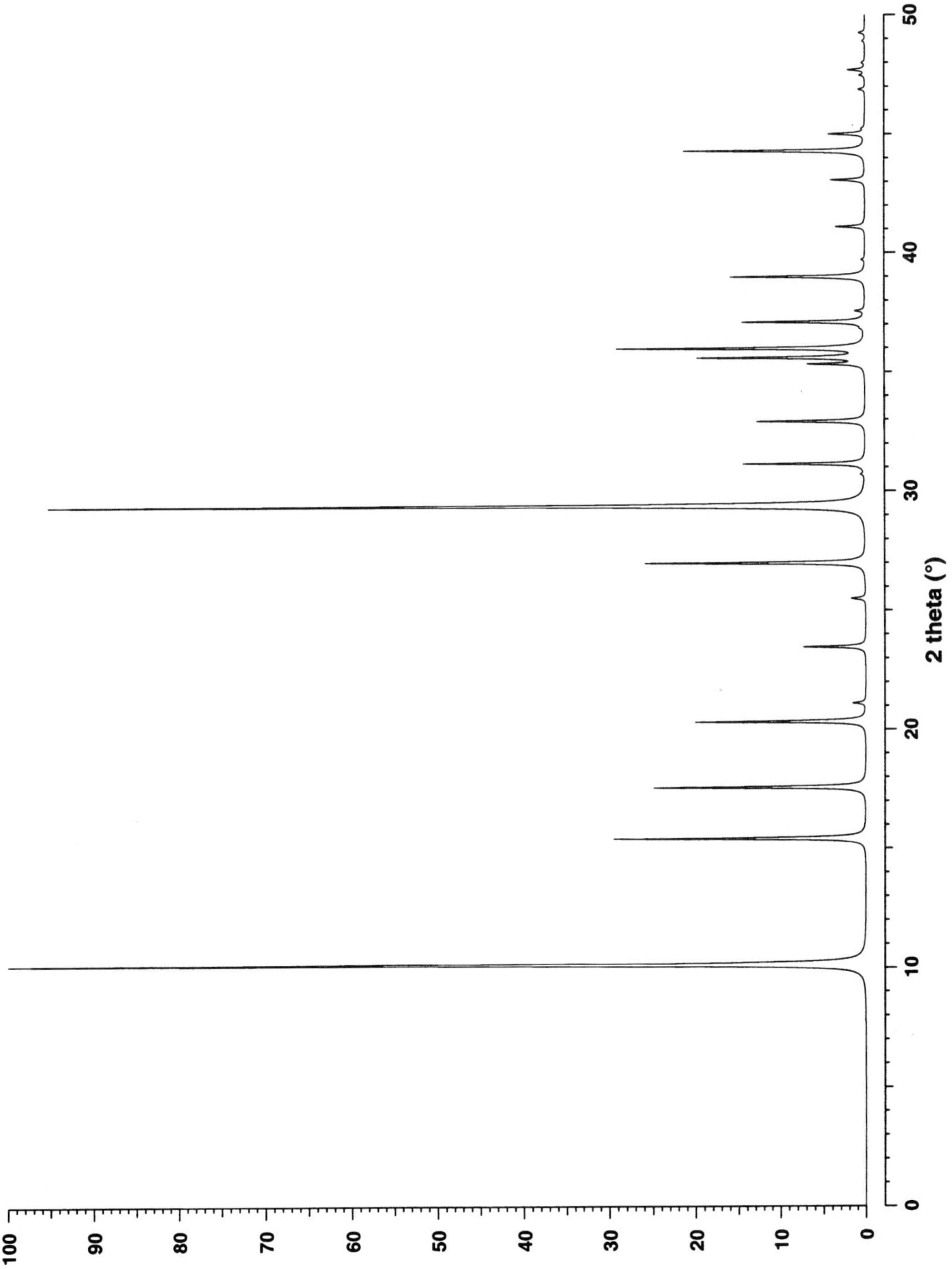
2 theta (°)
0
10
20
30
40
50
0
10
20
30
40
50
60
70
80
90
100

-PAR Partheite

CHEMICAL COMPOSITION: $|Ca_8(H_2O)_{16}|$ $[Si_{16}Al_{16}O_{60}(OH)_8]$
Taurus Mountains, Turkey

REFINED COMPOSITION: $|Ca_8(H_2O)_{16}|$ $[Si_{16}Al_{16}O_{60}(OH)_8]$

CRYSTAL DATA: $C1\,2/c\,1$ (No. 15) unique axis **b**, cell choice 1
$a = 21.555$ Å $b = 8.761$ Å $c = 9.304$ Å
$\alpha = 90°$ $\beta = 91.55°$ $\gamma = 90°$
X-ray single crystal refinement, $R = 0.07$

REFERENCE: N. Engel and K. Yvon,
Z. Kristallogr. **169** 165–175 (1984).

h	k	l	2θ	d	M	I_{rel}
2	0	0	8.21	10.774	2	100.0
1	1	0	10.90	8.116	4	57.5
-1	1	1	14.41	6.146	4	9.9
1	1	1	14.56	6.084	4	40.3
3	1	0	15.96	5.554	4	1.2
4	0	0	16.46	5.387	2	6.4
3	1	1	18.78	4.725	4	1.0
0	0	2	19.08	4.650	2	5.3
0	2	0	20.27	4.380	2	9.1
-2	0	2	20.60	4.312	2	10.3
2	0	2	21.01	4.228	2	2.0
2	2	0	21.90	4.058	4	9.4
-1	1	2	21.93	4.053	4	1.6
1	1	2	22.13	4.017	4	6.5
0	2	1	22.43	3.963	4	3.4
5	1	0	23.00	3.867	4	10.2
-2	2	1	23.83	3.733	4	43.2
2	2	1	24.02	3.705	4	2.7
-3	1	2	24.71	3.603	4	0.2
-5	1	1	24.72	3.602	4	27.6
6	0	0	24.79	3.591	2	15.5
-4	0	2	24.95	3.568	2	0.1
5	1	1	25.15	3.540	4	0.9
3	1	2	25.23	3.529	4	21.1
4	0	2	25.64	3.474	2	0.6
4	2	0	26.22	3.399	4	13.9
-4	2	1	27.79	3.210	4	0.6
0	2	2	27.98	3.189	4	35.1
2	2	2	29.36	3.042	4	32.1
-5	1	2	29.69	3.009	4	2.6
5	1	2	30.42	2.938	4	35.8
-1	1	3	30.77	2.906	4	59.2
7	1	0	30.79	2.904	4	0.4
1	3	0	30.90	2.894	4	0.9
1	1	3	30.98	2.886	4	1.1
-6	0	2	31.05	2.880	2	11.1
6	0	2	31.89	2.806	2	6.9
-7	1	1	32.05	2.792	4	32.9
6	2	0	32.23	2.777	4	12.1
-4	2	2	32.36	2.767	4	13.7
-1	3	1	32.37	2.766	4	0.7
1	3	1	32.43	2.760	4	6.6
7	1	1	32.53	2.752	4	8.8
-3	1	3	32.79	2.731	4	0.3
4	2	2	32.91	2.722	4	0.1
3	3	0	33.11	2.705	4	21.6
3	1	3	33.39	2.683	4	1.9
-6	2	1	33.48	2.676	4	11.8
6	2	1	33.88	2.646	4	0.7
-3	3	1	34.43	2.605	4	0.5
3	3	1	34.62	2.591	4	3.7
-7	1	2	36.04	2.492	4	2.8
-2	2	3	36.29	2.476	4	2.0
1	3	2	36.63	2.453	4	3.0
2	2	3	36.66	2.451	4	1.8
-5	1	3	36.71	2.448	4	4.8
5	3	0	37.19	2.418	4	4.0
-6	2	2	37.37	2.407	4	0.8
5	1	3	37.62	2.391	4	0.5
6	2	2	38.09	2.363	4	0.3
-8	0	2	38.16	2.359	2	1.0
-3	3	2	38.32	2.349	4	0.9
5	3	1	38.62	2.331	4	2.2
3	3	2	38.67	2.328	4	2.5
0	0	4	38.73	2.325	2	5.7
-4	2	3	38.99	2.310	4	0.7
9	1	0	39.00	2.309	4	0.3
8	0	2	39.10	2.304	2	1.0
8	2	0	39.27	2.294	4	0.4
2	0	4	39.88	2.260	2	3.5
-9	1	1	39.98	2.255	4	1.6
-8	2	1	40.27	2.240	4	2.7
1	1	4	40.46	2.229	4	0.1
9	1	1	40.49	2.228	4	13.4
8	2	1	40.72	2.216	4	0.1
0	4	0	41.22	2.190	2	4.2
-3	1	4	41.80	2.161	4	0.4
-5	3	2	41.85	2.158	4	6.5
-4	0	4	41.90	2.156	2	0.7
10	0	0	41.93	2.155	2	4.0
-7	1	3	42.09	2.147	4	2.4
5	3	2	42.40	2.132	4	5.9
3	1	4	42.46	2.129	4	0.4
-1	3	3	42.66	2.119	4	1.7
7	3	0	42.68	2.119	4	7.0
4	0	4	42.77	2.114	2	1.2
-2	4	1	43.21	2.094	4	6.8
-6	2	3	43.28	2.090	4	1.8
2	4	1	43.31	2.089	4	0.8
-8	2	2	43.58	2.077	4	0.7
-7	3	1	43.64	2.074	4	1.0
7	3	1	44.01	2.057	4	0.8
0	2	4	44.09	2.054	4	0.9
6	2	3	44.24	2.047	4	2.9
9	1	2	44.24	2.047	4	4.3
8	2	2	44.43	2.039	4	0.1
-2	2	4	44.72	2.026	4	1.2
-5	1	4	45.00	2.014	4	0.8
2	2	4	45.14	2.009	4	3.0
-4	4	1	45.67	1.987	4	2.2
0	4	2	45.79	1.981	4	0.5
4	4	1	45.87	1.978	4	0.1
-6	0	4	45.92	1.976	2	1.2
-10	0	2	45.94	1.976	2	1.8
5	1	4	46.03	1.972	4	9.2
-2	4	2	46.50	1.953	4	3.0
2	4	2	46.71	1.945	4	1.7
-7	3	2	46.79	1.942	4	3.7
10	0	2	46.95	1.935	2	2.8
-4	2	4	46.97	1.934	4	2.1
10	2	0	47.00	1.933	4	0.8
-5	3	3	47.33	1.921	4	2.1
7	3	2	47.49	1.915	4	3.6
4	2	4	47.77	1.904	4	2.4
-10	2	1	47.82	1.902	4	0.2
5	3	3	48.08	1.893	4	1.8
10	2	1	48.31	1.884	4	5.7
-11	1	1	48.35	1.882	4	1.2
-9	1	3	48.54	1.876	4	0.2
6	4	0	48.69	1.870	4	0.2
-4	4	2	48.79	1.867	4	2.3
-8	2	3	48.83	1.865	4	0.3
4	4	2	49.18	1.853	4	0.3
9	3	0	49.21	1.851	4	1.9
-6	4	1	49.59	1.838	4	1.4
-7	1	4	49.59	1.838	4	0.5
9	1	3	49.85	1.829	4	0.3
6	4	1	49.88	1.828	4	0.8

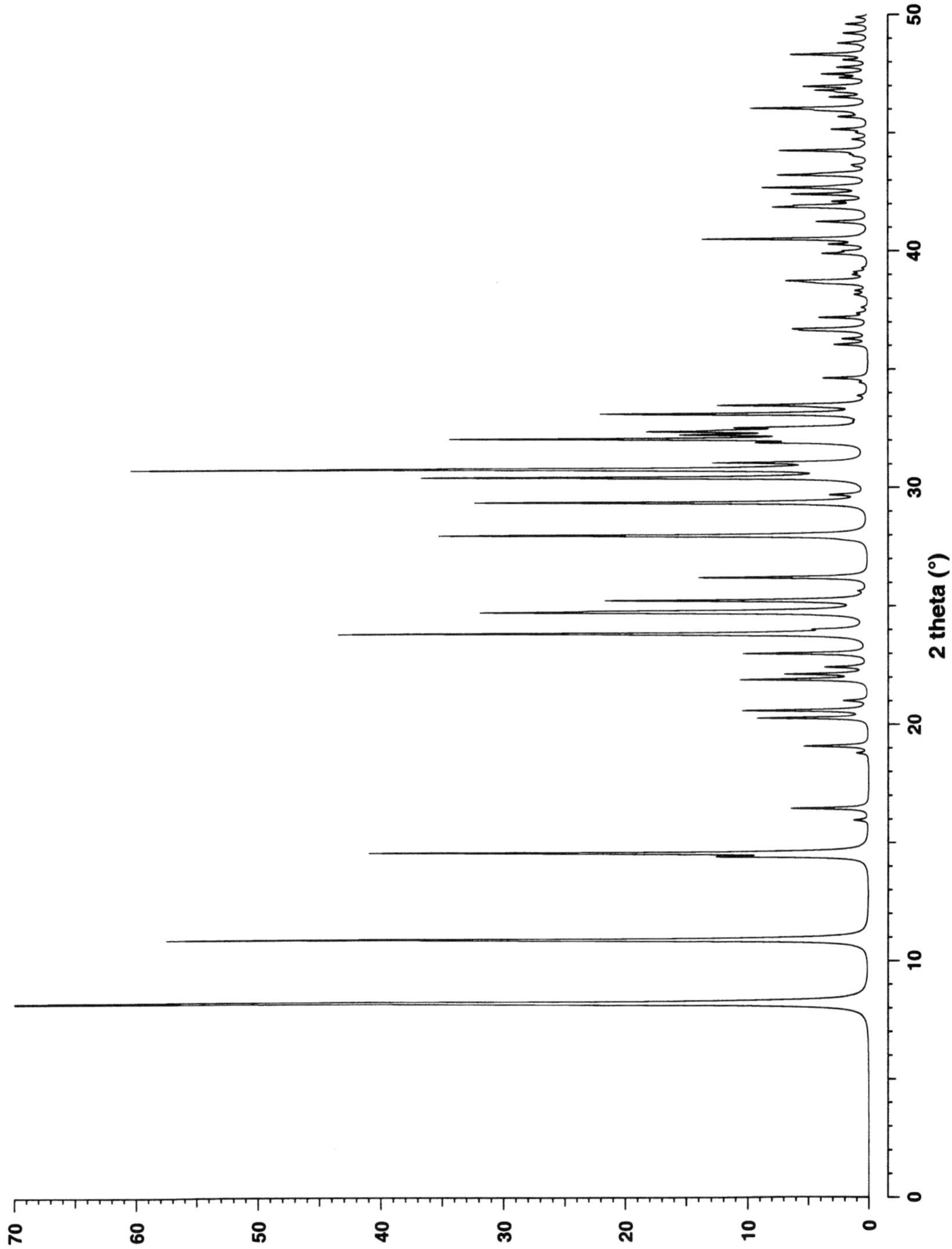
0
10
20
30
40
50
60
70
0
10
20
30
40
50
2 theta (°)

PAU Paulingite

CHEMICAL COMPOSITION: $|Na_{13}K_{68}Ca_{36}Ba_{1.5}(H_2O)_{705}|$ $[Si_{520}Al_{152}O_{1344}]$
Rock Island Dam, Washington, U.S.A.

REFINED COMPOSITION: $|Na_{9.68}K_{50.51}Ca_{26.75}Ba_{1.06}(H_2O)_{452}|$ $[Si_{520.12}Al_{151.88}O_{1344}]$

CRYSTAL DATA: $Im\bar{3}m$ (No. 229)
$a = 35.093$ Å $b = 35.093$ Å $c = 35.093$ Å
$\alpha = 90°$ $\beta = 90°$ $\gamma = 90°$
X-ray single crystal refinement, $R = 0.14$

REFERENCE: E. K. Gordon, S. Samson and W. B. Kamb, *Science* **154** 1004–1007 (1966).

h	k	l	2θ	d	M	I_{rel}
1	1	0	3.56	24.814	12	42.1
2	0	0	5.04	17.546	6	63.6
2	1	1	6.17	14.327	24	4.4
2	2	0	7.12	12.407	12	13.7
3	1	0	7.97	11.097	24	24.7
2	2	2	8.73	10.130	8	0.6
3	2	1	9.43	9.379	48	12.3
4	0	0	10.08	8.773	6	2.5
4	1	1	10.70	8.271	24	3.3
3	3	0	10.70	8.271	12	100.0
4	2	0	11.28	7.847	24	3.3
3	3	2	11.83	7.482	24	7.6
4	2	2	12.36	7.163	24	11.1
5	1	0	12.86	6.882	24	2.8
4	3	1	12.86	6.882	48	83.2
4	4	0	14.28	6.204	12	13.7
6	0	0	15.15	5.849	6	21.8
5	3	2	15.57	5.693	48	26.2
6	1	1	15.57	5.693	24	2.3
5	4	1	16.37	5.415	48	9.6
6	2	2	16.76	5.290	24	4.1
6	3	1	17.14	5.174	48	4.0
7	1	0	17.87	4.963	24	23.7
6	3	3	18.58	4.776	24	44.4
6	4	2	18.92	4.689	48	10.2
7	3	0	19.26	4.608	24	1.9
6	5	1	19.92	4.457	48	1.9
8	0	0	20.24	4.387	6	14.0
8	1	1	20.56	4.320	24	0.5
5	5	4	20.56	4.320	24	2.0
6	4	4	20.87	4.256	24	5.8
8	2	0	20.87	4.256	24	4.4
6	5	3	21.18	4.194	48	1.3
8	2	2	21.49	4.136	24	1.0
7	5	0	21.79	4.079	24	0.9
7	4	3	21.79	4.079	48	10.6
8	4	0	22.66	3.924	24	0.9
9	1	0	22.95	3.875	24	2.7
8	3	3	22.95	3.875	24	10.2
6	5	5	23.51	3.784	24	1.2
9	3	0	24.06	3.699	24	5.7
7	5	4	24.06	3.699	48	1.1
9	3	2	24.59	3.620	48	9.6
8	4	4	24.86	3.582	24	24.4
8	5	3	25.12	3.545	48	1.5
9	4	1	25.12	3.545	48	0.7
10	2	0	25.89	3.441	24	2.1
10	2	2	26.39	3.377	24	4.5
6	6	6	26.39	3.377	8	0.9
10	3	1	26.64	3.346	48	39.2
10	4	0	27.37	3.258	24	45.2
9	6	1	27.61	3.231	48	2.6
11	1	0	28.08	3.177	24	0.5
9	5	4	28.08	3.177	48	7.5
10	5	1	28.55	3.126	48	17.2
11	2	1	28.55	3.126	48	12.8
9	6	3	28.55	3.126	48	3.1
8	8	0	28.78	3.102	12	1.3
11	3	0	29.01	3.078	24	64.1
9	7	0	29.01	3.078	24	1.7
10	5	3	29.46	3.032	48	1.4
8	6	6	29.69	3.009	24	1.0
11	4	1	29.91	2.987	48	28.1
10	6	2	30.13	2.966	48	13.0
9	6	5	30.35	2.945	48	0.8
11	5	0	30.79	2.904	24	1.4
9	7	4	30.79	2.904	48	1.6
11	5	2	31.21	2.865	48	3.6
10	6	4	31.43	2.846	48	3.9
12	3	1	31.64	2.828	48	2.1
9	8	3	31.64	2.828	48	1.5
10	7	3	32.06	2.792	48	4.8
12	4	0	32.27	2.774	24	0.9
12	4	2	32.68	2.740	48	2.7
11	6	3	32.88	2.724	48	20.5
10	8	2	33.09	2.707	48	1.7
13	2	1	33.69	2.660	48	0.7
12	4	4	33.89	2.645	24	0.9
9	9	4	34.08	2.630	24	6.7
12	5	3	34.08	2.630	48	3.3
10	8	4	34.28	2.616	48	29.7
10	9	1	34.48	2.601	48	2.1
12	6	2	34.67	2.587	48	0.7
11	7	4	34.87	2.573	48	5.4
10	9	3	35.25	2.546	48	1.2
11	8	3	35.63	2.520	48	7.0
13	4	3	35.63	2.520	48	4.1
12	6	4	35.82	2.507	48	1.3
14	0	0	35.82	2.507	6	2.2
14	1	1	36.01	2.494	24	1.3
10	10	0	36.20	2.481	12	3.1
14	2	0	36.20	2.481	24	0.8
14	3	1	36.76	2.445	48	1.2
11	9	2	36.76	2.445	48	2.3
10	9	5	36.76	2.445	48	3.8
14	5	1	38.21	2.355	48	1.7
15	1	0	38.57	2.334	24	0.5
14	5	3	38.92	2.314	48	2.4
14	6	0	39.10	2.304	24	1.7
13	7	4	39.27	2.294	48	0.7
10	10	6	39.45	2.284	24	1.1
14	6	2	39.45	2.284	48	2.1
11	11	0	39.96	2.256	12	3.0
12	8	6	40.14	2.247	48	0.5
14	7	1	40.31	2.237	48	2.8
16	0	0	41.16	2.193	6	0.7
16	1	1	41.32	2.185	24	0.6
14	6	6	42.15	2.144	24	0.8
11	10	7	42.32	2.136	48	0.5
11	11	6	42.97	2.105	24	2.1
12	12	0	43.78	2.068	12	1.1
12	11	5	43.94	2.061	48	1.9
17	1	0	43.94	2.061	24	0.8
13	11	0	43.94	2.061	24	0.6
17	2	1	44.25	2.047	48	2.4
14	7	7	44.25	2.047	24	5.0
13	11	2	44.25	2.047	48	0.6
17	3	0	44.57	2.033	24	1.0
12	12	4	45.04	2.013	24	1.6
11	11	8	45.20	2.006	24	1.4
13	11	4	45.20	2.006	48	4.7
14	10	4	45.66	1.987	48	1.3
14	11	1	46.12	1.968	48	0.8
13	12	3	46.43	1.956	48	1.3
18	0	0	46.58	1.950	6	0.6
14	11	3	46.74	1.944	48	2.3
18	2	0	46.89	1.938	24	1.3
17	6	3	47.34	1.920	48	0.8
15	10	3	47.34	1.920	48	0.7
13	13	0	47.64	1.909	12	0.8
12	11	9	48.24	1.887	48	0.6
17	7	4	48.83	1.865	48	0.6
19	1	0	49.41	1.844	24	1.5
19	3	0	49.99	1.824	24	0.8
17	9	0	49.99	1.824	24	0.8

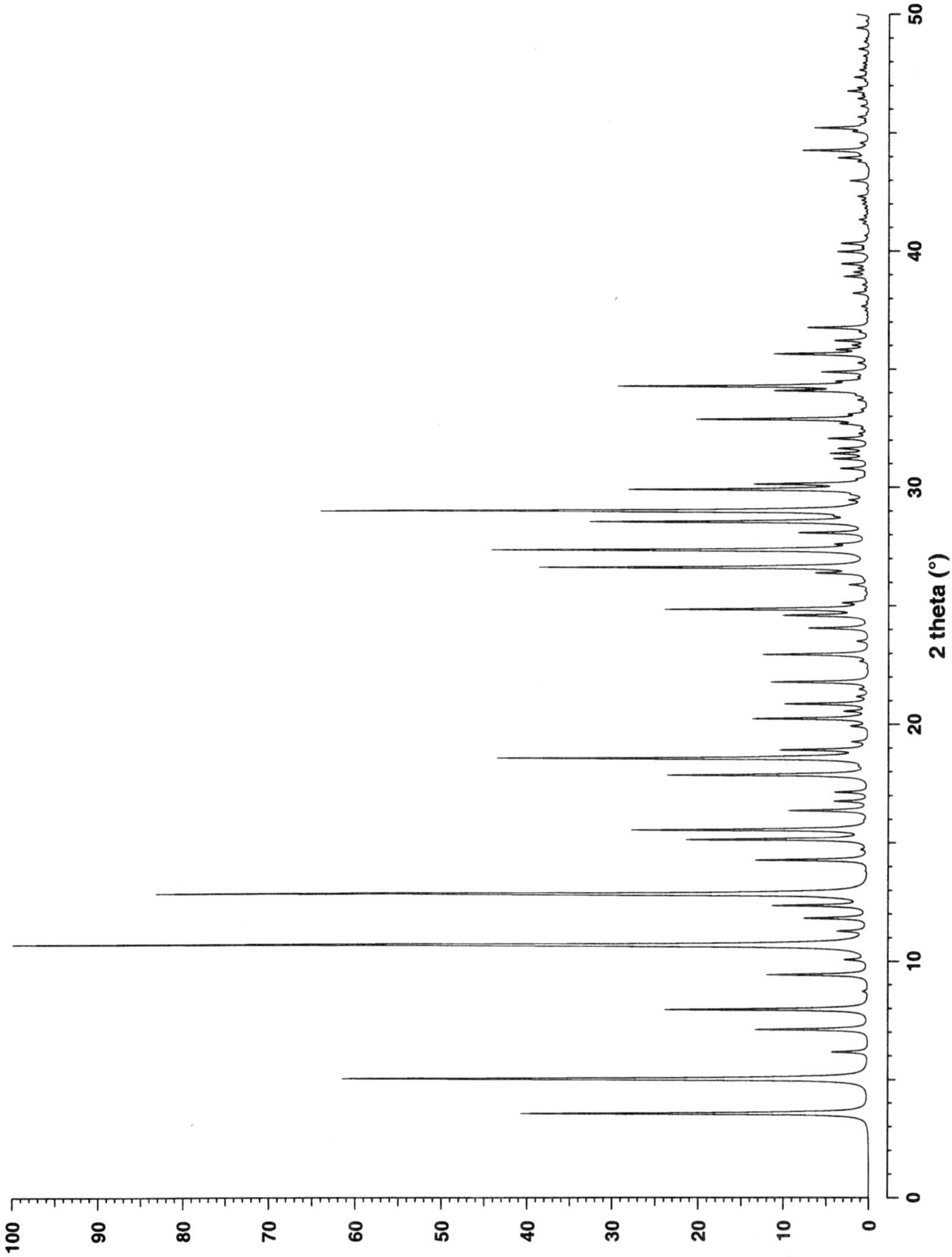
2 theta (°)
0
10
20
30
40
50
0
10
20
30
40
50
60
70
80
90
100

PHI Phillipsite

CHEMICAL COMPOSITION: $|K_2Ca_{1.7}Na_{0.4}(H_2O)_{13.5}|\ [Si_{10.6}Al_{5.3}O_{32}]$
Casal Brunori, Rome, Italy

REFINED COMPOSITION: $|K_2Ca_{1.648}(H_2O)_{12}|\ [Si_{10.7}Al_{5.3}O_{32}]$

CRYSTAL DATA: $P1\,2_1/m\,1$ (No. 11) unique axis **b**
$a = 9.865$ Å $b = 14.300$ Å $c = 8.668$ Å
$\alpha = 90°$ $\beta = 124.20°$ $\gamma = 90°$
X-ray single crystal refinement, $R_w = 0.057$

REFERENCE: R. Rinaldi, J. J. Pluth and J. V. Smith, *Acta Cryst.* **B30** 2426–2433 (1974).

h	k	l	2θ	d	M	I_{rel}
1	0	0	10.84	8.159	2	10.8
-1	0	1	10.93	8.095	2	24.3
0	0	1	12.35	7.169	2	82.5
0	2	0	12.38	7.150	2	64.3
1	1	0	12.49	7.087	4	30.5
-1	1	1	12.56	7.045	4	1.4
0	1	1	13.82	6.409	4	37.0
1	2	0	16.48	5.377	4	33.6
-1	2	1	16.54	5.359	4	18.6
0	2	1	17.52	5.063	4	45.8
-2	0	1	17.98	4.932	2	75.3
-2	1	1	19.03	4.663	4	11.7
1	0	1	20.58	4.315	2	10.6
-1	0	2	20.72	4.287	2	14.1
1	1	1	21.51	4.131	4	44.8
1	3	0	21.59	4.116	4	11.8
-1	3	1	21.63	4.107	4	6.8
-1	1	2	21.64	4.106	4	46.8
-2	2	1	21.89	4.060	4	27.1
-2	0	2	21.96	4.048	2	5.4
0	3	1	22.40	3.969	4	3.2
2	1	0	22.67	3.923	4	10.7
1	2	1	24.09	3.695	4	2.1
-1	2	2	24.21	3.677	4	3.6
0	0	2	24.84	3.585	2	0.3
0	1	2	25.62	3.477	4	6.8
-2	3	1	26.00	3.428	4	7.2
1	4	0	27.23	3.274	4	40.0
-1	4	1	27.27	3.270	4	21.5
-3	0	1	27.78	3.211	2	13.5
0	2	2	27.84	3.204	4	59.5
0	4	1	27.89	3.199	4	100.0
-3	0	2	27.89	3.199	2	17.9
1	3	1	27.89	3.199	4	7.9
-1	3	2	27.99	3.188	4	17.7
-3	1	1	28.49	3.133	4	62.1
-3	1	2	28.59	3.122	4	45.5
2	3	0	28.80	3.099	4	1.4
-2	3	2	28.94	3.085	4	10.5
-3	2	1	30.52	2.929	4	22.9
-3	2	2	30.61	2.920	4	14.5
2	0	1	30.71	2.911	2	0.8
-2	4	1	30.89	2.895	4	1.2
-2	0	3	30.96	2.888	2	12.7
0	3	2	31.22	2.865	4	0.8
1	0	2	32.44	2.760	2	22.1
1	4	1	32.52	2.753	4	13.3
-1	0	3	32.59	2.747	2	13.7
-1	4	2	32.61	2.746	4	18.2
1	1	2	33.06	2.710	4	26.8
1	5	0	33.19	2.699	4	27.1
-3	0	3	33.20	2.698	2	6.5
-1	1	3	33.20	2.698	4	11.0
-1	5	1	33.22	2.697	4	4.8
2	2	1	33.23	2.696	4	10.0
2	4	0	33.32	2.689	4	1.7
-2	4	2	33.44	2.679	4	6.5
-2	2	3	33.46	2.678	4	23.4
3	1	0	33.54	2.672	4	10.9
0	5	1	33.74	2.656	4	8.1
-3	3	2	33.74	2.656	4	2.4
-3	1	3	33.80	2.652	4	6.4
1	2	2	34.84	2.575	4	6.2
-1	2	3	34.99	2.565	4	4.6
3	2	0	35.31	2.542	4	6.8
0	4	2	35.46	2.531	4	16.1
-3	2	3	35.56	2.525	4	7.3
-4	0	2	36.43	2.466	2	3.1
-4	1	2	36.99	2.430	4	2.3
0	0	3	37.64	2.390	2	0.5
-3	4	1	37.65	2.389	4	14.3
1	3	2	37.66	2.388	4	4.5
-3	4	2	37.73	2.384	4	5.2
1	5	1	37.73	2.384	4	0.7
-1	5	2	37.81	2.379	4	3.0
3	3	0	38.09	2.362	4	3.9
0	1	3	38.18	2.357	4	1.9
-3	3	3	38.33	2.348	4	0.4
2	5	0	38.44	2.342	4	7.6
-2	5	2	38.54	2.336	4	11.4
-4	2	2	38.62	2.331	4	1.6
-4	0	3	38.71	2.326	2	0.7
-4	1	1	39.03	2.307	4	1.0
-4	1	3	39.24	2.296	4	3.9
0	6	1	39.86	2.262	4	0.7
2	4	1	39.93	2.257	4	6.3
-2	4	3	40.14	2.247	4	0.9
0	5	2	40.34	2.236	4	6.5
-4	3	2	41.21	2.190	4	2.2
1	4	2	41.33	2.185	4	0.7
-1	4	3	41.45	2.178	4	1.3
3	4	0	41.73	2.165	4	4.8
-3	4	3	41.95	2.154	4	9.0
2	1	2	42.36	2.134	4	1.4
-3	5	2	42.39	2.132	4	1.6
-3	1	4	42.47	2.128	4	0.3
1	6	1	43.37	2.086	4	3.3
-1	6	2	43.44	2.083	4	0.7
3	2	1	43.58	2.077	4	2.4
2	2	2	43.82	2.066	4	2.2
-3	2	4	43.93	2.061	4	1.9
-2	6	2	44.09	2.054	4	1.7
-2	2	4	44.10	2.053	4	5.1
2	5	1	44.40	2.040	4	2.9
4	0	0	44.41	2.040	2	5.3
-2	5	3	44.59	2.032	4	0.7
-4	4	2	44.64	2.030	4	0.5
1	0	3	45.13	2.009	2	1.0
-1	0	4	45.29	2.002	2	4.3
1	1	3	45.60	1.989	4	2.5
0	6	2	45.71	1.985	4	1.1
1	7	0	45.79	1.982	4	6.5
3	3	1	45.94	1.975	4	2.7
3	5	0	46.05	1.971	4	1.4
2	3	2	46.18	1.966	4	1.8
0	7	1	46.21	1.965	4	0.4
-3	5	3	46.25	1.963	4	3.7
-3	3	4	46.28	1.962	4	1.5
4	2	0	46.28	1.962	4	3.3
-5	0	2	46.40	1.957	2	3.9
-5	0	3	46.52	1.952	2	4.6
-4	4	3	46.58	1.950	4	1.5
-4	2	4	46.64	1.947	4	2.1
-5	1	2	46.86	1.939	4	0.4
-5	1	3	46.97	1.934	4	2.4
-3	6	1	47.51	1.914	4	1.3
-3	6	2	47.58	1.911	4	3.8
-5	2	2	48.22	1.887	4	2.0
4	3	0	48.55	1.875	4	1.1
3	4	1	49.11	1.855	4	1.2
1	3	3	49.22	1.851	4	0.4
1	7	1	49.35	1.846	4	0.6
-3	4	4	49.43	1.844	4	0.4
-2	4	4	49.59	1.838	4	1.1
0	5	3	49.72	1.834	4	8.5

Phillipsite PHI

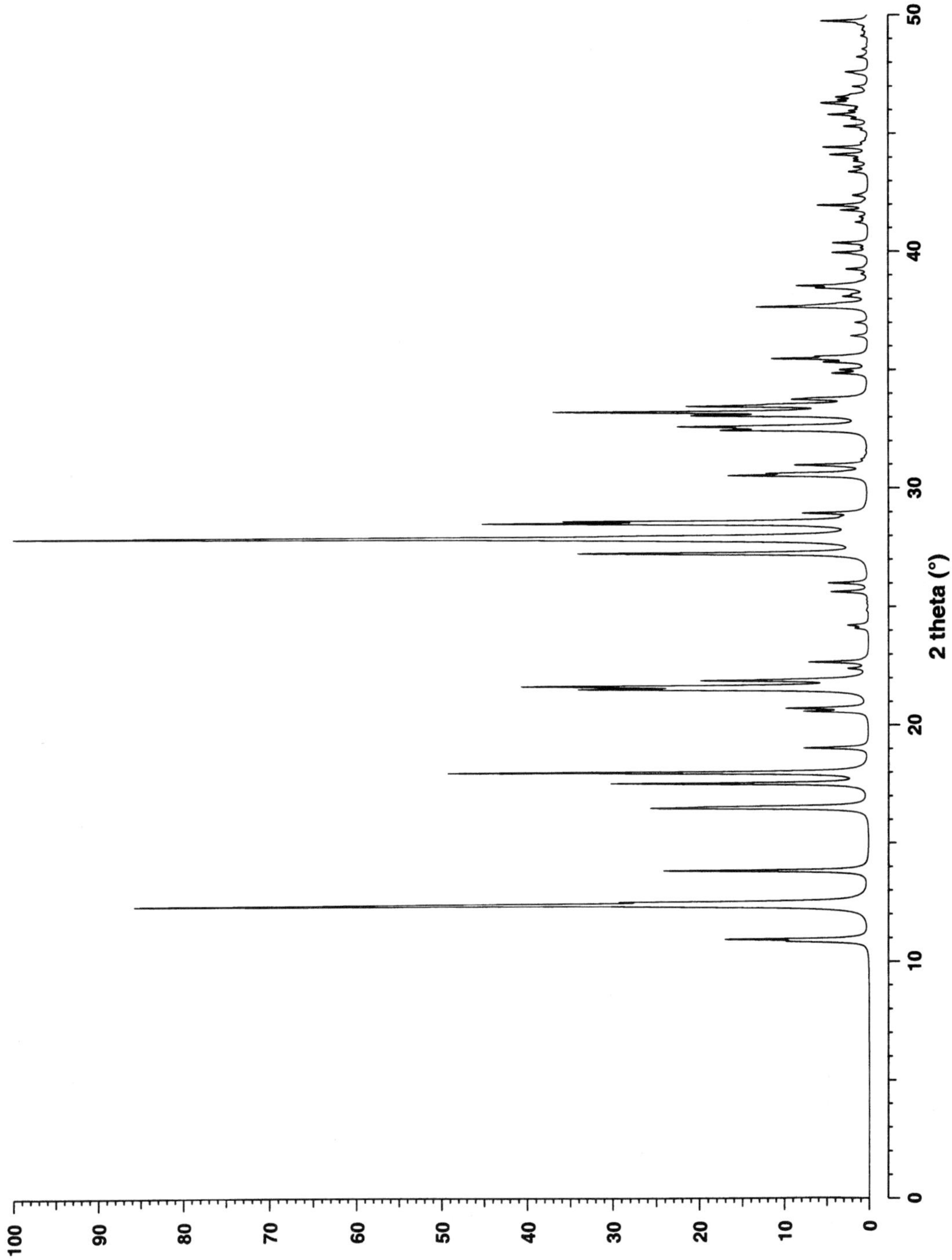

PHI Harmotome

CHEMICAL COMPOSITION: $|Ba_2Ca_{0.5}(H_2O)_{12}|$ $[Si_{12}Al_4O_{32}]$
Andreasberg, Hartz, Germany

REFINED COMPOSITION: $|Ba_2Ca_{0.6}(H_2O)_{12}|$ $[Si_{12}Al_4O_{32}]$

CRYSTAL DATA: $P1\,2_1/m\,1$ (No. 11) unique axis **b**
$a = 9.879$ Å $b = 14.139$ Å $c = 8.693$ Å
$\alpha = 90°$ $\beta = 124.81°$ $\gamma = 90°$
X-ray single crystal refinement, $R_w = 0.039$

REFERENCE: R. Rinaldi, J. J. Pluth and J. V. Smith, *Acta Cryst.* **B30** 2426–2433 (1974).

h	k	l	2θ	d	M	I_{rel}
-1	0	1	10.87	8.136	2	51.2
1	0	0	10.91	8.111	2	62.2
0	0	1	12.40	7.137	2	70.8
-1	1	1	12.55	7.052	4	12.0
1	1	0	12.58	7.036	4	5.8
0	1	1	13.90	6.372	4	100.0
0	2	1	17.66	5.023	4	33.7
-2	0	1	17.96	4.939	2	0.7
-1	0	2	20.69	4.292	2	18.6
1	0	1	20.75	4.282	2	10.0
-1	1	2	21.64	4.107	4	14.1
1	1	1	21.69	4.098	4	23.6
-1	3	1	21.79	4.078	4	29.2
1	3	0	21.81	4.075	4	14.5
-2	2	1	21.95	4.049	4	42.4
0	3	1	22.61	3.933	4	2.6
-2	1	2	22.75	3.909	4	8.6
2	1	0	22.81	3.898	4	18.6
-1	2	2	24.26	3.669	4	2.9
1	2	1	24.30	3.662	4	3.5
0	0	2	24.95	3.569	2	2.9
0	4	0	25.19	3.535	2	3.6
-2	2	2	25.26	3.526	4	1.4
0	1	2	25.75	3.460	4	10.6
-2	3	1	26.13	3.410	4	3.8
-1	4	1	27.51	3.242	4	16.4
1	4	0	27.53	3.240	4	27.6
-3	0	2	27.78	3.211	2	9.8
-3	0	1	27.82	3.206	2	1.0
0	2	2	28.01	3.186	4	8.0
-1	3	2	28.12	3.173	4	2.6
0	4	1	28.17	3.168	4	32.2
-3	1	2	28.51	3.131	4	42.8
-3	1	1	28.54	3.127	4	38.9
-2	3	2	29.00	3.079	4	11.2
2	3	0	29.05	3.074	4	12.0
-3	2	2	30.58	2.924	4	6.8
-3	2	1	30.61	2.920	4	20.5
-2	0	3	30.86	2.897	2	5.2
2	0	1	30.96	2.888	2	1.4
-2	4	1	31.11	2.875	4	3.4
0	3	2	31.44	2.845	4	6.3
-2	1	3	31.52	2.838	4	3.2
2	1	1	31.62	2.830	4	1.0
-1	0	3	32.62	2.745	2	4.5
1	0	2	32.68	2.740	2	6.7
-1	4	2	32.82	2.729	4	17.1
1	4	1	32.86	2.726	4	14.6
-1	1	3	33.25	2.695	4	11.7
1	1	2	33.30	2.690	4	22.6
-2	2	3	33.42	2.681	4	10.3
2	2	1	33.51	2.674	4	11.5
-1	5	1	33.55	2.671	4	14.0
1	5	0	33.56	2.670	4	14.3
-2	4	2	33.59	2.668	4	7.0
2	4	0	33.63	2.665	4	1.3
-3	1	3	33.65	2.663	4	4.2
-3	3	2	33.78	2.654	4	0.8
-3	3	1	33.81	2.651	4	0.8
0	5	1	34.10	2.629	4	9.9
-1	2	3	35.07	2.559	4	3.6
1	2	2	35.12	2.555	4	3.0
-3	2	3	35.45	2.532	4	16.2
3	2	0	35.55	2.525	4	10.8
0	4	2	35.75	2.511	4	11.4
-4	0	2	36.38	2.470	2	2.0
-2	3	3	36.40	2.468	4	6.1
2	3	1	36.48	2.463	4	1.4
-4	1	2	36.95	2.433	4	1.3
0	0	3	37.81	2.379	2	2.3
-1	3	3	37.93	2.372	4	4.1
1	3	2	37.98	2.369	4	6.1
1	5	1	38.14	2.360	4	2.9
0	6	0	38.19	2.357	2	1.5
0	1	3	38.37	2.346	4	1.3
3	3	0	38.38	2.345	4	6.4
-4	0	3	38.53	2.336	2	1.8
-4	2	2	38.62	2.332	4	4.0
-2	5	2	38.78	2.322	4	6.6
2	5	0	38.82	2.320	4	9.7
-4	1	3	39.08	2.305	4	3.6
-4	1	1	39.16	2.301	4	8.7
-1	6	1	39.82	2.263	4	0.9
1	6	0	39.83	2.263	4	2.8
0	2	3	39.98	2.255	4	3.4
-2	4	3	40.25	2.241	4	4.0
0	6	1	40.30	2.238	4	1.1
2	4	1	40.32	2.237	4	5.6
-4	2	3	40.67	2.218	4	2.8
-3	4	3	41.99	2.152	4	10.2
3	4	0	42.07	2.148	4	5.2
-2	0	4	42.10	2.146	2	1.3
-2	6	1	42.50	2.127	4	2.2
0	3	3	42.57	2.124	4	0.6
-3	5	1	42.63	2.121	4	0.7
2	1	2	42.72	2.117	4	0.7
-4	3	1	43.29	2.090	4	1.5
-1	6	2	43.83	2.066	4	3.6
1	6	1	43.85	2.064	4	5.0
3	2	1	43.95	2.060	4	2.1
-2	2	4	44.10	2.054	4	7.5
2	2	2	44.20	2.049	4	4.3
4	0	0	44.69	2.028	2	4.2
-4	4	2	44.76	2.025	4	1.2
-2	5	3	44.79	2.024	4	2.5
2	5	1	44.86	2.021	4	1.9
-4	1	4	45.03	2.013	4	4.4
-1	0	4	45.38	1.998	2	3.1
1	0	3	45.44	1.996	2	3.1
1	1	3	45.92	1.976	4	1.5
1	5	2	46.12	1.968	4	1.6
0	6	2	46.16	1.966	4	2.5
-3	3	4	46.22	1.964	4	2.9
-1	7	1	46.31	1.960	4	0.9
3	3	1	46.35	1.959	4	0.8
-5	0	3	46.37	1.958	2	1.3
-3	5	3	46.39	1.957	4	1.0
-5	0	2	46.42	1.956	2	3.9
-4	2	4	46.46	1.955	4	3.1
3	5	0	46.47	1.954	4	1.8
4	2	0	46.59	1.949	4	6.7
2	3	2	46.60	1.949	4	1.9
-4	4	3	46.60	1.949	4	1.3
-5	1	3	46.84	1.939	4	1.2
-1	2	4	47.27	1.923	4	1.9
1	2	3	47.32	1.921	4	3.0
-3	6	2	47.88	1.900	4	1.6
-3	6	1	47.91	1.899	4	2.5
-3	4	4	49.44	1.843	4	1.8
-1	3	4	49.54	1.840	4	1.0
1	3	3	49.60	1.838	4	0.9
-2	4	4	49.70	1.834	4	1.6
2	4	2	49.80	1.831	4	1.2
-5	0	4	49.92	1.827	2	1.2
2	6	1	49.95	1.826	4	1.4

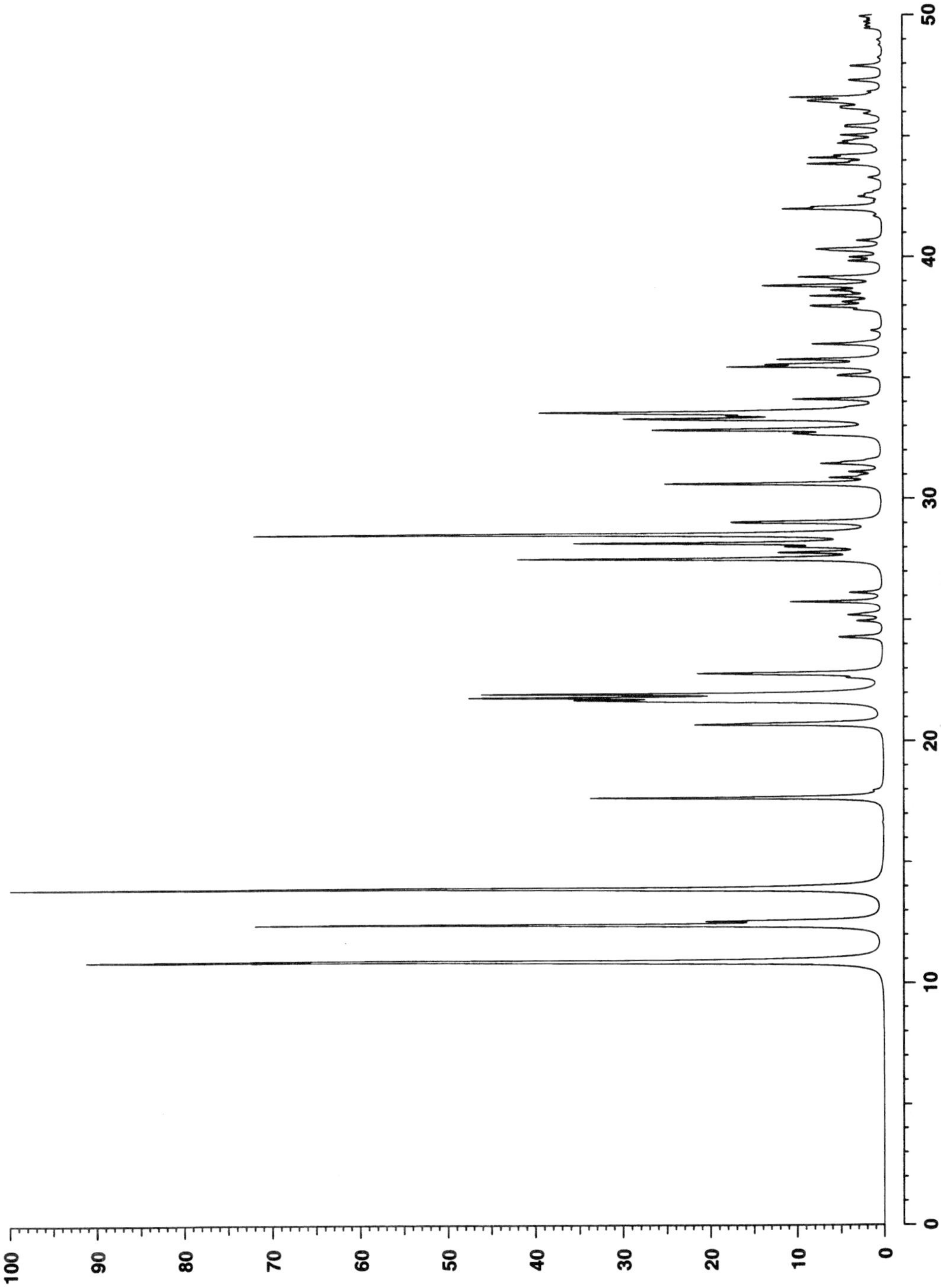
0
10
20
30
40
50
2 theta (°)
0
10
20
30
40
50
60
70
80
90
100

RHO Rho, Hydrated

CHEMICAL COMPOSITION: $|Na_3Cs_3(H_2O)_{73}|$ $[Si_{36}Al_{12}O_{96}]$

REFINED COMPOSITION: $|Na_{2.88}Cs_{3.12}(H_2O)_{73.27}|$ $[Si_{35.95}Al_{12.05}O_{96}]$

CRYSTAL DATA: $Im\bar{3}m$ (No. 229)
a = 15.031 Å b = 15.031 Å c = 15.031 Å
$\alpha = 90°$ $\beta = 90°$ $\gamma = 90°$
X-ray Rietveld refinement, $R_{wp} = 0.18$, $R_F = 0.134$

REFERENCE: L. B. McCusker and Ch. Baerlocher,
in *Proceedings of the 6th International Zeolite Conference, Reno 1983*,
Ed. by D. H. Olson and A. Bisio (Butterworth: Guildford) 812–822 (1984).

h	k	l	2θ	d	M	I_{rel}
1	1	0	8.32	10.629	12	70.2
2	0	0	11.77	7.516	6	1.9
2	1	1	14.43	6.136	24	5.0
2	2	0	16.68	5.314	12	18.2
3	1	0	18.67	4.753	24	49.0
2	2	2	20.47	4.339	8	13.0
3	2	1	22.13	4.017	48	16.2
4	0	0	23.68	3.758	6	7.8
3	3	0	25.14	3.543	12	2.6
4	1	1	25.14	3.543	24	100.0
4	2	0	26.52	3.361	24	49.5
3	3	2	27.84	3.205	24	11.4
4	2	2	29.10	3.068	24	8.4
4	3	1	30.32	2.948	48	7.1
5	1	0	30.32	2.948	24	26.1
5	2	1	32.63	2.744	48	33.5
4	4	0	33.73	2.657	12	3.0
4	3	3	34.80	2.578	24	3.5
5	3	0	34.80	2.578	24	3.5
6	0	0	35.84	2.505	6	0.2
4	4	2	35.84	2.505	24	32.1
5	3	2	36.86	2.438	48	2.9
6	1	1	36.86	2.438	24	0.1
6	2	0	37.85	2.377	24	1.3
5	4	1	38.83	2.319	48	0.5
6	2	2	39.78	2.266	24	2.1
6	3	1	40.71	2.216	48	4.7
7	1	0	42.53	2.126	24	2.9
5	5	0	42.53	2.126	12	3.8
6	4	0	43.41	2.084	24	0.1
7	2	1	44.28	2.045	48	0.3
5	5	2	44.28	2.045	24	5.5
6	4	2	45.14	2.009	48	0.3
7	3	0	45.98	1.974	24	1.4
7	3	2	47.64	1.909	48	1.5
6	5	1	47.64	1.909	48	6.8
8	0	0	48.45	1.879	6	8.5
7	4	1	49.25	1.850	48	2.5
8	1	1	49.25	1.850	24	1.1
5	5	4	49.25	1.850	24	4.0

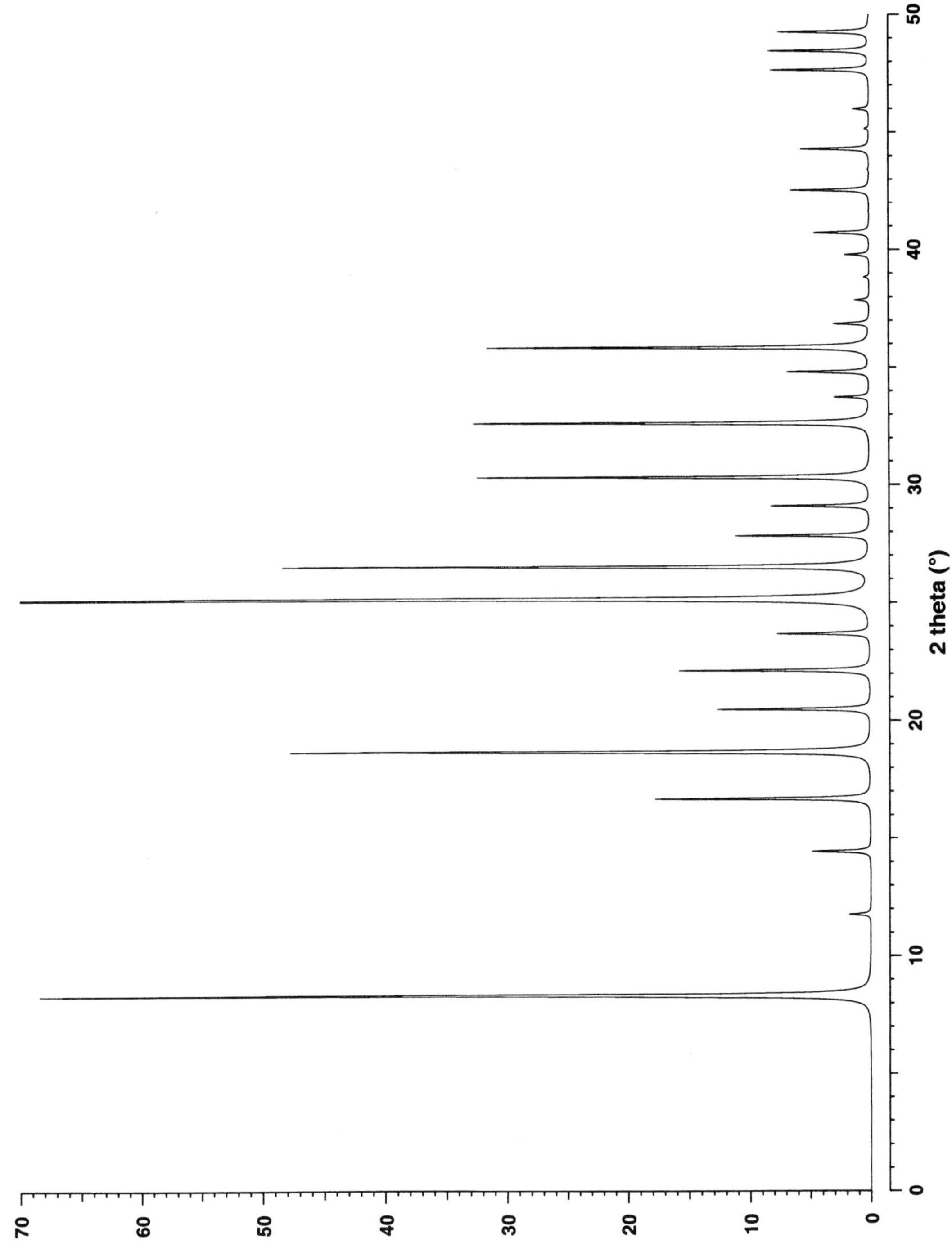
2 theta (°)
0
10
20
30
40
50
60
70

RHO Pahasapaite

CHEMICAL COMPOSITION: $|Li_{11.6}Na_{0.192}K_{1.2}Ca_{5.5}(H_2O)_{38}\square_{13.5}|$ $[Be_{24}P_{24}O_{96}]$
$\square$ = cation vacancy.
Custer, S. Dakota, U.S.A.

REFINED COMPOSITION: $|Li_{11.6}Na_{0.192}K_{1.2}Ca_{5.5}(H_2O)_{38}|$ $[Be_{24}P_{24}O_{96}]$

CRYSTAL DATA: $I23$ (No. 197)
$a = 13.781$ Å $b = 13.781$ Å $c = 13.781$ Å
$\alpha = 90°$ $\beta = 90°$ $\gamma = 90°$
X-ray single crystal refinement, $R_w = 0.035$

REFERENCE: R. C. Rouse, D. R. Peacor and S. Merlino,
American Mineralogist **74** 1195–1202 (1989).

h	k	l	2θ	d	M	I_{rel}	h	k	l	2θ	d	M	I_{rel}	h	k	l	2θ	d	M	I_{rel}
1	1	0	9.07	9.745	12	100.0	4	1	3	33.15	2.703	24	5.2	6	2	2	43.56	2.078	24	6.2
2	1	1	15.75	5.626	24	14.7	4	3	1	33.15	2.703	24	3.4	6	1	3	44.59	2.032	24	2.1
2	2	0	18.21	4.872	12	11.7	5	0	1	33.15	2.703	12	9.9	6	3	1	44.59	2.032	24	0.1
3	1	0	20.38	4.358	12	19.0	5	2	1	35.68	2.516	24	5.5	4	4	4	45.61	1.989	8	4.0
3	0	1	20.38	4.358	12	0.7	5	1	2	35.68	2.516	24	4.2	5	3	4	46.60	1.949	24	1.6
2	2	2	22.35	3.978	8	8.3	4	3	3	38.07	2.363	24	1.2	5	5	0	46.60	1.949	12	3.8
3	1	2	24.16	3.683	24	31.3	5	3	0	38.07	2.363	12	2.3	7	1	0	46.60	1.949	12	1.1
3	2	1	24.16	3.683	24	7.3	6	0	0	39.22	2.297	6	7.4	7	0	1	46.60	1.949	12	1.3
4	0	0	25.86	3.445	6	6.8	4	4	2	39.22	2.297	24	1.1	5	4	3	46.60	1.949	24	0.7
4	1	1	27.46	3.248	24	43.7	5	3	2	40.34	2.236	24	0.4	6	0	4	47.58	1.911	12	1.8
3	3	0	27.46	3.248	12	0.4	6	1	1	40.34	2.236	24	6.1	6	3	3	48.54	1.875	24	0.7
4	2	0	28.97	3.082	12	0.4	5	2	3	40.34	2.236	24	5.3	7	1	2	48.54	1.875	24	0.3
4	0	2	28.97	3.082	12	19.7	6	0	2	41.44	2.179	12	0.6	7	2	1	48.54	1.875	24	0.7
3	3	2	30.42	2.938	24	32.0	6	2	0	41.44	2.179	12	3.1	5	5	2	48.54	1.875	24	0.3
4	2	2	31.81	2.813	24	8.5	5	1	4	42.51	2.126	24	0.9	6	2	4	49.49	1.842	24	3.9
5	1	0	33.15	2.703	12	0.3	5	4	1	42.51	2.126	24	3.3	6	4	2	49.49	1.842	24	0.1

Pahasapaite RHO

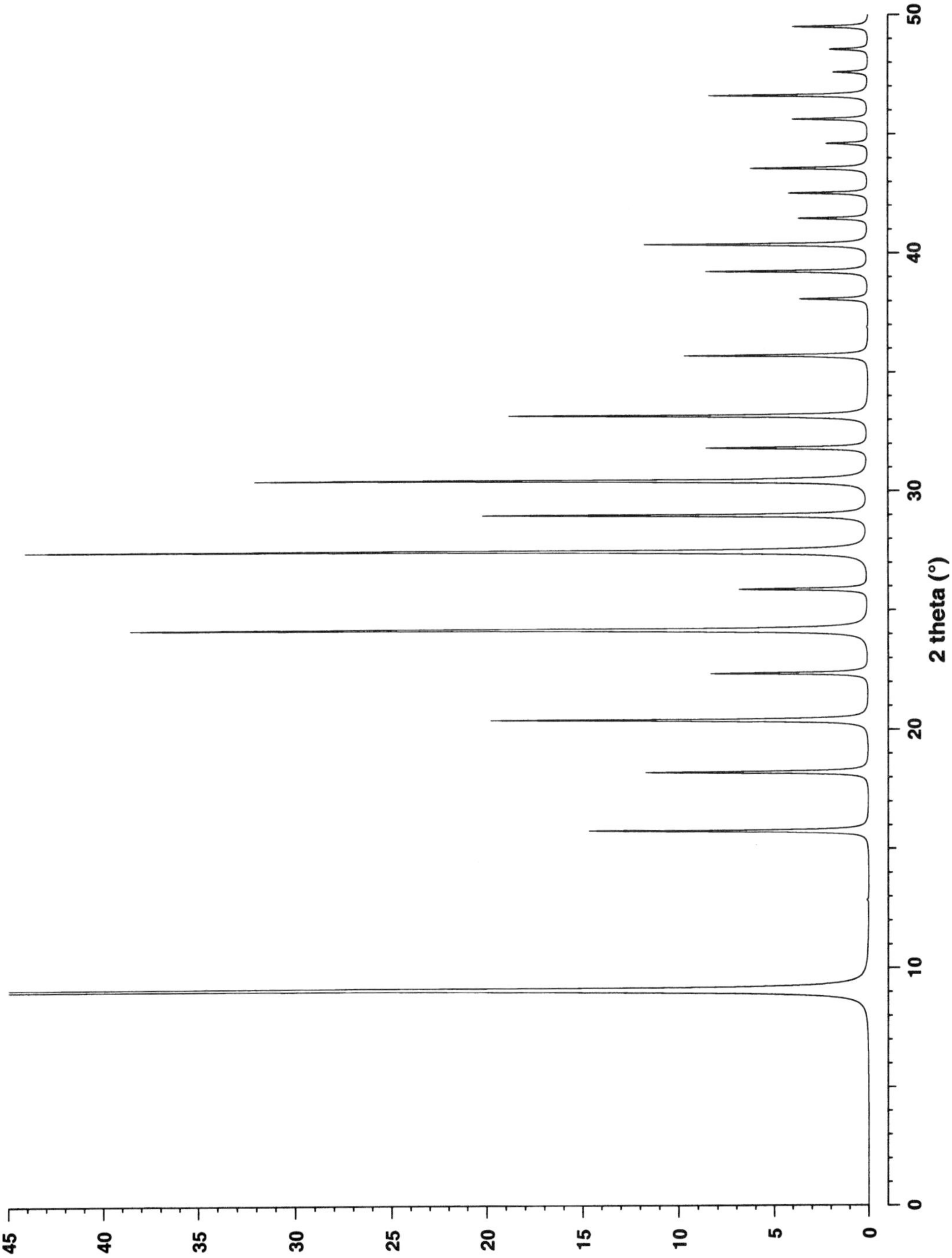

RHO Rho, Deuterated Berylloarsenate

CHEMICAL COMPOSITION: $|Rb_{24}(D_2O)_{3.2}|$ $[Be_{24}As_{24}O_{96}]$

REFINED COMPOSITION: $|Rb_{24}(D_2O)_{3.6}|$ $[Be_{24}As_{24}O_{96}]$

CRYSTAL DATA: $I23$ (No. 197)
a = 14.001 Å b = 14.001 Å c = 14.001 Å
α = 90° β = 90° γ = 90°
neutron R_{wp} = 0.041, X-ray R_{wp} = 0.145

REFERENCE: J. B. Parise, D. R. Corbin, T. E. Gier, R. L. Harlow, L. Abrams and R. B. Von Dreele, *Zeolites* **12** 360–368 (1992).
And J. B. Parise, Private communication (1994).

h	k	l	2θ	d	M	I_{rel}	h	k	l	2θ	d	M	I_{rel}	h	k	l	2θ	d	M	I_{rel}
1	1	0	8.93	9.900	12	50.6	5	0	1	32.61	2.746	12	10.3	6	3	1	43.86	2.064	24	14.9
2	0	0	12.64	7.001	6	9.6	5	1	0	32.61	2.746	12	0.5	6	1	3	43.86	2.064	24	3.0
2	1	1	15.50	5.716	24	17.6	5	1	2	35.10	2.556	24	31.7	4	4	4	44.85	2.021	8	0.7
2	2	0	17.92	4.950	12	4.4	5	2	1	35.10	2.556	24	18.4	5	4	3	45.83	1.980	24	4.6
3	1	0	20.05	4.428	12	1.6	4	4	0	36.30	2.475	12	6.9	7	1	0	45.83	1.980	12	0.2
3	0	1	20.05	4.428	12	22.9	5	0	3	37.45	2.401	12	4.7	7	0	1	45.83	1.980	12	0.3
2	2	2	21.99	4.042	8	5.0	4	3	3	37.45	2.401	24	3.5	5	3	4	45.83	1.980	24	5.8
3	2	1	23.78	3.742	24	84.6	5	3	0	37.45	2.401	12	0.9	5	5	0	45.83	1.980	12	2.0
3	1	2	23.78	3.742	24	23.3	6	0	0	38.58	2.334	6	2.0	6	0	4	46.79	1.942	12	7.6
4	0	0	25.45	3.500	6	1.7	4	4	2	38.58	2.334	24	1.8	6	4	0	46.79	1.942	12	1.1
4	1	1	27.02	3.300	24	100.0	6	1	1	39.68	2.271	24	4.6	6	3	3	47.73	1.905	24	3.4
3	3	0	27.02	3.300	12	2.6	5	3	2	39.68	2.271	24	0.5	7	1	2	47.73	1.905	24	5.9
4	2	0	28.51	3.131	12	4.0	5	2	3	39.68	2.271	24	6.9	5	5	2	47.73	1.905	24	1.0
4	0	2	28.51	3.131	12	53.3	6	2	0	40.76	2.214	12	5.3	7	2	1	47.73	1.905	24	7.5
3	3	2	29.93	2.985	24	65.5	6	0	2	40.76	2.214	12	0.2	6	4	2	48.67	1.871	24	8.7
4	2	2	31.30	2.858	24	27.0	5	1	4	41.81	2.160	24	7.6	6	2	4	48.67	1.871	24	2.9
4	3	1	32.61	2.746	24	22.7	5	4	1	41.81	2.160	24	8.9	7	3	0	49.58	1.838	12	0.8
4	1	3	32.61	2.746	24	6.3	6	2	2	42.84	2.111	24	4.2	7	0	3	49.58	1.838	12	2.3

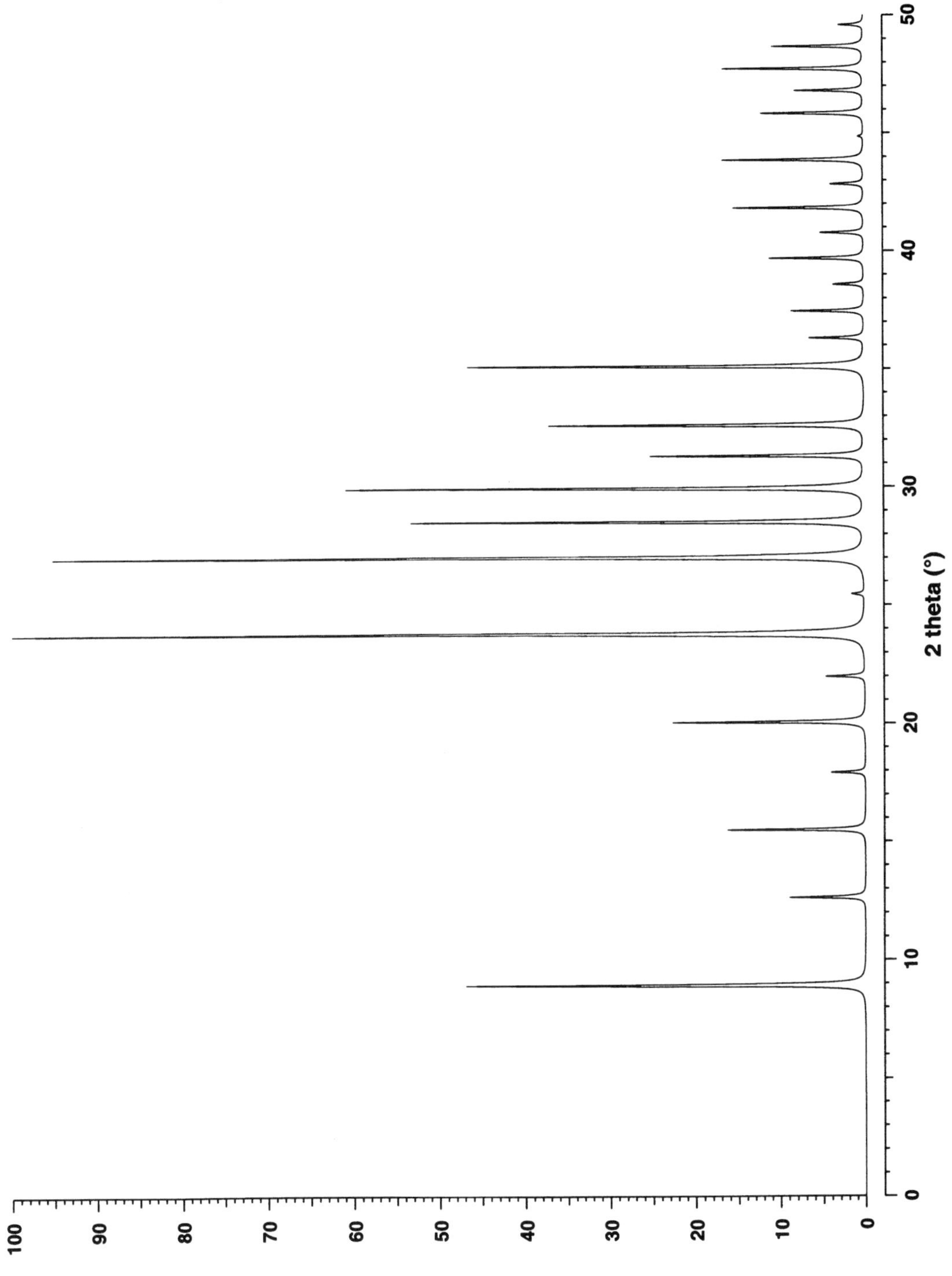
2 theta (°)
0
10
20
30
40
50
0
10
20
30
40
50
60
70
80
90
100

-RON Roggianite

CHEMICAL COMPOSITION: $|Ca_{16}(H_2O)_{<2.5}|$ $[(Be(OH)_2)_8Si_{32}Al_{16}O_{104}]$
Pizzo Marcio, Val Vigezzo, Italy

REFINED COMPOSITION: $|Ca_{16}(H_2O)_{18.72}|$ $[(Be(OH)_2)_8Si_{32}Al_{16}O_{104}]$

CRYSTAL DATA: $I\,4/m\,cm$ (No. 140)
$a = 18.33$ Å $b = 18.33$ Å $c = 9.16$ Å
$\alpha = 90°$ $\beta = 90°$ $\gamma = 90°$
X-ray single crystal refinement, $R_{obs} = 0.034$

REFERENCE: G. Giuseppetti, F. Mazzi, C. Tadini and E. Galli, *Neues Jahrb. Mineral. Monatsh.* **7** 307–314 (1991).

h	k	l	2θ	d	M	I_{rel}	h	k	l	2θ	d	M	I_{rel}	h	k	l	2θ	d	M	I_{rel}
1	1	0	6.82	12.961	4	100.0	2	1	3	31.26	2.861	16	6.3	6	6	0	41.82	2.160	4	0.8
2	0	0	9.65	9.165	4	35.0	5	1	2	31.64	2.828	16	12.8	2	2	4	41.84	2.159	8	0.8
2	2	0	13.66	6.481	4	2.8	5	4	1	32.78	2.732	16	0.7	6	1	3	42.13	2.145	16	0.3
2	1	1	14.50	6.109	16	25.5	4	4	2	33.89	2.645	8	8.5	7	3	2	42.43	2.131	16	2.6
3	1	0	15.29	5.796	8	6.0	6	3	1	34.24	2.618	16	0.7	3	1	4	42.44	2.130	16	0.3
4	0	0	19.37	4.582	4	0.6	3	2	3	34.26	2.618	16	1.0	8	3	1	43.31	2.089	16	0.5
0	0	2	19.38	4.580	2	0.2	7	1	0	34.60	2.592	8	0.5	8	4	0	44.19	2.049	8	0.6
3	2	1	19.97	4.445	16	0.6	5	5	0	34.60	2.592	4	2.5	8	0	2	44.20	2.049	8	3.1
3	3	0	20.56	4.320	4	1.8	5	3	2	34.61	2.592	16	0.5	4	0	4	44.21	2.048	8	0.2
1	1	2	20.57	4.318	8	3.0	6	4	0	35.31	2.542	8	0.2	9	1	0	44.77	2.024	8	2.0
4	2	0	21.68	4.099	8	0.6	6	0	2	35.32	2.541	8	11.4	3	3	4	44.79	2.023	8	0.5
2	0	2	21.69	4.097	8	5.0	4	1	3	35.67	2.517	16	1.4	8	2	2	45.35	2.000	16	0.3
4	1	1	22.23	4.000	16	0.4	6	2	2	36.69	2.449	16	5.0	6	6	2	46.48	1.954	8	0.5
2	2	2	23.79	3.740	8	5.2	7	2	1	37.03	2.428	16	0.3	7	6	1	46.75	1.943	16	1.9
5	1	0	24.77	3.595	8	1.1	7	3	0	37.36	2.407	8	0.1	9	2	1	46.75	1.943	16	0.3
3	1	2	24.77	3.594	16	21.7	4	3	3	38.37	2.346	16	2.8	9	3	0	47.03	1.932	8	5.5
4	3	1	26.18	3.404	16	28.0	8	0	0	39.32	2.291	4	2.2	7	5	2	47.03	1.932	16	0.1
4	0	2	27.53	3.239	8	5.3	0	0	4	39.34	2.290	2	4.8	5	1	4	47.05	1.931	16	0.6
5	2	1	27.96	3.191	16	14.9	6	5	1	39.64	2.273	16	1.5	4	4	4	48.69	1.870	8	0.2
5	3	0	28.39	3.144	8	9.7	5	2	3	39.65	2.273	16	0.7	6	5	3	48.95	1.861	16	0.9
3	3	2	28.40	3.143	8	6.2	5	5	2	39.96	2.256	8	0.8	7	7	0	49.21	1.852	4	3.2
4	2	2	29.24	3.054	16	0.4	1	1	4	39.98	2.255	8	1.5	5	3	4	49.23	1.851	16	7.7
6	2	0	30.85	2.898	8	5.0	8	2	0	40.58	2.223	8	3.1	8	6	0	49.74	1.833	8	1.0
6	1	1	31.25	2.863	16	5.2	2	0	4	40.61	2.222	8	0.3	6	0	4	49.76	1.832	8	0.9

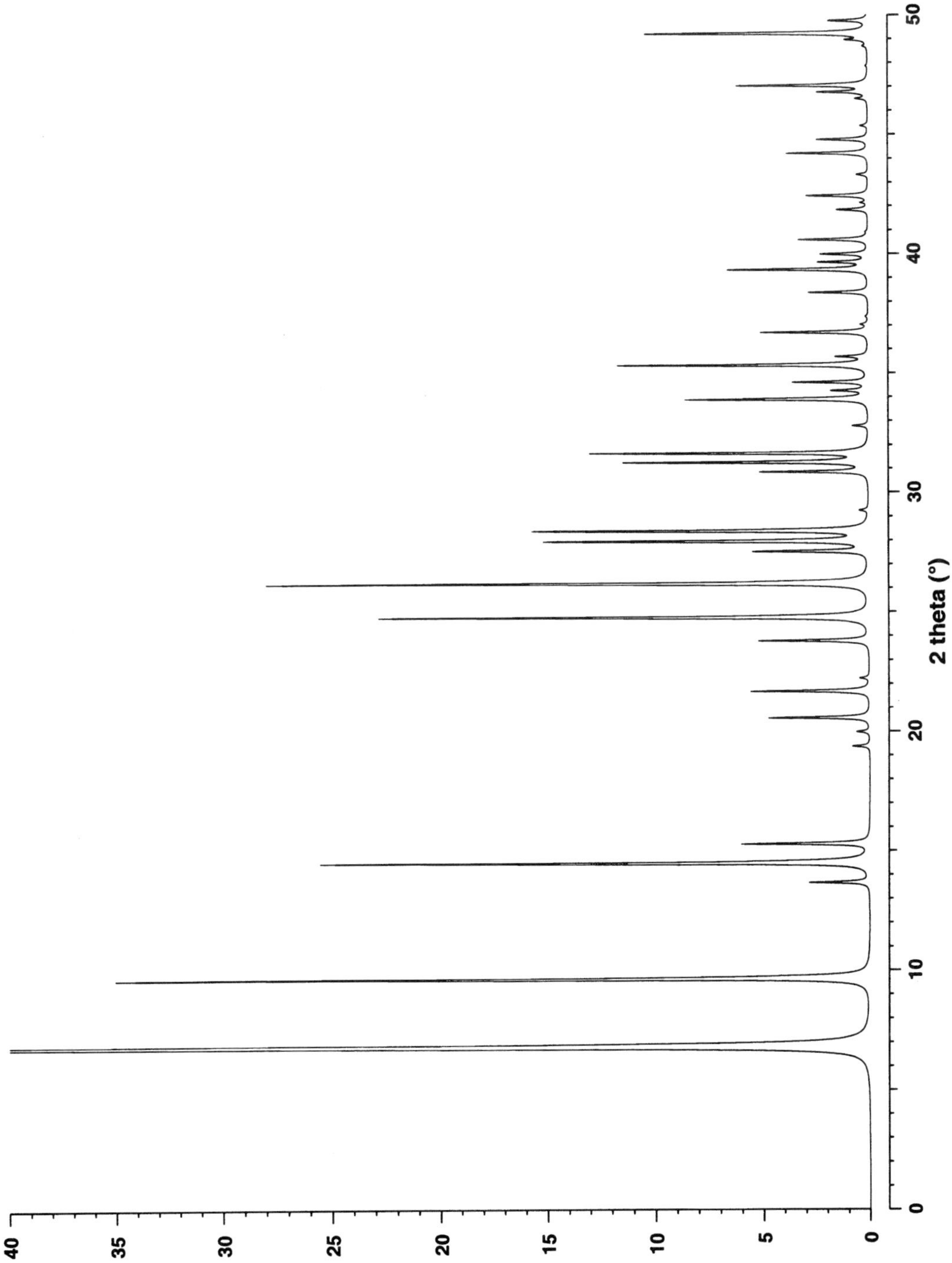
2 theta (°)
0
10
20
30
40
50
0
5
10
15
20
25
30
35
40

RSN RUB-17

CHEMICAL COMPOSITION: $|K_4Na_{12}(H_2O)_{18}|$ $[Si_{28}Zn_8O_{72}]$

REFINED COMPOSITION: $|K_4Na_{12}(H_2O)_{20.7}|$ $[Si_{28}Zn_8O_{72}]$

CRYSTAL DATA: $C1m1$ (No. 8) unique axis **b**, cell choice 1
$a = 7.238$ Å $b = 40.56$ Å $c = 7.308$ Å
$\alpha = 90°$ $\beta = 91.8°$ $\gamma = 90°$
X-ray Rietveld refinement, $R_{exp} = 0.238$, $R_{wp} = 0.164$, $R_I = 0.090$

REFERENCE: C. Röhrig and H. Gies,
Angew. Chem., Int. ed. **34** 63–65 (1995).

h	k	l	2θ	d	M	I_{rel}	h	k	l	2θ	d	M	I_{rel}	h	k	l	2θ	d	M	I_{rel}
0	4	0	8.72	10.140	2	16.5	2	6	1	30.90	2.894	4	2.2	3	3	1	40.31	2.238	4	1.9
0	0	1	12.12	7.304	2	3.5	1	13	0	31.22	2.865	4	1.6	3	7	0	40.52	2.226	4	2.5
1	1	0	12.43	7.122	4	6.0	1	7	2	31.79	2.814	4	8.9	-3	5	1	40.58	2.223	4	1.8
0	2	1	12.88	6.872	4	46.7	-2	8	1	32.49	2.756	4	5.4	1	5	3	41.02	2.200	4	2.7
0	6	0	13.10	6.760	2	1.4	2	8	1	33.09	2.707	4	23.6	0	8	3	41.13	2.195	4	4.2
1	3	0	13.88	6.379	4	100.0	2	10	0	33.18	2.700	4	2.3	-1	17	1	41.61	2.170	4	7.1
0	4	1	14.95	5.927	4	7.6	0	14	1	33.27	2.693	4	10.0	-1	7	3	41.79	2.162	4	1.4
1	5	0	16.42	5.399	4	4.3	-1	13	1	33.45	2.678	4	5.3	1	17	1	41.86	2.158	4	1.4
-1	1	1	17.12	5.180	4	24.4	1	13	1	33.75	2.656	4	2.7	0	18	1	41.96	2.153	4	4.4
0	8	0	17.49	5.070	2	28.4	1	9	2	34.23	2.620	4	5.7	-3	7	1	42.08	2.147	4	1.8
1	1	1	17.66	5.022	4	27.9	-2	0	2	34.34	2.611	2	16.6	3	9	0	42.52	2.126	4	2.6
0	6	1	17.88	4.961	4	31.0	-2	10	1	35.17	2.552	4	8.5	3	7	1	42.80	2.113	4	1.8
-1	3	1	18.21	4.872	4	5.4	0	16	0	35.41	2.535	2	10.9	-1	15	2	43.24	2.092	4	1.5
1	3	1	18.72	4.740	4	11.5	1	15	0	35.44	2.533	4	2.4	0	16	2	43.45	2.083	4	5.6
-1	5	1	20.22	4.391	4	1.7	2	0	2	35.47	2.531	2	7.8	2	16	0	43.60	2.076	4	1.6
1	5	1	20.68	4.294	4	5.1	-2	4	2	35.50	2.529	4	2.5	1	15	2	43.71	2.071	4	4.5
0	8	1	21.33	4.165	4	7.8	2	10	1	35.73	2.513	4	6.4	-1	9	3	43.74	2.070	4	6.0
1	9	0	23.25	3.825	4	2.2	2	2	2	35.76	2.511	4	5.2	-2	12	2	43.81	2.066	4	1.7
0	0	2	24.37	3.652	2	4.2	0	12	2	36.21	2.481	4	2.8	-3	1	2	44.42	2.040	4	2.9
2	0	0	24.61	3.617	2	1.6	2	12	0	36.38	2.470	4	1.8	-2	2	3	44.42	2.039	4	3.5
2	2	0	25.01	3.561	4	4.0	-1	11	2	36.53	2.460	4	8.7	1	9	3	44.44	2.039	4	1.7
0	10	1	25.11	3.546	4	15.5	2	4	2	36.60	2.455	4	1.9	3	11	0	44.91	2.018	4	1.3
0	4	2	25.93	3.436	4	18.7	0	0	3	36.92	2.435	2	3.0	-2	16	1	45.19	2.006	4	3.5
2	4	0	26.16	3.407	4	5.3	1	11	2	37.07	2.425	4	4.2	2	0	3	45.56	1.991	2	1.3
0	12	0	26.37	3.380	2	32.9	0	2	3	37.19	2.417	4	2.3	2	16	1	45.65	1.987	4	1.3
-1	1	2	27.09	3.291	4	24.8	3	1	0	37.36	2.407	4	2.3	2	2	3	45.79	1.982	4	1.9
1	11	0	27.14	3.285	4	37.8	0	16	1	37.56	2.395	4	6.3	-3	5	2	45.82	1.980	4	1.3
-2	0	1	27.16	3.283	2	5.4	1	15	1	37.72	2.385	4	1.7	0	12	3	45.94	1.976	4	1.7
-2	2	1	27.52	3.241	4	34.4	3	3	0	37.90	2.374	4	3.2	-1	11	3	46.08	1.970	4	1.6
0	6	2	27.76	3.213	4	38.5	0	4	3	38.01	2.368	4	1.5	1	19	1	46.15	1.967	4	6.4
1	1	2	27.79	3.210	4	17.2	-2	12	1	38.22	2.355	4	3.7	3	3	2	46.25	1.963	4	2.8
-1	3	2	27.81	3.208	4	14.8	2	12	1	38.74	2.324	4	4.7	-2	6	3	46.28	1.962	4	3.4
2	0	1	27.86	3.202	2	6.6	-2	8	2	38.79	2.322	4	2.2	0	20	1	46.47	1.954	4	1.4
2	6	0	27.98	3.189	4	58.7	3	5	0	38.96	2.312	4	5.3	0	18	2	47.41	1.918	4	3.1
2	2	1	28.22	3.163	4	19.7	-3	1	1	39.03	2.308	4	1.7	2	18	0	47.54	1.913	4	4.3
1	3	2	28.50	3.132	4	12.3	1	1	3	39.48	2.282	4	3.2	2	6	3	47.61	1.910	4	3.0
-2	4	1	28.58	3.123	4	1.9	-3	3	1	39.55	2.279	4	2.7	2	14	2	47.72	1.906	4	2.5
0	12	1	29.11	3.068	4	5.8	0	14	2	39.71	2.270	4	3.2	-2	8	3	47.86	1.900	4	1.6
-1	5	2	29.20	3.059	4	21.2	-1	13	2	39.74	2.268	4	4.6	3	7	2	48.50	1.877	4	3.4
2	4	1	29.25	3.053	4	4.0	1	17	0	39.78	2.266	4	4.2	-3	9	2	48.97	1.860	4	1.4
-1	11	1	29.66	3.012	4	9.2	3	1	1	39.79	2.265	4	2.0	-2	18	1	49.03	1.858	4	1.3
1	5	2	29.85	2.993	4	9.2	2	8	2	39.81	2.264	4	5.1	-3	13	1	49.04	1.857	4	1.5
1	11	1	29.98	2.980	4	11.6	2	14	0	39.87	2.261	4	3.3	1	13	3	49.42	1.844	4	2.5
0	8	2	30.16	2.963	4	12.4	1	3	3	40.00	2.254	4	4.9	2	18	1	49.46	1.843	4	1.5
-2	6	1	30.27	2.953	4	8.4	0	18	0	40.01	2.253	2	1.4	0	0	4	49.94	1.826	2	2.4

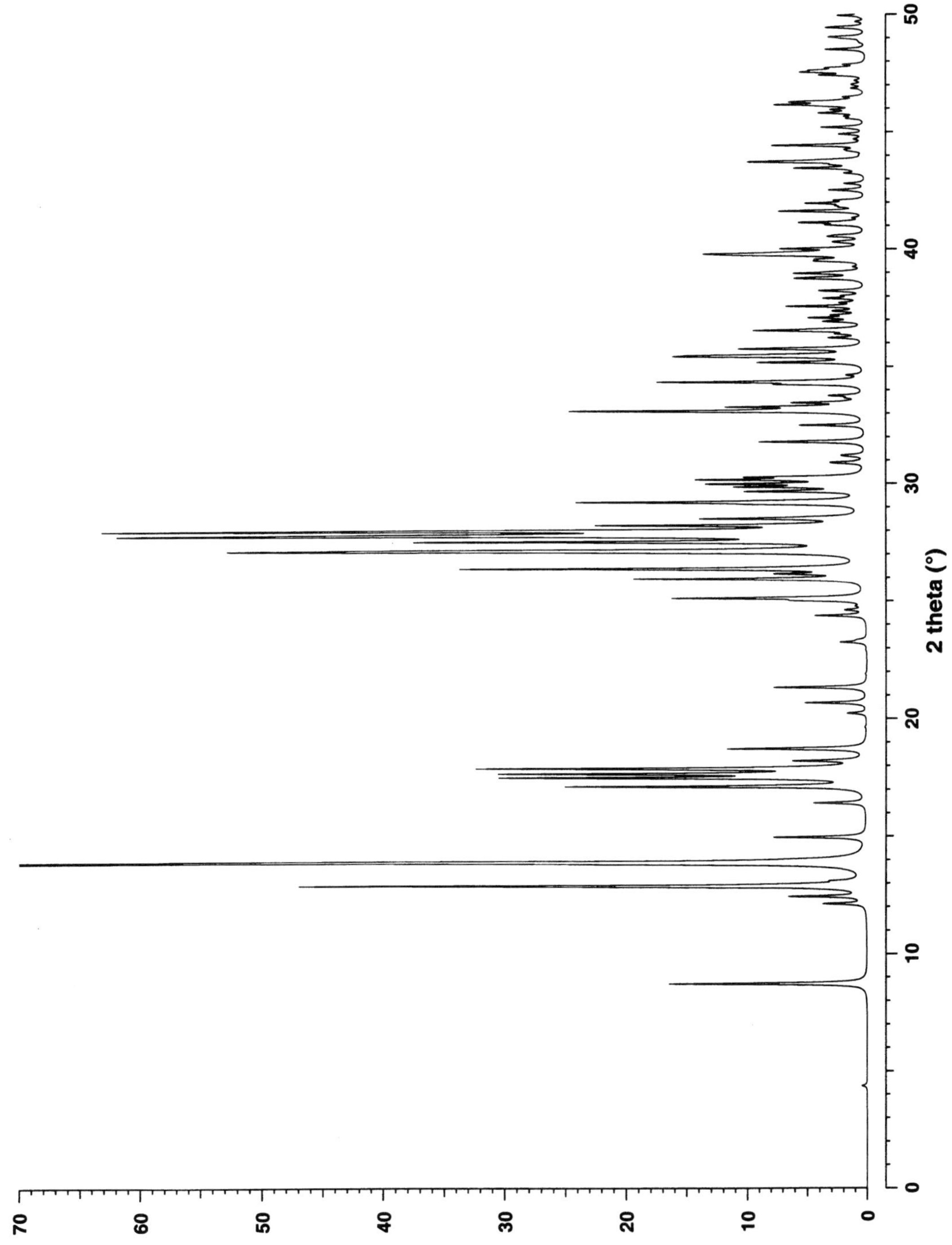
2 theta (°)
0
10
20
30
40
50
0
10
20
30
40
50
60
70

RTE RUB-3

CHEMICAL COMPOSITION: $|(C_7H_{13}N)_2|$ $[Si_{24}O_{48}]$
$C_7H_{13}N$ = (±)-exo-2-aminobicyclo[2.2.1]heptane

REFINED COMPOSITION: $|C_{17.6}|$ $[Si_{24}O_{48}]$

CRYSTAL DATA: $C1\,2/m\,1$ (No. 12) unique axis **b**, cell choice 1
a = 14.018 Å b = 13.612 Å c = 7.418 Å
α = 90° β = 102.12° γ = 90°
X-ray single crystal refinement, R_1 = 0.099

REFERENCE: B. Marler, A. Grünewald-Lüke and H. Gies, *Microporous and Mesoporous Materials* **26** 49–59 (1998).

h	k	l	2θ	d	M	I_{rel}	h	k	l	2θ	d	M	I_{rel}	h	k	l	2θ	d	M	I_{rel}
1	1	0	9.16	9.658	4	100.0	-1	3	2	31.17	2.869	4	0.8	-1	5	2	41.15	2.193	4	0.3
0	0	1	12.20	7.253	2	2.8	-4	0	2	31.95	2.801	2	4.6	-1	3	3	41.61	2.170	4	1.2
2	0	0	12.92	6.853	2	0.4	3	3	1	32.04	2.793	4	2.9	6	2	0	41.71	2.166	4	0.4
0	2	0	13.01	6.806	2	17.4	2	4	1	32.92	2.721	4	0.3	-4	4	2	41.76	2.163	4	3.9
-1	1	1	14.14	6.261	4	47.5	2	2	2	33.01	2.713	4	4.4	3	5	1	41.84	2.159	4	0.1
-2	0	1	15.82	5.603	2	0.6	-5	1	1	33.08	2.708	4	0.4	-5	3	2	41.93	2.154	4	0.2
1	1	1	16.33	5.427	4	1.7	1	3	2	33.33	2.688	4	4.9	4	2	2	42.05	2.149	4	0.5
0	2	1	17.87	4.963	4	90.2	5	1	0	33.34	2.687	4	0.2	-6	0	2	42.09	2.147	2	0.3
2	2	0	18.37	4.829	4	5.3	1	5	0	33.56	2.670	4	0.5	2	0	3	42.17	2.143	2	1.3
2	0	1	19.60	4.529	2	34.6	4	2	1	33.86	2.647	4	8.8	1	5	2	42.87	2.109	4	2.8
3	1	0	20.51	4.331	4	29.7	-4	2	2	34.63	2.591	4	0.9	-4	2	3	43.11	2.098	4	0.7
-2	2	1	20.53	4.326	4	56.0	-1	5	1	35.34	2.540	4	11.0	-3	3	3	43.35	2.087	4	0.5
1	3	0	20.62	4.307	4	40.8	0	4	2	36.20	2.481	4	0.4	6	0	1	43.99	2.058	2	2.5
-3	1	1	21.70	4.095	4	40.3	1	5	1	36.32	2.473	4	0.3	1	3	3	44.15	2.051	4	0.2
-1	3	1	23.31	3.816	4	13.5	-2	4	2	36.70	2.449	4	0.6	-6	2	2	44.24	2.047	4	0.9
2	2	1	23.60	3.770	4	0.2	-2	0	3	36.73	2.447	2	6.7	2	2	3	44.31	2.044	4	1.2
0	0	2	24.55	3.626	2	23.9	-1	1	3	36.97	2.431	4	0.5	2	6	1	44.67	2.028	4	0.3
-1	1	2	24.92	3.573	4	5.5	0	0	3	37.19	2.418	2	0.8	-2	4	3	45.67	1.987	4	1.2
-2	0	2	25.26	3.526	2	19.3	4	4	0	37.24	2.415	4	0.1	0	4	3	46.05	1.971	4	2.8
3	1	1	25.98	3.430	4	13.5	-5	1	2	37.32	2.409	4	4.9	3	1	3	46.52	1.952	4	1.8
4	0	0	26.00	3.426	2	15.1	-4	4	1	37.48	2.400	4	5.1	-5	5	1	46.85	1.939	4	0.5
0	4	0	26.19	3.403	2	16.8	5	1	1	38.03	2.366	4	3.7	7	1	0	46.88	1.938	4	0.4
-4	0	1	26.33	3.385	2	6.0	-5	3	1	38.13	2.360	4	0.2	5	5	0	47.04	1.932	4	0.4
1	1	2	27.52	3.241	4	29.9	5	3	0	38.36	2.346	4	0.5	1	7	0	47.21	1.925	4	0.2
3	3	0	27.71	3.219	4	6.9	3	5	0	38.49	2.339	4	1.4	-6	4	1	47.38	1.919	4	0.5
0	2	2	27.88	3.200	4	3.4	-6	0	1	38.76	2.323	2	5.4	-2	6	2	47.67	1.908	4	0.4
-2	2	2	28.51	3.131	4	15.9	-3	1	3	38.88	2.316	4	1.3	4	6	0	48.10	1.892	4	0.6
-3	1	2	28.79	3.101	4	2.2	-2	2	3	39.12	2.302	4	1.3	4	4	2	48.27	1.885	4	0.5
0	4	1	28.98	3.081	4	10.8	-3	5	1	39.18	2.299	4	1.9	-4	6	1	48.29	1.884	4	1.5
4	2	0	29.18	3.060	4	8.8	6	0	0	39.45	2.284	2	1.8	-7	1	2	48.44	1.879	4	4.8
2	4	0	29.30	3.048	4	11.5	0	2	3	39.56	2.278	4	0.4	-1	7	1	48.56	1.875	4	4.5
-4	2	1	29.47	3.031	4	10.1	4	0	2	39.81	2.264	2	2.1	-5	3	3	49.00	1.859	4	0.2
2	0	2	30.21	2.959	2	1.7	2	4	2	40.40	2.233	4	3.9	-2	0	4	49.21	1.852	2	1.1
-2	4	1	30.74	2.909	4	17.5	-4	0	3	40.92	2.205	2	0.6	-4	4	3	49.23	1.851	4	0.2
4	0	1	31.12	2.874	2	6.2	-6	2	1	41.05	2.198	4	1.7	-1	5	3	49.83	1.830	4	0.8

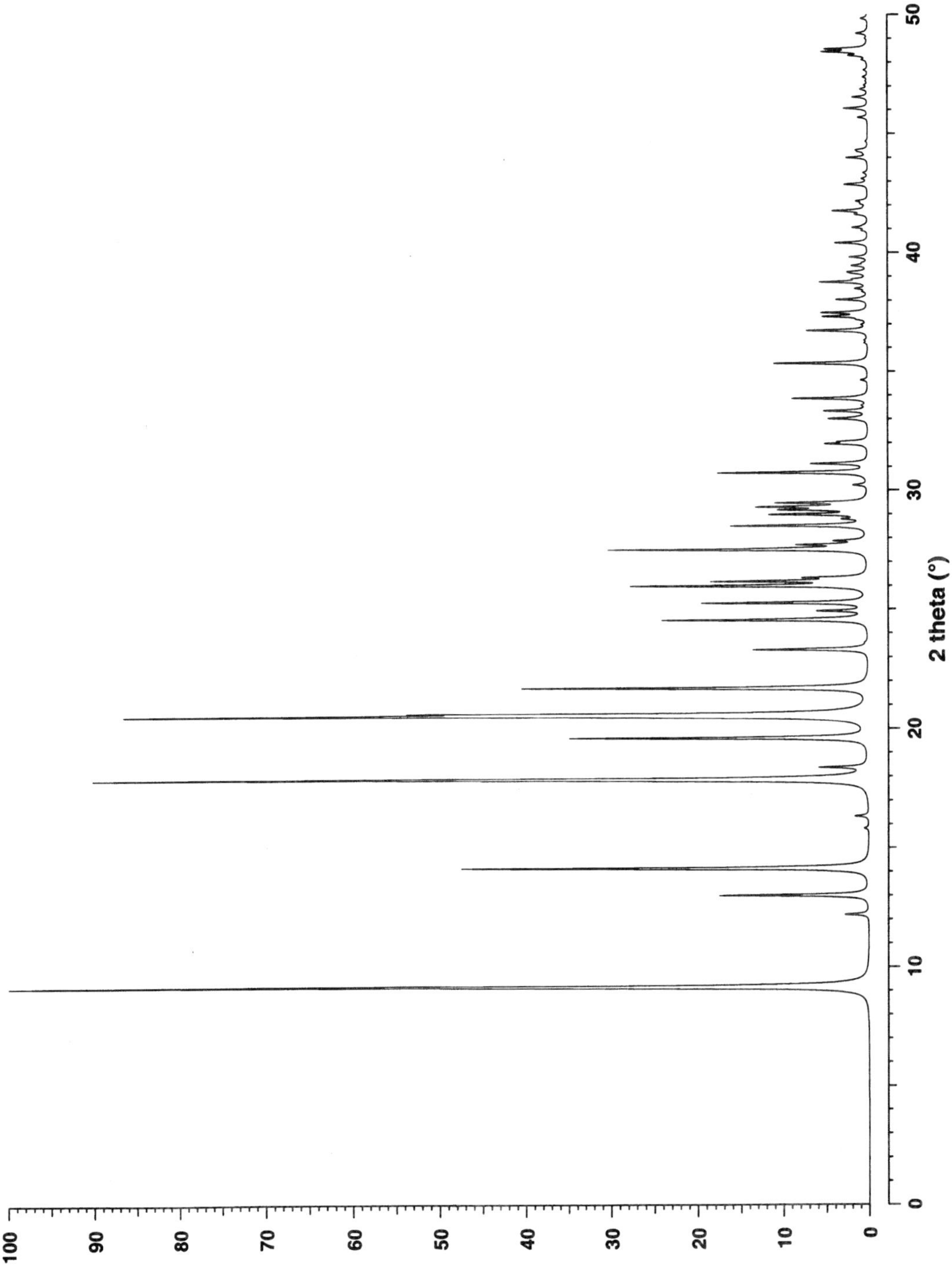
2 theta (°)
0
10
20
30
40
50
0
10
20
30
40
50
60
70
80
90
100

RTH RUB-13

CHEMICAL COMPOSITION: $|(C_{10}H_{21}N)_{1.7}|$ $[Si_{30.4}B_{1.6}O_{64}]$
$(C_{10}H_{21}N)^+$ = 1,2,2,6,6,-pentamethylpiperidine

REFINED COMPOSITION: $[Si_{31}O_{64}]$

CRYSTAL DATA: $C1\,2/m\,1$ (No. 12) unique axis **b**, cell choice 1
$a = 9.659$ Å $b = 20.461$ Å $c = 9.831$ Å
$\alpha = 90°$ $\beta = 96.58°$ $\gamma = 90°$
X-ray single crystal refinement, $R = 0.06$

REFERENCE: S. Vortmann, B. Marler, H. Gies and P. Daniels, *Microporous Materials* **4** 111–121 (1995).

h	*k*	*l*	*2θ*	*d*	*M*	I_{rel}	*h*	*k*	*l*	*2θ*	*d*	*M*	I_{rel}	*h*	*k*	*l*	*2θ*	*d*	*M*	I_{rel}
0	2	0	8.64	10.231	2	100.0	-2	0	3	31.39	2.850	2	1.1	2	8	1	41.55	2.173	4	0.4
0	0	1	9.05	9.766	2	91.5	-3	1	2	31.94	2.802	4	0.4	4	4	0	41.58	2.172	4	0.4
1	1	0	10.18	8.688	4	55.0	1	7	0	32.01	2.796	4	2.0	-1	9	1	41.62	2.170	4	0.2
0	2	1	12.53	7.064	4	3.8	0	6	2	32.01	2.796	4	1.7	3	5	2	41.79	2.161	4	0.4
-1	1	1	12.92	6.854	4	5.2	2	6	0	32.20	2.780	4	1.1	3	7	0	41.87	2.158	4	0.6
1	1	1	14.33	6.180	4	1.0	0	4	3	32.60	2.746	4	1.5	3	1	3	42.04	2.149	4	0.1
1	3	0	15.94	5.559	4	1.6	-2	2	3	32.62	2.745	4	5.0	1	9	1	42.12	2.145	4	0.2
0	4	0	17.34	5.115	2	1.7	2	4	2	32.70	2.738	4	0.4	-3	7	1	42.19	2.142	4	0.2
-1	3	1	17.83	4.975	4	12.0	1	3	3	32.79	2.731	4	1.5	-2	8	2	43.24	2.092	4	1.0
0	0	2	18.17	4.883	2	0.7	-2	6	1	32.91	2.721	4	0.2	1	7	3	43.36	2.087	4	2.7
2	0	0	18.49	4.798	2	0.1	-1	7	1	33.03	2.712	4	1.2	-2	4	4	43.38	2.086	4	0.8
1	3	1	18.88	4.699	4	34.4	3	3	1	33.17	2.701	4	0.9	4	4	1	43.62	2.075	4	0.1
0	4	1	19.59	4.531	4	7.0	1	7	1	33.63	2.665	4	0.9	3	7	1	43.65	2.073	4	0.4
-2	0	1	19.66	4.516	2	11.1	2	6	1	34.12	2.628	4	2.4	-3	1	4	44.24	2.047	4	0.3
-1	1	2	19.91	4.459	4	12.7	-3	3	2	34.32	2.613	4	0.6	-1	9	2	44.51	2.035	4	0.5
0	2	2	20.15	4.407	4	8.0	2	0	3	35.04	2.561	2	0.7	-3	7	2	44.58	2.033	4	0.2
2	2	0	20.45	4.344	4	6.1	3	1	2	35.54	2.526	4	1.0	4	2	2	44.85	2.021	4	0.8
-2	2	1	21.51	4.131	4	1.3	3	5	0	35.63	2.520	4	0.4	1	5	4	45.16	2.008	4	0.1
2	0	1	21.55	4.123	2	0.3	-1	5	3	35.64	2.519	4	1.7	-4	2	3	45.18	2.007	4	0.4
2	2	1	23.26	3.824	4	9.1	-2	4	3	36.08	2.490	4	0.2	-4	6	1	46.34	1.959	4	0.3
-1	3	2	23.44	3.796	4	3.1	-2	6	2	36.11	2.488	4	0.5	0	0	5	46.49	1.953	2	0.2
-2	0	2	24.47	3.637	2	0.6	2	2	3	36.16	2.484	4	0.1	-1	1	5	46.59	1.949	4	0.6
-1	5	1	24.97	3.567	4	0.1	0	8	1	36.31	2.474	4	0.1	2	4	4	47.13	1.928	4	0.3
1	3	2	25.06	3.554	4	3.8	-1	7	2	36.49	2.462	4	0.2	-5	1	1	47.43	1.917	4	1.2
0	4	2	25.21	3.532	4	12.5	-1	1	4	37.20	2.417	4	1.2	5	1	0	47.59	1.911	4	0.1
2	4	0	25.45	3.499	4	5.4	1	5	3	37.32	2.409	4	0.8	-2	8	3	47.78	1.903	4	0.2
1	5	1	25.74	3.461	4	12.5	-3	1	3	37.35	2.408	4	0.1	-4	4	3	47.87	1.900	4	0.5
-2	2	2	26.00	3.427	4	5.8	4	0	0	37.49	2.399	2	0.3	-2	6	4	47.93	1.898	4	0.2
0	6	0	26.13	3.410	2	0.2	-4	0	1	37.57	2.394	2	1.3	4	6	1	48.15	1.890	4	1.8
-2	4	1	26.33	3.385	4	0.3	1	7	2	37.60	2.392	4	1.3	0	10	2	48.22	1.887	4	0.1
0	0	3	27.40	3.255	2	0.9	3	5	1	37.66	2.388	4	0.4	-1	3	5	48.35	1.882	4	1.0
2	0	2	27.52	3.242	2	0.9	3	3	2	37.72	2.385	4	0.2	2	10	0	48.36	1.882	4	0.2
0	6	1	27.71	3.220	4	1.2	0	2	4	37.88	2.375	4	0.2	-3	7	3	48.80	1.866	4	0.2
3	1	0	28.24	3.160	4	3.7	0	6	3	38.22	2.355	4	0.8	1	1	5	48.84	1.865	4	0.2
-1	1	3	28.26	3.158	4	2.7	2	6	2	38.31	2.349	4	0.4	-2	10	1	48.87	1.864	4	0.2
-3	1	1	28.69	3.112	4	0.5	4	2	0	38.55	2.335	4	0.1	-2	2	5	49.12	1.855	4	0.5
0	2	3	28.78	3.102	4	1.7	-1	3	4	39.31	2.292	4	0.2	3	9	0	49.17	1.853	4	0.7
-2	4	2	30.15	2.964	4	3.0	-3	3	3	39.44	2.285	4	0.4	-5	3	1	49.17	1.853	4	0.3
1	1	3	30.30	2.950	4	0.3	4	0	1	39.70	2.270	2	0.2	-1	9	3	49.18	1.853	4	0.1
1	5	2	30.63	2.918	4	1.0	0	8	2	39.78	2.266	4	0.1	-5	1	2	49.18	1.852	4	0.1
3	1	1	30.71	2.912	4	2.4	-4	0	2	39.93	2.258	2	0.3	5	3	0	49.33	1.847	4	0.1
3	3	0	30.88	2.896	4	7.5	2	8	0	39.94	2.257	4	0.2	-3	5	4	49.58	1.838	4	0.5
-1	3	3	30.90	2.894	4	0.4	0	4	4	40.96	2.203	4	0.4	1	11	0	49.94	1.826	4	0.8
-3	3	1	31.29	2.858	4	0.2	1	3	4	41.36	2.183	4	0.1							

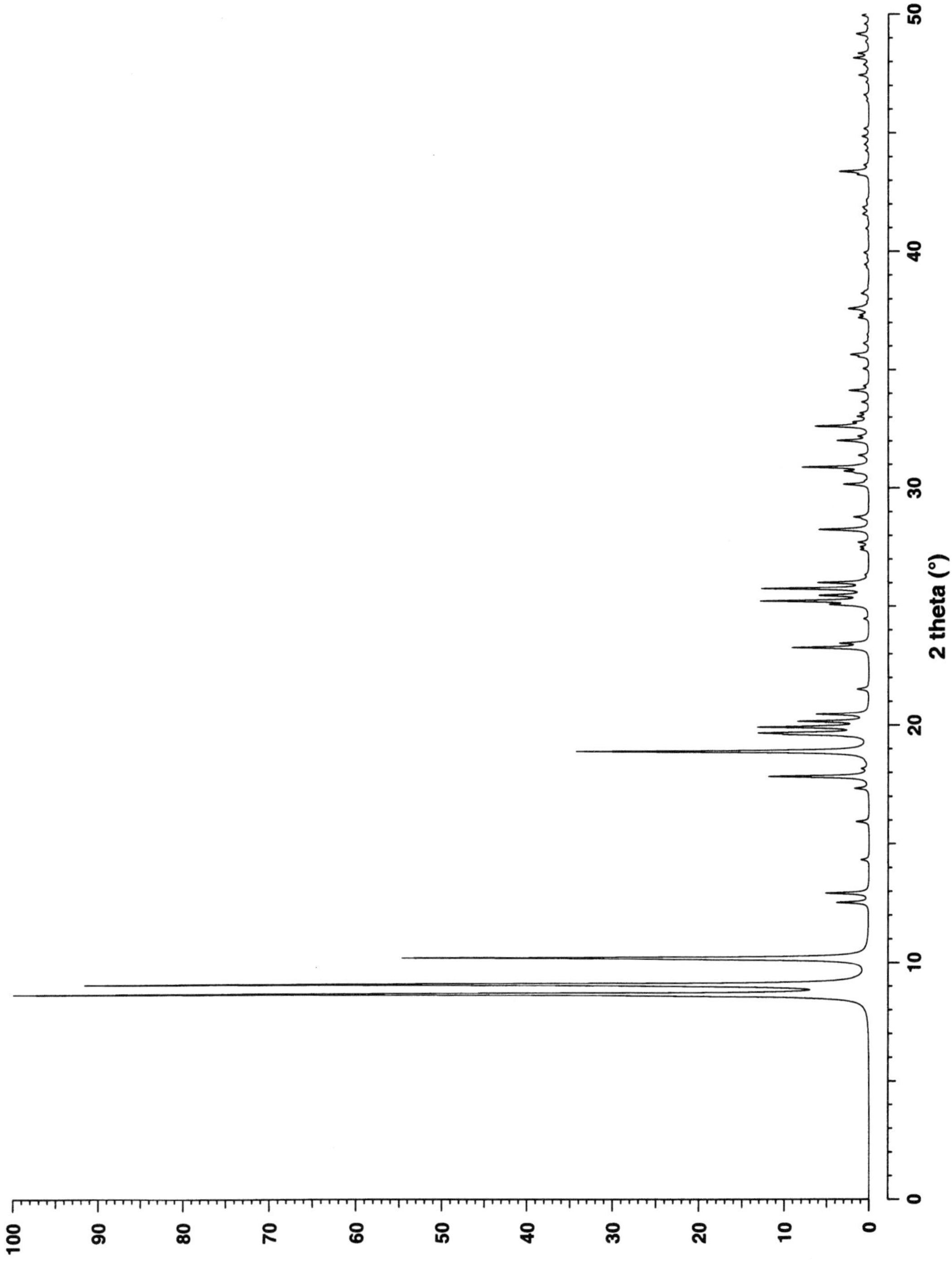
2 theta (°)
0
10
20
30
40
50
0
10
20
30
40
50
60
70
80
90
100

RUT RUB-10, SiO_2 Framework

CHEMICAL COMPOSITION: $[Si_{32}B_4O_{72}]$

REFINED COMPOSITION: $[Si_{36}O_{72}]$

CRYSTAL DATA: $P1\,2_1/a\,1$ (No. 14) unique axis **b**, cell choice 3
$a = 13.112$ Å $b = 12.903$ Å $c = 12.407$ Å
$\alpha = 90°$ $\beta = 113.50°$ $\gamma = 90°$
DLS refinement.

REFERENCE: H. Gies and J. Rius,
Z. Kristallogr. **210** 475–480 (1995).

h	k	l	2θ	d	M	I_{rel}
0	0	1	7.77	11.378	2	0.4
1	1	0	10.06	8.797	4	29.4
0	1	1	10.37	8.534	4	0.6
-1	1	1	10.77	8.215	4	100.0
-2	0	1	13.64	6.491	2	22.7
0	2	0	13.73	6.451	2	25.1
1	1	1	14.41	6.146	4	62.3
2	0	0	14.73	6.012	2	0.9
0	0	2	15.58	5.689	2	4.3
0	2	1	15.79	5.612	4	0.9
-1	1	2	15.89	5.576	4	9.0
-2	0	2	16.64	5.326	2	6.7
0	1	2	17.03	5.205	4	0.4
1	2	1	18.72	4.741	4	0.4
2	0	1	19.25	4.610	2	0.8
-2	2	1	19.40	4.576	4	0.5
-1	2	2	19.89	4.464	4	0.8
2	2	0	20.19	4.398	4	13.3
0	2	2	20.82	4.267	4	11.1
-3	1	1	21.49	4.135	4	1.7
-2	2	2	21.64	4.107	4	2.2
1	3	0	21.95	4.050	4	9.2
-2	0	3	22.17	4.009	2	13.0
-1	3	1	22.29	3.988	4	21.4
-3	1	2	22.52	3.947	4	8.0
-1	1	3	22.66	3.924	4	21.9
3	1	0	23.24	3.828	4	22.4
0	0	3	23.46	3.793	2	4.5
2	2	1	23.72	3.751	4	0.4
1	3	1	24.30	3.663	4	2.1
-1	3	2	25.22	3.531	4	1.3
2	0	2	25.49	3.495	2	10.7
-3	2	2	25.54	3.488	4	0.5
0	3	2	25.97	3.431	4	0.4
-3	1	3	26.03	3.424	4	0.8
-2	2	3	26.17	3.405	4	2.8
2	1	2	26.42	3.373	4	0.3
-2	3	2	26.64	3.346	4	0.9
0	2	3	27.28	3.270	4	0.7
-4	0	2	27.48	3.246	2	4.6
-4	0	1	27.50	3.243	2	0.9
0	4	0	27.65	3.226	2	12.1
-4	1	2	28.35	3.148	4	0.3
1	3	2	28.72	3.108	4	0.5
-2	0	4	28.88	3.092	2	1.0
2	2	2	29.06	3.073	4	6.0
-3	3	1	29.15	3.064	4	0.8
4	0	0	29.72	3.006	2	1.9
-3	3	2	29.93	2.985	4	2.8
-1	3	3	30.04	2.975	4	3.8
-1	1	4	30.07	2.972	4	0.6
3	3	0	30.48	2.932	4	4.0
-4	2	1	30.86	2.898	4	4.3
-3	1	4	31.23	2.864	4	4.7
0	0	4	31.45	2.844	2	4.5
2	4	0	31.47	2.842	4	5.7
0	4	2	31.89	2.806	4	2.6
-2	2	4	32.10	2.788	4	0.4
-1	2	4	32.43	2.761	4	0.3
-2	4	2	32.45	2.759	4	3.1
2	0	3	32.55	2.751	2	1.7
-3	3	3	32.70	2.738	4	1.3
-4	2	3	32.81	2.729	4	4.9
2	1	3	33.30	2.690	4	0.3
-4	0	4	33.65	2.663	2	0.3
3	3	1	33.72	2.658	4	0.8
4	0	1	33.74	2.657	2	5.9
0	2	4	34.46	2.603	4	1.1
1	3	3	34.50	2.600	4	0.6
-4	3	2	34.62	2.591	4	0.3
3	2	2	35.00	2.564	4	0.2
1	1	4	35.90	2.501	4	0.3
-5	1	3	36.04	2.492	4	0.6
-2	0	5	36.21	2.481	2	1.5
0	4	3	36.57	2.457	4	1.8
4	2	1	36.58	2.456	4	0.2
-5	2	2	37.01	2.429	4	0.4
-3	3	4	37.06	2.426	4	0.5
1	5	1	37.14	2.421	4	2.2
-3	1	5	37.51	2.398	4	0.4
-1	5	2	37.77	2.382	4	0.4
-1	1	5	37.84	2.378	4	0.8
2	4	2	37.96	2.370	4	0.3
5	1	0	38.06	2.364	4	0.5
-5	1	4	38.83	2.319	4	2.7
-2	2	5	38.89	2.316	4	2.0
-4	0	5	38.97	2.311	2	0.8
-4	4	2	39.38	2.288	4	0.9
-4	4	1	39.40	2.287	4	4.0
-3	2	5	39.48	2.282	4	0.3
0	0	5	39.60	2.276	2	0.3
4	1	2	39.72	2.269	4	0.5
2	0	4	40.11	2.248	2	0.4
-5	3	2	40.29	2.238	4	0.3
1	5	2	40.30	2.238	4	0.4
-2	4	4	40.41	2.232	4	0.8
-3	5	1	40.61	2.221	4	0.3
-5	3	1	40.91	2.206	4	0.3
4	4	0	41.04	2.199	4	0.5
3	2	3	41.19	2.191	4	0.3
-3	5	2	41.20	2.191	4	0.8
-5	3	3	41.28	2.187	4	0.4
-4	2	5	41.50	2.176	4	0.4
5	1	1	41.97	2.153	4	0.3
0	6	0	42.01	2.151	2	1.8
0	4	4	42.36	2.133	4	0.7
-6	0	1	42.57	2.124	2	0.5
-3	5	3	43.34	2.088	4	0.5
1	1	5	43.91	2.062	4	2.0
-2	0	6	43.97	2.059	2	0.5
-6	2	3	44.14	2.051	4	0.3
-3	1	6	44.52	2.035	4	0.4
-1	6	2	44.61	2.031	4	0.3
1	5	3	44.77	2.024	4	0.6
-4	5	2	44.87	2.020	4	0.6
-6	2	1	44.94	2.017	4	0.9
6	0	0	45.25	2.004	2	0.2
-1	1	6	45.92	1.976	4	0.5
-6	2	4	45.98	1.974	4	0.2
-1	5	4	46.04	1.971	4	1.2
3	1	4	46.46	1.954	4	1.5
2	6	1	46.60	1.949	4	0.7
5	3	1	46.65	1.947	4	0.5
1	6	2	46.83	1.940	4	0.3
-6	0	5	46.97	1.934	2	0.8
5	1	2	47.05	1.931	4	1.6
4	2	3	47.61	1.910	4	0.5
-5	3	5	47.63	1.909	4	0.2
3	5	2	47.99	1.896	4	0.5
1	3	5	48.44	1.879	4	0.5
4	4	2	48.54	1.875	4	0.8
-3	3	6	49.01	1.859	4	2.1
-6	2	5	49.17	1.853	4	0.3
-7	1	2	49.43	1.844	4	0.4
2	6	2	49.78	1.832	4	0.3

RUB-10, SiO_2 Framework RUT

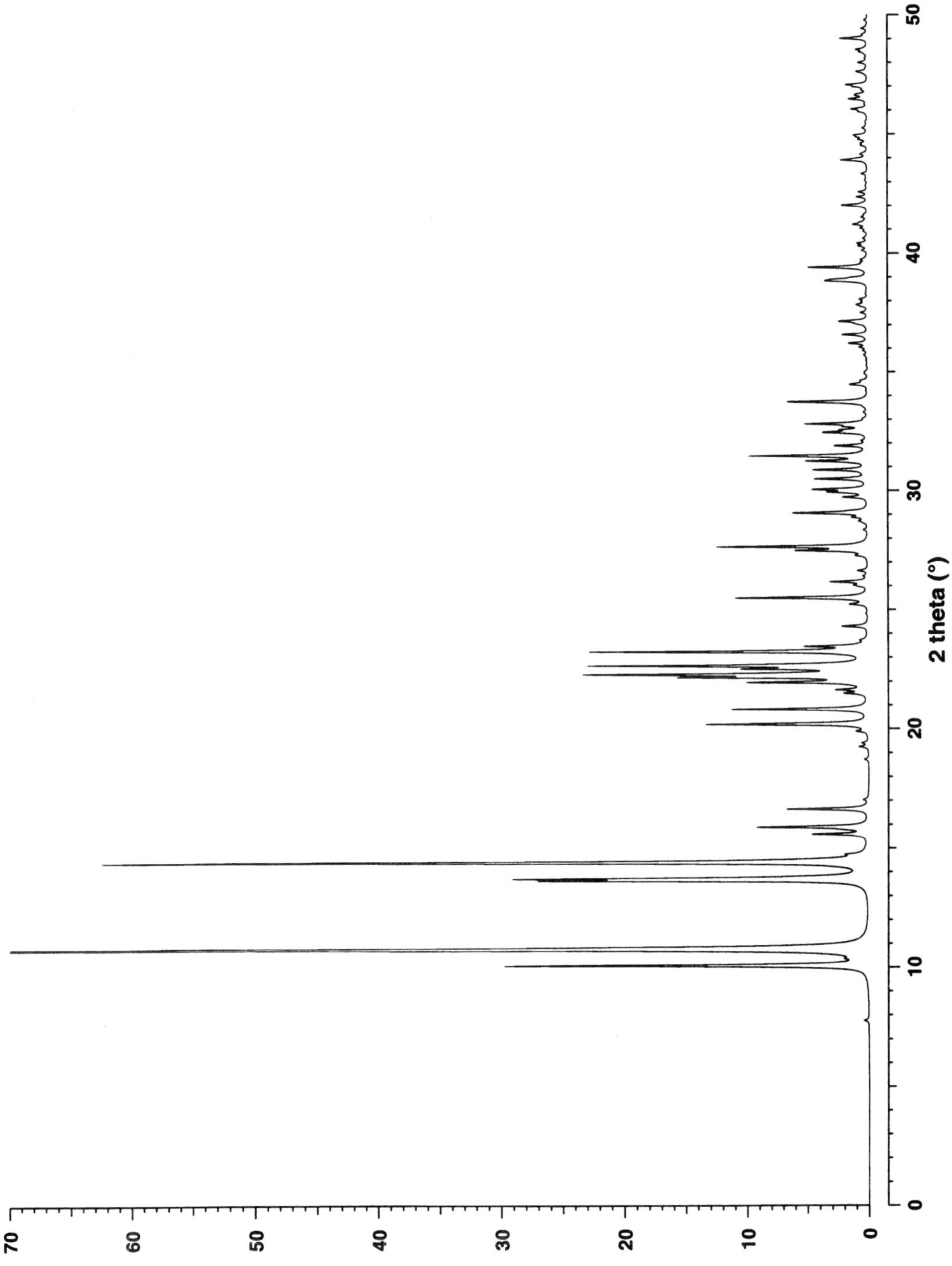

SAO STA-1, Magnesium Aluminophosphate

CHEMICAL COMPOSITION: $|((CH_2)_7(C_7H_{13}N)_2)_{2.74}(H_2O)_{6.16}|$ $[Mg_{5.04}Al_{22.96}P_{28}O_{112}]$
$((CH_2)_7(C_7H_{13}N)_2)^+$ = diquinuclidinium template

REFINED COMPOSITION: $[P_{28}Al_{28}O_{112}]$

CRYSTAL DATA: $P\bar{4}n2$ (No. 118)
$a = 13.620$ Å $b = 13.620$ Å $c = 21.649$ Å
$\alpha = 90°$ $\beta = 90°$ $\gamma = 90°$
X-ray single crystal refinement, $R = 0.059$, $R_w = 0.076$; At $T = 200$K.

REFERENCE: G. W. Noble, P. A. Wright, P. Lightfoot, R. E. Morris, K. J. Hudson, Å. Kvick and H. Graafsma, *Angew. Chem., Int. ed.* **36** 81–83 (1997).

h	k	l	2θ	d	M	I_{rel}
1	0	1	7.67	11.528	8	100.0
0	0	2	8.17	10.825	2	6.3
1	1	0	9.18	9.631	4	0.6
1	1	2	12.30	7.195	8	0.8
2	0	0	13.00	6.810	4	3.7
1	0	3	13.89	6.377	8	9.2
2	1	1	15.11	5.863	16	1.9
2	0	2	15.37	5.764	8	0.6
2	2	0	18.42	4.815	4	0.2
1	1	4	18.81	4.718	8	2.8
2	1	3	19.07	4.655	16	1.3
3	0	1	19.98	4.443	8	9.5
2	2	2	20.18	4.400	8	1.1
3	1	0	20.62	4.307	8	0.4
2	0	4	20.97	4.237	8	4.0
3	1	1	21.03	4.224	16	0.5
1	0	5	21.54	4.126	8	2.1
2	1	4	21.97	4.046	16	0.5
2	2	3	22.19	4.005	8	0.2
3	1	2	22.21	4.002	16	0.2
1	1	5	22.51	3.949	8	0.2
3	0	3	23.15	3.843	8	0.2
3	2	0	23.55	3.778	8	0.2
3	2	1	23.91	3.721	16	0.5
2	2	4	24.75	3.598	8	0.3

h	k	l	2θ	d	M	I_{rel}
3	2	2	24.97	3.567	16	0.1
2	1	5	25.24	3.529	16	1.9
4	0	0	26.17	3.405	4	4.7
3	1	4	26.45	3.370	16	0.8
3	2	3	26.63	3.347	16	0.2
4	1	1	27.31	3.266	16	1.0
4	0	2	27.46	3.248	8	0.5
2	0	6	27.98	3.188	8	1.4
3	3	2	29.01	3.078	8	0.2
1	0	7	29.62	3.016	8	0.9
4	1	3	29.74	3.004	16	2.1
4	2	2	30.49	2.932	16	0.9
4	0	4	31.03	2.882	8	2.9
3	2	5	31.43	2.846	16	1.0
2	1	7	32.47	2.758	16	0.2
4	3	1	33.15	2.703	16	0.2
4	2	4	33.77	2.654	16	0.2
4	1	5	34.14	2.626	16	1.5
1	1	8	34.42	2.605	8	0.6
5	1	2	34.59	2.593	16	0.6
3	0	7	35.11	2.556	8	0.4
4	3	3	35.21	2.548	16	0.2
5	0	3	35.21	2.548	8	1.4
2	0	8	35.70	2.515	8	0.1
5	2	1	35.74	2.512	16	0.3

h	k	l	2θ	d	M	I_{rel}
4	0	6	36.27	2.476	8	0.2
4	4	0	37.35	2.408	4	0.2
5	2	3	37.69	2.387	16	0.2
2	2	8	38.15	2.359	8	0.2
4	4	2	38.30	2.350	8	0.1
4	2	6	38.69	2.327	16	0.1
5	0	5	39.07	2.306	8	0.2
4	3	5	39.07	2.306	16	0.2
6	0	0	39.71	2.270	4	0.3
4	4	4	41.03	2.200	8	0.3
5	2	5	41.34	2.184	16	0.3
5	1	6	42.09	2.147	16	0.1
3	0	9	42.53	2.126	8	0.1
1	1	10	42.81	2.112	8	0.2
6	0	4	43.22	2.093	8	0.2
3	3	8	43.75	2.069	8	0.1
5	0	7	44.31	2.044	8	0.1
5	4	3	44.40	2.040	16	0.2
4	2	8	44.80	2.023	16	0.2
4	4	6	45.28	2.003	8	0.1
5	1	8	47.85	1.901	16	0.4
7	1	2	47.97	1.896	16	0.2
7	0	3	48.45	1.879	8	0.5

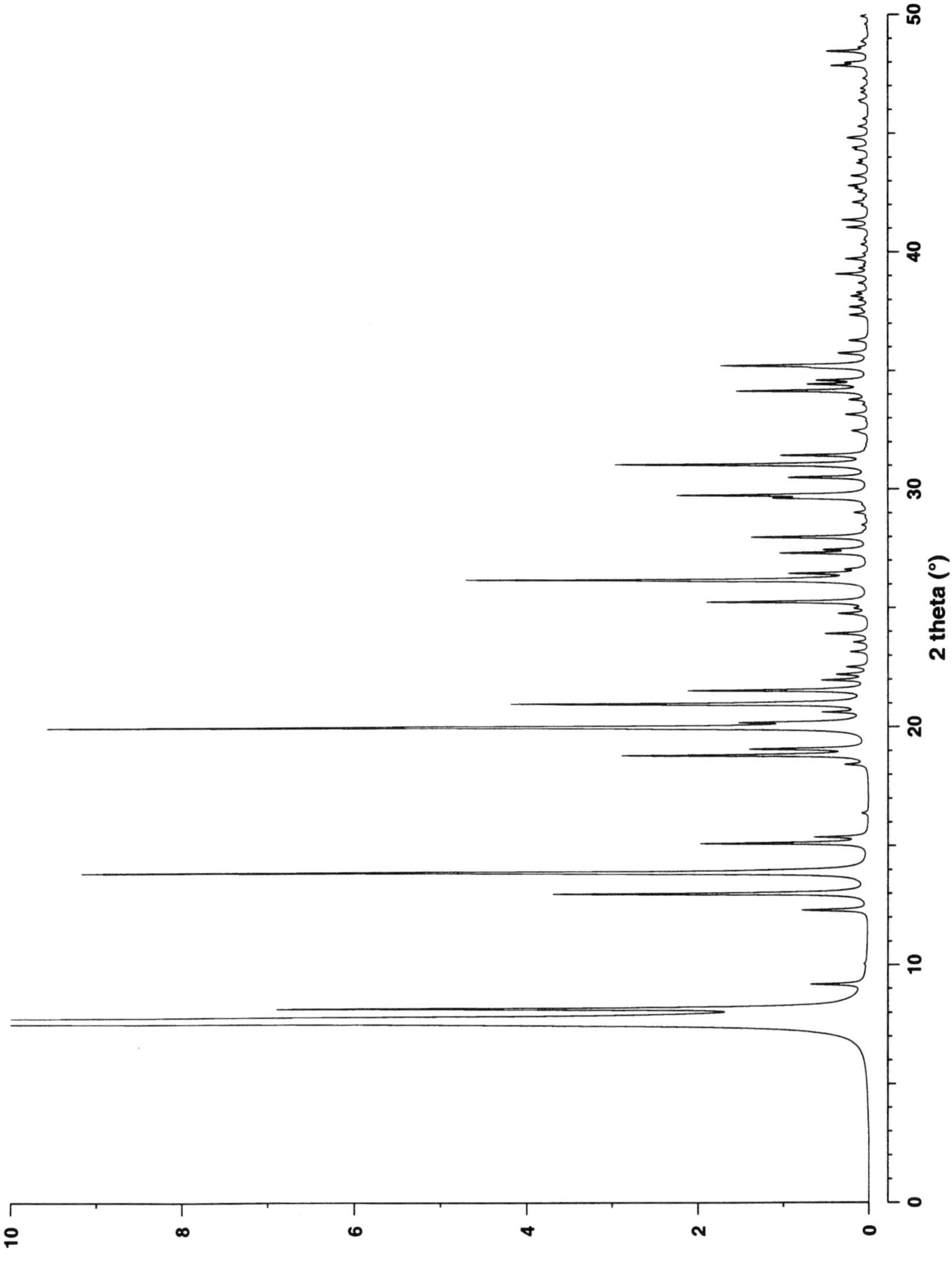
2 theta (°)
0
10
20
30
40
50
0
2
4
6
8
10

SAS STA-6

CHEMICAL COMPOSITION: $|(C_{14}H_{34}N_4)^{2+}_{1.5}(H_2O)_{2.5}|$ $[Mg_3Al_{13}P_{16}O_{64}]$
$C_{14}H_{32}N_4$ = 1,4,8,11-tetramethyl-1,4,8,11-tetraazacyclotetradecane

REFINED COMPOSITION: $|C_{19.44}N_{5.04}|$ $[Mg_{2.4}Al_{13.6}P_{16}O_{64}]$

CRYSTAL DATA: $P\,4/m\,n\,c$ (No. 128)
a = 14.322 Å b = 14.322 Å c = 10.424 Å
α = 90° β = 90° γ = 90°
X-ray synchrotron single crystal refinement, R_F = 0.094, wR_2 = 0.313

REFERENCE: V. Patinec, P. A. Wright, P. Lightfoot, R. A. Aitken and P. A. Cox, *J. Chem. Soc., Dalton Trans.* **22** 3909–3911 (1999).

h	k	l	2θ	d	M	I_{rel}
1	1	0	8.73	10.127	4	100.0
1	0	1	10.50	8.428	8	38.8
2	0	0	12.36	7.161	4	4.5
2	1	1	16.24	5.457	16	46.7
2	2	0	17.51	5.064	4	0.6
1	1	2	19.15	4.634	8	26.3
3	1	0	19.60	4.529	8	33.8
3	0	1	20.46	4.340	8	1.9
2	0	2	21.08	4.214	8	35.0
3	1	1	21.39	4.154	16	1.7
2	1	2	21.99	4.043	16	3.8
3	2	0	22.38	3.972	8	2.2
3	2	1	23.97	3.712	16	6.6
2	2	2	24.51	3.632	8	4.3
4	0	0	24.87	3.580	4	21.0
3	1	2	26.06	3.419	16	7.0
1	0	3	26.39	3.377	8	5.6
3	3	0	26.40	3.376	4	3.5
4	1	1	27.06	3.295	16	15.0
4	2	0	27.86	3.202	8	2.8
3	2	2	28.25	3.159	16	0.1
2	1	3	29.24	3.054	16	7.7
4	0	2	30.28	2.951	8	11.5
5	1	0	31.86	2.809	8	2.5
4	3	1	32.41	2.762	16	2.4
5	0	1	32.41	2.762	8	6.1
3	1	3	32.48	2.757	16	0.2
4	2	2	32.82	2.729	16	15.9
5	1	1	33.03	2.712	16	0.1
3	2	3	34.29	2.615	16	0.9
0	0	4	34.41	2.606	2	5.9
5	2	1	34.81	2.577	16	1.2
1	1	4	35.57	2.524	8	1.4
5	1	2	36.33	2.473	16	5.8
4	1	3	36.58	2.457	16	0.1
2	0	4	36.70	2.449	8	0.2
2	2	4	38.86	2.317	8	0.8
6	1	1	39.23	2.297	16	0.1
4	4	2	39.57	2.277	8	3.9
6	2	0	39.81	2.265	8	1.0
5	0	3	40.83	2.210	8	0.1
5	4	1	41.28	2.187	16	0.3
5	2	3	42.82	2.112	16	0.4
4	0	4	42.92	2.107	8	0.1
6	3	1	43.26	2.092	16	1.1
1	0	5	43.88	2.063	8	0.6
3	3	4	43.89	2.063	8	1.9
7	1	0	44.74	2.025	8	0.5
7	0	1	45.16	2.008	8	1.9
2	1	5	45.77	1.982	16	0.6
6	1	3	46.59	1.949	16	1.1
7	2	1	47.00	1.933	16	0.6
3	0	5	47.59	1.911	8	0.9
5	1	4	47.60	1.910	16	2.7
5	5	2	48.20	1.888	8	3.3
7	1	2	48.20	1.888	16	1.8
5	4	3	48.40	1.881	16	0.6
7	3	0	48.40	1.881	8	0.1
6	4	2	49.09	1.856	16	0.2

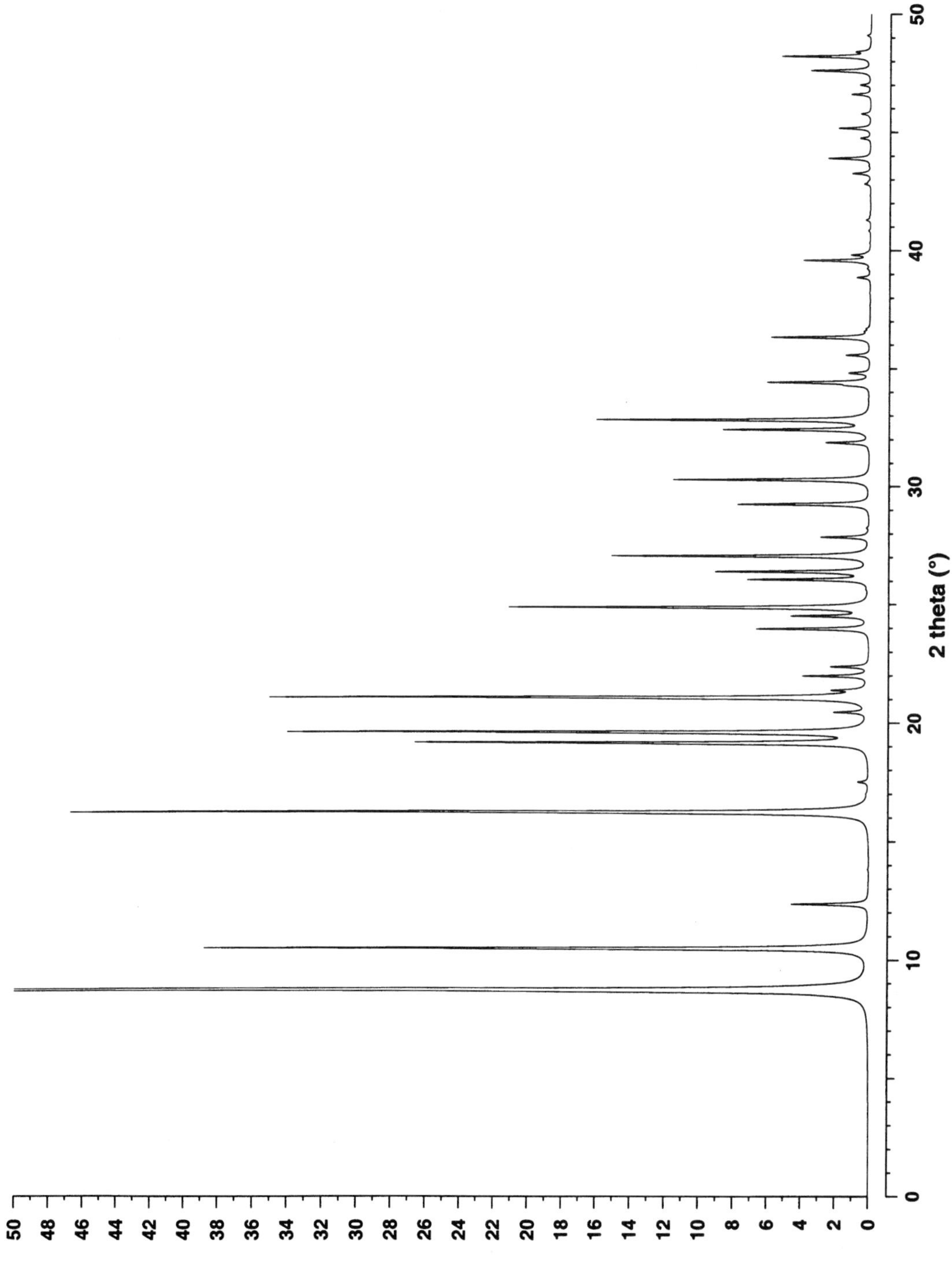
2 theta (°)
0
10
20
30
40
50
50 48 46 44 42 40 38 36 34 32 30 28 26 24 22 20 18 16 14 12 10 8 6 4 2 0

SAT

STA-2

CHEMICAL COMPOSITION: $|((CH_2)_4(C_7H_{13}N)_2)_3(H_2O)_{22.5}|\ [Mg_{5.4}Al_{30.6}P_{36}O_{144}]$
$((CH_2)_4(C_7H_{13}N)_2)^+$ = diquinuclidinium template

REFINED COMPOSITION: $|(C_{22}H_{50}N_2)_3|\ [Al_{36}P_{36}O_{144}]$

CRYSTAL DATA: $R\overline{3}$ (No. 148) hexagonal setting
$a = 12.726$ Å $b = 12.726$ Å $c = 30.939$ Å
$\alpha = 90°$ $\beta = 90°$ $\gamma = 120°$
X-ray single crystal refinement at 200 K. $R_F = 0.0757$

REFERENCE: G. W. Noble, P. A. Wright and Å. Kvick, *J. Chem. Soc., Dalton Trans.* 4485–4490 (1997).

h	k	l	2θ	d	M	I_{rel}
1	0	1	8.52	10.382	6	5.4
0	0	3	8.57	10.313	2	1.2
0	1	2	9.85	8.976	6	9.7
1	1	0	13.92	6.363	6	22.8
1	0	4	13.99	6.331	6	12.4
0	2	1	16.34	5.425	6	100.0
2	-1	3	16.37	5.415	6	2.2
1	1	3	16.37	5.415	6	2.4
0	1	5	16.43	5.396	6	31.9
2	0	2	17.08	5.191	6	3.9
0	0	6	17.20	5.156	2	5.8
0	2	4	19.78	4.488	6	11.5
2	1	1	21.52	4.128	6	75.3
3	-1	1	21.52	4.128	6	35.7
2	0	5	21.59	4.115	6	12.1
1	0	7	21.66	4.102	6	47.6
3	-2	2	22.10	4.022	6	4.5
2	-1	6	22.19	4.006	6	6.2
0	3	0	24.23	3.674	6	44.5
2	1	4	24.27	3.668	6	8.0
3	-1	4	24.27	3.668	6	5.2
0	1	8	24.39	3.649	6	33.6
3	0	3	25.74	3.461	6	0.8
3	-2	5	25.78	3.456	6	2.1
1	2	5	25.78	3.456	6	17.3
0	2	7	25.84	3.448	6	13.4
0	0	9	25.92	3.438	2	1.5
2	2	0	28.05	3.181	6	13.4
2	0	8	28.19	3.166	6	6.3
4	-3	1	29.36	3.042	6	0.3
1	3	1	29.36	3.042	6	3.0
4	-2	3	29.38	3.040	6	0.2
3	-1	7	29.46	3.031	6	12.1
2	1	7	29.46	3.031	6	11.4
2	-1	9	29.53	3.024	6	0.3
1	1	9	29.53	3.024	6	1.4
3	1	2	29.79	2.999	6	0.1
4	-1	2	29.79	2.999	6	2.6
4	-3	4	31.47	2.843	6	1.4
1	3	4	31.47	2.843	6	3.8

h	k	l	2θ	d	M	I_{rel}
1	2	8	31.57	2.834	6	7.0
3	-2	8	31.57	2.834	6	4.0
4	0	1	32.63	2.744	6	21.2
3	1	5	32.67	2.741	6	1.7
4	-1	5	32.67	2.741	6	4.6
0	1	11	32.86	2.725	6	3.0
0	4	2	33.02	2.713	6	2.8
2	2	6	33.08	2.708	6	0.1
4	0	4	34.56	2.596	6	13.3
0	0	12	34.80	2.578	2	5.7
5	-2	1	35.63	2.520	6	1.6
0	4	5	35.67	2.517	6	4.4
4	-3	7	35.71	2.514	6	0.3
1	3	7	35.71	2.514	6	1.0
0	3	9	35.77	2.510	6	0.2
2	0	11	35.84	2.505	6	0.5
3	-1	10	36.16	2.484	6	0.1
1	4	0	37.39	2.405	6	3.2
4	1	0	37.39	2.405	6	1.4
3	2	4	37.42	2.403	6	0.2
5	-2	4	37.42	2.403	6	0.5
4	-1	8	37.50	2.398	6	0.3
2	3	5	38.46	2.341	6	0.4
5	-3	5	38.46	2.341	6	1.4
4	0	7	38.50	2.338	6	4.1
3	-2	11	38.62	2.331	6	0.7
1	2	11	38.62	2.331	6	0.9
1	0	13	38.71	2.326	6	0.1
0	4	8	40.19	2.244	6	2.6
3	2	7	41.13	2.195	6	0.2
0	2	13	41.32	2.185	6	0.3
5	0	2	41.38	2.182	6	2.8
1	4	6	41.43	2.180	6	0.5
4	1	6	41.43	2.180	6	0.2
5	-4	6	41.43	2.180	6	0.5
5	-1	6	41.43	2.180	6	0.3
0	1	14	41.68	2.167	6	0.5
3	3	0	42.63	2.121	6	2.0
0	5	4	42.65	2.120	6	0.2
5	-3	8	42.73	2.116	6	0.1

h	k	l	2θ	d	M	I_{rel}
2	3	8	42.73	2.116	6	0.5
0	3	12	42.85	2.110	6	2.3
3	0	12	42.85	2.110	6	1.9
6	-3	3	43.56	2.078	6	0.4
3	3	3	43.56	2.078	6	0.4
5	0	5	43.59	2.076	6	1.2
4	-1	11	43.74	2.070	6	0.1
2	1	13	43.81	2.066	6	1.3
3	-1	13	43.81	2.066	6	0.3
0	0	15	43.89	2.063	2	6.8
2	0	14	44.15	2.051	6	0.7
6	-4	4	45.08	2.011	6	0.1
2	4	4	45.08	2.011	6	0.3
2	2	12	45.27	2.003	6	0.4
6	-1	1	45.94	1.975	6	1.6
5	1	1	45.94	1.975	6	2.0
4	2	5	45.98	1.974	6	0.4
6	-2	5	45.98	1.974	6	0.2
0	5	7	46.01	1.973	6	1.2
5	-1	9	46.06	1.971	6	0.7
4	1	9	46.06	1.971	6	0.2
1	4	9	46.06	1.971	6	0.1
0	4	11	46.12	1.968	6	2.1
1	1	15	46.27	1.962	6	0.6
2	-1	15	46.27	1.962	6	0.2
6	-3	6	46.28	1.962	6	0.3
5	-2	10	46.38	1.958	6	0.2
3	2	10	46.38	1.958	6	0.4
1	2	14	46.52	1.952	6	0.8
3	-2	14	46.52	1.952	6	0.9
6	-1	4	47.41	1.918	6	0.4
5	1	4	47.41	1.918	6	0.9
5	0	8	47.48	1.915	6	1.2
1	0	16	47.75	1.905	6	0.4
1	5	5	48.27	1.885	6	1.6
6	-5	5	48.27	1.885	6	0.3
2	4	7	48.31	1.884	6	0.2
1	3	13	48.48	1.878	6	0.2
0	6	0	49.63	1.837	6	0.3
6	-2	8	49.72	1.834	6	0.1

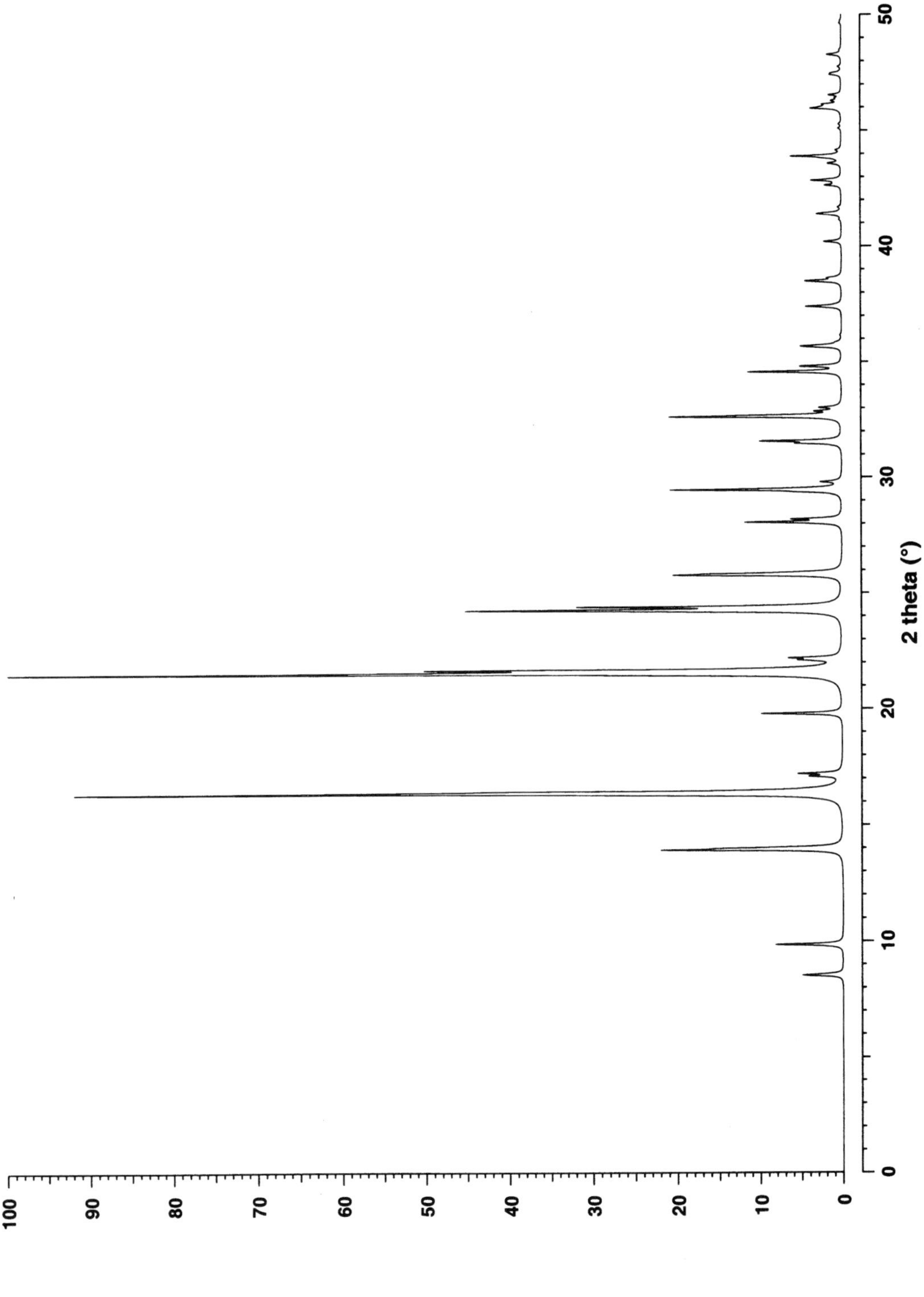
2 theta (°)
0
10
20
30
40
50
0
10
20
30
40
50
60
70
80
90
100

SAV Magnesium STA-7

CHEMICAL COMPOSITION: $|(C_{18}H_{42}N_6)_{1.96}(H_2O)_7|$ $[Mg_{4.8}Al_{19.2}P_{24}O_{96}]$
$C_{18}H_{42}N_6$=1,4,7,10,13,16-hexamethyl-1,4,7,10,13,16-hexaazacyclooctadecane

REFINED COMPOSITION: $[Mg_{4.8}Al_{19.2}P_{24}O_{96}]$

CRYSTAL DATA: $P4/n$ (No. 85) origin at centre $\overline{1}$
$a = 18.7732$ Å $b = 18.7732$ Å $c = 9.4537$ Å
$\alpha = 90°$ $\beta = 90°$ $\gamma = 90°$
X-ray synchrotron single crystal refinement, $R_p = 0.0539$, $wR_{wp} = 0.1422$

REFERENCE: P. A. Wright, M. J. Maple, A. M. Z. Slawin, V. Patinec, R. A. Aitken, S. Welsh and P. A. Cox, *J. Chem. Soc., Dalton Trans.* 1243–1248 (2000).

h	k	l	2θ	d	M	I_{rel}
1	1	0	6.66	13.275	4	11.9
0	0	1	9.35	9.454	2	38.2
2	0	0	9.42	9.387	4	100.0
1	0	1	10.48	8.444	8	25.6
1	1	1	11.49	7.701	8	0.7
2	0	1	13.29	6.661	8	5.0
2	2	0	13.34	6.637	4	2.6
1	3	0	14.92	5.937	4	5.8
3	1	0	14.92	5.937	4	4.5
2	2	1	16.32	5.432	8	27.8
3	0	1	16.99	5.218	8	17.1
1	3	1	17.64	5.028	8	0.2
0	0	2	18.77	4.727	2	2.1
4	0	0	18.91	4.693	4	2.2
1	0	2	19.36	4.584	8	2.0
2	3	1	19.46	4.561	8	2.3
3	2	1	19.46	4.561	8	2.7
1	1	2	19.94	4.453	8	1.3
3	3	0	20.07	4.425	4	8.2
2	0	2	21.04	4.222	8	3.2
4	0	1	21.13	4.204	8	2.8
4	2	0	21.16	4.198	4	0.6
2	4	0	21.16	4.198	4	4.7
1	2	2	21.57	4.119	8	3.1
1	4	1	21.66	4.102	8	1.5
4	1	1	21.66	4.102	8	0.4
3	3	1	22.18	4.008	8	4.0
2	2	2	23.10	3.850	8	0.6
2	4	1	23.18	3.837	8	2.7
4	2	1	23.18	3.837	8	0.4
3	0	2	23.59	3.772	8	7.9
3	1	2	24.07	3.698	8	0.5
1	5	0	24.17	3.682	4	0.4
5	1	0	24.17	3.682	4	0.3
2	3	2	25.45	3.500	8	1.8
3	2	2	25.45	3.500	8	2.5
3	4	1	25.53	3.490	8	0.3
4	3	1	25.53	3.490	8	0.6
5	0	1	25.53	3.490	8	1.8
5	1	1	25.97	3.431	8	0.4
1	5	1	25.97	3.431	8	0.6
4	0	2	26.77	3.330	8	0.4
4	4	0	26.86	3.319	4	0.3
2	5	1	27.26	3.271	8	2.2
5	2	1	27.26	3.271	8	2.0
3	3	2	27.61	3.230	8	1.0
3	5	0	27.71	3.220	4	1.8
5	3	0	27.71	3.220	4	1.8
1	0	3	28.73	3.108	8	0.3
1	1	3	29.12	3.066	8	0.9
3	5	1	29.30	3.048	8	1.9
5	3	1	29.30	3.048	8	1.5
2	0	3	29.91	2.987	8	2.8
6	0	1	30.08	2.970	8	2.8
2	6	0	30.11	2.968	4	0.9
6	2	0	30.11	2.968	4	1.3
2	1	3	30.29	2.950	8	0.4
1	2	3	30.29	2.950	8	0.9
5	0	2	30.40	2.940	8	2.8
4	3	2	30.40	2.940	8	0.9
3	4	2	30.40	2.940	8	0.8
6	1	1	30.47	2.934	8	0.6
1	6	1	30.47	2.934	8	1.2
2	2	3	31.42	2.847	8	3.3
2	6	1	31.59	2.832	8	2.8
6	2	1	31.59	2.832	8	3.3
3	0	3	31.79	2.815	8	0.8
5	2	2	31.90	2.806	8	0.3
2	5	2	31.90	2.806	8	0.9
4	5	1	31.96	2.800	8	0.4
5	4	1	31.96	2.800	8	0.9
1	3	3	32.16	2.783	8	1.1
3	1	3	32.16	2.783	8	1.0
4	4	2	32.98	2.716	8	0.2
6	3	1	33.39	2.683	8	1.1
3	6	1	33.39	2.683	8	0.8
3	5	2	33.68	2.661	8	0.9
5	3	2	33.68	2.661	8	0.9
5	5	0	33.76	2.655	4	0.5
4	0	3	34.27	2.616	8	0.4
6	0	2	34.37	2.609	8	0.2
4	6	0	34.45	2.603	4	0.3
1	6	2	34.71	2.584	8	0.2
6	1	2	34.71	2.584	8	0.4
3	3	3	34.95	2.567	8	0.8
5	5	1	35.11	2.556	8	0.3
7	2	1	36.10	2.488	8	0.2
3	7	0	36.45	2.465	4	0.3
5	0	3	37.25	2.414	8	0.3
6	3	2	37.34	2.408	8	0.3
3	6	2	37.34	2.408	8	0.2
8	0	0	38.36	2.347	4	0.6
1	0	4	38.39	2.345	8	0.7
7	0	2	38.60	2.333	8	0.7
1	2	4	39.61	2.275	8	0.2
8	1	1	39.87	2.261	8	0.3
1	8	1	39.87	2.261	8	0.2
6	0	3	40.63	2.220	8	0.3
2	8	1	40.77	2.213	8	0.2
8	2	1	40.77	2.213	8	0.2
6	6	0	40.78	2.212	4	0.3
1	6	3	40.93	2.205	8	0.2
6	2	3	41.81	2.161	8	0.2
2	6	3	41.81	2.161	8	0.3
6	6	1	41.94	2.154	8	0.3
6	5	2	42.18	2.143	8	0.2
5	6	2	42.18	2.143	8	0.3
3	8	1	42.23	2.140	8	0.2
8	3	1	42.23	2.140	8	0.4
3	3	4	43.41	2.085	8	0.2
2	4	4	43.96	2.059	8	0.2
4	2	4	43.96	2.059	8	0.3
2	8	2	44.15	2.051	8	0.2
4	8	1	44.20	2.049	8	0.2
5	5	3	44.63	2.030	8	0.2
7	1	3	44.63	2.030	8	0.2
1	7	3	44.63	2.030	8	0.2
3	7	3	46.79	1.942	8	0.2
6	5	3	47.58	1.911	8	0.2
5	6	3	47.58	1.911	8	0.2
3	5	4	47.74	1.905	8	0.2
1	0	5	48.38	1.881	8	0.3
10	0	0	48.49	1.877	4	1.6
6	8	0	48.49	1.877	4	0.2
1	6	4	48.51	1.876	8	0.4
1	1	5	48.64	1.872	8	0.2
6	7	2	48.69	1.870	8	0.2
7	6	2	48.69	1.870	8	0.2
2	0	5	49.15	1.854	8	0.2
2	8	3	49.38	1.845	8	0.3
8	6	1	49.50	1.841	8	0.2
6	8	1	49.50	1.841	8	0.4
10	0	1	49.50	1.841	8	0.4
10	2	0	49.51	1.841	4	0.2
2	10	0	49.51	1.841	4	0.2

Magnesium STA-7 SAV

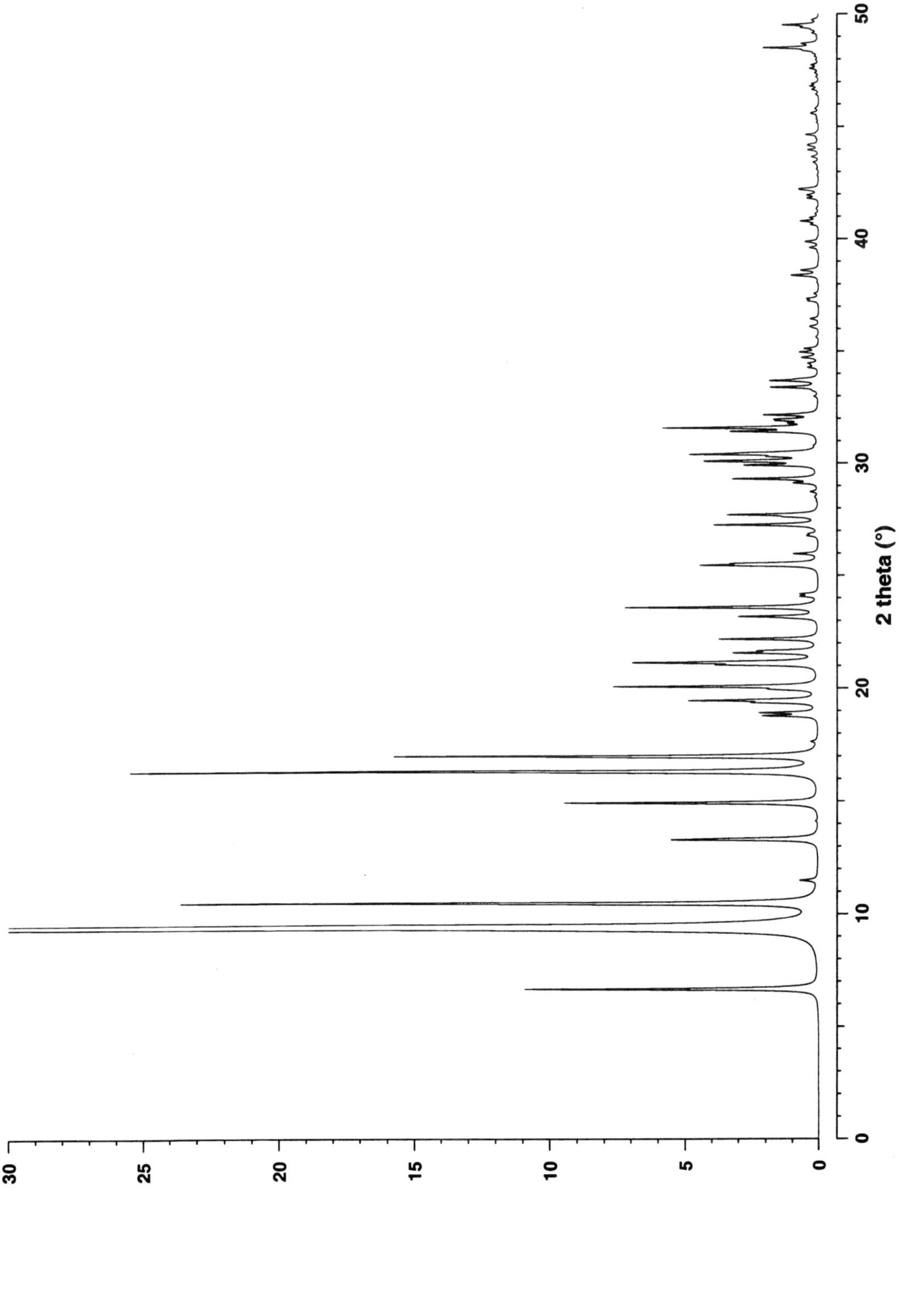

SBE UCSB–8Co

CHEMICAL COMPOSITION: $|(H_3N(CH_2)_9NH_3)^{2+}_{16}|$ $[Al_{32}Co_{32}P_{64}O_{256}]$
$H_2N(CH_2)_9NH_2$ = 1,9 diaminononane

REFINED COMPOSITION: $|C_{32}N_{16}|$ $[Al_{32}Co_{32}P_{64}O_{256}]$

CRYSTAL DATA: $P\,4/n\,n\,c$ (No. 126) origin at centre $\bar{1}$
a = 19.0654 Å b = 19.0654 Å c = 27.594 Å
$\alpha = 90°$ $\beta = 90°$ $\gamma = 90°$
X-ray single crystal refinement, $R_F = 0.098$, $R_w = 0.26$

REFERENCE: X. Bu, P. Feng and G. D. Stucky,
Science **278** 2080–2085 (1997).

h	k	l	2θ	d	M	I_{rel}
1	0	1	5.63	15.686	8	100.0
0	0	2	6.41	13.797	2	81.3
1	1	0	6.56	13.481	4	3.8
1	1	2	9.17	9.642	8	14.7
2	0	0	9.28	9.533	4	15.9
1	0	3	10.68	8.284	8	0.4
2	1	1	10.86	8.146	16	18.5
0	0	4	12.83	6.898	2	1.0
2	2	0	13.13	6.741	4	21.4
2	1	3	14.16	6.253	16	0.1
3	0	1	14.30	6.193	8	0.9
1	1	4	14.42	6.141	8	2.1
2	2	2	14.63	6.056	8	1.7
2	0	4	15.86	5.589	8	0.6
3	1	2	16.04	5.525	16	3.9
1	0	5	16.72	5.301	8	0.3
3	0	3	16.96	5.229	8	0.6
3	2	1	17.07	5.193	16	1.6
2	2	4	18.40	4.821	8	0.9
4	0	0	18.62	4.766	4	4.2
2	1	5	19.16	4.633	16	1.6
0	0	6	19.30	4.599	2	0.7
4	1	1	19.46	4.560	16	2.0
3	1	4	19.55	4.540	16	4.6
4	0	2	19.71	4.505	8	2.5
1	1	6	20.40	4.353	8	4.7
4	2	0	20.84	4.263	8	3.6
3	0	5	21.32	4.167	8	3.0
4	1	3	21.51	4.131	16	1.8
4	0	4	22.68	3.921	8	0.4
1	0	7	23.04	3.860	8	0.5
3	2	5	23.30	3.818	16	0.4
2	2	6	23.42	3.799	8	0.4
5	0	1	23.55	3.777	8	1.4
3	3	4	23.63	3.765	8	0.1
5	1	0	23.80	3.739	8	1.7
5	1	1	24.02	3.705	16	0.1
3	1	6	24.34	3.657	16	7.1
4	2	4	24.55	3.627	16	0.2
5	1	2	24.67	3.609	16	0.5
2	1	7	24.88	3.578	16	1.2
4	1	5	25.12	3.544	16	0.3
5	0	3	25.28	3.522	8	1.2
5	2	1	25.36	3.512	16	4.2
3	2	6	25.67	3.470	16	0.1
3	0	7	26.61	3.350	8	4.9
5	2	3	26.99	3.304	16	0.1
5	1	4	27.13	3.287	16	2.9
4	1	6	27.35	3.261	16	0.2
5	3	2	28.04	3.182	16	0.5
6	0	0	28.08	3.178	4	2.1
3	2	7	28.24	3.160	16	2.1
4	3	5	28.45	3.137	16	0.7
5	0	5	28.45	3.137	8	1.2
4	2	6	28.55	3.126	16	0.3
6	0	2	28.83	3.097	8	0.3
4	4	4	29.50	3.028	8	0.1
1	0	9	29.51	3.027	8	1.1
6	2	0	29.63	3.015	8	6.6
6	1	3	30.12	2.967	16	0.3
5	4	1	30.19	2.960	16	1.3
5	3	4	30.25	2.955	16	1.3
6	2	2	30.35	2.945	16	0.8
5	1	6	30.82	2.901	16	1.3
6	0	4	30.98	2.886	8	0.2
2	1	9	31.00	2.885	16	1.0
5	4	3	31.58	2.833	16	0.2
6	3	1	31.65	2.827	16	0.3
5	3	5	31.81	2.813	16	0.2
0	0	10	32.45	2.759	2	0.9
4	3	7	32.67	2.741	16	0.5
5	0	7	32.67	2.741	8	2.5
6	1	5	32.86	2.725	16	1.0
7	0	1	33.05	2.710	8	0.1
1	1	10	33.14	2.703	8	0.3
5	5	0	33.23	2.696	4	0.3
7	1	0	33.23	2.696	8	0.3
5	3	6	33.63	2.665	16	1.3
6	4	0	33.90	2.644	8	0.4
5	2	7	34.04	2.634	16	0.4
5	4	5	34.22	2.620	16	0.6
7	0	3	34.34	2.612	8	0.9
7	2	1	34.40	2.607	16	0.4
4	3	8	35.08	2.558	16	0.1
2	2	10	35.14	2.554	8	0.7
6	3	5	35.53	2.527	16	0.3
7	2	3	35.64	2.519	16	0.4
5	5	4	35.75	2.511	8	1.9
7	1	4	35.75	2.511	16	0.9
3	1	10	35.79	2.509	16	0.4
1	0	11	36.11	2.487	8	0.2
6	4	4	36.39	2.469	16	0.1
7	3	2	36.48	2.463	16	0.2
6	1	7	36.63	2.453	16	0.5
2	1	11	37.37	2.407	16	0.1
5	0	9	37.64	2.389	8	0.4
5	4	7	37.87	2.376	16	0.3
6	5	3	38.14	2.359	16	0.7
8	1	1	38.20	2.356	16	0.3
5	5	6	38.71	2.326	8	0.5
5	2	9	38.86	2.318	16	0.1
8	2	0	38.95	2.312	8	0.1
6	3	7	39.07	2.305	16	1.0
8	1	3	39.34	2.290	16	0.1
6	1	9	41.19	2.192	16	0.3
3	1	12	42.05	2.149	16	0.1
8	4	0	42.40	2.132	8	0.2
7	1	8	42.56	2.124	16	0.3
8	0	6	42.73	2.116	8	0.1
1	0	13	42.87	2.110	8	0.3
5	3	10	42.88	2.109	16	0.2
6	5	7	43.61	2.075	16	0.4
2	1	13	43.96	2.060	16	0.5
6	2	10	44.51	2.035	16	0.3
8	1	7	44.69	2.028	16	0.1
7	6	3	44.93	2.018	16	0.3
9	1	4	45.02	2.014	16	0.2
3	0	13	45.03	2.013	8	0.1
7	5	6	45.43	1.997	16	0.2
6	1	11	46.36	1.959	16	0.1
5	5	10	47.12	1.928	8	0.2
7	7	0	47.19	1.926	4	0.4
6	4	10	47.63	1.909	16	0.2
10	0	0	47.70	1.907	4	0.6
10	1	1	48.07	1.893	16	0.4
9	3	5	48.19	1.888	16	0.1
3	1	14	48.60	1.873	16	0.3
10	2	0	48.71	1.870	8	0.1
10	2	1	48.82	1.865	16	0.1
10	1	3	49.03	1.858	16	0.3
7	7	4	49.11	1.855	8	0.9
5	0	13	49.12	1.855	8	0.1
7	3	10	49.14	1.854	16	0.3
7	0	11	49.39	1.845	8	0.3
7	6	7	49.79	1.831	16	0.3

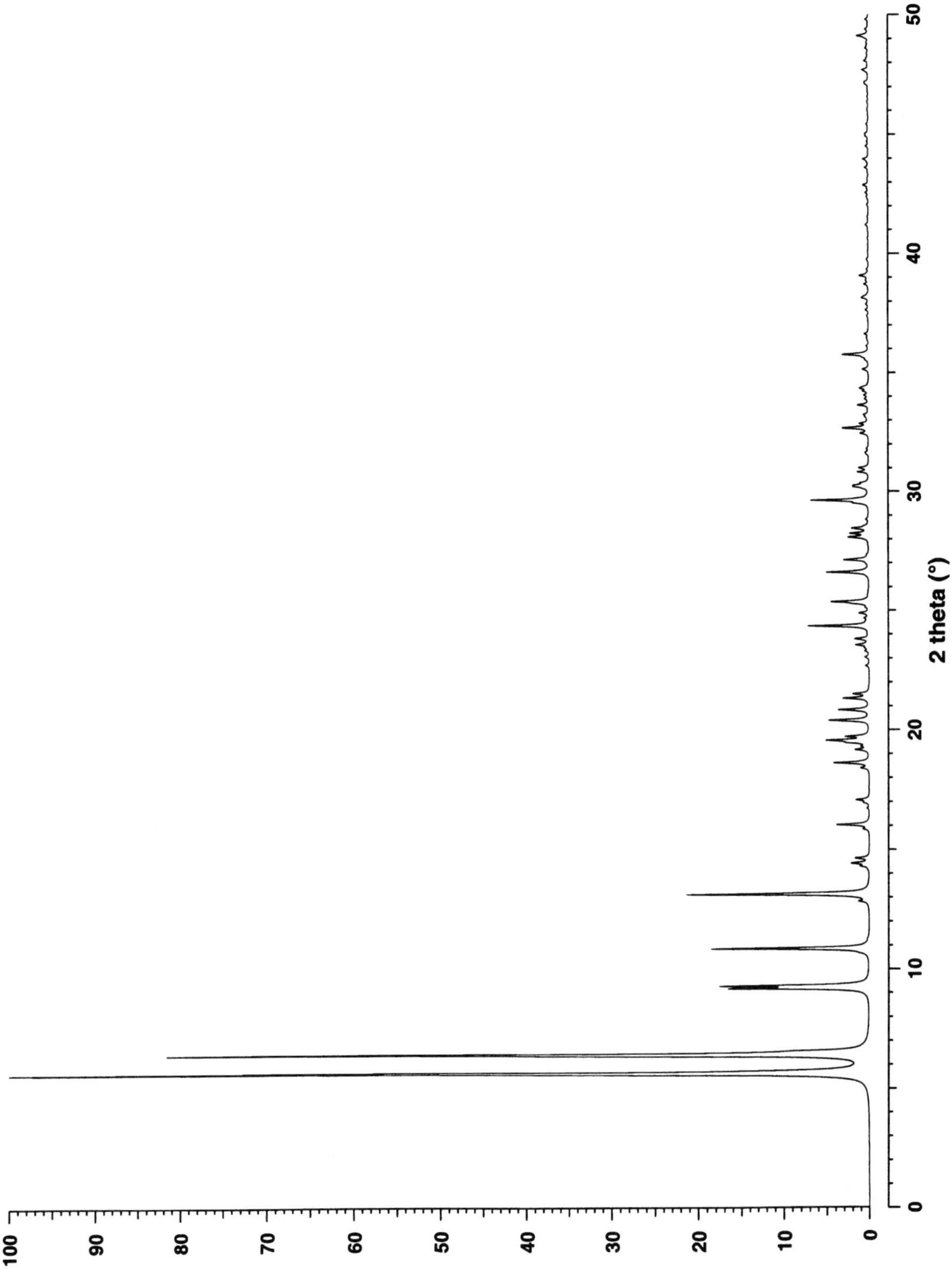

2 theta (°)
0
10
20
30
40
50
100
90
80
70
60
50
40
30
20
10
0

SBS UCSB-6GaCo

CHEMICAL COMPOSITION: $|(H_3N(CH_2)_7NH_3)^{2+}_{12}|$ $[Ga_{24}Co_{24}P_{48}O_{192}]$
$H_2N(CH_2)_7NH_2$ = 1,7 diaminoheptane

REFINED COMPOSITION: $|C_{16}N_{16}O_{12}|$ $[Ga_{24}Co_{24}P_{48}O_{192}]$

CRYSTAL DATA: $P\overline{3}1c$ (No. 163)
a = 17.8356 Å b = 17.8356 Å c = 27.1816 Å
α = 90° β = 90° γ = 120°
X-ray single crystal refinement, R_F = 0.070, R_w = 0.201

REFERENCE: X. Bu, P. Feng and G. D. Stucky,
Science **278** 2080–2085 (1997).

h	k	l	2θ	d	M	I_{rel}
1	0	0	5.72	15.446	6	32.2
0	0	2	6.50	13.591	2	99.2
1	0	1	6.58	13.429	12	100.0
1	0	2	8.67	10.203	12	1.4
1	1	0	9.92	8.918	6	21.7
2	0	0	11.46	7.723	6	7.7
1	1	2	11.87	7.456	6	6.2
2	-1	2	11.87	7.456	6	6.3
2	0	1	11.91	7.429	12	4.3
0	0	4	13.03	6.795	2	1.1
2	0	2	13.19	6.715	12	7.0
1	0	4	14.24	6.220	12	0.9
2	0	3	15.07	5.878	12	0.8
2	1	0	15.18	5.838	12	4.7
3	-1	1	15.52	5.708	12	5.1
2	1	1	15.52	5.708	12	1.7
1	1	4	16.40	5.405	6	0.4
2	-1	4	16.40	5.405	6	1.1
1	0	5	17.29	5.128	12	0.6
3	0	1	17.53	5.059	12	0.8
3	0	2	18.43	4.815	12	1.1
0	0	6	19.60	4.530	2	0.8
3	0	3	19.83	4.476	12	1.6
2	2	0	19.91	4.459	6	2.0
2	0	5	19.97	4.445	12	3.2
1	0	6	20.43	4.347	12	2.9
3	1	0	20.73	4.284	12	5.1
3	1	1	20.99	4.232	12	1.4
4	-1	1	20.99	4.232	12	0.9
3	0	4	21.65	4.104	12	2.2
3	1	2	21.75	4.086	12	1.0
4	-1	2	21.75	4.086	12	0.7
2	-1	6	22.01	4.039	6	0.4
1	1	6	22.01	4.039	6	2.3
2	1	5	22.35	3.979	12	1.6
2	0	6	22.76	3.908	12	0.3
4	-1	3	22.96	3.873	12	1.0
4	0	1	23.27	3.823	12	1.0
1	0	7	23.62	3.766	12	1.0
2	2	4	23.87	3.728	6	1.8
4	-2	4	23.87	3.728	6	0.8
4	0	2	23.96	3.714	12	0.3
3	1	4	24.56	3.624	12	0.3
3	-1	6	24.88	3.579	12	0.2
4	0	3	25.07	3.552	12	0.2
3	2	0	25.13	3.544	12	0.4
5	-2	1	25.35	3.514	12	0.8
3	2	1	25.35	3.514	12	1.0
2	0	7	25.68	3.469	12	4.7
3	2	2	25.98	3.429	12	0.6
3	0	6	26.20	3.401	12	2.4
4	1	0	26.44	3.371	12	1.5
3	1	5	26.49	3.365	12	0.9
4	-1	5	26.49	3.365	12	0.4
4	1	1	26.65	3.345	12	0.3
4	1	2	27.26	3.272	12	0.5
2	1	7	27.59	3.233	12	1.7
3	-1	7	27.59	3.233	12	0.9
4	-2	6	28.08	3.178	6	2.7
2	2	6	28.08	3.178	6	1.0
1	1	8	28.10	3.175	6	0.4
4	1	3	28.25	3.159	12	1.2
5	-2	4	28.41	3.142	12	0.6
3	2	4	28.41	3.142	12	0.3
4	-1	6	28.68	3.113	12	0.7
5	0	0	28.90	3.089	6	6.5
5	0	1	29.09	3.069	12	0.9
4	1	4	29.58	3.020	12	0.6
5	-1	4	29.58	3.020	12	0.3
5	0	2	29.66	3.012	12	1.0
3	3	0	30.06	2.973	6	2.2
5	-2	5	30.10	2.969	12	1.3
3	2	5	30.10	2.969	12	0.5
1	0	9	30.15	2.964	12	0.5
4	2	0	30.63	2.919	12	0.3
3	3	2	30.79	2.904	6	0.5
6	-2	1	30.81	2.902	12	0.7
4	-1	7	31.08	2.877	12	3.6
3	1	7	31.08	2.877	12	1.8
4	1	5	31.22	2.865	12	0.5
2	0	9	31.81	2.813	12	0.8
3	2	6	32.07	2.791	12	0.2
5	-2	6	32.07	2.791	12	0.3
6	-2	3	32.22	2.778	12	1.2
6	-1	1	32.44	2.760	12	2.0
0	0	10	32.95	2.718	2	0.8
5	-1	6	33.13	2.704	12	0.3
5	0	5	33.36	2.686	12	1.8
6	-2	4	33.41	2.682	12	0.3
4	2	4	33.41	2.682	12	0.3
1	0	10	33.47	2.677	12	0.4
3	1	8	33.67	2.662	12	0.3
5	1	3	33.79	2.653	12	0.2
3	2	7	34.26	2.617	12	0.4
1	1	10	34.49	2.600	6	0.3
6	0	0	34.85	2.574	6	0.6
6	-2	5	34.89	2.572	12	0.9
5	1	4	34.93	2.568	12	0.4
6	-1	4	34.93	2.568	12	0.7
6	0	1	35.01	2.563	12	1.0
5	-1	7	35.26	2.545	12	0.2
6	0	2	35.49	2.529	12	0.2
7	-3	1	35.51	2.528	12	0.2
6	0	3	36.28	2.476	12	1.0
5	2	0	36.32	2.473	12	1.0
3	1	9	36.40	2.468	12	0.4
5	-2	8	36.64	2.452	12	0.2
1	0	11	36.84	2.440	12	0.4
7	-2	2	36.94	2.433	12	0.4
5	0	7	37.19	2.418	12	3.0
6	0	4	37.35	2.407	12	0.8
6	1	0	38.21	2.356	12	0.2
6	-2	7	38.59	2.333	12	0.4
5	-2	9	39.19	2.299	12	0.4
3	2	9	39.19	2.299	12	0.2
6	0	6	40.29	2.238	12	0.7
8	-3	3	42.15	2.144	12	0.4
8	-3	4	43.10	2.099	12	0.4
6	1	6	43.29	2.090	12	0.2
1	0	13	43.69	2.072	12	1.0
8	-2	4	44.34	2.043	12	0.3
4	-2	12	44.88	2.019	6	0.3
2	0	13	44.91	2.018	12	0.5
4	4	6	45.34	2.000	6	0.3
7	1	3	45.45	1.996	12	0.4
6	3	0	46.67	1.946	12	0.7
0	0	14	46.79	1.942	2	0.4
7	0	7	47.39	1.918	12	1.1
5	4	4	47.90	1.899	12	0.3
2	-1	14	47.95	1.897	6	0.2
7	2	1	48.35	1.883	12	0.2
6	2	7	48.54	1.876	12	0.3
6	0	10	48.72	1.869	12	0.9
9	-4	5	49.01	1.859	12	0.2
6	-1	11	49.39	1.845	12	0.3

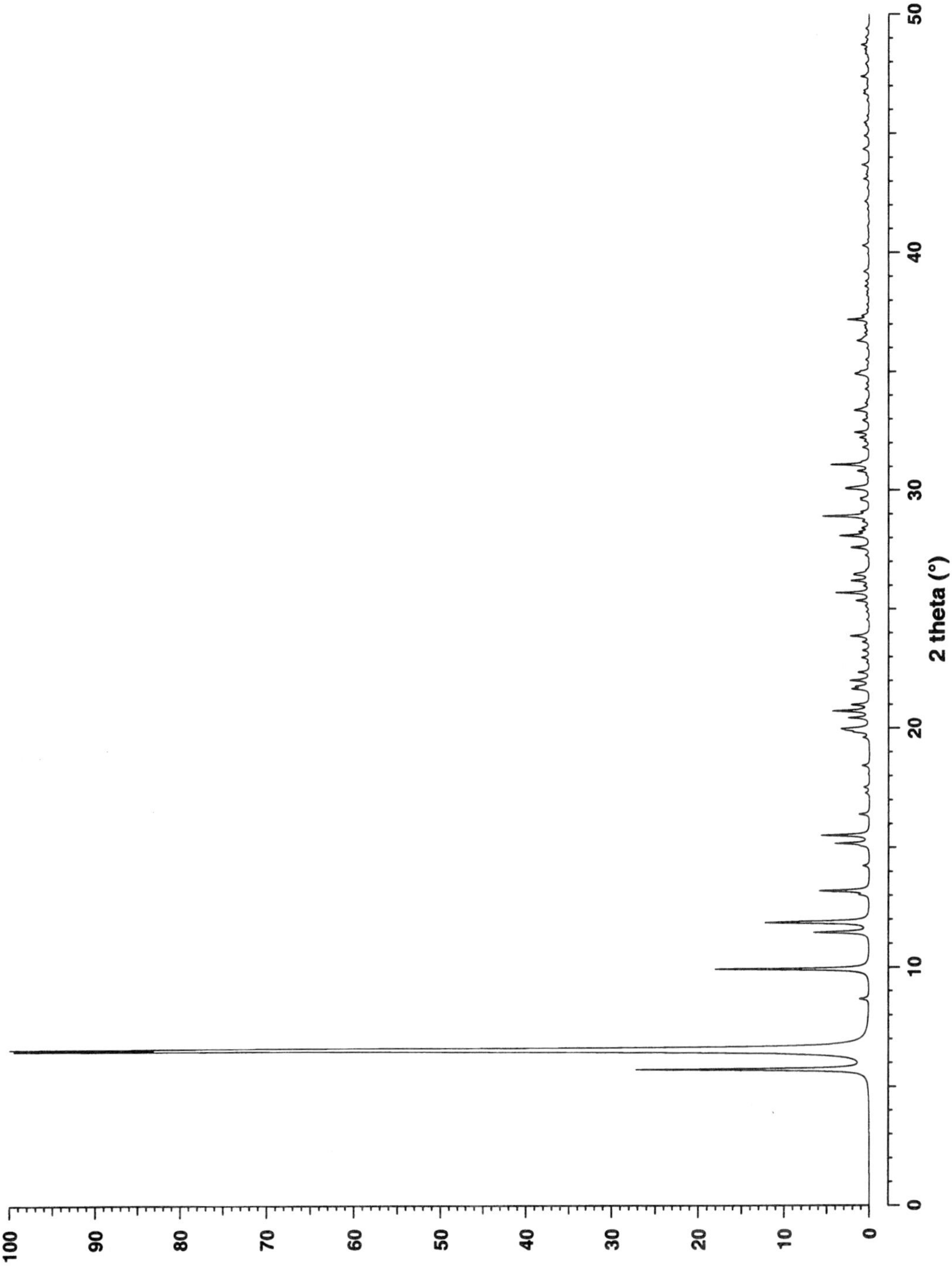
2 theta (°)
0
10
20
30
40
50
100
90
80
70
60
50
40
30
20
10
0

SBT UCSB–10GaZn

CHEMICAL COMPOSITION: $|(H_3N(CH_2)_3O((CH_2)_2O)_2(CH_2)_3NH_3)^{2+}_{18}|$ $[Ga_{36}Zn_{36}P_{72}O_{288}]$
$H_2N(CH_2)_3O((CH_2)_2O)_2(CH_2)_3NH_2$ = 4,7,10-trioxa-1,13-tridecanediamine

REFINED COMPOSITION: $|C_{30}N_{24}|$ $[Ga_{36}Zn_{36}P_{72}O_{288}]$

CRYSTAL DATA: $R\overline{3}$ (No. 148) hexagonal setting
a = 18.0804 Å b = 18.0804 Å c = 41.9511 Å
α = 90° β = 90° γ = 120°
X-ray single crystal refinement, R_F = 0.045, R_w = 0.140

REFERENCE: X. Bu, P. Feng and G. D. Stucky,
Science **278** 2080–2085 (1997).

h	k	l	2θ	d	M	I_{rel}
1	0	1	6.02	14.670	6	100.0
0	0	3	6.32	13.984	2	76.3
0	1	2	7.04	12.548	6	32.8
1	1	0	9.78	9.040	6	14.3
0	2	1	11.50	7.696	6	1.8
2	-1	3	11.66	7.592	6	5.0
1	1	3	11.66	7.592	6	4.7
0	1	5	11.97	7.395	6	0.4
2	0	2	12.07	7.335	6	10.4
0	0	6	12.66	6.992	2	0.9
3	-1	1	15.12	5.860	6	4.4
2	1	1	15.12	5.860	6	3.5
1	2	2	15.56	5.696	6	0.6
3	-2	2	15.56	5.696	6	0.8
2	-1	6	16.02	5.531	6	0.8
1	1	6	16.02	5.531	6	0.7
0	1	8	17.84	4.972	6	2.6
3	0	3	18.14	4.890	6	2.9
0	2	7	18.65	4.759	6	0.7
0	0	9	19.04	4.661	2	0.7
2	2	0	19.64	4.520	6	3.1
2	0	8	20.38	4.357	6	3.4
4	-3	1	20.56	4.320	6	0.5
1	3	1	20.56	4.320	6	4.3
4	-1	2	20.89	4.253	6	3.1
2	1	7	21.10	4.211	6	1.7
3	0	6	21.24	4.183	6	0.3
0	3	6	21.24	4.183	6	2.8
2	-1	9	21.45	4.143	6	0.3
1	1	9	21.45	4.143	6	3.3
1	0	10	21.93	4.052	6	2.4
1	2	8	22.65	3.925	6	0.8
4	0	1	22.82	3.898	6	0.2
3	1	5	23.06	3.857	6	0.4
4	-1	5	23.06	3.857	6	0.5
0	4	2	23.11	3.848	6	1.5
2	2	6	23.43	3.796	6	2.2
4	-2	6	23.43	3.796	6	1.1
0	2	10	24.07	3.698	6	1.7
0	4	5	25.10	3.547	6	0.7
5	-3	2	25.15	3.541	6	0.8
2	3	2	25.15	3.541	6	1.7
4	-3	7	25.33	3.517	6	0.4
1	3	7	25.33	3.517	6	0.6
0	3	9	25.62	3.477	6	1.4
3	0	9	25.62	3.477	6	1.1
2	0	11	25.99	3.429	6	3.3
3	-1	10	26.03	3.422	6	0.8
2	1	10	26.03	3.422	6	1.4
1	4	0	26.08	3.417	6	1.4
4	1	0	26.08	3.417	6	0.8
5	-2	4	26.22	3.398	6	0.2
3	1	8	26.65	3.345	6	0.9
1	4	3	26.86	3.319	6	0.6
5	-3	5	27.00	3.302	6	0.4
1	1	12	27.35	3.261	6	0.3
4	-2	9	27.49	3.245	6	2.8
2	2	9	27.49	3.245	6	0.9
3	-2	11	27.83	3.206	6	0.4
1	2	11	27.83	3.206	6	0.3
0	5	1	28.58	3.123	6	7.7
5	0	2	28.82	3.097	6	0.6
1	4	6	29.09	3.070	6	1.2
5	-4	6	29.09	3.070	6	0.3
5	-1	6	29.09	3.070	6	0.3
4	-3	10	29.61	3.017	6	3.8
1	3	10	29.61	3.017	6	0.5
3	3	0	29.64	3.013	6	1.9
0	2	13	29.95	2.984	6	0.7
5	-3	8	30.16	2.964	6	1.6
2	3	8	30.16	2.964	6	0.6
6	-4	1	30.28	2.952	6	0.9
3	3	3	30.34	2.946	6	0.3
0	1	14	30.37	2.943	6	0.7
4	-1	11	31.21	2.866	6	0.5
3	1	11	31.21	2.866	6	1.8
6	-1	1	31.89	2.806	6	1.0
2	0	14	31.98	2.799	6	0.4
0	0	15	32.00	2.797	2	0.7
6	-2	5	32.07	2.791	6	1.1
4	2	5	32.07	2.791	6	0.2
6	-5	2	32.11	2.787	6	0.3
1	5	2	32.11	2.787	6	0.3
0	5	7	32.25	2.776	6	1.0
5	-1	9	32.49	2.756	6	0.8
5	0	8	33.32	2.689	6	0.7
1	5	5	33.61	2.666	6	0.9
6	-5	5	33.61	2.666	6	0.6
0	6	0	34.36	2.610	6	0.6
4	-3	13	34.63	2.590	6	0.3
1	0	16	34.69	2.586	6	0.7
4	2	8	34.81	2.577	6	0.5
6	-2	8	34.81	2.577	6	1.1
6	0	3	34.98	2.565	6	0.4
0	6	3	34.98	2.565	6	1.7
7	-4	2	35.12	2.555	6	0.3
0	5	10	35.78	2.510	6	1.7
5	2	0	35.81	2.507	6	0.2
2	5	0	35.81	2.507	6	0.8
7	-5	3	36.40	2.468	6	0.3
3	1	14	36.43	2.466	6	0.3
6	0	6	36.76	2.445	6	1.0
5	0	11	37.15	2.420	6	1.7
3	2	13	37.46	2.401	6	0.3
5	-2	13	37.46	2.401	6	0.5
7	-6	1	37.73	2.384	6	0.3
1	6	1	37.73	2.384	6	0.3
4	-2	15	37.83	2.378	6	0.3
6	0	9	39.58	2.277	6	0.4
0	6	9	39.58	2.277	6	0.4
1	0	19	41.29	2.186	6	0.9
8	-3	5	41.79	2.161	6	0.4
0	2	19	42.54	2.125	6	0.6
8	-6	5	43.03	2.102	6	0.3
8	-5	7	43.17	2.096	6	0.3
0	1	20	43.53	2.079	6	0.3
1	6	10	43.61	2.075	6	0.4
4	-2	18	43.70	2.071	6	0.3
5	3	8	44.01	2.057	6	0.3
1	7	3	44.14	2.052	6	0.3
2	3	17	44.54	2.034	6	0.3
4	4	9	44.55	2.034	6	0.5
0	0	21	45.40	1.998	2	0.4
7	1	6	45.62	1.988	6	0.5
7	0	10	45.98	1.974	6	0.4
3	6	0	46.01	1.973	6	0.5
0	7	11	47.10	1.929	6	0.6
2	7	1	47.58	1.911	6	0.4
6	0	15	47.66	1.908	6	0.5
0	6	15	47.66	1.908	6	0.6
5	4	7	47.84	1.901	6	0.2
1	5	17	49.12	1.855	6	0.3
0	2	22	49.18	1.853	6	0.3
8	1	1	49.80	1.831	6	0.3
1	8	2	49.95	1.826	6	0.2

UCSB–10GaZn

SBT

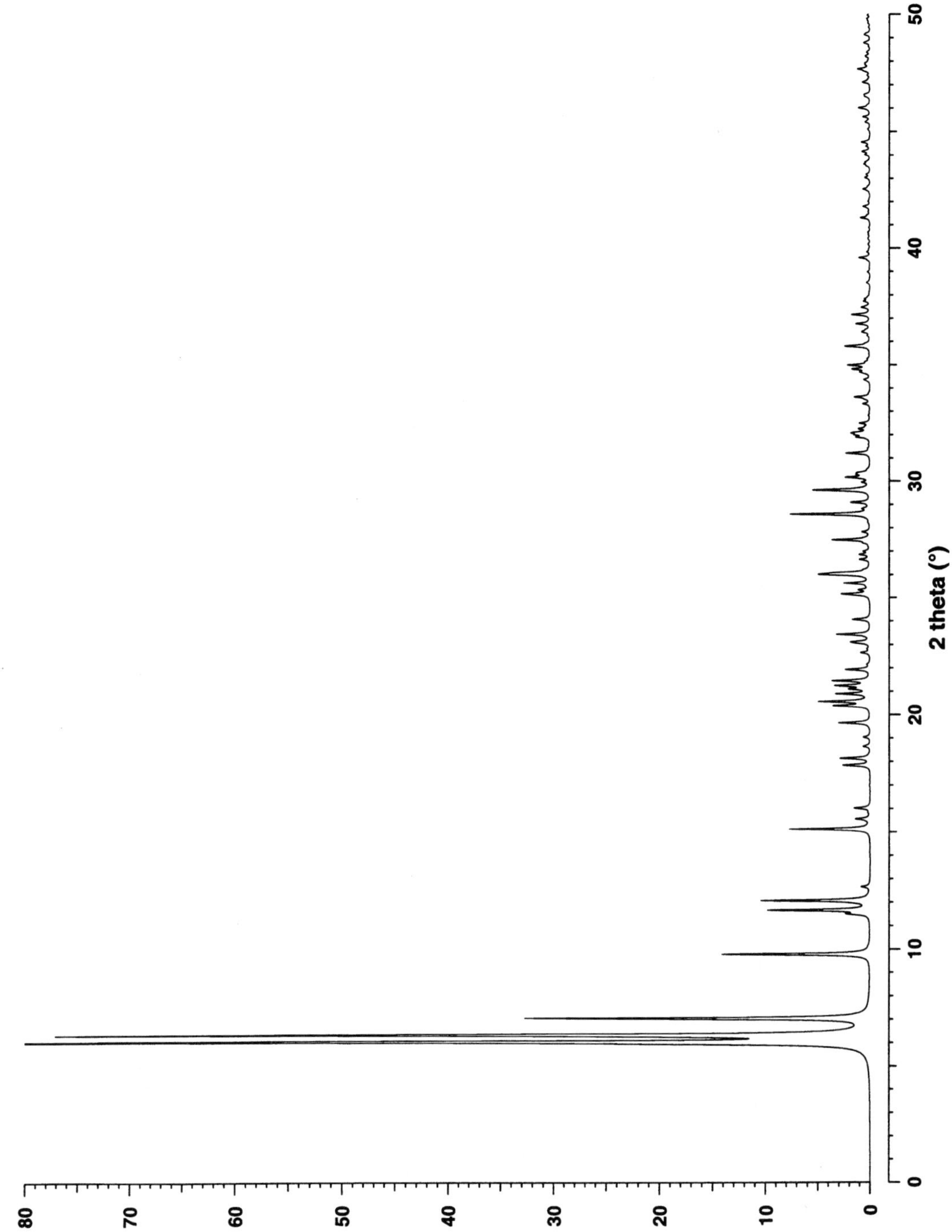

SFE SSZ–48, Calcined

CHEMICAL COMPOSITION: $[Si_{14}O_{28}]$

REFINED COMPOSITION: $[Si_{14}O_{28}]$

CRYSTAL DATA: $P2_1$ (No. 4) unique axis b
$a = 11.1527$ Å $b = 5.002$ Å $c = 13.667$ Å
$\alpha = 90°$ $\beta = 100.633°$ $\gamma = 90°$
Electron diffraction, and X-ray powder, $R_w = 0.0971$, $R_p = 0.0887$

REFERENCE: P. Wagner, O. Terasaki, S. Ritsch, J. G. Nery, S. I. Zones, M. E. Davis and K. Hiraga, *J. Phys. Chem. B* **103** 8245–8250 (1999).

h	k	l	2θ	d	M	I_{rel}
0	0	1	6.58	13.432	2	99.6
1	0	0	8.07	10.961	2	100.0
-1	0	1	9.43	9.383	2	40.0
1	0	1	11.32	7.815	2	21.8
0	0	2	13.18	6.716	2	0.3
-1	0	2	14.14	6.265	2	0.8
2	0	0	16.17	5.481	2	1.7
-2	0	1	16.30	5.438	2	2.7
1	0	2	16.70	5.307	2	2.8
2	0	1	18.58	4.776	2	2.7
-2	0	2	18.92	4.691	2	5.4
0	1	1	18.93	4.688	4	1.5
1	1	0	19.51	4.551	4	5.8
0	0	3	19.83	4.477	2	5.9
-1	0	3	19.99	4.442	2	0.2
-1	1	1	20.12	4.414	4	36.9
1	1	1	21.09	4.213	4	0.2
0	1	2	22.16	4.012	4	2.2
-1	1	2	22.75	3.909	4	24.5
2	0	2	22.76	3.908	2	6.0
-2	0	3	23.22	3.831	2	0.8
-3	0	1	24.03	3.703	2	3.2
2	1	0	24.09	3.695	4	4.2
-2	1	1	24.18	3.681	4	17.4
3	0	0	24.36	3.654	2	10.7
1	1	2	24.45	3.640	4	0.2
-3	0	2	25.51	3.491	2	0.6
2	1	1	25.79	3.454	4	1.1
-2	1	2	26.04	3.422	4	1.6
-1	0	4	26.28	3.391	2	2.2
3	0	1	26.43	3.372	2	3.1
0	0	4	26.54	3.358	2	10.0
0	1	3	26.72	3.336	4	2.6
-1	1	3	26.84	3.321	4	3.2
2	0	3	27.96	3.191	2	11.5
-2	0	4	28.49	3.132	2	1.2
-3	0	3	28.54	3.128	2	0.5
1	1	3	29.03	3.076	4	0.8
1	0	4	29.22	3.057	2	1.9
-2	1	3	29.37	3.041	4	0.4
-3	1	1	30.02	2.976	4	1.3
3	1	0	30.29	2.950	4	0.7
-3	1	2	31.24	2.863	4	1.3
-1	1	4	31.88	2.807	4	0.2
3	1	1	32.01	2.796	4	1.5
0	1	4	32.10	2.788	4	1.0
-4	0	1	32.11	2.788	2	0.4
-3	0	4	32.72	2.737	2	1.1
-1	0	5	32.83	2.728	2	1.3
2	1	3	33.30	2.690	4	0.1
0	0	5	33.35	2.686	2	0.2
-2	1	4	33.76	2.655	4	0.3
2	0	4	33.78	2.654	2	0.2
-3	1	3	33.80	2.652	4	2.1
-2	0	5	34.36	2.610	2	0.3
1	1	4	34.38	2.608	4	0.5
3	0	3	34.42	2.605	2	0.9
4	0	1	34.59	2.593	2	0.4
-4	0	3	35.10	2.557	2	0.2
1	0	5	35.85	2.505	2	0.3
0	2	0	35.91	2.501	2	8.7
0	2	1	36.54	2.459	4	1.2
1	2	0	36.86	2.438	4	0.8
-4	1	1	36.91	2.435	4	1.5
-1	2	1	37.20	2.417	4	0.3
4	1	0	37.42	2.403	4	0.2
-3	1	4	37.46	2.401	4	1.8
-1	1	5	37.55	2.395	4	2.0
-4	1	2	37.66	2.389	4	2.1
-3	0	5	37.72	2.385	2	0.2
1	2	1	37.77	2.382	4	0.3
2	1	4	38.40	2.344	4	0.6
-1	2	2	38.77	2.323	4	0.1
-2	1	5	38.92	2.314	4	1.2
3	1	3	38.98	2.311	4	0.2
4	1	1	39.13	2.302	4	0.5
-1	0	6	39.57	2.277	2	0.2
-4	1	3	39.59	2.277	4	0.1
2	2	0	39.61	2.275	4	0.2
3	0	4	39.66	2.272	2	0.1
-2	2	1	39.67	2.272	4	0.2
1	1	5	40.27	2.240	4	0.5
0	0	6	40.28	2.239	2	0.5
-5	0	2	40.88	2.208	2	0.1
-2	2	2	40.89	2.207	4	0.2
5	0	0	41.18	2.192	2	0.2
0	2	3	41.35	2.183	4	0.5
-3	1	5	41.97	2.152	4	0.1
-4	1	4	42.57	2.124	4	0.1
2	2	2	42.93	2.107	4	0.2
5	0	1	43.01	2.103	2	0.1
-2	2	3	43.20	2.094	4	0.4
-1	1	6	43.67	2.073	4	1.5
-3	2	1	43.67	2.073	4	0.1
3	1	4	43.75	2.069	4	0.2
3	2	0	43.87	2.064	4	1.1
2	1	5	44.07	2.055	4	0.2
0	1	6	44.33	2.043	4	0.2
-5	1	1	44.48	2.037	4	0.5
-3	2	2	44.56	2.033	4	0.5
-2	1	6	44.64	2.030	4	0.6
-5	1	2	44.88	2.020	4	0.2
3	2	1	45.14	2.009	4	0.2
5	1	0	45.16	2.008	4	0.6
0	2	4	45.20	2.006	4	0.7
5	0	2	45.85	1.979	2	0.9
2	2	3	46.11	1.968	4	0.2
-5	1	3	46.32	1.960	4	0.3
-4	1	5	46.44	1.955	4	0.8
-2	2	4	46.46	1.954	4	0.6
-3	2	3	46.49	1.953	4	1.2
1	1	6	46.56	1.950	4	0.3
2	0	6	46.57	1.950	2	1.4
-3	1	6	47.16	1.927	4	0.4
-2	0	7	47.21	1.925	2	0.1
-4	0	6	47.46	1.916	2	0.1
-5	1	4	48.75	1.868	4	0.3
-4	2	1	48.93	1.862	4	0.1
-6	0	1	49.07	1.857	2	0.1
3	1	5	49.15	1.854	4	0.3
-6	0	2	49.21	1.851	2	0.1
4	2	0	49.33	1.847	4	0.3
-3	2	4	49.36	1.846	4	0.4
-3	0	7	49.40	1.845	2	0.7
-1	2	5	49.44	1.844	4	0.3
-4	2	2	49.52	1.841	4	0.5
5	1	2	49.53	1.840	4	0.2
5	0	3	49.56	1.839	2	1.8
0	2	5	49.81	1.831	4	0.1
6	0	0	49.92	1.827	2	0.1

SSZ–48, Calcined **SFE**

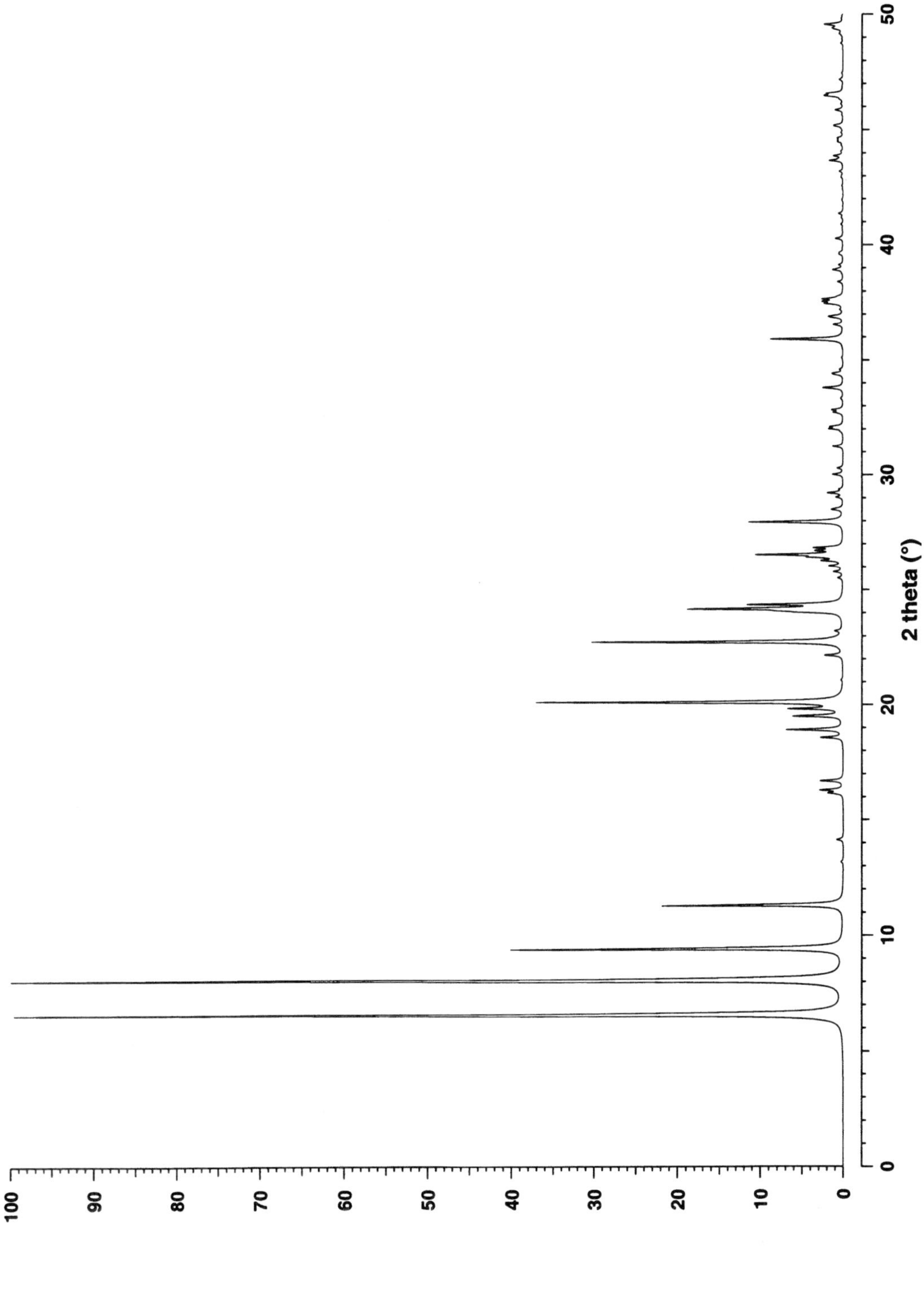

SFF SSZ-44, Calcined

CHEMICAL COMPOSITION: $[Si_{32}O_{64}]$

REFINED COMPOSITION: $[Si_{32}O_{64}]$

CRYSTAL DATA: $P1\,2_1/m\,1$ (No. 11) unique axis **b**
a = 11.4853 Å b = 21.9458 Å c = 7.3881 Å
α = 90.0° β = 94.702° γ = 90.0°
X-ray Rietveld refinement, R_p = 0.0892, R_{wp} = 0.1058

REFERENCE: P. Wagner, S. I. Zones, M. E. Davis and R. C. Medrud, *Angew. Chem., Int. ed.* **38** 1269–1272 (1999).

h	k	l	2θ	d	M	I_{rel}
1	0	0	7.72	11.447	2	50.8
0	2	0	8.06	10.973	2	100.0
1	1	0	8.71	10.149	4	55.9
1	2	0	11.17	7.921	4	5.6
0	0	1	12.02	7.363	2	4.5
0	1	1	12.68	6.981	4	7.7
-1	0	1	13.76	6.437	2	6.6
0	2	1	14.49	6.114	4	1.4
1	1	1	15.37	5.764	4	1.8
-1	2	1	15.96	5.552	4	0.5
2	1	0	16.00	5.538	4	0.4
0	4	0	16.15	5.486	2	0.3
0	3	1	17.09	5.190	4	1.3
1	4	0	17.93	4.947	4	2.8
-1	3	1	18.36	4.833	4	0.7
1	3	1	19.18	4.627	4	17.4
-2	1	1	19.27	4.605	4	10.0
2	3	0	19.69	4.508	4	6.3
0	4	1	20.18	4.399	4	0.4
-2	2	1	20.52	4.328	4	7.1
-1	4	1	21.28	4.176	4	1.8
1	5	0	21.68	4.098	4	5.7
2	2	1	21.98	4.043	4	2.3
2	4	0	22.45	3.961	4	0.2
-2	3	1	22.45	3.960	4	1.0
3	0	0	23.31	3.816	2	1.7
0	5	1	23.60	3.770	4	1.5
3	1	0	23.67	3.759	4	0.3
2	3	1	23.80	3.738	4	7.7
0	0	2	24.17	3.682	2	0.2
0	6	0	24.33	3.658	2	1.0
0	1	2	24.52	3.631	4	3.1
-1	5	1	24.55	3.626	4	0.5
3	2	0	24.70	3.604	4	2.3
-1	0	2	24.79	3.592	2	2.8
-2	4	1	24.92	3.574	4	4.9
-1	1	2	25.12	3.545	4	0.6
1	5	1	25.18	3.537	4	5.3
0	2	2	25.52	3.490	4	0.6
1	6	0	25.57	3.484	4	0.5
2	5	0	25.58	3.483	4	8.9
-3	1	1	25.72	3.463	4	3.5
1	0	2	26.02	3.424	2	0.4
-1	2	2	26.10	3.413	4	1.2
2	4	1	26.14	3.408	4	1.0
3	3	0	26.34	3.383	4	0.8
1	1	2	26.34	3.383	4	3.0
-3	2	1	26.68	3.341	4	0.7
0	3	2	27.11	3.289	4	5.5
0	6	1	27.22	3.276	4	0.8
1	2	2	27.28	3.269	4	0.7
-1	3	2	27.67	3.224	4	0.3
-2	0	2	27.71	3.219	2	3.7
-2	5	1	27.78	3.211	4	0.6
-2	1	2	28.02	3.185	4	0.3
-3	3	1	28.22	3.163	4	0.4
3	2	1	28.40	3.142	4	0.4
3	4	0	28.49	3.133	4	3.0
1	3	2	28.79	3.101	4	1.0
2	5	1	28.90	3.089	4	2.3
-2	2	2	28.91	3.089	4	0.2
2	6	0	28.97	3.082	4	0.7
0	4	2	29.21	3.057	4	0.3
1	7	0	29.54	3.024	4	0.2
-1	4	2	29.73	3.005	4	1.9
2	0	2	29.91	2.987	2	2.0
1	4	2	30.78	2.905	4	0.9
-2	6	1	30.95	2.889	4	0.2
0	7	1	31.00	2.885	4	1.0
3	5	0	31.06	2.880	4	1.8
4	0	0	31.26	2.862	2	0.3
-1	7	1	31.75	2.819	4	0.3
2	6	1	31.97	2.799	4	1.5
-1	5	2	32.20	2.780	4	0.4
-2	4	2	32.24	2.776	4	0.3
4	2	0	32.33	2.769	4	1.1
0	8	0	32.64	2.743	2	1.0
-3	5	1	32.68	2.740	4	0.9
4	3	0	33.63	2.665	4	0.4
-4	2	1	33.66	2.662	4	1.8
3	5	1	34.13	2.627	4	0.2
4	0	1	34.54	2.596	2	0.6
-3	3	2	34.68	2.586	4	0.5
0	8	1	34.90	2.571	4	2.3
-1	6	2	35.01	2.563	4	0.4
3	0	2	35.23	2.547	2	1.3
4	2	1	35.53	2.527	4	0.9
3	2	2	36.20	2.481	4	0.3
-3	4	2	36.38	2.469	4	0.3
0	0	3	36.61	2.454	2	0.4
4	3	1	36.73	2.447	4	0.4
-1	1	3	37.05	2.426	4	1.0
3	7	0	37.11	2.422	4	0.4
0	2	3	37.55	2.395	4	0.8
0	7	2	37.68	2.387	4	0.3
-1	2	3	37.75	2.383	4	0.5
-1	7	2	38.10	2.362	4	0.2
-4	1	2	38.45	2.341	4	0.3
-3	7	1	38.51	2.337	4	0.3
2	8	1	38.81	2.320	4	0.2
2	6	2	38.93	2.314	4	0.4
-2	2	3	39.61	2.275	4	0.3
3	7	1	39.77	2.266	4	0.5
-4	3	2	40.23	2.242	4	0.2
-4	6	1	41.12	2.195	4	0.8
0	10	0	41.13	2.195	2	0.4
5	3	0	41.32	2.185	4	0.3
-3	8	1	41.80	2.161	4	0.2
2	2	3	42.04	2.149	4	0.4
-2	4	3	42.20	2.141	4	0.2
3	8	1	42.98	2.104	4	1.9
0	10	1	43.01	2.103	4	0.7
2	3	3	43.09	2.099	4	0.2
3	9	0	44.07	2.055	4	0.3
5	3	1	44.15	2.051	4	0.6
2	4	3	44.52	2.035	4	0.2
-5	0	2	44.88	2.020	2	0.4
4	7	1	45.35	2.000	4	0.4
-3	4	3	45.38	1.998	4	0.4
-5	5	1	45.48	1.994	4	0.3
1	9	2	45.68	1.986	4	0.4
1	6	3	45.75	1.983	4	0.2
3	9	1	46.40	1.957	4	1.0
-4	1	3	46.93	1.936	4	0.2
3	3	3	47.31	1.921	4	0.6
-5	6	1	47.62	1.909	4	0.7
1	7	3	48.26	1.886	4	0.2
4	8	1	48.26	1.886	4	0.8
2	6	3	48.42	1.880	4	0.5
-3	6	3	49.23	1.851	4	0.5
-1	0	4	49.50	1.841	2	0.2
-2	11	1	49.62	1.837	4	0.5
-1	1	4	49.69	1.835	4	0.2
-3	9	2	49.86	1.829	4	0.6
-6	3	1	49.97	1.825	4	0.5

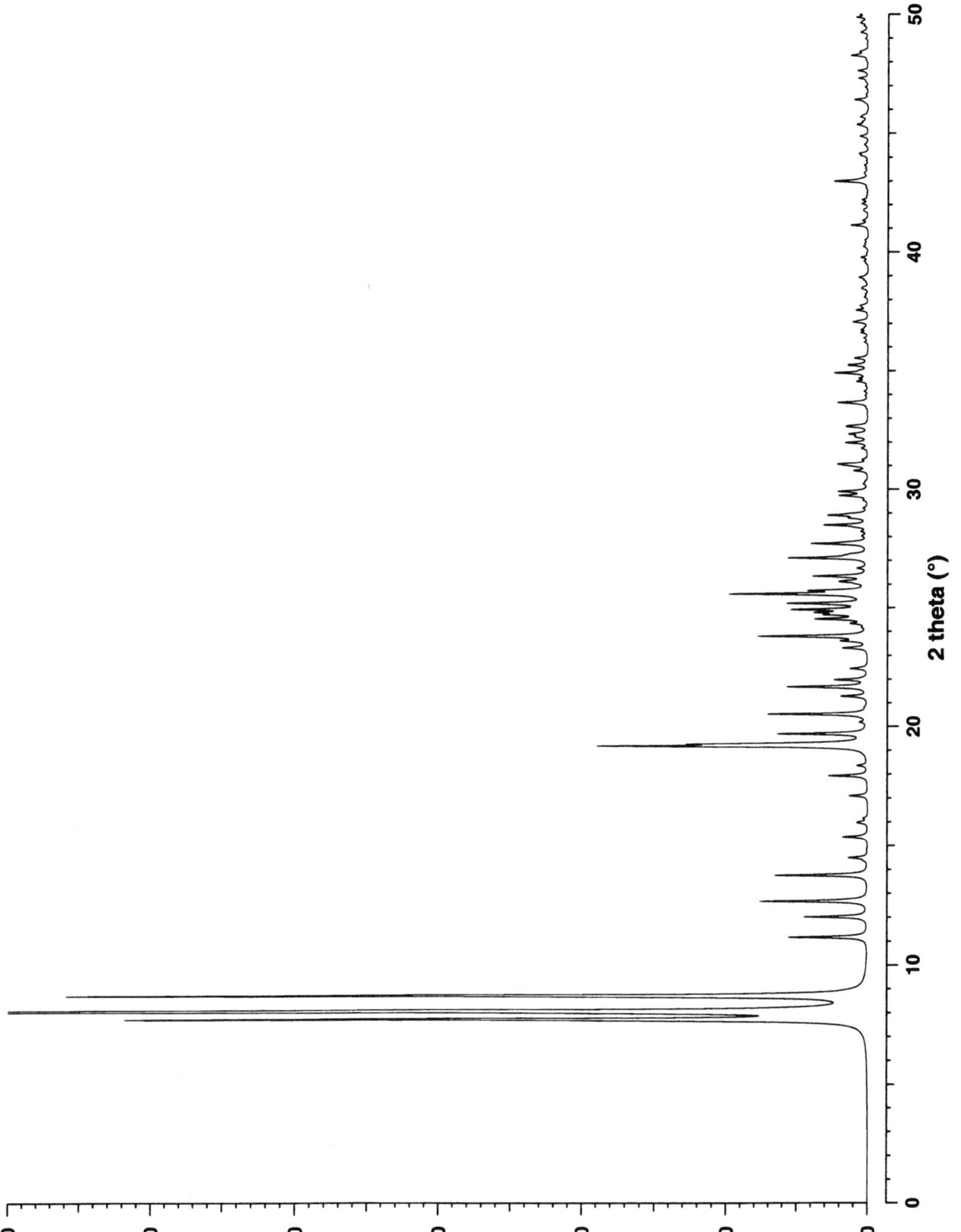
2 theta (°)
0
10
20
30
40
50
0
10
20
30
40
50
60

SGT 1-aminoadamantane Sigma-2

CHEMICAL COMPOSITION: $|(C_{10}H_{17}N)_4|$ $[Si_{64}O_{128}]$
$C_{10}H_{17}N$ = 1-aminoadamantane

REFINED COMPOSITION: $|C_{43.84}|$ $[Si_{64}O_{128}]$

CRYSTAL DATA: $I\,4_1/a\,m\,d$ (No. 141) origin at centre $(2/m)$
$a = 10.2387$ Å $b = 10.2387$ Å $c = 34.3829$ Å
$\alpha = 90°$ $\beta = 90°$ $\gamma = 90°$
X-ray Rietveld refinement, $R_{\rm wp} = 0.225$, $R_{\rm F} = 0.100$

REFERENCE: L. B. McCusker,
J. Appl. Cryst. **21** 305–310 (1988).

h	k	l	2θ	d	M	$I_{\rm rel}$
1	0	1	9.01	9.813	8	75.0
0	0	4	10.29	8.596	2	15.8
1	0	3	11.59	7.636	8	7.5
1	1	2	13.27	6.672	8	5.3
1	0	5	15.52	5.709	8	3.7
2	0	0	17.32	5.119	4	1.2
2	0	2	18.08	4.906	8	1.9
2	1	1	19.56	4.539	16	99.1
1	1	6	19.76	4.493	8	100.0
1	0	7	20.05	4.429	8	1.4
2	0	4	20.19	4.398	8	35.8
0	0	8	20.67	4.298	2	10.2
2	1	3	20.89	4.252	16	8.1
2	0	6	23.30	3.818	8	1.2
2	1	5	23.34	3.811	16	12.1
2	2	0	24.59	3.620	4	5.2
1	0	9	24.88	3.579	8	2.8
3	0	1	26.24	3.396	8	20.2
2	1	7	26.61	3.349	16	0.2
2	2	4	26.72	3.336	8	46.7
2	0	8	27.09	3.292	8	25.1
3	0	3	27.26	3.271	8	5.9
3	1	2	28.04	3.182	16	1.3
1	1	10	28.74	3.106	8	1.2
3	0	5	29.21	3.057	8	10.4
3	1	4	29.48	3.030	16	2.2
1	0	11	29.89	2.990	8	15.3
2	1	9	30.47	2.933	16	0.6

h	k	l	2θ	d	M	$I_{\rm rel}$
0	0	12	31.22	2.865	2	3.6
2	0	10	31.34	2.854	8	2.8
3	2	1	31.61	2.830	16	5.3
3	1	6	31.74	2.819	16	5.9
3	0	7	31.93	2.803	8	4.0
2	2	8	32.33	2.769	8	0.5
3	2	3	32.48	2.756	16	8.8
3	2	5	34.16	2.625	16	11.6
3	1	8	34.69	2.586	16	6.0
2	1	11	34.75	2.582	16	3.8
1	0	13	35.04	2.561	8	1.0
4	0	0	35.06	2.560	4	5.4
3	0	9	35.26	2.545	8	0.4
4	0	2	35.46	2.532	8	0.5
2	0	12	35.92	2.500	8	8.9
4	1	1	36.27	2.477	16	0.3
3	2	7	36.55	2.458	16	0.3
4	1	3	37.04	2.427	16	0.3
3	3	2	37.64	2.390	8	1.5
4	0	6	38.52	2.337	8	0.9
1	1	14	38.72	2.326	8	0.1
2	1	13	39.34	2.290	16	1.0
4	2	0	39.35	2.289	8	2.5
3	2	9	39.54	2.279	16	0.7
4	2	2	39.72	2.269	16	1.2
2	2	12	40.14	2.247	8	3.0
3	3	6	40.56	2.224	8	0.8
4	1	7	40.71	2.216	16	5.3

h	k	l	2θ	d	M	$I_{\rm rel}$
2	0	14	40.75	2.214	8	0.2
0	0	16	42.05	2.149	2	0.5
3	1	12	42.11	2.146	16	0.6
4	2	6	42.52	2.126	16	3.3
4	0	10	44.11	2.053	8	1.3
2	1	15	44.18	2.050	16	0.2
5	0	1	44.31	2.044	8	0.6
4	3	1	44.31	2.044	16	0.6
4	2	8	44.85	2.021	16	0.8
5	0	3	44.97	2.016	8	1.0
4	3	3	44.97	2.016	16	6.9
5	1	2	45.48	1.994	16	2.7
1	0	17	45.73	1.984	8	3.4
2	0	16	45.79	1.981	8	2.0
3	3	10	45.94	1.975	8	1.4
4	3	5	46.26	1.963	16	1.2
5	0	5	46.26	1.963	8	0.7
3	1	14	46.40	1.957	16	2.5
4	1	11	46.72	1.944	16	1.3
4	2	10	47.72	1.906	16	0.4
3	0	15	47.80	1.903	8	0.9
5	2	1	47.92	1.898	16	1.4
5	1	6	48.01	1.895	16	6.7
4	3	7	48.14	1.890	16	1.1
1	1	18	49.34	1.847	8	2.2
5	2	5	49.75	1.833	16	0.5

1-aminoadamantane Sigma-2 SGT

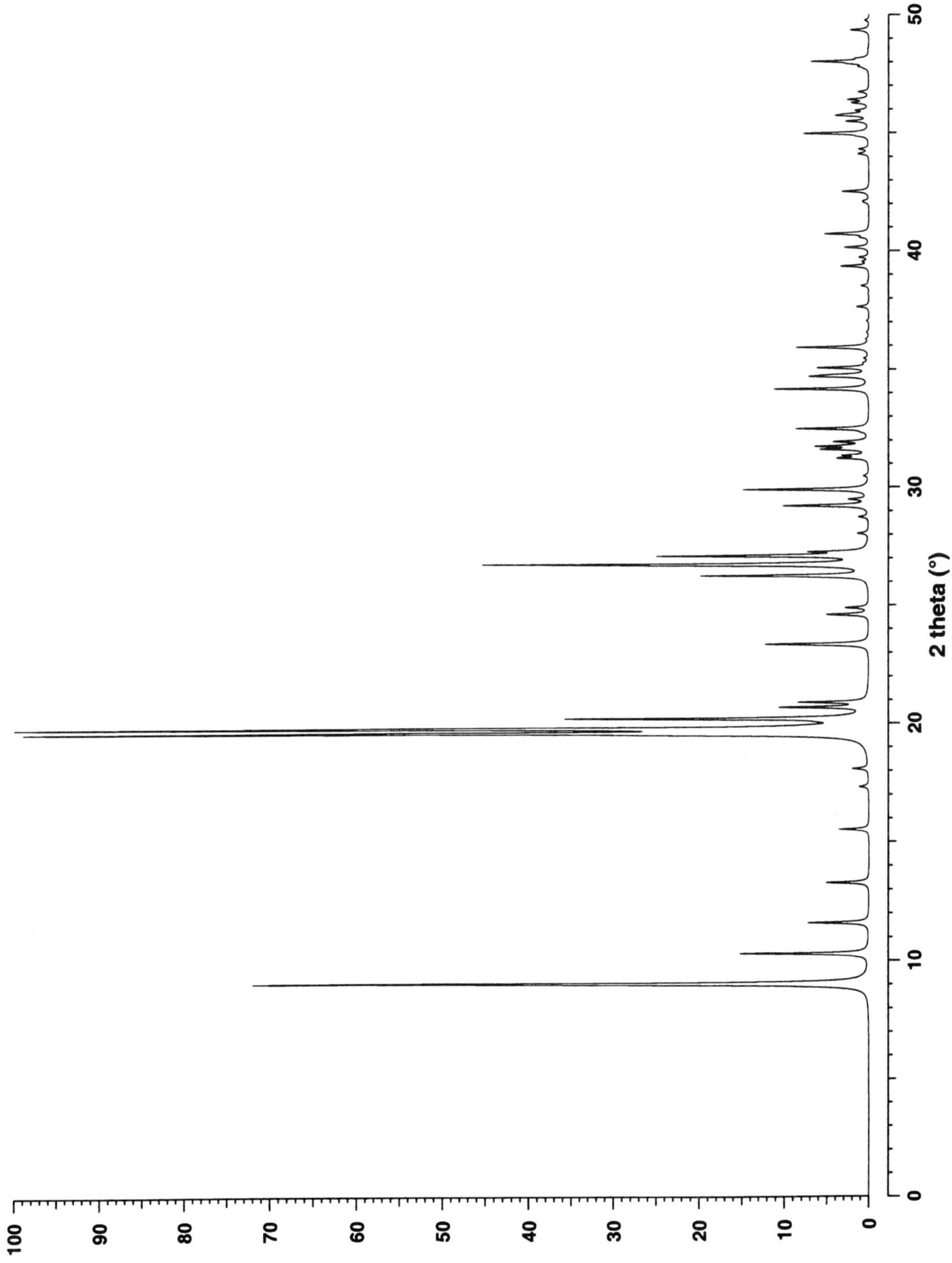

SOD Sodalite Octahydrate

CHEMICAL COMPOSITION: $|Na_6(H_2O)_8|\ [Si_6Al_6O_{24}]$

REFINED COMPOSITION: $|Na_6(H_2O)_8|\ [Si_6Al_6O_{24}]$

CRYSTAL DATA: $P\bar{4}3n$ (No. 218)
$a = 8.848$ Å $b = 8.848$ Å $c = 8.848$ Å
$\alpha = 90°$ $\beta = 90°$ $\gamma = 90°$
X-ray Rietveld refinement, $R_{wp} = 0.091$, $R_F = 0.047$

REFERENCE: J. Felsche, S. Luger and Ch. Baerlocher, *Zeolites* **6** 367–372 (1986).

h	k	l	2θ	d	M	I_{rel}	h	k	l	2θ	d	M	I_{rel}	h	k	l	2θ	d	M	I_{rel}
1	1	0	14.16	6.256	12	61.8	3	1	0	31.99	2.798	24	94.6	4	1	1	43.39	2.085	24	19.6
2	0	0	20.07	4.424	6	15.1	2	2	2	35.13	2.554	8	75.7	4	2	0	45.86	1.978	24	4.9
2	1	0	22.47	3.957	24	3.6	3	2	1	38.05	2.365	48	13.9	4	2	1	47.06	1.931	48	0.1
2	1	1	24.65	3.612	24	100.0	4	0	0	40.79	2.212	6	2.7	3	3	2	48.24	1.886	24	7.5
2	2	0	28.53	3.128	12	19.6	3	3	0	43.39	2.085	12	15.8							

Sodalite Octahydrate

SOD

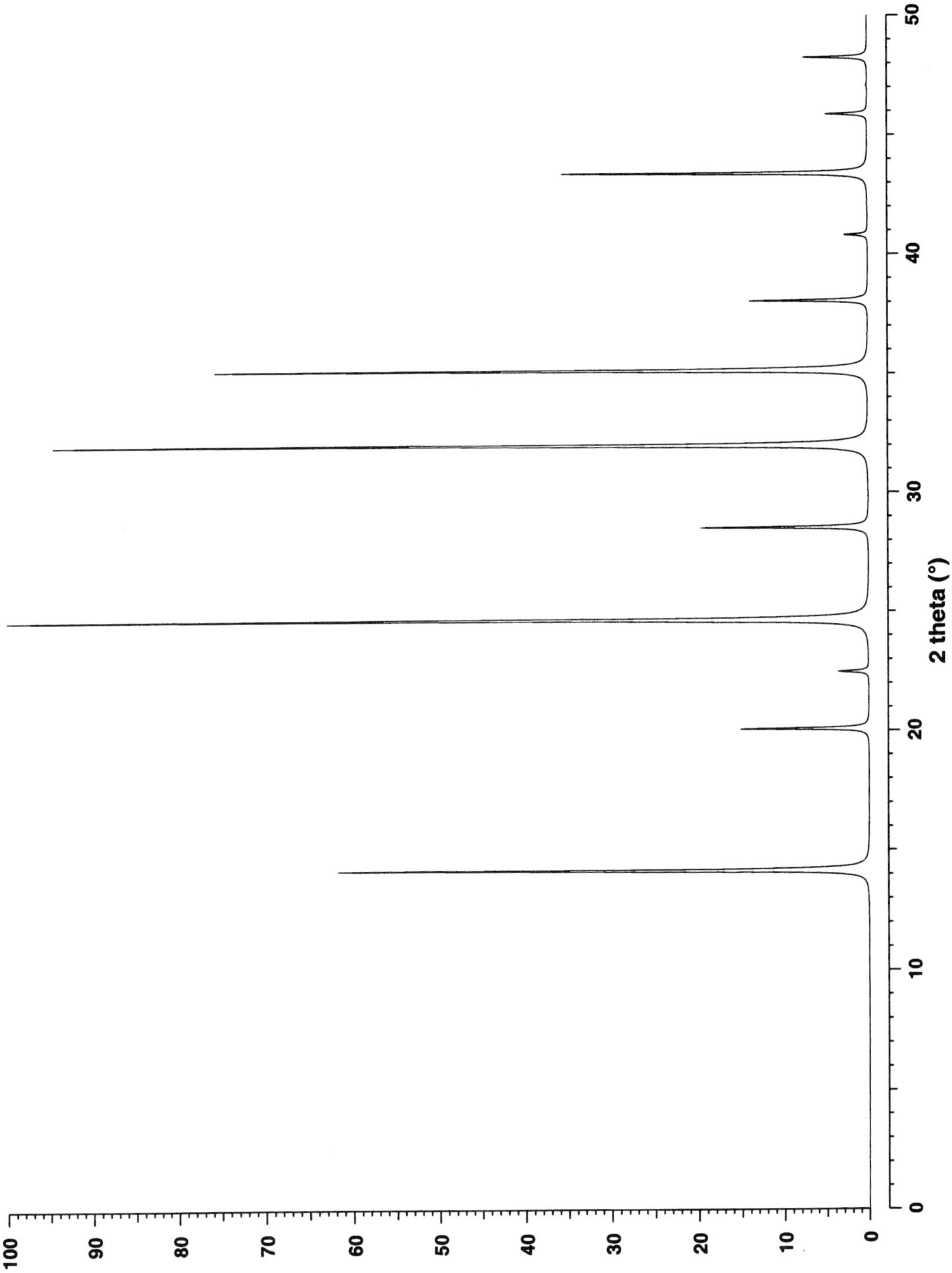

SOD Bicchulite

CHEMICAL COMPOSITION: $|Ca_8(H_2O)_8|$ $[Si_4Al_8O_{24}]$
Synthetic material

REFINED COMPOSITION: $|Ca_8(H_2O)_8|$ $[Si_4Al_8O_{24}]$

CRYSTAL DATA: $I\bar{4}3m$ (No. 217)
$a = 8.825$ Å $b = 8.825$ Å $c = 8.825$ Å
$\alpha = 90°$ $\beta = 90°$ $\gamma = 90°$
X-ray single crystal refinement, $R_w = 0.012$

REFERENCE: K. Sahl,
Z. Kristallogr. **152** 13–21 (1980).

h	k	l	2θ	d	M	I_{rel}	h	k	l	2θ	d	M	I_{rel}	h	k	l	2θ	d	M	I_{rel}
1	1	0	14.19	6.240	12	0.1	3	1	0	32.07	2.791	24	100.0	3	3	0	43.51	2.080	12	18.3
2	0	0	20.12	4.412	6	3.3	2	2	2	35.23	2.548	8	27.7	4	1	1	43.51	2.080	24	24.7
2	1	1	24.71	3.603	24	82.5	3	2	1	38.16	2.359	48	24.3	4	2	0	45.99	1.973	24	5.2
2	2	0	28.61	3.120	12	7.2	4	0	0	40.90	2.206	6	14.3	3	3	2	48.38	1.881	24	3.2

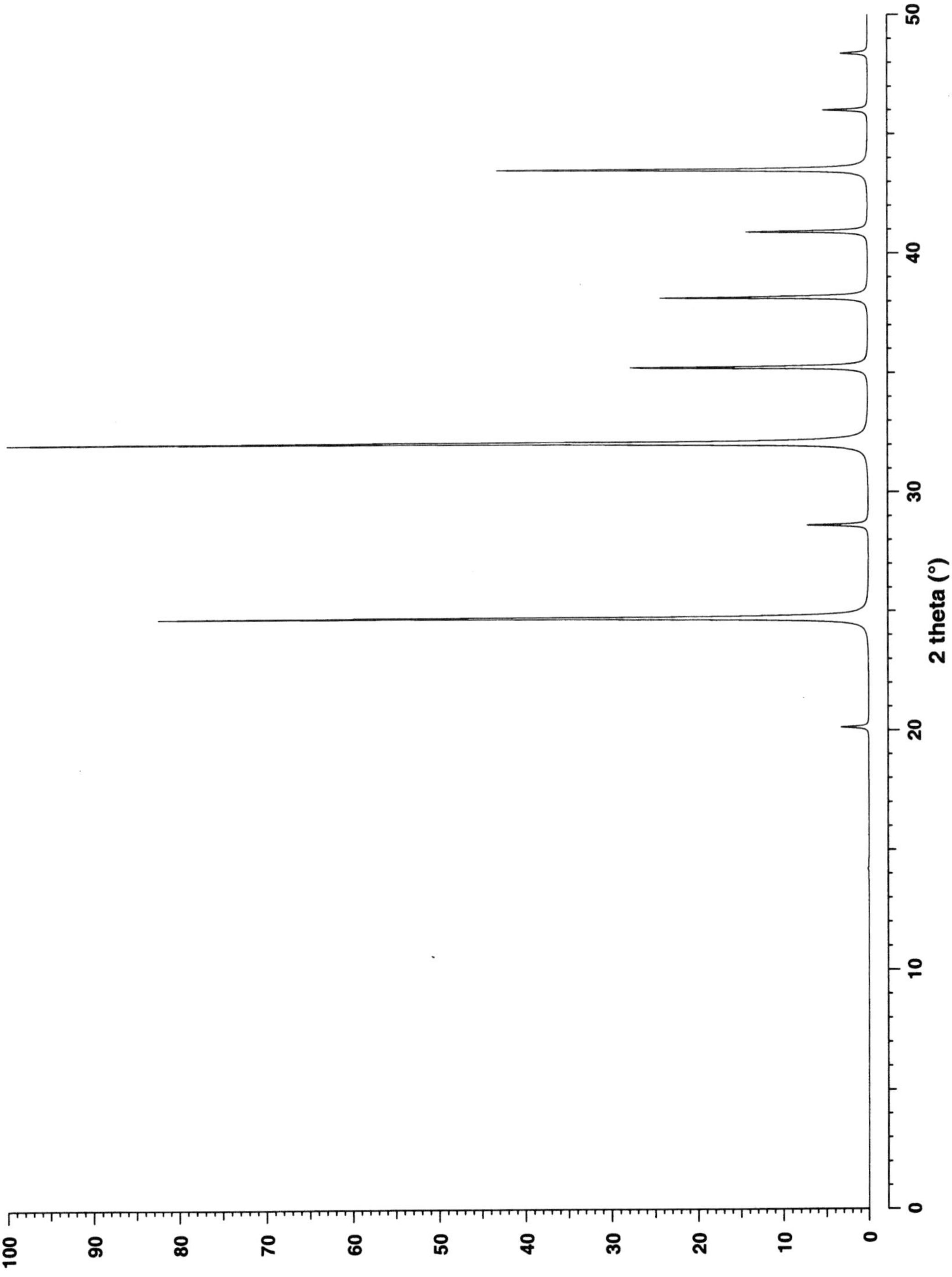
2 theta (°)
0
10
20
30
40
50
0
10
20
30
40
50
60
70
80
90
100

SOD Tugtupite

CHEMICAL COMPOSITION: $|Na_8Cl_2|\ [Si_8Al_2Be_2O_{24}]$
Ilimaussaq, South Greenland

REFINED COMPOSITION: $|Na_8Cl_2|\ [Si_8Al_2Be_2O_{24}]$

CRYSTAL DATA: $I\bar{4}$ (No. 82)
$a = 8.640$ Å $b = 8.640$ Å $c = 8.873$ Å
$\alpha = 90°$ $\beta = 90°$ $\gamma = 90°$
X-ray single crystal refinement, $R_w = 0.030$

REFERENCE: I. Hassan and H. D. Grundy, *Canadian Mineralogist* **29** 385–390 (1991).

h	k	l	2θ	d	M	I_{rel}	h	k	l	2θ	d	M	I_{rel}	h	k	l	2θ	d	M	I_{rel}
1	0	1	14.31	6.190	8	66.0	1	3	0	32.78	2.732	4	0.3	4	1	1	44.42	2.039	8	11.2
1	1	0	14.50	6.109	4	9.6	2	2	2	35.68	2.516	8	33.8	1	4	1	44.42	2.039	8	12.5
0	0	2	20.01	4.437	2	12.1	1	2	3	38.32	2.349	8	5.6	3	3	0	44.49	2.036	4	14.0
2	0	0	20.56	4.320	4	3.2	2	1	3	38.32	2.349	8	13.7	2	0	4	45.99	1.973	8	1.8
1	1	2	24.80	3.590	8	59.5	1	3	2	38.70	2.326	8	5.9	4	0	2	46.78	1.942	8	8.3
2	1	1	25.14	3.543	8	86.2	3	1	2	38.70	2.326	8	5.2	2	4	0	47.03	1.932	4	1.3
1	2	1	25.14	3.543	8	100.0	3	2	1	38.93	2.313	8	5.4	4	2	0	47.03	1.932	4	1.3
2	0	2	28.85	3.095	8	1.2	2	3	1	38.93	2.313	8	11.4	3	2	3	48.92	1.862	8	7.1
2	2	0	29.24	3.055	4	3.1	0	0	4	40.67	2.218	2	1.0	2	3	3	48.92	1.862	8	1.5
1	0	3	31.98	2.798	8	8.2	4	0	0	41.82	2.160	4	1.7	3	3	2	49.23	1.851	8	1.2
3	0	1	32.69	2.739	8	8.9	1	1	4	43.40	2.085	8	16.7							
3	1	0	32.78	2.732	4	5.9	3	0	3	43.88	2.063	8	22.6							

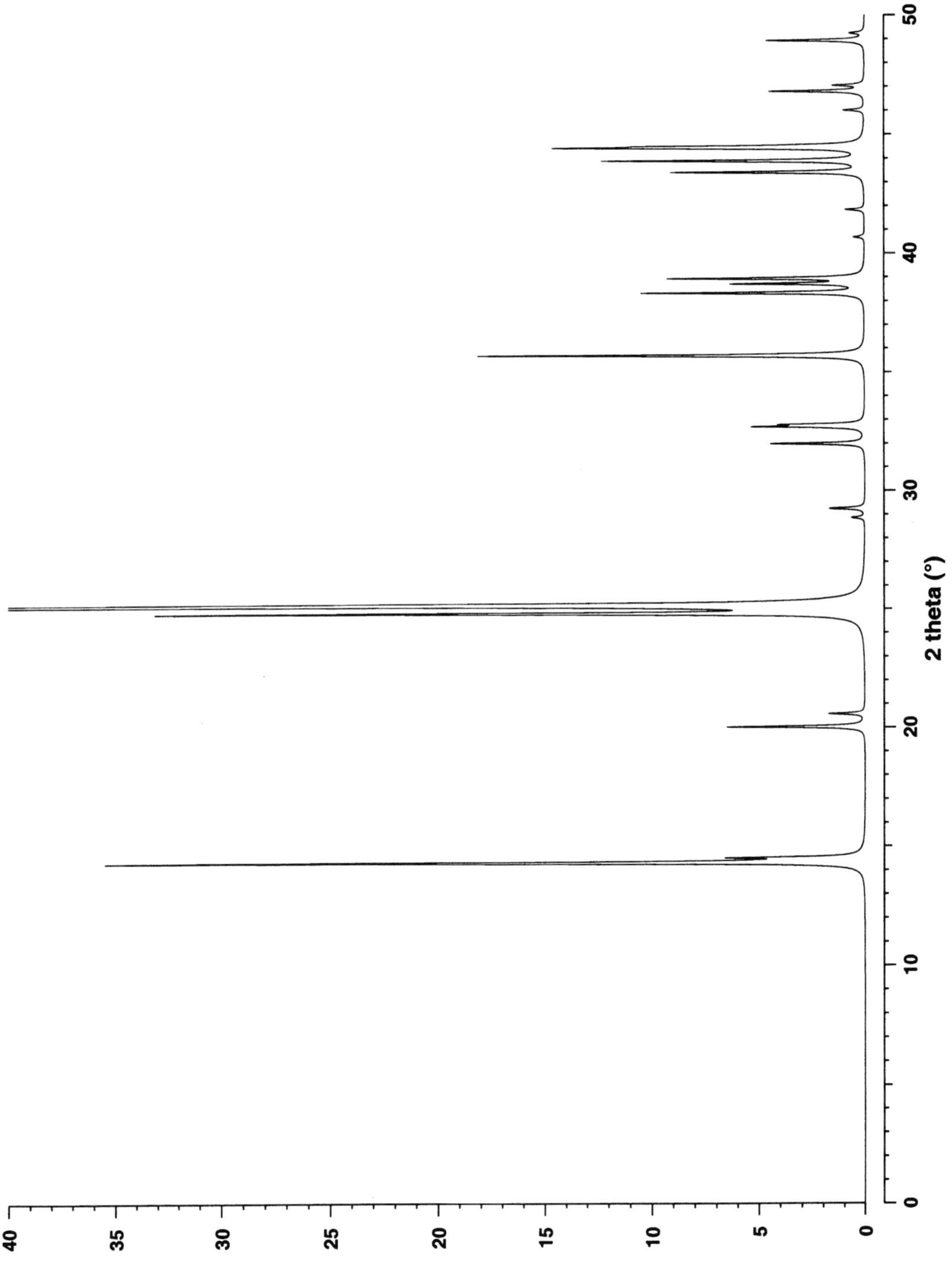
2 theta (°)
0
10
20
30
40
50
0
5
10
15
20
25
30
35
40

STF SSZ-35, Calcined

CHEMICAL COMPOSITION: $[Si_{16}O_{32}]$

REFINED COMPOSITION: $[Si_{16}O_{32}]$

CRYSTAL DATA: $P\bar{1}$ (No. 2)
$a = 11.4114$ Å $b = 11.5268$ Å $c = 7.3770$ Å
$\alpha = 94.661°$ $\beta = 96.206°$ $\gamma = 104.892°$
X-ray Rietveld refinement, $R_p = 0.0995$, $R_{wp} = 0.1188$

REFERENCE: P. Wagner, S. I. Zones, M. E. Davis and R. C. Medrud, *Angew. Chem., Int. ed.* **38** 1269–1272 (1999).

h	k	l	2θ	d	M	I_{rel}
0	1	0	7.99	11.068	2	98.4
1	0	0	8.09	10.929	2	100.0
1	-1	0	9.73	9.091	2	22.8
0	0	1	12.15	7.286	2	8.4
1	1	0	12.82	6.905	2	2.0
-1	0	1	13.67	6.476	2	6.3
0	-1	1	13.77	6.430	2	6.5
0	1	1	15.30	5.791	2	0.5
-1	1	1	15.45	5.737	2	1.6
1	0	1	15.49	5.719	2	0.2
1	-1	1	15.72	5.636	2	1.1
1	-2	0	15.89	5.577	2	0.5
0	2	0	16.02	5.534	2	0.2
2	-1	0	16.05	5.523	2	0.3
2	0	0	16.22	5.464	2	0.2
-2	0	1	18.95	4.683	2	1.6
0	-2	1	19.00	4.670	2	1.3
1	1	1	19.03	4.663	2	7.6
-2	1	1	19.39	4.578	2	9.3
2	-2	0	19.53	4.545	2	2.5
1	-2	1	19.60	4.530	2	14.2
1	2	0	19.83	4.477	2	4.2
2	1	0	19.96	4.449	2	6.8
-1	2	1	20.50	4.333	2	0.3
0	2	1	21.23	4.184	2	0.2
-1	-2	1	21.72	4.091	2	0.9
-2	-1	1	21.74	4.089	2	2.8
-2	2	1	22.87	3.888	2	6.7
2	-2	1	23.25	3.825	2	4.5
1	-3	0	23.29	3.820	2	2.3
3	-1	0	23.57	3.774	2	2.4
0	3	0	24.12	3.689	2	0.4
0	0	2	24.43	3.643	2	0.2
3	0	0	24.43	3.643	2	0.9
-1	0	2	24.70	3.604	2	4.0
1	2	1	24.82	3.587	2	4.5
0	-1	2	24.85	3.583	2	3.4
2	1	1	25.01	3.560	2	2.9
2	-3	0	25.24	3.528	2	3.2
3	-2	0	25.41	3.505	2	2.1
-3	1	1	25.47	3.497	2	2.1
1	-3	1	25.55	3.486	2	1.9
0	-3	1	25.80	3.453	2	2.2
2	2	0	25.80	3.453	2	6.8
-3	0	1	25.84	3.448	2	2.4
-1	1	2	26.19	3.403	2	1.6
1	-1	2	26.52	3.361	2	1.9
0	1	2	26.61	3.349	2	2.4
-2	-2	1	26.80	3.327	2	2.8
1	0	2	26.81	3.325	2	1.3
1	3	0	27.51	3.242	2	0.3
-2	0	2	27.55	3.238	2	4.4
-3	2	1	27.60	3.231	2	0.2
0	-2	2	27.75	3.215	2	4.1
2	-3	1	27.85	3.203	2	0.9
0	3	1	28.33	3.150	2	0.4
-3	-1	1	28.61	3.120	2	0.4
-2	-1	2	29.18	3.060	2	0.4
1	1	2	29.44	3.034	2	1.8
3	-3	0	29.48	3.030	2	3.9
-1	2	2	29.91	2.988	2	1.9
2	2	1	30.32	2.948	2	3.2
2	-1	2	30.44	2.937	2	1.6
0	2	2	30.88	2.896	2	0.3
1	-4	0	31.14	2.872	2	0.3
4	-1	0	31.55	2.835	2	0.4
3	-3	1	32.20	2.780	2	0.6
0	4	0	32.35	2.767	2	1.2
4	-2	0	32.42	2.761	2	0.4
4	0	0	32.78	2.732	2	0.9
-2	-2	2	32.85	2.727	2	0.5
3	2	0	32.86	2.726	2	0.2
-2	-3	1	33.19	2.699	2	0.3
0	-4	1	33.31	2.690	2	0.7
-4	0	1	33.45	2.679	2	2.0
-4	2	1	33.79	2.652	2	0.2
2	-4	1	33.89	2.645	2	1.2
-1	4	1	34.50	2.600	2	0.6
-2	4	1	34.98	2.565	2	0.2
2	-3	2	35.01	2.563	2	0.7
3	-4	0	35.07	2.559	2	0.4
4	-1	1	35.20	2.549	2	0.4
-1	3	2	35.22	2.548	2	1.5
4	-3	0	35.24	2.546	2	0.3
-2	3	2	35.82	2.507	2	0.2
3	-1	2	35.93	2.499	2	0.4
0	4	1	36.00	2.494	2	0.9
4	0	1	36.64	2.453	2	0.8
-1	0	3	36.82	2.441	2	0.3
-4	3	1	36.84	2.439	2	0.2
3	2	1	36.97	2.431	2	0.2
0	0	3	37.01	2.429	2	0.6
-1	-1	3	37.29	2.412	2	1.6
1	-4	2	38.27	2.352	2	0.2
-2	0	3	38.49	2.339	2	0.2
2	2	2	38.61	2.332	2	0.2
-3	-3	1	39.17	2.300	2	0.5
-4	2	2	39.36	2.289	2	0.2
-1	-2	3	39.53	2.280	2	0.3
-2	-4	1	40.38	2.234	2	0.2
-1	-4	2	40.63	2.220	2	0.2
0	5	0	40.76	2.214	2	0.2
-5	2	1	40.88	2.207	2	0.2
5	0	0	41.30	2.186	2	0.3
-4	4	1	41.40	2.181	2	0.3
-5	0	1	41.51	2.175	2	0.3
3	-5	0	41.67	2.167	2	0.4
-2	2	3	41.85	2.159	2	0.3
4	-4	1	41.85	2.158	2	0.5
5	-3	0	42.02	2.150	2	0.4
2	-2	3	42.52	2.126	2	0.2
3	3	1	43.01	2.103	2	2.5
3	-5	1	43.17	2.095	2	0.4
0	5	1	44.09	2.054	2	0.6
-3	2	3	44.14	2.052	2	0.4
-5	1	2	44.63	2.030	2	0.3
1	-5	2	44.79	2.024	2	0.4
2	-3	3	44.84	2.021	2	0.4
5	0	1	44.86	2.020	2	0.3
-2	3	3	45.82	1.980	2	0.2
-4	-3	1	46.07	1.970	2	0.5
-4	1	3	46.48	1.954	2	0.4
3	-2	3	46.70	1.945	2	0.2
-5	4	1	46.72	1.944	2	0.7
4	-5	1	47.00	1.933	2	0.7
0	3	3	47.07	1.930	2	0.2
-3	3	3	47.55	1.912	2	0.2
-4	2	3	47.89	1.899	2	0.3
-5	-2	1	48.40	1.881	2	0.3
2	-6	1	48.41	1.880	2	0.3
3	-3	3	48.46	1.879	2	0.3
2	2	3	49.10	1.855	2	0.2
-4	-3	2	49.25	1.850	2	0.4
-1	-1	4	49.74	1.833	2	0.6
3	4	1	49.90	1.828	2	0.3

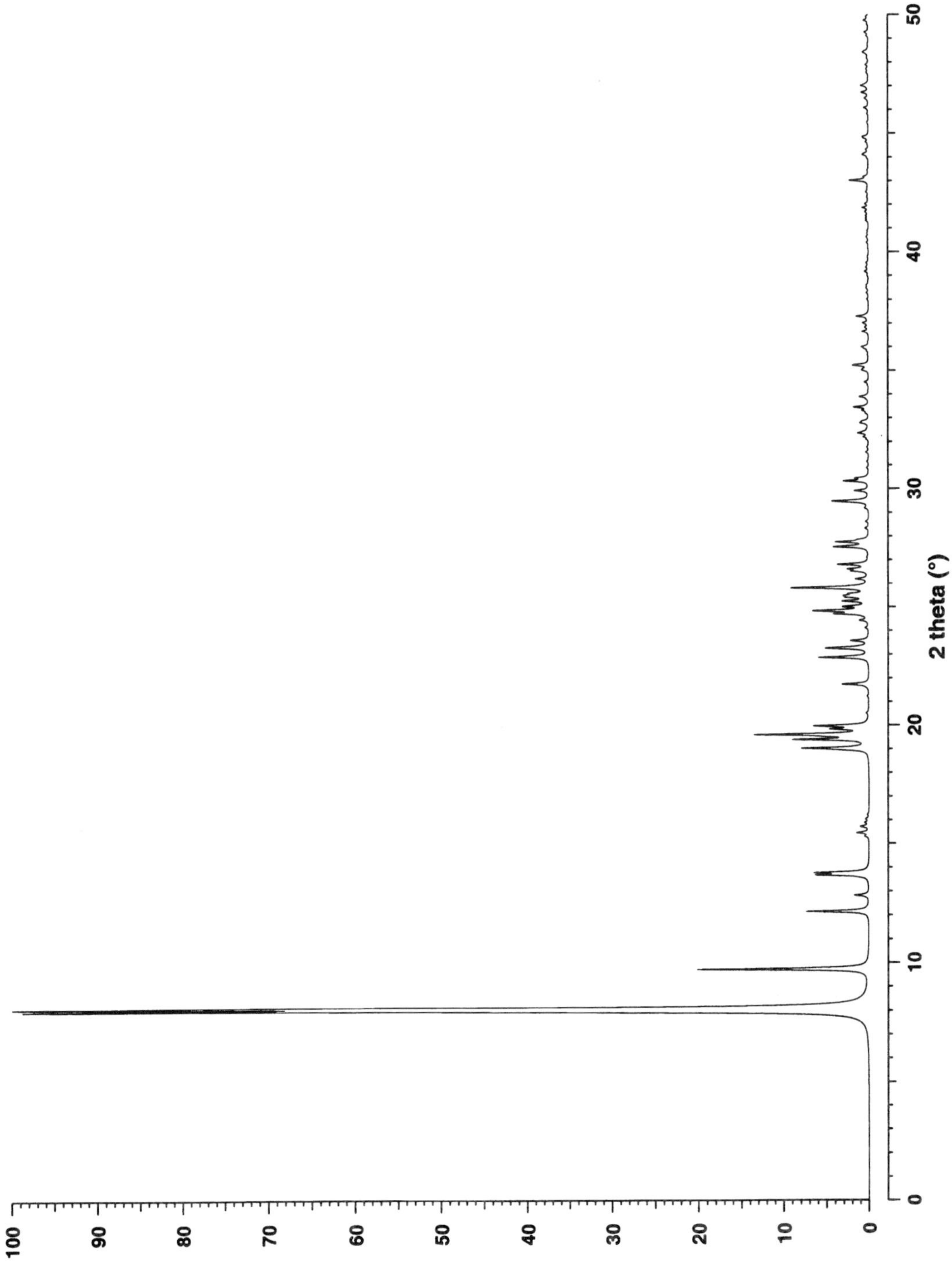
2 theta (°)
0
10
20
30
40
50
0
10
20
30
40
50
60
70
80
90
100

STI Stilbite

CHEMICAL COMPOSITION: $|Na_{1.28}Ca_{4.18}Mg_{0.18}(H_2O)_{34.1}|$ $[Si_{25.7}Al_{10.3}O_{72}]$
Iceland

REFINED COMPOSITION: $|Na_{1.76}Ca_4(H_2O)_{29.4}|$ $[Si_{25.68}Al_{10.32}O_{72}]$

CRYSTAL DATA: $C1\,2/m\,1$ (No. 12) unique axis **b**, cell choice 1
a = 13.64 Å b = 18.24 Å c = 11.27 Å
α = 90° β = 128.0° γ = 90°
X-ray single crystal refinement, R = 0.123

REFERENCE: E. Galli,
Acta Cryst. **B27** 833–841 (1971).

h	k	l	2θ	d	M	I_{rel}	h	k	l	2θ	d	M	I_{rel}	h	k	l	2θ	d	M	I_{rel}
-1	1	1	9.46	9.349	4	0.6	1	3	2	29.93	2.985	4	14.6	-6	2	3	40.91	2.206	4	1.6
0	2	0	9.70	9.120	2	100.0	0	0	3	30.19	2.960	2	5.5	3	1	2	41.27	2.187	4	1.1
0	0	1	9.96	8.881	2	24.7	0	6	1	31.09	2.876	4	2.3	-5	5	3	41.60	2.171	4	0.5
-2	0	1	12.98	6.819	2	4.2	0	2	3	31.78	2.816	4	4.4	-5	5	2	41.71	2.165	4	1.4
0	2	1	13.92	6.363	4	4.0	-3	5	1	31.89	2.806	4	6.0	0	2	4	41.88	2.157	4	2.7
-2	2	1	16.23	5.461	4	2.5	-3	1	4	32.15	2.784	4	1.2	0	6	3	42.63	2.121	4	5.7
-2	0	2	16.28	5.444	2	5.3	-2	6	1	32.24	2.777	4	26.1	-2	0	5	42.66	2.119	2	1.7
2	0	0	16.49	5.374	2	4.8	3	1	1	32.57	2.749	4	2.3	-2	8	2	43.01	2.103	4	4.1
-1	3	1	16.70	5.308	4	7.7	-2	0	4	32.82	2.729	2	1.6	2	8	0	43.10	2.099	4	0.6
1	3	0	16.75	5.292	4	3.1	-4	0	4	32.90	2.722	2	19.0	-3	3	5	43.24	2.092	4	0.6
-1	1	2	16.96	5.229	4	2.6	4	0	0	33.34	2.687	2	2.2	-1	7	3	43.53	2.079	4	1.4
-2	2	2	18.98	4.674	4	39.7	-4	4	1	34.46	2.602	4	1.8	1	7	2	43.64	2.074	4	0.5
2	2	0	19.17	4.630	4	41.4	2	2	2	34.62	2.591	4	9.6	-2	2	5	43.85	2.064	4	4.3
0	4	0	19.47	4.560	2	2.4	4	2	0	34.80	2.578	4	3.3	-1	5	4	43.90	2.063	4	2.0
0	0	2	20.00	4.440	2	5.8	-3	5	3	34.87	2.573	4	9.9	-5	5	4	44.03	2.057	4	1.0
-3	1	2	20.68	4.294	4	28.0	-3	3	4	35.10	2.556	4	5.1	1	5	3	44.05	2.056	4	2.9
-3	1	1	20.81	4.268	4	12.2	3	5	0	35.11	2.556	4	8.2	-5	3	5	44.18	2.050	4	6.3
-1	3	2	21.88	4.061	4	61.9	1	7	0	35.45	2.532	4	2.4	2	2	3	44.20	2.049	4	1.1
0	4	1	21.91	4.057	4	35.8	3	3	1	35.49	2.529	4	2.1	-5	5	1	44.35	2.042	4	2.0
1	3	1	22.00	4.039	4	60.4	0	6	2	35.80	2.508	4	4.9	-6	4	3	44.54	2.034	4	2.2
0	2	2	22.27	3.992	4	14.5	-1	5	3	35.83	2.506	4	0.5	-2	6	4	44.62	2.031	4	1.5
-2	4	1	23.47	3.791	4	2.4	0	4	3	36.18	2.483	4	4.6	0	8	2	44.68	2.028	4	1.9
-2	0	3	23.76	3.744	2	12.3	-1	1	4	36.26	2.477	4	0.5	5	3	0	44.71	2.027	4	7.7
2	0	1	24.06	3.699	2	13.2	-5	3	3	36.39	2.469	4	3.1	4	6	0	45.03	2.013	4	0.4
-3	1	3	25.02	3.560	4	0.5	-5	1	4	36.41	2.467	4	3.3	-6	0	1	45.04	2.013	2	1.2
-3	3	1	25.02	3.559	4	1.8	1	1	3	36.44	2.466	4	2.6	4	4	1	45.43	1.997	4	2.4
-2	4	2	25.48	3.496	4	2.2	-5	3	2	36.52	2.461	4	4.2	0	4	4	45.43	1.996	4	0.7
2	4	0	25.62	3.477	4	6.2	-5	1	1	36.79	2.443	4	9.2	-6	2	5	45.69	1.986	4	0.9
1	5	0	25.79	3.454	4	2.4	-2	6	3	38.13	2.360	4	3.9	-6	4	2	45.88	1.978	4	2.1
2	2	1	26.00	3.427	4	4.0	-1	7	2	38.28	2.351	4	2.1	-2	8	3	46.64	1.947	4	0.5
-4	0	2	26.13	3.410	2	22.4	2	6	1	38.33	2.349	4	1.8	-7	1	4	47.12	1.929	4	0.9
-1	1	3	26.30	3.389	4	11.3	1	7	1	38.35	2.347	4	2.1	-2	4	5	47.29	1.922	4	1.1
1	1	2	26.47	3.368	4	8.7	-2	4	4	38.44	2.342	4	0.7	-1	9	2	47.88	1.900	4	0.9
-4	0	3	27.90	3.197	2	16.4	2	4	2	38.73	2.325	4	0.5	1	9	1	47.94	1.898	4	1.8
-4	2	2	27.94	3.194	4	25.1	4	4	0	38.90	2.315	4	3.6	-4	8	2	48.00	1.895	4	2.6
0	4	2	28.05	3.181	4	4.7	-5	3	4	39.09	2.304	4	0.8	3	5	2	48.25	1.886	4	1.9
-4	0	1	28.16	3.169	2	14.6	1	3	3	39.11	2.303	4	2.3	-5	7	3	48.54	1.875	4	4.6
-3	3	3	28.64	3.116	4	9.3	-6	0	3	39.65	2.273	2	3.8	-4	0	6	48.64	1.872	2	0.6
3	3	0	28.92	3.087	4	3.7	-4	0	5	40.07	2.250	2	1.4	-5	5	5	48.70	1.870	4	3.6
0	6	0	29.38	3.040	2	4.8	-3	7	1	40.25	2.241	4	1.4	-5	1	6	48.95	1.861	4	0.6
-1	5	2	29.45	3.033	4	24.4	-3	5	4	40.45	2.230	4	2.4	5	5	0	49.20	1.852	4	1.3
1	5	1	29.54	3.024	4	28.1	4	0	1	40.62	2.221	2	1.0	-4	8	1	49.23	1.851	4	0.8
-4	2	3	29.60	3.017	4	10.2	0	0	4	40.63	2.220	2	2.0	-7	3	4	49.31	1.848	4	1.4
-1	3	3	29.78	3.000	4	10.4	-3	1	5	40.79	2.212	4	1.7	-3	9	2	49.48	1.842	4	0.8
-4	2	1	29.85	2.993	4	5.1	3	5	1	40.80	2.212	4	1.3	5	1	1	49.65	1.836	4	1.8

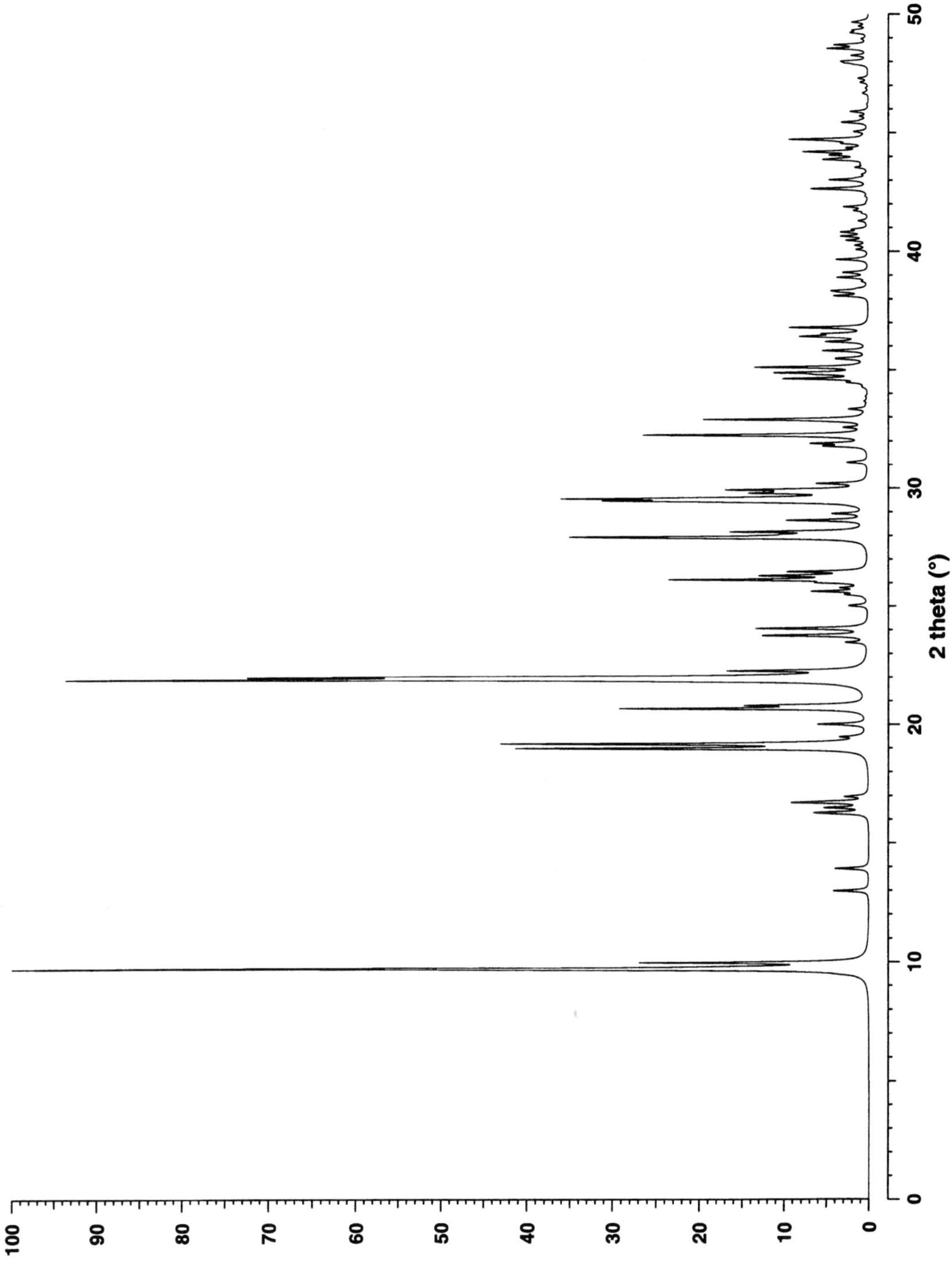
2 theta (°)
0
10
20
30
40
50
100
90
80
70
60
50
40
30
20
10
0

STI Stellerite

CHEMICAL COMPOSITION: $|Ca_{7.56}Mg_{0.06}Sr_{0.02}Na_{0.34}K_{0.28}(H_2O)_{56.2}|$ $[Si_{56.04}Al_{15.86}Fe_{0.12}O_{144}]$
Villanova-Monteleone, Sardinia, Italy

REFINED COMPOSITION: $|Ca_8(H_2O)_{55.52}|$ $[Si_{56}Al_{16}O_{144}]$

CRYSTAL DATA: $Fmmm$ (No. 69)
$a = 13.599$ Å $b = 18.222$ Å $c = 17.863$ Å
$\alpha = 90°$ $\beta = 90°$ $\gamma = 90°$
X-ray single crystal refinement, $R_w = 0.082$

REFERENCE: E. Galli and A. Alberti,
Bull. Soc. Fr. Minéral. Crystallogr. **98** 11–18 (1975).

h	k	l	2θ	d	M	I_{rel}
1	1	1	9.51	9.304	8	2.7
0	2	0	9.71	9.111	2	100.0
0	0	2	9.90	8.932	2	36.2
2	0	0	13.02	6.799	2	0.2
0	2	2	13.88	6.378	4	1.5
2	2	0	16.27	5.449	4	6.3
2	0	2	16.38	5.410	4	14.7
1	3	1	16.74	5.297	8	7.1
1	1	3	16.97	5.225	8	3.4
2	2	2	19.08	4.652	8	80.2
0	4	0	19.49	4.556	2	2.8
0	0	4	19.88	4.466	2	3.2
3	1	1	20.80	4.271	8	35.5
1	3	3	21.90	4.058	8	90.3
0	4	2	21.90	4.058	4	31.4
0	2	4	22.17	4.010	4	12.7
2	4	0	23.51	3.785	4	1.5
2	0	4	23.84	3.733	4	21.8
3	3	1	25.01	3.560	8	0.6
3	1	3	25.17	3.538	8	0.6
2	4	2	25.56	3.485	8	7.6
2	2	4	25.79	3.454	8	1.5
1	5	1	25.79	3.454	8	0.2
4	0	0	26.21	3.400	2	16.0
1	1	5	26.25	3.395	8	15.0
0	4	4	27.98	3.189	4	3.0
4	2	0	28.01	3.185	4	21.9
4	0	2	28.08	3.177	4	24.1
3	3	3	28.79	3.101	8	11.1
0	6	0	29.41	3.037	2	5.2
1	5	3	29.48	3.030	8	45.7
1	3	5	29.75	3.003	8	22.7
4	2	2	29.78	3.000	8	11.7
0	0	6	30.01	2.977	2	3.4
2	4	4	30.97	2.887	8	0.1
0	6	2	31.10	2.875	4	3.0
0	2	6	31.62	2.830	4	2.5
3	5	1	31.90	2.805	8	1.6
3	1	5	32.28	2.773	8	3.8
2	6	0	32.28	2.773	4	18.2
4	0	4	33.12	2.705	4	13.5
5	1	1	33.69	2.660	8	0.6
2	6	2	33.85	2.648	8	0.1
2	2	6	34.32	2.613	8	6.6
4	4	2	34.41	2.606	8	3.2
4	2	4	34.59	2.593	8	2.3
3	5	3	35.00	2.564	8	13.8
3	3	5	35.23	2.547	8	4.5
1	7	1	35.47	2.531	8	1.3
0	6	4	35.75	2.511	4	3.0
1	5	5	35.81	2.507	8	0.4
0	4	6	36.04	2.492	4	4.2
1	1	7	36.15	2.485	8	2.7
5	3	1	36.55	2.459	8	0.1
5	1	3	36.66	2.451	8	6.1
2	6	4	38.20	2.356	8	3.5
1	7	3	38.31	2.349	8	3.1
2	4	6	38.47	2.340	8	0.5
4	4	4	38.71	2.326	8	1.3
1	3	7	38.85	2.318	8	2.3
6	0	0	39.77	2.266	2	3.7
4	6	0	39.80	2.265	4	0.1
4	0	6	40.26	2.240	4	2.1
3	7	1	40.27	2.240	8	0.6
0	0	8	40.39	2.233	2	2.0
3	5	5	40.58	2.223	8	2.4
3	1	7	40.88	2.207	8	2.8
6	2	0	41.04	2.199	4	1.8
4	6	2	41.11	2.195	8	0.3
0	2	8	41.64	2.169	4	1.6
5	1	5	42.04	2.149	8	1.3
0	6	6	42.52	2.126	4	3.0
2	0	8	42.62	2.121	4	0.8
2	8	2	43.09	2.099	8	2.3
3	3	7	43.33	2.088	8	0.8
1	7	5	43.53	2.079	8	1.1
2	2	8	43.82	2.066	8	3.3
1	5	7	43.82	2.066	8	5.6
5	5	3	44.25	2.047	8	1.8
5	3	5	44.44	2.039	8	11.4
6	4	0	44.65	2.029	4	0.7
2	6	6	44.66	2.029	8	0.7
0	8	4	44.66	2.029	4	2.0
6	0	4	44.84	2.021	4	0.7
4	6	4	44.87	2.020	8	0.3
4	4	6	45.11	2.010	8	1.1
0	4	8	45.22	2.005	4	0.4
6	4	2	45.86	1.979	8	1.1
6	2	4	46.00	1.973	8	1.4
1	1	9	46.51	1.953	8	0.8
2	8	4	46.72	1.944	8	0.1
3	5	7	47.92	1.898	8	0.7
1	9	3	47.93	1.898	8	1.7
4	8	0	48.08	1.892	4	3.1
5	7	1	48.68	1.870	8	1.7
1	3	9	48.73	1.869	8	0.1
4	0	8	48.79	1.866	4	0.2
5	5	5	48.95	1.861	8	2.8
5	1	7	49.21	1.851	8	2.5
4	8	2	49.22	1.851	8	0.8
6	4	4	49.33	1.847	8	0.2
7	3	1	49.52	1.841	8	0.5
3	9	1	49.57	1.839	8	0.4
7	1	3	49.61	1.837	8	0.3
4	2	8	49.88	1.828	8	0.4

Stellerite STI

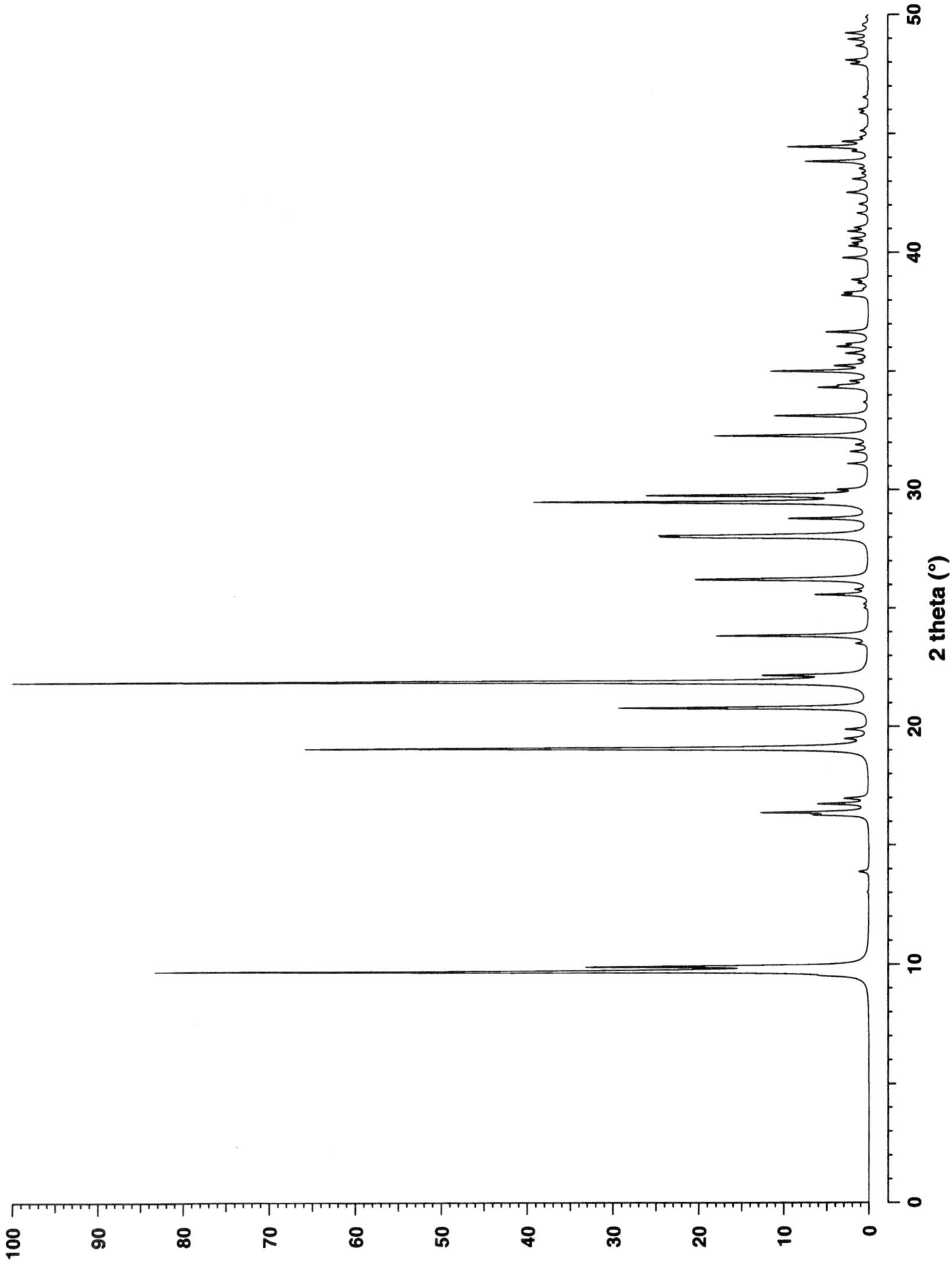

STI Barrerite

CHEMICAL COMPOSITION: $|Na_{10.9}K_{2.1}Mg_{0.3}Ca_{1.7}(H_2O)_{51.6}|$ $[Si_{55.6}Al_{16.4}O_{144}]$
Capo Pula, Sardinia, Italy

REFINED COMPOSITION: $|Na_{9.812}K_{1.912}Mg_{0.032}Ca_{1.52}(H_2O)_{58.56}|$ $[Si_{55.58}Al_{16.42}O_{144}]$

CRYSTAL DATA: *Amma* (No. 63) **cab** setting
a = 13.643 Å b = 18.200 Å c = 17.842 Å
α = 90° β = 90° γ = 90°
X-ray single crystal refinement, R_w = 0.067

REFERENCE: E. Galli and A. Alberti,
Bull. Soc. Fr. Minéral. Crystallogr. **98** 331–340 (1975).

h	k	l	2θ	d	M	I_{rel}
0	1	1	6.94	12.741	4	3.1
1	1	1	9.50	9.312	8	5.0
0	2	0	9.72	9.100	2	100.0
0	0	2	9.91	8.921	2	48.5
1	0	2	11.85	7.466	4	0.5
2	0	0	12.98	6.821	2	5.6
2	1	1	14.73	6.014	8	0.4
0	3	1	15.43	5.744	4	1.0
2	2	0	16.24	5.458	4	0.4
2	0	2	16.36	5.419	4	2.2
1	3	1	16.75	5.294	8	4.7
1	1	3	16.98	5.223	8	9.6
2	2	2	19.06	4.656	8	41.0
0	4	0	19.51	4.550	2	1.1
0	0	4	19.90	4.460	2	2.6
2	3	1	20.21	4.394	8	0.2
3	1	1	20.74	4.283	8	26.5
0	3	3	20.92	4.247	4	0.6
1	0	4	20.95	4.240	4	0.2
1	3	3	21.92	4.055	8	69.6
0	4	2	21.93	4.053	4	17.0
3	0	2	21.94	4.052	4	0.2
0	2	4	22.19	4.005	4	8.7
1	4	2	22.89	3.885	8	0.4
1	2	4	23.14	3.843	8	0.3
2	4	0	23.50	3.785	4	0.2
2	0	4	23.83	3.733	4	28.1
2	3	3	24.69	3.605	8	0.5
3	3	1	24.97	3.565	8	1.8
0	1	5	25.44	3.502	4	0.4
2	4	2	25.56	3.485	8	16.5
2	2	4	25.79	3.454	8	0.3
4	0	0	26.13	3.411	2	18.5
1	1	5	26.27	3.392	8	13.2
4	1	1	27.06	3.295	8	0.2
4	2	0	27.94	3.194	4	18.9
4	0	2	28.01	3.186	4	19.7
0	4	4	28.01	3.185	4	3.9
3	0	4	28.02	3.184	4	0.2
2	5	1	28.23	3.161	8	1.6
2	1	5	28.65	3.115	8	0.2
3	3	3	28.76	3.104	8	15.1
0	6	0	29.45	3.033	2	3.1
1	5	3	29.51	3.027	8	56.1
3	4	2	29.52	3.026	8	0.8
4	2	2	29.71	3.007	8	16.5
3	2	4	29.72	3.006	8	0.4
1	3	5	29.78	3.000	8	25.0
0	0	6	30.05	2.974	2	4.2
4	1	3	30.61	2.920	8	0.2
2	4	4	30.98	2.886	8	0.5
0	6	2	31.14	2.872	4	0.8
0	2	6	31.65	2.827	4	2.8
1	6	2	31.84	2.810	8	0.8
3	5	1	31.89	2.806	8	2.7
3	1	5	32.26	2.775	8	1.7
2	6	0	32.30	2.772	4	15.7
4	4	0	32.82	2.729	4	0.2
4	0	4	33.06	2.709	4	10.7
5	1	1	33.59	2.668	8	1.5
2	6	2	33.87	2.647	8	0.8
2	2	6	34.34	2.611	8	7.3
4	4	2	34.36	2.610	8	1.0
4	2	4	34.54	2.597	8	3.5
0	7	1	34.87	2.573	4	0.6
3	5	3	34.99	2.564	8	11.0
3	3	5	35.23	2.548	8	2.8
1	7	1	35.51	2.528	8	0.6
0	1	7	35.56	2.524	4	0.2
0	6	4	35.80	2.508	4	4.1
0	4	6	36.08	2.489	4	3.7
1	1	7	36.19	2.482	8	0.4
4	5	1	36.45	2.465	8	0.3
5	3	1	36.45	2.465	8	0.6
5	1	3	36.57	2.457	8	5.3
1	4	6	36.70	2.449	8	0.3
4	1	5	36.78	2.443	8	0.4
2	5	5	37.68	2.387	8	0.2
2	6	4	38.23	2.354	8	2.2
1	7	3	38.35	2.347	8	1.5
2	4	6	38.50	2.338	8	0.3
4	4	4	38.68	2.328	8	0.6
5	0	4	38.68	2.328	4	0.4
1	3	7	38.89	2.316	8	0.2
6	0	0	39.64	2.274	2	2.7
4	6	0	39.77	2.267	4	0.5
4	0	6	40.23	2.241	4	1.5
3	7	1	40.27	2.239	8	0.3
0	0	8	40.44	2.230	2	2.2
3	5	5	40.58	2.223	8	1.5
3	1	7	40.89	2.207	8	1.4
6	2	0	40.91	2.206	4	0.9
0	8	2	40.94	2.204	4	0.3
3	4	6	41.35	2.184	8	0.6
4	2	6	41.49	2.176	8	0.2
5	5	1	41.68	2.167	8	0.3
0	2	8	41.70	2.166	4	1.1
5	1	5	41.98	2.152	8	2.1
6	2	2	42.20	2.142	8	0.2
0	6	6	42.57	2.123	4	3.6
2	0	8	42.65	2.120	4	0.4
0	7	5	43.04	2.101	4	0.2
2	8	2	43.12	2.098	8	1.3
3	3	7	43.34	2.088	8	1.6
1	7	5	43.58	2.077	8	0.7
5	4	4	43.68	2.072	8	0.2
2	2	8	43.85	2.065	8	2.0
1	5	7	43.87	2.064	8	2.3
4	7	1	44.09	2.054	8	0.3
5	5	3	44.19	2.050	8	0.8
5	3	5	44.38	2.041	8	7.8
6	4	0	44.54	2.034	4	1.6
2	6	6	44.69	2.028	8	0.7
0	8	4	44.72	2.027	4	1.2
6	0	4	44.73	2.026	4	0.2
4	6	4	44.85	2.021	8	0.7
4	4	6	45.09	2.011	8	1.4
0	9	1	45.12	2.009	4	0.3
0	4	8	45.28	2.003	4	0.2
3	8	2	45.74	1.984	8	0.3
6	4	2	45.75	1.983	8	0.7
1	4	8	45.79	1.981	8	0.3
6	2	4	45.89	1.977	8	0.3
1	9	3	47.98	1.896	8	1.0
2	1	9	48.05	1.893	8	0.2
4	8	0	48.07	1.893	4	1.6
5	7	1	48.64	1.872	8	1.1
5	5	5	48.90	1.862	8	4.5
5	1	7	49.17	1.853	8	1.1
4	8	2	49.21	1.851	8	0.2
6	4	4	49.23	1.851	8	0.6
7	3	1	49.38	1.846	8	0.3
7	1	3	49.47	1.843	8	0.9
3	9	1	49.60	1.838	8	0.2
4	2	8	49.87	1.829	8	0.7

Barrerite STI

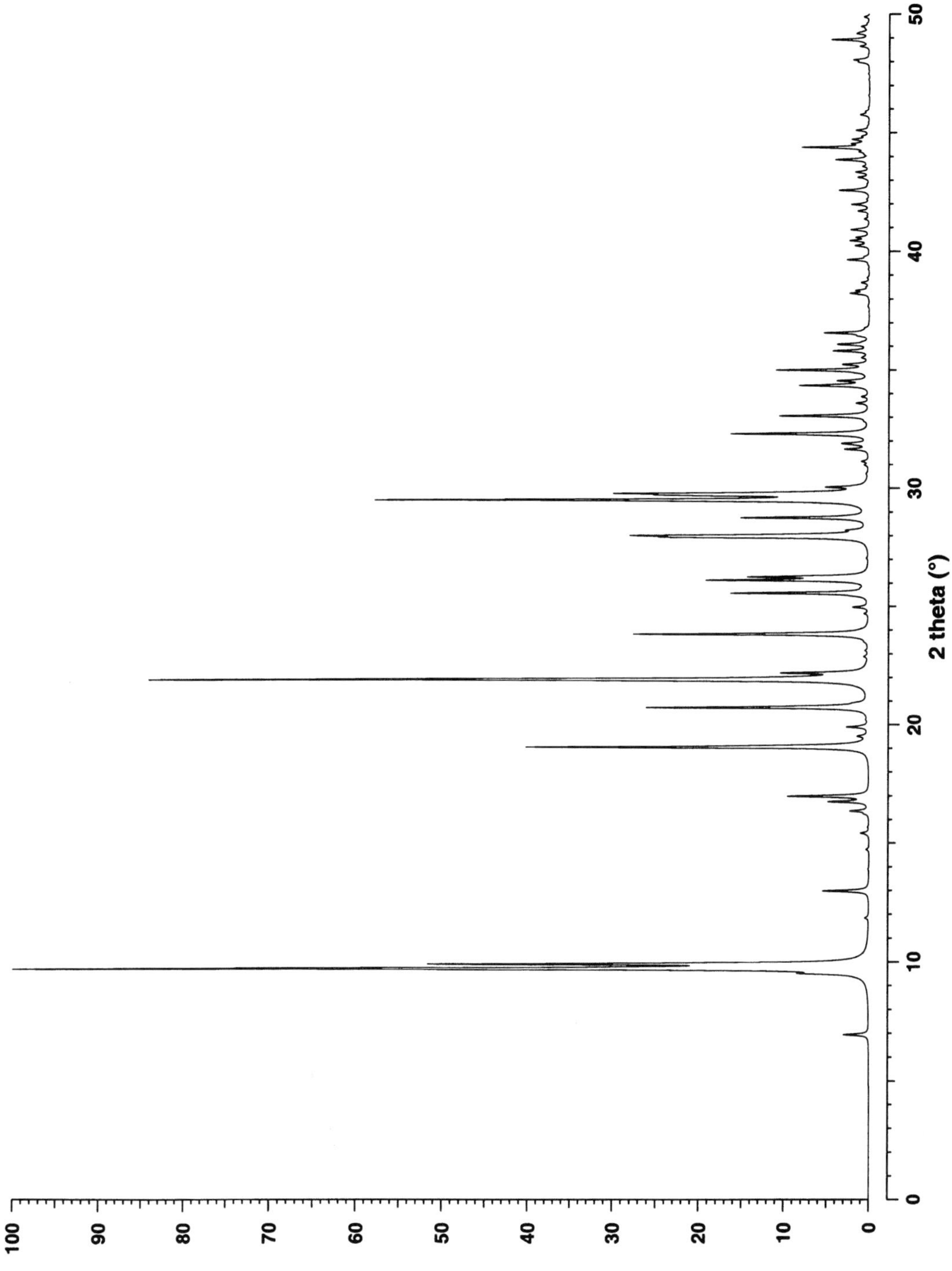

STT SSZ-23

CHEMICAL COMPOSITION: $|(C_{13}H_{24}N)_{4.1}F_{3.3}(OH)_{0.8}(H_2O)_{1.6}|$ $[Si_{64}O_{128}]$
$C_{13}H_{24}NF$ = *N,N,N*-trimethyl-1-adamantammonium fluoride

REFINED COMPOSITION: $|C_{52}N_4F_{2.84}|$ $[Si_{64}O_{128}]$

CRYSTAL DATA: $P1\,2_1/n\,1$ (No. 14) unique axis **b**, cell choice 2
a = 12.9594 Å b = 21.7919 Å c = 13.5980 Å
α = 90.0° β = 101.855° γ = 90.0°
X-ray single crystal refinement, $R_{F_{obs}}$ = 0.0811, wR_{F^2} = 0.1657

REFERENCE: M. A. Camblor, M.-J. Diaz-Cabanas, J. Perez-Pariente, S. J. Teat, W. Clegg, I. J. Shannon, P. Lightfoot, P. A. Wright and R. E. Morris, *Angew. Chem., Int. Ed.* **37** 2122–2126 (1998).

h	k	l	2θ	d	M	I_{rel}
1	1	0	8.07	10.962	4	11.0
0	2	0	8.11	10.896	2	100.0
-1	0	1	8.59	10.298	2	25.2
-1	1	1	9.50	9.311	4	70.5
0	2	1	10.49	8.431	4	17.5
1	0	1	10.58	8.363	2	44.6
1	2	0	10.70	8.265	4	7.1
0	0	2	13.31	6.654	2	2.0
0	3	1	13.89	6.376	4	1.4
2	0	0	13.96	6.341	2	1.6
2	1	0	14.55	6.089	4	11.6
0	2	2	15.60	5.679	4	16.7
2	1	1	17.17	5.165	4	3.2
-2	0	2	17.22	5.149	2	1.7
0	4	1	17.59	5.042	4	17.4
-2	1	2	17.70	5.011	4	7.5
1	4	0	17.72	5.006	4	12.3
0	3	2	18.08	4.907	4	12.1
1	2	2	18.20	4.874	4	7.2
-1	3	2	18.37	4.829	4	4.3
-1	4	1	18.42	4.816	4	2.2
2	3	0	18.57	4.777	4	38.5
-2	3	1	18.74	4.735	4	14.1
-2	2	2	19.06	4.656	4	12.4
1	4	1	19.45	4.565	4	9.3
-1	0	3	19.80	4.484	2	36.3
-1	1	3	20.22	4.392	4	35.5
1	3	2	20.37	4.359	4	12.3
2	3	1	20.70	4.290	4	23.3
0	4	2	21.08	4.215	4	2.4
-2	3	2	21.15	4.201	4	2.3
-1	4	2	21.33	4.166	4	4.8
3	1	0	21.41	4.150	4	2.4
-1	2	3	21.43	4.147	4	2.7
2	4	0	21.50	4.132	4	27.0
0	2	3	21.63	4.109	4	1.2
2	1	2	21.64	4.107	4	2.3
-2	4	1	21.65	4.105	4	33.0
-1	5	1	22.15	4.014	4	5.2
-3	2	1	22.26	3.993	4	13.9
-2	1	3	22.35	3.978	4	2.6
1	0	3	22.55	3.942	2	14.7
3	2	0	22.56	3.941	4	8.1
2	2	2	22.78	3.904	4	1.5
-3	1	2	22.85	3.891	4	18.1
1	1	3	22.92	3.879	4	12.3
2	4	1	23.38	3.805	4	15.4
3	1	1	23.71	3.753	4	6.4
-2	4	2	23.78	3.742	4	11.2
-3	3	1	24.08	3.695	4	3.7
2	3	2	24.56	3.624	4	19.4
-1	5	2	24.64	3.614	4	2.8
2	5	0	24.79	3.592	4	6.1
-2	5	1	24.91	3.574	4	19.5
1	6	0	25.51	3.492	4	1.9
-3	3	2	25.65	3.474	4	2.6
0	4	3	25.90	3.440	4	4.2
-3	1	3	26.28	3.391	4	3.0
2	5	1	26.44	3.371	4	2.2
-1	1	4	26.58	3.353	4	1.4
3	4	0	26.69	3.340	4	14.7
2	4	2	26.88	3.317	4	9.5
2	1	3	27.10	3.290	4	3.2
0	1	4	27.11	3.289	4	8.0
3	1	2	27.53	3.240	4	3.5
-2	0	4	27.60	3.232	2	5.3
2	2	3	28.03	3.183	4	2.9
0	2	4	28.04	3.182	4	7.3
4	0	0	28.14	3.171	2	1.5
-1	6	2	28.18	3.166	4	3.5
2	6	0	28.32	3.152	4	2.0
3	2	2	28.44	3.138	4	3.4
-3	3	3	28.76	3.104	4	4.3
-2	2	4	28.81	3.099	4	2.8
-1	3	4	29.04	3.075	4	1.5
1	1	4	29.41	3.037	4	3.6
3	5	0	29.43	3.035	4	2.0
0	7	1	29.47	3.031	4	1.9
2	3	3	29.52	3.026	4	2.1
1	7	0	29.54	3.023	4	7.3
2	5	2	29.60	3.017	4	3.1
1	2	4	30.27	2.952	4	4.0
-3	5	2	30.52	2.929	4	1.7
4	1	1	30.57	2.924	4	1.9
4	3	0	30.77	2.906	4	2.4
-3	4	3	30.79	2.904	4	7.8
-3	1	4	30.87	2.897	4	2.4
-4	1	3	31.40	2.849	4	2.0
4	2	1	31.41	2.848	4	2.8
0	4	4	31.51	2.839	4	4.1
1	3	4	31.66	2.826	4	3.2
-3	2	4	31.70	2.823	4	2.6
0	6	3	31.84	2.810	4	3.1
-1	7	2	31.91	2.805	4	3.6
2	7	0	32.03	2.795	4	4.9
-4	2	3	32.21	2.779	4	4.1
-3	6	1	32.29	2.772	4	1.7
4	4	0	32.68	2.740	4	2.0
2	0	4	32.87	2.725	2	4.9
-3	3	4	33.04	2.711	4	3.0
-3	5	3	33.22	2.697	4	1.6
0	8	1	33.58	2.669	4	2.8
1	8	0	33.65	2.663	4	2.1
-1	2	5	33.97	2.639	4	1.5
-4	5	1	34.51	2.599	4	3.0
1	8	1	34.63	2.590	4	3.2
-3	4	4	34.84	2.575	4	1.3
2	3	4	35.18	2.551	4	1.2
-4	4	3	35.31	2.542	4	2.4
-5	2	1	35.61	2.521	4	1.7
5	1	0	35.63	2.520	4	1.9
1	5	4	35.80	2.508	4	2.4
3	7	0	35.82	2.507	4	2.4
-2	8	1	35.97	2.497	4	1.3
-5	2	2	36.16	2.484	4	1.8
3	4	3	36.19	2.482	4	2.3
-3	1	5	36.23	2.479	4	3.8
2	6	3	36.62	2.454	4	1.6
2	4	4	36.88	2.437	4	1.2
3	7	1	37.30	2.411	4	1.3
4	6	0	37.66	2.389	4	3.1
4	7	0	40.61	2.221	4	1.2
-5	4	3	40.69	2.217	4	1.7
3	4	4	41.29	2.186	4	1.2
-5	6	2	43.34	2.088	4	2.9
4	8	0	43.81	2.066	4	2.6
1	3	6	44.67	2.028	4	1.4
4	7	2	44.83	2.022	4	2.1
-5	6	3	44.91	2.018	4	1.6
1	4	6	46.08	1.970	4	1.3
-1	0	7	46.84	1.939	2	1.5
-5	7	3	47.50	1.914	4	1.6
4	8	2	47.81	1.902	4	1.4
-4	4	6	47.97	1.897	4	1.2
-4	5	6	49.68	1.835	4	1.3

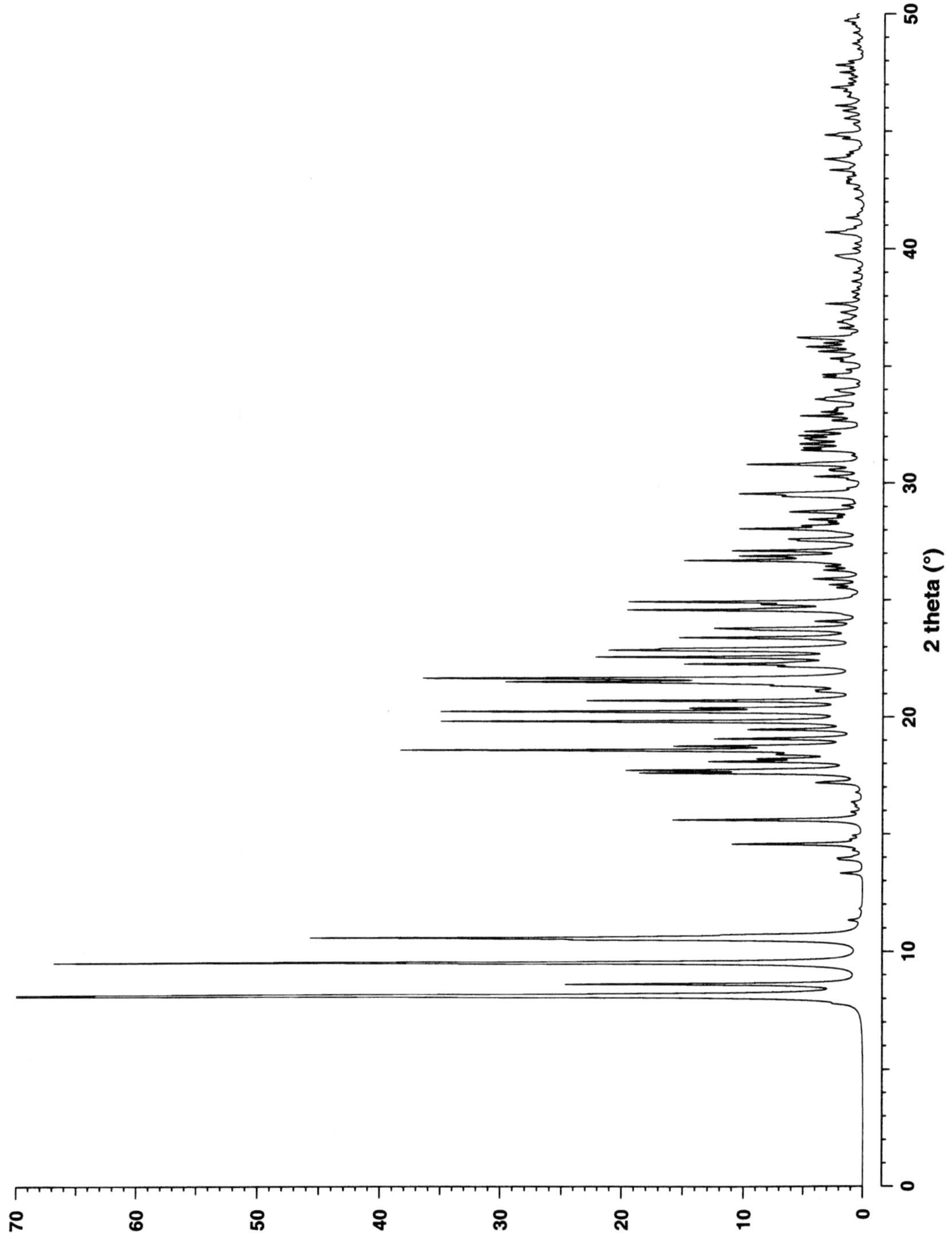
2 theta (°)
0
10
20
30
40
50
0
10
20
30
40
50
60
70

TER Terranovaite

CHEMICAL COMPOSITION: $|Na_{4.2}K_{0.2}Mg_{0.2}Ca_{3.7}(H_2O)_{>29}|\ [Al_{12.3}Si_{67.7}O_{160}]$
Mt. Adamson, N. Victoria Land, Antarctica

REFINED COMPOSITION: $|Na_{4.23}K_{0.167}Mg_{0.167}Ca_{3.74}|\ [Al_{12.32}Si_{67.68}O_{160}]$

CRYSTAL DATA: *Cmcm* (No. 63)
$a = 9.747$ Å $b = 23.880$ Å $c = 20.068$ Å
$\alpha = 90°$ $\beta = 90°$ $\gamma = 90°$
X-ray single crystal refinement, $R = 0.071$, $R_w = 0.068$

REFERENCE: E. Galli, S. Quartieri, G. Vezzalini, A. Alberti and M. Franzini, *American Mineralogist* **82** 423–429 (1997).

h	k	l	2θ	d	M	I_{rel}
0	2	0	7.40	11.940	2	82.9
0	2	1	8.62	10.261	4	100.0
0	0	2	8.81	10.034	2	63.4
1	1	0	9.80	9.024	4	64.1
1	1	1	10.75	8.230	8	5.3
0	2	2	11.52	7.682	4	2.4
1	1	2	13.19	6.710	8	3.4
1	3	0	14.37	6.165	4	23.7
0	4	0	14.84	5.970	2	8.8
1	3	1	15.03	5.893	8	0.3
0	2	3	15.18	5.836	4	15.2
0	4	1	15.49	5.722	4	2.3
1	1	3	16.50	5.374	8	1.8
1	3	2	16.88	5.253	8	0.4
0	4	2	17.28	5.131	4	12.4
0	0	4	17.68	5.017	2	7.3
2	0	0	18.20	4.873	2	14.4
0	2	4	19.19	4.625	4	0.5
2	2	0	19.67	4.512	4	2.8
0	4	3	19.93	4.454	4	0.3
2	2	1	20.17	4.402	8	0.2
1	1	4	20.25	4.385	8	0.8
2	0	2	20.26	4.384	4	4.8
1	5	0	20.71	4.289	4	0.1
1	5	1	21.18	4.194	8	5.3
2	2	2	21.59	4.115	8	0.2
0	6	0	22.34	3.980	2	2.9
1	5	2	22.55	3.944	8	0.1
1	3	4	22.85	3.891	8	0.3
0	4	4	23.16	3.841	4	0.2
0	2	5	23.38	3.804	4	71.6
2	4	0	23.56	3.775	4	60.7
2	2	3	23.79	3.741	8	8.9
0	6	2	24.05	3.700	4	1.5
1	1	5	24.27	3.667	8	4.2
1	5	3	24.66	3.610	8	22.4
2	4	2	25.20	3.533	8	8.4
2	0	4	25.48	3.496	4	0.1
0	6	3	26.05	3.420	4	8.9
0	0	6	26.65	3.345	2	2.1
0	4	5	26.76	3.331	4	3.6
2	4	3	27.12	3.288	8	1.2
0	2	6	27.70	3.221	4	0.2
1	7	0	27.70	3.220	4	2.6
3	1	0	27.71	3.219	4	0.7

h	k	l	2θ	d	M	I_{rel}
3	1	1	28.07	3.179	8	0.7
1	1	6	28.46	3.136	8	0.2
2	6	0	28.96	3.083	4	2.2
1	7	2	29.13	3.066	8	0.4
3	1	2	29.13	3.065	8	1.8
2	6	1	29.31	3.047	8	3.3
2	4	4	29.61	3.017	8	0.5
3	3	0	29.70	3.008	4	24.3
0	8	0	29.93	2.985	2	0.3
3	3	1	30.04	2.975	8	0.2
0	8	1	30.27	2.953	4	0.4
2	6	2	30.33	2.947	8	1.1
1	3	6	30.40	2.940	8	0.4
1	5	5	30.50	2.931	8	15.4
1	7	3	30.82	2.901	8	0.5
3	1	3	30.82	2.901	8	0.3
3	3	2	31.04	2.881	8	2.1
0	8	2	31.26	2.861	4	1.4
2	6	3	31.97	2.800	8	2.6
0	2	7	32.11	2.788	4	0.9
1	7	4	33.06	2.710	8	0.6
3	5	1	33.66	2.663	8	2.0
3	3	4	34.77	2.580	8	0.2
1	9	0	35.05	2.560	4	0.2
2	8	0	35.26	2.545	4	3.1
3	1	5	35.75	2.511	8	1.3
0	0	8	35.80	2.509	2	1.2
2	4	6	35.87	2.504	8	0.7
3	5	3	36.03	2.493	8	1.7
2	8	2	36.41	2.467	8	1.0
2	6	5	36.76	2.445	8	7.8
4	0	0	36.89	2.437	2	2.5
2	2	7	37.15	2.420	8	3.2
1	1	8	37.20	2.417	8	0.2
3	3	5	37.36	2.407	8	0.3
1	9	3	37.62	2.391	8	0.8
0	10	0	37.67	2.388	2	1.9
4	2	0	37.67	2.388	4	0.8
1	5	7	37.74	2.383	8	0.6
2	8	3	37.81	2.379	8	0.6
0	6	7	38.71	2.326	4	1.8
0	10	2	38.76	2.323	4	0.4
2	4	7	39.47	2.283	8	0.2
4	4	0	39.96	2.256	4	3.7
3	3	6	40.32	2.237	8	0.4

h	k	l	2θ	d	M	I_{rel}
3	5	5	40.40	2.232	8	1.3
2	0	8	40.44	2.230	4	1.3
3	7	3	40.65	2.219	8	0.5
0	2	9	41.18	2.192	4	0.4
1	5	8	41.71	2.165	8	0.2
1	9	5	41.85	2.158	8	1.1
2	10	0	42.14	2.144	4	0.1
3	1	7	42.21	2.141	8	0.3
2	10	1	42.39	2.132	8	0.2
3	7	4	42.43	2.130	8	0.2
1	11	0	42.67	2.119	4	0.4
2	6	7	43.09	2.099	8	0.3
2	10	2	43.14	2.097	8	0.9
2	4	8	43.30	2.089	8	0.4
0	8	7	43.78	2.068	4	0.5
4	4	4	44.01	2.058	8	0.2
3	9	0	44.06	2.055	4	0.2
4	2	5	44.13	2.052	8	0.1
2	10	3	44.36	2.042	8	0.2
3	7	5	44.64	2.030	8	0.7
0	0	10	45.18	2.007	2	6.2
0	12	0	45.58	1.990	2	0.4
4	6	3	45.71	1.985	8	1.4
0	12	1	45.82	1.980	4	0.4
0	2	10	45.85	1.979	4	0.2
1	7	8	45.86	1.979	8	5.5
3	9	3	46.21	1.964	8	5.5
1	1	10	46.35	1.959	8	0.2
5	1	0	46.75	1.943	4	1.7
5	1	1	46.98	1.934	8	0.2
0	8	8	47.33	1.920	4	1.6
1	3	10	47.65	1.908	8	0.3
5	1	2	47.67	1.908	8	0.4
0	12	3	47.68	1.907	4	0.1
2	8	7	47.78	1.903	8	0.3
0	4	10	47.82	1.902	4	0.5
2	10	5	48.11	1.891	8	0.2
4	8	0	48.21	1.888	4	0.2
4	8	1	48.43	1.879	8	0.2
5	3	2	48.95	1.861	8	0.5
2	0	10	49.09	1.856	4	0.2
4	6	5	49.38	1.845	8	0.8
2	12	0	49.47	1.842	4	0.2
0	10	7	49.69	1.835	4	0.2
3	9	5	49.85	1.829	8	2.1

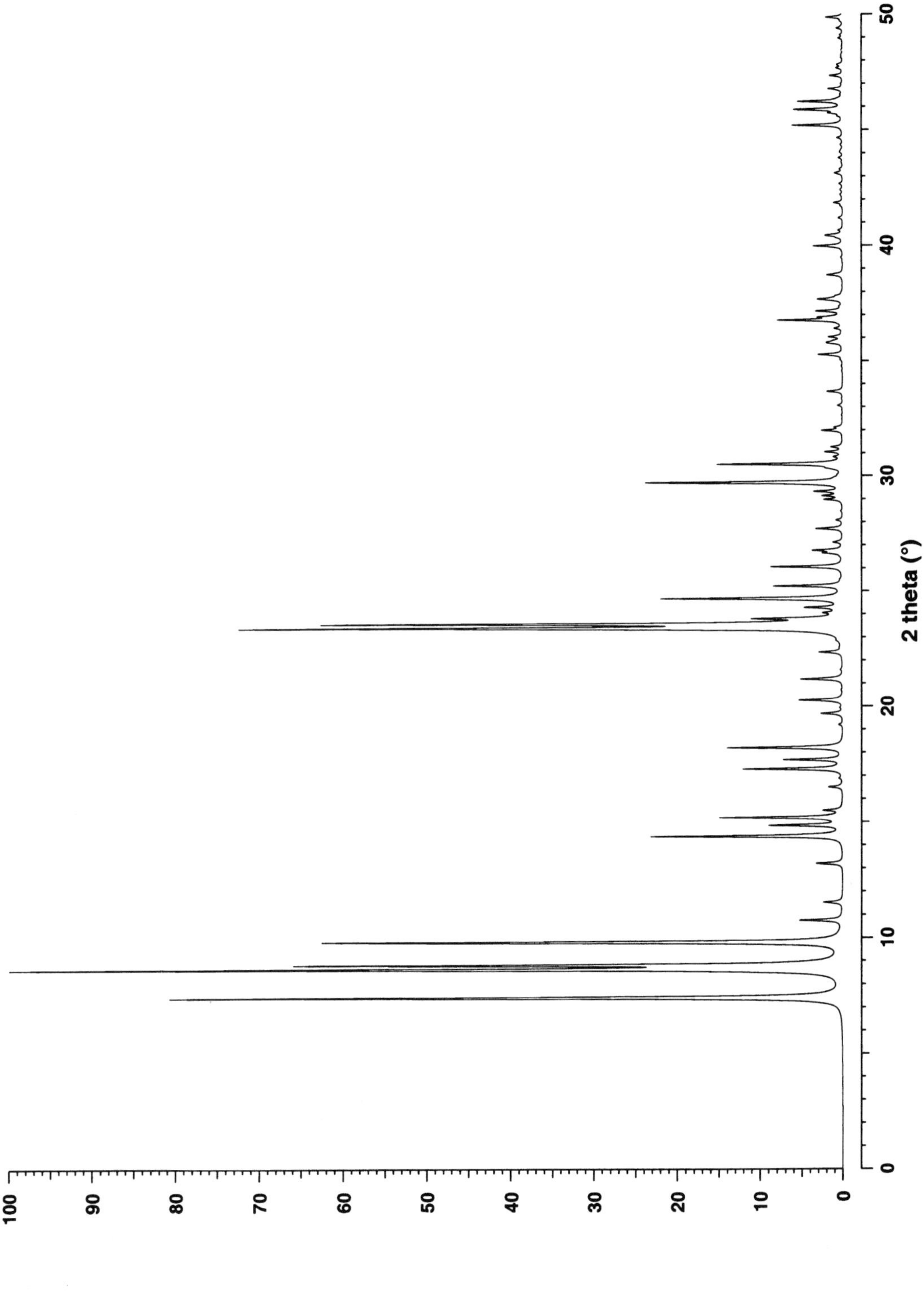
2 theta (°)
0
10
20
30
40
50
0
10
20
30
40
50
60
70
80
90
100

THO Thomsonite

CHEMICAL COMPOSITION: $|Na_4Ca_{7.5}Sr_{0.5}(H_2O)_{24}|\ [Si_{20}Al_{20}O_{80}]$
Death Valley, California, U.S.A.

REFINED COMPOSITION: $|Na_4Ca_{11.2}Sr_{0.8}(H_2O)_{24}|\ [Si_{20}Al_{20}O_{80}]$

CRYSTAL DATA: *Pncn* (No. 52) **bca** setting
$a = 13.088$ Å $b = 13.052$ Å $c = 13.229$ Å
$\alpha = 90°$ $\beta = 90°$ $\gamma = 90°$
Neutron single crystal refinement, $R_{F^2} = 0.048$

REFERENCE: J. J. Pluth, J. V. Smith and Å. Kvick,
Zeolites **5** 74–80 (1985).

h	k	l	2θ	d	M	I_{rel}
1	1	0	9.57	9.242	4	3.0
0	0	2	13.39	6.615	2	76.6
2	0	0	13.53	6.544	2	4.3
0	2	0	13.57	6.526	2	1.1
1	0	2	15.01	5.903	4	45.0
1	1	2	16.48	5.379	8	52.2
2	0	2	19.08	4.652	4	69.8
2	2	0	19.21	4.621	4	97.9
2	1	2	20.26	4.382	8	11.1
1	2	2	20.28	4.378	8	28.8
2	2	1	20.36	4.362	8	0.8
0	1	3	21.27	4.178	4	0.2
3	1	0	21.48	4.138	4	2.1
0	3	1	21.50	4.133	4	0.6
1	3	0	21.52	4.129	4	39.5
1	1	3	22.34	3.980	8	0.9
3	1	1	22.51	3.949	8	1.8
1	3	1	22.56	3.941	8	0.8
2	2	2	23.48	3.788	8	0.4
3	0	2	24.44	3.642	4	0.1
3	1	2	25.39	3.508	8	54.4
1	3	2	25.43	3.502	8	74.3
3	2	1	25.46	3.498	8	0.1
2	3	1	25.49	3.494	8	0.7
0	0	4	26.96	3.307	2	0.8
4	0	0	27.25	3.272	2	3.7
0	4	0	27.33	3.263	2	1.3
1	0	4	27.82	3.206	4	32.9
3	2	2	28.06	3.180	8	2.5
2	3	2	28.08	3.178	8	34.9
1	1	4	28.67	3.114	8	4.3
3	3	0	28.98	3.081	4	8.5
2	0	4	30.28	2.952	4	39.9
0	2	4	30.30	2.950	4	21.2
4	0	2	30.48	2.933	4	6.0
4	2	0	30.56	2.925	4	4.5
2	4	0	30.61	2.920	4	22.3
1	2	4	31.08	2.878	8	1.5
4	1	2	31.26	2.861	8	36.8
1	4	2	31.32	2.856	8	54.1
3	3	2	32.05	2.793	8	3.2
2	2	4	33.31	2.689	8	0.6
4	2	2	33.50	2.675	8	100.0
2	4	2	33.55	2.671	8	13.2
3	1	4	34.72	2.583	8	20.2
1	3	4	34.75	2.581	8	33.0
4	1	3	34.83	2.576	8	0.1
5	1	0	34.96	2.566	4	6.9
4	3	1	34.98	2.565	8	0.1
1	5	0	35.05	2.560	4	0.8
3	3	3	35.55	2.525	8	0.1
5	1	1	35.63	2.520	8	0.1
1	5	1	35.72	2.513	8	0.1
3	2	4	36.78	2.444	8	1.0
2	3	4	36.79	2.443	8	0.5
4	3	2	36.96	2.432	8	0.2
3	4	2	36.99	2.430	8	16.4
5	1	2	37.59	2.393	8	1.2
1	5	2	37.68	2.387	8	3.4
4	0	4	38.71	2.326	4	3.7
0	4	4	38.77	2.323	4	0.2
4	4	0	38.98	2.310	4	0.3
4	1	4	39.35	2.290	8	3.6
1	4	4	39.40	2.287	8	0.4
5	2	2	39.52	2.280	8	3.2
2	5	2	39.59	2.276	8	0.7
3	3	4	40.00	2.254	8	38.5
3	4	3	40.11	2.248	8	0.2
5	3	0	40.21	2.243	4	0.7
3	5	0	40.26	2.240	4	6.7
5	1	3	40.67	2.218	8	0.1
0	0	6	40.93	2.205	2	3.5
4	2	4	41.20	2.191	8	23.8
2	4	4	41.24	2.189	8	5.4
6	0	0	41.39	2.181	2	7.6
4	4	2	41.39	2.181	8	0.8
0	6	0	41.51	2.175	2	21.9
1	0	6	41.53	2.174	4	1.5
1	1	6	42.13	2.145	8	0.3
2	5	3	42.55	2.125	8	0.2
5	3	2	42.56	2.124	8	6.8
3	5	2	42.61	2.122	8	9.6
2	0	6	43.30	2.089	4	11.0
0	2	6	43.31	2.089	4	1.4
6	0	2	43.69	2.072	4	1.4
6	2	0	43.76	2.069	4	2.2
0	6	2	43.81	2.066	4	2.6
2	6	0	43.86	2.064	4	8.8
1	2	6	43.89	2.063	8	3.6
5	0	4	44.12	2.053	4	0.2
4	3	4	44.15	2.051	8	0.3
3	4	4	44.17	2.050	8	0.3
6	1	2	44.27	2.046	8	0.8
1	6	2	44.38	2.041	8	0.3
5	1	4	44.69	2.028	8	2.3
1	5	4	44.77	2.024	8	4.2
5	3	3	45.36	1.999	8	0.1
2	2	6	45.59	1.990	8	7.1
6	2	2	45.96	1.974	8	6.4
2	6	2	46.06	1.971	8	1.3
3	0	6	46.13	1.968	4	1.5
5	2	4	46.37	1.958	8	5.6
2	5	4	46.44	1.955	8	2.2
5	4	2	46.55	1.951	8	2.2
4	5	2	46.58	1.950	8	2.3
3	1	6	46.68	1.946	8	4.6
1	3	6	46.70	1.945	8	6.2
3	6	1	47.19	1.926	8	0.1
4	4	4	48.03	1.894	8	1.7
3	2	6	48.31	1.884	8	6.4
2	3	6	48.32	1.883	8	2.6
6	3	2	48.68	1.870	8	2.5
3	6	2	48.76	1.868	8	0.2
5	3	4	49.07	1.856	8	1.7
3	5	4	49.12	1.855	8	1.2
7	1	0	49.23	1.851	4	0.5
5	5	0	49.30	1.848	4	1.2
0	7	1	49.36	1.846	4	0.2
1	7	0	49.37	1.846	4	0.4
5	1	5	49.48	1.842	8	0.1
1	5	5	49.55	1.840	8	0.1
0	4	6	49.92	1.827	4	14.1

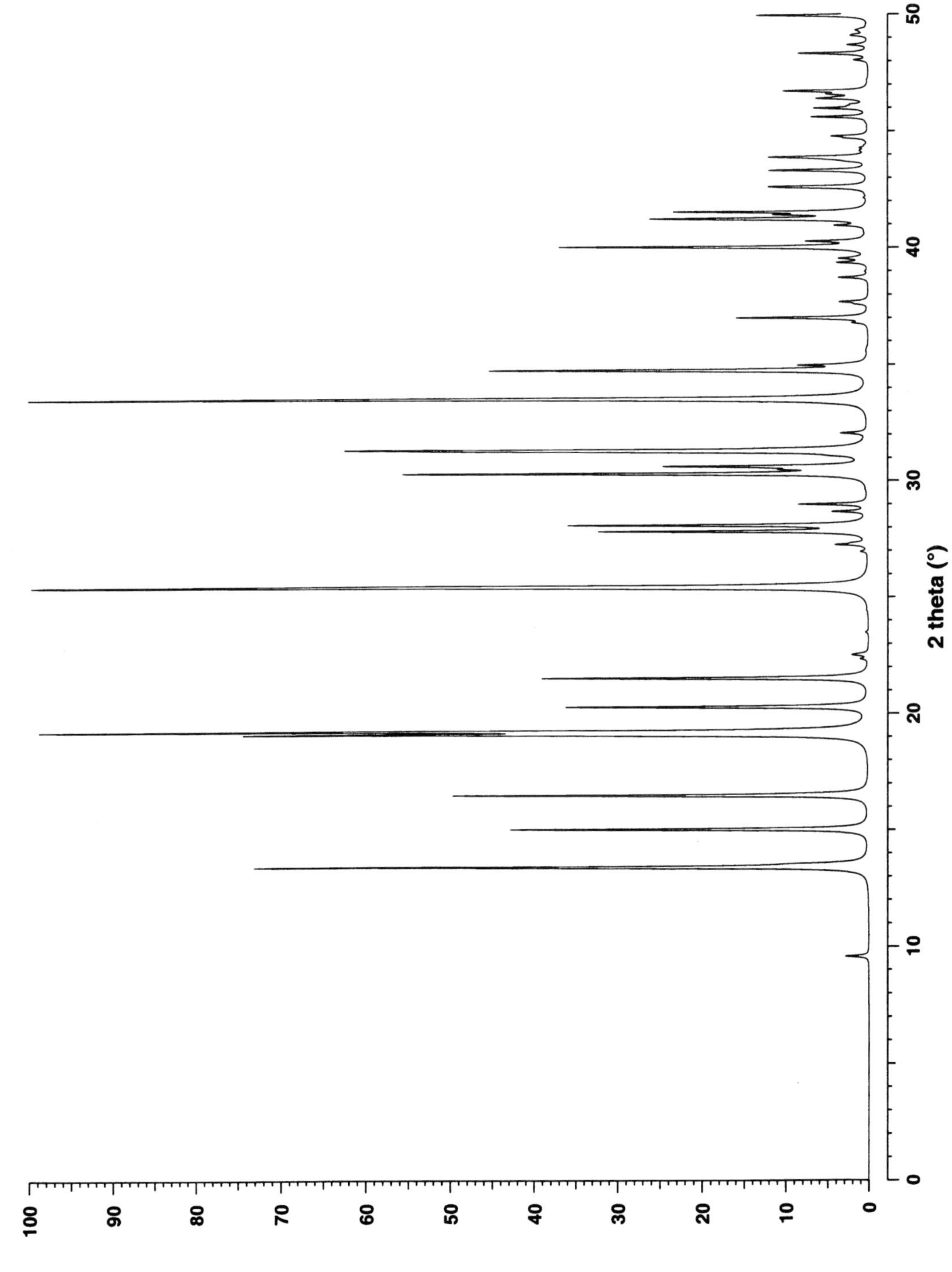
2 theta (°)
0
10
20
30
40
50
0
10
20
30
40
50
60
70
80
90
100

TON Diethylamine Theta-1 (Silica ZSM-22)

CHEMICAL COMPOSITION: $|(C_2H_5)_2NH|$ $[Si_{24}O_{48}]$
$(C_2H_5)_2NH$ = diethylamine

REFINED COMPOSITION: $|C_{5.1}|$ $[Si_{24}O_{48}]$

CRYSTAL DATA: $Cmc2_1$ (No. 36)
a = 13.859 Å b = 17.420 Å c = 5.038 Å
α = 90° β = 90° γ = 90°
X-ray single crystal refinement, R_w = 0.056

REFERENCE: B. Marler,
Zeolites **7** 393–397 (1987).

h	k	l	2θ	d	M	I_{rel}	h	k	l	2θ	d	M	I_{rel}	h	k	l	2θ	d	M	I_{rel}
1	1	0	8.15	10.845	4	100.0	5	1	0	32.71	2.737	4	2.7	2	4	2	43.51	2.080	8	0.2
0	2	0	10.16	8.710	2	19.1	4	2	1	33.02	2.713	8	2.7	3	3	2	43.80	2.067	8	3.9
2	0	0	12.77	6.929	2	22.3	4	4	0	33.04	2.711	4	0.7	6	2	1	44.38	2.041	8	0.5
2	2	0	16.35	5.423	4	7.3	2	6	0	33.46	2.678	4	0.3	6	4	0	44.39	2.041	4	0.5
1	3	0	16.55	5.356	4	4.6	0	0	2	35.64	2.519	2	13.8	4	0	2	44.47	2.037	4	0.5
1	1	1	19.43	4.569	8	9.0	0	6	1	35.69	2.516	4	3.5	4	6	1	44.51	2.036	8	2.2
3	1	0	19.88	4.465	4	0.3	5	3	0	35.90	2.501	4	0.2	1	5	2	44.88	2.020	8	2.1
0	2	1	20.36	4.361	4	69.6	1	1	2	36.62	2.454	8	2.1	3	7	1	45.13	2.009	8	1.1
0	4	0	20.39	4.355	2	15.6	3	5	1	36.91	2.435	8	9.6	0	8	1	45.37	1.999	4	3.1
2	2	1	24.11	3.691	8	8.7	0	2	2	37.15	2.420	4	0.4	5	5	1	45.53	1.992	8	0.4
2	4	0	24.14	3.687	4	11.5	5	1	1	37.39	2.405	8	2.9	4	2	2	45.73	1.984	8	0.4
1	3	1	24.25	3.670	8	49.5	4	4	1	37.67	2.388	8	1.9	7	1	0	46.14	1.967	4	0.1
3	3	0	24.62	3.615	4	50.0	2	0	2	38.01	2.367	4	0.7	2	8	1	47.33	1.920	8	0.7
4	0	0	25.71	3.465	2	33.5	2	6	1	38.06	2.365	8	7.2	1	9	0	47.43	1.917	4	0.3
1	5	0	26.38	3.379	4	2.6	2	2	2	39.44	2.285	8	0.4	0	6	2	47.80	1.903	4	3.7
3	1	1	26.68	3.342	8	6.1	1	3	2	39.53	2.279	8	0.3	7	3	0	48.58	1.874	4	8.6
0	4	1	27.06	3.295	4	6.6	5	3	1	40.25	2.240	8	1.5	3	5	2	48.77	1.867	8	0.8
4	2	0	27.71	3.219	4	1.4	6	2	0	40.40	2.233	4	0.5	5	1	2	49.15	1.854	8	0.3
2	4	1	30.03	2.975	8	2.7	4	6	0	40.54	2.225	4	0.2	5	7	0	49.20	1.852	4	0.3
3	3	1	30.43	2.937	8	3.2	1	7	1	40.97	2.203	8	0.1	4	4	2	49.38	1.845	8	0.8
0	6	0	30.80	2.903	2	2.6	3	1	2	41.14	2.194	8	0.2	4	8	0	49.43	1.844	4	1.4
1	5	1	31.89	2.806	8	0.9	3	7	0	41.20	2.191	4	0.4	2	6	2	49.69	1.835	8	0.2
3	5	0	32.18	2.782	4	1.5	5	5	0	41.64	2.169	4	0.4	7	1	1	49.76	1.832	8	1.8

Diethylamine Theta-1 (Silica ZSM-22) TON

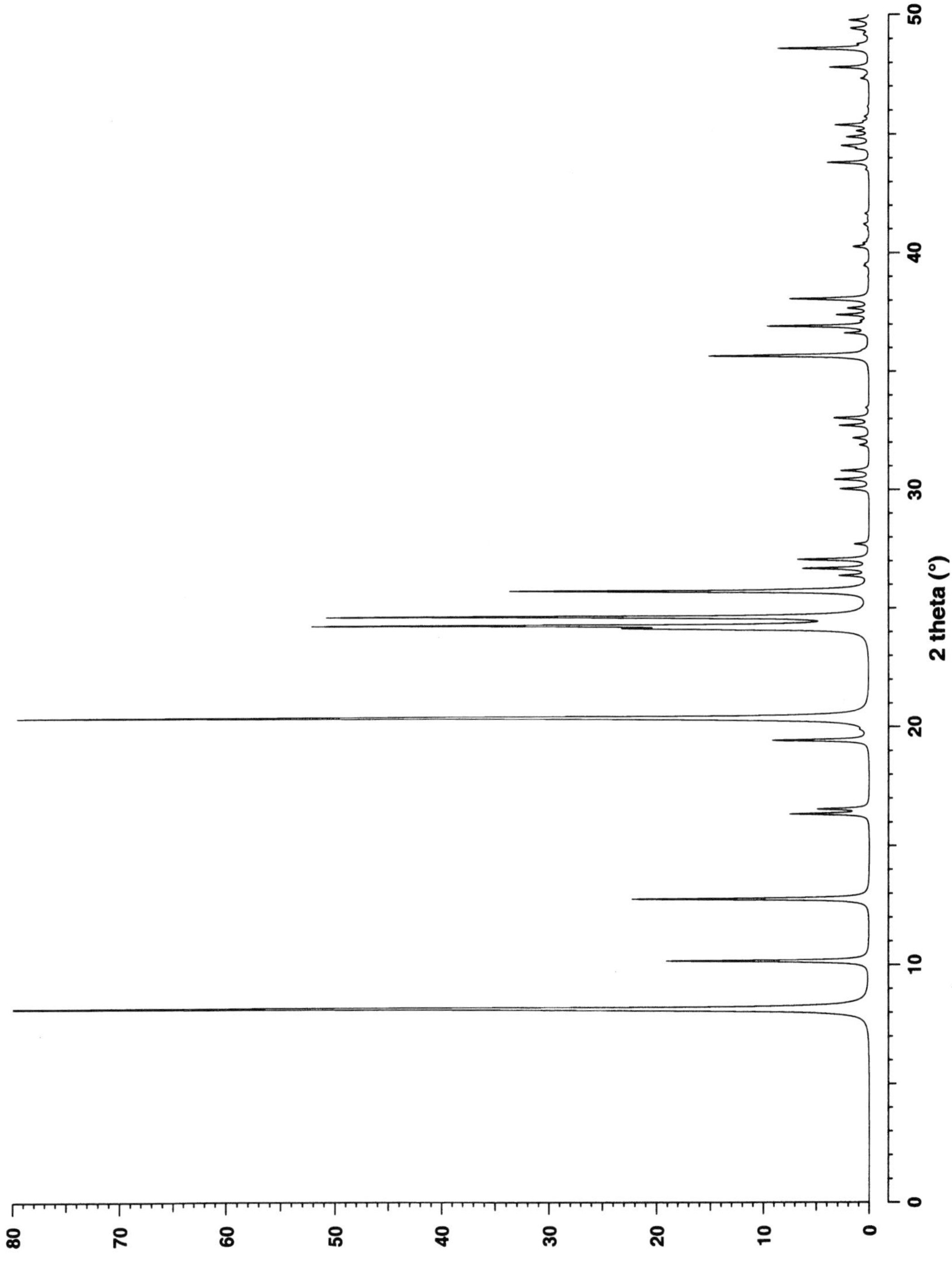

TSC Tschörtnerite

CHEMICAL COMPOSITION: $|K_{11.2}Ca_{89.6}Sr_{16.64}Ba_{4.8}Cu_{46.4}(H_2O)_{217}(OH)_{135.04}|$ $[Fe_{1.44}Al_{189.6}Si_{192.96}O_{768}]$
Bellberg volcano, near Mayen, Eifel, Germany

REFINED COMPOSITION: $|K_{12}Ca_{76}Sr_{12}Ba_{12}Cu_{48}Cl_{0.896}O_{327.448}(OH)_{128}|$ $[Si_{192}Al_{192}O_{768}]$

CRYSTAL DATA: $Fm\overline{3}m$ (No. 225)
$a = 31.62$ Å $b = 31.62$ Å $c = 31.62$ Å
$\alpha = 90°$ $\beta = 90°$ $\gamma = 90°$
Single Crystal X-ray Refinement. $R_p = 0.065$, $R_{wp} = 0.156$

REFERENCE: H. Effenberger, G. Giester, W. Krause and H.-J. Bernhardt, *American Mineralogist* **83** 607–617 (1998).

h	k	l	2θ	d	M	I_{rel}	h	k	l	2θ	d	M	I_{rel}	h	k	l	2θ	d	M	I_{rel}
1	1	1	4.84	18.256	8	100.0	10	0	0	28.22	3.162	6	0.5	13	3	1	38.07	2.363	48	0.1
2	0	0	5.59	15.810	6	14.5	8	6	2	28.79	3.101	48	0.4	12	6	0	38.18	2.357	24	0.2
2	2	0	7.91	11.179	12	1.4	9	5	1	29.21	3.057	48	0.2	10	8	4	38.18	2.357	48	0.1
3	1	1	9.28	9.534	24	0.5	7	7	3	29.21	3.057	24	0.2	11	7	5	39.81	2.264	48	0.5
2	2	2	9.69	9.128	8	5.4	10	2	2	29.35	3.043	24	0.8	14	0	0	39.91	2.259	6	0.2
3	3	1	12.20	7.254	24	0.8	6	6	6	29.35	3.043	8	0.9	10	10	0	40.34	2.236	12	0.5
4	2	2	13.72	6.454	24	0.4	9	5	3	30.31	2.949	48	0.3	13	5	3	40.65	2.219	48	0.2
5	1	1	14.56	6.085	24	0.4	10	4	0	30.45	2.936	24	1.3	14	2	2	40.76	2.214	24	0.3
4	4	0	15.85	5.590	12	0.6	8	6	4	30.45	2.936	48	0.8	10	10	2	40.76	2.214	24	0.1
5	3	1	16.59	5.345	48	0.3	10	4	2	30.98	2.886	48	1.6	9	9	7	41.48	2.177	24	0.2
6	0	0	16.82	5.270	6	0.2	11	1	1	31.37	2.851	24	0.1	11	9	3	41.48	2.177	48	0.1
6	2	2	18.61	4.767	24	0.8	7	7	5	31.37	2.851	24	0.1	11	7	7	42.30	2.137	24	0.1
5	5	1	20.05	4.428	24	1.0	8	8	0	32.02	2.795	12	0.9	13	5	5	42.30	2.137	24	0.3
7	1	1	20.05	4.428	24	0.9	11	3	1	32.41	2.763	48	0.8	11	9	5	43.10	2.099	48	0.5
6	4	0	20.25	4.385	24	0.5	9	5	5	32.41	2.763	24	1.1	15	1	1	43.10	2.099	24	0.3
5	5	3	21.59	4.117	24	0.6	9	7	1	32.41	2.763	48	0.6	10	8	8	43.20	2.094	24	0.2
7	3	1	21.59	4.117	48	0.3	10	6	0	33.04	2.711	24	0.8	14	4	4	43.20	2.094	24	0.1
7	3	3	23.02	3.863	24	1.4	11	3	3	33.41	2.682	24	0.1	10	10	6	43.99	2.058	24	0.2
6	4	4	23.20	3.834	24	0.8	10	6	2	33.53	2.672	48	0.3	9	9	9	44.67	2.028	8	0.4
8	2	0	23.20	3.834	24	1.2	8	8	4	34.02	2.635	24	0.3	13	7	5	44.67	2.028	48	0.2
6	6	0	23.88	3.726	12	0.7	11	5	1	34.39	2.608	48	0.4	11	11	1	44.67	2.028	24	0.2
6	6	2	24.54	3.627	24	0.4	7	7	7	34.39	2.608	8	1.2	16	0	0	45.92	1.976	6	0.2
7	5	3	25.67	3.471	48	2.1	12	2	0	34.51	2.599	24	0.2	15	5	3	46.20	1.965	48	0.2
8	4	2	25.82	3.450	48	0.7	10	6	4	34.98	2.565	48	1.4	14	8	0	46.30	1.961	24	0.2
6	6	4	26.44	3.371	24	0.5	12	4	2	36.39	2.469	48	0.5	13	9	5	47.69	1.907	48	0.1
8	4	4	27.64	3.227	24	1.7	8	8	6	36.39	2.469	24	0.7	14	8	4	47.79	1.903	48	0.2
9	3	3	28.08	3.178	24	0.5	10	8	0	36.39	2.469	24	0.2	12	12	0	48.88	1.863	12	0.5
7	5	5	28.08	3.178	24	0.5	10	6	6	37.29	2.411	24	0.2	13	11	3	49.87	1.829	48	0.1
8	6	0	28.22	3.162	24	0.5	11	7	3	38.07	2.363	48	0.1							

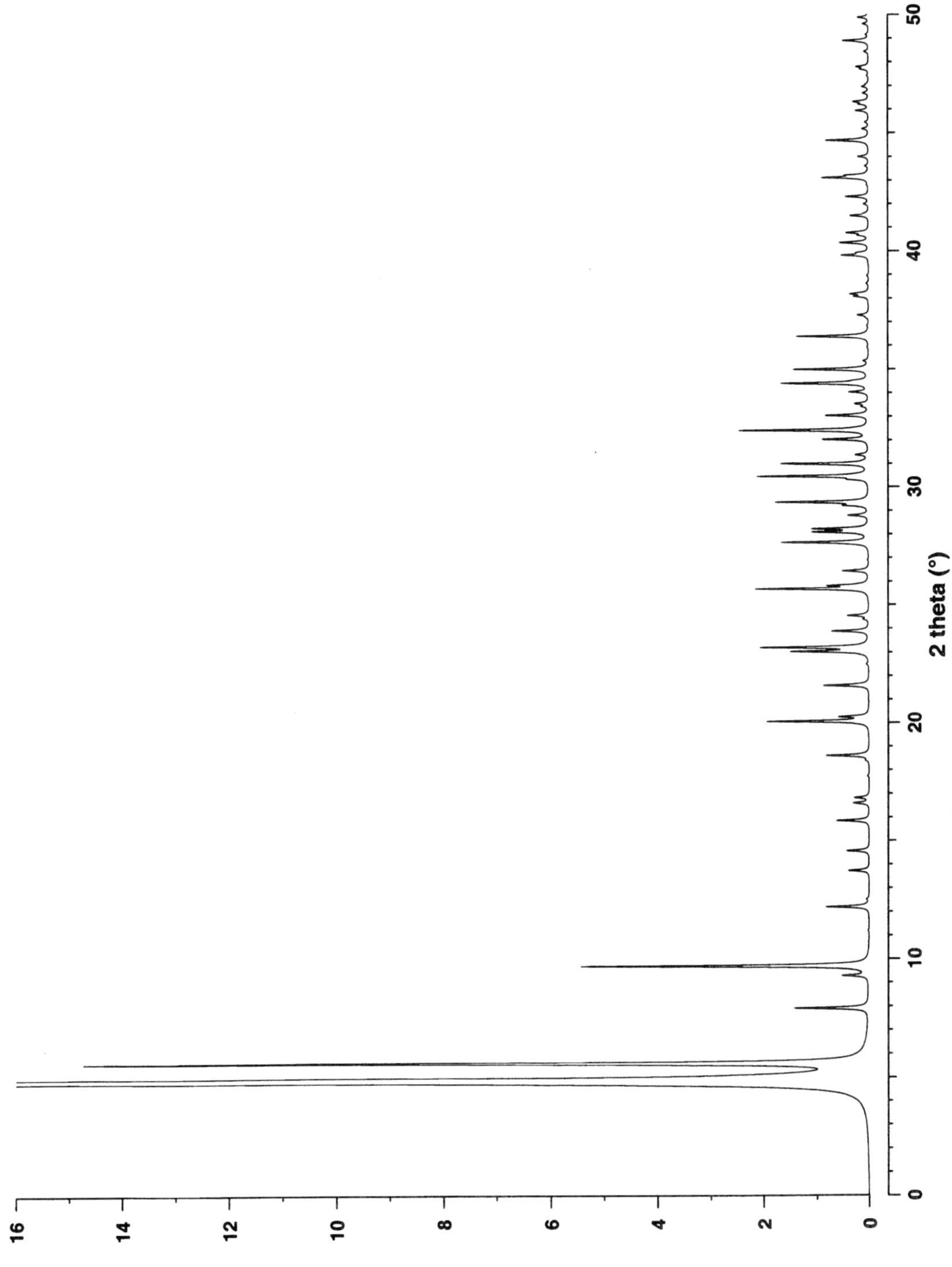
0
10
20
30
40
50
2 theta (°)
16
14
12
10
8
6
4
2
0

VET VPI-8

CHEMICAL COMPOSITION: $[Si_{17}O_{34}]$

REFINED COMPOSITION: $[Si_{17}O_{34}]$

CRYSTAL DATA: $P\overline{4}$ (No. 81)
$a = 13.045$ Å $b = 13.045$ Å $c = 5.034$ Å
$\alpha = 90°$ $\beta = 90°$ $\gamma = 90°$
X-ray Rietveld refinement, $R_{wp} = 0.125$, $R_p = 0.102$

REFERENCE: C. C. Freyhardt, R. F. Lobo, S. Khodabandeh, J. E. Lewis, Jr., M. Tsapatsis, M. Yoshikawa, M. A. Camblor, M. Pan, M. M. Helmkamp, S. I. Zones and M. E. Davis, *J. Am. Chem. Soc.* **118** 7299–7310 (1996).

h	k	l	2θ	d	M	I_{rel}	h	k	l	2θ	d	M	I_{rel}	h	k	l	2θ	d	M	I_{rel}
1	0	0	6.78	13.045	4	100.0	3	2	1	30.42	2.938	8	0.2	1	3	2	42.05	2.149	8	0.5
1	1	0	9.59	9.224	4	21.2	4	2	0	30.65	2.917	4	0.6	1	6	0	42.13	2.145	4	0.3
2	0	0	13.58	6.523	4	1.3	4	0	1	32.72	2.737	8	3.5	6	1	0	42.13	2.145	4	0.2
1	2	0	15.19	5.834	4	1.7	4	1	1	33.45	2.679	8	0.7	2	3	2	43.81	2.066	8	0.4
2	1	0	15.19	5.834	4	0.2	1	4	1	33.45	2.679	8	0.4	3	2	2	43.81	2.066	8	0.4
0	0	1	17.62	5.034	2	0.1	3	3	1	34.17	2.624	8	0.6	2	6	0	43.89	2.063	4	0.6
1	0	1	18.90	4.696	8	1.7	4	3	0	34.37	2.609	4	0.4	5	3	1	44.31	2.044	8	0.2
2	2	0	19.24	4.612	4	0.2	1	5	0	35.07	2.558	4	0.2	3	5	1	44.31	2.044	8	0.4
1	1	1	20.09	4.419	8	20.7	4	2	1	35.57	2.524	8	1.4	5	4	0	44.47	2.037	4	0.5
3	0	0	20.42	4.348	4	3.4	2	4	1	35.57	2.524	8	0.9	6	0	1	45.44	1.996	8	0.2
3	1	0	21.54	4.125	4	18.6	0	0	2	35.67	2.517	2	3.9	4	0	2	45.52	1.993	8	0.1
2	0	1	22.31	3.985	8	13.8	1	0	2	36.35	2.471	8	0.3	1	6	1	46.00	1.973	8	0.2
2	1	1	23.34	3.811	8	8.2	1	1	2	37.02	2.428	8	0.8	6	1	1	46.00	1.973	8	0.8
1	2	1	23.34	3.811	8	3.8	2	5	0	37.11	2.422	4	0.6	1	4	2	46.08	1.970	8	0.2
2	3	0	24.60	3.618	4	10.8	2	0	2	38.33	2.348	8	0.4	4	1	2	46.08	1.970	8	0.3
3	2	0	24.60	3.618	4	3.4	3	4	1	38.88	2.316	8	0.6	3	3	2	46.63	1.948	8	1.3
2	2	1	26.20	3.401	8	9.4	4	3	1	38.88	2.316	8	1.7	6	3	0	46.71	1.945	4	0.4
3	0	1	27.10	3.291	8	0.9	5	0	1	38.88	2.316	8	0.2	2	6	1	47.65	1.909	8	0.4
3	1	1	27.96	3.191	8	0.4	5	1	1	39.51	2.281	8	0.5	6	2	1	47.65	1.909	8	1.2
1	3	1	27.96	3.191	8	1.9	1	5	1	39.51	2.281	8	1.2	2	4	2	47.72	1.906	8	1.2
1	4	0	28.20	3.164	4	3.9	5	3	0	40.31	2.237	4	0.9	5	4	1	48.18	1.888	8	0.2
4	1	0	28.20	3.164	4	7.5	2	2	2	40.84	2.209	8	0.2	4	5	1	48.18	1.888	8	0.5
3	3	0	29.04	3.075	4	1.0	5	2	1	41.36	2.183	8	0.4	7	0	0	48.87	1.864	4	0.2
2	3	1	30.42	2.938	8	0.8	6	0	0	41.53	2.174	4	0.1	1	7	0	49.40	1.845	4	1.2

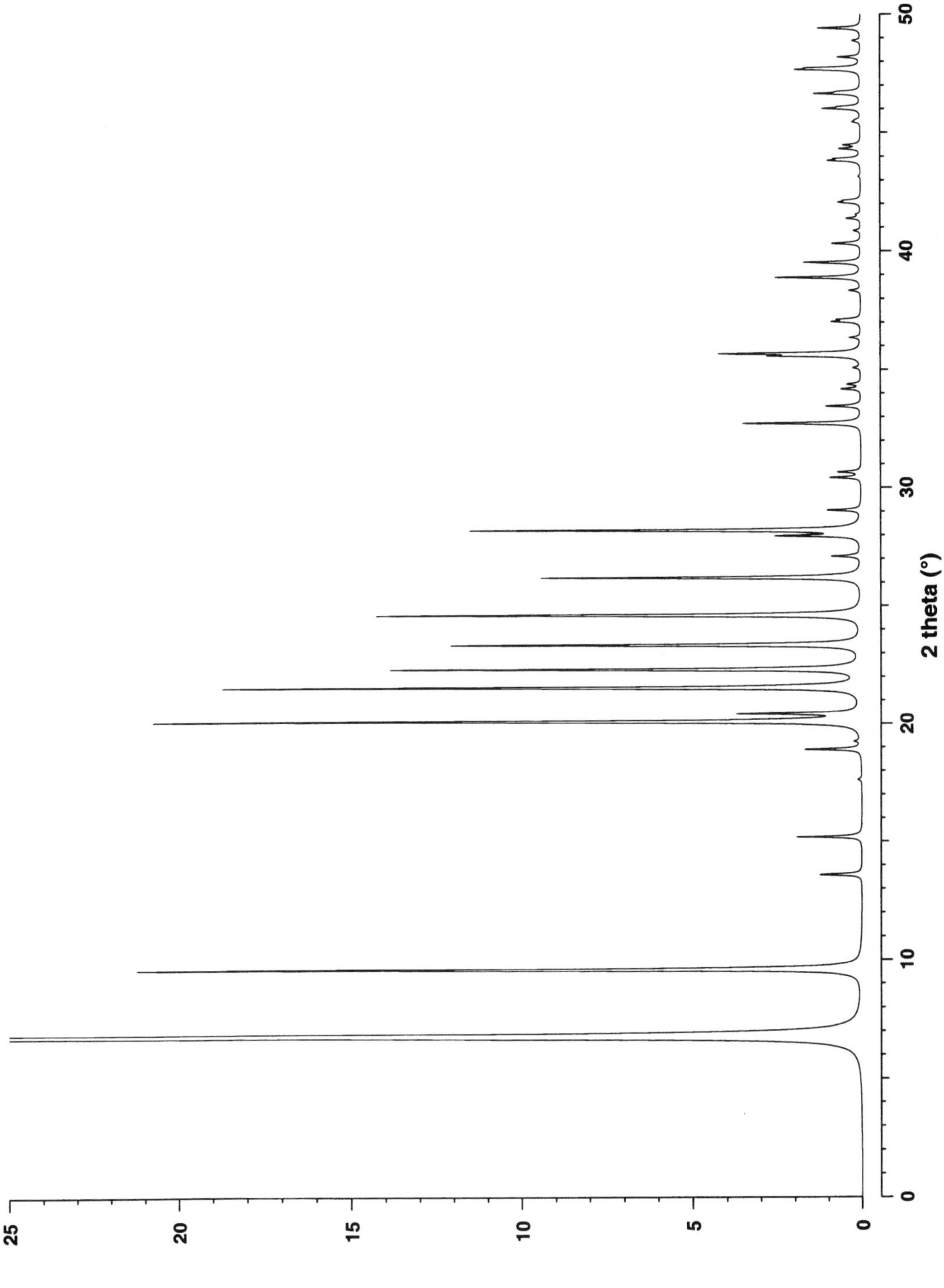
2 theta (°)
0
10
20
30
40
50
0
5
10
15
20
25

VFI VPI-5

CHEMICAL COMPOSITION: $|(H_2O)_{42}|$ $[Al_{18}P_{18}O_{72}]$

REFINED COMPOSITION: $|(H_2O)_{42}|$ $[Al_{18}P_{18}O_{72}]$

CRYSTAL DATA: $P6_3$ (No. 173)
$a = 18.9752$ Å $b = 18.9752$ Å $c = 8.1044$ Å
$\alpha = 90°$ $\beta = 90°$ $\gamma = 120°$
X-ray Rietveld refinement, $R_{exp} = 0.108$, $R_{wp} = 0.141$, $R_F = 0.086$

REFERENCE: L. B. McCusker, Ch. Baerlocher, E. Jahn and M. Bülow, *Zeolites* **11** 308–313 (1991).

h	k	l	2θ	d	M	I_{rel}
1	0	0	5.38	16.433	6	100.0
1	1	0	9.32	9.488	6	0.7
2	0	0	10.77	8.217	6	11.9
1	2	0	14.26	6.211	6	0.4
2	1	0	14.26	6.211	6	0.5
1	1	1	14.37	6.162	12	4.2
2	0	1	15.36	5.770	12	0.2
3	0	0	16.18	5.478	6	1.1
1	2	1	17.99	4.930	12	0.4
2	1	1	17.99	4.930	12	0.3
2	2	0	18.70	4.744	6	5.2
3	0	1	19.56	4.538	12	0.6
4	0	0	21.63	4.108	6	1.2
2	2	1	21.71	4.094	12	7.4
0	0	2	21.93	4.052	2	8.7
1	3	1	22.38	3.973	12	1.5
3	1	1	22.38	3.973	12	3.1
1	0	2	22.60	3.934	12	10.4
2	3	0	23.60	3.770	6	2.0
3	2	0	23.60	3.770	6	4.5
1	1	2	23.88	3.727	12	0.2
4	0	1	24.29	3.664	12	0.2
2	0	2	24.49	3.634	12	1.5
1	4	0	24.83	3.586	6	2.2
3	2	1	26.07	3.418	12	0.5
2	3	1	26.07	3.418	12	0.4
2	1	2	26.26	3.394	12	0.2
5	0	0	27.13	3.287	6	0.2
1	4	1	27.19	3.279	12	6.6
4	1	1	27.19	3.279	12	3.7
3	0	2	27.38	3.258	12	1.1
3	3	0	28.22	3.163	6	2.4
2	2	2	28.98	3.081	12	4.9
5	0	1	29.32	3.046	12	0.3
3	1	2	29.50	3.028	12	1.0
1	3	2	29.50	3.028	12	1.0
5	1	0	30.28	2.951	6	1.9
1	5	0	30.28	2.951	6	2.8
3	3	1	30.34	2.946	12	0.7
4	2	1	30.83	2.900	12	1.3
2	4	1	30.83	2.900	12	0.3
4	0	2	31.00	2.885	12	1.5
5	1	1	32.28	2.773	12	0.3
1	5	1	32.28	2.773	12	0.3
3	2	2	32.44	2.760	12	0.7
2	3	2	32.44	2.760	12	0.5
6	0	0	32.70	2.739	6	4.6
1	4	2	33.36	2.685	12	0.3
4	1	2	33.36	2.685	12	0.2
1	0	3	33.62	2.666	12	0.1
5	2	0	34.07	2.631	6	1.3
1	1	3	34.52	2.598	12	0.2
2	0	3	34.96	2.566	12	0.1
3	4	1	35.01	2.563	12	0.1
5	0	2	35.16	2.553	12	0.2
1	6	0	35.83	2.506	6	0.2
6	1	0	35.83	2.506	6	0.1
2	5	1	35.88	2.503	12	0.7
5	2	1	35.88	2.503	12	0.2
3	3	2	36.02	2.493	12	0.6
1	2	3	36.26	2.477	12	0.1
4	2	2	36.45	2.465	12	0.2
3	0	3	37.11	2.423	12	0.1
1	5	2	37.70	2.386	12	0.2
5	1	2	37.70	2.386	12	0.3
2	2	3	38.34	2.348	12	1.8
1	3	3	38.75	2.324	12	0.4
3	1	3	38.75	2.324	12	0.2
6	2	0	39.54	2.279	6	0.5
4	4	1	39.59	2.276	12	0.1
6	0	2	39.72	2.269	12	0.3
4	0	3	39.94	2.257	12	0.1
3	5	1	39.98	2.255	12	0.1
7	0	1	39.98	2.255	12	0.1
4	3	2	40.11	2.248	12	0.2
5	2	2	40.89	2.207	12	0.2
2	5	2	40.89	2.207	12	0.2
3	2	3	41.10	2.196	12	0.3
2	3	3	41.10	2.196	12	0.1
2	6	1	41.15	2.194	12	0.3
6	2	1	41.15	2.194	12	0.2
1	4	3	41.87	2.158	12	0.2
4	1	3	41.87	2.158	12	0.2
7	1	1	43.03	2.102	12	0.1
3	3	3	44.09	2.054	12	0.2
4	4	2	44.25	2.047	12	0.2
4	2	3	44.45	2.038	12	0.2
2	4	3	44.45	2.038	12	0.2
5	4	1	44.49	2.037	12	0.1
4	5	1	44.49	2.037	12	0.1
7	0	2	44.61	2.031	12	0.4
0	0	4	44.73	2.026	2	0.7
1	0	4	45.08	2.011	12	0.4
3	6	1	45.20	2.006	12	0.1
5	1	3	45.52	1.993	12	0.1
2	6	2	45.67	1.986	12	0.1
2	0	4	46.14	1.967	12	0.1
7	2	1	46.61	1.949	12	0.3
2	7	1	46.61	1.949	12	0.3
2	1	4	47.18	1.926	12	0.2
7	1	2	47.41	1.917	12	0.3
1	7	2	47.41	1.917	12	0.4
4	3	3	47.60	1.910	12	0.1
3	0	4	47.87	1.900	12	0.1
2	5	3	48.28	1.885	12	0.1
8	1	1	48.65	1.871	12	0.6
1	8	1	48.65	1.871	12	0.4
2	2	4	48.88	1.863	12	0.5
1	3	4	49.21	1.851	12	0.4
3	1	4	49.21	1.851	12	0.3
3	7	0	49.29	1.849	6	0.2
5	5	1	49.32	1.848	12	0.4
4	6	1	49.65	1.836	12	0.8
6	4	1	49.65	1.836	12	0.6
8	0	2	49.76	1.832	12	0.2

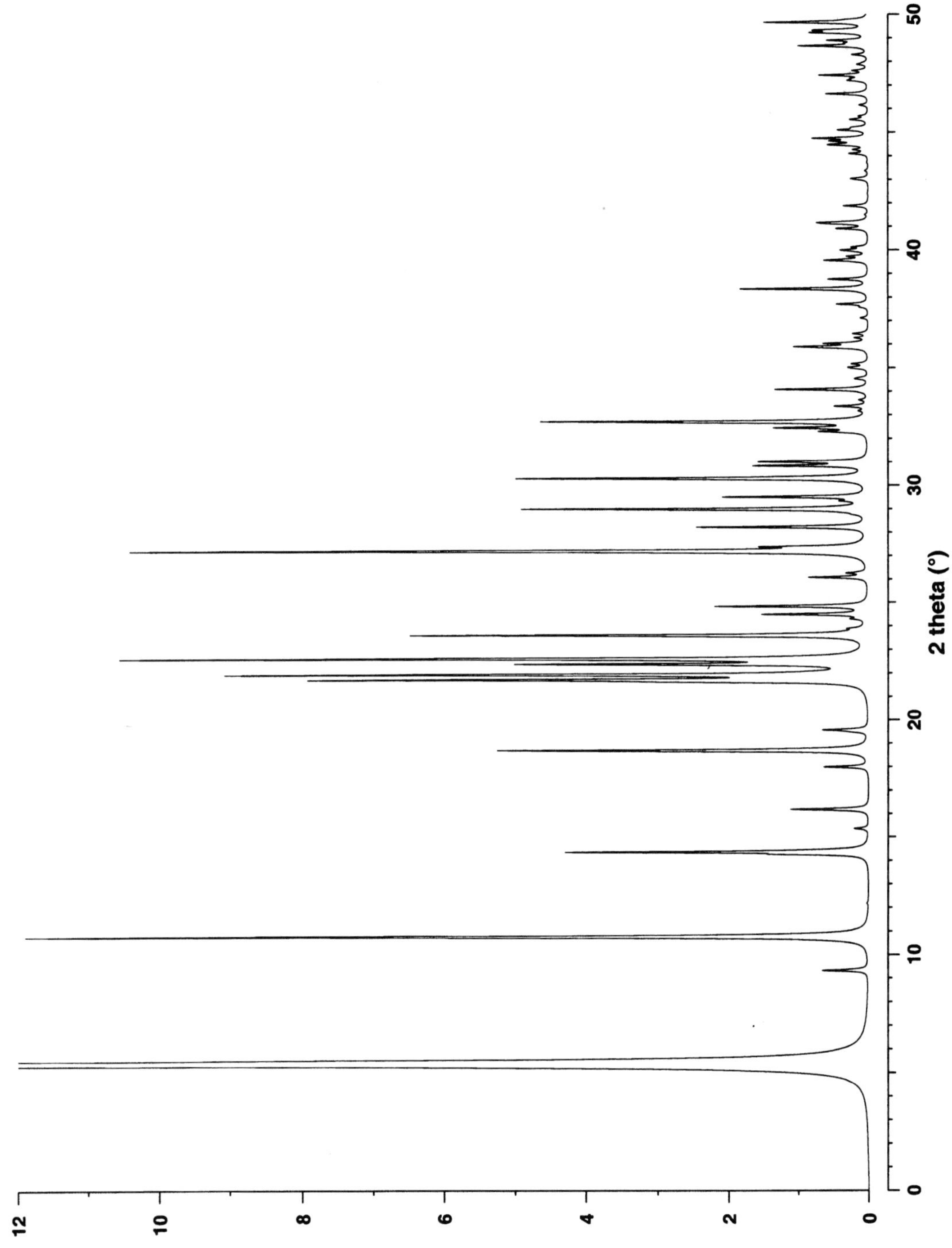
2 theta (°)
0
10
20
30
40
50
0
2
4
6
8
10
12

VNI — VPI-9

CHEMICAL COMPOSITION: $|Rb_{44}K_4(H_2O)_{48}|\ [Si_{96}Zn_{24}O_{240}]$

REFINED COMPOSITION: $|Rb_{44}K_4(H_2O)_{14}|\ [Si_{96}Zn_{24}O_{240}]$

CRYSTAL DATA: $P4_12_12$ (No. 92)
$a = 9.8837$ Å $b = 9.8837$ Å $c = 73.6505$ Å
$\alpha = 90°$ $\beta = 90°$ $\gamma = 90°$
X-ray Rietveld refinement, $R_{exp} = 0.099$, $R_{wp} = 0.147$, $R_F = 0.069$

REFERENCE: L. B. McCusker, R. W. Grosse-Kunstleve, Ch. Baerlocher, M. Yoshikawa, and M. E. Davis, *Microporous Materials* **6** 295–309 (1996).

h	k	l	2θ	d	M	I_{rel}
0	0	4	4.80	18.413	2	21.0
1	0	2	9.26	9.546	8	2.7
0	0	8	9.61	9.206	2	6.8
1	0	3	9.65	9.169	8	3.3
1	0	4	10.16	8.708	8	34.5
1	0	5	10.78	8.207	8	5.0
1	0	7	12.29	7.204	8	4.9
1	1	0	12.67	6.989	4	14.0
1	1	2	12.89	6.866	8	24.5
1	0	8	13.14	6.737	8	14.3
0	0	12	14.43	6.138	2	22.1
1	1	6	14.58	6.073	8	20.2
1	1	8	15.92	5.567	8	7.6
1	1	10	17.49	5.070	8	23.5
2	0	0	17.95	4.942	4	5.6
2	0	4	18.59	4.773	8	29.3
1	1	12	19.25	4.612	8	9.4
2	1	1	20.12	4.412	16	4.7
1	0	15	20.19	4.397	8	2.4
2	1	2	20.23	4.389	16	5.1
2	0	8	20.40	4.354	8	42.0
2	1	3	20.41	4.350	16	2.8
2	1	4	20.67	4.298	16	26.6
2	1	5	20.98	4.234	16	6.9
1	1	14	21.14	4.203	8	2.5
1	0	16	21.29	4.173	8	2.3
2	1	7	21.81	4.075	16	5.9
1	0	17	22.41	3.968	8	5.8
2	1	9	22.87	3.889	16	2.3
2	0	12	23.11	3.849	8	23.2
1	1	16	23.14	3.844	8	6.1
1	0	18	23.53	3.781	8	3.1
2	1	12	24.82	3.587	16	4.3
2	2	0	25.49	3.494	4	74.9
2	2	1	25.52	3.490	8	3.9
2	0	15	25.57	3.483	8	3.7
2	2	2	25.61	3.479	8	50.8
1	0	20	25.82	3.451	8	17.3
2	1	14	26.33	3.384	16	66.3
2	0	16	26.46	3.368	8	33.7
2	2	6	26.52	3.361	8	10.6
2	2	7	26.88	3.316	8	3.6
2	2	8	27.30	3.267	8	8.9
1	0	22	28.14	3.171	8	3.5
2	2	10	28.27	3.157	8	100.0

h	k	l	2θ	d	M	I_{rel}
1	1	21	28.47	3.135	8	3.7
3	1	0	28.56	3.126	8	37.8
3	1	1	28.58	3.123	16	3.1
3	1	2	28.66	3.114	16	55.2
3	0	8	28.78	3.102	8	64.5
3	1	4	28.98	3.081	16	89.8
0	0	24	29.10	3.069	2	96.8
3	1	5	29.21	3.057	16	6.4
3	0	9	29.22	3.056	8	3.9
2	2	12	29.41	3.037	8	29.8
1	1	22	29.59	3.019	8	46.0
2	1	18	29.75	3.003	16	10.1
3	1	8	30.20	2.960	16	3.6
2	0	20	30.27	2.953	8	53.3
1	1	23	30.71	2.911	8	3.7
3	1	10	31.08	2.877	16	5.6
2	0	21	31.27	2.860	8	3.9
3	1	11	31.59	2.832	16	2.7
3	1	12	32.14	2.785	16	18.2
3	1	13	32.72	2.737	16	2.3
3	0	15	32.73	2.736	8	2.6
1	0	26	32.89	2.723	8	2.5
3	2	4	33.04	2.711	16	7.8
3	1	14	33.34	2.687	16	21.4
3	0	16	33.45	2.679	8	49.4
2	1	22	33.58	2.669	16	5.0
2	2	18	33.73	2.657	8	17.7
3	2	7	33.79	2.653	16	3.6
3	2	8	34.13	2.627	16	8.8
3	1	16	34.69	2.586	16	4.2
2	2	20	35.41	2.535	8	5.9
2	1	24	35.61	2.521	16	3.8
3	2	12	35.88	2.503	16	4.1
3	1	18	36.16	2.484	16	14.7
4	0	0	36.36	2.471	4	4.5
3	1	19	36.94	2.433	16	2.4
3	2	14	36.98	2.431	16	5.0
2	2	22	37.19	2.417	8	6.6
3	0	21	37.45	2.401	8	4.4
3	2	15	37.58	2.393	16	4.3
4	0	8	37.69	2.386	8	3.1
2	1	26	37.72	2.385	16	3.3
3	1	20	37.75	2.383	16	12.6
4	1	5	38.03	2.366	16	2.3
4	1	7	38.52	2.337	16	4.3

h	k	l	2θ	d	M	I_{rel}
3	3	0	38.65	2.330	4	11.5
2	1	27	38.79	2.321	16	3.0
3	2	17	38.88	2.316	16	3.8
1	1	30	38.88	2.316	8	2.8
2	2	24	39.06	2.306	8	17.5
4	0	12	39.31	2.292	8	4.6
3	3	6	39.37	2.289	8	6.4
3	1	22	39.44	2.285	16	3.3
3	2	18	39.57	2.277	16	2.4
2	1	28	39.88	2.260	16	6.2
3	2	19	40.29	2.238	16	2.9
3	1	23	40.32	2.237	16	2.7
4	1	13	40.88	2.208	16	3.0
4	2	2	40.91	2.206	16	3.2
2	1	29	40.98	2.202	16	4.2
3	2	20	41.05	2.199	16	4.7
3	1	24	41.23	2.190	16	7.3
3	3	12	41.46	2.178	8	6.3
4	0	16	41.48	2.177	8	7.4
4	2	8	42.04	2.149	16	5.9
2	1	30	42.10	2.146	16	4.1
3	1	26	43.10	2.099	16	14.5
2	0	32	43.37	2.086	8	5.7
4	2	12	43.52	2.079	16	4.2
3	3	16	43.54	2.079	8	3.7
4	0	20	44.14	2.052	8	3.2
4	2	14	44.46	2.038	16	2.5
3	3	18	44.77	2.024	8	3.9
3	1	28	45.05	2.013	16	15.1
4	3	2	45.98	1.974	16	2.7
3	3	20	46.10	1.969	8	4.2
4	1	22	46.60	1.949	16	2.8
4	2	18	46.71	1.945	16	2.7
5	1	2	46.94	1.936	16	5.6
5	0	8	47.02	1.933	8	3.4
4	3	8	47.02	1.933	16	2.4
5	1	4	47.14	1.928	16	3.9
5	1	6	47.49	1.915	16	3.1
4	3	10	47.63	1.909	16	6.7
4	2	20	48.01	1.895	16	4.8
5	0	12	48.37	1.882	8	3.2
3	3	24	49.10	1.856	8	5.5
5	1	12	49.30	1.848	16	13.6
4	2	22	49.41	1.844	16	2.9
2	2	34	49.51	1.841	8	4.6

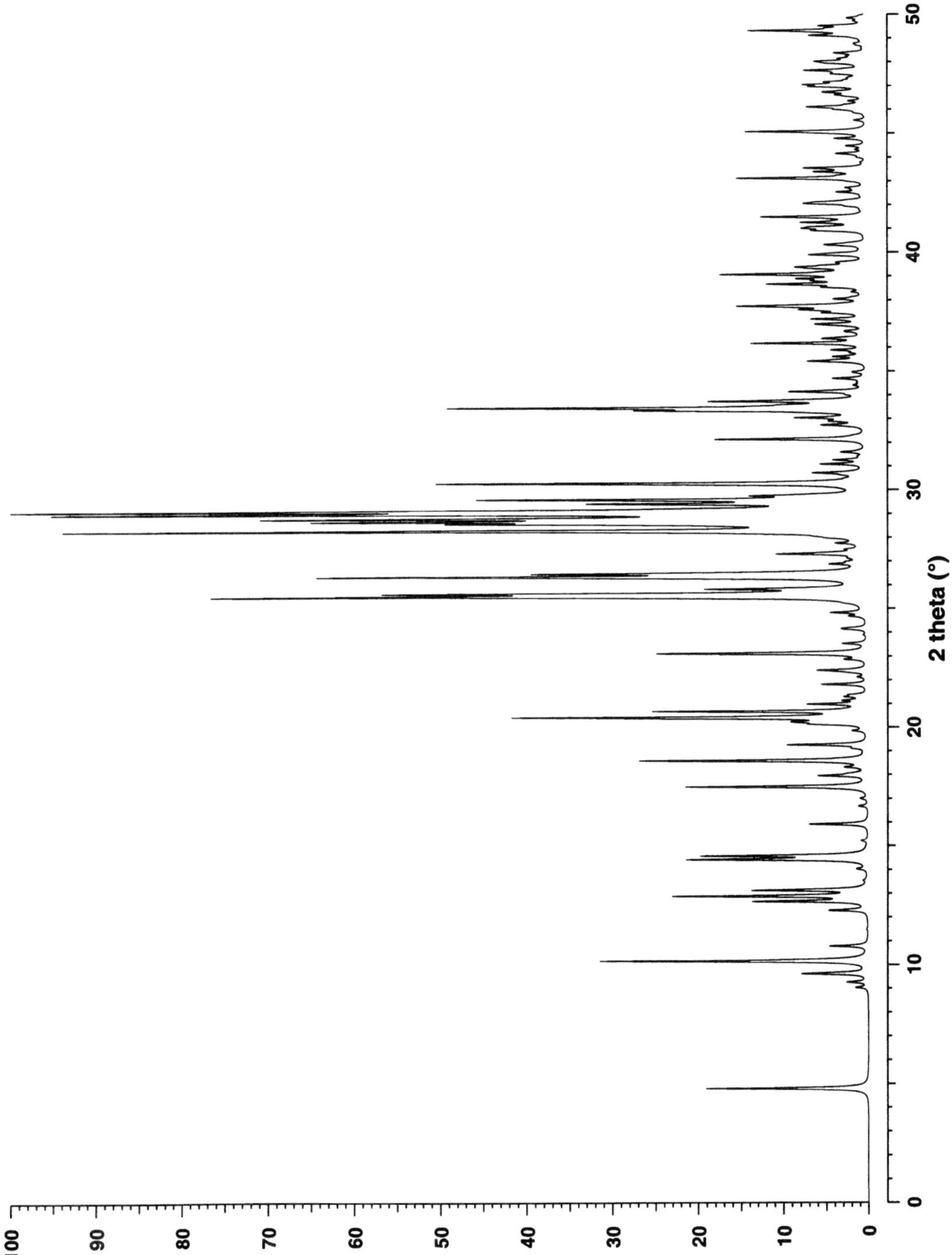
0
10
20
30
40
50
2 theta (°)
100
90
80
70
60
50
40
30
20
10
0

VSV VPI-7

CHEMICAL COMPOSITION: $|Na_{32}(H_2O)_{40}|\ [Si_{56}Zn_{16}O_{144}]$

REFINED COMPOSITION: $|Na_{28.48}(H_2O)_{49.76}|\ [Si_{56}Zn_{16}O_{144}]$

CRYSTAL DATA: $Fdd2$ (No. 43)
$a = 39.88$ Å $b = 10.326$ Å $c = 10.219$ Å
$\alpha = 90°$ $\beta = 90°$ $\gamma = 90°$
X-ray Rietveld refinement, $R_{exp} = 0.248$, $R_{wp} = 0.162$, $R_I = 0.100$

REFERENCE: C. Röhrig, H. Gies and B. Marler, *Zeolites* **14** 498–503 (1994).

h	k	l	2θ	d	M	I_{rel}
4	0	0	8.87	9.970	2	9.1
1	1	1	12.39	7.146	8	1.8
3	1	1	13.89	6.374	8	100.0
5	1	1	16.51	5.370	8	30.4
2	2	0	17.75	4.998	4	4.0
8	0	0	17.79	4.985	2	17.3
2	0	2	17.92	4.950	4	22.4
4	2	0	19.36	4.585	4	9.1
7	1	1	19.80	4.483	8	5.9
6	0	2	21.94	4.051	4	0.8
9	1	1	23.52	3.783	8	2.3
0	2	2	24.51	3.632	4	0.4
8	2	0	24.83	3.586	4	0.3
2	2	2	24.92	3.573	8	3.2
4	2	2	26.11	3.412	8	11.8
12	0	0	26.83	3.323	2	15.3
1	3	1	27.43	3.251	8	50.8
11	1	1	27.50	3.244	8	22.6
1	1	3	27.67	3.224	8	14.5
6	2	2	28.00	3.187	8	36.8
3	3	1	28.17	3.168	8	14.1
10	2	0	28.28	3.156	4	6.9
10	0	2	28.39	3.144	4	8.7
3	1	3	28.39	3.143	8	30.8
5	3	1	29.59	3.019	8	3.9
5	1	3	29.80	2.998	8	26.3
8	2	2	30.45	2.935	8	17.8
7	3	1	31.61	2.831	8	7.2
13	1	1	31.66	2.826	8	24.1
7	1	3	31.81	2.813	8	0.9
12	2	0	32.03	2.794	4	3.6
10	2	2	33.37	2.685	8	1.0
9	3	1	34.13	2.627	8	1.2
9	1	3	34.32	2.613	8	0.3
0	4	0	34.75	2.582	2	2.2
2	4	0	35.05	2.560	4	6.5
0	0	4	35.13	2.555	2	5.2
4	4	0	35.93	2.499	4	0.3
15	1	1	35.97	2.497	8	10.1
16	0	0	36.03	2.493	2	1.5
14	0	2	36.10	2.488	4	9.1
4	0	4	36.30	2.475	4	0.5
12	2	2	36.65	2.452	8	6.3
11	3	1	37.07	2.425	8	6.7
1	3	3	37.20	2.417	8	4.4
11	1	3	37.25	2.414	8	7.9
6	4	0	37.37	2.406	4	1.1
3	3	3	37.77	2.382	8	12.6
5	3	3	38.87	2.317	8	1.1
8	4	0	39.30	2.292	4	1.9
2	4	2	39.36	2.289	8	2.6
8	0	4	39.64	2.274	4	5.1
4	4	2	40.17	2.245	8	0.6
16	2	0	40.17	2.245	4	0.9
14	2	2	40.23	2.241	8	8.7
13	3	1	40.36	2.235	8	3.4
17	1	1	40.40	2.232	8	1.9
4	2	4	40.42	2.232	8	3.3
7	3	3	40.48	2.228	8	0.2
13	1	3	40.53	2.226	8	1.5
6	4	2	41.48	2.177	8	1.4
10	4	0	41.68	2.167	4	0.5
6	2	4	41.72	2.165	8	0.4
9	3	3	42.55	2.125	8	1.1
8	4	2	43.26	2.092	8	0.7
8	2	4	43.49	2.081	8	0.8
15	3	1	43.93	2.061	8	0.4
16	2	2	44.06	2.055	8	1.3
15	1	3	44.09	2.054	8	0.6
12	4	0	44.44	2.039	4	1.6
18	2	0	44.50	2.036	4	1.8
18	0	2	44.57	2.033	4	6.3
12	0	4	44.74	2.025	4	3.8
1	5	1	44.83	2.022	8	2.2
19	1	1	44.95	2.016	8	0.6
11	3	3	45.02	2.013	8	0.3
1	1	5	45.29	2.002	8	1.2
3	5	1	45.31	2.001	8	1.2
10	4	2	45.46	1.995	8	3.0
20	0	0	45.49	1.994	2	0.1
10	2	4	45.69	1.986	8	2.6
3	1	5	45.77	1.982	8	4.3
5	5	1	46.27	1.962	8	3.5
5	1	5	46.72	1.944	8	0.1
14	4	0	47.53	1.913	4	1.5
7	5	1	47.68	1.907	8	3.1
17	3	1	47.75	1.905	8	0.4
13	3	3	47.86	1.901	8	1.7
17	1	3	47.90	1.899	8	0.1
12	4	2	48.05	1.894	8	0.6
7	1	5	48.11	1.891	8	1.8
12	2	4	48.26	1.886	8	1.6
20	2	0	48.97	1.860	4	0.8
9	5	1	49.50	1.841	8	1.5
21	1	1	49.62	1.837	8	3.3
9	1	5	49.93	1.827	8	2.3

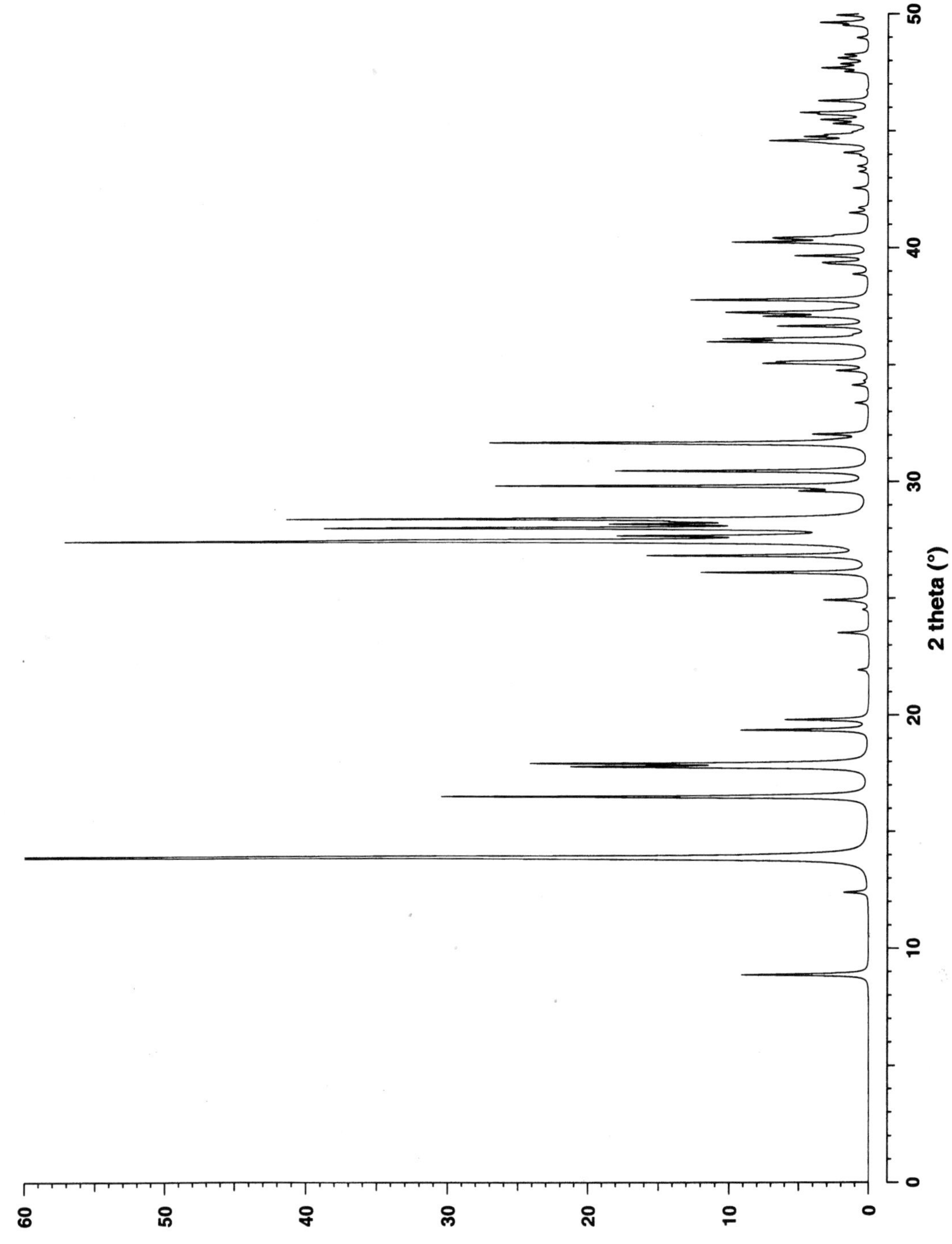
2 theta (°)
0
10
20
30
40
50
60

WEI Weinebeneite

CHEMICAL COMPOSITION: $|Ca_4(H_2O)_{16}|\ [Be_{12}P_8O_{32}(OH)_8]$
Weinebene Pass, Koralpe, Austria

REFINED COMPOSITION: $|Ca_4(H_2O)_{16}|\ [Be_{12}P_8O_{32}(OH)_8]$

CRYSTAL DATA: $C1c1$ (No. 9) unique axis **b**, cell choice 1
$a = 11.897$ Å $b = 9.707$ Å $c = 9.633$ Å
$\alpha = 90°$ $\beta = 95.76°$ $\gamma = 90°$
X-ray single crystal refinement, $R_w = 0.044$

REFERENCE: F. Walter,
Eur. J. Mineral. **4** 1275–1283 (1992).

h	k	l	2θ	d	M	I_{rel}
1	1	0	11.79	7.506	4	17.7
-1	1	1	14.52	6.101	4	19.0
2	0	0	14.97	5.918	2	100.0
1	1	1	15.45	5.735	4	69.7
0	2	0	18.28	4.853	2	35.3
0	0	2	18.51	4.792	2	34.2
0	2	1	20.51	4.330	4	52.2
-1	1	2	21.35	4.161	4	38.1
1	1	2	22.64	3.927	4	7.1
-2	0	2	22.67	3.922	2	8.6
2	2	0	23.71	3.753	4	1.4
3	1	0	24.35	3.655	4	7.4
-2	2	1	24.92	3.573	4	12.5
2	0	2	25.06	3.554	2	10.8
-3	1	1	25.26	3.526	4	7.1
2	2	1	26.04	3.421	4	63.8
0	2	2	26.13	3.410	4	7.4
3	1	1	26.90	3.314	4	8.3
1	3	0	28.60	3.121	4	5.0
-2	2	2	29.28	3.050	4	2.2
-3	1	2	29.32	3.046	4	30.2
-1	1	3	29.69	3.009	4	14.3
-1	3	1	29.87	2.991	4	19.3
4	0	0	30.20	2.959	2	38.3
1	3	1	30.35	2.945	4	34.3
1	1	3	31.11	2.874	4	27.5
2	2	2	31.19	2.867	4	2.2
3	1	2	32.15	2.784	4	33.0
0	2	3	33.58	2.669	4	0.3
-1	3	2	33.86	2.648	4	1.4
-4	0	2	33.97	2.639	2	3.5

h	k	l	2θ	d	M	I_{rel}
1	3	2	34.71	2.584	4	0.7
4	2	0	35.53	2.527	4	1.5
-3	1	3	35.56	2.525	4	3.8
-2	2	3	35.73	2.513	4	65.1
3	3	0	35.89	2.502	4	8.9
-4	2	1	35.97	2.496	4	1.8
-3	3	1	36.53	2.459	4	20.1
0	4	0	37.04	2.427	2	14.9
4	0	2	37.28	2.412	2	1.0
0	0	4	37.54	2.396	2	6.9
4	2	1	37.58	2.393	4	4.2
3	3	1	37.73	2.384	4	2.9
2	2	3	38.14	2.360	4	12.0
0	4	1	38.26	2.353	4	1.4
-1	1	4	38.71	2.326	4	5.7
-4	2	2	38.84	2.318	4	1.4
-2	0	4	39.12	2.303	2	1.9
3	1	3	39.13	2.302	4	4.5
5	1	0	39.17	2.300	4	1.3
-5	1	1	39.39	2.288	4	2.4
-1	3	3	39.84	2.263	4	1.4
2	4	0	40.16	2.245	4	4.3
1	1	4	40.23	2.242	4	7.6
-2	4	1	40.93	2.205	4	5.6
1	3	3	40.95	2.204	4	2.0
5	1	1	41.25	2.189	4	3.7
2	4	1	41.66	2.168	4	4.6
0	4	2	41.72	2.165	4	0.8
3	3	2	41.78	2.162	4	0.3
4	2	2	41.82	2.160	4	3.9
-5	1	2	41.88	2.157	4	2.5

h	k	l	2θ	d	M	I_{rel}
0	2	4	42.05	2.149	4	2.4
2	0	4	42.08	2.147	2	3.0
-3	1	4	43.18	2.095	4	2.2
-2	2	4	43.50	2.081	4	0.8
-4	2	3	43.73	2.070	4	5.8
-2	4	2	43.87	2.064	4	8.9
-3	3	3	44.55	2.034	4	0.7
2	4	2	45.25	2.004	4	1.5
5	1	2	45.37	1.999	4	2.3
6	0	0	46.00	1.973	2	8.9
2	2	4	46.23	1.964	4	0.8
-4	0	4	46.30	1.961	2	5.5
-5	1	3	46.34	1.959	4	3.2
0	4	3	47.02	1.932	4	3.1
3	1	4	47.25	1.924	4	6.2
1	5	0	47.46	1.916	4	0.6
3	3	3	47.57	1.912	4	7.2
5	3	0	47.59	1.911	4	1.4
4	2	3	47.77	1.904	4	4.7
-5	3	1	47.78	1.904	4	24.6
-6	0	2	48.08	1.892	2	1.9
-1	1	5	48.25	1.886	4	1.7
-1	5	1	48.29	1.885	4	1.0
1	3	4	48.50	1.877	4	0.3
1	5	1	48.61	1.873	4	1.1
-2	4	3	48.67	1.871	4	11.3
-4	4	1	48.86	1.864	4	2.5
5	3	1	49.39	1.845	4	2.1
1	1	5	49.84	1.830	4	0.4
6	2	0	49.90	1.828	4	3.8
-6	2	1	49.92	1.827	4	0.9

Weinebeneite

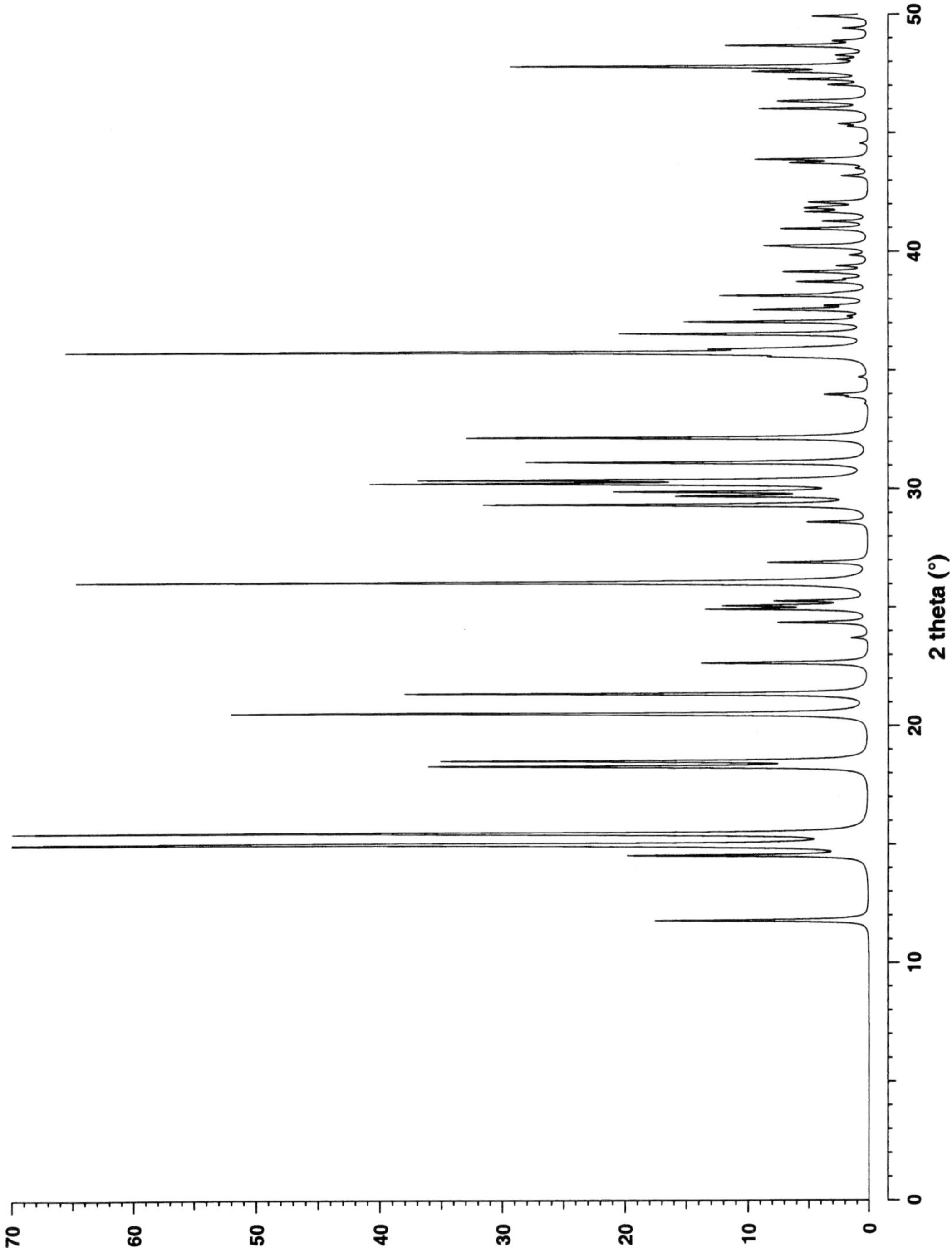

-WEN Wenkite

CHEMICAL COMPOSITION: $|Na_{0.5}K_{0.5}Ca_{5.5}Ba_{3.5}(SO_4)_3H_2O|$ $[Si_{11}Al_9O_{41}(OH)_2]$
Candoglia, Italy

REFINED COMPOSITION: $|Na_{0.5}K_{0.5}Ca_{5.5}Ba_{3.5}S_3(OH)_{12}|$ $[Si_{11}Al_9O_{41}(OH)_2]$

CRYSTAL DATA: $P\overline{6}2m$ (No. 189)
$a = 13.511$ Å $b = 13.511$ Å $c = 7.462$ Å
$\alpha = 90°$ $\beta = 90°$ $\gamma = 120°$
X-ray single crystal refinement, $R = 0.10$

REFERENCE: S. Merlino,
Acta Cryst. **B30** 1262–1266 (1974).

h	*k*	*l*	2θ	*d*	*M*	I_{rel}	*h*	*k*	*l*	2θ	*d*	*M*	I_{rel}	*h*	*k*	*l*	2θ	*d*	*M*	I_{rel}
1	0	0	7.56	11.701	6	0.1	2	1	2	31.37	2.852	24	37.4	4	2	1	42.64	2.120	24	23.8
0	0	1	11.86	7.462	2	24.9	4	0	1	32.89	2.723	12	15.0	4	1	2	42.92	2.107	24	1.7
1	1	0	13.11	6.755	6	27.7	3	0	2	33.23	2.696	12	32.7	5	1	0	43.04	2.102	12	2.5
1	0	1	14.08	6.292	12	12.0	3	2	0	33.38	2.684	12	68.0	3	0	3	43.13	2.097	12	21.9
2	0	0	15.14	5.850	6	3.9	4	1	0	35.15	2.553	12	3.6	5	1	1	44.80	2.023	24	0.5
1	1	1	17.71	5.008	12	17.5	3	2	1	35.54	2.526	24	2.4	2	2	3	45.27	2.003	12	2.3
2	0	1	19.28	4.604	12	32.1	2	2	2	35.86	2.504	12	16.3	5	0	2	45.77	1.982	12	8.6
2	1	0	20.08	4.423	12	4.3	0	0	3	36.11	2.487	2	3.1	3	1	3	45.97	1.974	24	8.8
3	0	0	22.80	3.900	6	0.1	3	1	2	36.70	2.449	24	30.9	6	0	0	46.57	1.950	6	3.7
2	1	1	23.38	3.805	24	31.6	4	1	1	37.22	2.416	24	18.6	3	3	2	47.14	1.928	12	3.9
0	0	2	23.85	3.731	2	32.5	5	0	0	38.47	2.340	6	19.1	4	3	0	47.25	1.924	12	2.5
1	0	2	25.05	3.555	12	56.8	1	1	3	38.57	2.334	12	5.5	4	2	2	47.81	1.902	24	0.9
3	0	1	25.77	3.457	12	100.0	4	0	2	39.13	2.302	12	7.7	4	0	3	48.01	1.895	12	10.6
2	2	0	26.39	3.378	6	62.5	2	0	3	39.36	2.289	12	15.1	5	2	0	48.59	1.874	12	1.6
1	1	2	27.31	3.266	12	4.2	3	3	0	40.04	2.252	6	7.1	0	0	4	48.82	1.865	2	19.5
3	1	0	27.48	3.245	12	63.2	5	0	1	40.39	2.233	12	65.5	4	3	1	48.89	1.863	24	10.2
2	0	2	28.37	3.146	12	83.0	4	2	0	40.81	2.211	12	1.1	1	0	4	49.47	1.842	12	0.7
2	2	1	29.02	3.077	12	20.1	3	2	2	41.44	2.179	24	19.0	5	1	2	49.80	1.831	24	1.6
3	1	1	30.03	2.976	24	26.9	2	1	3	41.66	2.168	24	5.1	3	2	3	49.99	1.824	24	22.6
4	0	0	30.56	2.925	6	0.1	3	3	1	41.90	2.156	12	3.0							

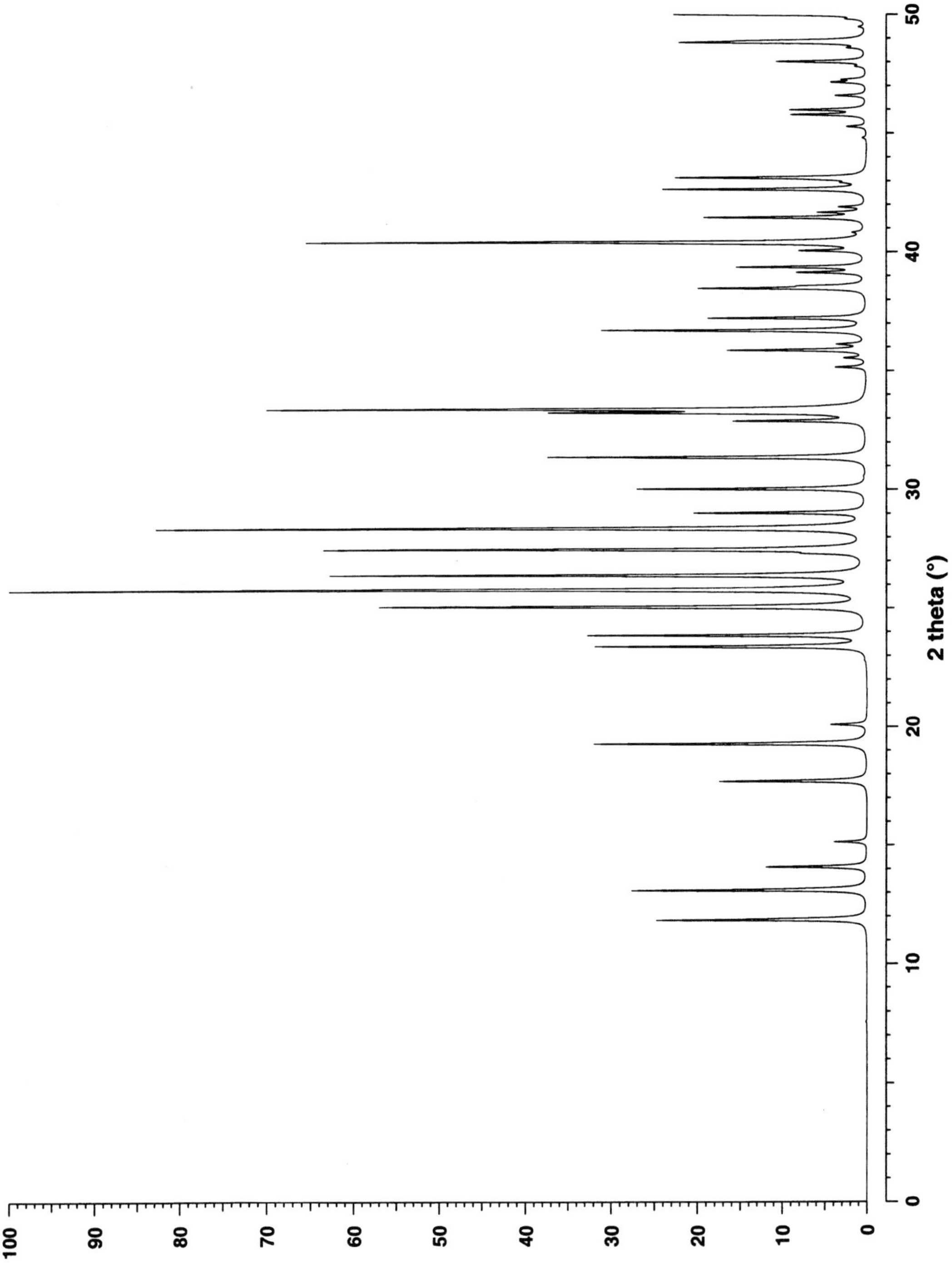
2 theta (°)
0
10
20
30
40
50
100
90
80
70
60
50
40
30
20
10
0

YUG Yugawaralite

CHEMICAL COMPOSITION: $|Ca_2(H_2O)_8|$ $[Si_{12}Al_4O_{32}]$
Hvalfjördur, Iceland

REFINED COMPOSITION: $|Ca_2H_{15.82}O_{8.136}|$ $[Si_{12}Al_4O_{32}]$

CRYSTAL DATA: $P1c1$ (No. 7) unique axis **b**, cell choice 1
$a = 6.700$ Å $b = 13.972$ Å $c = 10.039$ Å
$\alpha = 90°$ $\beta = 111.07°$ $\gamma = 90°$
Neutron single crystal refinement at $T = 13$K, $R_w = 0.045$

REFERENCE: Å. Kvick, G. Artioli and J. V. Smith,
Z. Kristallogr. **174** 265–281 (1986).

h	k	l	2θ	d	M	I_{rel}
0	1	0	6.33	13.972	2	4.8
0	1	1	11.37	7.781	4	8.1
0	2	0	12.67	6.986	2	30.6
1	0	0	14.17	6.252	2	6.4
-1	1	1	15.30	5.790	4	82.6
1	1	0	15.53	5.707	4	1.5
0	2	1	15.82	5.600	4	2.4
0	0	2	18.95	4.684	2	29.3
1	2	0	19.05	4.659	4	51.1
0	3	0	19.06	4.657	2	6.4
-1	0	2	19.16	4.632	2	71.5
0	1	2	19.99	4.441	4	5.9
-1	1	2	20.20	4.397	4	13.5
1	1	1	20.71	4.288	4	34.9
0	3	1	21.30	4.170	4	18.1
0	2	2	22.86	3.890	4	6.7
-1	2	2	23.04	3.860	4	2.2
1	2	1	23.49	3.787	4	1.9
-1	3	1	23.67	3.758	4	10.0
1	3	0	23.82	3.735	4	3.2
0	3	2	27.00	3.303	4	5.4
-1	3	2	27.15	3.284	4	3.5
0	4	1	27.25	3.273	4	2.1
-2	1	1	27.39	3.256	4	6.5
1	3	1	27.54	3.238	4	8.2
1	0	2	27.60	3.232	2	27.2
-1	1	3	27.70	3.220	4	11.1
-2	0	2	28.05	3.181	2	2.3
1	1	2	28.34	3.149	4	3.4
2	0	0	28.55	3.126	2	7.7
-2	1	2	28.78	3.102	4	5.2
2	1	0	29.28	3.051	4	1.6
1	4	0	29.29	3.049	4	100.0
0	1	3	29.31	3.047	4	16.9
-2	2	1	29.58	3.020	4	2.5
-1	2	3	29.87	2.991	4	3.9
1	2	2	30.47	2.933	4	14.9
-2	2	2	30.88	2.895	4	24.3
2	2	0	31.35	2.853	4	4.6
0	2	3	31.38	2.851	4	1.0
0	4	2	31.96	2.800	4	0.7
-1	4	2	32.09	2.789	4	0.5
1	4	1	32.43	2.761	4	14.7
-2	1	3	33.08	2.708	4	17.6
-1	3	3	33.21	2.698	4	4.3

h	k	l	2θ	d	M	I_{rel}
0	5	1	33.46	2.678	4	10.1
1	3	2	33.75	2.655	4	4.0
2	1	1	33.95	2.641	4	8.8
-2	3	2	34.13	2.627	4	8.3
2	3	0	34.56	2.596	4	2.6
-2	2	3	34.95	2.567	4	5.0
-1	5	1	35.07	2.559	4	1.7
1	5	0	35.18	2.551	4	3.2
2	2	1	35.78	2.510	4	1.0
-1	0	4	35.78	2.509	2	4.3
-1	1	4	36.37	2.470	4	5.5
1	1	3	37.08	2.425	4	5.9
-2	4	1	37.19	2.417	4	2.5
0	5	2	37.48	2.400	4	2.6
-1	5	2	37.59	2.393	4	0.7
-2	3	3	37.89	2.375	4	0.9
1	4	2	37.92	2.372	4	0.8
-1	2	4	38.10	2.362	4	9.4
-2	4	2	38.27	2.352	4	0.7
0	0	4	38.44	2.342	2	8.3
2	4	0	38.65	2.329	4	1.4
2	3	1	38.66	2.329	4	3.2
0	6	0	38.66	2.329	2	0.5
0	4	3	38.68	2.328	4	1.4
1	2	3	38.78	2.322	4	1.3
-2	0	4	38.89	2.316	2	2.1
0	1	4	38.99	2.310	4	1.0
-2	1	4	39.44	2.285	4	1.1
0	6	1	39.89	2.260	4	0.7
2	1	2	40.56	2.224	4	1.0
-3	0	2	40.56	2.224	2	1.2
0	2	4	40.63	2.221	4	1.1
-1	3	4	40.85	2.209	4	2.0
-2	2	4	41.06	2.198	4	2.5
-3	1	2	41.10	2.196	4	0.8
-1	6	1	41.28	2.187	4	3.3
-3	1	1	41.37	2.183	4	0.5
1	6	0	41.37	2.182	4	0.5
1	3	3	41.49	2.177	4	2.5
-2	4	3	41.70	2.166	4	1.6
-2	5	1	42.12	2.146	4	1.8
2	2	2	42.14	2.144	4	2.4
-1	5	3	42.33	2.135	4	1.9
2	4	1	42.42	2.131	4	4.2
1	5	2	42.78	2.114	4	1.3

h	k	l	2θ	d	M	I_{rel}
-2	5	2	43.08	2.099	4	1.3
-3	1	3	43.14	2.097	4	3.9
0	3	4	43.24	2.092	4	1.2
0	6	2	43.39	2.085	4	0.9
3	0	0	43.42	2.084	2	8.9
2	5	0	43.43	2.083	4	0.9
0	5	3	43.46	2.082	4	1.2
-1	6	2	43.50	2.081	4	0.8
-2	3	4	43.65	2.074	4	0.6
1	6	1	43.76	2.069	4	0.9
-1	4	4	44.45	2.038	4	2.0
-3	2	3	44.65	2.029	4	0.6
2	3	2	44.68	2.028	4	0.6
1	4	3	45.05	2.012	4	1.7
-3	3	2	45.18	2.007	4	2.5
3	2	0	45.41	1.997	4	0.9
-3	3	1	45.43	1.996	4	4.7
0	7	0	45.44	1.996	2	1.1
-1	1	5	45.74	1.983	4	1.3
1	0	4	46.01	1.973	2	3.3
-2	5	3	46.23	1.964	4	0.4
1	1	4	46.49	1.953	4	1.3
0	7	1	46.52	1.952	4	5.5
-3	0	4	46.79	1.942	2	2.3
2	5	1	46.89	1.938	4	0.6
-2	1	5	47.19	1.926	4	2.2
-1	2	5	47.19	1.926	4	0.7
-3	1	4	47.26	1.923	4	4.7
-1	6	3	47.76	1.904	4	2.1
3	3	0	47.81	1.902	4	0.7
1	7	0	47.84	1.901	4	1.8
1	2	4	47.92	1.898	4	0.8
2	4	2	48.06	1.893	4	9.1
3	1	1	48.48	1.878	4	0.5
2	1	3	48.48	1.878	4	1.9
-3	4	2	48.53	1.876	4	0.8
-2	2	5	48.60	1.873	4	1.0
2	6	0	48.76	1.867	4	1.6
-3	4	1	48.77	1.867	4	1.7
0	1	5	49.06	1.857	4	1.3
-1	3	5	49.52	1.841	4	1.0
0	7	2	49.65	1.836	4	0.6
-1	7	2	49.74	1.833	4	0.8
3	2	1	49.86	1.829	4	0.5
1	7	1	49.98	1.825	4	0.7

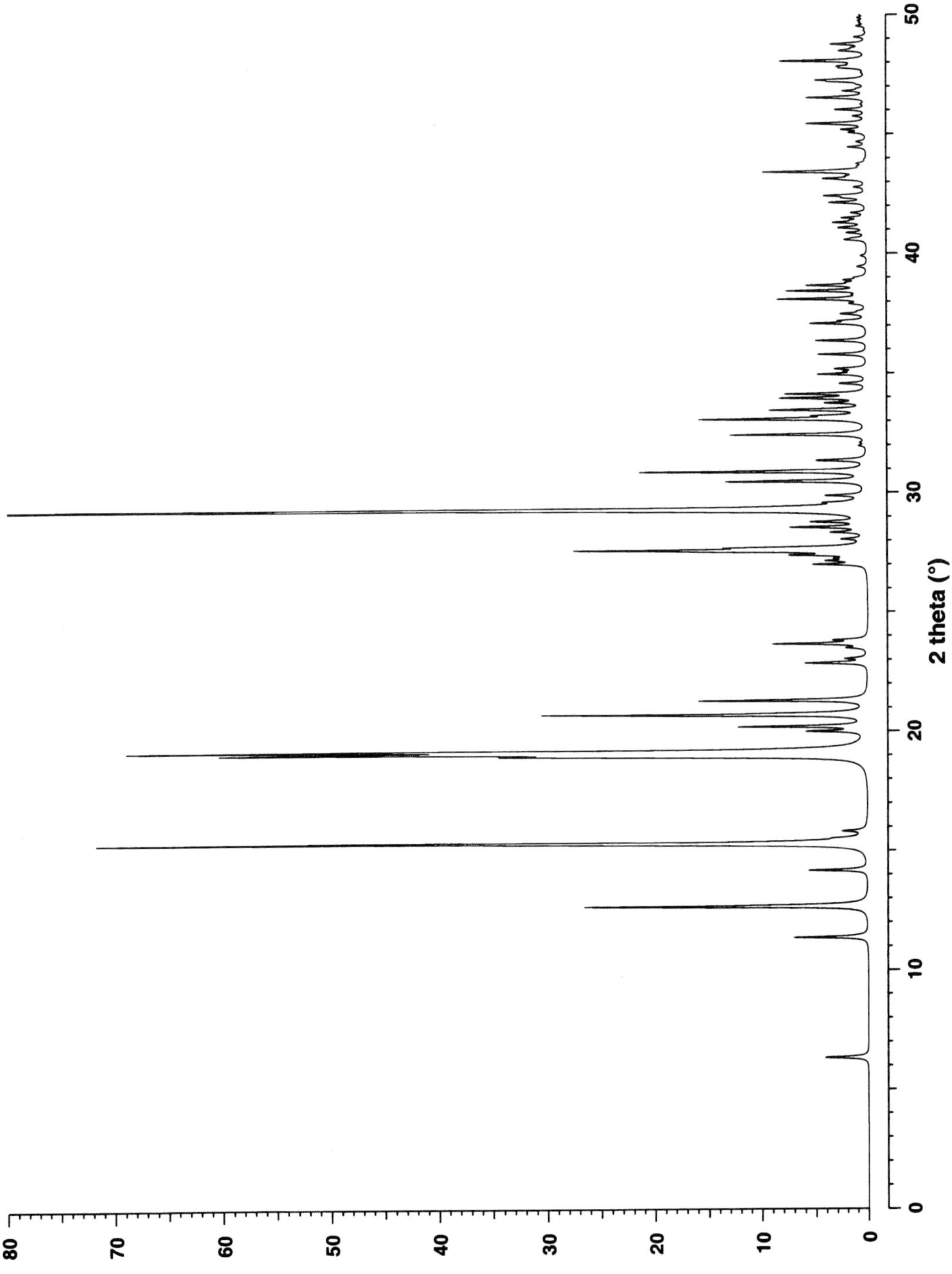
2 theta (°)
0
10
20
30
40
50
0
10
20
30
40
50
60
70
80

ZON Tetramethylammonium ZAPO-M1

CHEMICAL COMPOSITION: $|((CH_3)_4N)_8|\ [Al_{24}P_{32}Zn_8O_{128}]$
$((CH_3)_4N)^+$ = tetramethylammonium

REFINED COMPOSITION: $|N_8C_{42.656}|\ [Al_{25}Zn_7P_{32}O_{128}]$

CRYSTAL DATA: *Pbca* (No. 61)
$a = 14.226$ Å $b = 15.117$ Å $c = 17.557$ Å
$\alpha = 90°$ $\beta = 90°$ $\gamma = 90°$
X-ray single crystal refinement, $R_w = 0.069$

REFERENCE: B. Marler, J. Patarin and L. Sierra, *Microporous Materials* **5** 151–159 (1995).

h	*k*	*l*	2θ	*d*	*M*	I_{rel}
0	0	2	10.08	8.779	2	100.0
0	2	0	11.71	7.558	2	11.4
1	0	2	11.85	7.471	4	0.6
2	0	0	12.44	7.113	2	33.4
0	2	1	12.75	6.942	4	2.5
1	1	2	13.22	6.697	8	0.7
2	1	0	13.76	6.436	4	39.2
2	1	1	14.66	6.043	8	2.8
0	2	2	15.47	5.728	4	2.6
2	2	0	17.12	5.180	4	0.4
2	2	1	17.85	4.968	8	0.8
0	2	3	19.18	4.627	4	25.1
2	2	2	19.90	4.461	8	51.1
0	0	4	20.23	4.389	2	2.0
3	1	1	20.27	4.381	8	0.8
2	1	3	20.51	4.330	8	61.5
1	3	2	21.27	4.178	8	0.2
3	0	2	21.30	4.172	4	0.3
2	3	0	21.61	4.112	4	15.6
2	3	1	22.20	4.003	8	1.5
2	2	3	22.93	3.879	8	29.5
0	2	4	23.44	3.796	4	5.7
0	4	0	23.54	3.779	2	14.8
2	0	4	23.82	3.735	4	27.6
2	3	2	23.90	3.724	8	4.7
0	4	1	24.09	3.695	4	13.5
1	2	4	24.27	3.667	8	0.5
2	1	4	24.55	3.626	8	1.8
3	1	3	24.87	3.580	8	0.2
4	0	0	25.04	3.556	2	2.4
0	4	2	25.66	3.471	4	40.5
4	1	0	25.73	3.462	4	1.8
4	1	1	26.24	3.397	8	13.1
1	4	2	26.43	3.372	8	0.3
2	2	4	26.62	3.349	8	2.9
2	4	0	26.71	3.337	4	0.6
4	0	2	27.05	3.296	4	0.9
2	4	1	27.20	3.279	8	0.5
3	0	4	27.69	3.221	4	0.4
4	1	2	27.70	3.221	8	24.0
4	2	0	27.72	3.218	4	0.9
0	2	5	28.02	3.185	4	0.4
0	4	3	28.11	3.175	4	1.4
4	2	1	28.19	3.165	8	0.6
2	4	2	28.61	3.120	8	4.6
1	4	3	28.81	3.099	8	0.5
2	1	5	28.97	3.082	8	4.5
4	2	2	29.56	3.021	8	1.1
2	3	4	29.77	3.001	8	2.9
4	1	3	29.99	2.980	8	24.4
0	0	6	30.55	2.926	2	1.0
2	2	5	30.76	2.907	8	11.4
4	3	0	30.77	2.906	4	2.1
4	3	1	31.20	2.867	8	2.5
0	4	4	31.23	2.864	4	1.1
2	5	1	32.58	2.748	8	6.5
0	2	6	32.82	2.729	4	12.2
4	1	4	32.95	2.718	8	3.7
2	1	6	33.64	2.664	8	3.5
2	4	4	33.74	2.657	8	1.5
2	5	2	33.79	2.652	8	0.6
4	3	3	34.46	2.603	8	3.4
4	2	4	34.56	2.595	8	0.7
0	4	5	34.88	2.572	4	0.3
2	2	6	35.23	2.548	8	4.1
0	6	0	35.63	2.520	2	0.5
2	5	3	35.73	2.513	8	9.2
0	6	1	36.01	2.494	4	1.6
4	1	5	36.44	2.465	8	1.5
3	5	2	36.71	2.448	8	0.2
0	6	2	37.12	2.422	4	1.1
2	4	5	37.17	2.419	8	1.5
2	6	0	37.88	2.375	4	1.4
6	0	0	37.95	2.371	2	2.4
4	4	3	37.99	2.368	8	0.7
2	6	1	38.24	2.353	8	2.8
1	2	7	38.34	2.348	8	0.3
6	1	0	38.43	2.342	4	1.7
6	1	1	38.78	2.322	8	1.3
0	4	6	38.92	2.314	4	2.0
4	5	0	39.10	2.304	4	4.2
2	6	2	39.30	2.293	8	1.6
4	5	1	39.45	2.284	8	1.9
1	4	6	39.46	2.284	8	0.3
4	0	6	39.90	2.260	4	0.3
2	2	7	39.94	2.257	8	2.3
4	3	5	40.29	2.239	8	0.6
4	1	6	40.36	2.235	8	0.8
4	5	2	40.48	2.228	8	0.7
2	6	3	41.01	2.201	8	0.5
2	4	6	41.02	2.200	8	0.6
0	6	4	41.32	2.185	4	0.8
2	5	5	41.40	2.181	8	4.5
6	1	3	41.52	2.175	8	3.2
4	2	6	41.72	2.165	8	0.8
6	3	0	42.12	2.145	4	1.1
0	2	8	42.91	2.108	4	0.5
0	4	7	43.29	2.090	4	0.2
2	6	4	43.32	2.089	8	0.6
4	4	5	43.41	2.084	8	0.6
6	3	2	43.42	2.084	8	0.3
2	1	8	43.57	2.077	8	0.5
2	7	0	43.81	2.066	4	0.3
4	3	6	43.91	2.062	8	4.0
4	6	0	44.05	2.056	4	0.3
2	7	1	44.13	2.052	8	0.9
0	6	5	44.24	2.047	4	1.6
4	1	7	44.61	2.031	8	0.2
2	2	8	44.85	2.021	8	0.6
2	5	6	44.95	2.016	8	0.3
2	7	2	45.07	2.011	8	0.4
6	2	4	45.08	2.011	8	0.2
4	6	2	45.30	2.002	8	0.4
4	2	7	45.87	1.978	8	0.8
6	4	2	46.38	1.958	8	0.5
2	7	3	46.61	1.949	8	0.3
4	5	5	47.19	1.926	8	0.8
6	2	5	47.83	1.902	8	0.8
6	4	3	47.88	1.900	8	0.4
0	4	8	47.93	1.898	4	2.1
0	8	0	48.15	1.890	2	0.5
0	8	1	48.45	1.879	4	0.6
2	1	9	48.78	1.867	8	1.6
2	5	7	48.89	1.863	8	0.5
0	8	2	49.33	1.847	4	0.7
4	7	0	49.37	1.846	4	1.0
2	6	6	49.42	1.844	8	0.2
6	0	6	49.48	1.842	4	0.3
4	7	1	49.66	1.836	8	1.3
6	3	5	49.81	1.831	8	1.5
6	1	6	49.87	1.829	8	0.2
6	4	4	49.93	1.826	8	0.2
2	8	0	49.94	1.826	4	0.3
5	2	7	49.95	1.826	8	0.3
6	5	2	49.98	1.825	8	0.6

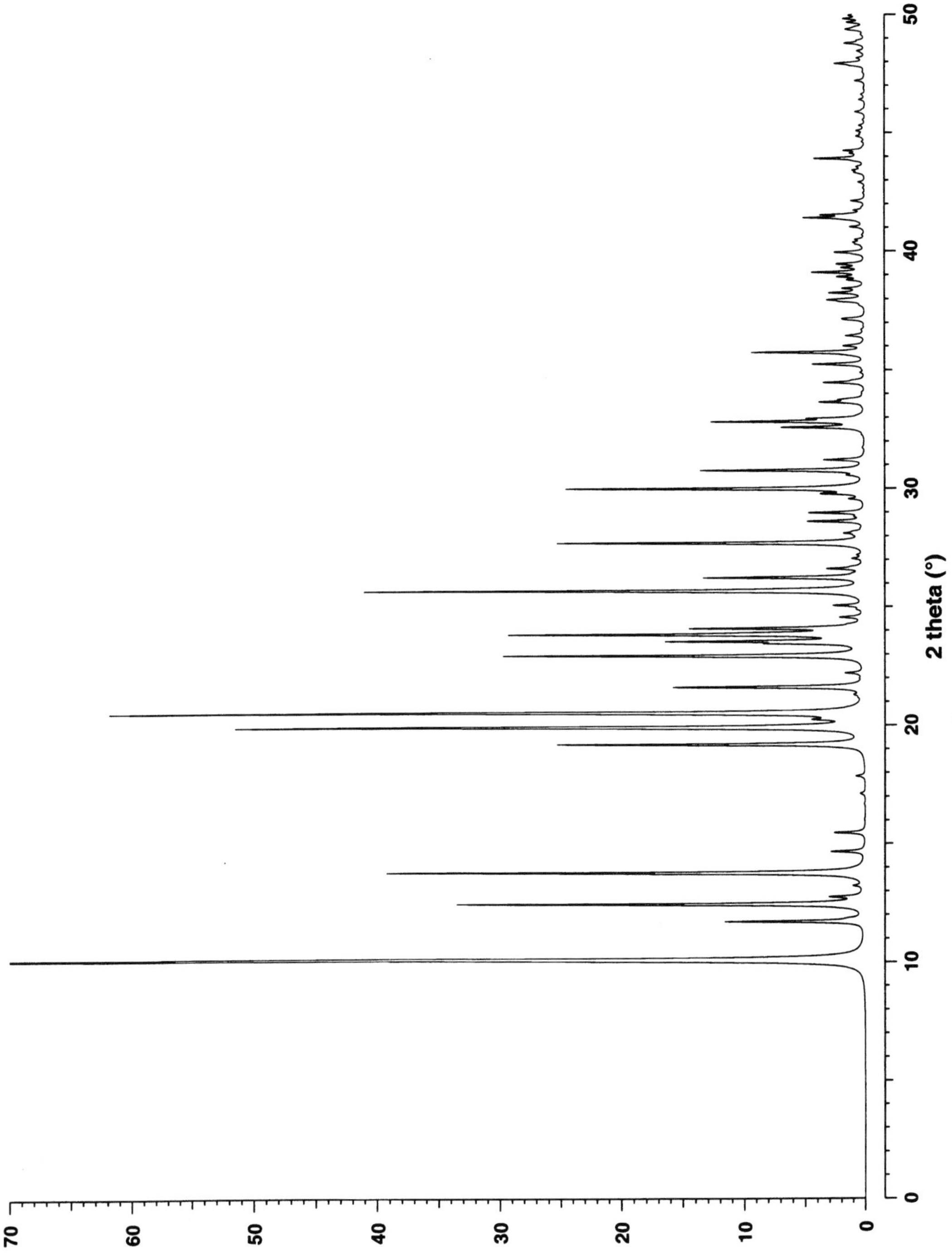
2 theta (°)
0
10
20
30
40
50
60
70

Alpha Cristobalite

CHEMICAL COMPOSITION: $[Si_4O_8]$
Ellora, Hyderabad, India

REFINED COMPOSITION: $[Si_4O_8]$

CRYSTAL DATA: $P4_12_12$ (No. 92)
$a = 4.97$ Å $b = 4.97$ Å $c = 6.94$ Å
$\alpha = 90°$ $\beta = 90°$ $\gamma = 90°$
X-ray single crystal refinement, $R_w = 0.044$

REFERENCE: W. A. Dollase,
Z. Kristallogr. **121** 369–377 (1965).

h	k	l	2θ	d	M	I_{rel}
1	0	1	22.00	4.041	8	100.0
1	1	0	25.34	3.514	4	0.7
1	1	1	28.47	3.135	8	9.0
1	0	2	31.44	2.845	8	11.2
2	0	0	36.15	2.485	4	14.0
1	1	2	36.38	2.469	8	4.3
2	1	1	42.72	2.117	16	2.7
1	0	3	43.13	2.097	8	0.2
2	0	2	44.86	2.020	8	2.6
1	1	3	47.03	1.932	8	5.2
2	1	2	48.65	1.872	16	5.4

Alpha Cristobalite

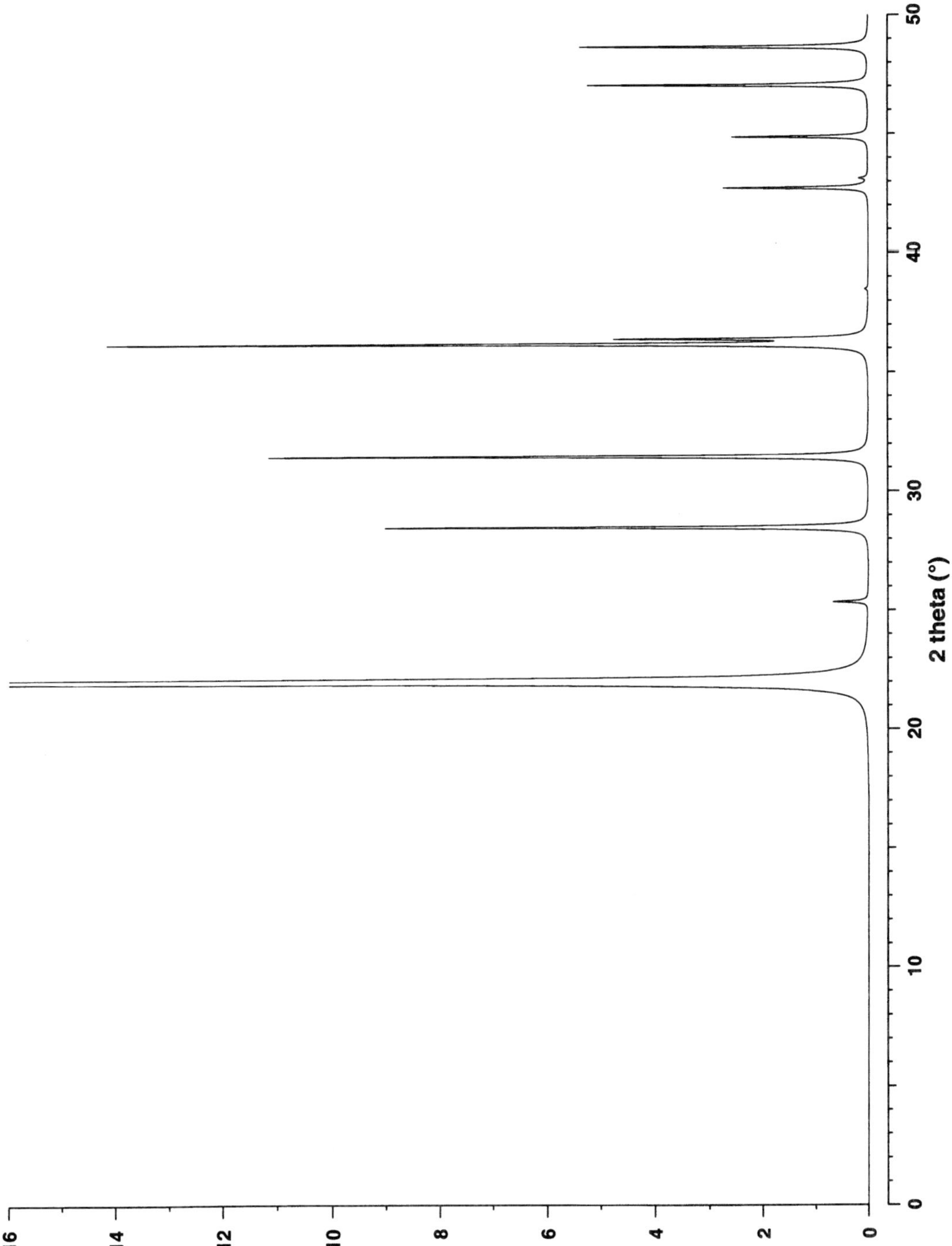

Keatite

CHEMICAL COMPOSITION: $[Si_{12}O_{24}]$
Synthetic material

REFINED COMPOSITION: $[Si_{12}O_{24}]$

CRYSTAL DATA: $P4_32_12$ (No. 96)
$a = 7.46$ Å $b = 7.46$ Å $c = 8.61$ Å
$\alpha = 90°$ $\beta = 90°$ $\gamma = 90°$
X-ray single crystal refinement, $R_{hk0} = 0.129$, $R_{hhl} = 0.098$

REFERENCE: J. Shropshire, P. P. Keat and P. A. Vaughan,
Z. Kristallogr. **112** 409–413 (1959).

h	k	l	2θ	d	M	I_{rel}
1	0	1	15.72	5.638	8	4.4
1	1	0	16.81	5.275	4	0.2
1	1	1	19.74	4.498	8	11.0
2	0	0	23.86	3.730	4	0.7
1	0	2	23.86	3.729	8	39.8
2	0	1	26.03	3.423	8	100.0
2	1	0	26.72	3.336	8	0.5
1	1	2	26.73	3.335	8	20.7
2	1	1	28.70	3.111	16	15.5
2	0	2	31.74	2.819	8	0.4
2	2	0	33.99	2.638	4	0.3
2	1	2	34.00	2.637	16	0.3
1	1	3	35.61	2.521	8	1.6
3	0	1	37.65	2.389	8	0.2
3	1	0	38.15	2.359	8	0.1
3	1	1	39.61	2.275	16	0.7
2	2	2	40.09	2.249	8	3.2
2	1	3	41.50	2.176	16	3.7
0	0	4	41.97	2.152	2	1.2
3	2	0	43.75	2.069	8	0.2
3	1	2	43.76	2.069	16	1.9
1	0	4	43.77	2.068	8	0.8
1	1	4	45.51	1.993	8	0.7
3	0	3	48.43	1.879	8	6.5
4	0	0	48.83	1.865	4	7.8
3	2	2	48.84	1.865	16	2.6

Keatite

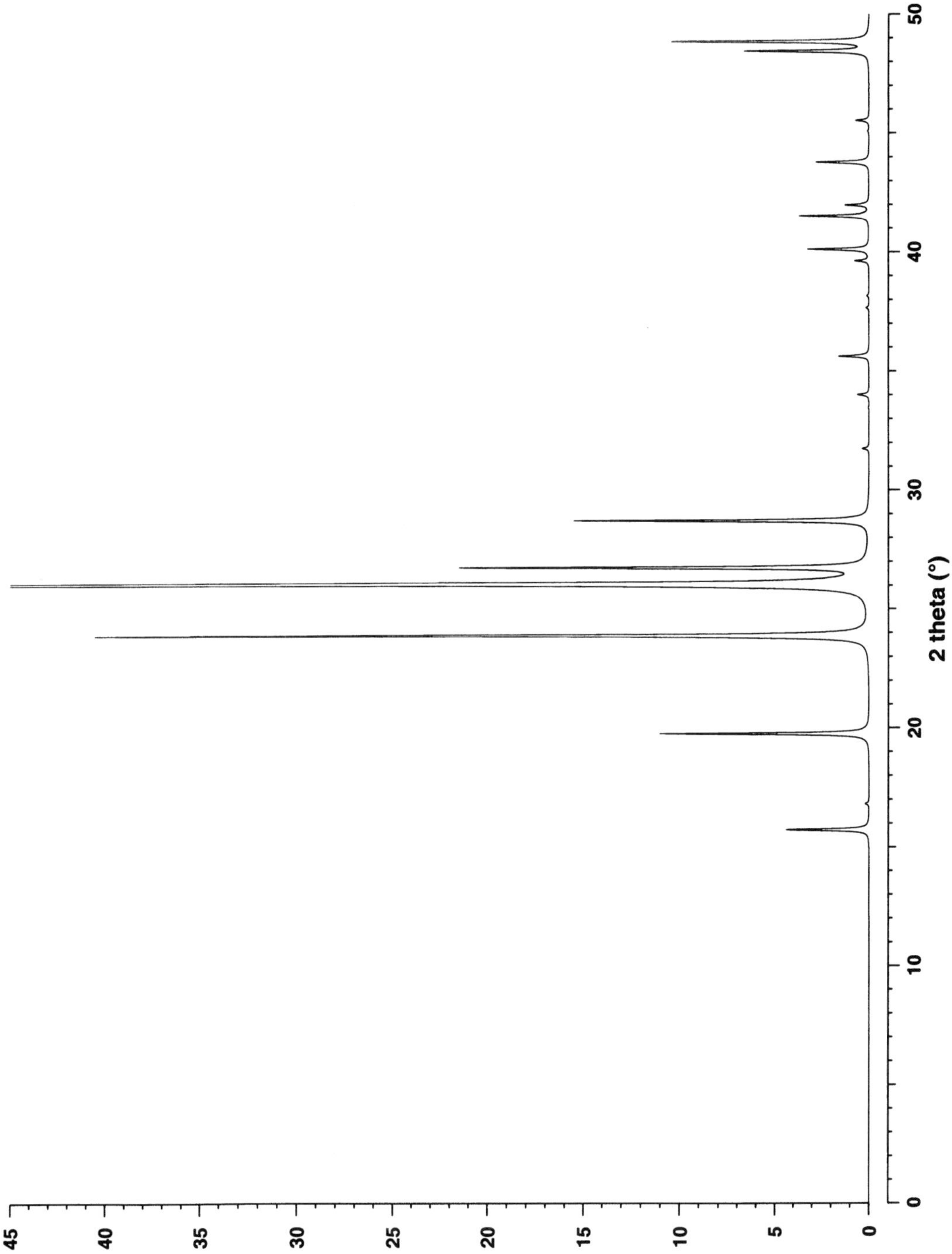

Moganite

CHEMICAL COMPOSITION: $[Si_{12}O_{24}]$
Mogan formation, Gran Canaria, Canary Islands

REFINED COMPOSITION: $[Si_{12}O_{24}]$

CRYSTAL DATA: $I1\,2/a\,1$ (No. 15) unique axis **b**, cell choice 3
$a = 8.758$ Å $b = 4.876$ Å $c = 10.715$ Å
$\alpha = 90°$ $\beta = 90.08°$ $\gamma = 90°$

REFERENCE: G. Miehe and H. Graetsch,
Eur. J. Mineral. **4** 693–706 (1992).

h	*k*	*l*	2*θ*	*d*	*M*	I_{rel}	*h*	*k*	*l*	2*θ*	*d*	*M*	I_{rel}	*h*	*k*	*l*	2*θ*	*d*	*M*	I_{rel}
0	0	2	16.55	5.357	2	0.2	0	2	0	36.87	2.438	2	9.2	2	2	0	42.43	2.130	4	9.5
0	1	1	20.01	4.438	4	45.7	-2	1	3	37.34	2.408	4	0.1	-4	0	2	44.69	2.028	2	1.9
2	0	0	20.28	4.379	2	22.3	2	1	3	37.38	2.406	4	1.3	4	0	2	44.73	2.026	2	3.8
-2	0	2	26.27	3.393	2	47.4	-1	2	1	39.27	2.294	4	3.3	2	2	2	45.85	1.979	4	0.5
2	0	2	26.30	3.388	2	75.5	1	2	1	39.27	2.294	4	5.0	-4	1	1	46.22	1.964	4	0.4
-1	1	2	26.73	3.336	4	100.0	-2	0	4	39.41	2.287	2	7.6	4	1	1	46.24	1.963	4	3.0
1	1	2	26.74	3.333	4	57.3	2	0	4	39.46	2.284	2	0.6	-1	2	3	46.25	1.963	4	0.5
-2	1	1	28.63	3.118	4	26.7	-3	1	2	39.70	2.270	4	0.4	1	2	3	46.27	1.962	4	0.4
2	1	1	28.65	3.116	4	12.1	3	1	2	39.74	2.268	4	0.4	0	1	5	46.28	1.962	4	0.6
0	1	3	31.04	2.881	4	16.4	1	1	4	39.76	2.267	4	4.2	-3	1	4	49.81	1.831	4	11.2
0	0	4	33.45	2.679	2	1.0	0	2	2	40.66	2.219	4	0.9	3	1	4	49.87	1.828	4	12.5
3	1	0	35.85	2.505	4	1.2	4	0	0	41.23	2.189	2	6.8							

Moganite

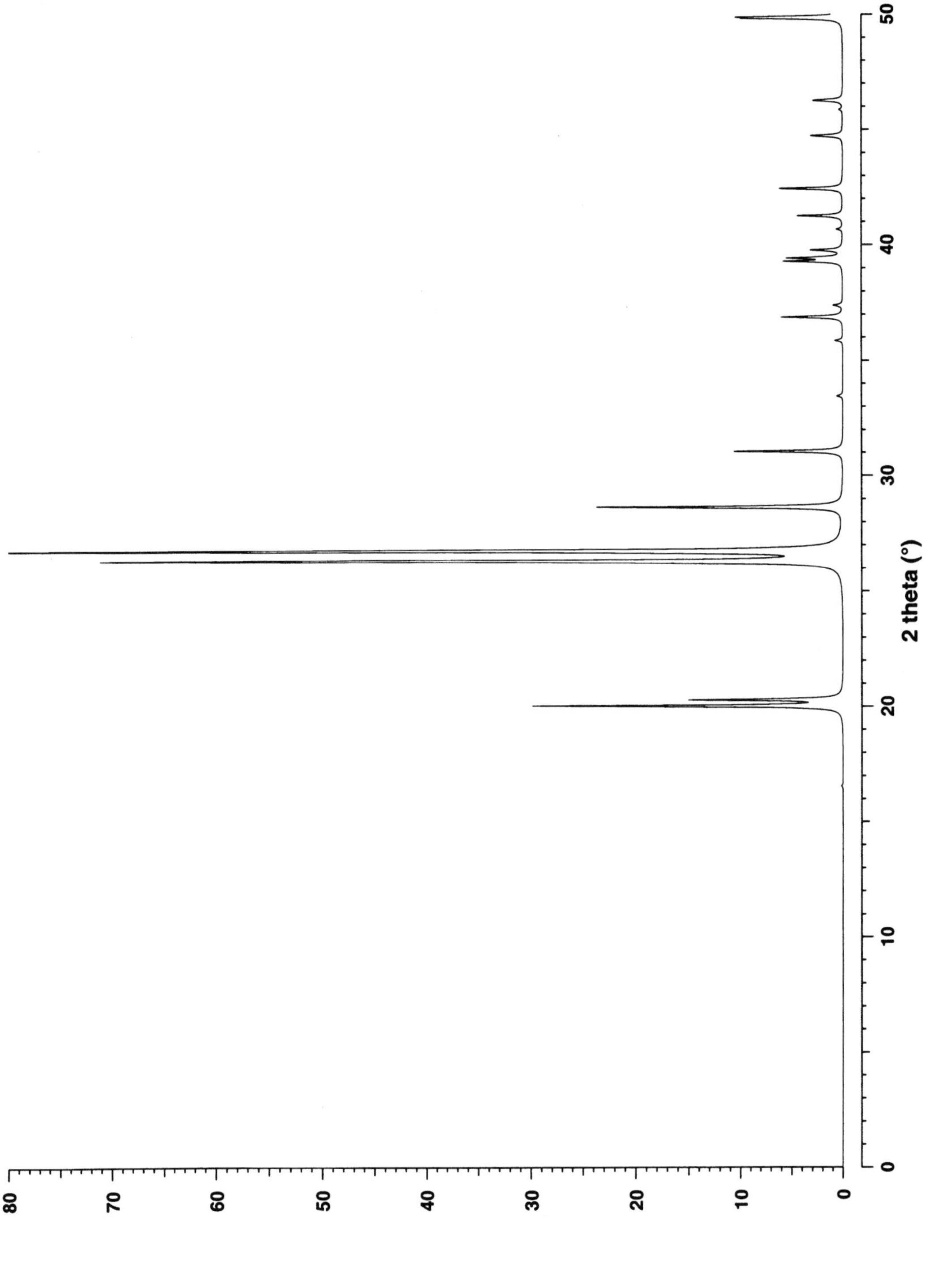

Alpha Quartz

CHEMICAL COMPOSITION: $[Si_3O_6]$
Dextro-rotatory synthetic material

REFINED COMPOSITION: $[Si_3O_6]$

CRYSTAL DATA: $P3_221$ (No. 154)
$a = 4.916$ Å $b = 4.916$ Å $c = 5.4054$ Å
$\alpha = 90°$ $\beta = 90°$ $\gamma = 120°$
X-ray single crystal refinement, $R_w = 0.016$

REFERENCE: L. Levien, C. T. Prewitt and D. J. Weidner,
American Mineralogist **65** 920–930 (1980).

h	k	l	2θ	d	M	I_{rel}
1	0	0	20.86	4.257	6	26.7
0	1	1	26.65	3.345	6	8.9
1	0	1	26.65	3.345	6	100.0
1	1	0	36.56	2.458	6	9.6
1	0	2	39.49	2.282	6	1.1
0	1	2	39.49	2.282	6	21.2
1	1	1	40.31	2.238	12	8.3
2	0	0	42.46	2.129	6	7.0
2	0	1	45.81	1.981	6	2.8
0	2	1	45.81	1.981	6	1.3

Alpha Quartz

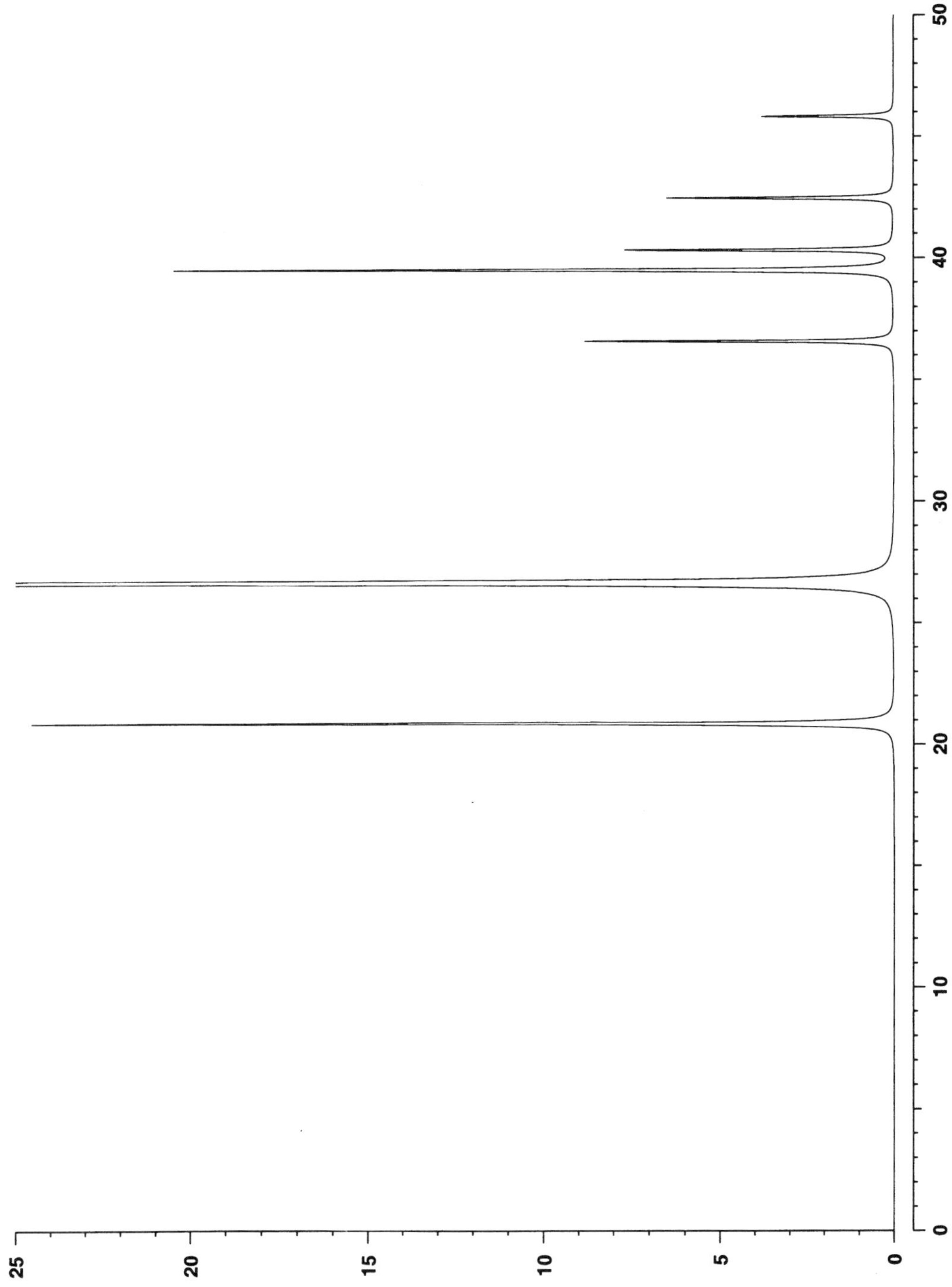

Tridymite

CHEMICAL COMPOSITION: $[Si_{48}O_{96}]$
From the Steinbach meteorite, Harvard Museum, U.S.A.

REFINED COMPOSITION: $[Si_{48}O_{96}]$

CRYSTAL DATA: *C*1*c*1 (No. 9) unique axis **b**, cell choice 1
$a = 18.524$ Å $b = 5.0032$ Å $c = 23.810$ Å
$\alpha = 90°$ $\beta = 105.82°$ $\gamma = 90°$
X-ray single crystal refinement, $R_w = 0.069$

REFERENCE: W. A. Dollase and W. H. Baur,
American Mineralogist **61** 971–978 (1976).

h	k	l	2θ	d	M	I_{rel}
0	0	2	7.72	11.454	2	0.4
2	0	0	9.93	8.911	2	0.2
2	0	2	14.16	6.255	2	0.2
1	1	0	18.42	4.817	4	0.6
-1	1	1	18.54	4.785	4	0.2
-4	0	2	19.31	4.597	2	2.4
-1	1	2	19.46	4.562	4	1.1
1	1	2	20.52	4.328	4	100.0
2	0	4	20.59	4.313	2	56.5
-1	1	3	21.07	4.217	4	1.7
-4	0	4	21.68	4.100	2	82.5
1	1	3	22.54	3.945	4	0.3
-2	0	6	22.70	3.917	2	0.6
-3	1	1	22.88	3.887	4	1.3
-1	1	4	23.24	3.828	4	27.0
3	1	0	23.24	3.827	4	26.8
0	0	6	23.30	3.818	2	15.7
4	0	2	23.31	3.816	2	15.0
-3	1	3	24.11	3.691	4	0.4
3	1	1	24.24	3.671	4	3.1
-3	1	4	25.62	3.476	4	0.5
3	1	2	25.81	3.452	4	2.5
-1	1	5	25.83	3.449	4	2.3
-4	0	6	26.27	3.392	2	4.5
-3	1	5	27.62	3.229	4	3.5
2	0	6	27.82	3.207	2	1.5
1	1	5	27.84	3.204	4	1.3
3	1	3	27.85	3.203	4	0.8
4	0	4	28.54	3.128	2	0.9
-6	0	2	28.92	3.087	2	1.1
-6	0	4	29.87	2.991	2	0.1
-3	1	6	30.02	2.977	4	6.1
-5	1	2	30.03	2.976	4	7.3
-2	0	8	30.06	2.972	2	4.5
6	0	0	30.08	2.970	2	5.1
-5	1	1	30.16	2.963	4	1.0
3	1	4	30.28	2.951	4	2.1
-5	1	3	30.41	2.939	4	2.1
1	1	6	30.94	2.890	4	0.5
-5	1	4	31.30	2.858	4	0.2
-1	1	7	31.91	2.805	4	0.5
5	1	1	31.92	2.803	4	1.9
-4	0	8	32.22	2.778	2	5.5
-5	1	5	32.64	2.743	4	0.4
-3	1	7	32.73	2.736	4	1.2
1	1	7	34.24	2.619	4	4.1
-5	1	6	34.41	2.607	4	2.6
4	0	6	34.87	2.573	2	1.1
5	1	3	35.43	2.533	4	3.4
2	0	8	35.44	2.533	2	0.1
-3	1	8	35.71	2.514	4	0.4
0	2	0	35.90	2.502	2	20.9
3	1	6	36.03	2.493	4	11.0
0	2	1	36.12	2.487	4	0.1
-5	1	7	36.53	2.460	4	0.4
-2	0	10	37.79	2.381	2	0.5
0	2	3	37.84	2.377	4	0.1
-1	1	9	38.78	2.322	4	0.6
-3	1	9	38.90	2.315	4	0.2
-7	1	1	38.92	2.314	4	0.3
-5	1	8	38.97	2.311	4	1.8
-7	1	4	38.98	2.311	4	4.6
-4	0	10	39.00	2.309	2	2.3
-8	0	2	39.03	2.308	2	6.7
3	1	7	39.24	2.296	4	0.4
7	1	0	39.72	2.269	4	0.2
-7	1	5	39.82	2.264	4	0.2
5	1	5	40.30	2.238	4	2.3
7	1	1	40.91	2.206	4	0.4
-8	0	6	40.96	2.203	2	0.6
-2	2	5	41.03	2.200	4	0.2
-7	1	6	41.04	2.199	4	0.1
0	2	5	41.11	2.196	4	0.1
1	1	9	41.32	2.185	4	1.6
2	2	4	41.74	2.164	4	0.3
4	0	8	41.89	2.156	2	0.3
4	2	1	42.13	2.145	4	0.2
-3	1	10	42.27	2.138	4	0.4
-4	2	4	42.32	2.135	4	2.4
-1	1	10	42.43	2.131	4	0.2
7	1	2	42.45	2.129	4	0.3
-7	1	7	42.61	2.122	4	0.6
3	1	8	42.63	2.121	4	1.9
-6	0	10	42.77	2.114	2	0.2
-2	2	6	42.89	2.108	4	0.1
5	1	6	43.13	2.097	4	0.4
0	2	6	43.24	2.092	4	3.0
4	2	2	43.24	2.092	4	2.7
2	0	10	43.38	2.086	2	2.0
6	0	6	43.40	2.085	2	1.3
-8	0	8	44.18	2.050	2	2.6
-5	1	10	44.61	2.031	4	0.7
4	2	3	44.70	2.027	4	0.2
-4	2	6	45.03	2.013	4	0.2
1	1	10	45.06	2.012	4	0.2
-2	2	7	45.06	2.012	4	0.6
0	2	7	45.65	1.987	4	0.3
-3	1	11	45.80	1.981	4	0.7
-2	0	12	45.81	1.981	2	0.3
2	2	6	46.01	1.973	4	0.6
3	1	9	46.17	1.966	4	0.3
5	1	7	46.18	1.966	4	0.1
-1	1	11	46.20	1.965	4	0.1
-4	0	12	46.36	1.959	2	0.7
4	2	4	46.48	1.954	4	0.3
7	1	4	46.49	1.953	4	0.2
-7	1	9	46.71	1.945	4	1.2
-6	2	3	46.88	1.938	4	0.1
-6	2	1	46.95	1.935	4	0.2
-6	2	4	47.37	1.919	4	0.3
-2	2	8	47.50	1.914	4	0.2
6	2	0	47.52	1.913	4	0.3
0	0	12	47.63	1.909	2	0.2
8	0	4	47.66	1.908	2	0.3
-5	1	11	47.76	1.904	4	0.1
-9	1	3	47.79	1.903	4	0.9
-9	1	2	48.02	1.895	4	0.1
-6	2	5	48.21	1.887	4	0.2
-9	1	5	48.38	1.881	4	0.6
6	2	1	48.43	1.880	4	0.2
2	2	7	48.55	1.875	4	0.1
-9	1	1	48.59	1.874	4	0.3
-8	0	10	48.62	1.872	2	0.6
1	1	11	48.91	1.862	4	0.2
7	1	5	48.93	1.861	4	0.6
-4	2	8	49.00	1.859	4	0.8
-7	1	10	49.17	1.853	4	0.2
-9	1	6	49.19	1.852	4	0.3
-6	0	12	49.20	1.852	2	0.3
-10	0	4	49.23	1.851	2	0.3
9	1	0	49.50	1.841	4	0.2
6	2	2	49.67	1.836	4	0.2
3	1	10	49.85	1.829	4	0.8
6	0	8	49.90	1.828	2	1.1

Tridymite

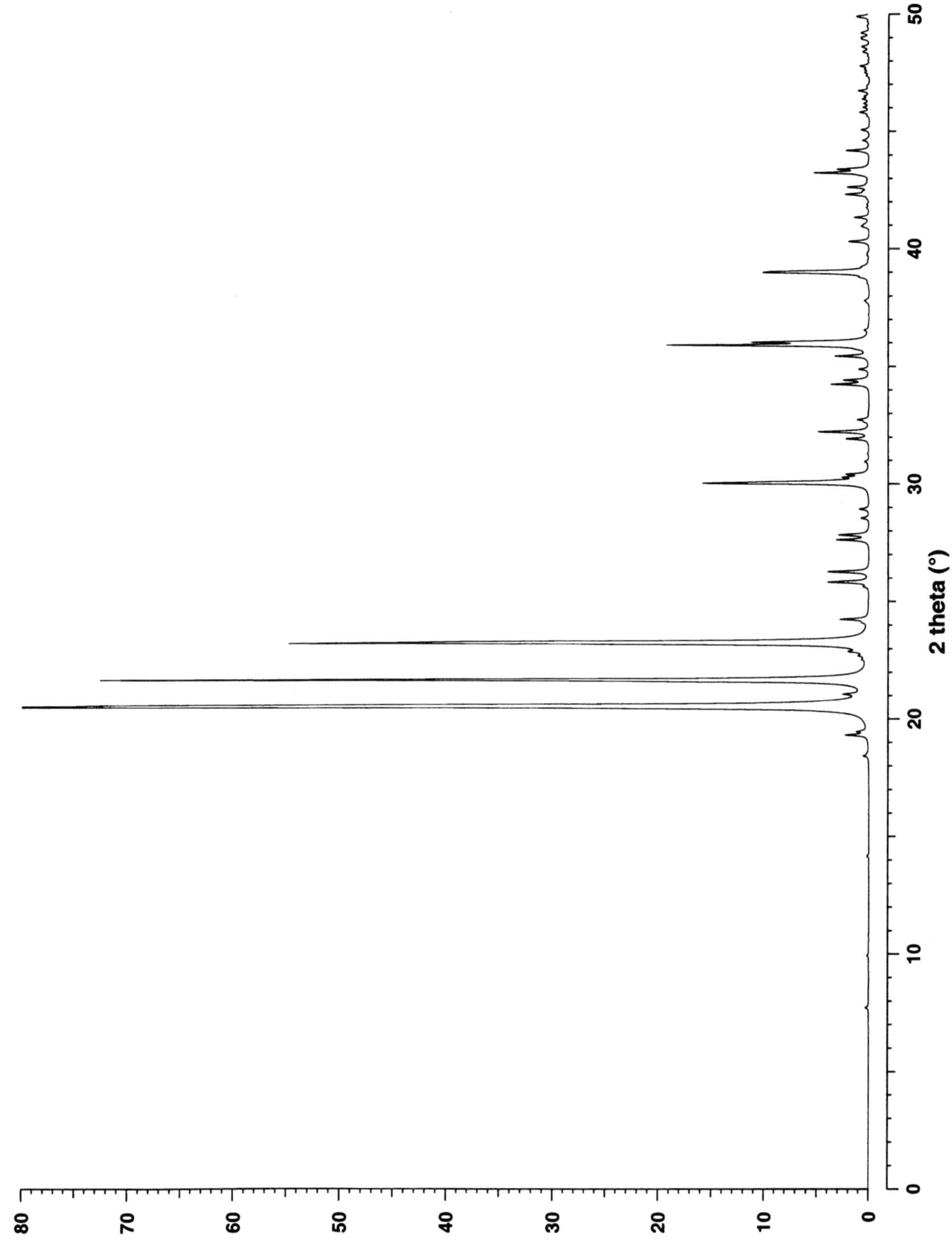

POWDER PATTERN SIMULATIONS OF DISORDERED INTERGROWTHS

Beta family, polymorph A (*BEA) – polymorph B

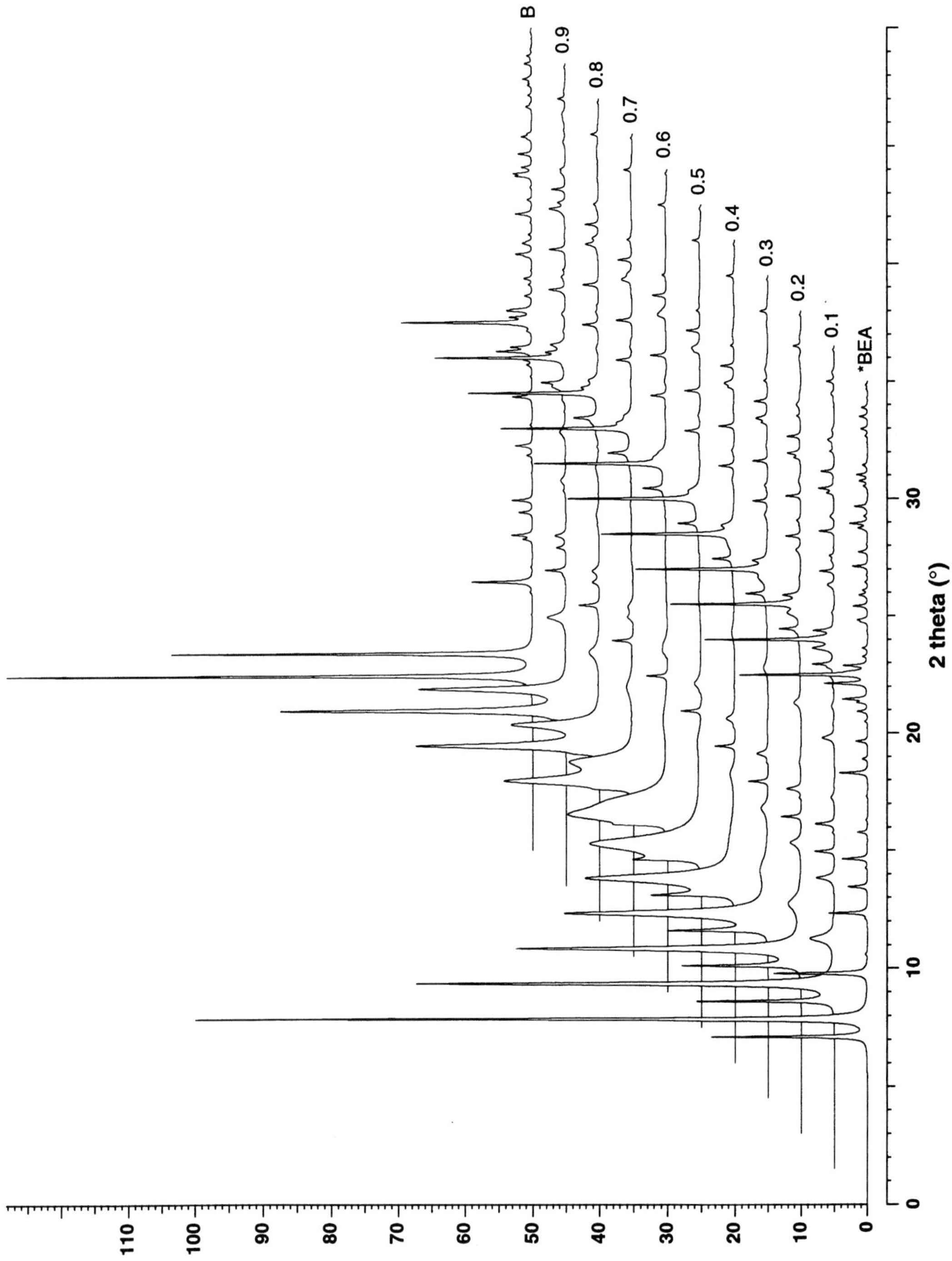

SSZ33/SSZ26 family, polymorph A – polymorph B

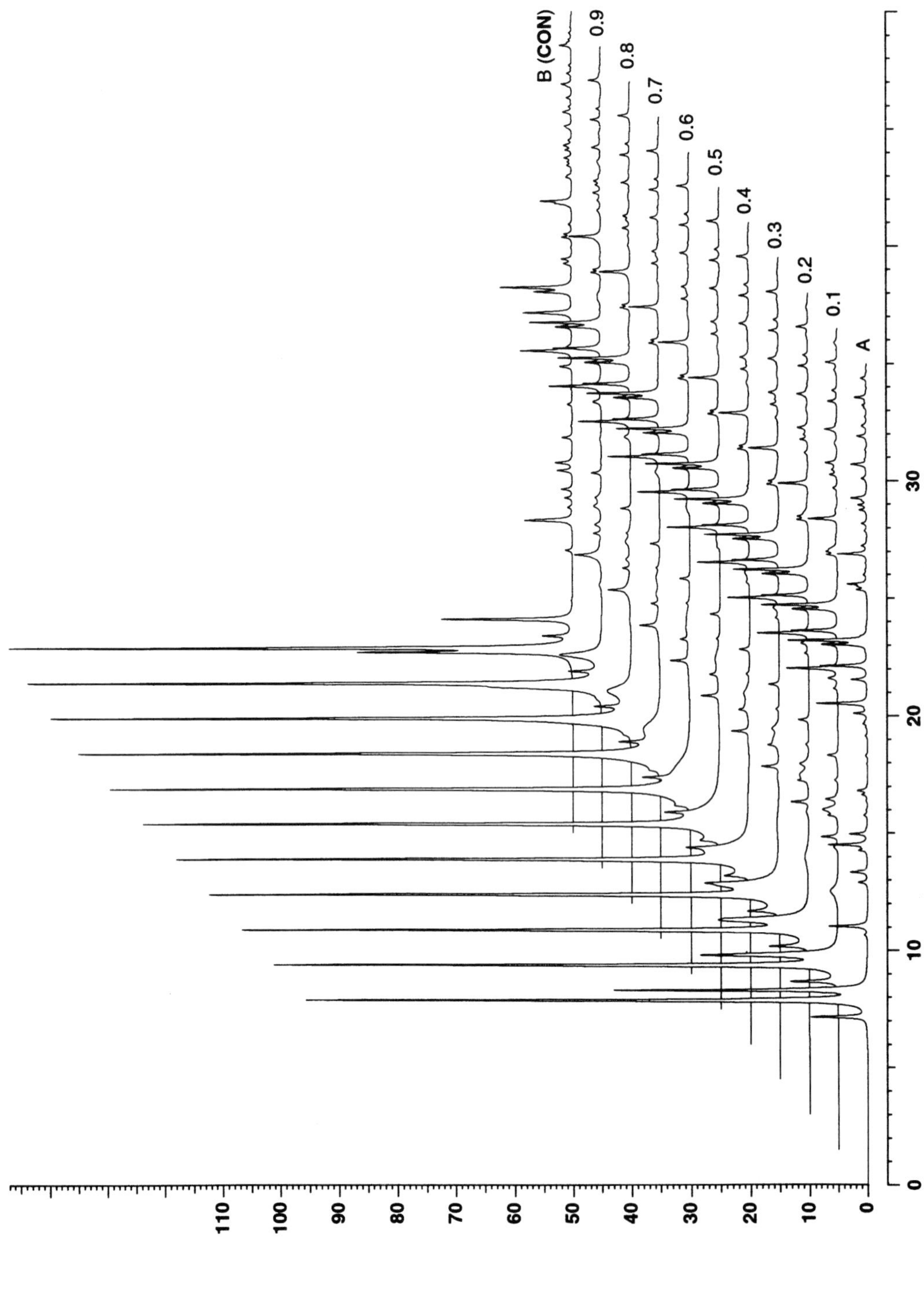

Faujasite family, FAU – EMT

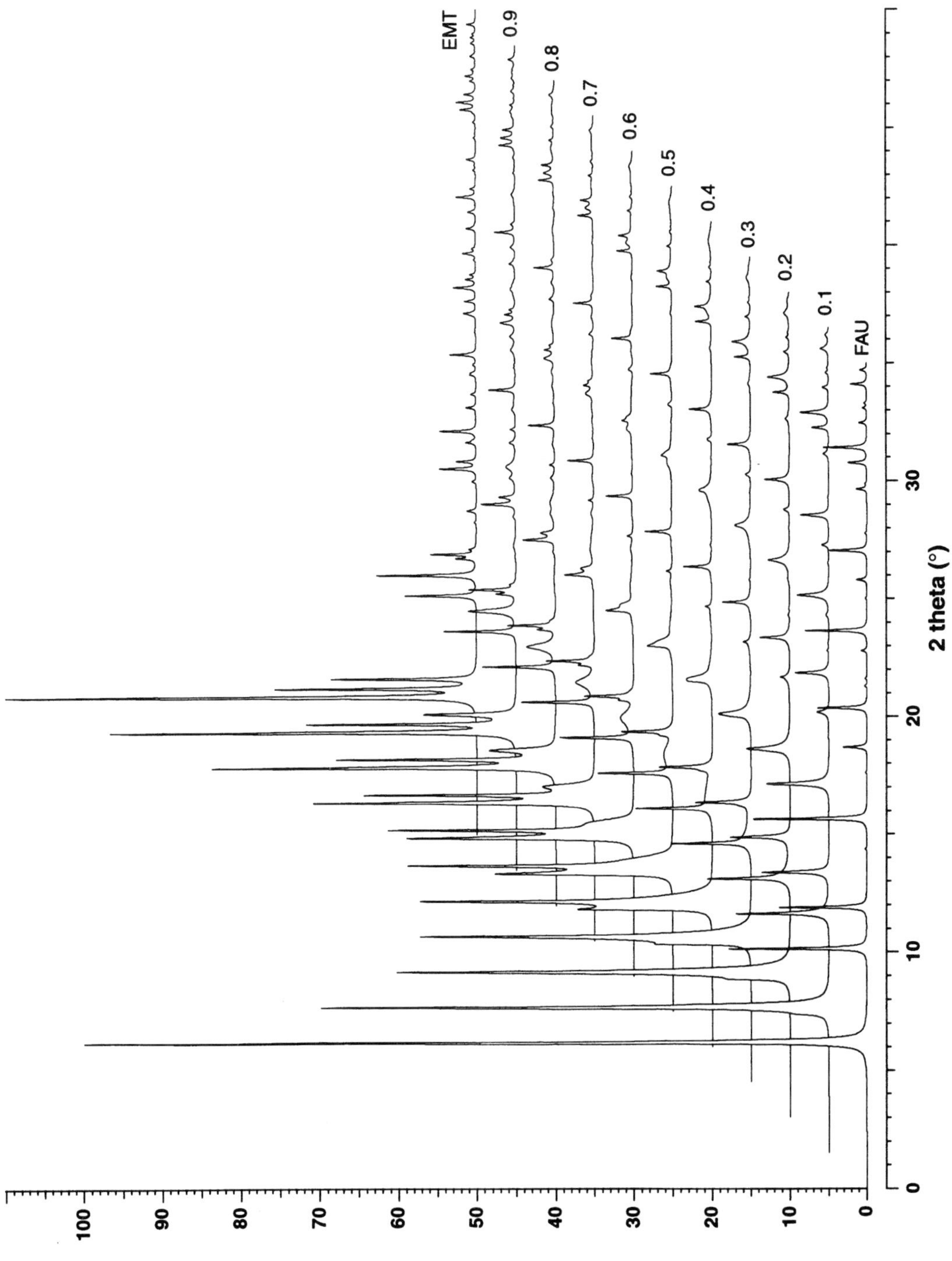

ZSM5/ZSM11 family, MFI – MEL

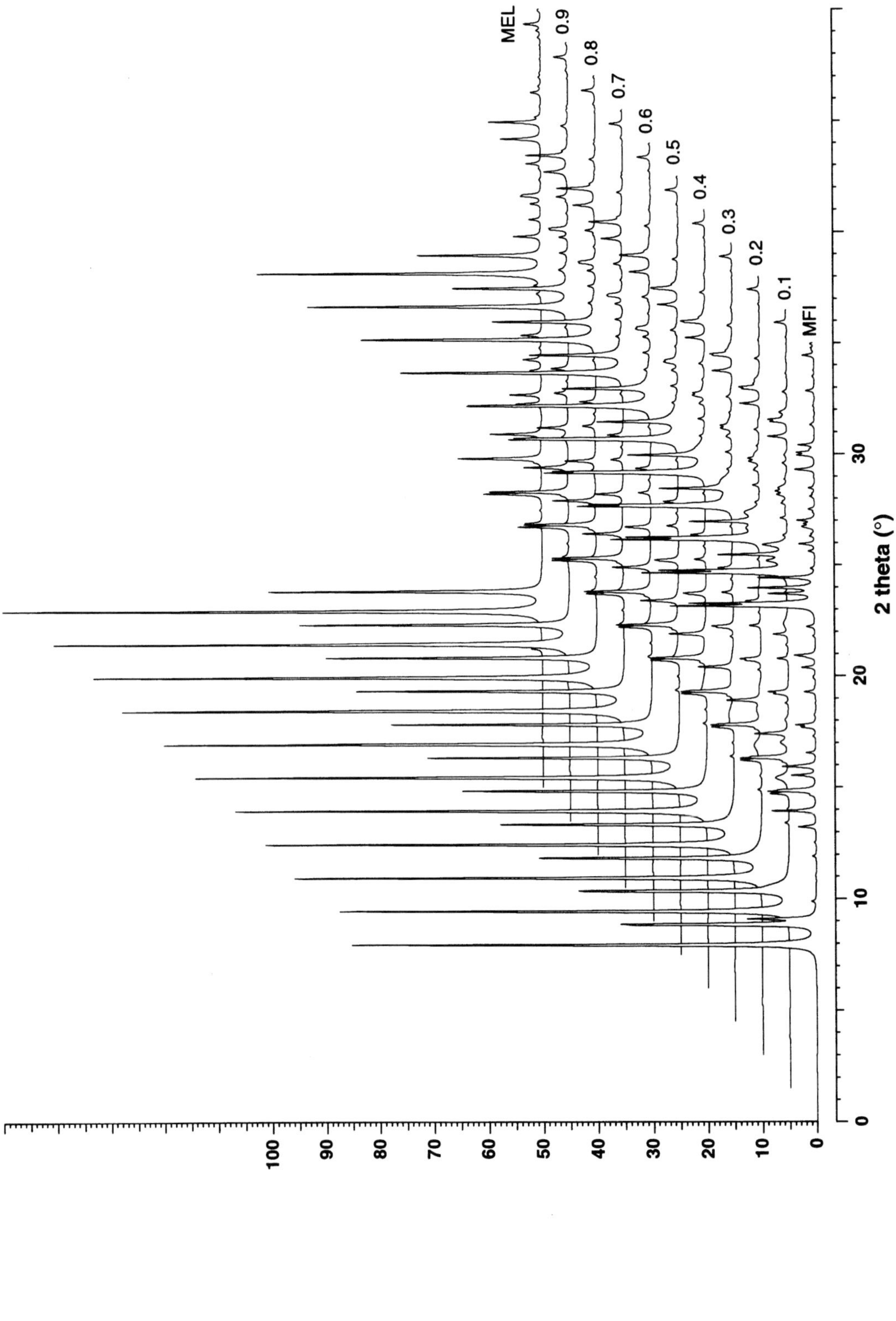

Offretite – Erionite family

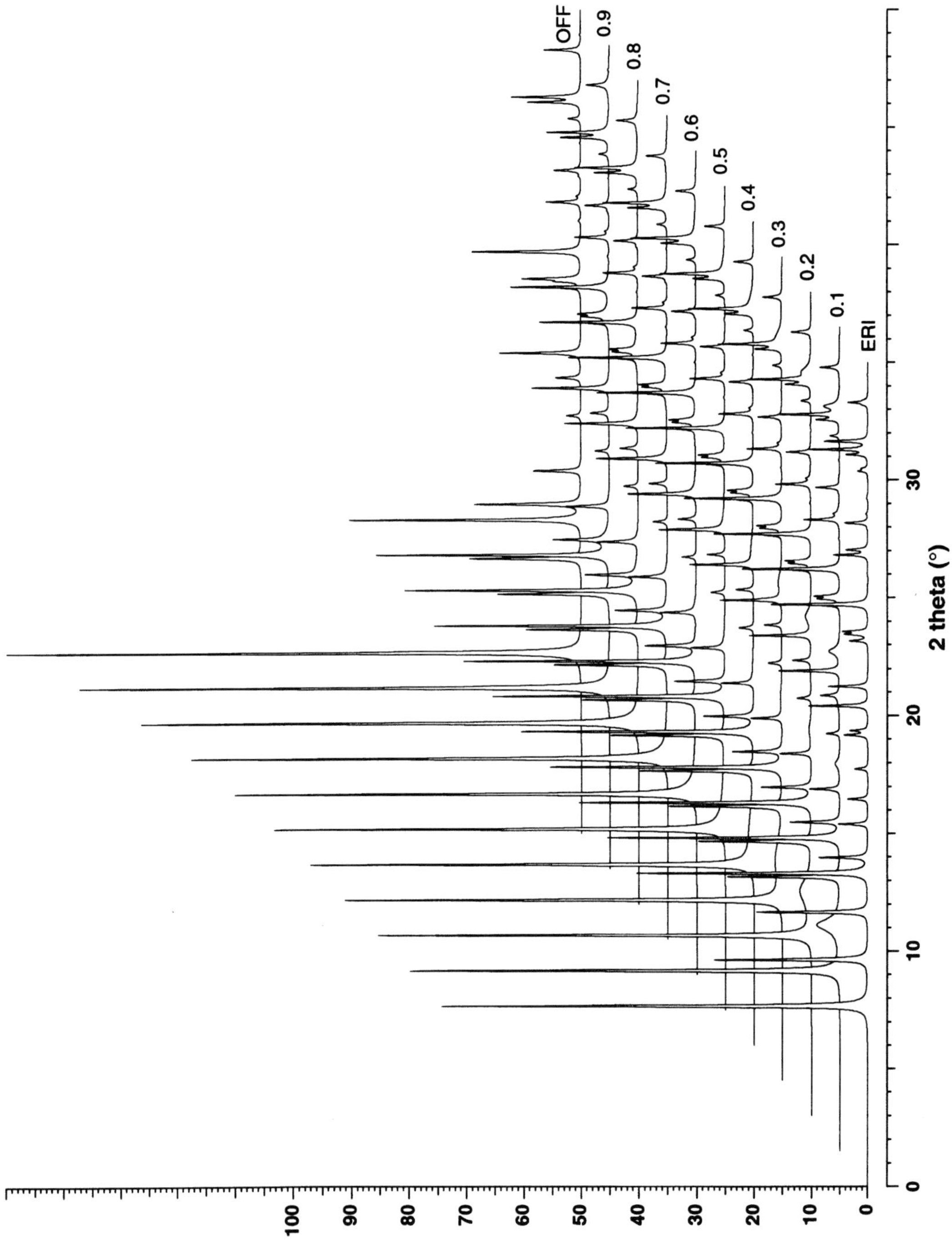